Biochemistry of Lipids, Lipoproteins and Membranes

Biochemistry of Lipids, Lipoproteins and Membranes

Sixth Edition

Edited by

Neale D. Ridgway, Ph.D.
Departments of Pediatrics and Biochemistry & Molecular Biology
Dalhousie University
Halifax, Nova Scotia, Canada

Roger S. McLeod, Ph.D.
Department of Biochemistry & Molecular Biology
Dalhousie University
Halifax, Nova Scotia, Canada

ELSEVIER

AMSTERDAM • BOSTON • HEIDELBERG • LONDON • NEW YORK • OXFORD • PARIS
SAN DIEGO • SAN FRANCISCO • SINGAPORE • SYDNEY • TOKYO

Elsevier
Radarweg 29, PO Box 211, 1000 AE Amsterdam, Netherlands
The Boulevard, Langford Lane, Kidlington, Oxford OX5 1GB, UK
225 Wyman Street, Waltham, MA 02451, USA

Notices
Knowledge and best practice in this field are constantly changing. As new research and experience
broaden our understanding, changes in research methods, professional practices, or medical treatment
may become necessary.

Practitioners and researchers may always rely on their own experience and knowledge in
evaluating and using any information, methods, compounds, or experiments described herein.
In using such information or methods they should be mindful of their own safety and the safety
of others, including parties for whom they have a professional responsibility.

To the fullest extent of the law, neither the Publisher nor the authors, contributors, or editors,
assume any liability for any injury and/or damage to persons or property as a matter of products
liability, negligence or otherwise, or from any use or operation of any methods, products,
instructions, or ideas contained in the material herein.

ISBN: 978-0-444-63438-2

British Library Cataloguing-in-Publication Data
A catalogue record for this book is available from the British Library

Library of Congress Cataloging-in-Publication Data
A catalog record for this book is available from the Library of Congress

For information on all Elsevier publications
visit our website at http://store.elsevier.com/

**Working together
to grow libraries in
developing countries**

www.elsevier.com • www.bookaid.org

Acquisition Editor: Jill Leonard
Editorial Project Manager: Pat Gonzalez
Production Project Manager: Karen East and Kirsty Halterman
Designer: Alan Studholme

Typeset by TNQ Books and Journals
www.tnq.co.in

Printed and bound in the United Kingdom

Contents

1. Functional Roles of Lipids in Membranes

William Dowhan, Mikhail Bogdanov and Eugenia Mileykovskaya

2. Approaches to Lipid Analysis

Jeff G. McDonald, Pavlina T. Ivanova and H. Alex Brown

3. Fatty Acid and Phospholipid Biosynthesis in Prokaryotes

Yong-Mei Zhang and Charles O. Rock

4. Lipid Metabolism in Plants

Katherine M. Schmid

5. **Fatty Acid Handling in Mammalian Cells**

Richard Lehner and Ariel D. Quiroga

6. Fatty Acid Desaturation and Elongation in Mammals

Laura M. Bond, Makoto Miyazaki, Lucas M. O'Neill, Fang Ding and James M. Ntambi

7. Phospholipid Synthesis in Mammalian Cells

Neale D. Ridgway

8. Phospholipid Catabolism

Robert V. Stahelin

11. Cholesterol Synthesis

Andrew J. Brown and Laura J. Sharpe

12. Bile Acid Metabolism

Paul A. Dawson

13. Lipid Modification of Proteins

Marilyn D. Resh

14. Intramembrane and Intermembrane Lipid Transport

Frederick R. Maxfield and Anant K. Menon

15. High-Density Lipoproteins: Metabolism and Protective Roles Against Atherosclerosis

Gordon A. Francis

16. Assembly and Secretion of Triglyceride-Rich Lipoproteins

Roger S. McLeod and Zemin Yao

17.   Lipoprotein Receptors

Wolfgang J. Schneider

18.   Atherosclerosis

Murray W. Huff, Alan Daugherty and Hong Lu

19. Diabetic Dyslipidaemia

Khosrow Adeli, Jennifer Taher, Sarah Farr, Changting Xiao and Gary F. Lewis

Contributors

Khosrow Adeli The Hospital for Sick Children, University of Toronto, Toronto, ON, Canada

Mikhail Bogdanov Department of Biochemistry and Molecular Biology, University of Texas–Houston, Medical School, Houston, TX, USA

Laura M. Bond Department of Biochemistry, University of Wisconsin-Madison, Madison, WI, USA

Andrew J. Brown School of Biotechnology and Biomolecular Sciences, The University of New South Wales (UNSW Australia), Sydney, NSW, Australia

H. Alex Brown Department of Pharmacology, Vanderbilt University School of Medicine, Nashville, TN, USA; Department of Biochemistry and The Vanderbilt Institute of Chemical Biology, Vanderbilt University School of Medicine, Nashville, TN, USA

Alan Daugherty Saha Cardiovascular Research Center, University of Kentucky, Lexington, KY, USA

Paul A. Dawson Division of Pediatric Gastroenterology, Hepatology and Nutrition, Department of Pediatrics, School of Medicine, Emory University, Atlanta, GA, USA

Fang Ding Institute of Animal Genetics and Breeding, College of Animal Science and Technology, Sichuan Agricultural University, Wenjiang, Sichuan, PR, China

William Dowhan Department of Biochemistry and Molecular Biology, University of Texas–Houston, Medical School, Houston, TX, USA

Sarah Farr The Hospital for Sick Children, University of Toronto, Toronto, ON, Canada

Gordon A. Francis Department of Medicine, Centre for Heart Lung Innovation, University of British Columbia, Vancouver, BC, Canada

Anthony H. Futerman Department of Biological Chemistry, Weizmann Institute of Science, Rehovot, Israel

Murray W. Huff Departments of Medicine and Biochemistry, Robarts Research Institute, The University of Western Ontario, London, ON, Canada

Pavlina T. Ivanova Department of Pharmacology, Vanderbilt University School of Medicine, Nashville, TN, USA

Richard Lehner Group on Molecular and Cell Biology of Lipids, University of Alberta, Edmonton, AB, Canada; Departments of Pediatrics and Cell Biology, University of Alberta, Edmonton, AB, Canada

Gary F. Lewis University Health Network, University of Toronto, Toronto, ON, Canada

Hong Lu Saha Cardiovascular Research Center, University of Kentucky, Lexington, KY, USA

Frederick R. Maxfield Department of Biochemistry, Weill Cornell Medical College, New York, NY, USA

Jeff G. McDonald Department of Molecular Genetics, UT Southwestern Medical Center, Dallas, TX, USA

Roger S. McLeod Department of Biochemistry & Molecular Biology, Dalhousie University, Halifax, NS, Canada

Anant K. Menon Department of Biochemistry, Weill Cornell Medical College, New York, NY, USA

Eugenia Mileykovskaya Department of Biochemistry and Molecular Biology, University of Texas–Houston, Medical School, Houston, TX, USA

Makoto Miyazaki Department of Medicine, Division of Renal Diseases and Hypertension, University of Colorado, Anschutz Medical Campus, Aurora, CO, USA

Robert C. Murphy Department of Pharmacology, University of Colorado-Denver, Aurora, CO, USA

James M. Ntambi Department of Biochemistry, University of Wisconsin-Madison, Madison, WI, USA; Department of Nutritional Science, University of Wisconsin-Madison, Madison, WI, USA

Lucas M. O'Neill Department of Biochemistry, University of Wisconsin-Madison, Madison, WI, USA

Ariel D. Quiroga Group on Molecular and Cell Biology of Lipids, University of Alberta, Edmonton, AB, Canada; Institute of Experimental Physiology (IFISE-CONICET), Faculty of Biochemical and Pharmacological Sciences, National University of Rosario, Suipacha, Rosario, Argentina

Marilyn D. Resh Cell Biology Program, Memorial Sloan Kettering Cancer Center, New York, NY, USA

Neale D. Ridgway Departments of Pediatrics and Biochemistry & Molecular Biology, Dalhousie University, Halifax, NS, Canada

Charles O. Rock Department of Infectious Diseases, St. Jude Children's Research Hospital, Memphis, TN, USA

Katherine M. Schmid Department of Biology, Butler University, Indianapolis, IN, USA

Wolfgang J. Schneider Department of Medical Biochemistry, Medical University of Vienna, Vienna, Austria

Laura J. Sharpe School of Biotechnology and Biomolecular Sciences, The University of New South Wales (UNSW Australia), Sydney, NSW, Australia

William L. Smith Department of Biological Chemistry, University of Michigan Medical School, Ann Arbor, MI, USA

Robert V. Stahelin Department of Biochemistry & Molecular Biology, Indiana University School of Medicine-South Bend, South Bend, IN, USA; Department of Chemistry & Biochemistry, University of Notre Dame, Notre Dame, IN, USA

Jennifer Taher The Hospital for Sick Children, University of Toronto, Toronto, ON, Canada

Changting Xiao University Health Network, University of Toronto, Toronto, ON, Canada

Zemin Yao Department of Biochemistry, Microbiology and Immunology, University of Ottawa, Ottawa, ON, Canada

Yong-Mei Zhang Department of Biochemistry and Molecular Biology, Medical University of South Carolina, Charleston, SC, USA

Preface

This represents the sixth edition of the textbook first edited by Drs. Dennis and Jean Vance in 1985. In taking on the editorship, we are revising the textbook that we used as an essential resource early in our academic careers. With this in mind we have strived to assemble a text that will inspire students to embrace the challenges of the future in lipid, lipoprotein and membrane biology.

Since the last edition that appeared in 2008 there have been remarkable advances in lipid and membrane biology in terms of identification of fundamental metabolic processes and their relationship to a broad spectrum of human diseases. The authors of this edition are at the forefront of these discoveries, and have provided chapters with a basic knowledge component coupled with unique insights into their respective fields of research. This edition also has a more defined focus on the impact on new technologies and relevance to chronic disease with a view to future studies of lipid metabolism. As with previous editions, the content is easily accessible to a broad spectrum of learners with a basic understanding of biochemistry and metabolism. The text also serves as a gateway to further exploration of topics, and provides a bridge between basic concepts and the current research literature. Thus undergraduate and graduate students will find this book to be an essential resource for course and research-related studies, while experienced researchers can use it as a reference guide to the lipid field.

All of the chapters have been revised from the fifth edition and new authors have taken on the task for many. We asked the authors to resist the temptation to be comprehensive; that we were not seeking to assemble a compendium of reviews. In addition, we limited the number of citations, attempting to glean classic and exceptional recent studies in each area. We are grateful that all the authors have complied. This edition of the book has full-color figures embedded in the text and standardized tables, which add visual appeal and clarity.

The contributors and editors assume full responsibility for the content and would appreciate any and all feedback for refinement of future editions.

Neale D. Ridgway and Roger S. McLeod
Halifax, Nova Scotia, Canada
June 2015

Chapter 1

Functional Roles of Lipids in Membranes

William Dowhan, Mikhail Bogdanov, Eugenia Mileykovskaya
Department of Biochemistry and Molecular Biology, University of Texas–Houston, Medical School, Houston, TX, USA

ABBREVIATIONS
CL Cardiolipin
DAG Diacylglycerol
EMD Extramembrane domain
GlcDAG Monoglucosyl diacylglycerol
GlcGlcDAG DIglucosyl diacylglycerol
NAO 10-N-nonyl acridine orange
PA Phosphatidic acid
PC Phosphatidylcholine
PE Phosphatidylethanolamine
PG Phosphatidylglycerol
PI Phosphatidylinositol
PS Phosphatidylserine
T_m Midpoint temperature
TMD Transmembrane domain

1. INTRODUCTION AND OVERVIEW

Lipids as a class of molecules display a wide diversity in structure and biological function. A primary role of lipids is to form the membrane bilayer permeability barrier of cells and organelles (Figure 1). Glycerophospholipids (termed phospholipids hereafter) make up about 75% of total membrane lipids of prokaryotic and eukaryotic cells, but other lipids are important components. Table 1 shows the major lipids found in the membranes of various cells and organelles but does not include the minor lipids, many of which are functionally important. Sterols are present in all eukaryotic cells and in a few bacterial membranes. The major sterol of mammalian cells is cholesterol whereas yeast contain ergosterol. Bacteria do not make sterols but some species incorporate sterols from the growth medium. Interestingly *Drosophila* also must acquire cholesterol from exogenous sources. The ceramide-based sphingolipids are present in the membranes of all eukaryotes. Neutral diacylglycerol (DAG) glycans are major membrane-forming

Biochemistry of Lipids, Lipoproteins and Membranes. http://dx.doi.org/10.1016/B978-0-444-63438-2.00001-8

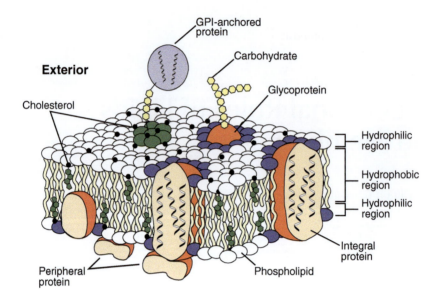

FIGURE 1   Model for membrane structure. This model of the plasma membrane of a eukaryotic cell is an adaptation of the original model proposed by Singer and Nicholson (1972). The phospholipid bilayer is shown with integral membrane proteins largely containing α-helical transmembrane domains (TMDs). Peripheral membrane proteins associate either with the lipid surface or with other membrane proteins. Lipid rafts (dark *green* head groups) are enriched in cholesterol and contain a PI glycan-linked (GPI) protein. The *purple* head groups depict lipids in close association with protein. The irregular surface and wavy acyl chains denote the fluid nature of the bilayer.

components in many Gram-positive bacteria and in the membranes of plants, while Gram-negative bacteria utilise a saccharolipid (Lipid A) as a major structural component of the outer leaflet of the outer membrane. The variety of hydrophobic domains of lipids results in additional diversity. In eukaryotes and eubacteria these domains are saturated and unsaturated fatty acids or lesser amounts of fatty alcohols; many Gram-positive bacteria also contain branched chain fatty acids. Instead of esterified fatty acids, *Archaea* contain long chain reduced polyisoprene moieties in ether linkage to glycerol. Such hydrophobic domains are highly resistant to the harsh environment of these organisms. Further stability of the lipid bilayer of *Archaea* comes from many of the hydrocarbon chains spanning the membrane with covalently linked head groups at each end. If one considers a simple organism such as *Escherichia coli* with three major phospholipids and several different fatty acids along with many minor precursors and modified lipid products, the number of individual phospholipid species ranges in the hundreds. In more complex eukaryotic organisms with greater diversity in both phospholipids and fatty acids, the number of individual species is in the thousands. Sphingolipids also show a similar degree of diversity and

TABLE 1 Lipid Composition of Various Biological Membranes

Lipid	Erythrocyte[a]	CHO Cells[b]	Mitochondria[c]		Endoplasmic Reticulum[d]	*Escherichia coli*[e]
			Outer	Inner		
Cholesterol	25	–	N.D.	N.D.	20	N.D.
PE	18	21	33	24	21	75
PC	19	51	46	38	46	N.D.
Sphingomyelin	18	9	–	–	9	N.D.
PS	9	7	1	4	2	<1
PG	0	1	N.D.	N.D.	–	20
CL	0	2.3	6	16	–	5
PI	1	8	10	16	2	N.D.
Glycosphingolipid	10	–	–	–	–	N.D.
PA	–	1	4	2	–	<2

The data are expressed as mol% of total lipid. N.D. indicates not detected and blank indicates not analysed.
[a]*Human (Tanford, 1980).*
[b]*Chinese hamster cells (Ohtsuka et al., 1993).*
[c]Saccharomyces cerevisiae *inner and outer mitochondrial membrane (Zinser et al., 1991).*
[d]*Murine L cells (Murphy et al., 2000).*
[e]*Inner and outer membrane excluding Lipid A (Raetz, 1990).*

when added to the steroids the size of the eukaryotic lipidome dwarfs that of the proteome.

Lipids provide the solvent within which integral membrane proteins (those whose transmembrane domains (TMDs) span the bilayer) are integrated. Peripheral proteins also interact with the membrane surface and are even found partially inserted into the lipid bilayer. These amphitropic proteins are found in the aqueous compartments of cells and interact with the membrane surface in a reversible manner. The lipid bilayer provides a rich and varied environment for proteins, which includes a highly hydrophobic interior bounded by the hydrophilic and/or charged lipid head groups. The latter organises water and counterions in a manner significantly differently from that of the cell aqueous phase, which imparts distinct properties to the aqueous layer in close contact with the membrane surface. Each lipid molecular class is made up of a wide spectrum of chemical and structural variants, which as an ensemble determine membrane fluidity, lateral pressure, permeability and surface charge. The lipid and protein components of the membrane are not held together by covalent interactions and therefore are in dynamic equilibrium undergoing transient interactions organised into the supermolecular structure of the lipid bilayer.

In this chapter, the diversity in structure, chemical properties and physical properties of lipids will be outlined. The various genetic approaches available for studying lipid function in vivo will be summarised. Finally, how the physical and chemical properties of lipids relate to their multiple functions in living systems will be reviewed to provide a molecular basis for the diversity of lipid structures in natural membranes. Due to space limitations, recent review articles and research articles, which contain the primary background references supporting the summaries in the text, are cited.

2. DIVERSITY IN LIPID STRUCTURE

Lipids are defined as those biological molecules readily soluble in organic solvents such as chloroform, ether or toluene. However, many peptides and some very hydrophobic proteins are soluble in organic solvents, and lipids with large hydrophilic domains such as saccharolipids are not soluble in these solvents. Here we will consider those lipids that contribute significantly to membrane structure or have a role in determining protein structure or function. The LIPID MAPS consortium (http://www.lipidmaps.org) in the United States, Lipid Bank (http://www.lipidbank.jp) in Japan and the LipidomicNet (http://www.lipidomicnet.org) in Europe have cooperated to devise classification systems, methodology and forums for the benefit of researchers.

2.1 Glycerolipids

The DAG backbone in eubacteria and eukaryotes is *sn*-3-glycerol (L-glycerol) esterified at the 1- and 2-position with long chain fatty acids (Figure 2) (Chapters 3 and 7). In *Archae* (Figure 3) *sn*-1-glycerol (D-glycerol) forms the backbone and the hydrophobic domain is composed of phytanyl (saturated isoprenyl) groups in ether linkage at the 2- and 3-positions (an archaeol) (Koga and Morii, 2007). In addition, two *sn*-1-glycerol groups are connected in ether linkage by two biphytanyl groups (dibiphytanyldiglycerophosphatetetraether) to form a covalently linked bilayer. Some eubacteria (mainly hyperthermophiles) have dialkyl (long chain alcohols in ether linkage) phospholipids and similar ether linkages are found in the plasmalogens of eukaryotes. The head groups of the phospholipids (boxed area of Figure 2) extend the diversity of lipids defining phosphatidic acid (PA, with OH), phosphatidylcholine (PC), phosphatidylserine (PS), phosphatidylglycerol (PG), phosphatidylinositol (PI) and cardiolipin (CL). *Archae* analogues exist with head groups of glycerol and glyceromethylphosphate as well as all of the above except PC. *Archae* also have neutral glycan lipid derivatives in which mono- and disaccharides (glucose or galactose) are directly linked to the *sn*-1 position of archaeol (Figure 3). Plants (mainly in the thylakoid membrane) and many Gram-positive bacteria also have high levels of neutral DAG glycans

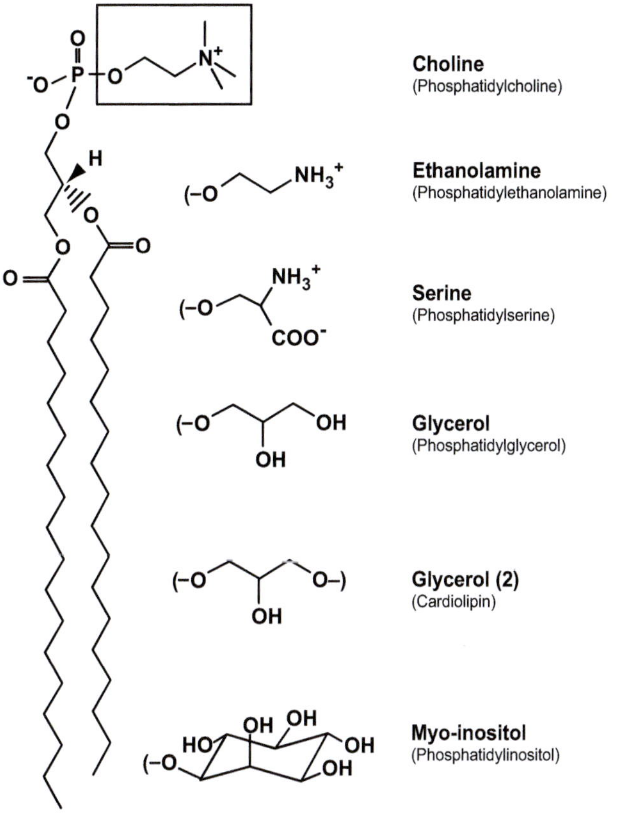

FIGURE 2 Structure of glycerophosphate-based lipids. The lipid structure shown is 1,2 distearoyl-*sn*-glycerol-3-phosphocholine or phosphatidylcholine (PC). Substitution of choline in the box with the head groups shown on the right forms the other phospholipid structures. Cardiolipin (CL) is also referred to as diphosphatidylglycerol since it contains two PAs joined by a glycerol.

with mono- or disaccharides linked to the 3-carbon of *sn*-3-glycerol (Chapter 4). In addition to head group diversity, a range of alkyl chains are attached to the glycerol moiety. In eubacteria, fatty acid chain lengths vary from 12 to 18 carbons and can be fully saturated or contain double bonds. Some Gram-positive bacteria contain odd-numbered, branched chain fatty acids rather than unsaturated fatty acids. Eukaryotic lipids contain fatty acid chains up to 26 carbons in length with multiple or no double bonds. Therefore, the diversity of glycerol-based lipids in a single organism is significant, but the diversity throughout nature is enormous.

The majority of information on the chemical and physical properties of lipids comes from studies on the major phospholipid classes of eubacteria and eukaryotes with only limited information on the lipids from *Archae*. The biosynthetic pathways and the genetics of lipid metabolism have also been extensively studied in eubacteria (Chapter 3) and eukaryotes (Chapter 7). Clearly the archaeol lipids confer some advantage with respect to the environment of

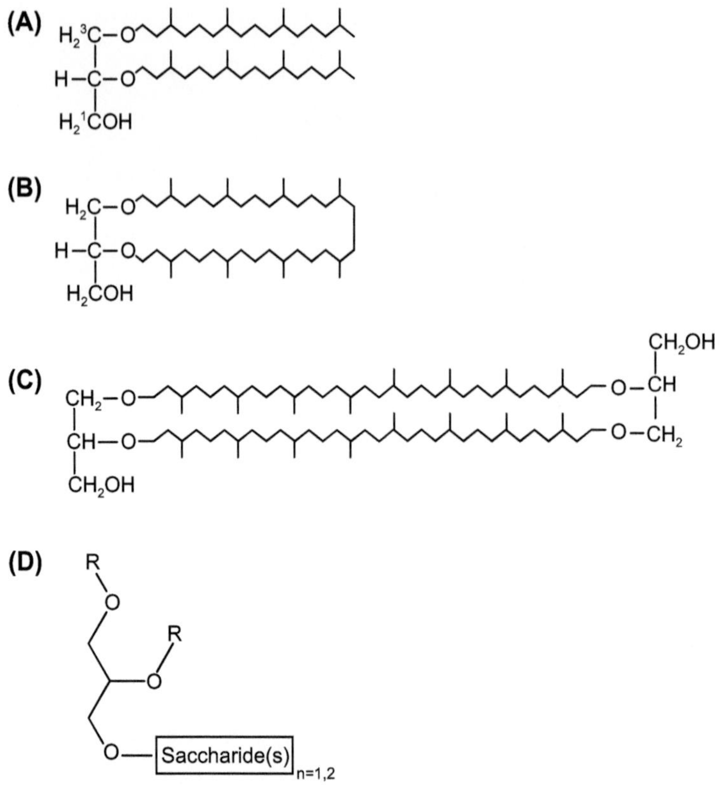

FIGURE 3 Structure of dialkylglycerols in *Archae*. *Archae* have phytanyl chains in ether linkage to the 2- and 3-positions of *sn*-1-glycerol (archaeol). The 1-position can be derivatised with phosphodiesters. (A) Diphytanylglycerol (C20–C20 diether) with the stereochemistry of glycerol indicated. (B) Cyclic biphytanyl (C40) diether. (C) Biphytanyl diglycerol diether. (D) A glyceroglycan with either a mono or a disaccharide (glucose or galactose) at the 1-position of *sn*-1-glycerol. The R groups are ether-linked phytanyl chains. Similar glyceroglycans are found in eubacteria and plants with a *sn*-3-glycerol backbone and ester-linked fatty acid chains at the 1- and 2-positions.

archaebacteria. Interestingly, the pathways for phospholipid biosynthesis in eubacteria and *Archaea* are very similar even though their lipids differ in chirality of the glycerol backbone. How the physical properties of the more commonly studied lipids change with environment will be discussed later.

2.2 Saccharolipids

The outer membrane of Gram-negative bacteria (Figure 4) contains lipopolysaccharide or endotoxin, which is a lipid made up of a backbone derived from glucosamine phosphate (Whitfield and Trent, 2014) rather than glycerophosphate. The core lipid (Lipid A, see Figure 5) in *E. coli* contains two glucosamine groups in β 1–6 linkage that are decorated at positions 2, 3, 2′ and 3′ with *R*-3-hydroxymyristic

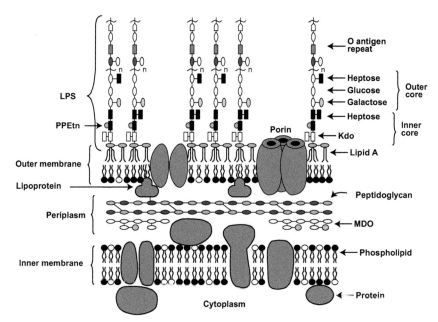

FIGURE 4 *Escherichia coli* cell envelope. The complete cell envelope of Gram-negative bacteria contains an inner phospholipid bilayer and is the permeability barrier of the cell. The outer membrane is composed of an inner monolayer of phospholipid and an outer monolayer of the Lipid A portion of lipopolysaccharide (LPS). The structure of KDO_2–Lipid A is shown in Figure 5 and is connected to a polysaccharide to build up the inner core, outer core and the O antigen repeat of LPS. The outer membrane is a permeability barrier allowing molecules of less than 750–1000 Da to pass through various pores in the outer membrane. The periplasmic space contains many proteins and the membrane-derived oligosaccharide (MDO), which is one component of the osmolarity regulatory system. MDO is decorated with *sn*-glycerol-1-phosphate and ethanolamine phosphate derived from PG and PE, respectively. The amino acid–sugar cross-linked peptidoglycan gives structural rigidity to the cell envelope. One-third of the lipoprotein (*lpp* gene product) pool is covalently linked via its carboxyl terminus to the peptidoglycan and, in complex with the remaining lipoproteins as trimers, associates with the outer membrane via covalently linked fatty acids at its amino terminus. The amino terminal cysteine is blocked with a fatty acid, derived from membrane phospholipids, in amide linkage and is derivatised with diacylglycerol, derived from PG, in thioether linkage. *Figure provided by and reproduced with permission from C.R. Raetz.*

acid (C14) and at positions 1 and 4′ with phosphates. Further modification at position 6′ with a KDO disaccharide (two 3-deoxy-D-*manno*-octulosonic acids in a 1–3 linkage) results in KDO_2–Lipid A that is further modified by an inner core, an outer core and the O antigen (Figure 4). Laboratory strains of *E. coli* such as K-12 and *Salmonella typhimurium* lack the O antigen found in the wild-type and clinically important strains.

The core Lipid A forms the outer monolayer of the outer membrane bilayer of Gram-negative bacteria with the inner monolayer (Figure 4), which is made up of phospholipids (about 90% phosphatidylethanolamine (PE)). Lipopolysaccharide is modified postassembly in response to the environment including growth media,

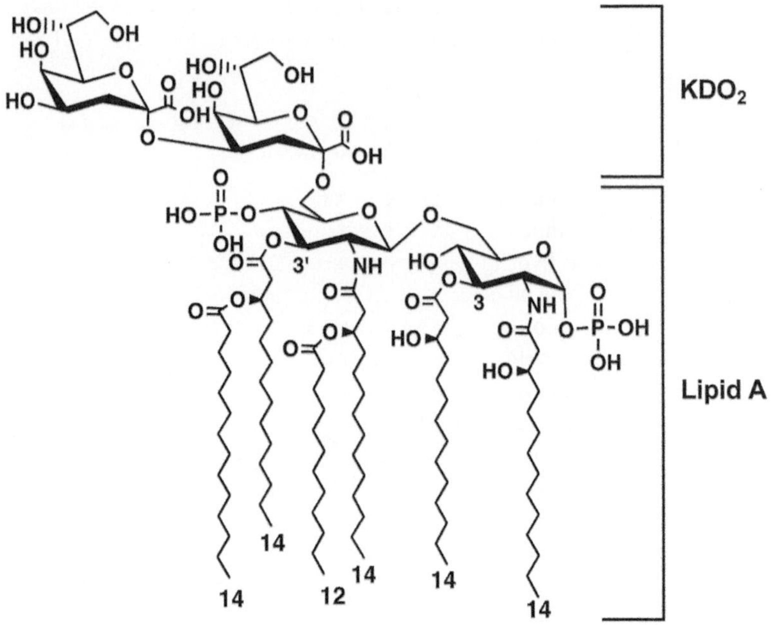

FIGURE 5   Structure of KDO$_2$–Lipid A. Lipid A is a disaccharide of glucosamine phosphate that is multiply acylated with both amide and ester linkages to fatty acids of the chain lengths indicated. As illustrated in Figure 4, Lipid A is attached to KDO$_2$ that is then elongated with the remainder of the lipopolysaccharide structure. *Figure provided by and reproduced with permission from C.R. Raetz.*

temperature, ionic properties and antimicrobial agents and displays additional diversity among enteric and nonenteric Gram-negative bacteria. Studies on Lipid A are of clinical importance because it is the primary antigen responsible for toxic shock syndrome.

2.3 Sphingolipids

All eukaryotic cells contain sphingolipids derived from the condensation of pal-mitoyl-coenzyme A and serine followed by slightly different species-specific con-version to the core ceramide molecule (Chapter 10). In higher eukaryotes there is additional diversity of the long chain base derived from palmitate with additional double bonds and hydroxyl groups as well as considerable diversity in the fatty acid in amide linkage, which can range up to 26 carbons in length. Yeast con-tain mainly derivatives of phytoceramide (4-hydroxy ceramide) and C26 fatty acid chains in amide linkage. The major classes of sphingolipids are grouped according to what is esterified at the primary hydroxyl β to the amide carbon of ceramide. Sphingomyelin has choline phosphate at this position whereas the glycosphingolipids have various lengths of oligosaccharides. The acidic glyco-sphingolipids, found primarily in mammalian cells, contain either sulphated sug-ars (sulphatide) or sialic acid (gangliosides) in the terminal sugar position. Yeast sphingolipids contain inositol phosphate and mannose inositol phosphate linked at this hydroxyl. Although the synthesis of sphingolipids occurs in the endoplasmic reticulum and the Golgi, they are primarily found in the outer leaflet of the plasma membrane.

3. PROPERTIES OF LIPIDS IN SOLUTION

The matrix that defines a biological membrane is a lipid bilayer composed of a hydrophobic core excluded from water and an ionic surface that interacts with water and defines the hydrophobic–hydrophilic interface (Figure 1). Much of our understanding of the physical properties of lipids in solution and the driving force for the formation of lipid bilayers comes from the concept of the 'hydrophobic bond' as described by Walter Kauzmann (Kauzmann, 1959) in the context of the forces driving protein folding and later extended as the 'hydrophobic effect' by Charles Tanford (Tanford, 1980) to explain self-association of lipids within biological systems. The Tanford book is a must read for anyone wishing to work with membrane components. The 'fluid mosaic' model for membrane structure further popularised these concepts (Singer and Nicolson, 1972). This model envisioned membrane proteins as undefined globular structures freely moving in a homogeneous sea of lipids. Although this model stimulated research in the area of membrane proteins, it relegated lipids to a monolithic role as a fluid matrix within which membrane proteins reside and function. As will be become apparent, our current understanding of the role of lipids in cell function is as specific and dynamic as that of proteins, which are now more precisely defined with respect to their structure and interaction with lipids.

3.1 Why Do Polar Lipids Self-Associate?

Polar lipids are amphipathic in nature containing both hydrophobic domains that do not interact with water and hydrophilic domains that readily interact with water. The basic premise of the hydrophobic effect (Tanford, 1980) is that the hydrocarbon domains of polar lipids distort the stable hydrogen bonded structure of water by inducing ordered cage-like structures around the apolar domains. Self-association of the hydrophobic domains minimises the total surface area in contact with water, resulting in entropy-driven relaxation of the water structure and an energy minimum for the final self-associated molecular organisation. The polar domains of lipids interact through either hydrogen bonding or ionic interaction with water or other lipid head groups and therefore are energetically stable in an aqueous environment. The structural organisation that a polar lipid assumes in water is determined by its concentration and the law of opposing forces, that is, hydrophobic forces driving self-association of hydrophobic domains versus steric and ionic repulsive forces of the closely associated polar domains opposing self-association. At low concentrations, amphipathic molecules exist as monomers in solution. As the concentration of the molecule increases, its stability in solution as a monomer decreases until the unfavourable repulsive forces of closely packed polar domains are outweighed by the favourable self-association of the hydrophobic domains. At this point, a further increase in concentration results in the formation of increasing amounts of self-associated monomers in equilibrium with a constant amount of free monomer. This point of self-association and the remaining constant free monomer concentration is the critical micelle concentration. Due to the increased hydrophobic effect, a larger hydrophobic domain results in a lower critical micelle concentration. However, the larger the polar domain, because of either the size of neutral domains or the charge repulsion for like-charged ionic domains, the higher the critical micelle concentration due to the unfavourable steric hindrance or charge repulsion in bringing these domains into close proximity. The critical micelle concentration of amphipathic molecules with a net charge is lowered by increasing ionic strength of the medium due to dampening of the charge repulsion effect. Addition of chaotropic agents, such as urea, that disrupt water structure or organic solvents that lower the dielectric constant of water raises the critical micelle concentration by stabilising the hydrophobic domain in an aqueous environment. Therefore, the critical micelle concentration of the detergent sodium dodecyl sulphate is reduced 10-fold when the NaCl concentration is raised from 0 to 0.5 M but is increased on addition of urea or ethanol.

These physical properties and the shape of amphipathic molecules define three supermolecular structural organisations of polar lipids and detergents in solution (Figure 6). Detergents, lysophospholipids (containing only one alkyl chain) and phospholipids with short alkyl chains (eight or fewer carbons) have an inverted cone shape (large head group relative to a small hydrophobic domain) and self-associate above the critical micelle concentration with a small radius of curvature to form micellar structures with a hydrophobic core, excluding water. The micelle surface, rather than being a smooth spherical or elliptical structure with the hydrophobic domains completely sequestered inside a shell of polar residues that interact with water, is a very rough surface with many of the hydrophobic domains exposed to water. The overall structure reflects the optimal packing of amphipathic molecules at an energy minimum by balancing the attractive force of the hydrophobic effect and the repulsive force of close head group association. The critical micelle concentration for most detergents ranges from micromolar to millimolar. Lysophospholipids also form micelles with critical micelle concentrations in the micromolar range. However, phospholipids with chain lengths of 16 self-associate at a concentration around 10^{-10} M due to the hydrophobic driving force contributed by two alkyl chains. Phospholipids with long alkyl chains do not form micelles but organise into bilayer structures, which allow tight packing of adjacent side chains with the maximum exclusion of water from the hydrophobic domain. Due to repulsive forces of the head groups, significant lengths of the hydrocarbon chain near the glycerol backbone are exposed to water. In living cells phospholipids are not found free as monomers in solution but are organised into membrane bilayers or found complexed with proteins.

When long chain phospholipids are first dried to a solid from organic solvent and then hydrated, they spontaneously form large multilamellar bilayer sheets

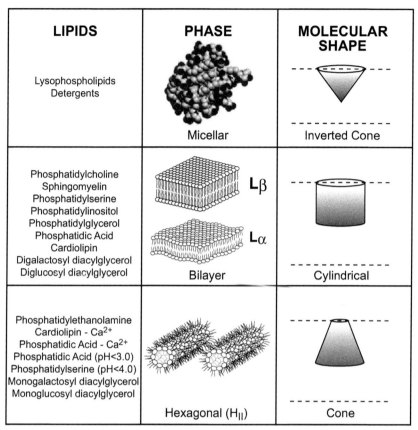

LIPIDS	PHASE	MOLECULAR SHAPE
Lysophospholipids Detergents	Micellar	Inverted Cone
Phosphatidylcholine Sphingomyelin Phosphatidylserine Phosphatidylinositol Phosphatidylglycerol Phosphatidic Acid Cardiolipin Digalactosyl diacylglycerol Diglucosyl diacylglycerol	Lβ Lα Bilayer	Cylindrical
Phosphatidylethanolamine Cardiolipin - Ca^{2+} Phosphatidic Acid - Ca^{2+} Phosphatidic Acid (pH<3.0) Phosphatidylserine (pH<4.0) Monogalactosyl diacylglycerol Monoglucosyl diacylglycerol	Hexagonal (H$_{II}$)	Cone

FIGURE 6   Polymorphic phases and molecular shapes exhibited by lipids. The space-filling model for the micellar phase is of the β-D-octyl glucoside micelle (50 monomers). The polar portions of the detergent molecules (oxygen atoms are *black*) do not completely cover the micelle surface (hydrocarbons in *grey*), leaving substantial portions of the core exposed to bulk solvent. Inverted cone-shaped molecules form micelles. Polar lipids with two long alkyl chains adopt a bilayer or nonbilayer (H$_{II}$) structure, depending on the geometry of the molecule (cylindrical or cone shaped, respectively) and environmental conditions. The L$_\beta$ (ordered gel) and L$_\alpha$ (liquid crystalline) bilayer phases differ in the order within the hydrophobic domain and in mobility of the individual molecules.

separated by water. Sonication disperses these sheets into smaller unilamellar bilayer closed structures that satisfy the hydrophobic nature of the ends of the bilayer sheets by forming sealed vesicles (also termed liposomes) defined by a continuous single bilayer and an aqueous core much like the membrane surrounding cells. Liposomes can also be made by physical extrusion of multilamellar structures through a small orifice or by dilution of a detergent–lipid mixture below the critical micelle concentration of the detergent (Patil and Jadhav, 2014).

Cylindrical-shaped lipids (head group and hydrophobic domains of similar diameter), such as PC, form lipid bilayers. Cone-shaped lipids (small head groups relative to a large hydrophobic domain) such as PE containing at least one unsaturated fatty acid or CL in the presence of divalent cations favour an inverted micellar structure where the head groups sequester an internal aqueous core and the hydrophobic domains are oriented outwards and self-associate in nonbilayer structures. These are denoted as the hexagonal II (H_{II}) and cubic phases (a more complex organisation similar to the H_{II} phase). The ability of lipids to form multiple structural associations is referred to as lipid polymorphism. Lipids such as PE, PA, CL and monosaccharide derivatives of DAG can exist in either a bilayer or a nonbilayer phase depending on solvent conditions, alkyl chain composition and temperature. These phases are governed by the packing geometry of the hydrophilic and hydrophobic domains on self-association as discussed below.

Both cone-shaped and inverted cone-shaped lipids are considered nonbilayer-forming lipids and when mixed with bilayer-forming lipids change the physical properties of the bilayer by introducing lateral stress or strain within the bilayer structure. When bilayer-forming lipids are spread as a monolayer at an aqueous–air interface they orient with the hydrophobic domain facing air, and they have no tendency to bend away from or towards the aqueous phase due to their cylindrical symmetry. Monolayers of the asymmetric cone-shaped lipids (H_{II} forming) tend to bend towards the aqueous interface (negative radius of curvature) while monolayers of asymmetric inverted cone-shaped lipids (micelle-forming) tend to bend away from the aqueous phase (positive radius of curvature). The distribution of lipids with different geometrical shapes between monolayers may determine the overall curvature of a bilayer membrane. The significance of shape mismatch in lipid mixtures is covered below.

3.2 Physical Properties of Membrane Bilayers

The organisation of DAG-containing polar lipids in solution is dependent on the nature of the alkyl chains, the head groups and the solvent conditions (i.e. ion content, pH and temperature). The transition between these phases for pure lipids in solution can be measured by various physical techniques such as ^{31}P NMR and microcalorimetry. The difference between the ordered gel (L_β) and the liquid crystalline (L_α) phases (Figure 6) is the viscosity or fluidity of the hydrophobic domains of the lipids, which is a function of temperature and the alkyl chain structure. At any given temperature, the 'fluidity' (the inverse of the viscosity) of the hydrocarbon core of the bilayer increases with increasing content of unsaturated or branched alkyl chain or with decreasing alkyl chain length. Due to the increased mobility of the fatty acid chains with increasing temperature, the fluidity and space occupied by the hydrophobic domain of lipids increases, which also tends to move the head groups apart. A bilayer-forming lipid such as PC assumes a cylindrical shape over a broad temperature range and with different alkyl chain compositions. When analysed in pure form, PC exists in either the L_β or the L_α phase, mainly dependent on its alkyl chain composition and the temperature (Figure 6). Nonbilayer-forming lipids such as

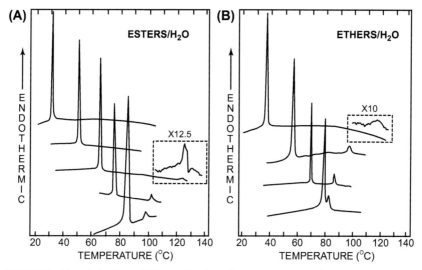

FIGURE 7 Phase behaviour of PE as a function of temperature and chain length. As hydrated lipids pass through a phase transition heat is absorbed as indicated by the peaks. The large peaks at the lower temperatures are due to the L_β to L_α transition and the smaller peaks at higher temperatures are due to the L_α to H_{II} transition. (A) Even numbered diacyl-PEs ranging from C12 to C20 (top to bottom). (B) Even numbered dialkyl-PEs in ether linkage ranging from C12 to C18 (top to bottom). The insets indicate an expanded scale for the transition to H_{II}. *Reprinted with permission from Seddon et al. (1983). Copyright 1983 American Chemical Society.*

PE exist at low temperatures in the L_β phase, at intermediate temperatures in the L_α phase and at elevated temperatures in the H_{II} or cubic phase (Figure 7). The last transition is temperature dependent but also dependent on the shape of the lipid. The supermolecular organisation of lipids with relatively small head groups can change from cylindrical to conical (H_{II} phase) with increasing unsaturation or length of the alkyl chains or with increasing temperature. As can be seen from Figure 7, the midpoint temperature (T_m) of the transition from the L_β to L_α phase increases with an increase in the length of the fatty acids, but the midpoint of the transition temperature (T_{LH}) between the L_α and the H_{II} phases decreases with increasing chain length (or increasing unsaturation, not shown).

Similar transition plots as well as complex phase diagrams have been generated for mixtures of lipids. The physical property of a lipid mixture is collectively determined by each of the component lipids. T_m of biological membranes depends on its lipid composition. Liposomes composed of a single lipid exhibit a sharp phase transition while complex biological membranes display a broader transition with lipid composition finely tuned to reduce the sharpness of the transition. A large number of studies indicate that the L_α state of the membrane bilayer is required for cell viability, and cells adjust their lipid composition in response to many environmental factors so that the collective property of the membrane exhibits the L_α state. Cells regulate either their lipid fatty acid composition and/or their hydrophilic domain composition to maintain overall constant bilayer properties. Addition of nonbilayer-forming lipids to bilayer-forming lipids can result in nonbilayer formation but at a higher temperature than that for the pure nonbilayer-forming lipid. Addition of nonbilayer-forming lipids also adds other parameters of tension between the two monolayers and lateral stress within each monolayer. These lipids in each leaflet of the bilayer tend to reduce the radius of curvature of each monolayer that results in a tendency to pull the bilayer apart by curving the monolayers away from each other. This process results in potential energy residing in the bilayer that is a function of the presence of nonbilayer lipids. Forcing nonbilayer-forming lipids into a bilayer structure also exposes the hydrophobic core of the nonbilayer-forming lipids to the aqueous phase due to increasing lateral stress, which when relieved by insertion of proteins into the bilayer results in a release of free energy. Mixtures of lipids with dissimilar phase properties can also generate phase separations with local domain formation. Such discontinuities in the bilayer structure may be required for many structural organisations and cellular processes such as accommodation of proteins into the bilayer, movement of macromolecules across the bilayer, cell division and membrane fusion and fission events. The need for bilayer discontinuity may be the reason that all natural membranes contain a significant proportion of nonbilayer-forming lipids even though the membrane under physiological conditions is in the L_α phase.

Addition of cholesterol to lipid mixtures has a profound effect on the physical properties of a bilayer. Increasing amounts of cholesterol inhibit the organisation of lipids into the L_β phase and favour a less fluid but more ordered structure than that of the L_α phase resulting in the lack of a phase transition normally observed in the absence of cholesterol. The solvent surrounding the lipid bilayer also influences these transitions, primarily by affecting the size of the head group relative to the hydrophobic domain (Figure 6). Ca^{2+} and other divalent cations (Mg^{2+}, Sr^{2+} but not Ba^{2+}) reduce the effective size of the negatively charged head groups of CL and PA, thus endowing nonbilayer properties. Low pH has a similar effect on the head group of PS. Since Ca^{2+} is an important signalling molecule that elicits many cellular responses, it is possible that part of its effect is transmitted through changes in the physical properties of membranes. In eukaryotes CL is found exclusively in the inner membrane (and to lesser extent in the outer membrane) of the mitochondria where Ca^{2+} fluxes play important regulatory roles.

3.3 What Does the Membrane Bilayer Look Like?

A primary role of biological membranes is to define the limits of a cell or organelle by maintaining a controlled permeability barrier to small polar and charged molecules (O_2, CO_2, H_2O, H^+, K^+, HCO_3^-, Mg^{2+}, Ca^{2+}, etc.) as well as macromolecules. A second role is to provide the solvent and surface in which many essential biological processes are organised. Amphipathic lipid molecules are

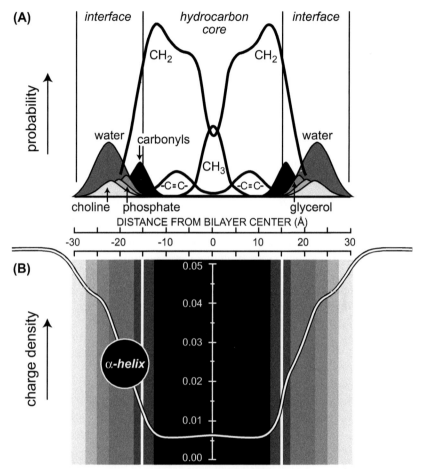

(A)

interface   *hydrocarbon core*   *interface*

probability

CH$_2$ CH$_2$

water   carbonyls   water

CH$_3$

-C=C-   -C=C-

choline   phosphate   glycerol

DISTANCE FROM BILAYER CENTER (Å)

-30   -20   -10   0   10   20   30

(B)

charge density

0.05

0.04

0.03

α-*helix*   0.02

0.01

0.00

FIGURE 8   The probability distribution for chemical constituents across a bilayer of PC. (A) The diagram was generated from X-ray and neutron diffraction data. The interface region between the hydrocarbon core and the free solvent region extends for approximately 15 Å on either side of the 30-Å-thick bilayer. The width of each peak defines the mobility of each constituent of PC. (B) As an α-helical peptide moves from either side of the bilayer towards the centre, the charge density of the environment steeply declines as indicated by the line. *Figure adapted from White et al. (2001). Copyright 2001 The American Society for Biochemistry and Molecular Biology.*

organised into a flexible noncovalently associated supermolecular structure that optimally satisfies these requirements. The functional properties of natural fluid bilayers are not only influenced by the hydrophobic core and the hydrophilic surface but by also the interface region containing bound water and ions. Figure 8(A) shows the distribution of the component parts of dioleoyl-PC across the bilayer (White et al., 2001) and illustrates the dynamic rather than static nature of the membrane. The length of the fatty acid chains defines the bilayer thickness of 30 Å for the above phospholipid. However, the thickness is not a static number as indicated by the probability of finding CH_2 residues randomly distributed over a range of distances. Bilayer thickness can vary over the surface of a membrane if microdomains of lipids are formed with different alkyl chain lengths. The width ($15\,\text{Å}$ on either side of the bilayer) of the interface region between the hydrocarbon core and the free water phase (Figure 8(B)) is generally not appreciated. This region contains a complex mixture of chemical species defined by the ester linked glycerophosphate moiety, the variable head groups, bound water and ions. The many biological processes that occur within this interface region are dependent on its unique properties, including the steep polarity gradient within which surface-bound cellular processes occur.

4. ENGINEERING OF MEMBRANE LIPID COMPOSITION

Given the diversity in both lipid structure and function, how can the role of a given lipid be defined at the molecular level? Unlike proteins, lipids have neither inherent catalytic activity nor obvious functions in isolation (except for ligand–receptor interactions). Many potential functions of lipids have been uncovered serendipitously based on their effect on catalytic processes or biological functions studied in vitro. Although considerable information has accumulated with this approach, such studies are prone to artefacts. In addition, many of these functions have not been verified in living cells. The physical properties of lipids are as important as their chemical properties in determining function. Yet there is little understanding of how the physical properties of lipids measured in vitro relate to their in vivo function. Genetic approaches are generally the most useful for identifying in vivo function, but this approach has considerable limitations when applied to lipids. First, genes do not encode lipids, and in order to make mutants with altered lipid composition, the genes encoding enzymes along a biosynthetic pathway must be targeted. Therefore, the results of genetic mutation are indirect and many times far removed from the primary lesion. Second, a primary function of major membrane lipids is to provide the permeability barrier of the cell. Therefore, alterations in lipid composition may compromise cell permeability before other functions of a particular lipid are uncovered. Genetic approaches might strongly suggest that a lipid is essential for cell viability, but due to compromising cell viability the precise molecular basis for the requirement might be difficult to assess. The challenge is to use genetic information to manipulate the lipid composition of cells without severely compromising cell viability. In cases where this has been possible, the combination of the genetic approach to uncover phenotypes of cells with altered lipid composition, and the dissection in vitro of the molecular basis for the phenotype, has proven to be a powerful approach for defining lipid function. The most useful information to date has come from genetic manipulation of prokaryotic cells and eukaryotic microorganisms. However, the basic molecular principles underlying lipid function are generally applicable to more complex mammalian systems.

4.1 Alteration of Lipid Composition in Bacteria

The pathways for formation of the major phospholipids (PE, PG and CL) of *E. coli* (Chapter 3) were biochemically established mainly by Eugene Kennedy and coworkers (Figure 9) and subsequently verified using genetic approaches. The design of strains in which lipid composition can be genetically altered in a systematic manner has been very important in defining new roles for lipids in cell function (see Dowhan (2013) for a summary).

Surprisingly, *E. coli* mutants completely lacking either PE and PS or PG and CL are viable. Null mutants in the *pgsA* gene that cannot synthesise PG and CL are lethal, but suppressors of this mutation have been identified. In such mutants the major outer membrane lipoprotein precursor (Figure 4), which depends on PG for its lipid modification, accumulates in the inner membrane and apparently kills the cell. Mutants lacking this lipoprotein are viable without PG and CL but are temperature sensitive for growth at 42 °C, indicating that PG and CL are not absolutely required for viability, only for optimal growth. However,

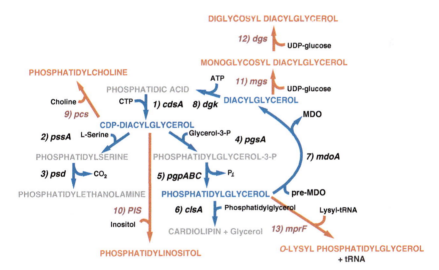

FIGURE 9 Synthesis of native and foreign lipids in *Escherichia coli*. The native pathways (*blue arrows*), native lipids (*blue and grey*) and the respective gene names by the numbers are shown. 1) CDP-diacylglycerol synthase; 2) phosphatidylserine synthase; 3) phosphatidylserine decarboxylase; 4) phosphatidylglycerophosphate synthase; 5) phosphatidylglycerophosphate phosphatase encoded by three genes; 6) cardiolipin synthase (also encoded by *clsB* and *clsC*, which utilise PE and an unknown second substrate, respectively, rather than a second PG); 7) phosphatidylglycerol:premembrane-derived oligosaccharide (MDO) *sn*-glycerol-1-*P* transferase; 8) diacylglycerol kinase. The following enzymes and their respective gene names and sources noted have been expressed in *E. coli* and synthesise the indicated products (*red arrows*); 9) phosphatidylcholine synthase (*Legionella pneumophila*); 10) phosphatidylinositol synthase (*Saccharomyces cerevisiae*); 11) monoglucosyl diacylglycerol synthase (*Acholeplasma laidlawii*); 12) diglucosyl diacylglycerol synthase (*Acholeplasma laidlawii*); 13) lysyl t-RNA:phosphatidylglycerol lysine transferase (*Staphococcus aureus*). *Figure adapted from Dowhan (2013). Copyright 2013 Elsevier Press.*

the anionic nature of these lipids (apparently substituted by increased levels of PA) is necessary for the proper membrane association and function of peripheral membrane proteins as discussed later. Mutants of *E. coli* lacking the amine containing lipids PE (*psd* null) or PS and PE (*pssA* null) are viable when grown in the presence of millimolar concentrations of Ca^{2+}, Mg^{2+} and Sr^{2+} but have a complex mixture of defects in cell division, growth rate, outer membrane barrier function, energy metabolism and assembly of some membrane proteins (mainly including sugar and amino acid transporters).

The plasticity in specific lipid requirements for viability of *E. coli* has been utilised to design strains in which phospholipid composition can be varied at steady state as well as temporally over the growth cycle of a culture. Such variability is attained through a combination of null mutants and placing specific genes under the control of regulated promoters. In addition, incorporation of genes from other organisms has made possible the expression of foreign lipids in *E. coli* in place of the naturally occurring phospholipids. This collection of mutants (Figure 9) has been instrumental in defining specific roles for some of the phospholipids of *E. coli*. Since the foreign lipids display an array of physical and chemical properties (Figure 10), their ability to suppress the phenotype of mutants lacking specific natural lipids has been used to establish which

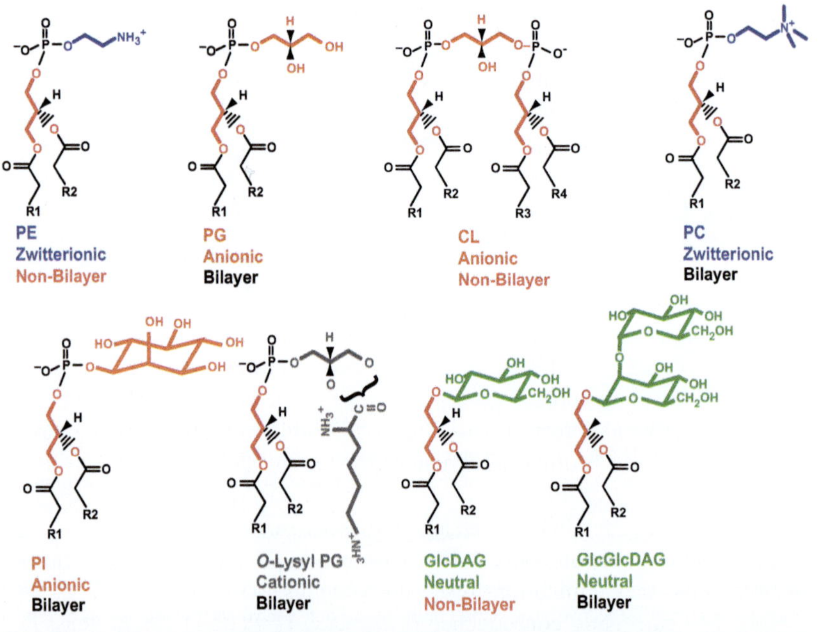

FIGURE 10   Structure and physical properties of lipids. The glycerol backbone (*red*) shown in ester linkage to fatty acids (aliphatic chains R_1–R_4) at the *sn*-1 and *sn*-2 positions and either in phosphodiester linkage for phospholipids or in a glycosidic linkage for glycolipids at the *sn*-3 position. Head groups are colour coded to indicate the charge nature of each head group. *Figure and legend taken from Dowhan (2013). Copyright 2013 Elsevier Press.*

property of a given lipid is necessary to support a particular cellular function (Dowhan, 2013).

4.2 Alteration of Lipid Composition in Yeast

The pathways of phospholipid synthesis and the genetics of lipid metabolism in the yeast *Saccharomyces cerevisiae* (Henry et al., 2012) are as well understood as those in *E. coli*. *S. cerevisiae* has pathways similar to those of *E. coli* for CDP-DAG-dependent synthesis of PE and PG. However, CL synthesis in all eukaryotes involves transfer of a phosphatidyl moiety from CDP-DAG to PG rather than from one PG to another PG as in bacteria. In addition, yeast also have the mammalian pathways for synthesis of PI, PE and PC, including the methylation of PE to form PC (Chapter 7). Unlike mammalian cells, yeast possess a bacterial-type PS synthase as well as a mitochondrial PS decarboxylase; the latter is also found in mammalian cells. All gene products necessary for the synthesis of DAG, CDP-DAG and PI in yeast are essential for viability. PS synthesis is not essential if growth medium is supplemented with ethanolamine in order to make PE and PC. However, PE is definitely required whereas PC is only required for optimal growth.

No gene products involved in lipid metabolism are encoded by mitochondrial DNA, which in *S. cerevisiae* encodes eight proteins primarily required for oxidative phosphorylation. Mitochondrial PS is imported from its site of synthesis in the endoplasmic reticulum and converted to PE by the mitochondrial-localised PS decarboxylase (*PSD1* gene product). Yeast express nuclear genes that encode a CDP-DAG synthase targeted to the endoplasmic reticulum (*CDS1*) and one targeted to the mitochondria (*TAM41*). PG and CL (localised solely to the mitochondria) are synthesised from CDP-DAG via nuclear gene products imported into mitochondria. Null mutants of *CRD1* (encodes CL synthase) lack CL but accumulate the immediate precursor PG (normally 10-fold lower than CL) to levels approaching that of CL (ca. 20% of the inner mitochondrial membrane phospholipids). These mutants grow normally on glucose for which mitochondrial ATP formation is not required. However, growth on nonfermentable carbon sources such as glycerol or lactate is considerably slower, indicating a partial defect in oxidative phosphorylation. Therefore, CL appears to be required for optimal mitochondrial function but is not essential for viability. However, lack of PG and CL synthesis due to a null mutation in the *PGS1* gene (encodes PG phosphate synthase) results in the inability to utilise nonfermentable carbon sources for growth. Similar effects are seen in mammalian cells with a mutation in the homologous *PGS1* gene (Ohtsuka et al., 1993). The surprising consequence of the lack of PG and CL in yeast is the lack of translation of mRNAs of four mitochondria-encoded proteins (cytochrome *b* and cytochrome *c* oxidase subunits I–III) as well as cytochrome *c* oxidase subunit IV (Su and Dowhan, 2006) that is nuclear encoded. These results indicate that some aspects of translation of a subset of mitochondrial proteins (those associated with electron transport complexes in the inner membrane but not ATP metabolism) require PG and/or CL.

5. ROLE OF LIPIDS IN CELL FUNCTION

There are at least two ways (White et al., 2001) by which lipids can affect protein structure and thereby cell function. Protein structure is influenced by specific protein–lipid interactions that depend on the chemical and structural anatomy of lipids (head group, backbone, alkyl chain length, degree of unsaturation, chirality, ionisation and chelating properties). However, protein structure is also influenced by the unique self-association properties of lipids that result from the collective properties (bilayer fluidity, thickness, shape, packing properties) of the lipids organised into membranes.

5.1 The Bilayer as a Supermolecular Lipid Matrix

Biophysical studies on membrane lipids coupled with biochemical and genetic manipulation of membrane lipid composition have established that the L_α state of the membrane bilayer is essential for cell viability. However, membranes are made up of a vast array of lipids that have different physical properties (Figure 10), can assume individually different physical arrangements and contribute collectively to the final physical properties of the membrane. Animal cell membranes are exposed to a constant temperature, pressure and solvent environment and therefore do not change their lipid make up dramatically in response to external conditions. The complex membrane lipid composition that includes cholesterol stabilises mammalian cell membranes in the L_α phase over the variation in conditions they encounter. However, there are considerable differences in lipid composition (Table 1) between the various membranes that define the multiple organelles of eukaryotic cells (Drin, 2014). Microorganisms are exposed to a broad range of environmental conditions so have developed systems for changing membrane lipid composition in order to maintain the L_α phase.

5.2 Physical Organisation of the Bilayer

As the growth temperature of E. coli is lowered, the content of unsaturated fatty acids in phospholipids increases to maintain the L_α state and membrane fluidity. Genetic manipulation of phospholipid fatty acid composition in E. coli is possible by introducing mutations in genes required for the synthesis of unsaturated fatty acids (Parsons and Rock, 2013). The mutants require supplementation with unsaturated fatty acids in the growth medium and incorporate these fatty acids to adjust membrane fluidity in response to growth temperature. When mutants with membranes containing a high content of unsaturated fatty acids are grown at low temperature they lyse when raised rapidly to high temperature, probably due to the increased membrane permeability of fluid membranes and a transition from the L_α to the H_{II} phase by PE and/or CL. Conversely, when mutants with membranes containing a high content of saturated fatty acids are grown at high temperatures, growth arrest occurs after a shift to low temperature, due to the reduced fluidity of the membrane. Wild-type cells (not requiring fatty acid supplementation) arrest growth after a temperature shift until fatty acid composition is adjusted to provide favourable membrane fluidity.

Bacterial cells also regulate the ratio of bilayer- to nonbilayer-forming lipids in response to growth conditions (Dowhan, 1997). Bacterial nonbilayer-forming lipids are PE with unsaturated alkyl chains, CL in the presence of divalent cations, and monoglucosyl diacylglycerol (GlcDAG). Extensive studies of lipid polymorphism have been carried out on *Acholeplasma laidlawii* because this organism alters its ratio of GlcDAG (capable of assuming the H_{II} phase) to GlcGlcDAG (diglucosyl diacylglycerol, which only assumes the L_α phase) in response to growth conditions. High temperature and unsaturation of fatty acids favour the H_{II} phase for GlcDAG. At a given growth temperature the GlcDAG to GlcGlcDAG ratio is inversely proportional to the unsaturated fatty acid content of GlcDAG. As growth temperature is lowered, *A. laidlawii* either increases the incorporation of unsaturated fatty acids from the medium into GlcDAG or increases the ratio of GlcDAG to GlcGlcDAG to adjust the H_{II} phase potential of its lipids to remain just below the transition from bilayer to nonbilayer. Therefore, the cell maintains the physical properties of the membrane well within that of the L_α phase but with a constant potential to undergo transition to the H_{II} phase.

In contrast to *A. laidlawii*, *E. coli* maintains its nonbilayer lipids, CL (in the presence of divalent cations) and PE, within a narrow range and in wild-type cells adjusts the fatty acid content of PE to increase or decrease its nonbilayer potential. In mutants completely lacking PE, the role of nonbilayer lipid appears to be filled by CL, the levels of which rise from <10% to nearly 50% of total phospholipid. The viability of mutants lacking PE is maintained by divalent cations in the same order of effectiveness ($Ca^{2+} > Mg^{2+} > Sr^{2+}$, but not Ba^{2+}) for these ions to induce the formation of the nonbilayer phase of CL. The CL content of these mutants varies with the divalent cation used during growth. However, the L_α to H_{II} transition for the extracted lipids (in the presence of the divalent cation in which they were grown) is always the same as that of lipids from wild-type cells (containing PE) grown in the absence of divalent cations. Therefore, even though *E. coli* normally does not alter its PE or CL content to adjust the physical properties of the membrane, these mutants are able to adjust CL levels to maintain the optimal physical properties of the membrane bilayer (Dowhan, 1997).

5.3 Selectivity of Protein–Lipid Interactions

The folding and interaction of integral membrane proteins occur via multiple modes of interaction with their environment. Extramembrane domains (EMDs) are exposed to the aqueous milieu, where they interact with freely diffusing water, solutes, ions and water-soluble proteins. The interface region at the hydrophobic–aqueous interface (Figure 8) defines a unique aqueous environment of structured water, ions and lipid head groups that provide a unique environment for EMDs. TMDs composed of a high content of hydrophobic amino acids are buried within the approximately 30-Å-thick hydrophobic interior of the membrane and interact directly with the hydrophobic domain of lipids. TMDs also interact with each other and also with lipids immobilised within the interior of proteins. The driving force for the overall assembly of a biological membrane is the favourable thermodynamics determined by net energetics of the above interactions of a membrane protein with its diverse environment. The resulting lipid–protein supermolecular structure, although highly mobile, resides at a free energy minimum. Amphitropic proteins may spend part of their time completely in the cytosol and are recruited to the membrane surface, or even partially inserted into the membrane, in response to various signals.

Much of what is known about protein–lipid interactions has come from protein purification and reconstitution of function in different lipid environments. Genetic approaches coupled with in vitro verification of function have uncovered new roles for lipids. Results from X-ray crystallographic analysis of membrane proteins have revealed lipids in specific and tight association with proteins. The predominant structural motif for the membrane-spanning domain of membrane proteins is an α-helix of 20–25 mostly hydrophobic amino acids, which is sufficient to span the 30 Å core of the bilayer. A β-barrel motif is also found to a lesser extent (see below).

5.3.1 Lipid Association with α-Helical Proteins

CL is aligned with a high degree of structural complementarity within a high-resolution structure of the light harvesting photosynthetic reaction centre from *Rhodobacter sphaeroides* (Figure 11(A)). The head group of CL is located on the surface of the reaction centre, is in close contact with residues from all three of the reaction centre subunits and is engaged in hydrogen bond interactions with polar residues in the interface region (at the cytoplasmic side of the membrane). A striking observation was that the acyl chains of CL lie along grooves in the α-helices that form the hydrophobic surface of the protein and are restricted in movement by van der Waals interactions. A PE molecule was resolved in the X-ray structure of the photosynthetic reaction centre from *Thermochromatium tepidum*. The phosphate group of PE is bound to Arg and Lys by electrostatic interaction and to Tyr and Gly by hydrogen bonds. PE acyl chains fit into the hydrophobic clefts formed between α-helices of three different subunits of the complex. The fatty acid chains of these phospholipids are unsaturated to allow bending of the chains to fit the grooves on the surface of these proteins.

Bacteriorhodopsin is a light-driven ion pump that is found in the purple membrane of the archaebacterium *Halobacterium salinarum*. Bacteriorhodopsin monomers consist of a bundle of seven TMD α-helices that are connected by short EMDs and enclose a molecule of retinal that is buried in the protein

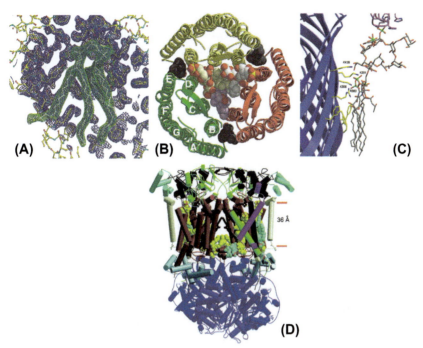

FIGURE 11   Structure of membrane proteins with bound lipids. (A) Model of CL (*green*) tightly bound to the surface of the photosynthetic reaction centre (*blue*) from *Rhodobacter sphaeroides*. The space-filling model was derived from X-ray crystallographic data that resolved between 9 and 15 carbons of the acyl chains of CL. (*Figure adapted from McAuley et al. (1999). Copyright 1999 National Academy of Sciences, USA.*) (B) Lipid packing in crystals of bacteriorhodopsin. Top view of the trimer in three different colours (domains A–E noted in one of the trimers) in complex with lipid (space-filling models) viewed from the extracellular side. Three phytanyl chains of lipid (*grey*) lie in the crevices formed between the A–B domain of one monomer and the D–E domain of the adjacent monomer. The central core of the trimer is filled with a sulphated triglycoside attached to archaeol. *Red* denotes the oxygen atoms of the sugars in *white*. (*Figure adapted from Essen et al. (1998). Copyright 1998 National Academy of Sciences, USA.*) (C) Crystal structure of FhuA complexed with lipopolysaccharide. The ribbon structure (*blue*) represents the outside surface of the β-barrel of FhuA with extended chains (*yellow*) of amino acids. The amino acids of the aromatic belt interact with the acyl chains (*grey*) and the basic amino acids interact with the phosphate (*green* and *red*) groups of Lipid A. (*Figure adapted from Ferguson et al. (2000). Copyright 2000 Elsevier Science, Ltd.*) (D) Crystal structure of yeast Complex III dimer with the interface between monomers in the centre. The *cyan* CL and *yellow* PE on the front right side and left back side are near the proposed surface that interacts with Complex IV (Pfeiffer et al., 2003). Bottom faces the mitochondrial matrix. (*Figure D was adapted from Hunte (2005). Copyright 2005 Biochemical Society, London.*)

interior, approximately half way across the membrane (Essen et al., 1998). Proton pumping by bacteriorhodopsin is linked to photoisomerisation of the retinal and conformational changes in the protein, in a series of changes called a photochemical cycle. Specific lipids can influence the steps in this cycle. A combination of squalene (an isoprenoid) and the methyl ester of PG phosphate is required to maintain normal photochemical cycle behaviour. In a high-resolution structure of bacteriorhodopsin, 18 full or partial lipid acyl chains per monomer were resolved (Figure 11(B)), four pairs of which are linked with a glycerol backbone to form diether lipids identified as native archaeol-based lipids. One of the lipid alkyl chains buried in the centre of the membrane appears to be squalene. Lipid chains were also observed in the hydrophobic crevices between the ends of the monomers in the trimeric structure and probably hold the complex together. This organisation explains the requirement for the natural archaeol lipids to maintain structure and function of the protein. The two glycolipid molecules are positioned at the extracellular side (top) of the molecule leaving a central hole in the bacteriorhodopsin trimer facing the cytoplasm resulting in a 5 Å 'membrane thinning' relative to the surrounding bilayer. This may cause a steeper electric gradient across the central core than in the bulk lipid phase.

5.3.2 Lipid Association with β-Barrel Proteins

The pore-forming proteins of the outer membrane of *E. coli* are organised as antiparallel β-chains forming a barrel structure with an internal aqueous pore and an exterior hydrophobic interface with the membrane bilayer (Figure 11(C)). The X-ray crystal structure of *E. coli* outer membrane ferric hydroxamate uptake receptor (FhuA) contains bound lipopolysaccharide in 1:1 stoichiometry (Ferguson et al., 2000). The acyl chains of the lipopolysaccharide are ordered on the protein surface approximately parallel to the axis of the β-barrel along the half of the hydrophobic belt oriented towards the extracellular surface of the outer membrane. Numerous van der Waals interactions with surface-exposed hydrophobic residues are observed. The large polar head group of lipopolysaccharide makes extensive interactions with a cluster of eight positively charged residues on the surface of the barrel. In the interface region of the membrane there are clusters of aromatic amino acid residues positioned as belts around the protein. Similar organisation of aromatic amino acids of α-helical proteins near the membrane interface region has been observed in other membrane proteins and may be involved in π-bonding interactions with the head groups of lipids.

5.3.3 Organisation of Protein Complexes

Rather than being associated with the exterior surface of membrane proteins, many phospholipids are found wedged between the subunits of oligomeric complexes. Anionic phospholipids have a particularly important function in energy-transducing membranes, such as the bacterial cytoplasmic membrane and the inner mitochondrial membrane. In particular, CL is a key factor in the maintenance of the optimal activity of the major integral proteins of the inner mitochondrial membrane, including NADH dehydrogenase, the cytochrome bc_1 complex, ATP synthase, cytochrome c oxidase and the ATP/ADP translocase (Mileykovskaya and Dowhan, 2014). CL is an integral part of the structure of *E. coli* succinate dehydrogenase and formate dehydrogenase-N.

The yeast ubiquinol:cytochrome c oxidoreductase is a membrane protein complex (cytochrome bc_1 or Complex III) residing in the inner mitochondrial membrane.

The catalytic core is composed of the b subunit, which is encoded by mitochondrial DNA, and the c_1 and Rieske iron–sulphur protein subunits, which are encoded by nuclear DNA. The complex also contains an additional seven nonidentical, noncatalytic nuclear encoded subunits, three haem groups and two quinones. Fourteen phospholipid molecules have been identified in the 2.3 Å crystal structure (Figure 11(D)) of the dimer from yeast (Palsdottir and Hunte, 2004). Four are CL, two are PI, six are PE and two are PC. Six of the phospholipids per dimer are integral to the structure. Two PE molecules (one per monomer, Figure 11(D), yellow lipid at the centre) contact the b subunit from both monomers of the dimer and lie at the interface between the two monomers. Two CL molecules (one per monomer, cyan lipid near the centre) are located near the PEs. Each PI is intercalated between the three catalytic subunits of each monomer. The acyl chains fit tightly in grooves between the helices and are fixed by hydrophobic interactions with residues of all catalytic subunits. Importantly, PI is wrapped around a TMD of the Rieske protein, which may stabilise the helix packing between the transmembrane anchor of the Rieske protein and the core of the complex. Since PI binds close to the point of movement of the extrinsic domain of the Rieske protein, it could dissipate torsional forces generated by the fast movement of this domain in the process of catalysis. The remaining phospholipids are immobilised annular lipids on the surface of the complex that define the TMD of Complex III (red bars). Among these annular lipids are the CL (cyan) and PE (yellow) shown on the right front side and left rear side of Figure 11(D). These faces form a cleft containing CL and PE that has been postulated to be the interface with cytochrome oxidase (Complex IV) in formation of a supercomplex (Pfeiffer et al., 2003). This CL molecule lies at the entrance to one of the proton uptake sites associated with quinone reduction.

In the above cases, specific lipids mediate protein–protein contacts within a multimeric complex and are very important for structural and functional integrity of complex membrane proteins. The advantage of using lipid molecules to form a significant part of the contact surface between adjacent protein subunits is that they have a high degree of conformational flexibility, and are usually available in a range of molecular shapes and sizes. The use of lipids as interface material reduces the need for highly complementary protein–protein interactions and provides flexible and deformable interactions between subunits.

5.3.4 Supercomplex Formation

Kinetic and structural analysis of the mammalian mitochondrial respiratory chain suggested that its individual Complexes I–IV (NADH dehydrogenase, succinate dehydrogenase, bc_1 complex and cytochrome c oxidase, respectively) physiologically exist in equilibrium with supercomplexes or 'respirasomes' composed of the individual complexes (see Mileykovskaya and Dowhan (2014) for a review). Cytochrome c and CoQ are the low molecular weight substrates that mediate electron transfer between the individual complexes. Thus, electron transfer in the respiratory chain would be through either substrate channelling between associated individual complexes within the supercomplexes or random collision between cytochrome c/CoQ and individual respiratory complexes depending on metabolic conditions. In *S. cerevisiae* mitochondria equilibrium between individual Complexes III and IV is shifted to a supercomplex composed of a Complex III (which is a dimer) flanked by Complex IV monomers on each side. The three-dimensional density map of this supercomplex (IV_1–III_2–IV_1) determined by cyro-electron microscopy supports this arrangement of individual complexes (Mileykovskaya et al., 2012). Biochemical, structural and genetic evidence supports an essential role for CL in 'gluing' components of the mitochondrial respirasome together in a functional manner (Mileykovskaya and Dowhan, 2014). In intact yeast mitochondria or extracts of mitochondria prepared by the mild detergent digitonin, Complexes III and IV behave kinetically or on gel electrophoresis as an associated supercomplex. However, in mutants of yeast lacking the ability to make CL (even with accumulation of PG), Complexes III and IV behave kinetically or are displayed on gel electrophoresis as individual complexes. A cavity in the surface of Complex III formed by membrane-embedded TMDs of cytochrome c_1 and cytochrome b with a lid on top of this cavity formed by subunits Qcr8 and Qcr6p has been suggested as a possible site of interaction between Complexes III and IV (Figure 11(D)). CL together with PE fills this cavity and might provide a flexible amphipathic linkage between the complexes. Mutagenesis of the lysine residues of cytochrome c_1 participating in CL binding in this cavity leads to the conclusion that neutralisation of the positive charges at this site by CL also is important for interaction of CIII with CIV and stabilisation of the SC.

Although the structure of the yeast supercomplex (IV_1–III_2–IV_1) solved by cryo-electron microscopy supports the above site of Complex III as the interface with Complex IV, the CL integral to Complex III is not sufficient to stabilise supercomplex formation. The supercomplex can be isolated by solubilisation with digitonin. In this case there are about 50 molecules of CL associated with the supercomplex. This is in large excess over the 8–10 integral CL molecules found in the individual Complexes III and IV after solubilisation by more stringent detergents used for crystallographic structural determination. The supercomplex can be reconstituted from purified Complexes III and IV (stripped of all phospholipids except those integral to the structure) after digitonin removal and incorporation into liposomes (Mileykovskaya and Dowhan, 2014). However, the native supercomplex (IV_1–III_2–IV_1) only forms if CL is present in the liposome, consistent with a requirement for additional CL molecules over and above those integral to the individual complexes.

CL in all mitochondria has a unique species-specific fatty acid composition highly enriched in unsaturated fatty acids and largely restricted to chain lengths of 16 and 18 carbons. Nascent CL (with fatty acid composition similar to bulk phospholipids) is remodelled through the action of a CL-specific deacylase (encoded in yeast by the *CLD1* gene) followed by an acyltransfer (the *TAZ1* gene product) that enriches CL with unsaturated fatty acids derived from other phospholipids. In heart muscle over 80% of the CL species are tetralinoleic (C18 with two double bonds) CL. Mutations in the human *TAZ* gene result in Barth syndrome, an X-linked genetic disorder characterised by cardiomyopathy, skeletal myopathy, neutropenia and growth retardation. Lymphoblasts from patients with Barth syndrome display decreased stability of the respiratory supercomplexes, suggesting that loss of mature CL species in Barth syndrome results in an altered organisation of the respiratory chain. Reduced supercomplex formation is also seen in yeast *TAZ1* gene mutants. However, yeast mutants in the deacylase that initiates CL remodelling do not display a phenotype even though their CL pool has not been remodelled. This result would suggest that the level of CL (also decreased in *TAZ* mutants) is more critical in regulating supercomplex formation than the degree of its remodelling.

5.3.5 Binding Sites for Peripheral Membrane Proteins

A common mechanism of cellular regulation is to organise functional complexes at the membrane on demand from existing components. Peripheral membrane proteins interact with or are partially embedded in the membrane surface, and are easily released from the membrane by chelating agents, high salt, high pH or chaotropic agents that do not disrupt the membrane bilayer. Integral membrane proteins with at least one domain that spans the bilayer are generally not released by such treatments and require detergents or other agents that disrupt the bilayer in order to be rendered in a 'soluble' form. Amphitropic proteins are a subclass of peripheral membrane proteins that transiently associate with the membrane usually in response to a metabolic signal. Peripheral membrane proteins can interact with the membrane lipids in at least four modes (Cho, 2001): coulombic interactions between a positively charged domain on the protein surface and a membrane domain of anionic phospholipids (PA, PG, PI, CL or PS); aromatic residues exposed on the protein surface interacting through π bonding with the positive head group of phospholipids (PC or PE); interaction of specific binding sites on the protein with lipid second messengers (polyphosphorylated PI or DAG); partial insertion (without crossing the lipid bilayer) of hydrophobic domains of the protein into the membrane bilayer, for example an amphipathic α-helix. In many cases, the hydrophobic face of an amphipathic helix inserts parallel to the bilayer with the cationic face residing in the interface region and interacting with anionic lipid head groups.

Numerous structure-specific domains, mostly found in eukaryotic cells, have been identified for membrane association and activation of amphitropic proteins. For example, the C1 lipid clamp is a conserved cysteine-rich protein domain that binds lipids and is found in protein kinases C and other enzymes regulated by the second messenger DAG. This receptor domain interacts with one molecule of DAG and recruits protein kinase C to specific membrane sites. The C1 domain adopts a β-sheet structure with an open cavity. The C2 domain generally binds anionic phospholipids such as PS in a Ca^{2+}-dependent manner and is conserved among phospholipases C, phospholipases A_2, PI-3-phosphate kinases and calcium-dependent protein kinases C. The crystal structure of the C2 domain of protein kinase Cα in complex with PS reveals that the recognition of PS involves a direct interaction with two Ca^{2+} ions. The pleckstrin homology (PH) domain is shared by protein kinase Cβ and some phospholipases C. This domain is responsible for association of peripheral membrane proteins with the membrane via the phosphoinositide head group of polyphosphorylated PIs in an enantiomer-specific manner. PH domains consist of 7-stranded β-sheets with positively charged loops that attract the negatively charged PI head group.

In prokaryotic cells, the protein structural features defining lipid-binding domains are less well conserved than in eukaryotes, and the membrane ligand appears to be an anionic lipid-rich domain with little selectivity for the chemical species of lipid (Mileykovskaya and Dowhan, 2009). DnaA, MinD and MinE are amphitropic proteins in *E. coli* that perform different functions but become membrane associated and activated by similar mechanisms. The involvement of anionic lipids in the function of these proteins was discovered through the use of *E. coli* mutants in which the anionic lipid content could be controlled (see Section 4.1). DnaA is required for initiation of DNA replication and is active in its ATP- but not ADP-bound form. In vitro, the exchange of ADP for ATP in the complex is greatly stimulated by almost any anionic phospholipid including non-*E. coli* lipids like PI. An anionic phospholipid-specific positively charged amphipathic domain composed of distant residues along the polypeptide chain has been identified in DnaA that directs membrane association followed by conformational changes that alter the ATP/ADP-binding properties (Regev et al., 2012).

MinD is also an ATP/ADP-binding protein containing a similar highly conserved carboxyl terminal amphipathic motif that binds anionic phospholipids. This motif is not structured in crystals of MinD but is predicted to be an amphipathic α-helix, with one side of the helix containing mainly hydrophobic amino acids and the other side containing mainly positively charged amino acids. ATP binding to MinD induces a conformational change in the protein that exposes the C-terminal hydrophobic motif followed by binding to phospholipid bilayers with the induction of α-helix formation. The ATP-bound form of MinD has high affinity for anionic phospholipids. Interaction of MinD with MinE results in ATP hydrolysis and release of MinD from the membrane. MinE also interacts with membrane phospholipids through its N-terminal domain, which can form an amphipathic α-helix. ATP binding and hydrolysis cause the rapid movement of MinD from one cell pole to the opposite cell pole, by alternately forming membrane-associated zones of coiled oligomeric MinD structures extending from each pole. This cycle restricts MinD and its tightly associated partner MinC to the membrane surface at the poles. MinC strongly inhibits formation of the Z-ring that initiates cell septum formation in preparation for cell division. This mechanism appears to be critical for positioning of the Z-ring at the centre of cell and not at the poles.

What was once thought to be a specific interaction of DnaA, MinD or MinE with PG or CL is actually an interaction with an anionic surface charge on the membrane. Mutants completely lacking PG and CL but with elevated levels of PA still initiate DNA replication (DnaA) and divide properly (MinD). These proteins can be activated in vitro in reconstituted systems with a wide range of anionic lipids including those not found in *E. coli*. It appears that these proteins recognise, via positively charged amphipathic helices, clusters or domains of negative charge rather than specific lipids on the membrane surface.

5.4 Assembly of Integral Membrane Proteins

Much less is known about the role of phospholipids in insertion and organisation of integral membrane proteins than about the protein machinery required for protein insertion. Most integral membrane proteins contain one or more α-helical TMDs spanning the membrane bilayer. Hydrogen bonding is satisfied within the α-helical structure, which allows energetically favourable insertion into the lipid bilayer of protein domains rich in hydrophobic amino acids. These helices are connected by EMDs alternately exposed on either side of the membrane. A subset of integral membrane proteins form a barrel structure surrounding an open pore (Figure 11(C)). The barrel wall is made up of β-sheets stacked in alternating directions with hydrogen bonding between the alternating strands. How do lipids act in specific ways to guide and determine final membrane protein structure and organisation?

5.4.1 Lipid-Assisted Folding of Membrane Proteins

The membrane clearly serves as the solvent within which integral membrane proteins fold and function. Since membrane lipid composition and membrane protein structure have coevolved for optimal function of the latter, a specific role for lipids outside of providing a hydrophobic environment for folding of membrane proteins only became strikingly evident through the phenotype of cells with genetically altered membrane lipid composition. The major evidence for lipid-assisted folding and topological organisation of membrane proteins comes from studies on the requirement for PE in the assembly and function of three secondary transport proteins of *E. coli*, lactose permease (LacY), phenylalanine permease (PheP) and γ-aminobutyrate permease (GapP) (see Bogdanov et al. (2014) and Dowhan (2013) and references within). LacY is a polytopic membrane protein with 12 α-helical TMDs (Figure 12(A)). LacY transports lactose either in an energy-independent mode to equilibrate lactose across the membrane (downhill transport) or by coupling uphill movement of lactose against a concentration gradient with downhill movement of a proton coupled to the proton electrochemical gradient across the membrane (uphill transport). Uncovering a role for PE in the assembly of LacY came about by the fortuitous availability of reagents and techniques. The availability of viable *E. coli* strains in which the level of PE can be regulated provided a reagent for studying the steady state and temporal requirement for PE in the assembly of LacY in vivo. The development of a blotting technique termed 'Eastern–Western' made

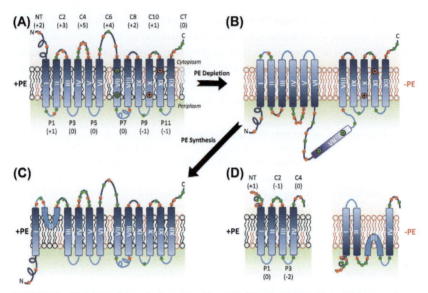

FIGURE 12   Topological organisation of LacY and PheP in the *Escherichia coli* inner membrane. (A) The topological organisation of LacY when assembled in membranes containing PE (normal configuration). The 12 hydrophobic membrane spanning α-helices are numbered in Roman numerals from the amino (N) to the carboxyl (C) terminus. The even-numbered hydrophilic EMDs ('C') connect the TMDs on the cytoplasmic side of the membrane. The odd-numbered EMDs ('P') connect the TMDs on the periplasmic side of the membrane. The presence of negative (*green* dot) and positive (*red* dot) amino acids in each EMD is indicated with net charge of each EMD in parentheses. PE is coloured *grey* and PG/CL are coloured *red*. (B) The topological organisation of LacY when assembled in the absence of PE. Note the first six TMDs along with their connecting EMDs are inverted with respect to the plane of the membrane when compared to the normal conformation. TMD VII is very hydrophilic and resides outside of the membrane in the altered structure. (C) The topological organisation of LacY after initial assembly in PE-lacking cells followed by induction of PE synthesis postassembly of LacY. (D) Topological organisation of TMDs I–IV of PheP assembled in the presence and absence of PE. Topology of the remaining eight TMDs is insensitive to lipid composition. *Figure adapted from Bogdanov et al. (2014). Copyright 2014 Elsevier Press.*

possible the screening for lipids affecting the refolding of LacY in vitro or the conformation of LacY made in vivo using a conformation-sensitive monoclonal antibody. Finally, reconstitution of purified LacY into proteoliposomes of varied lipid composition provided the evidence for a direct lipid–protein interaction as a determinant of membrane protein topological organisation.

In the Eastern–Western procedure, lipids are first applied to a solid support such as nitrocellulose. Next, proteins subjected to sodium dodecyl sulphate polyacrylamide gel electrophoresis are transferred by standard Western blotting techniques to the solid support in such a manner that the protein of interest is transferred to the lipid patch. During electrotransfer of protein to the solid support, protein, lipid and sodium dodecyl sulphate mix and as transfer continues the sodium dodecyl sulphate is removed leaving behind the protein to refold in the presence of lipid. Attachment of the refolded protein to a solid support allows one to probe protein structure using conformation-sensitive antibodies or protein function by direct assay. This combined blotting technique makes possible the detection of membrane protein conformational changes as influenced by individual lipids during refolding (Dowhan and Bogdanov, 2009).

The initial observation that PE was required for LacY function was concluded from studies of reconstitution of transport function in sealed vesicles made of purified LacY and lipid during the 1970s by the Wilson and Kaback laboratories. When reconstituted in lipid vesicles containing PE, both uphill and downhill transport was observed. In vesicles lacking PE and containing PG, CL or PC only downhill transport occurred. The physiological importance of PE for LacY function was established using mutants lacking PE. LacY expressed in PE-containing cells displayed uphill transport of substrate, but in cells lacking PE only downhill transport was observed (Dowhan, 2013). Using Western and Eastern–Western blotting techniques and a monoclonal antibody sensitive to the conformation of the EMD P7 (Figure 12(A)), it was established that LacY assembled in the presence, but not in the absence, of PE displays 'native' structure. LacY maintained its native structure even when PE was completely removed. LacY originally assembled in the absence of PE was restored to the native conformation of P7 by partial denaturation in sodium dodecyl sulphate followed by renaturation in the presence of PE (Eastern–Western blot). LacY assembled (either in vivo or in vitro) in membranes lacking PE was restored to native structure and function by postassembly induction of PE synthesis in cells or addition of PE to proteoliposomes. Taken together these data strongly indicate that PE assists in the folding of LacY by a transient noncovalent interaction with a late folding and nonnative intermediate, thereby fulfilling the minimum requirements of a lipid-assisted folding mechanism analogous to that propagated by protein molecular chaperones (Dowhan and Bogdanov, 2009).

The initial observation that PC in place of PE does not support an uphill transport function of LacY after reconstitution into liposomes illustrates the need to use native lipids to study protein structure and function in vitro and to verify in vitro results with in vivo studies. Earlier reconstitutions protocols of LacY employed either commercially available dioleoyl-PC or naturally occurring PC containing a high content of unsaturated fatty acids, neither of which supported uphill transport or recognition by the conformation-sensitive antibody. Surprisingly, when LacY was expressed in a PE-lacking strain of *E. coli* that was capable of making PC (Figure 10), full uphill transport function and recognition by the conformation-sensitive antibody were observed. PE and PC derived from *E. coli* contain primarily one saturated and one unsaturated fatty acid. Use of *E. coli*-derived PC as well as synthetic PC's containing at least one saturated fatty acid in proteoliposomes was found to support uphill transport and recognition by the antibody. Therefore, a combination of head group and fatty acid composition is required for full LacY function. By combining in vivo results with in vitro results, the molecular basis for the lipid requirement of LacY function was established.

The molecular basis for the loss of native structure and function of LacY (as well as PheP and GabP) assembled in the absence of PE is a topological misassembly of the protein. In the absence of PE, the N-terminal six TMD helical bundle and the connecting EMDs of LacY are inverted with respect to the plane of the membrane bilayer (Figure 12(A) vs (B)). Cytoplasmic loops become periplasmic, and periplasmic loops become cytoplasmic. TMD VII, a helical domain of low hydrophobicity normally stabilised in the membrane bilayer by intramolecular salt bridges, stably resides outside of the membrane in the absence of PE. The ability of LacY to adopt alternative conformations in different lipid environments is consistent with the existence of two separately folded subdomains (N-terminal and C-terminal halves) that can be coexpressed independently within the cell and associate into a functional protein. Therefore each domain can display considerable conformational flexibility and respond independently to the lipid environment. However, in order for these domains to respond to the lipid environment independent of each other while tethered, the interface between these domains must possess considerable flexibility so that the intrinsic preferred orientation in the lipid bilayer is governed by their respective thermodynamic minima. Indeed the hydrophobicity of TMD VII influences the operation of the molecular hinge within LacY; mutation of the Asp 240 to Ile within this domain, which decreases it stability when exposed to the aqueous phase, prevents inversion of the N-terminal helix bundle in cells lacking PE. Therefore, TMDs on either side of a flexible hinge region can organise independently of each other in response to a lipid environment, while proteins without such a hinge region either cannot assume different topologies or cannot fold and are degraded.

If LacY is first assembled in the absence of PE and then postassembly PE synthesis is initiated, there is correction of topology (Figure 12(C)), that is, a reorientation of TMDs with respect to the plane of the bilayer accompanied by regain of uphill transport. Thus membrane protein topology can be rearranged after membrane insertion, challenging the dogma that once TMD orientation is established during assembly is static and not subject to postinsertional topological editing or alteration. The ability of LacY to adopt different topologies dependent on the presence or absence of PE raises the question of whether protein topogenesis is determined primarily by protein–lipid interactions or depends on the lipid requirements of the components of the membrane protein insertion and assembly machinery or other cellular factors. To address this question, misassembled LacY or properly assembled LacY was purified from PE-lacking or PE-containing cells, respectively, followed by reconstitution into liposomes of various lipid compositions. Irrespective of the source of LacY, the final lipid composition of proteoliposomes determined both topology and function. Therefore, both the topological organisation and the transport function of LacY are determined primarily by the phospholipid composition independent of cellular protein assembly machinery.

Several secondary transport proteins with high overall structural homology to LacY are also dependent on PE for assembly and function. PE-lacking cells are unable to carry out uphill transport of a wide variety of amino acid and sugar substrates. In the case of PheP and GapP (Figure 12(D)) only the N-terminal helical hairpin (two TMDs) is inverted in cells lacking PE. Therefore, the role of PE in determining topological orientation of TMDs may be a general requirement for proper assembly of this family of secondary transport proteins. These results dramatically illustrate the specific effects of membrane lipid composition on structure, function and dynamics of membrane proteins. The ability of changes in lipid composition to effect such large alterations in protein structure has important implications for regulatory roles of lipids in cell processes. For example, as eukaryotic proteins move through the secretory pathway, they encounter different membrane lipid compositions (Table 1) that might affect protein structure in dramatic ways to turn on or turn off function. Local changes in lipid composition may also result in similar changes in structure and function. Incompatibility of lipids involved in correct topogenesis of membrane proteins from different sources may explain the unsuccessful heterologous expression of some membrane proteins in a foreign host.

5.4.2. Molecular Determinants of Protein Topology

Based on statistical analysis and experimental determination, cytoplasmic EMDs connecting TMDs predominantly carry a net positive charge in contrast to the remaining EMDs that are either neutral or negative in their overall charge (Positive Inside Rule). LacY faithfully follows this rule (Figure 12(A)). Although amino acid sequence determines membrane protein topology, the sequence is encoded for a specific membrane lipid environment as has been demonstrated for the secondary transport proteins of *E. coli*. Coordinated manipulation of the net charge of EMDs and the negative charge density of the membrane surface contributed by anionic lipid head groups has provided new insight into the rules that determine membrane protein topological organisation. EMDs facing the cytoplasm generally follow the Positive Inside Rule but many of these contain a mixture of positively and negatively charged amino acids. Some of these mixed charged EMDs (see Figure 12(D)) have a net negative charge, indicating that positive residues are dominant over negative residues in determining cytoplasmic orientation. Increasing the net positive charge of the cytoplasmic EMDs (containing both negative and positive residues) of the N-terminal six TMD bundle of LacY in a position-independent manner prevented its inversion in *E. coli* cells lacking PE. However, inversion in PE-containing cells occurred when the above EMD surface was made net negative, indicating that the presence of PE dampens or attenuates the translocation potential of negative residues in favour of positively charged residues acting as retention signals. Increasing the hydrophobicity of marginally hydrophobic TMD VII, which is an EMD in PE-lacking cells, prevented inversion of LacY in PE-lacking cells. However, significantly increasing the net negative charge of the EMDs of the N-terminal bundle overcame the thermodynamic barrier to exposure of the now highly hydrophobic TMD VII to an aqueous environment and allowed inversion of the N-terminal bundle. Furthermore, several net neutral lipids (PC, GlcDAG or GlcGlcDAG), which dilute the high negative surface charge contributed by PG and CL, are substituted for PE in preventing topological inversion. These results indicate that opposing thermodynamic forces between the lipid-driven inversion and exposure of a TMD to an aqueous environment are balanced to determine topological organisation.

Most dramatic is that reversible interconversion between topological conformers can be induced after protein folding in living cells or after reconstitution of LacY into liposomes by supplying PE or diluting PE in either cells or proteoliposomes after stable membrane insertion, respectively. Therefore, simply changing membrane lipid composition independent of other cellular factors (Bogdanov et al., 2014) can change transmembrane protein topology. These results demonstrate that membrane protein topogenesis is a primarily thermodynamically driven process that can occur in any cell membrane at any time. Thus, the final topology of membrane proteins results from a finely tuned interaction between topogenic signals on the protein and topological determinants within the membrane that are influenced by the net charge of protein EMDs and lipid head groups through direct lipid–protein interactions. These effects of lipid environment on membrane protein structure have been consolidated into the Charge Balance Rule (Figure 13), which takes into account the effect of lipid–protein interactions on the potency of charged residues as topological signals (Bogdanov et al., 2014) and is an extension of the Positive Inside Rule.

5.5 Heterologous Organisation of Membrane Components

Compartmentalisation of biological processes such as biosynthesis, degradation, energy production and metabolic signalling plays an important role in cell function.

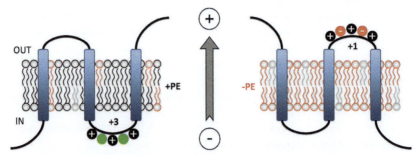

FIGURE 13 Charge Balance Rule. An EMD facing the cytoplasm in the presence of PE is shown containing a mixture of positive and negative charges. Zwitterionic PE (*black*) and anionic lipids (*red*) are indicated in each bilayer. *Left panel*. In the presence of net neutral lipids (PE) the EMD retention potential of positive residues (*black*) is dominant over the translocation potential of negative residues (*green*). *Right Panel*. In the absence of net neutral lipids the translocation potential of negative residues (*red*) is increased to equal or dominate the effect of positive residues. The arrow indicates the membrane potential (positive outwards). *Reproduced from Bogdanov et al. (2014). Copyright 2014 Elsevier Press.*

These functions are compartmentalised in organelles and in membrane structures and by partitioning between cytosol and membrane. The original fluid mosaic model envisioned the membrane bilayer as a homogeneous sea of lipid into which proteins are dispersed (Figure 1). The current view of biological membranes is that they contain microdomains of different lipid and protein compositions and that these domains serve to further compartmentalise cellular processes.

Defined lipid mixtures undergo phase separations due to differences in steric packing of the acyl chains, length of acyl chains and steric/charge differences between lipid head groups. Mixtures of bilayer and nonbilayer lipids undergo multiple phase transitions as a function of temperature, supporting the existence of segregated domains within the bilayer. In model systems, amphipathic polar lipid analogues with similar hydrophobic domains will self-associate even if their polar regions carry the same net charge. Therefore, head group repulsive forces can be overcome by orderly packing of the hydrophobic domains. There is considerable acyl chain mismatch between phospholipids and sphingolipids; that is, phospholipids tend to have shorter acyl chains (16–18) with higher degrees of unsaturation compared to the longer (20–24 for the acyl group) saturated chains of sphingolipids. Naturally occurring sphingolipids undergo the L_β to L_α transition near the physiological temperature of 37 °C while the transition for naturally occurring phospholipids is near or below 0 °C. Therefore, the more laterally compact hydrophobic domains of sphingolipids can readily segregate from the more disordered domains of unsaturated acyl chains of phospholipids. Lipid segregation can also be facilitated by specific polar head group interactions, particularly intermolecular hydrogen bonding to other lipids and to protein networks involving hydroxyls, phosphates, amines, carbohydrates and alcohols.

The most widely studied compartmentalisation of lipids and proteins in eukaryotic cell membranes occurs through the formation of lipid rafts (Lingwood and Simons, 2010). Lipid rafts are liquid ordered phases of lipids and proteins that exist as microdomains within the more dispersed L_α bilayer. Lipid rafts are operationally defined as the membrane fraction of eukaryotic cells that is resistant to solubilisation by the detergent Triton X-100 (detergent-resistant membrane fraction) near 4 °C. This fraction is greatly enriched relative to the total membrane in cholesterol, glycosphingolipids, sphingomyelin and a subset of membrane proteins (Figure 14). The liquid-ordered- to liquid-disordered-phase transition temperature of rafts is up to 15 °C above the transition temperature of the surrounding lipid bilayer because of the high cholesterol and sphingolipid content. Many cell surface receptors colocalise to lipid rafts by virtue of the apparent specificity of their TMDs for the raft environment. Also coclustered with lipid rafts are soluble globular protein domains tethered to the raft lipids via covalent linkage to fatty acids, cholesterol, isoprenoid compounds or PI (Chapter 13). The PI-linked proteins are attached via their carboxyl termini directly to the amino group of ethanolamine phosphate, which in turn is linked to a trisaccharide and then to the inositol of PI (Figure 14).

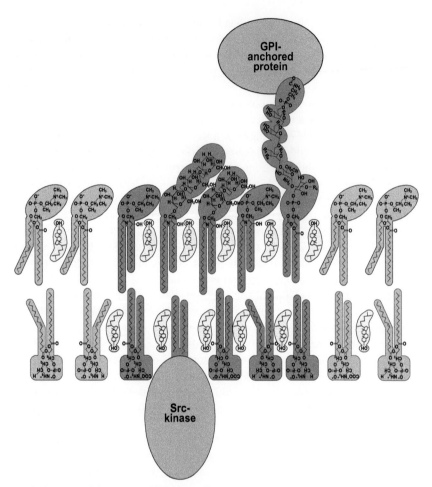

FIGURE 14   Model of lipid raft. A PI glycan (GPI)-linked protein is attached to the exterior mono-layer of the membrane and Src-kinase to the interior monolayer of the membrane by their respective covalently attached lipids. The mechanism for clustering and coupling Src-kinase with a PI glycan-linked protein is hypothetical. Clustered (dark *grey*) around the PI glycan are ordered (straight alkyl chains) glycosphingolipids, sphingomyelin and PC with intercalated cholesterol. The phospholipids with kinked (unsaturated) chains indicate the more disordered L$_\alpha$ state of the surrounding bilayer. *Reprinted (abstracted/excerpted) with permission from Simons and Ikonen (2000). Copyright 2000 American Association for the Advancement of Science.*

The sphingolipids and PI glycan-linked proteins occupy the outer surface monolayer of the plasma membrane bilayer, and the acyl chains of these lipids are generally more saturated and longer than those of the plasma membrane phospholipids. The similarity in the structure of the more ordered hydrophobic domains of the raft lipids and their dissimilarity with the surrounding more fluid phospholipids favour a self-association of the raft lipids and the PI glycan-linked proteins. The hydrogen-bonding properties of the glycosphingolipids with themselves and with the constituents of the PI glycan-linked proteins stabilise the complexes. Finally, the planar shape of cholesterol favours its intercalation parallel to the ordered acyl chains of the raft lipids with its single hydroxyl group facing the surface. The stability of this structure appears to explain why it is not dissipated by detergent extraction.

Lipid rafts appear to be a mechanism to compartmentalise processes on the cell surface by bringing together various receptor-mediated and signal transduction processes. A general phenomenon is that when PI glycan-linked proteins aggregate on the membrane surface, they also become enriched in the detergent-resistant membrane fraction and are phosphorylated by kinases believed to localise to lipid rafts from the cytosolic side of the membrane via covalently attached fatty acids that insert into the membrane. The existence of lipid rafts and their function are still an evolving area of research.

Isolation and characterisation of detergent-resistant membrane fractions, and studies in model systems and whole cells, all support the concept of lipid rafts. Lipid rafts are currently envisioned as highly dynamic domains of lipids and proteins that vary from 10 to 120 nm in diameter. However, it is now clear that other subdomain arrangements of lipids and proteins exist at membrane–membrane junctions, through interaction of proteins with specific lipids such as polyphosphorylated PI, at the sight of cell division, and within the cell surface of specialised cells such as spermatozoa, mucosal cells and neurons (Trimble and Grinstein, 2015). Each of these subdomain structures is characterised by the presence of specific lipids and proteins.

Bacteria also contain membrane domains with lipid and protein compositions that differ from the surrounding membrane area primarily at the septal region and cell poles, which are derived from the septal region. Lipid domains enriched in CL and possibly other anionic lipids (Oliver et al., 2014) were observed in *E. coli* or *Bacillus subtilis* using the fluorescent dye 10-*N*-nonyl acridine orange (NAO), which has affinity for CL and other anionic lipids (Mileykovskaya and Dowhan, 2009). Association with CL induces a shift from green to red in the emission spectrum of the compound. The spectral shift on NAO binding to CL is due to π–π bond stacking induced by neighbouring CL molecules, analogous to acridine orange binding to DNA, and is indicative of CL microdomains in the membrane. In wild-type *E. coli* the fluorescence is localised at the cell poles. During cell division, fluorescence is observed between the nuclei at the septal region. Time-lapse microscopy of NAO-stained *E. coli* cells showed positioning of anionic phospholipid domains at nascent division sites, then their gradual development into septa, and finally after cell constriction into cell pole domains. In filamentous *E. coli* with multiple genomes distributed along their length, the fluorescence is localised between the genomes. Visualisation of CL in *B. subtilis* in different phases of sporulation revealed specific targeting of CL into the engulfment and forespore membranes.

In *E. coli,* cell division is initiated after genome duplication by oligomerisation of FtsZ, a prokaryotic analogue of tubulin, in a ring-structure midway between the poles of the cell. This protein ring is the scaffold that recruits a series of proteins to the division site that brings about constriction and eventually cell division. The heterogeneous distribution of phospholipids in bacterial membranes is an essential factor in the bacterial cell cycle. First, it is important for initiation of DNA replication and division site selection, since it involves amphitropic proteins DnaA, MinD and MinE, whose functions are directly controlled by membrane anionic phospholipids through the interaction with amphipathic helices. Next, protein FtsA (contains a C-terminal amphipathic motif) at least partially controls recognition of the division site and tethers the cytoskeletal protein FtsZ to the division site. Thus, the mid-cell domain provides optimal phospholipid composition first for initiation of DNA replication and then for Z-ring positioning. The concentration of nonbilayer phospholipids such as CL at the division site is consistent with this requirement.

Other proteins are also localised to specific membrane sites enriched in specific lipids. The osmosensory transporter ProP also localises to the cell poles. Signal transduction cascades are organised into functional membrane microdomains that are enriched in polyisoprenoid lipids and/or CL. The insolubility of these microdomains in Triton X-100 and the presence of polyisoprenoid lipids (rather than cholesterol) suggest a similar organisation of proteins and lipid as seen in eukaryotic lipid rafts (Bramkamp and Lopez, 2015).

6. SUMMARY AND FUTURE DIRECTIONS

Biological membranes due to their lipid component are flexible self-sealing boundaries that form the permeability barrier for cells and organelles and provide the means to compartmentalise functions; at the same time they perform many other duties. As a support for both integral and peripheral membrane processes, the physical properties of the lipid component directly affect these processes in ways that are often difficult to assess. Each specialised membrane has a unique structure, composition and function. Also within each membrane exist microdomains such as lipid rafts, lipid domains and organisations of membrane-associated complexes with their own unique lipid composition. Lipids provide the complex hydrophobic–hydrophilic solvent within which membrane proteins fold and function, but they can also act in a more specific manner in determining final membrane protein organisation and orientation in the membrane. These diverse functions of lipids are made possible by a family of low molecular weight molecules that are physically fluid and deformable to enable interaction in a flexible and specific manner with other macromolecules. At the same time they can organise into the very stable but highly dynamic supermolecular structures we know as membranes.

Defining lipid function is a challenging undertaking because of the diversity of chemical and physical properties of lipids and the fact that each lipid type potentially is involved at various levels of cellular function. The challenge for the future will be to determine the function, at the molecular level, of the many lipid species already discovered and yet to be discovered. Coupling genetic and biochemical approaches has been historically a very powerful approach to defining structure–function relationships of physiological importance. Using this approach in microorganisms has proven to be very fruitful. As the sophistication of mammalian cell and whole animal genetics evolves, genetic manipulation coupled with biochemical characterisation will begin to yield new and useful information on the function of lipids in more complex organisms. Interest in understanding biodiversity through detailed characterisation of the vast number of microorganisms will yield additional novel lipids that must be characterised structurally and functionally. As we discover more about the role of lipids in normal cell function, the role lipids play in disease will become more evident. How lipids influence the folding, assembly and function of proteins will be useful in developing novel therapeutic approaches for treatment of conformational and topological membrane protein disorders.

REFERENCES

Bogdanov, M., Dowhan, W., Vitrac, H., 2014. Lipids and topological rules governing membrane protein assembly. Biochim. Biophys. Acta 1843, 1475–1488.

Bramkamp, M., Lopez, D., 2015. Exploring the existence of lipid rafts in bacteria. Microbiol. Mol. Biol. Rev. 79, 81–100.

Cho, W., 2001. Membrane targeting by C1 and C2 domains. J. Biol. Chem. 276, 32407–32410.

Dowhan, W., 1997. Molecular basis for membrane phospholipid diversity: why are there so many phospholipids. Annu. Rev. Biochem. 66, 199–232.

Dowhan, W., Bogdanov, M., 2009. Lipid-dependent membrane protein topogenesis. Annu. Rev. Biochem. 78, 515–540.

Dowhan, W., 2013. A retrospective: use of *Escherichia coli* as a vehicle to study phospholipid synthesis and function. Biochim. Biophys. Acta 1831, 471–494.

Drin, G., 2014. Topological regulation of lipid balance in cells. Annu. Rev. Biochem. 83, 51–77.

Essen, L., Siegert, R., Lehmann, W.D., Oesterhelt, D., 1998. Lipid patches in membrane protein oligomers: crystal structure of the bacteriorhodopsin-lipid complex. Proc. Natl. Acad. Sci. U.S.A. 95, 11673–11678.

Ferguson, A.D., Welte, W., Hofmann, E., Lindner, B., Holst, O., Coulton, J.W., Diederichs, K., 2000. A conserved structural motif for lipopolysaccharide recognition by procaryotic and eucaryotic proteins. Struct. Fold. Des. 8, 585–592.

Henry, S.A., Kohlwein, S.D., Carman, G.M., 2012. Metabolism and regulation of glycerolipids in the yeast *Saccharomyces cerevisiae*. Genetics 190, 317–349.

Hunte, C., 2005. Specific protein-lipid interactions in membrane proteins. Biochem. Soc. Trans. 33, 938–942.

Kauzmann, W., 1959. Some factors in the interpretation of protein denaturation. Adv. Protein Chem. 14, 1–63.

Koga, Y., Morii, H., 2007. Biosynthesis of ether-type polar lipids in archaea and evolutionary considerations. Microbiol. Mol. Biol. Rev. 71, 97–120.

Lingwood, D., Simons, K., 2010. Lipid rafts as a membrane-organizing principle. Science 327, 46–50.

McAuley, K.E., Fyfe, P.K., Ridge, J.P., Isaacs, N.W., Cogdell, R.J., Jones, M.R., 1999. Structural details of an interaction between cardiolipin and an integral membrane protein. Proc. Natl. Acad. Sci. U. S. A. 96, 14706–14711.

Mileykovskaya, E., Dowhan, W., 2009. Cardiolipin membrane domains in prokaryotes and eukaryotes. Biochim. Biophys. Acta 1788, 2084–2091.

Mileykovskaya, E., Penczek, P.A., Fang, J., Mallampalli, V.K., Sparagna, G.C., Dowhan, W., 2012. Arrangement of the respiratory chain complexes in *Saccharomyces cerevisiae* supercomplex III_2IV_2 revealed by single particle cryo-electron microscopy. J. Biol. Chem. 287, 23095–23103.

Mileykovskaya, E., Dowhan, W., 2014. Cardiolipin-dependent formation of mitochondrial respiratory supercomplexes. Chem. Phys. Lipids 179, 42–48.

Murphy, E.J., Stiles, T., Schroeder, F., 2000. Sterol carrier protein-2 expression alters phospholipid content and fatty acyl composition in L-cell fibroblasts. J. Lipid Res. 41, 788–796.

Ohtsuka, T., Nishijima, M., Akamatsu, Y., 1993. A somatic cell mutant defective in phosphatidylglycerophosphate synthase, with impaired phosphatidylglycerol and cardiolipin biosynthesis. J. Biol. Chem. 268, 22908–22913.

Oliver, P.M., Crooks, J.A., Leidl, M., Yoon, E.J., Saghatelian, A., Weibel, D.B., 2014. Localization of anionic phospholipids in *Escherichia coli* cells. J. Bacteriol. 196, 3386–3398.

Palsdottir, H., Hunte, C., 2004. Lipids in membrane protein structures. Biochim. Biophys. Acta 1666, 2–18.

Parsons, J.B., Rock, C.O., 2013. Bacterial lipids: metabolism and membrane homeostasis. Prog. Lipid Res. 52, 249–276.

Patil, Y.P., Jadhav, S., 2014. Novel methods for liposome preparation. Chem. Phys. Lipids 177, 8–18.

Pfeiffer, K., Gohil, V., Stuart, R.A., Hunte, C., Brandt, U., Greenberg, M.L., Schagger, H., 2003. Cardiolipin stabilizes respiratory chain supercomplexes. J. Biol. Chem. 278, 52873–52880.

Raetz, C.R., 1990. Biochemistry of endotoxins. Annu. Rev. Biochem. 59, 129–170.

Regev, T., Myers, N., Zarivach, R., Fishov, I., 2012. Association of the chromosome replication initiator DnaA with the *Escherichia coli* inner membrane in vivo: quantity and mode of binding. PLoS One 7, e36441.

Seddon, J.M., Cevc, G., Marsh, D., 1983. Calorimetric studies of the gel-fluid (Lß-La) and lamellar-inverted hexagonal (La-HII) phase transitions in dialkyl- and diacylphosphatidylethanolamines. Biochemistry 22, 1280–1289.

Simons, K., Ikonen, E., 2000. How cells handle cholesterol. Science 290, 1721–1726.

Singer, S.J., Nicolson, G.L., 1972. The fluid mosaic model of the structure of cell membranes. Science 175, 720–731.

Su, X., Dowhan, W., 2006. Translational regulation of nuclear gene *COX4* expression by mitochondrial content of phosphatidylglycerol and cardiolipin in *Saccharomyces cerevisiae*. Mol. Cell. Biol. 26, 743–753.

Tanford, C., 1980. The Hydrophobic Effect: Formation of Micelles and Biological Membranes, second ed. Wiley, New York. 233.

Trimble, W.S., Grinstein, S., 2015. Barriers to the free diffusion of proteins and lipids in the plasma membrane. J. Cell Biol. 208, 259–271.

White, S.H., Ladokhin, A.S., Jayasinghe, S., Hristova, K., 2001. How membranes shape protein structure. J. Biol. Chem. 276, 32395–32398.

Whitfield, C., Trent, M.S., 2014. Biosynthesis and export of bacterial lipopolysaccharides. Annu. Rev. Biochem. 83, 99–128.

Zinser, E., Sperka-Gottlieb, C.D., Fasch, E.V., Kohlwein, S.D., Paltauf, F., Daum, G., 1991. Phospholipid synthesis and lipid composition of subcellular membranes in the unicellular eukaryote *Saccharomyces cerevisiae*. J. Bacteriol. 173, 2026–2034.

Chapter 2

Approaches to Lipid Analysis

Jeff G. McDonald,[1] Pavlina T. Ivanova,[2] H. Alex Brown[2,3]

[1]*Department of Molecular Genetics, UT Southwestern Medical Center, Dallas, TX, USA;*

[2]*Department of Pharmacology, Vanderbilt University School of Medicine, Nashville, TN, USA;*

[3]*Department of Biochemistry and The Vanderbilt Institute of Chemical Biology, Vanderbilt University School of Medicine, Nashville, TN, USA*

ABBREVIATIONS

EI Electron ionisation
GC Gas chromatography
HPLC High-performance liquid chromatography
LLE Liquid–liquid extraction
MS Mass spectrometry
NIST National Institute of Standards and Technology
TAG Triacylglycerol
TLC Thin-layer chromatography
RRF Relative response factor
TMAO Trimethylamine-*N*-oxide

1. INTRODUCTION AND OVERVIEW

Lipids are among the fundamental building blocks of life. Simple ancient organisms living in the preoxidising environment likely used hydrocarbon chains and pentacyclic ring sterol-like compounds to evolve the huge diversity of complex structures that constitute the rich array of fats found throughout the tree of life. These molecules have remarkable chemical and structural heterogeneity that contributes to innumerable biochemical and biophysical processes.

That organic oils (a mixture of chemicals composed largely of fatty acids and triacylglycerols (TAGs)) are precious and valuable commodities have been recognised by humans for thousands of years. This value justifies the classification and careful measurement of these compounds. The analysis of lipids by mass spectrometry (MS) has been performed for decades, but a systems biology-based analysis that monitors changes in many lipid analytes in parallel has its beginnings in the early part of the current century. As elegantly defined by McLafferty and Tureček (1993), 'The mass spectrum shows the mass of the molecule and the masses of the pieces from it'. This is the fundamental basis for the analysis of lipids by MS just as for other small molecule analytes. The mass spectrum of a lipid does not reveal the precise chemical structure of the

Biochemistry of Lipids, Lipoproteins and Membranes. http://dx.doi.org/10.1016/B978-0-444-63438-2.00002-X

molecular species, but its systematic fragmentation gives information about the chemical arrangement. For precise determination of the structures of unknown lipids, a combination of MS with other techniques, such as nuclear magnetic resonance, is required.

2. LIPID DIVERSITY

Lipids have many definitions but, in simplest terms, are biologically relevant molecules that are soluble in organic solvents (major lipid classes and their metabolic origins are shown in Figure 1). They include a wide variety of chemical entities such as fatty acids and their derivatives, sterols and steroids and isoprenoids. Fatty acid moieties can be esterified to alcohols, forming glycerol-based lipids like TAGs (Chapters 5 and 6), or be linked by an amide bond to long-chain bases, forming sphingolipids (Chapter 10). Each of these lipids can also contain phosphoric acid, forming the phospholipid category (Chapters 3 and 7). Some lipids are a source of energy in biological systems, while others serve as building blocks and major participants in cellular recognition and signal transduction across the cell membrane. This vast diversity in structure reflects the unique amphiphilic property of lipids imparted by their hydrophobic acyl chains and hydrophilic head groups. Lipids can generally be divided into

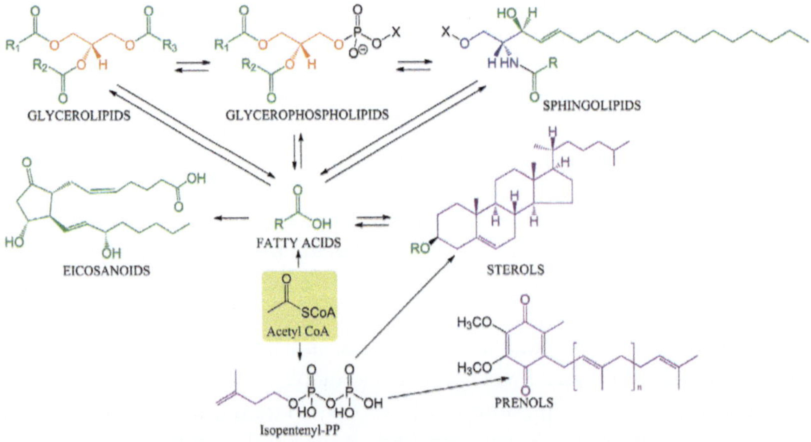

FIGURE 1 **Human lipid diversity.** Relationships among the major mammalian lipid categories are shown starting with the 2-carbon precursor acetyl-CoA, which is the building block in the biosynthesis of fatty acids. Fatty acyl substituents in turn are transferred to be part of the complex lipids, namely sphingolipids, glycerolipids, glycerophospholipids and sterols (as sterol esters). Fatty acids are also converted to eicosanoids. A second major biosynthetic route from acetyl-CoA generates the 5-carbon isoprene precursor isopentenyl pyrophosphate, which provides the building blocks for the prenol and sterol lipids. Fatty acyl-derived substituents are coloured green; isoprene-derived atoms are coloured purple; glycerol and serine-derived groups are coloured orange and blue, respectively. Arrows denote multistep transformations among the major lipid categories starting with acetyl-CoA. *This figure and legend are modified from the original that appeared in Quehenberger et al. (2010), © the American Society for Biochemistry and Molecular Biology.*

three categories – fatty acids, simple lipids and complex lipids. Fatty acids from plant and animal origin have different properties. Fatty acids of animal origin are summarised in several subcategories – *cis/trans*, saturated, mono- or poly-unsaturated. Plant fatty acids are more complex and may contain different functionalities, including epoxy-, hydroxy- and keto- groups. Bacterial fatty acids can contain branched, odd-carbon or cyclopropane groups (the synthesis and metabolism of these fatty acids is described in Chapters 3–6). Simple lipids only contain fatty acids and alcohol constituents. The alcohol can be glycerol, esterified on one, two or all three hydroxyl groups, sterols or a long-chain alcohol. They are represented in triacyl-, diacyl- or monoacylglycerols, cholesterols and cholesteryl esters and wax esters. Complex lipids include glycoglycerolipids, glycerophospholipids and sphingolipids that are the major phospholipids present in mammalian cells. The comprehensive analysis of lipids generally involves separation into simpler categories, according to their chemical nature and identification and eventual quantitative measurement of a specific class, subclass or individual lipid (Hanahan, 1997).

3. CHROMATOGRAPHIC-BASED ANALYSIS OF LIPIDS

3.1 Historical Perspective

The identification of lipids by chemical analysis dates back more than two centuries. Excellent timelines and historical perspectives of lipid science can be found throughout this book and elsewhere on the development of MS-based techniques (Brügger et al., 1997; Ivanova et al., 2001; Murphy et al., 2001; Han and Gross, 1995; Brown and Murphy, 2009). Detailed reports of methodologies for lipid measurements can be found dating back more than a century. These include the first mention of colourimetric assays, column chromatography and thin-layer chromatography (TLC). The introduction of spectrophotometric techniques provided additional insight for lipid researchers. Although modern versions of these tools are still employed in today's laboratories, there have been considerable advances made in the quantitative measurement of lipids. Advances in chromatography, the advent of modern electronics (especially computers) and breakthroughs in gas-phase ionisation have dramatically expanded the tools for lipid analysis. A detailed explanation of the technical aspects of lipid MS from one of the pioneers of the lipid biochemistry field, Robert Murphy (1993), provides insightful explanations of collision-induced dissociation, ionisation techniques and other important details for the analysis of major lipid classes as well as interesting comments regarding the history and development of lipid analysis by MS.

3.2 Lipid Extraction from Biological Sources

Extraction of lipids from biological samples (e.g. cells, tissues, fluids) is the first step in any lipid analysis. An extraction begins when organic solvents are used to solubilise and separate the lipids from proteins, which are largely insoluble in these solvents. If necessary, the choice of extraction solvent can be tailored to a specific class of lipids (e.g. sphingolipids, phospholipids, lysophospholipids, phosphoinositides). In addition to solubilising lipids, solvents can serve to inactivate enzymes and denature proteins that would otherwise degrade or modify the extracted lipids. Solvent extraction will also help minimise the oxidation of polyunsaturated fatty acids. For these reasons, lipids are typically extracted from fluids, tissues or cultured cells soon after their removal or isolation. Antioxidants such as butylated hydroxytoluene or ascorbic acid can be added during the extraction procedures to reduce oxidation.

There is no single ideal organic solvent suitable for comprehensive lipid extraction due to the different characteristics of each lipid class. The precise details of optimal conditions for extraction depend on the sample origin and type of lipid being analysed, which is summarised in a series of excellent reviews on methodology (Methods in Enzymology, 2007; Biochimica et Biophysica Acta, 2011). Properties such as polarity and solubility vary tremendously between lipid classes (Christie and Han, 2010), necessitating the use of mixtures of polar solvents, such as alcohols, and a nonpolar solvent, such as isooctane. Halogenated solvents such as chloroform or dichloromethane are also commonly used due to their unique properties. Two of the most popular methods used to extract lipids from biological samples are those developed by Folch et al. (1951) and Bligh and Dyer (1959). The Folch method uses chloroform and methanol in a 2:1 ratio and large volume of water to remove nonlipid, aqueous-soluble components. The Bligh and Dyer method uses a similar approach and is designed for lipid extraction from tissues and fluids with a high percentage of endogenous water. The extraction begins with creation of azeotrop (single phase) composed of chloroform:methanol:water followed by the addition of equal volumes of water and chloroform to generate a two-phase system. The lower layer contains the lipids, whereas the majority of the proteins and aqueous-soluble components partition in the upper layer or the interface. The lower layer is collected and subjected to additional processing steps. Methods for phospholipid extraction from plant tissues involving initial isopropanol extraction for enzyme inactivation followed by a 'Folch-like' procedure developed in the 1960s are still widely employed (Nichols, 1963).

3.3 Classic Chromatographic Techniques

Chromatography is a general term for separating mixtures using a variety of scientific techniques. Liquid–liquid extraction (LLE), TLC, high-performance liquid chromatography (HPLC) and gas chromatography (GC) are the four basic methods of chromatography. Using any of these techniques, a complex mixture may be separated into simpler groups, or individual components, depending on the desired outcome.

3.3.1 Liquid–Liquid Extraction

LLE is a useful partitioning technique when isolating and purifying lipids from biological samples. The chromatographic phase extraction system is established by combining two immiscible solvents, such as methanol and hexane. Introduction of a biological sample (e.g. serum) to the two-phase system results in the aqueous components of the sample partitioning into the methanol layer and the hydrophobic components partitioning to the hexane layer. The partitioning of components into phases takes time but can be facilitated by aggressive mixing and subsequent centrifugation. The most frequently used method for separating lipids is the Bligh/Dyer extraction process (Bligh and Dyer, 1959) (see Section 3.2). Further fractionation of the lipid extract can be performed using additional two-phase solvent systems. For example, addition of an aqueous lipid sample to an ethyl acetate and isooctane mixture results in the partitioning of the polar, charged lipids (phospholipids) into the ethyl acetate layer while the neutral species (tri-, di- and monoacylglycerol, cholesteryl esters) partition to the isooctane layer (Leiker et al., 2011).

3.3.2 Thin-Layer Chromatography

TLC involves the separation of a mixture by partitioning between a mobile liquid phase and a solid absorbent phase. A lipid sample is deposited on a solid surface (typically a glass plate) coated with an absorbent material (usually silica) and the bottom portion of the plate is submerged in the organic solvent, which is the mobile phase. As the solvent begins migrating up the plate, the mixture of lipids separates based on affinity for the solvent or stationary phase; lipids preferring the solvent phase will migrate further up the plate than those that prefer the stationary phase. The separated lipids are then visualised either by chemically treating or by exposing the plate to high temperatures to visualise the separated lipid groups. TLC can resolve lipids by class (e.g. TAGs, cholesteryl esters, phosphatidylcholines (PCs)) or by a feature, like acyl-chain length or unsaturation. Although TLC is a very effective tool for the basic assessment of lipid composition, it generally cannot separate individual lipid species and has limited utility for advanced lipid separation.

3.3.3 Gas–Liquid Chromatography

Gas–liquid chromatography, more commonly referred to as gas chromatography, is a more advanced form of separation that offers a resolving power not often matched by other chromatographic techniques. GC is a mature, well-developed technology used to perform quantitative analyses in virtually every aspect of science in laboratories around the world. A gas chromatograph is relatively easy to use and affordable, compared to other advanced analytical instruments. GC uses a carrier gas to move analytes through a capillary. Separation occurs based on the differential affinity of the lipid for the gas phase or stationary phase (coating on the column wall). This method requires that the lipid be volatile, or amenable to chemical derivatisation to render it volatile, in order to migrate through the column in the gas phase. Analysing lipids by GC involves injecting a lipid extract (1–2 µl) into a heated injection port (200–300 °C), where the sample is transitioned to the gas phase and swept into a coated silica column via a carrier gas, such as helium. Modern GC columns range from 10 to 100 m in length and 0.1–0.5 mm in diameter with an interior coated with modified polysiloxane. The oven containing the column precisely regulates the temperature. Once the sample has transitioned to the gas phase, it begins migrating through the column. The individual components migrate through the column based on a combination of physical properties, such as boiling point, and their affinity for the stationary phase. The resolving capability of GC is affected by temperature, flow rate, column dimensions and composition of the stationary phase.

In order to measure the compounds as they elute from the column, GC instruments are coupled to a detector, sometimes including a mass spectrometer, which will be addressed below. The simplest GC detectors use flame ionisation where the column eluate passes through a hydrogen-fuelled flame and the resultant ions produce an electronic signal measured as they pass between two charged plates. There are numerous other detectors that can be used with GC, some broadly applicable and others specific to a class of compounds.

While GC offers exceptional resolving capabilities, it has limited applicability. The lipid of interest must transition to the gas phase without undergoing thermal decomposition. Compounds that contain charged sites, such as acids, bases or compounds containing hydrogen bonding moieties (–OH, –NH₂), cannot readily make this transition without decomposition. This can be overcome by chemically altering the reactive lipid to prevent thermal degradation. One of the most well-known examples of this type of modification is the esterification of fatty acids with a methyl group (fatty acid methyl ester or FAME) to make a neutral derivative, suitable for GC analysis. Compounds like cholesterol can transition to the gas phase without thermal degradation. However, the hydrogens on alcohol groups bond to alcohols on other sterol molecules, reducing the efficiency of the gas phase transition. The efficiency of the transition can be improved through derivatisation, such as trimethylsilylation, which replaces the reactive hydrogens with a trimethylsilyl group. Other neutral lipids, such as TAGs and cholesteryl esters, cannot effectively transition to the gas phase due to their high molecular weight. Both may partially enter the gas phase, but not quantitatively, and not without some portion of the molecule undergoing thermal degradation. As a rule of thumb, GC is best suited for neutral lipids that are thermally stable and less than 500 Da.

3.3.4 High-Performance Liquid Chromatography

HPLC is a chromatographic technique similar to GC that involves the migration of a lipid mixture through a column containing a stationary phase. However, in HPLC the mobile phase is a liquid instead of a gas. Analysing a lipid extract by HPLC typically involves injecting a sample (20–200 µl) into an HPLC column while a mobile phase (solvent) is flowing through the column. The mobile phase can have many solvent combinations, but it typically contains water and an organic component. The HPLC column is usually a stainless-steel tube ranging from 50 to 250 mm in length and 1–4.6 mm in diameter, packed with chemically modified silica particles (<1–5 μm in diameter) with a consistency of very fine sand. The smaller the particle, the better the resolution of the mixture. There are numerous stationary phases available for HPLC, the most common being silica particles modified with C_{18} groups. Because of the numerous choices in column diameter, length and particle size, one can quickly appreciate the vast number of options afforded to the researcher. Furthermore, there are almost endless solvent choices. HPLC analyses fall into four different categories: (1) reverse phase chromatography where the stationary phase is hydrophobic (C_{18}-modified silica), (2) normal phase chromatography where the stationary phase is hydrophilic (silica), (3) hydrophilic interaction chromatography, a hybrid of techniques 1 and 2, and (4) ion chromatography where the stationary phase is an ion exchange material for cationic or anionic analytes. Additional references and information on lipid analysis can be found in two excellent handbooks (Mukherjee and Weber, 1993; Cevc, 1993).

HPLC is a versatile chromatographic technique that is amenable to resolution of most lipids of interest. After an aliquot of sample is injected into the HPLC column, the solvent composition may be held constant (isocratic elution) or the organic component may be increased (gradient elution) depending on the desired outcome of the separation. Components of the lipid sample will flow through the column and will elute at different times depending on affinity for the stationary and mobile phase. Recent advances in HPLC stationary phase technology have dramatically increased the resolving capabilities of HPLC. The utility of HPLC has dramatically increased due to the ability to couple HPLC to MS via electrospray and other atmospheric-pressure ionisation techniques.

4. BASIC CONCEPTS OF ANALYTICAL BIOCHEMISTRY

4.1 What is Quantitative Measurement?

One of the most fundamental principles in science is the ability to make quantitative measurements. In the case of lipids, the quantity of a specific species in a sample is expressed relative to a unit volume, mass or other denominators. Often, the goal of quantitative analysis is to evaluate the result of a treatment. Measurements are made prior to and after treatment to determine the effect of the treatment on the chemical/substance of interest. Clinicians use quantitative measurements to determine the best course of treatment for their patients. A well-known example of quantitative lipid analysis is the measurement of an individual's total serum cholesterol level. In the United States this is reported as mass per unit volume, or milligrams of cholesterol per decilitre of serum (abbreviated mg/dl). A total cholesterol level of less than 200 mg/dl is generally considered desirable. Because lipid measurements in biological fluids are reported as mass per unit volume, they are easily comparable between laboratories.

Lipid measurements in tissues, such as the amount of cholesterol in mouse brain, are reported as weight of lipid per weight of tissue. When measuring lipid levels in tissue, it is important to specify the condition of the tissue (e.g. wet vs lyophilised) at the time of the extraction. For example, the cholesterol level in mouse brain is reported to contain ~15 mg cholesterol/g tissue (wet weight) (Russell et al., 2009). Alternatively, another laboratory may report the cholesterol level to be ~4 mg cholesterol/g tissue (lyophilised weight). Reporting of lipid levels in cultured cells also requires specification of how the measurement was normalised. The data may be normalised to the number of cells (mg lipid/10^6 cells), amount of protein or DNA in the sample (μg lipid/μg DNA or protein) or per weight of the cell pellet (μg lipid/mg cells). In order to ensure quality lipid measurements, it is important to select the appropriate extraction method, to employ the best laboratory technique and to ensure the accuracy and precision of the tools used to perform the steps leading to and including the quantitation. As such, it is critical to use properly calibrated equipment, such as pipettes or scales, when making quantitative measurements.

4.2 Standard Calibration Curve

A standard calibration curve is the basis for making quantitative measurements. It is a method by which the amount of an analyte (e.g. lipid) in a test sample is derived by comparing the test sample to a set of samples of known concentration. A calibration curve may also be used to (1) establish a working concentration range for the intended analysis, (2) determine the linearity of the working concentration range and (3) establish instrument and method detection limits. A basic calibration curve is shown in Figure 2(A). The x-axis represents known concentration of an analyte (lipid). The y-axis represents a measure of signal intensity such as absorbance (UV–VIS), spectral intensity (fluorescence) or counts per second (MS). The slope of the regression line for the known concentrations is then used to determine analyte levels present in the test samples. This type of calibration curve is referred to as an external standard calibration curve, because the calibration is separate from the measurement of the sample.

The advantage of performing an external standard calibration is its simplicity and applicability to a wide variety of quantitation methods. However, external standard calibration curves are influenced by the stability of the detector and the presence of interfering factors in the sample. Additionally, the accuracy of the method depends on knowledge of the standard concentrations used to produce the curve. These issues make it difficult to use external standard calibration curves for MS-based quantitation. As a result, more advanced calibration techniques, such as internal calibration curves, are employed. An internal standard calibration involves adding a chemical substance at a constant

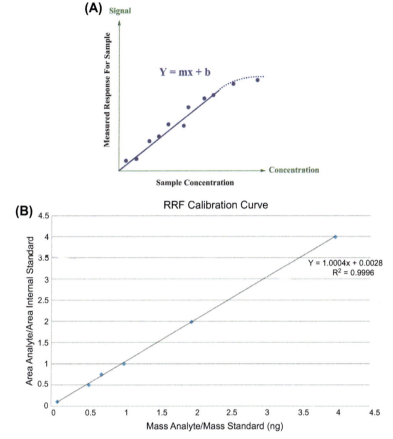

FIGURE 2 **Schematic of a basic calibration curve.** Calibration curves are used in assigning a quantitative value with a scientific measurement. (A) Slope of curve is represented in $Y = mx + b$ format. Linear (dynamic) range is depicted in solid, while nonlinear response is shown with a dashed line. (B) A calibration curve used specifically for isotope dilution mass spectrometry. The slope of the regression is plotted as a ratio of chromatographic peak area versus ratio of analyte mass per sample. The resulting slope of the regression is the relative response factor (RRF).

amount to the samples, blank and calibration standards. An internal standard calibration employs plotting the ratio of the analyte signal as a function of the standard. The standard may be added at the beginning of the sample extraction and carried through to the final extract (surrogate) or added to the sample following extraction and reconstitution (internal).

4.3 Surrogate versus Internal Standard

The terms 'surrogate standard' and 'internal standard' are interchangeable in the modern language of analytical chemistry, yet are distinctly different in the purposes they serve in the quantitation process. A known amount of surrogate standard is added at *the beginning* of an extraction and carried through to the final extract. The surrogate standard mimics the lipid of interest and is lost at a similar rate during the extraction process. This allows the researcher to calculate the amount of analyte present in the test sample by correcting for loss of surrogate during extraction. On the other hand, an internal standard serves as a marker of instrumental consistency across the analysis of a sample set. A known amount of internal standard is added *after* the extraction of the test sample. Therefore, there is no expected loss of the internal standard as a result of sample extraction.

Current quantitation methods frequently use a combination of a surrogate standard and an internal standard in a single assay. While the use of a surrogate standard in combination with an internal standard is not critical for the quantitative analysis of lipids, it does allow for both a quantitative assessment of extraction efficiency and a quality control check of instrument performance. A marked decrease in either the extraction efficiency and/or the instrument performance would indicate the need to evaluate all aspects of the extraction and instrumental method being employed.

4.4 Choice of Surrogate and Internal Standards

When choosing a surrogate and/or internal standard for quantitation, it is important to select one that is not endogenous to the test sample. The use of MS for quantitation dramatically increases the options for surrogate and internal standards. Unlike most spectroscopic techniques, MS can differentiate an endogenous lipid from one that contains an isotopically labelled analogue such as ^{13}C or ^{2}H containing. This allows for the use of a labelled version of the analyte of interest for calibration and quantitation. For example, when quantitating cholesterol by MS, synthetically prepared cholesterol in which seven hydrogens were replaced by seven deuteriums can be used as a surrogate standard. Using spectroscopic techniques, the two molecules could not be differentiated but can be uniquely separated from one another by MS due to the difference in monoisotopic weight (endogenous cholesterol is 386.3 Da and synthetically prepared d7-cholesterol is 393.3 Da). Most isotopically labelled standards contain deuterium since it is simpler to synthesise and cost-effective. However, there are other stable isotope labels such as ^{13}C, ^{18}O and ^{15}N that may be appropriate for use in certain quantitative analyses.

Odd-carbon acyl chains are a suitable choice as a surrogate and/or internal standard when quantifying lipids containing endogenous even-carbon acyl chains. Lipids containing odd-carbon fatty acyl chains are utilised because they do not exist naturally in the test samples from human materials and when there are no available stable isotope-labelled compounds for each analyte. An advantage of this approach is that spectral density of some classes of lipids can be quite high in regions of interest and odd-carbon numbered lipid chains fit into natural spaces in the spectrum. Regardless of the standards chosen, one should ensure that there are no isobaric interferences at or near the mass-to-charge ratio (*m/z*) of the surrogate or internal standard.

4.5 The Advantage of Surrogate Standards through Isotope Dilution

Quantitation by internal standard using isotope dilution offers several advantages over external calibration. Isotope dilution is simply quantitation using an internal standard that contains stable isotope labels, most often deuterium or ^{13}C. The first advantage is that it corrects for the variation observed in MS over relatively short periods of time. The variability of electrospray (most widely used ionisation source) can be dependent on many factors and is subject to a phenomenon called ion suppression. This results in part from high concentrations of charged or neutral species and can cause changes in the electrospray process, leading to sample-to-sample variability. When a known amount of surrogate standard is added to each sample and a ratio is used for quantitation, the variability of electrospray is reduced. Factors related to temperature and chromatography conditions can also affect signal intensities but provided quantitation is normalised against a surrogate standard, these effects are minimised. The second advantage is that it negates the need to know exact volumes during the sample extraction process. Extracting and isolating lipids from biological matrices are complex, time-consuming processes involving many steps. Each time a sample is transferred, processed or otherwise manipulated, a portion of the sample is lost. A surrogate standard added at the beginning of the extraction process will mirror its representative analyte(s) and account for losses during sample workup. This is a self-correcting calculation as one assumes that loss of standard and surrogate occurs at the same rate. Another benefit of internal calibration is that it removes the need to precisely control volumes during the extraction process. Due to the nature of extractions, the multiple steps, and the use of volatile organic solvents, controlling most volumes during extraction are challenging. By adding a known amount of surrogate standard to each sample and using the relative response factor (RRF) calculation, the final value will be reported as mass per sample and not concentration, which is dependent on precise volumes as would be done in most spectroscopic measurements. It should be noted that the surrogate and internal standards should be added with the highest precision and accuracy possible, as all quantitative values will derive from these volumes. Likewise, it is critical that volume of liquid sample (e.g. plasma or serum) or weight of tissue (wet or dry) be measured with the highest rigor.

4.6 The Relative Response Factor

RRF is the equation used for isotope dilution quantitation via MS. The equation is given as follows:

$$\text{Area}_{is}/\text{Mass}_{is} = \text{RRF} \left(\text{Area}_{analyte}/\text{Mass}_{analyte}\right) \tag{1}$$

where

Area_{is} = area of the chromatographic peak or MS signal for the isotopic standard

$Area_{analyte}$ = area of the chromatographic peak or MS signal for the primary standard

$Mass_{is}$ = mass of the isotopic standard

$Mass_{analyte}$ = mass of the primary standard.

The RRF is generated from a single-point calibration or can be calculated by the regression of a calibration curve. In the single-point calibration, a mixture of deuterated and primary standard is prepared that represents the levels expected in an actual sample. The mixture is analysed by the appropriate chromatographic and MS techniques and peak areas are obtained. With the masses of primary and labelled standard known, the RRF can be calculated using Eqn (1). The single point approach is commonly used when the range of concentrations for a lipid or lipids is small, the sample type remains constant and the linearity of the working range has been evaluated. Single-point calibration also minimises use of labelled standards, which can be very expensive. It is advisable to run the RRF mixture at the beginning and end of the sample batch and depending on the batch size, in the middle of the batch. This allows evaluation of the stability of the RRF and allows for the calculation of an average RRF value over the duration of the analysis.

For samples with a wide range of concentrations, or other variables that might complicate quantitation, generating the RRF from a calibration curve may be a better choice. In this example, a range of concentrations for a primary analyte is chosen that covers the expected or known range that will be observed in a sample set. In each sample, a fixed amount of isotope-labelled standard is added to each primary standard and peak values are obtained by integration. Plotting $Area_{analyte}$/$Area_{is}$ on the y-axis and $Mass_{analyte}$/$Mass_{is}$ on the x-axis generates a linear plot with the slope equal to the RRF (Figure 2(B)). This RRF can be averaged over the given concentration range, or the calibration curve can be used to calculate individual concentrations by solving for mass/analyte. Modern mass spectrometers typically have at least four orders of magnitude linearity, so the single-point calibration is often a valid choice. Additional information on quantitation with MS can be found in Duncan et al. (2006) and Boyd et al. (2008).

4.7 Single Standard, Single Analyte

The most robust quantitative approach involves the use of isotope dilution with a primary standard for lipid of interest, and a stable isotope-labelled analogue. This ensures that the RRF will be tailored to the lipid of interest and any interferences or anomalies occurring in the mass spectrometer for the specific lipid can be adjusted. There are few lipid classes that have complete or semicomplete coverage for primary and stable isotope-labelled analogues. The eicosanoids and the sterols have an extensive array of standards available, due to their relative simplicity compared to the other fatty acyl-containing lipid classes and reasonable synthetic routes for manufacture. It is impractical, if not impossible, to synthesise a comprehensive set of labelled analogues and primary standards for all diacylglycerols, as an example. This and other complex classes require a different analytical and quantitative approach, which will be addressed in the next section.

4.8 Single Standard, Multiple Analytes

In most lipid classes, such as glycerophospholipids and glycerolipids, there are many thousands of possible species. It is not feasible to synthesise a primary and stable isotope-labelled analogue for each compound. In this case, several standards that represent the class are chosen based on four factors: head group, acyl chain length, type of linkage (ester vs. ether) and level of unsaturation. For example, the standard for cholesteryl esters could be a C18:1 fatty acyl chain species that represents the entire class. Alternatively, a more in-depth quantitative scheme can be designed with C16:0, C16:1, C22:2, C22:5 fatty acyl chains that represent the range of cholesteryl ester species. During quantitation, limits would be set such that C16:0 would be used to determine levels of any acyl chain between C12 and C18 with one, two or three double bonds. Likewise, the C22 acyl chain standards would cover any cholesteryl esters with an acyl chain over C19 and the two levels of unsaturation would cover mono- and polyunsaturated species. In cases that involve complex lipid classes, a wide range of primary standards is available. Access to stable isotope-labelled or odd-chain fatty acids is limited and synthesis of novel standards can be prohibitively expensive. The most common compromise is to select one or two representative labelled analogues and a representative set of primary standards that cover the most abundant and common lipid compositions. In addition, it is important to have representation of different types of linkages between the fatty acid and the glycerol backbone. Most lipid species have two ester linkages, but in some species the *sn-1* attachment may have an ether linkage as either plasmanyl or plasmenyl, alkyl and alkenyl ethers, respectively. Differences in the stability of these different types of ether linkages contribute to considerations of extraction and chromatography conditions and inclusion of ether-linked lipids among the standards can provide important compositional information.

4.9 Relative versus Absolute Quantitation

Relative and absolute quantitations share several features, and ultimately differ only in the units in which the data are reported. Relative quantitation is a simple approach that typically reports the fold change between a lipid in two samples, or groups of samples. Fold changes can be calculated based on either chromatographic peak area or mass spectral signal intensity. A more advanced relative quantitation is done by adding a known amount of a stable isotope-labelled or odd-carbon acyl-chain surrogate standard to each sample before the sample is extracted and lipids are isolated. The area or intensity of a lipid signal in each sample is then divided by the area or signal of the internal standard and the ratio is calculated. One should demonstrate that the working range is linear through a basic calibration curve, but for relative quantitation, the advanced calibration steps are bypassed. Depending on the required data output, relative quantitation can offer significant information, especially for preliminary investigations or in untargeted analyses, and is much simpler and less expensive to perform compared to a full quantitative analysis. Considering the complexities and challenges associated with full quantitative analysis, relative quantitation affords the researcher a reasonable approach for comparing the levels of lipids between samples of interest.

4.10 Bulk versus Species Analysis of Lipids

Until recent advances in chromatography and MS, lipids were often measured as bulk species. This is still frequently done today. It is important to understand the range of information that is achieved by the different levels of analysis. For example, plasma cholesterol is reported as total cholesterol and also as the amount of cholesterol in various lipoprotein particles. This assay uses a colourimetric, spectroscopic measurement and a commercial kit or clinical chemistry analyser. What is not evident from this analysis is that cholesterol exists in both free and esterified forms. Furthermore, the numerous synthetic precursors and metabolites of cholesterol present in human serum are captured in a 'total cholesterol' assay. Cholesterol is three- to sixfold more abundant than any of its analogues, so this will not affect the cholesterol measurement. However, if these cholesterol intermediates and precursors need to be quantified, specialised and advanced analytical procedures involving chromatography and MS are necessary (McDonald et al., 2012). Likewise, TAGs are commonly measured as part of a standard lipid panel. TAGs are a complex mixture of species with various combinations of acyl chain lengths and levels of unsaturation. There are hundreds and potentially even thousands of TAG species within an organism, but they are reduced to a 'bulk' measurement for clinical diagnosis. Like the cholesterol example, unique TAGs are measured with advanced chromatography and MS, but the effort and expertise are substantially greater than the clinical analyser that measures total TAG. For TAG and other glycerolipids, this issue is further complicated as current analytical technology cannot precisely assign double bond or acyl chain positions.

5. LIPID MASS SPECTROMETRY

5.1 Lipidomics as a Branch of Metabolomics

Advances in genomics and proteomics and the necessity to define the functional processes encoded within these databases, at a systems level, led to small molecule analysis or metabolomics. While metabolomics analysis captures some lipids, the diversity of lipid species and the multitude of roles and functions they mediate in biological systems required the development of a comprehensive and quantitative field of study termed lipidomics. Ideally, biological samples are analysed by parallel measurements of lipid molecular species and aqueous metabolites. The integration of metabolomics and lipidomics provides a more comprehensive picture of biological changes induced by a stimulus or other perturbations. Of course, changes in highly interconnected lipid pathways may cause a ripple effect throughout the network, much like a pebble perturbing the water in a pond.

As illustrated in Figure 1, most of the major classes of lipids intersect at nodes, such that the components of intermediary metabolism and fatty acid synthesis influence compositional changes in multiple categories. A careful analytical biochemist orders the sequence of these events with a well-conducted time course, and differentiates acute changes in lipid composition as either those restoring baseline steady-state lipid levels or those resulting from the rapid consumption of substrates to generate new products. This is frequently accomplished by the measurement of the rates of flux of a precursor component through successive branch points in a pathway (see Section 6).

5.2 Classification of Lipids

Previously used lipid nomenclature schemes are not directly amenable to classification of the large number of lipid species identified by lipidomic analysis. The development of a new comprehensive systems biology classification scheme was driven by the necessity to integrate lipid precursor and product analysis with genomic and proteomic information on enzymes, transporters and receptors that modify lipid metabolism and function. The new classification is based on the chemical structure of the lipid and has been widely adopted in the scientific literature to present the complexity and composition of lipid species within a cell-type, tissue or biological fluid (Fahy et al., 2005). In essence, it includes a two-letter abbreviation of the head group from the eight lipid classes (fatty acids, glycerolipids, glycerophospholipids, sphingolipids, sterol and prenol lipids, saccharolipids and polyketides) followed by numbers indicating the structure of the containing fatty acyl chains (number of C atoms and the number of double bonds), including geometry of the double bonds (*E* vs *Z*) were known. The diversity and quantity of lipid species present important information for comparison of different disease conditions and results after treatments, and in the plant world for genetic mutations and perturbations. As an example, a human plasma standard reference material analysed by different MS methods was determined to contain almost 600 different types of lipids of varying concentrations (Table 1). The plasma sample was a pool collected from multiple individuals in a narrow age range and ethnic composition of the United States. It was provided by the National Institute of Standards and Technology (NIST) using procedures certified from the Center for Disease Control (NIST, 2009) intending to provide reference material for clinical researchers to identify

TABLE 1 Lipid Categories and Species Quantified in Human Plasma

Lipid Category	Number of Species	Sum (nmol/ml)	Sum (mg/dl)
Fatty acyls			
Fatty acids	31	214	5.82
Eicosanoids	76	0.071	0.002
Total	**107**	**214**	**5.82**
Glycerolipids			
Triacylglycerols	18	1058	90.6
1,2-Diacylglycerols	28	39	2.36
1,3-Diacylglycerols	27	13	0.805
Total	**73**	**1110**	**93.7**
Glycerophospholipids			
Phosphatidylethanolamine (PE)	38	435	32.7
Lysophosphatidylethanolamine (LPE)	7	36.6	1.78
Phosphatidylcholine (PC)	31	1974	157
Lysophosphatidylcholine (LPC)	12	103	5.25
Phosphatidylserine (PS)	20	7.00	0.559
Phosphatidylglycerol (PG)	16	6.12	0.480
Phosphatidic acid (PA)	15	2.50	0.173
Phosphatidylinositol (PI)	19	31.5	2.74
N-acyl-PS	2	0.013	0.001
Total	**160**	**2596**	**201**
Sphingolipids			
Sphingomyelins	101	303.468	22.817
Monohexosylceramides	56	2.3135	0.180
Ceramides	41	11.586	0.732
Sphingoid bases	6	0.5678	0.02029
Total	**204**	**318**	**23.7**

TABLE 1 Lipid Categories and Species Quantified in Human Plasma—cont'd

Lipid Category	Number of Species	Sum (nmol/ml)	Sum (mg/dl)
Sterol lipids*			
Free sterols	14	826	31.8
Esterified sterols	22	2954	114
Total	**36**	**3780**	**146**
Prenol lipids			
Dolichols	6	0.025	0.003
Coenzyme-Q	2	4.59	0.394
Total	**8**	**4.62**	**0.397**
Total	**588**	**8023**	**471**

*Sterols in plasma (nmol/ml) were determined as described in Quehenberger et al. (2010). Weight calculations (mg/dl) were based on the sterol backbone for sterol esters. Total cholesterols were also measured by clinical laboratories, following procedures certified by the Center for Disease Control and they reported similar plasma levels.
Reprinted with permission from Quehenberger et al. (2010), © the American Society for Biochemistry and Molecular Biology.

plasma metabolites useful for diagnostic purposes. By having a highly standardised, population-based plasma preparation in combination with a numerical classification system, a truly systems approach can be applied to lipid analysis of biological tissues and fluids. This dataset serves as a kind of cornerstone of basal lipidomic analyses.

Several useful online resources provide detailed information on lipid molecule structures, classification and functions, including LIPID MAPS (LIPID Metabolites And Pathways Strategy; http://www.lipidmaps.org), Lipid Library (http://lipidlibrary.co.uk), Lipid Bank (http://lipidbank.jp), LIPIDAT (http://www.lipidat.chemistry.ohio-state.edu) and Cyberlipids (http://www.cyberlipid.org). In addition, the recent contribution by Robert Murphy (Murphy, 2015) is an excellent resource on the mechanism of product ion formation for a breadth of lipid species.

5.3 Mass Spectrometry Principles and Instrumentation

Lipid analysis was one of the first applications of the mass spectrometer, but the technique has been continuously developed and improved. The combination of selectivity, specificity, sensitivity and speed makes MS an ideal technique for lipid analysis. Different ionisation sources are available starting with electron ionisation (EI) or chemical ionisation, but these have presented challenges in analysis of labile biomolecules such as phospholipids. The 'soft' ionisation introduced in electrospray ionisation (ESI) is now the preferred option. ESI does not cause extensive fragmentation and, combined with high accuracy, reproducibility and applicability to complex extraction mixtures, presents advantageous opportunities in these analyses. ESI can be applied in positive- or negative-ionisation mode depending on the analyte. The choice of polarity is determined by the possible charge state of the phospholipid class in solution. All zwitterionic phospholipids PC, lysophosphatidylcholine (LPC), sphingomyelin (SM) and phosphatidylethanolamine (PE)) can ionise in both positive- and negative-ion modes. With the exception of PE, they are more efficiently analysed in positive mode. Because of their negative charge at neutral pH, anionic phospholipids (phosphatidylinositol (PI), phosphatidic acid (PA), phosphatidylglycerol (PG) and phosphatidylserine (PS)) produce primarily $[M-H]^-$ (deprotonated molecule retaining a negative charge) as the molecular ion peak and are detected in negative-ion mode.

ESI-MS was first developed for the analysis of biomolecules (Fenn et al., 1989) and its ability to produce ions from highly polar and mostly nonvolatile molecules made it the method of choice for lipid analysis. The analyte solution is passed through a metal needle on which a high voltage is applied, desolvating and ionising the molecules at the same time.

ESI sources combined with different mass analysers such as ion trap, quadrupole, time of flight (TOF) and Fourier transform ion cyclotron offer a variety of configurations that excel at different analyses and for different purposes. The minimal fragmentation during ESI ionisation makes it possible to detect most lipid species as molecular ions or molecular adduct ions when analysed in single stage mode using Q1 or Q3 quadrupoles as mass analysers (Figure 3).

Fragmentation is an essential tool in the identification and structural characterisation of lipid molecules. Often, when the ionisation energy is high the resulting internal energy is sufficient to fragment ions within the mass spectrometer, causing in-source fragmentation. Each ion of interest is subjected to collision-induced dissociation (CID) with a neutral atom or molecule in the gas phase, resulting in a product ion spectrum (Figure 3). While detection and quantitation of lipid species are accomplished during single stage MS, identification and structural information about specific peaks is obtained by collision induced dissociation or fragmentation, termed tandem MS. Tandem MS is performed either in space or in time. Space tandem MS requires physical separation of the instrument components into multiple quadrupoles (e.g. triple quadrupole) and TOF mass analysers, which can serve as mass analysers and collision chambers. Mass analysis separation can also be achieved using a single mass spectrometer with MS stages separated in time. FTMS and hybrid triple quadrupole ion trap instruments are used for this purpose. Tandem MS allows four types of experiments to identify key structural fragments: precursor ion scan, product ion scan, neutral loss scan (NL) and selected or multiple reaction monitoring (MRM). All of these scans are performed on tandem MS instruments. In a precursor ion

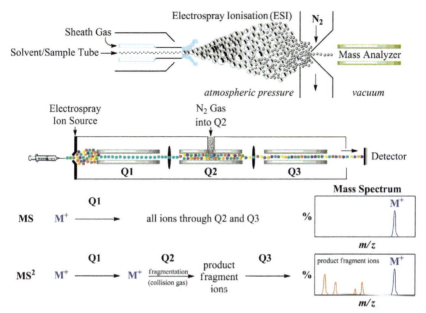

FIGURE 3 **Basics of an electrospray ionisation and triple quadrupole mass spectrometer.** ESI-MS is unsurpassed for complete analysis of phospholipid mixtures. The 'soft' electrospray ionisation causes little or no fragmentation and virtually all phospholipid species can be detected as molecular ion species (denoted as M+), when analysed in a single-stage mode using either Q1 or Q3 quadrupoles operating as mass analysers. The choice of polarity is determined by the possible charge state of the phospholipid class in solution. All zwitterionic phospholipids (PC, LPC, SM and PE) can ionise in both positive- and negative-ion modes. With the exception of PE, they are more efficiently analysed in positive mode. Because of their negative charge at neutral pH, anionic phospholipids (PI, PA, PG and PS) produce primarily [M–H]⁻ (deprotonated molecule retaining a negative charge) as the molecular ion peak and are detected in negative-ion mode. Thus, two sets of mass spectra are obtained for each sample – one in positive- and one in negative-ionisation mode, with all detected phospholipid species presented as peaks at their specific *m/z* value. The identification and structural information about a particular peak is acquired by tandem MS (MS²). The relevant ion is subjected to CID by interaction with a collision gas (usually nitrogen). A fragmentation spectrum (product or 'daughter' spectrum is generated as a result, revealing structural information (acyl chain or head group chemistry, depending on the ionisation mode). On that basis, 'fragmentation libraries' can be created for the phospholipid species composition of any cell type. *Figure adapted with permission from Ivanova et al. (2004).*

scan the precursor masses are scanned in the first mass analyser, whereas the product ion is selected in the second mass analyser. In product ion scan, a selected precursor ion is allowed to fragment and resulting masses are scanned and detected in the second mass analyser (typically used for identifying transitions for tandem MS). In a neutral loss scan, all the masses are scanned in the first mass analyser, while the second mass analyser scans at a different *m/z*. The difference in mass is the neutral loss commonly observed during fragmentation of lipids. Precursor and/or neutral loss scans in positive- or negative-ion mode are used in lipid head group identification, whereas fatty acid composition is identified by precursor ion scan in nega-tive-ion mode. Both mass analysers are set to a selected mass during MRM. For a more detailed explanation of these processes the reader is referred to Murphy (1993). Product ions resulting from both positive- and negative-ion mode fragmentation processes provide structural information on fatty acid,

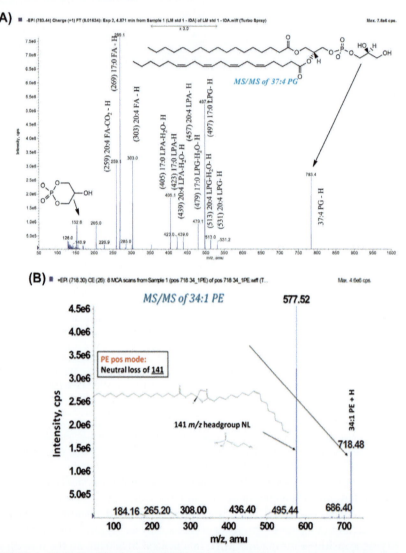

FIGURE 4 **Lipid identification by MS/MS.** (A) Fragmentation mass spectrum of phosphati-dylglycerol in negative-ionisation mode reveals structural information about head group and fatty acid content. (B) Fragmentation mass spectrum of phosphatidylethanolamine in positive-ionisation mode. The consistent neutral loss of 141 Da has been used as a diagnostic tool for the determination of phosphatidylethanolamine species in complex lipid mixtures and class identification.

lyso-lipid species and head group-related fragments expected for each lipid class (Figure 4). Typical fragmentation patterns for PG and PE are shown in Figure 4(A) and (B), respectively.

Other intriguing techniques are rapidly evolving that will improve MS-based lipid analysis. One such technique is ion-mobility MS in which the cross-sectional area of an analyte is used to add a further dimension to the analysis of molecular species of lipids. It relies on the mobility of molecules through a pressurised chamber filled with neutral gas based on their size and structure. Even isobaric species show different transition times through the chamber, allowing isobaric separation as well as increased peak capacity. John McLean (Kliman et al., 2011) have recently demonstrated the utility of this application in the development of larger databases of lipids and aqueous metabolites present in a variety of organisms.

5.4 Analytical Techniques to Enhance Class Separation by GC and HPLC

The coupling of gas chromatographic systems to mass spectrometers has been routine since before 1965. The nature and simplicity of the GC made coupling the two instruments achievable. A GC typically operates with a carrier gas flow rate of 1 ml/min. This gas flow is a reasonable load for the vacuum system that keeps the components of the mass spectrometer under high vacuum (10^{-4} to 10^{-6} Torr). The most widely used ionisation source for GC–MS is EI, which requires vacuum conditions and is an ideal ionisation source for most compounds analysed by GC–MS.

The coupling of HPLC to MS presents greater challenges with regard to engineering and ionisation. Solvent flow rates for most common HPLC methods range from 0.2 to 1.0 ml/min. This volume of solvent cannot be put directly into the high vacuum required by a mass spectrometer. Early coupling of HPLC to MS relied on heated interfaces to evaporate the solvent prior to entry into the MS. Hybrid chemical ionisation techniques were often used to charge the molecules. The use of heated interfaces was not ideal and compounds that are thermally labile were difficult to measure. Development of ESI (Fenn et al., 1989) greatly advanced HPLC-MS analysis of lipids. A key feature of this technique, which involves application of a high voltage to the eluate to create a fine aerosol, is that it occurs at atmospheric pressure prior to the vacuum region of a mass spectrometer. Another important feature of electrospray is that it is a 'soft' ionisation source that results in limited molecular fragmentation, and generation of protonated or deprotonated molecular ions. In this way, the effluent from the HPLC can be desolvated, analytes can be ionised and the charged molecules can be directed into the MS via a small orifice that separates atmospheric from vacuum pressure regions (Figure 3). Electrospray was a significant revolution in HPLC-MS and is now a common tool in many laboratories throughout the world. The importance of coupling modern chromatographic techniques and mass spectrometers cannot be overstated in its importance in almost every aspect of chemistry and biology and the interfaces between them.

5.5 Other MS Instruments and Techniques, MS/MS[ALL]

Starting in the early 2000s, researchers began pioneering a technique called 'shotgun lipidomics' as an alternative to chromatography-based MS techniques. Direct infusion of the lipid extracts in the mass spectrometer allowed for a fast profiling of abundant lipid species (Fridricsson et al., 1999; Ivanova et al., 2001). In 2003, Han and Gross published on the comprehensive analysis of cellular lipids and in 2005 they coined the phrase 'shotgun lipidomics' (Han and Gross, 2003, 2005). Much like shotgun sequencing transformed the field of genomics, shotgun lipidomics is now poised to transform the field of lipidomics. By extracting the lipids from a biological sample and infusing it directly into a mass spectrometer without chromatographic separation, they were able to demonstrate the utility of infusion-based analysis of lipids in biological extracts. However, this technique has limitations in that many lipids are isobaric (identical mass). Several investigators have employed LiOH and other salts to induce adduct formation to help deconvolute the complex mass spectrum and enhance the detectability of trace-level and neutral lipids. Dedicated precursor and neutral loss scans were used to differentiate some isobaric species and gain additional information.

Recent advances in instrument control software have allowed the progression of shotgun lipidomics to a more comprehensive analysis termed MS/MS[ALL]. With this technique, a lipid extract is infused into a Q-TOF mass spectrometer for several minutes. During the infusion, product-ion spectra are collected at each unit mass over a wide mass range (e.g. 200–1200 Da). By acquiring a product-ion spectrum of 'everything', a comprehensive dataset is available that can be queried postacquisition for any product-ion or neutral loss value. In previous shotgun analyses, a dedicated acquisition for each precursor or product-ion scan would have been done independently. These recent advances have made 'shotgun' analyses an increasingly useful tool in the field of lipidomics. It should be made clear that infusion-based lipidomics is complementary to more rigorous chromatographic-based approaches like those used for the NIST plasma lipidome (Quehenberger et al., 2010). Chromatography is a powerful tool for separating isobaric species, reducing interferences and detecting trace-level lipids. Ultimately the relationship between cost/effort and data/speed must be considered when determining which MS techniques to use for lipid measurements.

Performing an infusion-based lipidomic analysis is a relatively simple experiment. A bulk lipid extract is generated with a Bligh/Dyer extraction and the sample is infused into the mass spectrometer for approximately 5 min while the instrument performs the acquisition. While this is relatively simple compared to HPLC-MS, it can generate a data output with 50–100 K lines of MS data. Software tools to perform basic data analysis, including identifying lipid species, statistical comparisons and relative comparisons between sample groups have been unavailable until recently. Several researchers have put forth custom software (Herzog et al., 2012; Husen et al., 2013) that is capable of deconvoluting data, but is not in a widely usable software platform that would be accessible to the nonexpert, and is still not fully applicable for the evaluation of the more complex lipid analysis (e.g. phospholipids).

5.6 Glycerophospholipid Analysis and Quantitation by MS

Tandem mass spectrometry (MS/MS) is well suited for structural characterisation of lipids. All six classes of glycerophospholipids (PA, PC, PE, PS, PG, PI) can be analysed in ESI negative-ionisation mode for collection of structural information. Each lipid class except PA has a characteristic head group fragment (Figure 4 and Table 2). Utilising information-dependent analysis (IDA) along with class separation (LC-MS/MS) unambiguous identification is achieved by analysis of retention time and fragmentation pattern and comparison with spectra from chemically defined standards.

MS analysis of underivatised total lipid extract relies on the ability of each lipid class to acquire charge during ESI. Each molecular species is represented as a single molecular ion with an m/z characteristic for its monoisotopic weight. PC is less ionisable in negative mode but adducts formed with volatile salts of chloride, formate or acetate enhance ionisation and can provide additional mass spectral information. Employing LC-MS with ESI analysis allows quantitation of phospholipid molecular species due to the ability of NP HPLC to separate head group classes by polarity, while also decreasing ion suppression caused by the presence of multiple classes. Detailed descriptions of LC methods can be found elsewhere (Ivanova et al., 2007; Myers et al., 2011). Due to the high sensitivity and accuracy of the MS, instrument responses are sensitive to head group structure, acyl chain length and unsaturation (Figure 5).

Different molecular species are ionised with different efficiency. Accurate quantification of phospholipids by MS requires the use of internal standards with chemistry similar to that of the analytes. Several standards for each lipid class are necessary in order to reduce the need to control the phospholipid concentration (a single standard can be used at low total lipid concentration since the responses are linear). Inclusion of multiple standards allows quantitation of low abundance species at high total lipid concentration (Brügger et al., 1997; Hermansson et al., 2005). For glycerophospholipids, calibration curves were generated using four odd-carbon fatty acid-containing standards per class, covering a large acyl chain length-unsaturation range (Myers et al., 2011). The

TABLE 2 Summary of MS/MS Methods for Glycerophospholipid Head Group Analysis: Characteristic Fragments from the Major Classes of Glycerophospholipids (GPLs)

Lipid Class	Precursor Ion	MS/MS Mode	Fragment
Phosphatidic acid (PA)	[M–H]⁻	PIS, 153 m/z	Glycerol phosphate –H_2O
Phosphatidylcholine (PC)	[M+H]⁺	PIS, 184 m/z	Phosphocholine
	[M+Li]⁺	NL, 189 m/z	Li choline phosphate
	[M+Na]⁺	NL, 205 m/z	Na choline phosphate
	[M+Li/Na]⁺	NL, 59 m/z	Trimethylamine
	[M+Li/Na]⁺	NL, 183 m/z	Phosphocholine
	[M+Cl]⁻	NL, 50 m/z	Methyl chloride
Phosphatidylethanolamine (PE)	[M–H]⁻	PIS, 196 m/z	Glycerol phosphoethanolamine –H_2O
Phosphatidylglycerol (PG)	[M–H]⁻	PIS, 153 m/z	Glycerol phosphate –H_2O
		PIS, 227 m/z	Glycerol phosphoglycerol –H_2O
Phosphatidylinositol (PI)	[M–H]⁻	PIS, 153 m/z	Glycerol phosphate –H_2O
		PIS, 241 m/z	Cyclic inositol phosphate
Phosphatidylserine (PS)	[M–H]⁻	PIS, 153 m/z	Glycerol phosphate –H_2O
		NL, 87 amu	Serine

Reprinted with permission from Ivanova et al. (2009).

average integrations of the peaks from two odd-carbon acyl-chain species standards encompassing the acyl-chain lengths of the species of interest are used for internal standardisation. Using this strategy, standard curves have been developed and validated in 'spiking' experiments to verify that recoveries of the added species are sufficient (example for PA is presented on Figure 6).

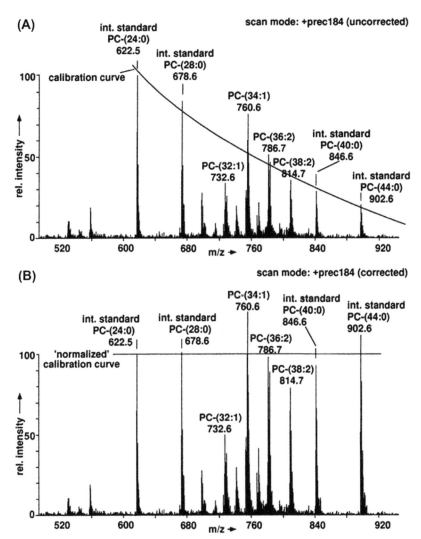

FIGURE 5 **Example of phosphatidylcholine quantitation using internal standards to determine the calibration curves for molecular species.** An unprocessed total lipid extract of 5000 CHO cells containing equimolar amounts of PC (24:0, 28:0, 40:0 and 44:0) was analysed by parent ion scanning for *m/z* 184. Internal standards and calibration curves for each class of phospholipids analysed are required for quantitation. (A) Uncorrected ion intensities. The signal intensities of the internal standards were used for generation of a calibration plot. (B) Corrected ion intensities of the phosphatidylcholine (PC) signals so that the monoisotopic signals represent the true molar abundance of the corresponding PC molecular species. *Figure from Brügger et al. (1997). Reprinted with kind permission of PNAS and the author.*

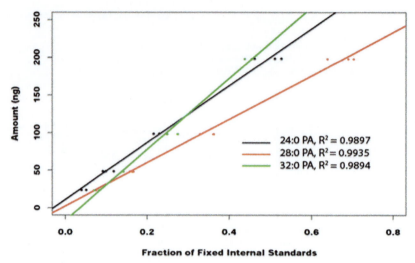

FIGURE 6 **Example of standard curves.** Standard curves of three saturated phosphatidic acid (PA) lipid species. The curves were generated using even number carbon phosphatidic acid lipid standards and fixed amounts of four odd number carbon phosphatidic acid internal standards. *Figure from Ivanova et al. (2007). Reprinted with permission of Elsevier Academic Press.*

6. FUTURE DIRECTIONS

Modern MS-based lipidomics methods have transformed the way in which lipids are studied. It is now feasible to quantitate hundreds or even thousands of species in parallel to determine how stress or an extracellular ligand perturbs the lipid species inside a cell or even at the whole organism level. Most of these analyses are typically a snapshot taken at a single defined time. A detailed analysis of lipid precursors and products over a time course can provide a clearer picture of changes, but limitations remain. For example, if the net biosynthesis of a substrate is roughly equivalent to the rate at which it is being catabolised by an enzyme, the initial conclusion erroneously drawn may be that the species is not changing (Figure 7). Strategic use of stable isotope precursors allows one to measure the relative flux of substrates and products. Each of the successful substrate–product relationships can be regarded as a distinct unit in the pathway that is likely to have one or more regulated steps. Looking at the sequential accumulation of a distinct isotope in the pathway can be used to calculate rates of flux. An excellent example on enzyme kinetics and flux through a pathway is found in Mullen et al. (2011) and an introduction to flux balance analysis is provided by Orth et al. (2010).

Alternatively, tracking changes in cellular composition can also be achieved through the use of molecules that can be covalently modified by affinity tagging. The use of an alkyne tag reacting with an azide by joining small units together has become commonplace with recent advances in 'Click Chemistry'.

$$E_1 \quad E_2 \quad E_3 \quad E_4 \quad E_5 \quad E_6$$
$$A \rightleftharpoons B \rightleftharpoons C \rightleftharpoons D \rightleftharpoons E \rightleftharpoons F \rightleftharpoons G$$

FIGURE 7 **Schematic flux representation.** Multistep enzymatic reactions with defined reaction rates can be tracked by measurement of analyte flux through the pathway.

Alkyne-modified phospholipids can be identified and separated from native lipid counterparts through the use of capture and release approaches, such as the formation of dicolbalthexacarbonyl complexes (Milne et al., 2010). Such approaches can be used to follow substrate–product relationships through many steps in a lipid signalling or metabolic pathway as the synthesis of subsequent products can be captured by formation of covalent bonds, essentially achieving a type of affinity chromatography of small molecules. This methodology allows the investigator to unambiguously chart the steps in a pathway in a time-dependent sequence.

The initial decade of lipidomics was empowered by technological innovations in MS and focused on the development of analytical methods and work flows. Recently, we have begun to realise the dividends of these investments. Biomarker signatures, defined patterns of lipid and metabolite changes, are becoming increasingly associated with disease progression and, in some cases, with clinical outcomes (Manna et al., 2014). An example is the quantification of sterols, oxysterols and vitamin D in plasma from >3200 patients in the Dallas Heart Study II (Stiles et al., 2014). A comparison of the quantitative level of each sterol or vitamin D against the patient phenotype database revealed shared metabolic pathways, new enzyme substrates and regulation of lipid metabolic networks at the transcription level. Figure 8 shows a plot correlating sterol and vitamin D levels against gender, age, ethnicity and numerous clinical parameters (blood glucose, insulin, plasma lipid levels) for this cohort. These findings would have been difficult, if not impossible, to uncover if not for the integration of high-quality analytical measurement of sterols and vitamin D coupled with a large well-documented patient data base.

This approach illustrates the power of comprehensive lipid measurements and their ability to make novel scientific discoveries, but also illustrates a shortcoming. The effort that went into this work far exceeds the capacity for all but the most well-staffed, equipped and financially viable laboratories. The project was successful in large part because scientists from many disciplines (biology, genetics, analytical chemistry, statistics) came together in a collaborative manner. The infusion-based lipidomics techniques described earlier offer the possibility of measuring lipids in mammalian samples with reduced effort and expertise, but these measurements lack the comprehensiveness that can be achieved with targeted, specialised approaches.

The implications of the maturing field of lipidomics to biomedical research since 1994 have been dramatic. Some important examples include a report from Eggert and colleagues that used lipidomics to provide a detailed, systematic analysis of lipid changes during cell cycle progression (Atilla-Gokcumen et al., 2014). Changes associated with pathophysiological alterations in tumours and

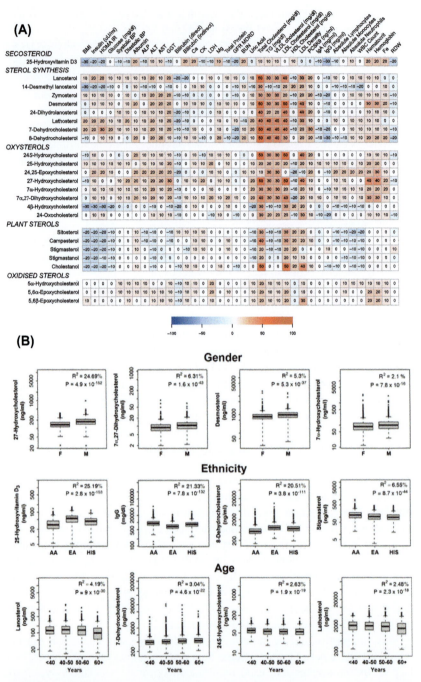

FIGURE 8   (A) Sterol and secosteroid levels in serum from a large human cohort compared against clinical phenotypes. (B) Distribution of lanosterol, 27- and 25-hydroxycholesterol measured in human serum based on gender, ethnicity and age. AA, African American; EA, European American; HIS, hispanic. *Figure from Stiles et al. (2014). Reprinted with permission of PNAS.*

cancer cells, including glioblastomas (Bruntz et al., 2014), colon cancers (Fhaner et al., 2012) and metabolic diseases (Kumari et al., 2012; Li et al., 2012) have been revealed. Similarly, lipidomic profiling has revealed roles for lipid species in various aspects of viral entry, egress, coat composition and innate immune response pathways during infection of host cells (Oguin et al., 2014; Liu et al., 2011; Wenk, 2006). Changes in plant lipids as a result of adaptations to cold, heat and other stress stimulations as well as changes associated with genetic alterations have been identified using a systems-based lipidomics approach (Welti et al., 2007). The change in a lipid species across a breadth of classes can be used to differentiate benign steatosis from the potentially precancerous nonalcoholic steatohepatitis (Gorden et al., 2015). Intriguing work by Hazen and colleagues has shown that the intestinal microbiota metabolism of choline, PC, or L-carnitine is ultimately metabolised in the body to trimethylamine-*N*-oxide (TMAO), which is linked to accelerated atherosclerosis in mouse models (Tang et al., 2013; Koeth et al., 2013; Wang et al., 2011). Similarly, lipidomic analysis showed that elevations in endogenous branched fatty acid esters of hydroxyl fatty acids have positive effects on type 2 diabetes (Yore et al., 2014). These are but a few of the exciting examples of identification of new pathways and biomarkers achieved through profiling in clinical diagnostics. This technology will increasingly be used to decide when to treat a patient and the selection of the best pharmacological agent as part of an evolving trend toward personalised medicine. The emerging field of *omics* will undoubtedly have an increasingly greater role in clinical diagnostics and therapeutic development over the coming decades.

REFERENCES

Atilla-Gokcumen, G.E., Muro, E., Relat-Goberna, J., Sasse, S., Bedigian, A., Coughlin, M.L., Garcia-Manyes, S., Eggert, U.S., 2014. Dividing cells regulate their lipid composition and localization. Cell 156 (3), 428–439.

Biochimica et Biophysica Acta (BBA)-Molecular and Cell Biology of Lipids, 2011. Vol. 1811, Issue 11: Lipidomics and Imaging Mass Spectrometry. In: Merrill, A.H., Murphy, R.C. (Eds.), pp. 635–1000.

Bligh, E.G., Dyer, W.J., 1959. A rapid method of total lipid extraction and purification. Can. J. Biochem. Physiol. 37, 911–917.

Boyd, R.K., Basic, C., Bethem, R.A., 2008. Trace Quantitative Analysis by Mass Spectrometry. John Wiley & Sons, England.

Brügger, B., Erben, G., Sandhoff, R., Wieland, F.T., Lehmann, W.D., 1997. Quantitative analysis of biological membrane lipids at the low picomole level by nano-electrospray ionization mass spectrometry. Proc. Natl. Acad. Sci. USA 94, 2339–2344.

Brown, H.A., Murphy, R.C., 2009. Working towards an exegesis for lipids in biology. Nat. Chem. Biol. 5 (9), 602–606.

Bruntz, R.C., Taylor, H.E., Lindsley, C.W., Brown, H.A., 2014. Phospholipase D2 mediates survival signaling through direct regulation of Akt in glioblastomas cells. J. Biol. Chem. 289 (2), 600–616.

Cevc, G., 1993. Phospholipids Handbook. Marcel Dekker, New York.

Christie, W.W., Han, X., 2010. Lipid Analysis, Fourth Edition: Isolation, Separation, Identification and Lipidomic Analysis. The Oily Press.

Duncan, M.W., Gale, P.J., Yergey, A.L., 2006. The Principals of Quantitative Mass Spectrometry, first ed. Rockpool Productions LLC, Denver, CO.

Fahy, E., Subramaniam, S., Brown, H.A., Glass, C.K., Merrill, A.H., Murphy, R.C., et al., 2005. A comprehensive classification system for lipids. J. Lipid Res. 46 (5), 839–862.

Fenn, J.B., Mann, M., Meng, C.K., Wong, S.F., Whitehouse, C.M., 1989. Electrospray ionization for mass spectrometry of large molecules. Science 246, 64–71.

Fhaner, C.J., Liu, S., Ji, H., Simpson, R.J., Reid, G.E., 2012. Comprehensive lipidome profiling of isogenic primary and metastatic colon adenocarcinoma cell lines. Anal. Chem. 84 (21), 8917–8926.

Folch, J., Ascoli, I., Lees, M., Meath, J.A., LeBaron, N., 1951. Preparation of lipide extracts from brain tissue. J. Biol. Chem. 191, 833–841.

Fridricsson, E.K., Shipkova, P.K., Sheets, E.D., Holowka, D., Baird, B., McLafferty, F.W., 1999. Quantitative analysis of phospholipids in functionally important membrane domains from RBL-2H3 mast cells using tandem high-resolution mass spectrometry. Biochemistry 38 (25), 8056–8063.

Gorden, L.D., Myers, D.S., Ivanova, P.T., Fahy, E., Maurya, M.R., Gupta, S., et al., 2015. Biomarkers of NAFLD progression – an *omics* approach to an epidemic. J. Lipid Res. 56 (3), 722–736.

Han, X., Gross, R.W., 1995. Structural determination of picomole amounts of phospholipids via electrospray ionization tandem mass spectrometry. J. Am. Soc. Mass Spectrom. 6 (12), 1202–1210.

Han, X., Gross, R.W., 2003. Global analyses of cellular lipidomes directly from crude extracts of biological samples by ESI-mass spectrometry: a bridge to lipidomics. J. Lipid Res. 44, 1071–1079.

Han, X., Gross, R.W., 2005. Shotgun lipidomics: electrospray ionization mass spectrometric analysis and quantitation of cellular lipidomes directly from crude extracts of biological samples. Mass Spectrom. Rev. 24 (3), 367–412.

Hanahan, D.J., 1997. A Guide to Phospholipid Chemistry. Oxford University Press, New York. pp. 3–24.

Hermansson, M., Uphoff, A., Käkelä, R., Somerharju, P., 2005. Automated quantitative analysis of complex lipidomes by liquid chromatography/mass spectrometry. Anal. Chem. 77, 2166–2175.

Herzog, R., Schuhmann, K., Schwudke, D., Sampaio, J.L., Bornstein, S.R., Schroeder, M., Schevchenko, A., 2012. LipidXplorer: a software for consensual cross-platform lipidomics. PLoS One 7 (1), e29851.

Husen, P., Tarasov, K., Katafiasz, M., Sokol, E., Vogj, J., Baumgart, J., Nitsch, R., Ekroos, K., Ejsing, C.S., 2013. Analysis of Lipid Experiments (ALEX): a software framework for analysis of high-resolution shotgun lipidomics data. PLoS One 8 (11), e79736.

Ivanova, P.T., Cerda, B.A., Horn, D.M., Cohen, J.S., McLafferty, F.W., Brown, H.A., 2001. Electrospray ionization mass spectrometry analysis of changes in phospholipids in RBL-2H3 mastocytoma cells during degranulation. Proc. Natl. Acad. Sci. USA 98 (13), 7152–7157.

Ivanova, P.T., Milne, S.B., Forrester, J.S., Brown, H.A., 2004. Lipid arrays: new tools in the understanding of membrane dynamics and lipid signaling. Mol. Interv. 4, 86–96.

Ivanova, P.T., Milne, S.B., Byrne, M.O., Xiang, Y., Brown, H.A., 2007. Glycerophospholipid identification and quantitation by electrospray ionization mass spectrometry. Meth. Enzymol. 432, 21–57.

Ivanova, P.T., Milne, S.B., Myers, D.S., Brown, H.A., 2009. Lipidomics: a mass spectrometry based, systems level analysis of cellular lipids. Curr. Opin. Chem. Biol. 13, 526–531.

Kliman, M., May, J.C., McLean, J.A., 2011. Lipid analysis and lipidomics by structurally selective ion mobility – mass spectrometry. Biochim. Biophys. Acta. Mol. Cell Biol. Lipids 1811, 935–945.

Koeth, R.A., Wang, Z., Levison, B.S., Buffa, J.A., Org, E., Sheehy, B.T., Britt, E.B., Fu, X., Wu, Y., Li, L., Smith, J.D., DiDonato, J.A., Chen, J., Li, H., Wu, G.D., Lewis, J.D., Warrier, M., Brown, J.M., Krauss, R.M., Tang, W.H., Bushman, F.D., Lusis, A.J., Hazen, S.L., 2013. Intestinal microbiota metabolism of L-carnitine, a nutrient in red meat, promotes atherosclerosis. Nat. Med. 19 (5), 576–585.

Kumari, M., Schoiswohl, G., Chitraju, C., Paar, M., Cornaciu, I., Rangrez, A.Y., et al., 2012. Adiponutrin (PNPLA3) functions as a nutritionally regulated lysophosphatidic acid acyltransferase. Cell Metab. 15, 691–702.

Leiker, T.J., Barkley, R.M., Murphy, R.C., 2011. Analysis of diacylglycerol molecular species in cellular lipid extracts by normal-phase LC-electrospray mass spectrometry. Inter. J. Mass Spectrom. 305, 103–108.

Li, J.Z., Huang, Y., Karaman, R., Ivanova, P.T., Brown, H.A., Roddy, T., Castro-Perez, J., Cohen, J.C., Hobbs, H.H., 2012. Chronic expression of PNPLA3I148M in mouse liver causes hepatic steatosis. J. Clin. Invest. 122 (11), 4130–4144.

Liu, S.T.H., Sharon-Friling, R., Ivanova, P.T., Milne, S.B., Myers, D.S., Rabinowitz, J., Brown, H.A., Shenk, T., 2011. Synaptic vesicle-like lipidome of human cytomegalovirus reveals a role for SNARE machinery in virion egress. Proc. Natl. Acad. Sci. USA 108 (31), 12869–12874.

Manna, J.D., Wepy, J.A., Hsu, K.-L., Chang, J.W., Cravatt, B.F., Marnett, L.J., 2014. Identification of the major prostaglandin glycerol ester hydrolase in human cancer cells. J. Biol. Chem. 289, 33741–33753.

McDonald, J.G., Smith, D.D., Stiles, A.R., Russell, D.W., 2012. A comprehensive method for extraction and quantitative analysis of sterols and secosteroids from human plasma. J. Lipid Res. 53 (7), 1399–1409.

McLafferty, F.W., Tureček, F., 1993. Interpretation of Mass Spectra, fourth ed. University Science Books, Sausalito, CA.

Methods in Enzymology. In: Alex Brown, H. (Ed.), 2007. Lipidomics and Bioactive Lipids: Mass Spectrometry-Based Lipid Analysis, vol. 432. Elsevier, pp. 1–170. Chapters 1–6.

Milne, S.B., Tallman, K.A., Serwa, R., Rouzer, C.A., Armstrong, M.D., Marnett, L.J., Lukehart, C.M., Porter, N.A., Brown, H.A., 2010. Capture and release of alkyne-derivatized glycerophospholipids using cobalt chemistry. Nat. Chem. Biol. 6 (3), 205–207.

Mukherjee, K.D., Weber, N., 1993. CRC Handbook of Chromatography: Analysis of Lipids. CRC, Boca Raton, FL.

Mullen, A.R., Wheaton, W.W., Jin, E.S., Chen, P., Sullivan, L.B., Cheng, T., Yang, Y., Linehan, W.M., Chandel, N.S., DeBerardinis, R.J., 2011. Reductive carboxylation supports growth in tumour cells with defective mitochondria. Nature 481, 385–388.

Murphy, R.C., 1993. Handbook of Lipid Research. Mass Spectrometry of Lipids. Plenum Press, New York.

Murphy, R.C., Fiedler, J., Hevko, J., 2001. Analysis of nonvolatile lipids by mass spectrometry. Chem. Rev. 101 (2), 479–526.

Murphy, R.C., 2015. Tandem Mass Spectrometry of Lipids: Molecular Analysis of Complex Lipids. Royal Society of Chemistry.

Myers, D.S., Ivanova, P.T., Milne, S.B., Brown, H.A., 2011. Quantitative analysis of glycerophospholipids by LC-MS: acquisition, data handling, and interpretation. Biochim. Biophys. Acta. Mol. Cell Biol. Lipids 1811, 748–757.

Nichols, B.W., 1963. Separation of the lipids of photosynthetic tissues: improvements in analysis by thin-layer chromatography. Biochim. Biophys. Acta 70, 417–422.

NIST, 2009. Development of a Standard Reference Material for Metabolites in Plasma. http://www.nist.gov/mml/csd/organic/metabolitesinserum.cfm.

Oguin, T.H., Sharma, S., Stuart, A.D., Duan, S., Scott, S., Jones, C.K., Daniels, J.S., Lindsley, C.W., Brown, H.A., 2014. Phospholipase D facilitates efficient entry of influenza virus allowing escape from innate immune inhibition. J. Biol. Chem. 289 (37), 25405–25417.

Orth, J.D., Thiele, I., Palsson, B.Ø., 2010. What is flux balance analysis? Nat. Biotechnol. 28 (3), 245–248.

Quehenberger, O., Armando, A.M., Brown, H.A., Milne, S.B., Myers, D.S., Merrill, A.H., et al., 2010. Lipidomics reveals remarkable diversity of lipids in human plasma. J. Lipid Res. 51 (11), 3299–3305.

Russell, D.W., Halford, R.W., Ramirez, D.M.O., Shah, R., Kotti, T., 2009. Cholesterol 24-hydroxylase: an enzyme of cholesterol turnover in the brain. Annu. Rev. Biochem. 78, 1017–1040.

Stiles, A.R., Kozlitina, J., Thompson, B.M., McDonald, J.G., King, K.S., Russell, D.W., 2014. Genetic, anatomic, and clinical determinants of human serum sterol and vitamin D levels. Proc. Natl. Acad. Sci. USA 111 (38), E4006–E4014.

Tang, W.H., Wang, Z., Levison, B.S., Koeth, R.A., Britt, E.B., Fu, X., Wu, Y., Hazen, S.L., 2013. Intestinal microbial metabolism of phosphatidylcholine and cardiovascular risk. N. Engl. J. Med. 368 (17), 1575–1584.

Wang, Z., Klipfell, E., Bennett, B.J., Koeth, R.A., Levison, B.S., Dugar, B., Feldstein, A.E., Britt, E.B., Fu, X., Chung, Y.M., Wu, Y., Schauer, P., Smith, J.D., Allayee, H., Tang, W.H., DiDonato, J.A., Lusis, A.J., Hazen, S.L., 2011. Gut flora metabolism of phosphatidylcholine promotes cardiovascular disease. Nature 472 (7341), 57–63.

Welti, R., Shah, J., Li, W., Li, M., Chen, J., Burke, J.J., Fauconnier, M.-L., Chapman, K., Chye, M.-L., Wang, X., 2007. Plant lipidomics: discerning biological function by profiling plant complex lipids using mass spectrometry. Front. Biosci. 12, 2494–2506.

Wenk, M., 2006. Lipidomics of host-pathogen interactions. FEBS Lett. 580 (23), 5541–5551.

Yore, M.M., Syed, I., Moraes-Vieira, P.M., Zhang, T., Herman, M.A., Homan, E.A., Patel, R.T., Lee, J., Chen, S., Peroni, O.D., Dhaneshwar, A.S., Hammarstedt, A., Smith, U., McGraw, T.E., Saghatelian, A., Kahn, B.B., 2014. Discovery of a class of endogenous mammalian lipids with anti-diabetic and anti-inflammatory effects. Cell 159 (2), 318–332.

Chapter 3

Fatty Acid and Phospholipid Biosynthesis in Prokaryotes

Yong-Mei Zhang,[1] Charles O. Rock[2]

[1]*Department of Biochemistry and Molecular Biology, Medical University of South Carolina, Charleston, SC, USA;* [2]*Department of Infectious Diseases, St. Jude Children's Research Hospital, Memphis, TN, USA*

ABBREVIATIONS

ACP Acyl carrier protein
CoA Coenzyme A
fabA, **FabA** lowercase italics indicates gene, while uppercase Roman type indicates the protein product of the gene
PC Phosphatidylcholine
PE Phosphatidylethanolamine
PG Phosphatidylglycerol
PS Phosphatidylserine
PGP Phosphatidylglycerolphosphate
CL Cardiolipin
LPS Lipopolysaccharides
ABC ATP-binding cassette
MDO Membrane-derived oligosaccharides

1. OVERVIEW OF BACTERIAL LIPID METABOLISM

Bacteria are a versatile tool for the study of metabolic pathways. This is especially true for the Gram-negative *Escherichia coli*. These bacteria are easy to grow, large amounts of material can be obtained and the growth conditions can be controlled and manipulated by the investigator. Most important, they are suitable for genetic manipulation and their genome sequences are available. *Escherichia coli* is the most extensively studied bacterium, and the genes and enzymes of fatty acid and phospholipid metabolism were first delineated using this organism (Rock and Cronan, 1996; Cronan et al., 1996). Although the *E. coli* paradigm provides a solid core of information that is shared by all bacteria, the biochemical details of lipid metabolic pathways in other bacterial cannot be inferred from the *E. coli* model. The explosion of genomic information in the last several years has allowed detailed comparisons of lipid metabolism in various bacteria based on their complement of genes.

Biochemistry of Lipids, Lipoproteins and Membranes. http://dx.doi.org/10.1016/B978-0-444-63438-2.00003-1

The goal of this chapter is to concisely layout the membrane phospholipid biosynthetic pathways as they are known in most bacteria. This means that a lot of interesting and important biochemistry in prokaryotes is not explored in detail. Interested readers should refer to an extensively referenced recent review of this area, which cites the original experimental work that underpin many of the statements of fact made in this chapter (Parsons and Rock, 2013). The structures and biosynthesis of the unique lipids in the Archaea are not covered, leaving out metabolism in an entire domain of life. The role of isoprenoids, carotenoids, and other hydrophobic carriers and membrane constituents in membrane lipid metabolism and structure are not well known. Likewise, the fascinating lipid biochemistry that is involved in the synthesis of co-factors (biotin, lipoic acids, etc.), signalling molecules (homoserinelactones, quinolones, etc.) and storage lipids (polyhydroxyalkanoates), which are important cellular components derived from lipid metabolism that are required for intermediary metabolism, survival and social behaviour in bacteria, is beyond the scope of this chapter. The pathways and regulation discussed in this chapter, while common to the bacteria being investigated most thoroughly, may not apply to a specific bacterial species. There is a large diversity in the structures and metabolism of membrane lipids, and not a single paradigm can be considered typical.

The study of phospholipid enzymology in *E. coli* dates back to the early 1960s, when work in the laboratory of Vagelos discovered that the intermediates in fatty acid synthesis are bound to a heat stable cofactor named acyl carrier protein (ACP) (Rock and Cronan, 1996; Cronan et al., 1996). The enzymes of bacterial type II fatty acid synthesis (FASII) are soluble proteins, the individual activities of which can be purified and studied independently from the other components. This is markedly different from the mammalian fatty acid synthase, a large multifunctional polypeptide with intermediates covalently attached (Chapter 5). While the mammalian system offers some efficiency in catalysis, it only produces a single fatty acid, whereas FASII is much more flexible and capable of producing a range of fatty acid structures and other intermediates essential for the synthesis of co-factors and signalling molecules. Thus, the FASII enzymes became a focus of intensive study in the laboratories of Vagelos, Bloch and Wakil. The structures of all of the intermediates are known, and the basic chemical reactions are described (Bloch and Vance, 1977). During the late 1960s, work on the enzymes of phospholipid synthesis in bacteria flourished, and, mainly through classical identification experiments in the Kennedy laboratory, the intermediates in the pathway were established (Raetz, 1982).

A second phase of bacterial phospholipid research, during the 1970s and early 1980s, was the identification of mutants in the pathway (Clark and Cronan, 1981). Mutants in many specific enzymes were generated by using mutagens in combination with a battery of clever selection and screening techniques (Bloch and Vance, 1977; Clark and Cronan, 1981). Such mutations generally fall into one of two classes. First, they may confer an auxotrophy on a strain, such as a requirement for unsaturated fatty acids or glycerol phosphate. Such mutants have generally lost the ability to produce a key biosynthetic enzyme (e.g. *fabA* mutants, which require supplementation with unsaturated fatty acids), or they may be more complex (e.g. *plsB* mutants require high glycerol phosphate concentrations due to a K_m defect in the enzyme and a second site mutation). Second, they may be conditionally defective, usually at elevated temperatures. For example, strains with the *fabI*(Ts) mutation grow at 30 °C, but are not viable at 42 °C. This type of defect is usually ascribed to an amino acid change in an essential enzyme that renders it unstable at higher temperatures. Techniques using the bacteriophage P1 are available for the movement of these alleles into other host strains, thus allowing for the mapping of the genes to specific regions of the chromosome, or the generation of strains with particular combinations of mutations. Regulatory mutants were identified, affecting multiple enzymes with a single mutation that allowed for regulatory networks to be investigated. The membrane-bound enzymes of phospholipid synthesis were largely not amenable to identification by protein purification, and the genetic approach allowed these enzymes to be identified.

Next, came the cloning and detailed study of the enzymes of lipid metabolism during the late 1980s and 1990s (Cronan et al., 1996). Plasmid-based expression systems were used to examine overexpression of enzymes on pathway regulation, and purified enzymes could be more easily obtained for biochemical analysis. With the availability of the sequence data generated by these clones, more precise methods for the construction of specific mutations also became available. Specific genes, or portions of genes, could be 'knocked-out' by targeted replacement based on sequence information, as opposed to random insertion of phage DNA. This period also saw an explosion of structural information on the enzymes of FASII that revealed in detail the biochemical mechanisms of these enzymes (White et al., 2005). In the late 2000s, the genomic sequences of a broad spectrum of bacteria became available on the Web, complemented by user-friendly bioinformatic tools. The explosion of bacterial genomic sequences continues to this day, and there is now an enormous compendium of complete bacterial genome sequences available that provide a solid bioinfomatic framework for comparative lipid metabolism. Research after 2005 has largely shifted to delineating the fine details of pathway enzymology and the characterisation of human pathogens, with an eye toward the development of therapeutics that target FASII.

This chapter on bacterial lipid metabolism is divided into sections corresponding to the stages of lipid synthesis that are common to all bacteria. The initiation module is responsible for assembling the precursors for pathway operation and the first condensation step in FASII. Each time the initiation module fires, a new fatty acid is made; thus this pathway component is responsible for determining the amount and type of fatty acids made. The elongation module is the central engine of FASII that determines the length and degree of unsaturation of fatty acids produced and, importantly, the rate at which they are made. The acyltranfer module is the interface between FASII and phospholipid synthesis creating the key intermediate, phosphatidic acid, from which all membrane lipids are derived. The phospholipid module is responsible for introducing the diversity of lipid structures found in the membrane. A diverse collection of enzymes give each bacterial species its unique membrane lipid composition. A list of the most relevant lipid metabolism genes and a description of their protein products is found in Table 1.

2. MEMBRANE SYSTEMS OF BACTERIA

Phospholipids in *E. coli* and other Gram-negative bacteria are used in the construction of the inner and outer membranes. The inner membrane (IM) is impermeable to solutes unless specific transport systems are present and corresponds to the cytoplasmic membrane of mammalian cells. The outer membrane (OM) is a rigid structure that is rich in structural lipoproteins and proteins involved in the transport of high molecular weight compounds. Specific pores in the outer membrane allow the passage of molecules having a molecular weight less than 600. The outer layer of the outer membrane is composed primarily of lipopolysaccharides rather than phospholipid. Between the inner and outer membranes is an osmotically active compartment called the periplasmic space PS. Membrane-derived oligosaccharides (MDO), peptidoglycan, and binding proteins involved with metabolite transport are found in this compartment. Gram-positive bacteria do not possess an outer membrane. Instead, they have a membrane bilayer surrounded by a thick layer of peptidoglycan (PG) decorated with proteins, carbohydrates, and often, teichoic and lipoteichoic acid.

3. THE INITIATION MODULE

3.1 Acyl Carrier Protein

A central feature of FASII is that the intermediates are carried from enzyme to enzyme by a small (8.86 kDa), acidic, and highly soluble protein, ACP, the product of the *acpP* gene. ACP is one of the most abundant proteins in *E. coli*, constituting about 0.25% of the total soluble protein ($\sim 6 \times 10^4$ molecules/cell). The acyl intermediates of fatty acid biosynthesis are bound to the protein through a thioester linkage to the terminal sulphhydryl of the 4′-phosphopantetheine prosthetic group. The prosthetic group contains the only thiol group of ACP and is attached to the protein via a phosphodiester linkage to Ser-36. ACP must interact specifically and transiently with all of the enzymes of FASII and does so through interactions with exposed negatively charged residues on ACP with a patch of positive residues on the surfaces of the *fab* enzymes (White et al., 2005).

The ACP pool in normally growing cells is approximately one-eighth the coenzyme A (CoA) pool, the other acyl group carrier in cells. The prosthetic group of ACP is derived from CoA, and the common feature of both is the

TABLE 1 Abbreviated List of Genes in Bacterial Lipid Metabolism

Gene	Protein/Enzyme
aas	2-Acyl-GPE acyltransferase
aasS	Acyl-ACP synthetase
accA	Carboxyltransferase subunit
accB	Biotin carboxy carrier protein
accC	Biotin carboxylase
accD	Carboxyl transferase subunit
acpP	Acyl carrier protein
acpH	Acyl carrier protein hydrolase
acpS	Acyl carrier protein synthase
cdh	CDP-diacylglycerol hydrolase
cdsA	CDP-diacylglycerol synthase
cdsS	Stabilises mutant CDP-diacylglycerol synthase
cfa	Cyclopropane fatty acid synthase
cls	Cardiolipin synthase
desA	Phospholipid desaturase
desB	Acyl-CoA desaturase
desT	Transcriptional regulator
desR	Transcriptional regulator
dgkA	Diacylglycerol kinase (Gram-negative)
dgkB	Diacylglycerol kinase (Gram-positive)
fabA	β-Hydroxydecanoyl-ACP dehydratase I
fabB	β-Ketoacyl-ACP synthase I
fabD	Malonyl-CoA:ACP transacylase
fabF	β-Ketoacyl-ACP synthase II
fabG	β-Ketoacyl-ACP reductase
fabH	β-Ketoacyl-ACP-synthase III
fabI	Enoyl-ACP reductase I
fabK	Enoyl-ACP reductase II
fabL	Enoyl-ACP reductase III

Continued

TABLE 1 Abbreviated List of Genes in Bacterial Lipid Metabolism—cont'd

Gene	Protein/Enzyme
fabM	*trans*-2-Enoyl-ACP isomerase
fabN	β-Hydroxydecanoyl-ACP dehydrase/isomerase II
fabR	Transcriptional regulator (Gram-negative)
fabT	Transcriptional regulator (Gram-positive)
fabZ	β-Hydroxyacyl-ACP dehydratase
fadA	β–Ketoacyl-CoA thiolase
fadB	4-Function enzyme of β-oxidation: β–hydroxyacyl-CoA dehydrogenase and epimerase; *cis*-β–*trans*-2-enoyl-CoA isomerase and enoyl-CoA hydratase
fadD	Acyl-CoA synthetase
fadE	Electron transferring flavoprotein
fadF	Acyl-CoA dehydrogenase
fadG	Acyl-CoA dehydrogenase?
fadH	2,4-Dienoyl-CoA reductase
fadL	Long-chain fatty acid transport protein precursor
fadR	Transcriptional regulator
fakA	The ATP-binding component of fatty acid kinase
fakB	The fatty acid binding component of fatty acid kinase
fapR	Transcriptional regulator
fatA	Unknown, possible transcription factor
gpsA	Glycerol phosphate synthase
htrB	KDO_2-lipid IV_A acyloxy lauroyltransferase
kdtA	KDO transferase
lplT	Lysophospholipid flippase
lpxA	UDP-GlcNAc β-hydroxymyristoyl-ACP acyltransferase
lpxB	Disaccharide-1-P synthase
lpxC	UDP-β–*O*-hydroxymyristoyl-GlcNAc deacetylase
lpxD	UDP-β–*O*-hydroxymyristoyl-GlcN N-acyltransferase
lpxK	Disaccharide-1-P 4′-kinase
lpxP	KDO_2-lipid IV_A acyloxy palmitoyltransferase

Gene	Protein/Enzyme
mdoB	Phosphatidylglycerol:membrane-oligosaccharide glycerophosphotransferase
msbA	Lipid flippase
msbB	KDO$_2$-Lipid IV$_A$ acyloxy myristoyltransferase
pgpA	PGP phosphatase
pgpB	PGP phosphatase
pgsA	PGP synthase
pldA	Detergent-resistant phospholipase A
pldB	Inner membrane lysophospholipase
plsB	Glycerol phosphate acyltransferase
plsC	1-Acylglycerol phosphate acyltransferase
plsX	Acyl-ACP phosphotransferase
plsY	Acylphosphate:glycerolphosphate acyltransferase
psd	PS decarboxylase
pssA	PS synthase
tesA	Thioesterase I
tesB	Thioesterase II

pantetheine arm for thioester formation. ACP is maintained in the active, holo-form in vivo, indicating that the supply of prosthetic group does not limit fatty acid biosynthesis. The 4′-phosphopantetheine prosthetic group is transferred from CoA to apo-ACP by the 14 kDa trimeric [ACP]synthase (Figure 1). The [ACP]synthase from *Bacillus subtilis* was crystallised in complex with ACP to give the first detailed look at ACP-protein interactions.

ACP is one of the most interactive proteins in bacterial physiology (Butland et al., 2005), and the physiological significance of its recognition by all its binding partners continues to be the subject of research. The ACPs from one bacterial species readily substitutes for other ACPs in FASII due to the conservation of residues along helix-2, termed the recognition helix (Zhang et al., 2003). The conserved negatively charged and hydrophobic residues of ACP interact with a complementary constellation of basic and hydrophobic residues on the target protein that are adjacent to the active site tunnel. Whereas the *fab* enzymes do not have a conserved primary sequence that can be recognised as an ACP recognition motif, their 3-dimensional structures illustrate that the basic hydrophobic

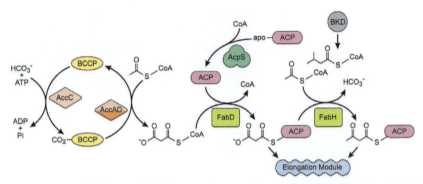

FIGURE 1   The initiation module. This group of enzymes is responsible for the initiation of FASII. Acetyl-CoA carboxylase is a 4-protein complex that produces malonyl-CoA by a two-step reaction. AccC carboxylates biotin attached to BCCP, and then AccAD transfers the carboxyl group to acetyl-CoA. Acyl carrier protein (ACP) is the key acyl group carrier in FASII. The primary gene product is an apoprotein that is converted to ACP by the transfer of the 4′-phosphopantetheine moiety of CoA to ACP catalysed by ACP synthase (AcpS). Malonyl-CoA:ACP transacylase (FabD) maintains an equilibrium between the malonyl-CoA and malonyl-ACP pools. FabH is the initiation-condensing enzyme (3-ketoacyl-ACP synthase III) that condenses acetyl-CoA to malonyl-ACP to form the first 3-ketoacyl-ACP intermediate in FASII. Each time FabH fires, a new acyl chain is made. Bacteria that produce branched-chain fatty acids use a branched-chain acyl-CoA primer derived from amino acid metabolism via branched-chain ketoacid dehydrogenase (BKD).

recognition region at the active site entrance is a common structural feature critically important for ACP binding (White et al., 2005; Zhang et al., 2003).

3.2 Acetyl-Coenzyme A Carboxylase

Acetyl-CoA carboxylase catalyses formation of the key precursor that fuels FASII, the conversion of acetyl-CoA to malonyl-CoA. Acetyl-CoA is a key intermediate in many pathways, and forms most of the CoA species within the cell at concentrations of about 0.5–1.0 mM during logarithmic growth of *E. coli* on glucose. Malonyl-CoA is normally present at 0.5% of this level and is used almost exclusively by FASII. The overall carboxylation reaction consists of two distinct half reactions: the ATP-dependent carboxylation of biotin with bicarbonate to form carboxybiotin; and transfer of the carboxyl group from carboxybiotin to acetyl-CoA, forming malonyl-CoA (Figure 1) (Cronan and Waldrop, 2002). Each acetyl-CoA carboxylase half reaction is catalysed by a distinct protein subcomplex. The vitamin biotin is covalently coupled through an amide bond to a lysine residue on biotin carboxyl carrier protein (BCCP; a homodimer of 16.7 kDa subunits encoded by *accB*) by a specific enzyme, biotin-apoprotein ligase (encoded by *birA*). The crystal and solution structures of the biotinyl domain of BCCP have been determined, and reveal a unique 'thumb' required for activity. Carboxylation of biotin is catalysed by biotin carboxylase (encoded by *accC*), a homodimeric enzyme composed of 55 kDa subunits. The *accB* and *accC* genes form an operon in most bacteria. The three-dimensional structure of the biotin carboxylase (AccC) subunit has been solved by X-ray diffraction, revealing an 'ATP-grasp' motif for nucleotide binding. The mechanism of biotin carboxylation involves the reaction of ATP and bicarbonate to form the short-lived carboxyphosphate, which then carboxylates the 1'-nitrogen of biotin on BCCP. The carboxyltransferase enzyme that transfers the carboxy group from the biotin moiety of BCCP to acetyl-CoA is a heterotetramer (dimer of dimers) composed of two copies of two dissimilar subunits encoded by *accA* and *accD*. Strains with mutations in *accB* and *accD* have been obtained that are temperature-sensitive for growth, illustrating that this reaction is essential for FASII. It is thought that the enzyme present in vivo is composed of one copy of each subcomplex, with a combined molecular weight of 280 kDa. However, on cell lysis, the AccC, AccAD, and AccB components dissociate and the entire complex cannot be purified as a single entity.

3.3 Malonyl Transacylase

For the malonate group to be used for fatty acid synthesis, it must first be transferred from malonyl-CoA to ACP by the 32.4 kDa monomeric malonyl-CoA:ACP transacylase, the product of the *fabD* gene (Figure 1). A malonyl-serine enzyme intermediate is formed during the course of the FabD reaction and subsequent nucleophilic attack on this ester by the sulfhydryl of ACP yields malonyl-ACP. FabD catalyses a rapid equilibrium transacylation reaction. The high reactivity of the serine in malonyl-CoA:ACP transacylase is due to the active site being composed of a nucleophilic elbow, as observed in α/β hydrolases. The serine is hydrogen bonded to His-201 in a fashion similar to serine hydrolases.

3.4 3-Ketoacyl-Acyl Carrier Protein Synthase III

The last two carbons of the fatty acid chain (i.e. those most distal from the carboxylate group) are the first introduced into the nascent chain, and acetyl-CoA can be thought of as the 'primer' molecule of fatty acid synthesis in *E. coli*. The initial condensation reaction, catalysed by 3-ketoacyl-ACP synthase III (FabH), uses acetyl-CoA and malonyl-ACP to form the four carbon acetoacetyl-ACP with concomitant loss of bicarbonate (Figure 1). FabH is a ping-pong enzyme that contains a catalytic triad composed of Cys-His-Asn. Every FabH reaction leads to a new fatty acid being formed by FASII, making FabH the focal point for regulating the amount of fatty acids produced. The idea that FabH is the universal initiator of FASII was recently shattered by the discovery of FabY, which contains a Cys-His-His active site triad and initiates FASII in *Pseudomonas aeruginosa*. This discovery is one of the many examples highlighting the diversity of bacterial lipid metabolism.

The FabH proteins play a major role in specifying product diversity. *Escherichia coli* FabH is specific for acetyl-CoA as the primer, and this organism makes straight-chain, even numbered fatty acids. However, many Gram-positive bacteria produce branched-chain fatty acids. The branched-chain primers are derived from amino acid metabolism via a branched-chain ketoacid dehydrogenase (BKD, Figure 1), which when present forms another component of the initiation module. The FabH proteins from these organisms prefer 5-carbon branched-chain acyl-CoAs, resulting in the production of *anteiso* and *iso* branched-chain fatty acid products. *Mycobacterium tuberculosis* FabH prefers long-chain acyl-CoAs and initiates the production of very long-chain mycolic acids.

3.5 Regulation in the Initiation Module

The initiation module is an ideal point of regulation in membrane lipid biosynthesis. The two most significant enzymes in the initiation module are ACC and FabH (Figure 1). ACC performs the first committed step in FASII, and every FabH reaction results in the formation of a new fatty acid. In *E. coli*, the ACC is feedback inhibited by long-chain acyl-ACP representing an end-product feedback mechanism (Figure 2). Unacylated ACP shows no inhibition. The biotin carboxylase (AccC) and carboxyltransferase (AccAD) reactions individually are not inhibited by acyl-ACP, pointing to regulation occurring in the ACC multiprotein complex (Figure 2). The importance of long-chain acyl-ACP is highlighted by experiments in vivo, in which blocking the elongation module with specific drugs leads to the continued synthesis of malonyl-CoA. Unlike *E. coli*, *Streptococcus pneumoniae* can transfer exogenous fatty acids to ACP via fatty acid kinase and PlsX. Following treatment with exogenous oleate, intracellular malonyl-CoA pool falls to less than 5% of the untreated cells. Although a biochemical link between acyl-ACP in this physiological response has not been established, it seems reasonable to postulate that acyl-ACP may

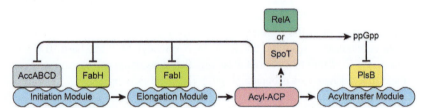

FIGURE 2 Regulation of lipid synthesis by acyl-ACP and ppGpp. The long-chain acyl-ACP end products of the elongation module act as feedback regulators of the elongation module at the enoyl-ACP reductase (FabI) step to slow the rate of fatty acid synthesis. Acyl-ACP also bichemically inhibit 3-ketoacyl-ACP synthase III (FabH) to suppress the intiation of FASII and acetyl-CoA carboxylase (AccABCD) to block both inititation and elongation. The regulatory nucleotide, ppGpp is produced in resonse to amino acid starvation and inhibits the glycerol-phosphate acyltransferase (PlsB) in the acyltransfer module to block phospholipid synthesis and trigger the accumulation of acyl-ACP. ACP also interacts with SpoT and alternate ppGpp synthase. How this protein–protein interaction biochemically regulates SpoT to impact ppGpp levels remains to be elucidated.

repress ACC activity (Figure 2). Thus, inhibition of the ACC by acyl-ACP is a logical feedback system linking the end product of the pathway with the initial reaction and is likely to be widespread in bacteria.

Overproduction of the ACC in *E. coli* does not result in a significant increase in the rate of fatty acid synthesis, unless a soluble acyl-ACP thioesterase (TesA) is co-expressed to uncouple acyl-ACP use from phospholipid synthesis. These data confirm the extremely tight regulation involving acyl-ACP in *E. coli* and also point to another control point in FASII besides the ACC. This additional control point is FabH (Figure 2). Unlike the acyl-chain length independent effect of acyl-ACP on the ACC, the potency of acyl-ACP inhibition of FabH increases with the length of the fatty acid. This graded response helps to tune the activity of the initiation module to the elongation and acyltransfer modules, ensuring fatty acids of the desired length are synthesised. The mode of inhibition is competitive with respect to malonyl-ACP and mixed inhibition with acetyl-CoA, indicating that acyl-ACP binds to the free-enzyme and the acyl-enzyme intermediate. The combinatorial regulation of ACC and FabH by acyl-ACP, the product of the elongation module, allows for synergistic feedback regulation of the initiation cycle to control the quantity of fatty acid produced (Figure 2). The inhibition of FabH by acyl-ACP has only been characterised in *E. coli*, but may be present in other organisms. Oleic acid treatment of *Staphylococcus aureus* reduces lipid synthesis by 50%, although malonyl-CoA levels remain unchanged, suggesting FabH is a target for regulation of FASII at the FabH step by oleoyl-ACP. The lack of repression of ACC activity combined with the reduction of lipid synthesis points to FabH as the regulatory point and potentially acyl-ACP as the regulatory ligand (Figure 2), but further studies are needed to confirm this.

4. THE ELONGATION MODULE

Four enzymatic reactions participate in each iterative cycle of chain elongation (Figure 3). First, 3-ketoacyl-ACP synthase I or II (FabB or FabF) adds a two-carbon unit from malonyl-ACP to the growing acyl-ACP. The resulting ketoester is reduced by a NADPH-dependent 3-ketoacyl-ACP reductase (FabG), and a water molecule is then removed by a 3-hydroxyacyl-ACP dehydratase (FabA or FabZ). The last step is catalysed by enoyl-ACP reductase (FabI or FabK) to form a saturated acyl-ACP.

Historically, scientists in the field have debated whether the enzymes of bacterial fatty acid synthase system are 'dissociable' or 'dissociated'. This seemingly minor semantic distinction has larger ramifications for the in vivo physiology of the cell. The implication of the word dissociable is that the enzymes form a complex in vivo and only became separated (or dissociated) on cell disruption, whereas in a dissociated system, the enzymes do not form a complex in vivo. The concept of a large complex mimicking the multifunctional type I enzyme would support the notion of substrate channelling between the

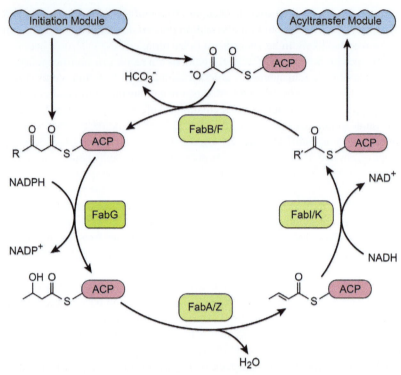

FIGURE 3 The elongation module. This module consists of four steps. The initiation module supplies the first 3-ketoacyl-ACP and malonyl-ACP to the elongation module. The 3-ketoacyl-ACP intermediate is reduced by 3-ketoacyl-ACP reductase (FabG). The hydroxyl intermediate is dehydrated by an isoform of 3-hydroxyl-acyl-ACP dehydratase. FabZ is the most widely distributed in bacteria, and FabA is also important in unsaturated fatty acid synthesis (Figure 4). The last step is catalysed by enoyl-ACP reductase. The FabI and FabK isoforms are the most widely distributed in bacteria.

active sites, increasing the catalytic efficiency of the pathway as a whole. Evidence for protein–protein interactions in the *M. tuberculosis* FASII derived from yeast 2-hybrid screens indicates the existence of potentially three protein complexes, each anchored by the initiation and elongation condensing enzymes in the pathway. The *M. tuberculosis* FASII system is distinct from other bacterial FASIIs in that it initiates with a long-chain acyl-CoA and produces very long-chain mycolic acids. However, it does suggest that prototypical FASII systems should be analysed to determine whether similar complexes exist. There is one important feature of the FASII system that is consistent with its configuration as a dissociated system. Although a 'hardwired' type I mechanism may seem more efficient, it is essentially a one trick pony that produces only a single product. In contrast, the type II system is not only responsible for producing all the diversity of fatty acids in membrane, the intermediates of the pathway are diverted for the synthesis of other key molecules. These include biotin, lipoic acid and the quorum-sensing acylhomoserine lactones. These molecules are produced in relatively low abundance, but in the case of *P. aeruginosa* the intracellular storage lipids and the extracellular rhamnolipids are produced in quantities that exceed the number of acyl groups in the membrane. The existence of a dissociated pathway allows these ancillary metabolic pathways to compete with the FASII enzymes for the acyl-ACP intermediates in the elongation module to produce a diversity of products that are critical to cellular metabolism. Although it is clear that each pathway enzyme strongly interacts with ACP, there is no compelling evidence for multimolecular complexes in vivo.

4.1 3-Ketoacyl-Acyl Carrier Protein Synthases I and II

There are two isoforms of the elongation condensing enzymes. Almost all bacteria have the 3-ketoacyl-ACP synthase II (FabF). γ-Proteobacteria that produce unsaturated fatty acids also contain the 3-ketoacyl-ACP synthase I (FabB) isoform, which, in addition to elongation, plays a special role in introducing the double bond (see below). These enzymes are dimers of identical subunits, and use a ping-pong reaction mechanism. In the ping step, the acyl group is transferred from acyl-ACP to the active site cysteine, and the free ACP departs. The acyl-enzyme undergoes a conformational change to allow the entry of malonyl-ACP into the active site, which is condensed with the acyl-group releasing 3-ketoacyl-ACP, HCO_3^-, and the free enzyme. 3-Ketoacyl-ACP synthase II (FabF) is similar to FabB (38% identical at the amino acid level in *E. coli*). FabF is not essential to growth in *E. coli,* but is by far the most common and essential condensing enzyme in bacteria. FabF is the only elongation condensing enzyme present in most bacteria, including the Gram-positive pathogens. The crystal structures of many examples of FabB and FabF have been determined, and they are virtually identical with a common thiolase fold. Both use a conserved catalytic triad of Cys-His-His. The elongation condensing enzymes are responsible in large part for the structure of fatty acids produced, and the chain-length specificity of the condensing enzyme places an upper limit on the length of acyl chains produced by FASII.

4.2 3-Ketoacyl-Acyl Carrier Protein Reductase

The 3-ketoacyl-ACP reductase gene (*fabG*) is located within the *fab* gene cluster between the *fabD* and *acpP* genes and is co-transcribed with *acpP*. Insertional mutants that prevent *fabG* transcription while allowing ACP to be produced were generated in the Cronan laboratory and suggest that *fabG*-encoded reductase activity is essential in *E. coli*. The *fabG*-encoded NADPH-specific 3-ketoacyl-ACP reductase is a homotetrameric protein of 25.6 kDa subunits. The enzyme uses all chain lengths in vitro and exhibits cooperative binding of NADPH. A dramatic conformational change occurs on cofactor binding, as evidenced by the crystal structures of the free and NADPH-bound protein. FabG is the only protein in the elongation cycle that exists as a single isoform and is universally distributed in bacteria. There is no known regulatory role for FabG in the Elongation Module.

4.3 3-Hydroxyacyl-Acyl Carrier Protein Dehydratases

Escherichia coli possesses two 3-hydroxyacyl-ACP dehydratases. One is encoded by *fabZ*. FabZ is the most widely distributed dehydratase and is active on all chain lengths of saturated and unsaturated intermediates. This enzyme is distinct from the dual-function 3-hydroxydecanoyl-ACP dehydratase/isomerase (encoded by *fabA*), which was first described by Bloch and coworkers. The FabA enzyme also catalyses a key isomerisation reaction at the point where the biosynthesis of unsaturated fatty acids diverges from saturated fatty acids (Figure 4). FabA always occurs with FabB, and their expression is largely restricted to the γ-proteobacteria. Both FabA and FabZ enzymes share weak overall homology, with 28% identity and 50% similarity at the amino acid level. The monofunctional FabZ protein (subunit MW of 17 kDa) is a hexamer (a trimer of dimers), whereas the *fabA*-encoded bifunctional enzyme is a dimer (subunit MW of 19 kDa). The crystal structures of dehydratases show that the dimers of both FabA and FabZ adopt a hot-dog fold. The active sites are nearly identical, and it is thought that the shape of the acyl chain binding tunnel determines whether the proteins can isomerise substrates or not. Supporting this concept, some Gram-positive bacteria that make unsaturated fatty acids express a FabZ-like protein that, like FabA, possess dehydratase/isomerase activity and can accommodate the kink that arises through introduction of the *cis*-3 double bond.

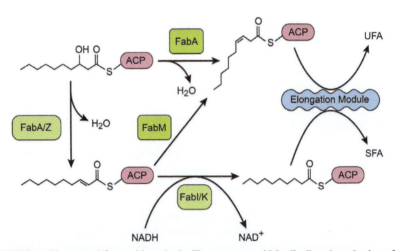

FIGURE 4 Unsaturated fatty acid synthesis. The two most widely distributed mechanisms for unsaturated fatty acid synthesis are the isomerisation of the *trans*-2-decenoyl-ACP intermediate to the *cis*-3-decenoyl-ACP by FabA in Gram-negative and FabM in Gram-positive bacteria.

A second way to introduce a double bond into the growing acyl chain is carried out by an isomerase called FabM that is expressed in Gram-positive bacteria like *S. pneumoniae*. FabM converts *trans*-2-decenoyl-ACP to *cis*-decenoyl-ACP and is structurally related to the isomerase in fatty acid oxidation (Figure 4). Other Gram-positive bacteria, such as *Clostridium*, produce copious amounts of unsaturated fatty acids, but do not contain the isomerases shown in Figure 4, leaving the enzyme(s) involved a mystery.

4.4 Enoyl-Acyl Carrier Protein Reductase

The final step in each round of fatty acyl elongation in *E. coli* is the NADH-dependent reduction of the *trans* double bond, catalysed by the homotetrameric (subunit mass of 29 kDa) NADH-dependent enoyl-ACP reductase I (encoded by *fabI*). The FabI amino acid sequence is similar (34% identical) to the product of a gene called *inhA* from mycobacteria. InhA is involved in mycolic acid biosynthesis. The synthesis of these unusual 70–80 carbon mycobacterial acids requires a pathway composed of enzymes essentially identical to those of fatty acid synthesis. Missense mutations within the *inhA* gene result in resistance to the antituberculosis drugs isoniazid and ethionamide. The crystal structures of FabI and InhA have been solved and are virtually superimposable for most of the protein. FabI has a flexible substrate binding loop that becomes ordered on binding of specific small molecule inhibitors (diazaborine, triclosan), which occupy the substrate binding site. A novel enoyl-ACP reductase II (FabK) is found in Gram-positive bacteria. FabK is an NADH-dependent enzyme, but also has a flavin mononucleotide cofactor and is structurally unrelated to FabI. In addition to these two major enoyl-ACP reductases, there are other enoyl-ACP reductases with structures related to FabI (FabL and FabV), but distinct enough to be clearly separate entities. Why some bacteria express multiple enoyl-ACP reductases remains to be elucidated.

Experiments primarily with the *E. coli* system indicate that FabI plays an important role in determining the rate of the elongation module. The dehydratases (FabA and FabZ) catalyse a rapid equilibrium reaction that favours the 3-hydroxyacyl-ACP over enoyl-ACP intermediate. Thus, the activity of FabI is responsible for pulling cycles of elongation to completion and is thought to be a pacemaker of FASII that is feedback regulated by acyl-ACP (Figure 2). However, it is not clear whether the *E. coli* paradigm can be extended to other groups of bacteria, particularly the many organisms that lack FabI or have multiple enoyl-ACP reductases.

4.5 Regulation in the Elongation Module

The activities and substrate specificities of the elongation enzymes are finely balanced to produce fatty acids of the desired length with the correct ratio of unsaturated:saturated fatty acids. The acyl chain length of the fatty acids generated by the cycle is determined by the competition for the acyl-ACP between the glycerolphosphate acyltransferases (such as PlsB) and the elongation condensing enzymes (such as FabB and FabF). The longer chain acyl-ACPs are poor substrates for FabB/FabF and ideal substrates for the acyltransferases. Either the overexpression of an elongation condensing enzyme or the inhibition of PlsB alters the kinetic balance between the elongation and acyltransfer modules, leading to abnormally long fatty acids in the membrane phospholipids. The acyl-chain length specificity of FabB/FabF is explained by the size of substrate binding tunnel adjacent to the active site. Many crystal structures illustrate that the pocket accommodates acyl-chains up to 16 carbons and limits the enzyme to producing 18- or 20-carbon fatty acids as an upper limit. The trans-2-enoyl-ACP reductase (FabI) is also feedback inhibited by acyl-ACP to slow the rate at which the elongation cycle turns (Figure 2). This effect is attributed to product inhibition. In summary, biochemical regulation at the FabB/F and FabI steps in the elongation module pathway work in concert to provide a tight control over the rate of FASII, and the interplay between the elongation and acyltransfer modules determines the fatty acid chain length. The roles of FabA from E. coli and FabM from S. pneumoniae in regulating the unsaturated fatty acid content are also relevant. FabA works in concert with FabB, and these two genes are transcriptionally linked. How unsaturated fatty acid synthesis is controlled in other systems is a mystery.

4.6 Bacteria With a Type I Synthase

A general distinction between prokaryotic and eukaryotic fatty acid synthases is that bacteria possess the dissociated enzymes described above (type II), while higher organisms have a single, multifunctional protein (type I) that catalyses all of the reactions. There are exceptions to this rule. Mycobacteria, for example, possess a type I fatty acid synthase for the production of their membrane fatty acids. This enzyme is a homohexamer of 290 kDa subunits. Each subunit possesses the six different active sites required to generate a fatty acid. Unlike the type II system, the products of a type I enzyme are acyl-CoAs. For the mycobacterial enzyme, the saturated acyl chains produced are between 16 and 24 carbons in length. Even more unusual is that the mycobacterium possesses a type II synthase system for the further elongation of the fatty acyl products of the type I system into the 70–80 carbon mycolic acids. Brevibacterium ammoniagenes, a highly developed bacterium thought to be a progenitor of the fungi, possesses a type I fatty acid synthase that is capable of producing both saturated and unsaturated fatty acids anaerobically.

5. THE ACYLTRANSFER MODULE

5.1 The PlsB/PlsC System

Fatty acid biosynthesis in E. coli normally ends when the acyl chain is 16 or 18 carbons in length. This is due in part to the substrate specificities of the elongation condensing enzymes, but the substrate preferences of the acyltransferases also impact the fatty acid chain length. The acyl-ACP end products are substrates for the acyltransferases that will transfer the fatty acyl chain into the membrane phospholipids (Figure 5). The first enzyme (PlsB) was discovered in *E. coli* and transfers fatty acids from acyl-ACP to the 1-position of glycerol phosphate. The PlsB protein is an integral inner membrane protein of 91 kDa and has a preference for saturated fatty acids. The product of the reaction, 1-acylglycerol phosphate, is the first membrane-associated intermediate in the pathway. The second acyltransferase (PlsC), a membrane protein of 27 kDa, esterifies the 2-position of the glycerol backbone and prefers unsaturated acyl chains. Thus, bacterial phospholipids usually have an asymmetric distribution of fatty acids between the 1- and 2-position of the glycerol phosphate backbone that arise from the substrate specificity of the two acyltransferases.

The isolation of *E. coli plsB* mutants with defective acyltransferase activity by Bell's laboratory heralded a major advance in the study of the acyltransferases. These mutants were glycerol phosphate auxotrophs and exhibited an increased Michaelis constant for glycerol phosphate in acyltransferase assays

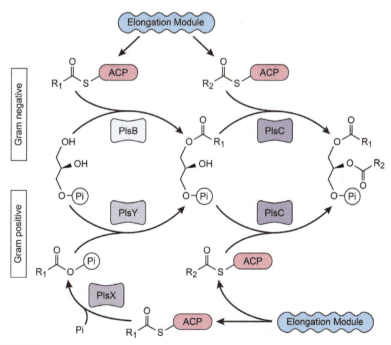

FIGURE 5 The acyltransfer module. Gram-negative bacteria use the PlsB isoform of the glycerol-phosphate acyltransferase to initiate phospholipid synthesis, followed by the acylation of the 2-position by PlsC. These enzymes often use either acyl-CoA or acyl-ACP as the acyl donor. Gram-positive bacteria first convert the acyl-ACP end product of FASII to the unique acyl-phosphate intermediate, which is then used by PlsY to acylate the 1-position of glycerol-phosphate. The second step is catalysed in PlsC. The enzymes of Gram-positive bacteria do not use acyl-CoA substrates.

in vitro. The increased K_m arises from a single missense mutation in the open reading frame. Therefore, *plsB* mutants require an artificially high intracellular concentration of glycerol phosphate for activity. Complementation of these mutants facilitated the cloning of PlsB, which possesses a catalytic His–Asp dyad common to glycerolipid acyltransferases. PlsB is a monomer and specifically activated by acidic phospholipids, phosphatidylglycerol and cardiolipin, as shown by mixed micelle assays containing detergents and phospholipid. PlsB also exhibits negative cooperativity with respect to glycerol phosphate binding, a property that may account in part for the finding that dramatic increases in the intracellular glycerol phosphate concentration do not increase the amount of phospholipids in *E. coli*.

The next step in phospholipid biosynthesis is catalysed by 1-acylglycerol phosphate acyltransferase (PlsC), which acylates the product of the PlsB step to form phosphatidic acid (Figure 5). PlsC is an integral membrane protein. Phosphatidic acid comprises only about 0.1% of the total phospholipids in *E. coli* and turns over rapidly, a property consistent with its role as an intermediate in phospholipid synthesis. The *plsC* gene is universally expressed in bacteria, but not all PlsCs have the same substrate preference for acyl chains or acyl donors. The PlsC of *E. coli* prefers unsaturated fatty acids and is capable of using both acyl-ACP and acyl-CoA thioesters. In contrast, the PlsC of *S. aureus* prefers 15-carbon branched-chain fatty acids and only accepts acyl-ACP thioesters.

5.2 The PlsX/PlsY/PlsC System

PlsB homologues are found in mice and humans, so early in bacterial lipid research it was thought that PlsB was a universal acyltransferase. However, PlsB is not widely distributed in bacteria, and most organisms use the PlsX/PlsY system to initiate membrane phospholipid synthesis (Figure 5) (Lu et al., 2006). This pathway begins by the conversion of acyl-ACP end products to acylphosphate by PlsX to create a novel activated fatty acid intermediate. Next, an acylphosphate-specific glycerolphosphate acyltransferase (PlsY) transfers the acyl chain to the 1-position of the glycerol-phosphate backbone. The acylphosphate intermediate is not physiologically stable, and a blockade of the PlsY step leads to the accumulation of its breakdown product, free fatty acid, inside the cell. PlsX is purified as a soluble protein, but associates with the membrane in vivo. PlsY is smaller than PlsB and is an integral membrane protein. The *plsX* gene is widely distributed in bacteria and is present even in those bacteria that use the PlsB acyltransferase. The role of PlsX in these organisms remains a mystery, but it is clear that it plays some role because the auxotrophic phenotype of the *plsB* mutant in *E. coli* is only evident in a *plsX* knockout background. The PlsC fulfills the same role in this pathway as in the PlsB/PlsC pathway. Usually, the PlsCs in this pathway use only acyl-ACP thioesters as substrates. Some bacteria using the PlsX/PlsY/PlsC pathway has a positional distribution of fatty acid chains between the 1- and 2-positions like *E. coli*; however, other bacteria in this group make only branched-chain fatty acids, and their phospholipid fatty acids are also asymmetrically distributed. Thus, positional asymmetry is a common property of bacterial phospholipids that is controlled by the specificity of the acyltransferases.

5.3 Regulation in the Acyltransfer Module

The synchronisation of lipid synthesis with DNA, RNA and protein synthesis must occur to maintain the correct proportion of membrane lipids and proteins. This is a key area of bacterial physiology that requires additional research to clarify. Many of the early experiments did not reveal coupling between membrane and macromolecular synthesis; however, these studies used *E. coli* strains that possess lesions in the synthesis of the signalling molecule, ppGpp, an established regulator of protein translation. Most mutants were defective in both *relA1* and *spoT1* alleles that result in the absence of ppGpp. The RelA protein catalyses the synthesis of ppGpp from GTP and GDP, whereas SpoT catalyses its synthesis and degradation. Multiple studies suggest that ppGpp can inhibit PlsB activity in vitro when palmitoyl-CoA is used as the acyl-donor, and the inhibition of the elongation module triggers a SpoT-dependent increase in ppGpp. These experiments suggest that SpoT activity is responsible for the coordination of fatty acid and protein synthesis. Recent research showed that an interaction between SpoT and ACP plays a role in triggering SpoT-dependent ppGpp formation on inhibition of fatty acid synthesis (Battesti and Bouveret, 2006, 2009). Deletion of *fabH* is possible in wild-type strains, but these deletions are not viable in an *E. coli recA1 spoT1* strain lacking ppGpp. Collectively, these experiments indicate the involvement of ppGpp in the control of lipid metabolism and its coordination with macromolecular synthesis, but considerably more work needs to be done to clarify the detailed biochemical mechanism.

5.4 Use of Extracellular Fatty Acids

Most of the ATP and reducing equivalents required for the synthesis of membrane phospholipids goes into the production of the fatty acids. Thus, cells can spare a significant amount of energy if they are able to assimilate fatty acids encountered in the environment into their phospholipids. The ability of bacteria to use effectively an extracellular fatty acid resource is highly variable. Some intracellular pathogens have lost the FASII genes and only use fatty acids derived from the host. Some bacteria have a β-oxidation pathway and use extracellular fatty acids as a carbon source. However, the ability to use fatty acids as a carbon source is not widely distributed in bacteria. Some bacteria appear uninterested in this extracellular resource, whereas others shut down de novo synthesis and robustly use the fatty acid provided in the environment for phospholipid synthesis. The key step in the incorporation of fatty acids into phospholipids is activation to allow their use. There are three different activation pathways known that couple to the two different acyltransferase systems of bacteria. One of the unanswered questions about fatty acid uptake in bacteria is the mechanism used to traverse the inner membrane. Despite extensive mutagenesis studies and bioinformatic analyses, there are no proteins known to catalyse the membrane transport of fatty acids. Current thinking is that fatty acids are protonated on the outer leaflet by the proton/pH gradient and the protonated fatty acids spontaneously flip from the outer to the inner leaflet (Figure 6). Gram-negative bacteria have outer membranes that protect them from environmental insults like detergents. In these organisms the movement of fatty acids across the outer membrane is catalysed by the FadL transporter that was first described in *E. coli*. Once inside the cell, the fatty acids are activated by one of three enzyme systems (Figure 5).

5.4.1 Acyl-Coenzyme A Synthetase

The first enzyme discovered to activate extracellular fatty acids was the acyl-CoA synthetase (FadD) of *E. coli* (Figure 6). FadD was discovered as an essential enzyme for the use of extracellular fatty acids as a carbon source via β-oxidation, and most bacteria with a FadD also use fatty acids as a carbon source. FadD is a peripheral membrane protein and produces acyl-CoA for use by both PlsB and PlsC. The enzyme requires ATP for activation of the fatty acid to an enzyme-bound acyl-AMP intermediate, which then transfers the acyl chain to the sulfhydryl of CoA.

5.4.2 Acyl-Acyl Carrier Protein Synthetase

Acyl-ACP synthetase (AasS) is related to acyl-CoA synthetase by using the same mechanism involving an acyl-AMP intermediate (Figure 6). This enzyme was first described in *Vibrio harveyi,* where it was implicated in light production via luciferase. AasS also permits *V. harveyi* to incorporate extracellular fatty acids into phospholipids and introduce them into FASII. Like FadD, AasS is a soluble protein that associates with the membrane to extract the fatty acid and transfer it to ACP via an acyl-AMP intermediate. Accordingly, FadD and AasS are related enzymes; however, bioinformatics parameters to distinguish the two activities are not robust enough to reach conclusions about their phylogenetic distribution in bacteria based on sequence alone. The expression of AasS in *E. coil* allows extracellular fatty acids to be converted to acyl-ACP and enter the FASII cycle for elongation. Some confusion arose from the discovery of an acyl-ACP synthetase activity in vitro in *E. coli* extracts. However, this enzyme does not release acyl-ACP for use by the acyltransferases (PlsB and PlsC), but is rather hardwired to a secondary acyltransferase involved in the reacylation of lysophosphlipids (Section 5.6.2).

5.4.3 Fatty Acid Kinase

A soluble two-component enzyme called fatty acid kinase (Fak) was discovered in Gram-positive bacteria. The first component is the ATP-binding protein

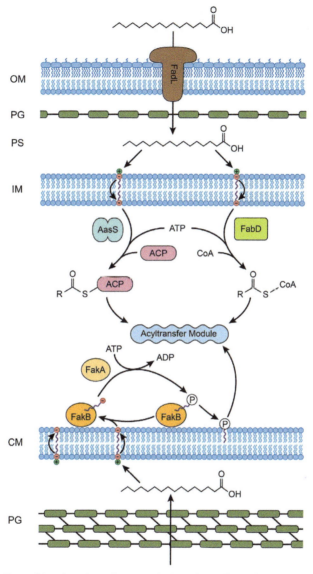

FIGURE 6 Fatty acid uptake and use. Gram-negative bacteria pass fatty acids through the outer membrane via the FadL porin. The acyl chain intercalates into the outer leaflet of the inner membrane, where it becomes protonated by the electrochemical gradient. The protonated fatty acid spontaneously flips to the inner leaflet. There the fatty acid is activated by either an acyl-CoA synthetase (FadD) or acyl-ACP synthetase (Aas). Both acyl-CoA and acyl-ACP are used by the acyltransfer module, and acyl-CoA can be funnelled to β-oxidation for energy generation. In Gram-positive bacteria, there is no known enzyme needed for passage through the cell wall, and the fatty acid moves to the inner membrane leaflet by flipping. The fatty acid associates with the FakB component of fatty acid kinase, which is phosphorylated by the FakA component. The FakB exchanges the acyl-phosphate for a fatty acid making the activated fatty acid intermediate available to the acyltransfer module. OM, outer membrane; PG, peptidoglycan; PS, periplasmic space; IM, inner membrane; and CM, cytoplasmic membrane.

called FakA. The second protein required is a fatty acid binding protein called FakB. This system is restricted to the Gram-positive pathogens, which use the PlsX/PlsY/PlsC acyltransferase system. These bacteria express a single *fakA* gene, but express two to six *fakB* genes. FakA phosphorylates the fatty acid bound to FakB to form FakB-acylphosphate. This intermediate then exchanges with a membrane-bound fatty acid to deposit the acylphosphate in the membrane, where it is used by PlsY (Figure 6). The released FakB-fatty acid may then undergo another round of phosphorylation by FakA. Fatty acid kinase is also involved in virulence factor production in *S. aureus* and is essential in other bacteria, indicating that it has important unknown functions outside of lipid metabolism.

5.4.4 Thioesterases

Thioesterases preferentially cleave the thioester bond of an acyl-CoA or acyl-ACP to liberate the cofactor and free fatty acid. *Escherichia coli* contains two well characterised thioesterases. Thioesterase I (encoded by *tesA*) is a periplasmic enzyme of 20.5 kDa, with a substrate specificity for acyl chains >12 carbon atoms. TesA hydrolyses synthetic substrates used in the assay of chymotrypsin, which led to the initial conclusion that TesA was a protease ('protease I'). However, the purified protein does not cleave peptide bonds. Thioesterase I also appears to possess lysophospholipase activity. Thioesterase II is a cytosolic tetrameric protein composed of 32 kDa subunits encoded by the *tesB* gene. There are a number of other thioesterases found in bacteria, and they possess activity against both acyl-ACP and acyl-CoA in all cases tested. The physiological function of thioesterases is unknown, and they may have no specific role in lipid metabolism. Many of these enzymes have other hydrolytic activities, and those located in the periplasm are likely to be involved in nutrient acquisition. Null mutants of *tesA* and *tesB* have been constructed, as have double mutant strains. None of these strains has an observable lipid phenotype. Nonetheless, genetically engineered strains expressing thioesterases to liberate free fatty acids from the biosynthetic pathway have proven key to biofuel research.

5.4.5 Fatty Acids as a Carbon Source

Degradation of fatty acids proceeds via an inducible set of enzymes that catalyse the pathway of β-oxidation. β-Oxidation occurs via repeated cycles of reactions that are essentially the reverse of the reactions of fatty acid synthesis. However, three major differences distinguish the two pathways. First, β-oxidation uses acyl-CoA thioesters, and not acyl-ACP. Second, the β-hydroxy intermediates have the opposite stereochemistry (L in β-oxidation and D in synthesis). Finally, the enzymes of β-oxidation share no homology with those of synthesis.

The first step in the pathway is the dehydrogenation of acyl-CoA by the enzyme acyl-CoA dehydrogenase. While other organisms have several dehydrogenases with different chain length specificities, *E. coli* has one enzyme active on all chain lengths. The acyl-CoA dehydrogenase is a flavoprotein and is linked to an electron transferring flavoprotein (*fadE*). Mutant strains of *E. coli* blocked in β-oxidation with *fadE* (also known as *fadF or yahF*) mutations can accumulate acyl-CoA species, but cannot degrade them. The second step in the cycle is enoyl-CoA hydratase, an activity commonly referred to as 'crotonase'. Traditionally, in vitro measurements of this activity use crotonoyl-CoA (*trans*-2-butenoyl-CoA) as the substrate. Crotonase activity in *E. coli* is one function present in a multifunctional protein encoded by *fadB*. The next step in the cycle is β-hydroxyacyl-CoA dehydrogenase, another function of the FadB enzyme. The β-ketoacyl-CoA produced in this reaction is a substrate for the monofunctional *fadA*-encoded β-ketoacyl thiolase, which cleaves acetyl-CoA from the acyl-CoA to produce an acyl-chain two carbons shorter than when it enters the cycle. The cycle is then repeated until the fatty acid is converted to acetyl-CoA. The FadB protein is a homodimer of 78 kDa subunits and is purified in complex with the homodimeric 42 kDa *fadA* gene product. The total complex is thus an $\alpha_2\beta_2$ heterotetramer with an apparent mass of about 260 kDa. The *fadA* gene encodes the β-subunit, while the *fadB* gene gives the α-subunit, the confusing nomenclature a remnant of the days of classical genetics.

Unsaturated fatty acids can also be degraded by the β-oxidation pathway. The FadB protein possesses *cis*-β-enoyl-CoA isomerase activity, which converts *cis*-3 double bonds to *trans*-2. A 2,4-dienoyl-CoA reductase encoded by *fadH* is also required for the metabolism of polyunsaturated fatty acids. An epimerase activity of FadB allows for the use of D-hydroxy fatty acids. The epimerase is actually a combination of a D-β–hydroxyacyl-CoA dehydratase and the crotonase (hydratase) activities, resulting in the conversion of the D to the L enantiomer. The substrate specificities of the enzyme complex in vitro suggest that all the enzymes can use all chain lengths of substrates, with the possible exception of the crotonase activity. This function of FadB appears somewhat limited to short chain substrates, and it has been suggested that a separate long-chain enoyl-CoA hydratase may exist in *E. coli*. Two open reading frames in the *E. coli* genome are predicted to encode homologues of FadA and FadB. Thus, two complexes may be present with preferences for long- or short-chain acyl-CoAs. Finally, there is also an anaerobic pathway for fatty acid degradation in *E. coli* that is shared by other Gram-negative bacteria. This system allows the use of fatty acids in the anaerobic intestinal environment.

6. THE PHOSPHOLIPID MODULE

Escherichia coli possesses only three major phospholipid species in its membranes, making it one of the simplest organisms to study with regard to phospholipid biosynthesis. This chapter reviews these pathways with the recognition that this module is diverse in bacteria. Phosphatidylethanolamine (PE) comprises the bulk of the phospholipids (75%), with PG and CL forming the remainder (15–20% and 5–10%, respectively). The scheme for the synthesis of these membrane phospholipids

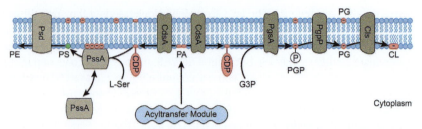

FIGURE 7   The phospholipid module. The acyltransfer module deposits phosphatidic acid in the membrane, where it is activated to CDP-diacylglycerol by CDP-diacylglycerol synthase (CdsA). This intermediate is used for both phosphatidylethanolamine (PE) synthesis via phosphatidylserine synthase (PssA) and phosphatidylserine decaroboxylase (Psd). Phosphatidylglycerol (PG) is formed from the same intermediate by phosphatidylglycerolphosphate synthase (PgsA), and the phosphorylated intermediate is dephosphorylated by phosphatidylglycerolphosphate phosphatase (PgpP). Cardiolipin (CL) is produced by the condensation of two PG molecules by cardiolipin synthase (Cls).

follows the classic Kennedy pathway (Figure 7). For interested readers, there is an excellent review of the early years of research in this area, which highlights the seminal discoveries and discusses the contributions of individual scientists who established the field of bacterial phospholipid biochemistry and genetics (Dowhan, 2012).

6.1 Phosphatidate Cytidylyltransferase

The key activated intermediate in bacterial phospholipid synthesis, CDP-diacylglycerol, comprises only 0.05% of the total phospholipid pool and undergoes rapid turnover. The 27.6 kDa enzyme phosphatidate cytidylyltransferase (or CDP-diacylglycerol synthase) catalyses the conversion of phosphatidic acid to a mixture of CDP-diacylglycerol and dCDP-diacylglycerol. Normally, CDP-diacylglycerol synthase (CdsA) is present in large excess of the minimum amount of enzyme required to sustain phospholipid synthesis. Strains of *E. coli* with mutations in the *cds* gene that only retain 5% of the normal levels of CDP-diacylglycerol synthase are known, but grow normally under standard laboratory conditions. These mutants accumulate substantial amounts of phosphatidic acid (up to 5% of the total phospholipid). Null mutations in *cds* have not been reported and would presumably be nonviable due to the complete lack of phospholipid synthesis. CDP-diacylglycerol stands at the branch point between PE synthesis and PG and CL synthesis (Figure 7).

6.2 Phosphatidylethanolamine Production

6.2.1 Phosphatidylserine Synthase

The first step in the synthesis of PE is the condensation of CDP-diacylglycerol with serine catalysed by PS synthase to form PS (Figure 7). During cell disruption, the 58 kDa PS synthase (PssA) appears associated with ribosomes, but reattaches to membrane vesicles once substrate is added. PS is a minor membrane constituent of *E. coli* because it is rapidly converted to PE by PS decarboxylase. Mutants in the *pss* gene encoding PS synthase contain no PS or PE in their membranes, but are viable when supplemented with divalent metal ions. PE is capable of forming the hexagonal (nonbilayer) H_{II} lipid phase, and Dowhan has suggested that the divalent cations interact with CL to replace the function of PE in the formation of an H_{II} phase. There cells also have perturbations in the function of permeases, electron transport, motility, and chemotaxis. The defect appears to arise from the improper folding of membrane proteins, leading to the idea that specific phospholipids act as chaperones to guide the folding of membrane proteins (Chapter 1). The physiological processes dependent on the formation of local regions of nonbilayer structure or that specifically require PE remain to be elucidated, but the process of cell division, the formation of contacts between inner and outer membranes, and the translocation of molecules across the membrane are viable candidates.

6.3 Phosphatidylserine Decarboxylase

PS is decarboxylated by PS decarboxylase (Psd) to yield the zwitterionic PE (Figure 7). This inner membrane enzyme has a subunit molecular mass of 36 kDa. PS decarboxylase has a pyruvate prosthetic group that participates in the reaction by forming a Schiff base with PS. Overproduction of the enzyme 30- to 50-fold by plasmid-borne copies of the *psd* gene has no effect on membrane phospholipid composition, indicating that the level of this enzyme does not regulate the amount of PE in the membrane. Most of the PE is found in the periplasmic leaflet of the inner membrane, and there is a rapid flipping from the inner to outer leaflet that is thought to be catalysed by the MsbA lipid flippase. However, there is much to be learnt about how phospholipids are trafficked from the inner leaflet of the inner membrane to the outer leaflet and outer membrane of *E. coli*.

Mutants with a temperature-sensitive decarboxylase accumulate PS at the nonpermissive temperature. The mutants continue to grow for several hours after the shift to the nonpermissive temperature, despite the reduced levels of PE and the concomitant increase in PS. Complete inactivation of *psd* by insertional mutagenesis has the same divalent cation requiring-phenotype as the *pss* mutants described above. The requirement for CL is consistent with the inability to introduce a null *cls* allele into *psd* strains. Thus, PE appears essential for the polymorphic regulation of lipid structure. However, some bacteria like *S. aureus* do not contain PE making this function irrelevant to many bacteria.

6.4 Phosphatidylglycerol Synthesis

6.4.1 Phosphatidylglycerolphosphate Synthase

CDP-diacylglycerol is condensed with glycerol phosphate to form phosphatidylglycerolphosphate (PGP), an intermediate in the production of the acidic phospholipids PG and CL (Figure 7). The reaction is analogous to the synthesis of PS, with the product CMP being released. Mutants (*pgsA*) defective in PGP synthesis contain less than 5% of normal PGP synthase activity in vitro; however, there is no growth phenotype associated with these mutants. The PgsA protein is predicted to be a 20.7 kDa integral membrane protein. It has been thought that PGP synthase (PgsA) is essential and that cells cannot survive without acidic phospholipids. There are many important cellular functions that are affected by reduced PG and/or CL content of the membrane. PG is required for protein translocation across the membrane, and acidic phospholipids are required for channel activity of bacterial colicins and the interaction of antibiotics with the membrane. Cell division proteins such as FtsY also apparently require acidic phospholipids for activity, as does the DnaA protein involved in chromosome segregation. However, Matsumoto has recently inactivated the *pgsA* gene with a kanamycin cassette. This *pgsA::kan* strain has no detectable PG or CL, is not viable above 40°C, and contains increased concentrations of PA, which may at least partially compensate for the absence of PG and CL.

6.4.2 Phosphatidylglycerolphosphate Phosphatases

The second step in the synthesis of PG is the dephosphorylation of PGP (Figure 7). Three independent genes have been identified that encode PGP phosphatases (PgpP) based on in vitro assays (PgpA, PgpB, and PgpC), and the triple knockout strain is not viable unless rescued by one of the phosphatases. All three are inner membrane proteins, and it is not immediately clear which phosphatase is designed to function in the biosynthetic pathway. However, all three are capable of hydrolysing a number of different phosphorylated membrane lipids, but the biological significance of the three phosphatases in phospholipid metabolism remains unclear.

6.5 Cardiolipin Biosynthesis

Unlike mammalian mitochondria, in which CL is synthesised by the reaction of CDP-diacylglycerol with PG (Chapter 7), the major pathway to CL in bacteria involves the condensation of two PG molecules (Figure 7). CL accumulates as the cells enter the stationary phase of growth and is required for prolonged survival of the bacteria. There are three CL synthases (Cls). ClsA was the first isoform discovered, and *clsA* knockout mutants do not have detectable CL during log phase growth, but stationary phase cells still accumulated CL. A second gene product, *clsB*, catalyses CL synthesis in vitro. Finally, a third gene, *clsC*, was identified recently that differs from the others by using PE and PG as substrates instead of PG. Multiple *cls* genes are a common theme in bacteria, but considerably more work is needed to sort out the specific physiological roles for each of these enzymes.

6.6 Phospholipids as Precursors

Phospholipids are used as substrates in the synthesis of other molecules and are particularly useful donors when the biochemical reactions occur outside the cell. This compartment is devoid of the normal forms of activated fatty acids, or other intermediates that are used in the biosynthesis of proteins, lipids and polysaccharides inside the cell. The pieces of the phospholipid that remain are usually salvaged and reintroduced into the phospholipid module after they are translocated back to the interior of the cell.

6.6.1 Use of Phospholipid Headgroups

The polar headgroup of PG is rapidly lost in a pulse-chase experiment, whereas that of PE is stable. The conversion of PG to CL, catalysed by CL synthase, does not account for all the loss of ^{32}P-labelled PG observed in the pulse-chase experiments. A phosphate-containing nonlipid compound derived from the head group of PG was sought, which led to the discovery of the MDO by Kennedy's group. These compounds are composed of *sn*-glycerol-1-phosphate (derived from PG), glucose and (usually) succinate moieties and have molecular weights in the range of 4000–5000 Da. These oligosaccharides are found in the periplasm of Gram-negative bacteria, an osmotically sensitive compartment. The synthesis of the MDO compounds is regulated by the osmotic pressure of the growth medium and decreased osmotic pressure activates MDO synthesis. Thus, MDO compounds are involved in osmotic regulation.

In the synthesis of MDO, the *sn*-glycerol-1-phosphate polar headgroup of PG is transferred to the oligosaccharide, with 1,2-diacylglycerol as the other product. The 85 kDa transmembrane protein that catalyses this reaction is encoded by *mdoB*. Diacylglycerol kinase phosphorylates the diacylglycerol to phosphatidic acid, which re-enters the phospholipid biosynthetic pathway to complete the diacylglycerol cycle. In the overall reaction, only the *sn*-glycerol-1-phosphate portion of the PG molecule is consumed; the lipid portion of the molecule is recycled back into phospholipids. MDO synthesis is responsible for most of the metabolic instability of the polar group of PG, because blocking MDO synthesis at the level of oligosaccharide synthesis by lack of UDP-glucose greatly reduces PG turnover. Moreover, the rate of accumulation of diacylglycerol in strains lacking diacylglycerol kinase *(dgkA)* correlates with the presence of both the oligosaccharide acceptor and the osmolarity of the growth medium. DgkA is a trimer of identical 13 kDa subunits, each with three predicted trans-membrane helices. Activity appears to be limited by diffusion of substrate across the membrane to the cytoplasmic active site. It should be noted that some species of MDO contain phosphoethanolamine. Although direct proof is lacking, it is likely that the ethanolamine moiety is derived from PE, the only known source of ethanolamine.

There is an analogous diacylglycerol cycle operating in Gram-positive bacteria that is constitutively active. These organisms have a molecule called lipoteichoic acid that is formed by the addition of about 25 *sn*-1-glycerolphosphate moieties to the headgroup of glucosyldiacylglycerols. Like MDO synthesis, the glycerol-phosphate is transferred from PG and the resulting diacylglycerol is reintroduced into the phospholipid biosynthetic pathway. This group of bacteria has a different type of diacylglycerol kinase called DgkB. DgkB is a peripheral membrane protein that, unlike the integral membrane DgkA, its association with anionic phospholipids activates ATP binding and efficiently reintroduces diacylglycerol into the phospholipid biosynthetic pathway. In both schemes, diacylglycerol is thought to spontaneously flip from the outer to the inner leaflet.

6.6.2 Use of Phospholipid Fatty Acids

Both of the two routes for the use of phospholipid fatty acids involve the maturation of bacterial lipoproteins. The first step is transfer of the diacylglycerol portion of PG to a cysteine residue in the amino terminal domain of the lipoprotein. This modification creates a substrate for a protease that cleaves the pro-protein at the modified cysteine residue to create a free amino terminus. An acyltransferase then transfers the 1-position fatty acid from PE to the amino terminus of the lipoprotein to complete its synthesis. All these reactions occur outside the cell on the outer leaflet of the inner membrane. 2-Acylglycerolphosphoethanolamine is generated by transfer of the acyl moiety. 2-Acylglycerolphosphoethanolamine acyltransferase (encoded by *aas*) is an inner membrane enzyme that esterifies the 1-position of 2-acylglycerolphosphoethanolamine using acyl-ACP (but not acyl-CoA) as the acyl donor to regenerate PE. The acyltransferase was first recognised as a protein called acyl-ACP synthetase that catalyses the ligation of fatty acids to ACP, hence the gene designation *aas*. ACP acts as a bound protein subunit for accepting the acyl intermediate in the acyltransferase reaction that occurs at nonphysiological concentrations of chaotropic salts that are required to dissociate the acyl-ACP intermediate from the enzyme in vitro. This means that 2-acylglycerolphosphoethanolamine acyltransferase is the only activity of Aas in vivo. Mutants lacking the *aas* gene do not accumulate 2-acylglycerolphosphoethanolamine in vivo unless they are also defective in the *pldB* gene, which encodes a lysophospholipase that degrades 2-acylglycerolphosphoethanolamine to glycerophosphoethanolamine and fatty acid. The 2-acylglycerolphosphoethanolamine is transported into the cell by an inner membrane lysoPtdEtn filppase (*lplT*) that is transcribed in the same operon as *aas*. In some bacteria, the three activities are fused together and translated as a single protein.

6.7 Modification of Phospholipids

6.7.1 Origin of Cyclopropane Fatty Acids

Fatty acids attached to membrane phospholipids can be postbiosynthetically converted to their cyclopropane derivatives during the stationary phase of bacterial growth. Their biosynthesis and function have been elucidated by the Cronan laboratory. The conversion of the double bond of unsaturated fatty acids to a cyclopropane ring occurs on the intact phospholipid molecule. Cyclopropane fatty acid synthesis begins as the cultures enter the stationary phase of growth. *Escherichia coli* mutants that completely lack cyclopropane fatty acid synthase activity (owing to null mutations in the *cfa* gene) grow and survive normally under virtually all conditions, except that *cfa* mutant strains are more sensitive to freeze-thaw treatment and acid shock than the isogenic wild-type strains. Thus, the stable cyclopropane derivative appears to protect the reactive double bond in unsaturated phospholipid molecular species during stationary phase. Cyclopropanation involves a significant energy commitment by the cell because the reaction uses *S*-adenosylmethionine, which requires three molecules of ATP for regeneration.

The 44 kDa cylcopropane synthase protein is metabolically unstable, but protein levels peak sharply due to increased *cfa* transcription as cultures enter the stationary phase. Cfa levels drop in late stationary phase cultures as the enzyme is destroyed by proteolysis, probably by a protease of the heat shock response. The *cfa* gene possesses two promoters of approximately equal strengths, with the more distal promoter functioning throughout the growth cycle. The proximal promoter requires the specialised sigma factor RpoS for transcription and is thus active only as cultures enter stationary phase. The cyclopropane fatty acid content of *rpoS* strains is low, and transcription does not increase in stationary phase. As wild-type strains remain in stationary phase and phospholipid biosynthesis ceases, the low levels of Cfa that persist no longer encounter an expanding substrate pool. Thus, an increasing amount of the fatty acyl chains are converted to their cyclopropane derivatives over time. The instability of Cfa results in little carryover of synthetic capacity when exponential growth resumes, and the existing cyclopropane fatty acids are quickly diluted by de novo phospholipid synthesis.

6.7.2 Phospholipid Desaturases

In addition to controlling the rate of fatty acid synthesis in response to a shift in the composition of pathway intermediates, *B. subtilis* has developed regulatory mechanisms to detect environmental changes in temperature and adjust the expression of desaturating enzymes accordingly (Mansilla and de Mendoza, 2005). The ability to adjust the unsaturation of membrane phospholipids is crucial to counteract the increase in rigidity that arises when the temperature is below the phase transition of biological membranes. The Des protein from *B. subtilis* modifies membrane structure by inserting a *cis*-double bond into the $\Delta 5$ position of existing phospholipids. How this system operates is described in more detail in Section 6.3.2.

6.7.3 Modification of Polar Headgroups

Some bacteria, including *S. aureus* and *B. subtilis* synthesise the unique lipid, lysyl-phosphatidylglycerol (Lys-PG). This positively charged phospholipid is synthesised by the aminoacylation of PG by the transmembrane MprF enzyme using Lys-tRNA as a lysine doner (Ernst and Peschel, 2011). The lysinylation of the anionic PG changes the net charge of the phospholipid from −1 to +1. The enzyme was named MprF as a notation for multiple peptide resistance factor in reference to the increased susceptibility of an *S. aureus mprF* null mutant to cationic antimicrobial peptides (CAMPs) and lipopeptide antibiotics such as defensins and daptomycin. The positive net charge on membranes due to the production of Lys-PG perturbs this electrostatic attraction and mitigates the antimicrobial effects. Although Lys-PG does not have a defined structural role, it clearly has an integral role in pathogenesis of *S. aureus* because *mprF* deletion mutants have attenuated virulence. Why a nonpathogenic soil-dwelling organism like *B. subtilis* would produce Lys-PG is less obvious, but it has been hypothesised that many soil-dwelling organisms actively mitigate the effect of antimicrobial peptides secreted by themselves and other soil bacteria. MprF from *S. aureus* produces strictly Lys-PG, whereas an MprF-homologue from *Pseudomonas aeruginosa* produces only Ala-PG. The MprF enzyme from *Enterococcus faecium* and *B. subtilis* have less strict substrate specificities and are able to use both Lys and Ala to modify PG.

The MprF enzyme consists of two different catalytic domains. The C-terminal domain has Lys-PG synthase activity that catalyses the transfer of the lysine from the aminoacyl-tRNA to the hydroxyl group of terminal glycerol of PG. The amino terminal domain is the hydrophobic flippase component that catalyses the translocation of Lys-PG from the inner leaflet of the cell membrane where it is produced, to the outer leaflet where it can repel CAMPs. Production of Lys-PG alone on the inner leaflet is not sufficient to induce resistance to CAMPs. In *P. aeruginosa*, Ala-PG synthesis is induced by acidic growth conditions and Lys-PG production in *S. aureus* varies throughout the stage of growth. The *mprF* gene and other resistance mechanisms are upregulated in response to defensins by the ApsRSX three component regulator. ApsS is the membrane-bound sensor kinase that interacts with CAMPs and activates ApsR by phosphorylation, inducing transcription of *mprF* and genes involved in lysine biosynthesis.

6.7.4 Phospholipases

Based mainly on cell free assays, multiple enzymatic activities that degrade phospholipids, intermediates in the phospholipid biosynthetic pathway, or triacylglycerol have been reported. The detergent-resistant phospholipase A_1 (encoded by *pldA*) of the outer membrane, characterised by Nojima and colleagues, is the most studied of these enzymes. This enzyme is unusually resistant to inactivation by heat and ionic detergents and requires calcium for maximal activity. The mature phospholipase has a subunit molecular mass of 31 kDa. Hydrolysis of fatty acids from the 1-position of phospholipids is the most rapid reaction, but the enzyme will also hydrolyse 2-position fatty acids, as well as both isomeric forms of lysophosphatides and mono-acylglycerols and diacylglycerols. A detergent-sensitive phospholipase A_1 has also been described, although this activity has not been assigned to a gene. This enzyme differs from the detergent-resistant protein in that it is located in the soluble fraction of the cell, is inactivated by heat and ionic detergents, and has a high degree of specificity for PG. The cytoplasmic phospholipase A_1 requires calcium for activity. There are also inner membrane and cytoplasmic lysophospholipases. The best characterised of these is the inner membrane lysophospholipase L2 (*pldB*), which hydrolyses 2-acylglycerophosphoethanolamine efficiently, but is barely active on the 1-acyl isomer. This lysophospholipase also catalyses the transfer of fatty acids from 2-acylglycerophosphoethanolamine to the terminal hydroxyl of the headgroup of PG to form acyl-PG.

The physiological role of these degradative enzymes remains largely unknown. Mutants lacking the detergent-resistant phospholipase (*pldA*), lysophospholipase L2 (*pldB*), or both enzymes do not have any obvious defects in growth, phospholipid composition, or lipid turnover. Moreover, strains that overproduce the detergent-resistant enzyme also grow normally. It has been established that the detergent-resistant phospholipase is responsible for the release of fatty acids from phospholipids during infection with T4 and λ phages. However, phospholipid hydrolysis is not essential for the life cycle of these bacteriophages. One possible function for the hydrolytic activities with unassigned genes is that they are actually biosynthetic proteins (acyltransferases) that act as lipases in the absence of suitable acceptor molecules in the assay systems used. An example of such an enzyme is PS synthase, which catalyses both phospholipase D and CDP-diacylglycerol hydrolase reactions. PS synthase appears to function via a phosphatidyl-enzyme intermediate, and in the absence of a suitable acceptor such as serine or CMP, the phosphatidyl-enzyme complex is hydrolysed by water to release phosphatidic acid; thus, the enzyme acts as a phospholipase D. Also, some of these enzyme activities may reflect the broad substrate specificity of a single enzyme rather than the presence of several distinct ennzyme species. For example, the observed lipase activity that cleaves the 1-position fatty acids from triacylglycerols (a lipid usually not found in *E. coli*) may arise from the detergent-resistant phospholipase A_1 acting on triacylglycerol as an alternate substrate. Further, a lysophospholipase L(1) activity has been attributed to thioesterase I. Three open reading frames (*ybaC*, *yhjY*, and *yiaL*) present in the *E. coli* genome potentially encode proteins with lipase activity, but have not been studied to date. Lipases and phospholipases secreted by many pathogenic bacteria and are responsible for tissue damage and inflammation. In these instances, the biological significance of the lipases and phospholipases are obvious.

6.8 Phospholipid Diversity in Bacteria

Escherichia coli have a simple phospholipid composition of primarily PE, PG, and CL. However, the prokaryotic kingdom possesses a wide array of headgroups that defy adequate description in this short space; hence the reader is referred to Goldfine's review (Goldfine, 1982) for a more comprehensive treatment of bacterial phospholipid structures. The phosphocholine headgroup stands worthy of mention for its uniqueness and distinct mechanisms of synthesis. Phosphatidylcholine (PC) had long been considered a eukaryotic phospholipid, when it is synthesised by transfer of the choline from CDP-choline to diacylglycerol or by methylation of PE (Chapter 7).

Rhodospseudomonas spheroides, *Bradyrhizobium japonicum* and a few other specialised photosynthetic or nitrogen fixing bacteria synthesise PC by three successive methylation reactions starting with PE and using *S*-adenosylmethionine as the methyl donor. The first methyltransferase, encoded by *pmtA*, has been disrupted in *B. japonicum* and the mutants containing significantly reduced PC content and are less able to fix nitrogen. Thus, PC seems to be involved in host:bacteria interactions to establish symbiosis. The prokaryotic PE methyltransferases share weak homology, but have no homology with their eukaryotic counterparts. *R. spheroides* and other photosynthetic bacteria are also somewhat unique among bacteria in that they contain intracellular membranes, which hold the photosynthetic machinery. The amount of intracellular membrane correlates with the amount of incident light, indicating a light-specific regulation of phospholipid synthesis in these organisms.

Sinorhizobium meliloti synthesises PC by direct condensation of choline with CDP-diacylglycerol, as well as by the methyltransferase pathway. The *pcs* gene was identified and expression in *E. coli* confirmed that it encodes a PC synthase. The genomes of *P. aeruginosa* and *Borrelia burgdorferi* contain similar genes and have been reported to contain PC in their membranes. The PC synthase protein shares weak homology with PS synthase (a CDP-diacylglycerol:serine O-phosphatidyltransferase) from other bacteria, but not to any eukaryotic proteins.

Sphingobacterium and Bacteriodes species produce sphingolipids by a pathway similar to that in mammals (Chapter 10). Clostridia produce plasmalogens (1-alk-1'-enyl lipids) by an anaerobic pathway clearly different to the O_2-dependent pathway in mammals. Branched-chain fatty acids are also found in which the methyl group is inserted post synthetically into the middle of the chain, in a manner analogous to cyclopropane fatty acid synthesis. *S*-Adenosylmethionine is also the methyl donor for these reactions. The biochemistry surrounding the formation of these and many other bacterial phospholipids remains to be elucidated.

6.9 Membrane Lipids Lacking Phosphorus

Phosphorus is often limiting in the environment. Many bacteria can switch to the synthesis of lipids that do not contain phosphorus and conserve this valuable resource. The two major lipids classes are ornithine lipids and glycolipids. Ornithine lipids are produced by the successive acylation of ornithine by enzymes that resemble the PlsC acyltransferases of phospholipid synthesis, but clearly form their own subgroup. The common theme is that the key biosynthetic genes in ornithine lipid synthesis are induced by the *pho* regulon, which coordinates the whole cell response to phosphate starvation. Glycolipids are abundant in many bacteria. In Gram-positive organisms, glycolipids are major membrane constituents and also the anchor of the lipoteichoic acids, which are key components of the cell surface. Mono- and di-glycolipids are most prevalent and produced by the stepwise glycosylation of diacylglycerol by glycosyltransferases that use UDP-sugars as substrates.

6.10 Regulation in the Phospholipid Module

6.10.1 Regulation of PS Synthesis

The PE to PG content of membranes is maintained by a series of strict biochemical regulatory controls. The branch point between zwitterionic phospholipids (PE) and anionic phospholipids (PG and cardiolipin) arises after phosphatidic acid has been converted to CDP-DAG. The PS synthase (Pss) and the PG-phosphate synthase (PgsA) enzymes compete for the CDP-DAG. Overexpression of Pss or PgsA individually in *E. coli* has minimal effect on membrane phospholipid composition, indicating both enzymes are tightly controlled by independent feedback regulatory loops. Treatment of *E. coli* with the artificial glycerol-1-phosphate acceptor arbutin stimulates the transfer of glycerol-1-phosphate from PG to form arbutin-phosphoglycerol and diacylglyerol. The arbutin-mediated removal of glycerol-1-phosphate from PG does not significantly reduce the PG content of the cell membrane, although a huge increase in PgsA activity is required to maintain the PG content under these conditions. The mechanism for this biochemical feedback regulation has not been explored. Phosphatidylserine decarboxylase (Psd) is an intrinsic membrane protein that catalyses the formation of PE from PS. Only trace amounts of PS are detected in *E. coli* cells, illustrating that Psd is extremely efficient.

PssA is a key enzyme controlling PE content. PssA is an interfacial enzyme that binds to anionic lipids on the membrane surface. This requirement for anionic phospholipids for PssA activity results in a clever detection mechanism that responds to an increase in anionic membrane phospholipids by stimulating PS (and consequently PE) synthesis. This also provides an explanation for the lack of increased PE synthesis following Pss overexpression. In contrast, PssA from *B. subtilis* is an integral membrane protein. The replacement of *E. coli* PssA with *B. subtilis* PssA triggers a dose-dependent increase in the PE content. Thus, regulation of the membrane-bound *B. subtilis* PssA is different than the *E. coli* enzyme.

7. GENETIC REGULATION OF LIPID METABOLISM

There is a sophisticated group of diverse transcriptional regulators of lipid metabolism in bacteria that work in concert with biochemical regulation to control the pathway (Zhang and Rock, 2009). The expression of different fatty acid biosynthetic genes are coordinated with growth rate, nutrient availability and environmental stimuli. In organisms possessing a fatty acid β-oxidation pathway in addition to biosynthesis, the expression of the degradation machinery is coordinated with fatty acid synthetic enzymes.

7.1 Gram-Negative Bacteria

7.1.1 Coordination of Fatty Acid Synthesis and Degradation by FadR

The FadR transcription factor has been most thoroughly studied in *E. coli*, but homologues are present in Gram-positive and Gram-negative bacteria. *Escherichia coli* FadR is a member of the GntR family of transcription factors and functions as a classical repressor of genes involved in fatty acid β-oxidation. FadR is not the sole transcriptional regulator of fatty acid degradation genes; the ArcAB system also negatively regulates the pathway, and the global cyclic-AMP receptor protein-cAMP complex permits the activation the *fad* genes.

The binding of long-chain acyl-CoA thioesters by the FadR repressor induces a conformational change that reduces it DNA binding affinity. Structural data have been obtained for the free protein, the R-DNA complex, and the FadR-acyl-CoA complex. The structures illustrate that FadR contains a DNA binding winged-helix at the amino terminus and an α-helix bundle in the carboxy-terminal domain that binds the acyl-CoA regulatory ligand. The free protein and DNA-bound structures are almost identical, whereas the acyl-CoA-bound structure shows a conformational change to accommodate the acyl-CoA that causes the helices in the DNA binding domain to separate. This conformational change reduces the affinity of FadR for the DNA recognition sequence. Biochemical studies have concluded that FadR has the highest affinity for long-chain acyl-CoAs (C16–C18) and low affinity for < C10-CoA or free fatty acids. The dissociation constants are in the nanomolar range underlining the high sensitivity of FadR to cellular long-chain acyl-CoA concentrations.

FadR is also an activator of two genes involved in unsaturated fatty acid synthesis (*fabA* and *fabB*) and *iclR*. The IclR protein is a repressor of the *aceAB* genes, which encode enzymes involved in the glyoxylate shunt pathway that is required to use the acetyl-CoA derived from β-oxidation. The activation of *fabA* and *fabB* by FadR is abolished on binding of FadR to acyl-CoA. The simultaneous activation of fatty acid degradation and repression of unsaturated fatty acid synthesis in response to exogenous fatty acids coordinates the two pathways. FadR is present in a subset of bacteria with a β-oxidation pathway. A FadR homologue was discovered in *B. subtilis* that also represses the genes of β-oxidation but shows no evidence of the activator function observed in *E. coli*, most likely due to the lack of a FabB/FabA mediated unsaturated fatty acid synthesis pathway. Although *B. subtilis* and *E. coli* FadR proteins share sequence homology, they are structurally distinct; the *B. subtilis* protein belongs to the TetR family, while the *E. coli* enzyme is a GntR-like repressor. In *Vibrio cholerae*, FadR also binds to the promoter region and represses transcription of *plsB*. This is understood based on lipid metabolism in *V. cholerae*, which uses the PlsX/PlsY/PlsC system for the acylation of glycerol-phosphate by de novo synthesised fatty acids and PlsB for acylation by exogenous fatty acids.

7.1.2 FabR/DesT Control of Unsaturated Fatty Acid Synthesis

In addition to the influence of *fabA* and *fabB* on unsaturated fatty acid biosynthesis, *P. aeruginosa* also possesses Δ9-desaturases that insert a double bond into the acyl-chains of existing phospholipids (DesA) or saturated acyl-CoAs (DesBC). The expression of *fabAB* and *desBC* is coordinately controlled by the DesT repressor. DesT is a TetR family transcription factor that senses the overall fatty acid composition of the acyl-CoA pool. DesT binds saturated and unsaturated acyl-CoAs with equal affinity, but binding to DNA is enhanced when DesT is bound to an unsaturated acyl-CoA and inhibited when DesT is bound to a saturated acyl-CoA (Subramanian et al., 2010; Zhang et al., 2007). This allows *P. aeruginosa* to respond to the availability of exogenous saturated and unsaturated fatty acids and adjust gene expression to maintain membrane fluidity. Growth of *P. aeruginosa* on oleate represses the expression of *desBC* and *fabAB*, whereas growth on stearic acid stimulated *deBC* and *fabAB* transcription. The differential influence of DesT binding on *fabAB* and *desBC* gene expression was attributed to the slightly different recognition palindromes in the promoters (Subramanian et al., 2010; Zhang et al., 2007). X-ray crystallography determined that DesT adopts two conformations: a 'relaxed' conformation when bound to an unsaturated acyl-CoA that facilitates DNA binding, and a 'tense' conformation when bound to a saturated acyl-CoA species that reduces affinity for the DNA target sequence. Thus, DesT is a lipid compositional sensor.

A second transcription factor, designated FabR, also contributes to fine tuning the biophysical properties of the membrane. FabR binds to regions of the *fabA* and *fabB* promoter downstream of the FadR recognition sequence to repress transcription. Deletion of *fabR* in *E. coli* causes a two- to four-fold increase in *fabA* and *fabB* messenger RNA (mRNA), coupled with an increase in the unsaturated to saturated fatty acid ratio from approximately one to two–. FabR DNA binding is regulated by unsaturated acyl-CoA and acyl-ACP thioesters, thus allowing FabR to tune the expression of *fabA* and *fabB* according to the unsaturated fatty acid content of the acyl-ACP pool. The ability of FabR to bind both CoA and ACP thioesters indicates FabR is a sensor of both de novo and exogenously obtained fatty acids, in contrast to FadR that only senses the exogenously derived acyl-CoA pool. Both FadR and FabR can simultaneously bind to the *fabB* promoter sequence, demonstrating combinatorial transcriptional regulation. Although not all FabR homologues may function in this manner, the ability of these transcription factors to sample the composition of the intermediate pools provides an elegant mode of regulation, compared with classical repressors that respond to ligand concentration rather than ligand composition.

7.2 Gram-Positive Bacteria

Gram-positive bacteria lack the FabB and FabA enzymes for synthesising unsaturated fatty acids. One group, exemplified by *S. aureus* and *B. subtilis*, produce primarily branched-chain saturated fatty acids (Figure 1), whereas *S. pneumoniae* uses the FabM enzyme for the *trans–cis* isomerisation reaction to initiate unsaturated fatty acid synthesis (Figure 4). This difference in the fatty acid biosynthetic machinery is reflected by the differences in transcriptional regulation. Two different systems for the transcriptional control of FASII gene expression have been described in Gram-positive bacteria, although there remains much to be learned about this diverse group of pathogens.

7.2.1 The FapR System

The first transcription factor discovered in Gram-positive bacteria was the FapR repressor of *B. subtilis*. FapR is a global regulator of FASII genes plus the acyltransferases of phospholipid biosynthesis. FapR is a repressor that, when bound to malonyl-CoA, releases from its DNA binding site. Structural studies have revealed FapR as a homodimer consisting of a 'hot dog' fold. Interestingly, strains lacking FapR contain no more phospholipid than wild-type strains, despite the overexpression of the pathway enzymes. However, a fatty acid compositional defect likely underlies the cold intolerance of *B. subtilis* *fapR* mutants.

7.2.2 The FabT System

Streptococcus pneumoniae use FabT as a transcriptional repressor that governs expression of the FASII gene cluster. FASII genes in *S. pneumoniae* reside in a single locus on the chromosome that includes the *fabT* gene. The genes are divided into two operons; the first contains *fabT-fabH-acpP*, and the second includes *fabK-fabD-fabG-fabF-accD-fabZ-accC-accD-accA*. A FabT operator sequence is located in the promoter region of *fabT* and *fabK*. Disruption of the *fabT* gene distinctly alters the fatty acid composition of membrane phospholipids by reducing the unsaturated:saturated fatty acid ratio and increasing fatty acid chain length. Long-chain acyl-ACP enhances FabT binding to DNA. On binding to acyl-ACP, the affinity of FabT for its DNA binding site is significantly increased, and the FabT-acyl-ACP complex docks on the DNA and represses transcription of the FASII cluster. FabT binds acyl-ACPs of all chain lengths, but only the long-chain acyl-ACPs induce the conformational change that activates DNA binding. The transcriptional feedback regulation involving long-chain acyl-ACPs links accumulation of the end product of the pathway with expression of the genes encoding the pathway enzymes. This regulatory circuit results in the stringent repression of FASII genes in the presence of exogenous fatty acids.

7.3 Stress Response Regulators

7.3.1 Two-Component Systems

In addition to controlling the rate of fatty acid synthesis in response to a shift in the composition of pathway intermediates, *B. subtilis* has developed regulatory mechanisms to detect changes in temperature and adjust the expression of desaturating enzymes accordingly (for review see Mansilla and de Mendoza, 2005). The ability to adjust the unsaturation of membrane phospholipids is crucial to counteract the increase in rigidity that arises when biological membranes are below their phase transition temperature. The *des* mRNA is virtually undetectable in *B. subtilis* cultures grown at 37 °C, but is dramatically induced on cold shock. This regulation is accomplished by the DesKR two-component regulator consisting of a sensor histidine kinase/phosphatase component, which phosphorylates/dephosphorylates a response regulator that exerts its effect, usually through activation of a specific target gene. The *B. subtilis* system consists of the membrane-associated kinase, DesK, which phosphorylates DesR and drives transcription of the *des* gene by recruiting RNA polymerase to the promoter. DesK acts as a thermometer, detecting the ambient temperature by adopting different signaling states depending on membrane fluidity. The ability of DesK to detect perturbations in membrane fluidity is demonstrated by experiments that show no *des* induction during cold shock when *B. subtilis* is treated with unsaturated fatty acids. An increased proportion of branched-chain *anteiso* fatty acids in the *B. subtilis* membrane also increase membrane fluidity, and DesK is activated in cells starved for the *anteiso* precursor isoleucine.

Streptococcus pneumoniae uses the two-component regulator WalRK (previously designated YycFG or VicRK) to increase acyl chain length by upregulating *fabF* and downregulating *fabH*. However, the WalRK-mediated change in gene expression is modest (1.5–2×) and not specific to *fabH* and *fabF* because all the pathway enzymes except *fabH*, *fabM* and *acpP* are upregulated. Activation of the WalRK regulator causes an increase in 18-carbon fatty acid species compared with 16-carbon. It is not known whether WalRK is activated in response to increased membrane fluidity, but it participates in a cell wall damage response pathway.

7.3.2 Alternate Sigma Factors

Transcriptional control by the extracytoplasmic function (ECF) sigma factors is highly complex, with many being allosterically regulated by antisigma factors, which in turn can be controlled by anti-antisigma factors. Cross regulation is also evident. An example is the σ^w promoter region that also contains a σ^a binding site. This paints a complex picture of the mechanisms that control each particular ECF sigma factor; however, their roles in the transcription of several genes controlling phospholipid biosynthesis are evident through genetic and physiologic studies.

Bacillus subtilis expresses seven ECF σ factors. Several of the σ factors detect changes in membrane integrity and elicit a response to counteract the problem. A prime example is σ^w, which, in contrast to DesKR, responds to an increase in membrane fluidity. σ^w is activated under conditions of membrane perturbation, such as high concentrations of detergent or when the lipid:protein ratio is disrupted. σ^w reacts to an increase in membrane fluidity by elevating straight-chain fatty acid synthesis by the downregulation of *fabHa* and increasing the average chain length by increased *fabF* expression. There is a σ^w binding site within the *fabHa* coding sequence in the *B. subtilis fabHa-fabF* operon that when occupied upregulates transcription of *fabF* and downregulates *fabHa*. Unlike most bacteria discussed in this chapter, the *B. subtilis* chromosome encodes two initiation-condensing enzymes termed FabHa and FabHb, which differ in their substrate specificities. FabHa has a preference for the branched-chain precursors isobutyryl, isovaleryl, and 2-methylbutyryl-CoA, whereas FabHb shows maximum activity with the straight-chain precursor acetyl-CoA. Plasmid-based overexpression of σ^w increased transcription of *fabF* and decreased *fabHa* transcripts. Deletion of σ^w is associated with increased susceptibility to environmental hazards such as detergents and membrane active compounds; however, many of these physiological effects are likely due to the plethora of efflux pumps and detoxification enzymes that are concurrently induced as opposed to specific modification of membrane structure. Whether σ^w is induced under other conditions that increase membrane fluidity, such as growth under high temperature or excessive unsaturated fatty acid production remains undetermined. Other ECFs also regulate membrane lipid composition in *B. subtilis*. For example, σ^x senses membrane damage from antimicrobial peptides and responds by inducing the synthesis of the zwitterionic lipid PE, thereby decreasing the net anionic charge of the membrane. Promoter regions of the *pss* and *psd* genes in *B. subtilis* also contain σ^x binding sites. The principal behind σ^x-mediated resistance to antimicrobial peptides is similar to that of MprF present in several Gram-positive bacteria that synthesises the positively charged Lys-PG to repel the peptides electrostatically.

8. FUTURE DIRECTIONS

There remain many unanswered questions in bacterial lipid metabolism. The most daunting is grasping the diversity in the structures and biochemical pathways that support the extremely varied lifestyles in the bacterial kingdom. Many membrane lipids have been structurally identified, the biosynthetic pathways of which are not known, illustrating that there is unique biochemistry to be discovered. Production of membrane lipids is an essential facet of bacterial physiology. Thus, defining the biochemical and regulatory pathways in pathogens will be important to determine whether these can be targeted for developmental therapeutics. In some bacteria with unsaturated fatty acids, the mechanisms described in this chapter are not present, and how these bacteria produce these fatty acids remains unknown. PlsX remains an enigma. It has a clear role in phospholipid synthesis in Gram-positive bacteria, but *plsX* is widely conserved, and its role is not clear in bacteria that do not contain *plsY*. Why do these bacteria generate acyl-phosphate from acyl-ACP? A broader issue is the biochemical mechanism for the acute regulation of fatty acid synthesis. In *E. coli*, acyl-ACP clearly plays a key role as a feedback regulator, but whether this is a universal paradigm or an oddity in a subset of bacteria is unknown. It seems that these biochemical mechanisms trump genetic regulation of the pathway, because manipulation of the levels of biosynthetic enzymes does not alter the final amount of phospholipid produced. Fatty acid synthesis is the most energy intensive process in phospholipid synthesis, and these biochemical mechanisms are key elements of bacterial physiology. Finally, this chapter did not cover lipid metabolism in obligate or facilitative intracellular bacteria. This group of bacteria likely steals lipid components or precursors from the host cell using unique biochemistry, which remains to be discovered.

Therapeutic applications are another important reason to study bacterial lipid metabolism. The importance of FASII to bacterial physiology has made it an attractive target for the development of antibiotics, and interested readers should see recent reviews to follow the rapidly evolving opportunities for the development of therapies based on attacking FASII (Parsons and Rock, 2011).

REFERENCES

Bloch, K., Vance, D.E., 1977. Control mechanisms in the synthesis of saturated fatty acids. Annu. Rev. Biochem. 46, 263–298.

Butland, G., Peregrin-Alvarez, J.M., Li, J., Yang, W., Yang, X., Canadien, V., Starostine, A., Richards, D., Beattie, B., Krogan, N., Davey, M., Parkinson, J., Greenblatt, J., Emili, A., 2005. Interaction network containing conserved and essential protein complexes in *Escherichia coli*. Nature (London) 433, 531–537.

Battesti, A., Bouveret, E., 2009. Bacteria possessing two RelA/SpoT-like proteins have evolved a specific stringent response involving the acyl carrier protein-SpoT interaction. J. Bacteriol. 191, 616–624.

Battesti, A., Bouveret, E., 2006. Acyl carrier protein/SpoT interaction, the switch linking SpoT-dependent stress response to fatty acid metabolism. Mol. Microbiol. 62, 1048–1063.

Cronan Jr., J.E., Rock, C.O., 1996. Biosynthesis of membrane lipids. In: Neidhardt, F.C., Curtis, R., Gross, C.A., Ingraham, J.L., Lin, E.C.C., Low, K.B., Magasanik, B., Reznikoff, W., Riley, M., Schaechter, M., Umbarger, H.E. (Eds.), *Escherichia coli* and *Salmonella typhimurium*: Cellular and Molecular Biology. American Society for Microbiology, Washington, D.C.

Clark, D.P., Cronan Jr., J.E., 1981. Bacterial mutants for the study of lipid metabolism. Methods Enzymol. 72, 693–707.

Cronan Jr., J.E., Waldrop, G.L., 2002. Multi-subunit acetyl-CoA carboxylases. Prog. Lipid Res. 41, 407–435.

Dowhan, W., 2012. A retrospective: use of *Escherichia coli* as a vehicle to study phospholipid synthesis and function. Biochim. Biophys. Acta 1831, 471–494.

Ernst, C.M., Peschel, A., 2011. Broad-spectrum antimicrobial peptide resistance by MprF-mediated aminoacylation and flipping of phospholipids. Mol. Microbiol. 80, 290–299.

Goldfine, H., 1982. Lipids of prokaryotes: structure and distribution. Curr. Top. Membr. Transp. 17, 1–43.

Lu, Y.-J., Zhang, Y.-M., Grimes, K.D., Qi, J., Lee, R.E., Rock, C.O., 2006. Acyl-phosphates initiate membrane phospholipid synthesis in Gram-positive pathogens. Mol. Cell 23, 765–772.

Mansilla, M.C., de Mendoza, D., 2005. The *Bacillus subtilis* desaturase: a model to understand phospholipid modification and temperature sensing. Arch. Microbiol. 183, 229–235.

Parsons, J.B., Rock, C.O., 2013. Bacterial lipids: metabolism and membrane homeostasis. Prog. Lipid Res. 52, 249–276.

Parsons, J.B., Rock, C.O., 2011. Is bacterial fatty acid synthesis a valid target for antibacterial drug discovery? Curr. Opin. Microbiol. 14, 544–549.

Rock, C.O., Cronan Jr., J.E., 1996. *Escherichia coli* as a model for the regulation of dissociable (type II) fatty acid biosynthesis. Biochim. Biophys. Acta 1302, 1–16.

Raetz, C.R.H., 1982. In: Hawthorne, Ansell (Eds.), Genetic Control of Phospholipid Bilayer Assembly. Elsevier Biomedical Press.

Subramanian, C., Zhang, Y.-M., Rock, C.O., 2010. DesT coordinates the expression of anaerobic and aerobic pathways for unsaturated fatty acid biosynthesis in *Pseudomonas aeruginosa*. J. Bacteriol. 192, 280–285.

White, S.W., Zheng, J., Zhang, Y.-M., Rock, C.O., 2005. The structural biology of type II fatty acid biosynthesis. Annu. Rev. Biochem. 74, 791–831.

Zhang, Y.-M., Marrakchi, H., White, S.W., Rock, C.O., 2003. The application of computational methods to explore the diversity and structure of bacterial fatty acid synthase. J. Lipid Res. 44, 1–10.

Zhang, Y.-M., Rock, C.O., 2009. Transcriptional regulation in bacterial membrane lipid synthesis. J. Lipid Res. 50, S115–S119.

Zhang, Y.-M., Zhu, K., Frank, M.W., Rock, C.O., 2007. A *Pseudomonas aeruginosa* transcription factor that senses fatty acid structure. Mol. Microbiol. 66, 622–632.

Chapter 4

Lipid Metabolism in Plants

Katherine M. Schmid
Department of Biology, Butler University, Indianapolis, IN, USA

ABBREVIATIONS

ACCase Acetyl-CoA carboxylase
ACP Acyl carrier protein
DAG Diacylglycerol
DGAT Diacylglycerol acyltransferase
ER Endoplasmic reticulum
GPAT Glycerol-3-phosphate acyltransferase
IPP Isopentenyl pyrophosphate
KAS 3-Ketoacyl-ACP synthase
KCS β-Ketoacyl-CoA synthase
LPAAT Lyso-phosphatidic acid acyltransferase
LPCAT Lyso-phosphatidylcholine acyltransferase
MEP 2-*C*-methyl-D-erythritol 4-phosphate
MVA Mevalonate
OPDA Oxophytodienoic acid
PA Phosphatidic acid
PC Phosphatidylcholine
PDAT Phospholipid:diacylglycerol acyltransferase
PDCT Phosphatidylcholine:diacylglycerol cholinephosphotransferase
PE Phosphatidylethanolamine
PG Phosphatidylglycerol
TAG Triacylglycerol

1. INTRODUCTION

Plants produce the majority of the world's lipids, and most animals, including humans, depend on these lipids as a major source of calories and essential fatty acids. Like other eukaryotes, plants require lipids for membrane biogenesis, as signal molecules and as a form of stored carbon and energy. In addition, bark, herbaceous shoots and roots each have distinctive protective lipids that help prevent desiccation and infection. To what extent does the biochemistry of plant lipid metabolism resemble that in other organisms? This chapter notes a number of similarities, but emphasises aspects unique to plants.

Biochemistry of Lipids, Lipoproteins and Membranes. http://dx.doi.org/10.1016/B978-0-444-63438-2.00004-3

TABLE 1 Comparison of Plant, Mammalian and Bacterial Lipid Metabolism

	Higher Plants	Mammals	*Escherichia coli*
Fatty Acid Synthase			
Structure	Type II (multicomponent)	Type I (multifunctional)	Type II (multicomponent)
Location	Plastids	Cytosol	Cytosol
Acetyl-CoA carboxylase(s)	Multisubunit and multifunctional	Multifunctional	Multisubunit
Primary Desaturase Substrates			
Δ^9	18:0-ACP	18:1-CoA	None
ω-6	18:1 on glycerolipids	None	None
ω-3	18:2 on glycerolipids	None	None
Isoprenoid Synthesis			
Pathways	Mevalonate and MEP pathways	Mevalonate pathway	MEP pathway
Prominent membrane sterols	Sitosterol and many others	Cholesterol	None
Primary substrate(s) for phosphatidic acid synthesis	Acyl-ACP and acyl-CoA	Acyl-CoA	Acyl-ACP
Prominent bilayer lipids	Galactolipid> phospholipid	Phospholipid	Phospholipid
Main β-oxidation function	Provides acetyl-CoA for glyoxylate cycle	Provides acetyl-CoA for TCA cycle	Provides acetyl-CoA for TCA cycle

Major differences between lipid metabolism in plants and other organisms are summarised in Table 1.

The presence of chloroplasts and related organelles in plants has a profound effect on both the gross lipid composition and the flow of lipid within the cell. Fatty acid synthesis occurs not in the cytosol as in animals and fungi, but in the chloroplast and other plastids. Acyl groups must then be distributed to multiple compartments, and the complex interactions between alternative pathways are a major focus of plant lipid biochemists. It is also significant that the lipid bilayers of chloroplasts are largely composed of galactolipids rather than phospholipids.

As a result, galactolipids are the most abundant acyl lipids in green tissues and probably on earth.

Plant lipids also have a substantial impact on the world economy and human nutrition. More than three-quarters of the edible and industrial oils marketed annually are derived from seed and fruit triacylglycerol (TAG). These figures are particularly impressive given that, on a whole organism basis, plants store more carbon as carbohydrate than as lipid. Since plants are not mobile, and since photosynthesis provides fixed carbon on a regular basis, plant require-ments for storage lipid as an efficient, lightweight energy reserve are less acute than those of animals.

Information on the pathways responsible for plant lipid metabolism has exploded in recent years, particularly for the model organism *Arabidopsis*, a small weed of the mustard family that was the first target of the plant genome project (Li-Beisson et al., 2013). Hundreds of genes required for plant lipid biosynthesis, utilisation and turnover have now been cloned. In addition to pro-viding valuable information on enzyme structure and function, these genes are being exploited to improve understanding of factors underlying oilseed yields and to design new, more valuable plant oils. Questions about coordination of lipid metabolic genes with each other and with their potential regulators are also becoming more tractable, as data from microarrays, RNAseq and lipidomics technology become more available.

2. PLANT LIPID GEOGRAPHY

Although all eukaryotic cells have much in common, the ultrastructure of a plant cell differs from that of the typical mammalian cell in three major ways. First, all living plant cells contain plastids. Second, the plasma membrane of plant cells is shielded by the cellulosic cell wall, preventing lysis in the naturally hypotonic environment but making preparation of cell fractions more difficult. Finally, the nucleus, cytosol and organelles are pressed against the cell wall by the tonoplast, the membrane of the large, central vacuole that can occupy 80% or more of the cell's volume.

2.1 Plastids

The plastids are a family of organelles containing a circular chromosome pres-ent in multiple copies. Young or undifferentiated cells contain tiny proplastids that, depending on the tissue, may differentiate into photosynthetic chloro-plasts, carotenoid-rich chromoplasts, or any of several varieties of colourless leucoplasts, including plastids specialised for starch storage. These different types of plastids, which may be interconverted in vivo, have varying amounts of internal membrane but, except in some algae, are bounded by two mem-branes. The internal structure of chloroplasts is dominated by the flattened green membrane sacks known as thylakoids. The thylakoid membranes house

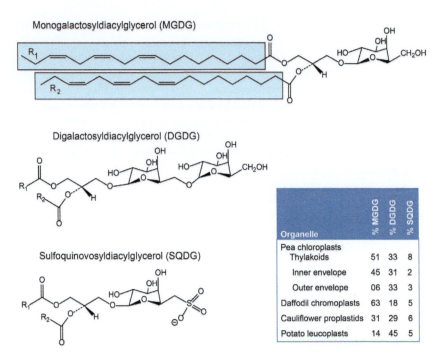

Monogalactosyldiacylglycerol (MGDG)

Digalactosyldiacylglycerol (DGDG)

Sulfoquinovosyldiacylglycerol (SQDG)

Organelle	% MGDG	% DGDG	% SQDG
Pea chloroplasts Thylakoids	51	33	8
Inner envelope	45	31	2
Outer envelope	06	33	3
Daffodil chromoplasts	63	18	5
Cauliflower proplastids	31	29	6
Potato leucoplasts	14	45	5

FIGURE 1 Structures and distribution of plant glycolipids. Except during phosphate deficiency, the pictured galacto- and sulpholipids are located primarily in plastids. Mol% of components are based on total acyl lipid in the designated organelle or membrane. *Data from Sparace and Kleppinger-Sparace (1993) and Andersson and Dörmann (2009).*

the light harvesting pigments and the rest of the apparatus for photosynthetic ATP and NADPH synthesis.

As noted above, chloroplasts and other plastids are enriched in galactolipids (Figure 1). They also contain a unique sulpholipid, sulphoquinovosyldiacylglycerol, whose head group is a modified galactose. The phospholipid components of plastids are less abundant. Phosphatidylglycerol (PG), the most prominent phospholipid contributor to the thylakoid membrane system, composes less than 10% of chloroplast glycerolipids, whereas plastidial phosphatidylcholine (PC) is limited primarily to the organelle's outer membrane.

Plastids are at the heart of plant acyl lipid metabolism, since they supply the 16- and 18-carbon fatty acid backbones used throughout the plant cell. All classes of plastid incorporate acetate into long-chain fatty acids, desaturate 18:0 to 18:1, and assemble phosphatidic acid (PA) and galactolipids. Chloroplasts also synthesise PG, including molecular species containing the unusual *trans*-3-hexadecenoic acid at the *sn*-2 position, and produce polyunsaturated fatty acids on galactolipid and sulpholipid substrates. It should also be noted that, although net lipid traffic is out of the plastids, they also import glycerolipids manufactured elsewhere. The quantitative significance of this backflow depends on the plant species (Section 5.3).

Plastids also house one of two isoprenoid synthesis pathways in plant cells (Section 11). Their isoprenoid products range from monoterpenes to the carotenoids and plastoquinone necessary for photosynthesis. Chlorophylls, by virtue of their phytol tails, also rank as chloroplast lipids.

2.2 Endoplasmic Reticulum and Lipid Bodies

The endoplasmic reticulum (ER) has traditionally been viewed as the primary source of phospholipids in plant cells. With the exception of cardiolipin, all of the common phospholipids can be produced by microsomal fractions. The ER also serves as the major site of fatty acid diversification. Since the polyunsaturated fatty acids synthesised in plastids are typically retained by the galacto- and sulpholipids on which they are made, ER desaturation pathways are of particular importance for developing seeds that store large quantities of 18:2 and 18:3. Many of the unusual fatty acids found primarily in seed oils are also products of microsomal enzymes. Not surprisingly, the ER is instrumental in the formation of TAG and probably in the differentiation of the lipid bodies in which they are stored (Section 7). Other functions of the plant ER include production of sphingolipids, some isoprenoids including the sterols, and many components of the cuticle (Sections 8–9, 11).

2.3 Mitochondria

Next to plastids and the ER, the mitochondrion is probably the plant organelle investigated the most thoroughly with respect to lipid metabolism. Synthesis of PG and cardiolipin by mitochondria is well established, although in Arabidopsis a microsomal phosphatidylglycerophosphate synthase can provide adequate precursor for cardiolipin synthesis in the absence of the dually located plastid/mitochondrial isozyme (Tanoue et al., 2014). While most fatty acids for mitochondrial membranes are imported from the plastids or the ER, mitochondria synthesise low levels of fatty acids from malonate. Octanoate is a major product of this pathway and serves as a precursor for the lipoic acid cofactor needed by glycine decarboxylase and pyruvate dehydrogenase. Arabidopsis mitochondria also express six homologs to bacterial lipid A pathway genes, although the function of the resulting lipopolysaccharides in plants is unknown (Li-Beisson et al., 2013). Plant mitochondria contain enzymes of β-oxidation, but these appear to be associated primarily with branched-chain amino acid catabolism rather than with oxidation of long-chain fatty acids.

2.4 Peroxisomes and Glyoxysomes

A discussion of the compartmentation of lipids and their metabolism would be incomplete without reference to the organelles responsible for fatty acid oxidation. Unlike mammals, plants use peroxisomal β-oxidation to catabolise long-chain fatty acids all the way to acetyl-coenzyme A (CoA). Moreover, plants also contain differentiated, specialised peroxisomes called glyoxysomes. Glyoxysomes oxidise fatty acids and feed the resulting acetyl-CoA into carbohydrate via the glyoxylate cycle, a pathway absent from animals. Since plants cannot transport fatty acids over long distances, this conversion of acetate to sucrose, which can be transported by the plant vascular system, makes lipid a practical carbon reserve for germinating seeds and pollen tubes (Graham, 2008). Peroxisomal β-oxidation is also crucial in the production of jasmonate and related plant hormones (Section 10).

3. ACYL-ACYL CARRIER PROTEIN SYNTHESIS IN PLANTS

Fatty acid synthases may be classified into two groups. 'Type I' fatty acid synthases are characterised by the large, multifunctional proteins typical of yeast and mammals (Chapter 5), while 'Type II' synthases of most prokaryotes are dissociable into components that catalyse individual reactions (Chapter 3). Plants, while certainly themselves eukaryotic, appear to have inherited a Type II fatty acid synthase from the photosynthetic prokaryotes from which plastids originated.

Early studies by Overath and Stumpf (1964) established not only that the constituents of the avocado fatty acid synthesis system could be dissociated and reconstituted but also that the heat stable fraction from *Escherichia coli* known as acyl carrier protein (ACP) could replace the corresponding fraction from avocado. Plant ACPs share both extensive sequence homology and significant elements of three-dimensional structure with their bacterial counterparts. Typically, plants contain multiple isoforms of this small, acidic protein. In addition to their classic functions in fatty acid synthesis, plant acyl-ACPs are substrates for synthesis of monounsaturated fatty acids and plastidial glycerolipids.

3.1 Components of Plant Fatty Acid Synthase

Fatty acid synthase is generally defined as including all polypeptides required for the conversion of acetyl and malonyl-CoA to the corresponding ACP derivatives, the acyl-ACP elongation cycle diagrammed in Chapter 3, and the cleavage of ACP from completed fatty acids by enzymes termed thioesterases or acyl-ACP hydrolases. All components of fatty acid synthase occur in plastids, although they are encoded in the nuclear genome and synthesised on cytosolic ribosomes. Most of the 8–10 enzymes of the pathway are soluble when isolated from homogenates. Nevertheless, at least ACP and some subunits of acetyl-CoA carboxylase (ACCase) may be associated with the plastid membranes.

Although plant fatty acid synthase preparations have both malonyl-CoA:ACP and acetyl-CoA:ACP acyltransferase activities, acetyl-ACP does not appear to play a major role in plant fatty acid synthesis. Instead, the first condensation takes place between acetyl-CoA and malonyl-ACP. This reaction is catalysed by ketoacyl-ACP synthase III, one of three required 3-ketoacyl-ACP synthase

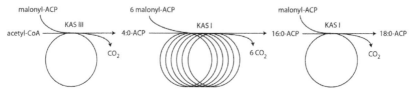

FIGURE 2 Contribution of the three ketoacyl-ACP synthases (KAS I, KAS II and KAS III) to fatty acid synthesis in plastids. Each circle represents addition of two carbons from malonyl-ACP by one cycle of the fatty acid synthase reactions catalysed by the appropriate KAS, enoyl-ACP reductase, hydroxyacyl-ACP dehydrase and acyl-ACP reductase.

(KAS) isoforms in plant systems (Figure 2). The acetoacetyl-ACP product then undergoes the standard reduction–dehydration–reduction sequence to produce 4:0-ACP, the initial substrate of KAS I. KAS I is responsible for the condensations in each elongation cycle up to the production of 16:0-ACP. The third ketoacyl synthase, KAS II, is dedicated to the final plastidial elongation, that of 16:0-ACP to 18:0-ACP (Li-Beisson et al., 2013).

3.2 The First Double Bond Is Introduced by Soluble Acyl-Acyl Carrier Protein Desaturases

The major components of the long-chain acyl-ACP pool in most plant tissues are 16:0-ACP, 18:0-ACP and 18:1-ACP. This finding highlights the importance of stearoyl-ACP desaturase, the plastidial enzyme responsible for Δ^9-desaturation in plants. In contrast to the desaturation system of *E. coli* (Chapter 3), the plant enzyme introduces the double bond directly to the Δ^9 position. Unlike yeast and mammalian Δ^9-desaturases, it is a soluble enzyme and is specific for acyl-ACPs rather than acyl-CoAs. Genes for acyl-ACP desaturases have been cloned from a number of plant species, and crystal structures have been determined for both the castor bean Δ^9-18:0-ACP desaturase and an unusual English ivy isoform that can synthesise Δ^9-18:0, Δ^4-16:1 or even $\Delta^{4,9}$-dienes. The two acyl-ACP desaturases are homodimers with very similar structures, including active sites featuring a diiron-oxo cluster (Shanklin et al., 2009). Reduction of the iron by ferredoxin leads to its binding of molecular oxygen. The resulting complex ultimately removes electrons at the Δ^9 position, resulting in double bond formation.

Although the most common unsaturated fatty acids in plants are derived from oleic acid, a wide range of unusual fatty acids are found in the seed oils of different species. The English ivy Δ^4/Δ^9-desaturase noted above is one of several acyl-ACP desaturase isoforms that account for some of this diversity. Coriander houses a more substrate specific Δ^4-16:0-ACP-desaturase compared to ivy, but elongates most of its Δ^4-16:1-ACP product to Δ^6-18:1. *Thunbergia alata*, another producer of Δ^6-18:1-rich seed oil, produces the fatty acid directly via a Δ^6-18:1-ACP-desaturase.

Several structure–function relationships suggested by unusual desaturases have been tested in the acyl-ACP desaturase system. For example, shortening the acyl-binding pocket of castor Δ^9-18:0-ACP desaturase to a length suggested by the cat's claw Δ^9-16:0 desaturase, an operation requiring only a single amino acid change, allowed the castor enzyme to produce oil enriched in Δ^9-16:1. Regiospecificity has been more difficult to model, although the castor Δ^9-desaturase can be converted to a Δ^6-desaturase using a set of five specific amino acids suggested by the *Thunbergia* Δ^6-18:0 desaturase. Most recently, site-directed mutagenesis based on crystal structures for the castor enzyme complexed with 18:0-ACP confirmed that the interaction between the phosphate of ACP's phoshopantetheine ring and a charged amino acid near the entrance to the active site is critical in determining Δ^4 versus Δ^9 desaturation (Guy et al., 2011).

3.3 Acyl-Acyl Carrier Protein Thioesterases Release Fatty Acids for Export

Among most prokaryotes, all acyl groups exiting the fatty acid synthase are transferred directly from ACP to glycerol 3-phosphate to form polar lipids. However, plants must also release large quantities of fatty acid from ACP to accommodate membrane lipid assembly outside the plastids. Since the typical chloroplast exports primarily 18:1 and 16:0, the same fatty acids that compose the greatest fraction of long-chain acyl-ACPs, it might be assumed that a relatively nonspecific thioesterase releases fatty acids from ACP. However, molecular and biochemical analyses of cloned plant thioesterases have revealed one family of acyl-ACP thioesterases with a strong preference for 18:1-ACP, and another specialising in 16:0-ACP or other saturated fatty acids. Altered thioesterase expression has therefore been a popular tool for altering the ratio of saturated to unsaturated fatty acids in oilseeds (Moreno-Pérez et al., 2014).

Plants whose plastids export unusual fatty acids often have additional or modified thioesterases. For example, several plant species that produce storage oils containing large proportions of 8- to 14-carbon acyl chains contain thioesterases specific for those chain lengths. By removing acyl groups from ACP prematurely, the medium-chain thioesterases simultaneously prevent their further elongation and release them for TAG synthesis outside the plastids. As a result of medium-chain thioesterases, coconut and palm kernel oils accumulate 45% or more of 12:0, providing feedstocks for such ubiquitous products as sodium dodecyl sulphate, the detergent most familiar to scientists for its role in sodium dodecyl sulphate–polyacrylamide gel electrophoresis. At least one plant genus, *Cuphea*, combines a special medium-chain keto-acyl synthase (KAS IV) with a specialised thioesterase to maximise 10:0 production (Cahoon and Schmid, 2008).

Plants with unusual acyl-ACP desaturases may also express unusual thioesterases (Moreno-Pérez et al., 2014). For example, macadamia nut and coriander, whose seed oils accumulate Δ^9-16:1 and Δ^6-18:1 respectively, have acyl-ACP thioesterases with corresponding activities. By regulating expression of different thioesterases, plants can both fine-tune and radically modify the exported fatty acid pool.

4. ACETYL-COENZYME A CARBOXYLASE AND CONTROL OF FATTY ACID SYNTHESIS

4.1 Most Plants Have Two Acetyl-Coenzyme A Carboxylases

The malonyl-CoA that supplies two-carbon units for fatty acid synthesis is produced from acetyl-CoA and bicarbonate by ACCase. In plants, malonyl-CoA for fatty acid synthesis is provided by a plastid-localised ACCase, while a cytosolic ACCase contributes malonyl units for fatty acid elongation beyond C18 as well as synthesis of flavonoids, polyketides and other metabolites. As with fatty acid synthases, ACCase forms may be categorised as either 'eukaryotic' enzymes, which are dimers of a multifunctional polypeptide (Chapter 5), or 'prokaryotic' enzymes, which are heteromers of four subunits: biotin carboxyl carrier protein, biotin carboxylase and two subunits of carboxyltransferase (Chapter 3). In the grass family, both plastids and cytosol house eukaryotic enzymes. However, dicots and monocots other than grasses appear to have both forms, with the eukaryotic form limited primarily to the cytosol, and prokaryotic enzymes dominating in the plastids (Sasaki and Nagano, 2004). Assembly of the prokaryotic form requires participation of both the nuclear genome, which encodes biotin carboxyl carrier protein, biotin carboxylase, and the alpha subunit of carboxyltransferase, and the plastid genome, which has retained the gene for the carboxyltransferase beta subunit, perhaps due to its requirement for RNA editing.

4.2 Acetyl-Coenzyme A Carboxylase is a Control Point for Fatty Acid Synthesis

In other kingdoms, ACCase is a major control point for fatty acid biosynthesis. Although the mechanisms acting in plants are incompletely characterised, there is evidence that plant ACCases are also tightly regulated. For example, both redox regulation via thioredoxin and phosphorylation of the carboxyltransferase have been implicated in upregulation of the chloroplast ACCase by light (Sasaki and Nagano, 2004). Conversely, feedback inhibition of ACCase has been observed in both rapeseed and tobacco (Andre et al., 2012). Due to its impact on the rate of fatty acid synthesis, ACCase is considered a promising target in oilseed improvement programmes. Although results have been variable, some increases in oil content have been reported following overexpression of the plastid-encoded subunit or introduction of the cytosolic ACCase to plastids.

5. PHOSPHATIDIC ACID SYNTHESIS OCCURS VIA PROKARYOTIC AND EUKARYOTIC ACYLTRANSFERASES

Since PA serves as a precursor of phospholipids, galactolipids and TAG, it is not surprising that its assembly via acylation of glycerol 3-phosphate is dispersed among the plastids, endomembranes and mitochondria. Plants have multiple families of glycerol-3-phosphate acyltransferases (GPAT), some of which esterify at the *sn*-2 position of glycerol 3-phosphate rather than the usual *sn*-1 position. However, the *sn*-2 variants appear to be associated with production of protective lipids (Section 8) rather than membrane lipids and TAG (Yang et al., 2012). Thus, it is *sn*-1-lyso-PA that serves as substrate for the lyso-PA acyltransferases (LPAAT) that generate PA.

Despite the consistency of their acylation sequences, plastidial and extraplastidial acyltransferases use different thioesters and acyl chains for PA synthesis. As a result, pools of glycerolipid backbones derived from the two compartments have distinctive fatty acid profiles that have proved very useful in dissecting the relationships between plastid and ER glycerolipids, as summarised in Figure 3.

5.1 Plastidial Acyltransferases Use Acyl-Acyl Carrier Protein Substrates

In the plastids, acyltransferases provide a direct route for acyl groups from ACP to enter membrane lipids. Since this is the standard pathway in *E. coli* and cyanobacteria, both the enzymes of PA synthesis in plastids and the glycerolipid backbones they produce are termed prokaryotic. In both chloroplasts and non-green plastids, GPAT is a soluble enzyme that, unlike the *E. coli* enzyme, shows preference for 18:1-ACP over 16:0-ACP. Plastidial LPAAT, which is a component of the inner envelope, is extremely selective for 16:0-ACP. The presence of a 16-carbon fatty acid at *sn*-2 is therefore considered diagnostic for lipids synthesised in the plastids (Li-Beisson et al., 2013).

5.2 Extraplastidial Acyltransferases Use Acyl-Coenzyme A Substrates

Both mitochondrial and endomembrane acyltransferases preferentially employ acyl-CoA substrates. In the ER, which is quantitatively the most significant of the extraplastidial sites for PA synthesis, saturated fatty acids are almost entirely excluded from the *sn*-2 position. The GPAT is less selective but, due to substrate availability, more often fills the *sn*-1 position with 18:1 than with 16:0. It is therefore possible to judge the relative contributions of the prokaryotic and eukaryotic pathways by comparing the proportions of eukaryotic 18/18 or 16/18 glycerolipids with prokaryotic 18/16 or 16/16 glycerolipids (Li-Beisson et al., 2013).

5.3 The 16:3 and 18:3 Plants Have Different Proportions of Prokaryotic Flux

Relative fluxes through the prokaryotic and eukaryotic pathways vary between plant species and among tissues. Plastids have the potential to use PA from the prokaryotic pathway to build all of their characteristic glycerolipids. However, not all plants do so; in some cases, the prokaryotic acyl chain arrangement is

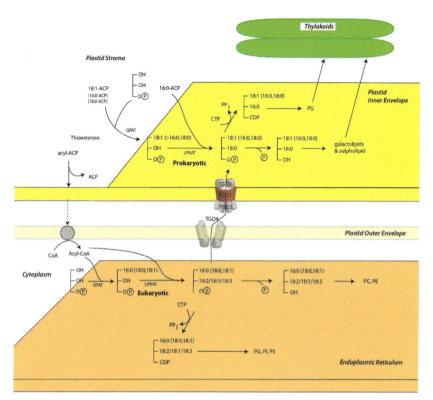

FIGURE 3 The prokaryotic and eukaryotic pathways of plant glycerolipid synthesis. The pro-karyotic pathway uses plastidial glycerol-3-phosphate acyltransferase (GPAT) and lysophospha-tidic acid acyltransferase (LPAAT) to assemble acyl-ACP substrates into phosphatidic acid (PA) enriched in *sn*-2 16:0. The eukaryotic pathway relies on GPAT and LPAAT at the endoplasmic reticulum to generate PA unusual at the *sn*-2 position. In both com-partments, PA is shuttled to either the CDP–diacylglycerol (DAG) or DAG route for glycerolipid synthesis. Prokaryotic PA provides substrate for plastidial phosphatidylglycerol (PG) in all plants, while eukaryotic PA feeds into PC, PE, phosphatidylserine (PS), phosphatidylinositol (PI), TAG and extraplastidial PG. Plastids typically export fatty acyl units, but import glycerolipid via TGD4 at the outer envelope and the TGD1–3 complex at the inner envelope. Plants differ in the propor-tions of prokaryotic and eukaryotic lipids used to assemble galactolipids and sulpholipids in the plastidial inner envelope. The 16:3 plants use primarily DAG derived from prokaryotic, plastidial PA, whereas 18:3 plants use mainly imported DAG with eukaryotic acyl chains. In their respective compartments, 18:1 or 16:1 esterified to galactolipid and 18:1 esterified phosphatidylcholine are major substrates for further desaturation.

found only in plastidial PG, whereas galactolipids are derived from diacylglycerol (DAG) imported to the plastids from the ER. As indicated above, the eukaryotic acyltransferases of the ER produce substantially more 18/18 than 16/18 lipids, and it is chiefly the 18/18 units that are assembled into galactolipids by plants with a minor prokaryotic pathway. Because galactolipids become highly unsatu-rated, plants that import DAG for galactolipids are rich in 18:3 and are called

18:3 plants. Species in which most galactolipid is derived from the prokaryotic 18/16 or 16/16 DAG contain substantial 16:3 and are known as 16:3 plants.

Kunst et al. (1988) first demonstrated that a 16:3 plant, *Arabidopsis thaliana*, may be converted to a de facto 18:3 plant by a single mutation in plastidial GPAT. Under these conditions, 16:3 content is reduced dramatically, and when isolated chloroplasts are labelled with glycerol 3-phosphate, only PG is labelled. Nevertheless, the percentage of galactolipids in mutant plants is nearly identical to that in wild-type plants, emphasising the ability of plants to compensate for reduction of the prokaryotic pathway. Other mutant studies have confirmed that plants have an amazing capacity to adapt to many, though not all, perturbations of lipid metabolism.

6. MEMBRANE GLYCEROLIPID SYNTHESIS

Prokaryotic and eukaryotic PA feeds into a complex web of reactions distributed among multiple compartments (Figure 3). As in mammals, the synthesis of other glycerolipids is initiated either by the formation of CDP–DAG from PA and CTP or by cleavage of phosphate from PA to produce DAG (Chapter 7).

CTP:phosphatidate cytidyltransferase activity has been observed in plastids, mitochondria and ER. In all three compartments, the CDP–DAG derived from PA is used in the synthesis of PG. Mitochondria also combine a second CDP–DAG with PG to produce cardiolipin, while the ER directs part of its CDP–DAG into phosphatidylinositol (PI) and, in some plants, phosphatidylserine. However, plastids appear to dedicate the CDP–DAG pathway solely to PG, which is required for thylakoid development and as a component of the photosynthetic apparatus. Assembly of PG on the plastid inner envelope employs mainly prokaryotic CDP–DAG, although plants engineered to have large proportions of eukaryotic PG in their plastids are able to photosynthesise normally. Plastidial PG is also the substrate for an unusual desaturase that inserts a Δ^3-*trans*-double bond into its 16:0. Curiously, although this desaturase is evolutionarily conserved and Δ^3-*trans*-16:1 production is typically coordinated with thylakoid development, mutants lacking detectable Δ^3-*trans*-16:1 grow normally under laboratory conditions (Boudière et al., 2014).

PA phosphatase is likewise present in multiple compartments, including the ER, mitochondria and plastids. In the ER, a large proportion of the resulting DAG reacts with CDP–choline or CDP–ethanolamine to produce PC or phosphatidylethanolamine (PE), respectively. The enzymes catalysing these reactions are nonspecific aminoalcohol phosphotransferases that can use either substrate (Li-Beisson et al., 2013). However, affinities for the two substrates are not necessarily equal. For example, wheat has both a constitutive enzyme favouring CDP–choline and a cold-induced variant favouring CDP–ethanolamine (Sutoh et al., 2010). It should be noted that plants also produce a fraction of their PE via phosphatidylserine decarboxylase, a mitochondrial enzyme in *Arabidopsis*.

Plants synthesise the phosphocholine precursor of CDP–choline by sequential methylation of phosphoethanolamine to monomethylphosphoethanolamine, dimethylphosphoethanolamine and phosphocholine. They can also incorporate any of these intermediates into PC via aminoalcohol phosphotransferases. However, unlike yeast and mammals, most plants are unable to methylate PE to PC. They lack the first methyltransferase of this pathway, although they can methylate N-methyl-PE the rest of the way to PC (Li-Beisson et al., 2013). The relative contributions of these pathways to PC may vary depending on the plant species, tissue and environmental conditions. One factor in some plants is the use of choline as a precursor of the glycine betaine that helps protect them against freezing and drought stress, and which, in lower plants, may even give rise to betaine lipids. PC is also at the centre of a web of acyltransferase, phospholipase and acyl modifications that can make analysing its metabolism daunting. Many of these are under investigation for their contributions to the composition of TAG, which are also derived from the phosphatidate phosphatase branch of phospholipid metabolism (Section 7).

In plastids, DAG supplies the lipid backbone for sulpholipids and galactolipids. The head groups of sulpholipid and monogalactosyldiacylglycerol are transferred to DAG from UDP-sulphoquinovose and UDP-galactose, respectively. A second UDP-galactose converts monogalactosyldiacylglycerol to digalactosyldiacylglycerol. Under normal growth conditions, the inner envelope of chloroplasts is the primary site of galactolipid and sulpholipid synthesis, and the molecules are incorporated into the thylakoid membranes and photosynthetic complexes. However, plants have two sets of monogalactosyl- and digalactosyldiacylglycerol synthase genes. The second set is strongly expressed when phosphate is limiting, a common situation for plants. These additional galactolipid synthases, which are especially prominent in nongreen tissues, are localised to the outer envelope of plastid. They are especially valuable in conversion of eukaryotic DAG to digalactosyldiacylglycerol. Export of this galactolipid from the plastids helps to make up for the shortfall of membrane phospholipids. Sulpholipid production is also upregulated when phosphate is limiting and recruited to substitute for PG (Boudière et al., 2014). Finally, a plastid envelope enzyme can add up to three galactose units to monogalactosyldiacylglycerol, using additional monogalactosyldiacylglycerol molecules as galactose donors. Disruption of this enzyme activity increases sensitivity to freezing and osmotic stresses, perhaps because the plant loses the ability to shunt nonbilayer-forming monogalactosyldiacylglycerol to TAG. It has also been suggested that the polygalactosyl species themselves may prevent freezing-induced membrane fusions (Moellering et al., 2010).

6.1 Lipid Trafficking between Plastids and Endomembranes

The interactions between the plastids and the ER described above require both export and import of lipids by plastids. The unraveling of the multiple mechanisms facilitating lipid trafficking in plants remains a major challenge. Even the necessary export of fatty acids from the plastids is not well understood. Transport is known to involve a nonesterified fatty acid intermediate, as expected given the activities of plastidial acyl-ACP thioesterases, and probably is a channelled or facilitated process rather than free diffusion because only a tiny pool of free fatty acid is ever detected. An acyl-CoA synthetase on the outer envelope of plastids has the capacity to quickly convert exported fatty acids to diffusible acyl-CoA thioesters. However, deletion of this isoform in *Arabidopsis*, although it reduces plastidial acyl-CoA synthesis, does not affect the plant's capacity for glycerolipid synthesis unless coupled with deletion of an ER isoform. There are also kinetic data suggesting that acyl groups leaving the plastid reach PC before the acyl-ACP pool, and the outer envelope has acyl exchange activity that could account for this flow. In addition, at least one ABC-binding cassette transporter is a strong candidate for export of TAG building blocks from *Arabidopsis* plastids, although it is uncertain whether it carries individual fatty acids, PA and/or DAG. There is clear evidence for direct membrane contact sites between the plastids and the ER, and between inner and outer plastid envelopes. The contributions of both these membrane contact sites and the many acyl-CoA-binding proteins identified in plants to lipid trafficking remain unclear (Hurlock et al., 2014; Andersson, 2013).

Extensive backflow of lipids from the ER to the plastids, particularly among 18:3 plants, must also be accounted for. Even in a 16:3 plant such as *Arabidiopsis*, about a third of the galactolipids are derived from eukaryotic PA and carry acyl lipids that have been desaturated on PC substrates. The best understood mechanism for glycerolipid import by *Arabidopsis* plastids involves a dimeric outer envelope beta-barrel protein (TGD4) that binds PA, coupled with an inner envelope ABC-binding cassette protein assembled from the TGD1, 2 and 3 proteins (Hurlock et al., 2014) (Figure 3).

6.2 Glycerolipids Are Substrates for Polyunsaturated Fatty Acid Synthesis

In addition to soluble acyl-ACP desaturases and the PG desaturase that inserts the Δ^3-*trans*-16:1 bond, plants contain a number of membrane-bound enzymes that desaturate fatty acids esterified within glycerolipids. The products of these enzymes include $\Delta^{9,12}$-18:2 and $\Delta^{9,12,15}$-18:3, fatty acids essential to the human diet that may have significant effects on human health and disease.

Once again, separate pathways for synthesis of polyunsaturated fatty acids occur in plastids and ER. Clarification of the number of desaturases involved in plant lipid metabolism by brute force screening for mutants, followed by cloning and expression of the relevant genes, was an early success of *Arabidopsis* as a model organism. Many other plants have now been shown to share homologues of the gene families described below.

Plastids are able to convert 18:1 to 18:3 and 16:0 to 16:3 using a combination of three membrane-bound desaturases (Li-Beisson et al., 2013). One of them, encoded by *FAD5 (ADS3)*, is relatively specific for the conversion of 16:0 on monogalactosyldiacylglycerol to Δ^7-16:1. This 16:1 and Δ^9-18:1 may then have a second and third double bond inserted by the *FAD6* and the *FAD7* or *FAD8* gene products, respectively. The latter two desaturases are less selective in their choice of glycerolipid substrate, and will accept appropriate fatty acids on PG, sulpholipid, or either of the major galactolipids. Whereas *FAD7* is expressed constitutively, *Arabidopsis FAD8* is upregulated by cold, and FAD8 protein is destabilised at higher temperatures. In the ER, 18:1 esterified to PC or occasionally PE can be desaturated to 18:2 by FAD2 and to 18:3 by FAD3.

Fatty acids entering one of the multistep desaturation pathways are not necessarily committed to completing that set of reactions. It is particularly common for 18:2 to be an end product of ER desaturation that can be incorporated into TAG or even enter the galactolipid pathway and receive a third double bond in the chloroplast. However, heterodimers between FAD2 and FAD3, and between FAD6 and FAD7 or 8, allow channeling of fatty acids from 18:1 to 18:3 (Lou et al., 2014).

Recently, another *Arabidopsis* desaturase, ADS2, was localised to ER and, to a lesser extent, Golgi and plastids. This enzyme appears to desaturate 16:0 on both PG and monogalactosyldiacylglycerol. At 23 °C, *Arabidopsis* lacking functional *ADS2* resemble wild-type plants, but when moved to 6 °C they become male sterile and grow poorly compared to controls (Chen and Thelen, 2013). These findings are consistent with a wide array of mutant studies highlighting the importance of unsaturated fatty acids during cold stress (Table 2). For example, *Arabidopsis* unable to desaturate 18:1 in the ER survives well at 22 °C but not when transferred to 6 °C. If both plastidial and ER 18:1 desaturase activities are lost, temperature is no longer the primary issue: the combination is lethal. The double mutants can be recovered only on sucrose-supplemented media. The sucrose grown plants, which contain less than 6% polyunsaturated fatty acids, are chlorotic and unable to carry out photosynthesis but otherwise remarkably normal (McConn and Browse, 1998). These results, while confirming the significance of polyunsaturated fatty acids to photosynthesis, indicate that the vast majority of membrane functions can proceed despite drastically reduced levels of polyunsaturated fatty acids.

Triunsaturated fatty acids normally dominate chloroplast membranes and thus are the most abundant fatty acids in plants. Surprisingly, *Arabidopsis* triple mutants lacking all 18:2 desaturation activity are still able to grow, photosynthesise, and even flower. However, they are male sterile and therefore cannot produce seeds unless the anthers are rescued by treatment with 18:3 or jasmonate, a plant hormone derived from 18:3 (Section 10). This result is a dramatic reminder that the most severe phenotypes of plants impaired in desaturation and other aspects of lipid metabolism are not necessarily mediated by bulk changes in membrane composition.

TABLE 2 Biochemical and Physiological Responses of Selected Arabidopsis Lipid Mutants

Mutant	Enzyme Blocked	Fatty Acid or Lipid Phenotype	Physiological Response
fab1	3-Ketoacyl-ACP synthase II	↑16:0	Death of plants after prolonged exposure to 2 °C
fab2	18:0-ACP Δ^9 desaturase	↑18:0	Dwarf at 22 °C
fad4	*Trans*-Δ^3 16:0 desaturase, plastid PG *sn-2*	↓*Trans*-Δ^3-16:1	No obvious phenotype; slight photosystem instability noted during gel electrophoresis.
fad5 (*ads3*)	16:0 Δ^7 desaturase (plastid galactolipid *sn2*)	↑16:0; ↓16:3	Enhanced growth rate at high temperatures. Leaf chlorosis, reduced growth and impaired chloroplast development at low temperature.
fad6	Plastid ω^6 desaturase (16:1→ 16:2, 18:1→ 18:2, *sn1* and *sn2*, all plastid lipids)	↑16:1; ↓16:3, ↓18:3	Leaf chlorosis, reduced growth and impaired chloroplast development at low temperature. Enhanced thermotolerance of photosynthetic electron transport at high temperatures.
fad7	Plastid ω^3 desaturase (16:2→ 16:3, 18:2→ 18:3, *sn1* and *sn2*, all plastid lipids)	↑16:2; ↑18:2 ↓16:3; ↓18:3	Reduced chloroplast size and altered chloroplast ultrastructure.
fad2	ER ω^6 desaturase	↑18:1; ↓18:2	Greatly reduced stem elongation at 12 °C. Death at 6 °C.

TABLE 2 Biochemical and Physiological Responses of Selected Arabidopsis Lipid Mutants—cont'd

Mutant	Enzyme Blocked	Fatty Acid or Lipid Phenotype	Physiological Response
fad2/fad6	Plastid and ER ω^6 desaturases	<6% polyunsaturated	Loss of photosynthesis
fad3/ fad7/ fad8	Plastid and ER ω^3 desaturases	<1% trienoic	Male sterile, insect resistance decreased
ads2	Primarily ER 16:0 ω^9 desaturase; cold inducible	↑16:0	Dwarf and male sterile at 6°C; reduced freezing tolerance
dgd1	Digalactosyldiacyl-glycerol synthase	↓Digalactosyldiacyl-glycerol	Dwarfism, abnormal chloroplast size
act1	Plastid acyl-ACP:G3P acyltransferase	↓Phosphatidylglycerol, ↓16:3	Altered chloroplast structure
pgp1	Plastid and mitochondrial phosphatidylglycerol phosphate synthase	↓Phosphatidylglycerol	Loss of photosynthesis
pah1/ pah2	ER phosphatidic acid phosphatase	↑Phosphatidic acid, phosphatidycholine, phosphatidylethanolamine	Giant ER cisternae
mgd1	Diacylglycerol glycosyltransferase	↓Monogalactosyldiacyl-glycerol	Abnormal chloroplast development
AS11	Diacylglycerol acyltransferase	50% reduction in seed TAG	Slow germination
wri1	Transcription factor; glycolysis impaired	80% reduction in seed TAG	Slow germination

6.3 Some Plants Use Endoplasmic Reticulum Glycerolipids as Substrates for Production of Unusual Fatty Acids

Although the membrane lipids of plants have a very limited number of fatty acid species, the fatty acid composition of storage oils is notoriously variable, with more than 300 types of fatty acids known to occur in plant TAGs. As described earlier, some of these are produced by nonstandard acyl-ACP desaturases and acyl-ACP thioesterases. A third major source of fatty acid diversity is modification of fatty acids esterified to ER glycerolipids (Cahoon and Schmid, 2008) (Table 3). Some representatives of this category are desaturase variants that merely insert double bonds at unusual positions. For example, members of the borage family have a Δ^6-desaturase distantly related to *FAD2* and *FAD3*. By combining this enzyme with the usual ER desaturases, borage produces γ-linolenic acid, and its oil is marketed as a nutraceutical. At least two classes of enzyme from the *FAD2* family tree act as conjugases, which convert a normal *cis*-double bond to a pair of double bonds, for example, *cis*-Δ^{12}→*trans* Δ^{11}+*trans* Δ^{13}. Still another *FAD2* variant can convert a double bond to a triple bond. However, the activities of such enzymes are not limited to desaturation. The first divergent *FAD2* cloned from plants inserts a hydroxyl group rather than a double bond. Specifically, it converts the ordinary Δ^9-18:1 in PC to 12-OH-Δ^9-18:1, better known as the ricinoleic acid, and ultimately composes 85–95% of castor oil. Production of such hydroxy-fatty acids actually requires relatively little modification of desaturase enzymes. In the laboratory, as few as four amino acid changes can convert a desaturase into a hydroxylase. Certain plants also use membrane-bound substrates to produce epoxy- or cyclopropane fatty acids.

Some unusual plant desaturases are homologous to animal acyl-CoA desaturases, although not all of them use acyl-CoA substrates (Cahoon and Schmid, 2008). Note that the *ACL2* and *FAD5* discussed above also belong to this clade. However, some variants, such as the enzyme that produces Δ^5-20:1 in meadowfoam, do seem to work with acyl-CoAs rather than glycerolipid-bound substrates.

7. LIPID STORAGE IN PLANTS

A plant stores reserve material in its seeds in order to allow seedling growth of the next generation until photosynthetic capacity can be established. The three major storage materials are oil, protein and carbohydrate, and almost all seeds contain some of each. However, their proportions vary greatly. For example, the amount of oil in different species may range from as little as 1–2% in grasses such as wheat, to as much as 60% of the total dry weight of the castor seed. With the exception of the jojoba plant, which accumulates wax esters in seeds, seeds store oil as TAG.

7.1 Lipid Body Structure and Biogenesis

In the mature seed, TAG is stored in densely packed lipid bodies, which are roughly spherical in shape with an average diameter of 1 μm (Figure 4) (Chapman et al., 2012). This size does not change during seed development, and accumulation of oil is accompanied by an increase in the number of lipid bodies. The very large number of lipid bodies in an oilseed cell (often >1000) contrasts strikingly with animal adipose tissue where oil droplets produced in the cytosol can coalesce into one or a few droplets. Plant lipid bodies are

TABLE 3 Some Unusual Fatty Acids Resulting from Modifications in the Endoplasmic Reticulum

Fatty Acid Type	Example	Biosynthesis	Source	Applications
γ-Linolenic		Δ^6-desaturase = cytochrome b5-desaturase fusion	*Borago officinalis*	Nutraceutical
Ricinoleic (hydroxy)		18:1 Δ^{12}-hydroxylase, divergent FAD2	*Ricinus communis* (castor bean)	Coatings, lubricants
Vernolic (epoxy)		Divergent FAD2 or cytochrome P450 monoxygenase	*Vernonia fordii*	Plasticiser
Crepenynic (acetylenic)		18:2 acetylenase, divergent FAD2	*Crepis foetida*	Polymers
Parinaric (conjugated)		18:3 Δ^{12}-conjugase, divergent FAD2	*Impatiens balsamina*	Fluorescent probes, coatings
Sterculic (cyclopropene)		ER cyclopropane fatty acid synthase + unknown desaturase	*Stercuia foetida*	Lubricants, polymers

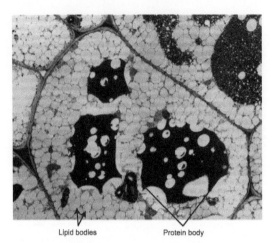

Lipid bodies   Protein body

FIGURE 4   Thin-sectional view of cells in a cotyledon of a developing cotton embryo harvested 42 days after anthesis, magnification 9000×. The cells are densely packed with lipid bodies. The large, dark objects are protein bodies containing crystals of mineral-rich phytin. *Photo courtesy of Richard Trelease, Arizona State University.*

bounded by a phospholipid monolayer in which the polar head groups face the cytosol, while the nonpolar acyl groups are associated with the nonpolar TAG within. The membranes of isolated lipid bodies, which compose less than 5% of a lipid body's weight, contain both phospholipids and characteristic proteins. The best studied of these are the oleosins, small (15–26 kDa) proteins that help to preserve individual lipid bodies as discrete entities during desiccation. Several lines of evidence show that loss of oleosins is associated with lipid body fusion and cell disruption (Schmidt and Herman, 2008). Each of the many oleosins cloned to date has an extraordinarily long hydrophobic sequence of 68–74 amino acids. Structurally, oleosins are roughly analogous to the animal apolipoproteins that coat the surface of lipoproteins during their transport between tissues (Chapters 15 and 16). At least one oleosin, peanut OLE3, also can switch between monoacylglycerol acyltransferase and phospholipase A2 activity based on its phosphorylation state (Parthibane et al., 2012). Other lipid body proteins include caleosins, calcium-binding proteins that stabilise lipid bodies and participate in oxylipid metabolism, steroleosins that may have roles in plant steroid hormone signalling (Chapman et al., 2012), and LDAPs (lipid droplet-associated proteins) that also appear to participate in formation of large, nondesiccation-resistant lipid bodies such as those of avocado fruits (Horn et al., 2013).

When a seed germinates, the TAG stored in the lipid bodies becomes the substrate for lipases. In some plants, this is preceded by peroxidation of polyunsaturated fatty acids by a lipid body lipoxygenase. Typically lipases and lipid body lipoxygenase are active only after germination is triggered by imbibition and other environmental signals. Fatty acids released by the lipid bodies are further metabolised through the β-oxidation pathway and glyoxylate cycle in the glyoxysomes (Section 2.4).

7.2 The Pathways of Triacylglycerol Biosynthesis

As in animal tissues, TAG can be produced by a relatively simple four reaction pathway. The first three steps, as discussed earlier, involve production of DAG by sequential acylation of glycerol 3-phosphate, followed by cleavage of the phosphate (Figure 3). The last and committed step is the transfer of a fatty acid from CoA to the vacant third hydroxyl of the DAG. This reaction, which directly produces TAG, is catalysed by diacylglycerol acyltransferase (DGAT). However, TAG assembly is not always straightforward. In many oilseeds, pulse–chase labelling has revealed that fatty acids reach TAG only after passing through PC or, to a lesser extent, PE. As described in Section 6, both polyunsaturated fatty acids and various unusual fatty acids are produced on ER glycerolipids, especially PC. Since these fatty acids often comprise large proportions of TAG acyl units, transit through PC helps to explain some of the fatty acid diversity in TAG.

Some potential routes by which a modified fatty acid could move between PC and TAG are diagrammed in Figure 5. The relative contributions of these pathways probably differ among plants, but the clearest picture comes from work with *Arabidopsis*, in which flow of polyunsaturated fatty acids from PC to TAG can be eliminated by inactivation of the genes for two enzymes (Bates et al., 2013).

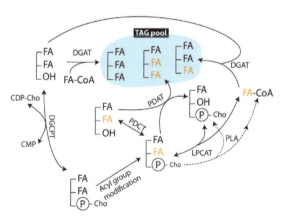

FIGURE 5 Flux through phosphatidylcholine (PC) promotes acyl group diversity in plant triacylglycerols (TAG). Fatty acids desaturated, hydroxylated or otherwise modified at the *sn-2* position of PC may ultimately reach any position on TAG. Enzymes involved can include diacylglycerol acyltransferase (DGAT), phosphlipid:diacylglycerol acyltransferase (PDAT), PC:diacylglycerol cholinephosphotransferase (PDCT), lysophophosphatidylcholine acyltransferase (LPCAT) in both directions, diacylglycerol cholinephosphotransferase (DGCPT) and phospholipases coupled with acyl-CoA synthetases.

1. Lyso-PC acyltransferase (LPCAT) transfers acyl groups from CoA to lyso-PC. Since the reaction is reversible, it can both release modified fatty acids to the CoA pool and replace them with others. Although similar effects could be achieved by a combination of phospholipase A and acyl-CoA synthase activities, this does not seem to be a major pathway in *Arabidopsis*.
2. PC:diacylglycerol cholinephosphotransferase (PDCT) shunts the choline head group to a new DAG, leaving the DAG backbone of the original molecule free to enter the TAG pool. This objective could also be obtained by other mechanisms. For example, in some plants the synthesis of PC from DAG and CDP–choline appears to be rapidly reversible.

Of course, *Arabidopsis* is not faced with many of the unusual fatty acids populating the plant kingdom. One enzyme that may be especially significant for incorporation of unusual fatty acids into TAG is phospholipid:diacylglycerol acyltransferase (PDAT), which synthesises TAG by transferring a fatty acid from a phospholipid substrate rather than coenzyme A. PDATs with high specificity for hydroxy- and epoxy-fatty acids have been identified in plants whose seed oils accumulate the corresponding acyl groups. It is more difficult to evaluate the proportions of TAG produced by PDAT vs. DGAT for ordinary fatty acids, since a plant in which one is compromised may compensate with the other (Bates et al., 2013).

7.3 Control of Triacylglycerol Yield

A major goal in the oilseed industry is to increase yields of TAG without compromising other agronomic factors. For example, there would not be much market for a high oilseed that germinated poorly, produced plants that were more susceptible to drought, or reduced profits on desirable by-products. Nevertheless, there are hopes that better understanding of the genetic and metabolic factors underlying TAG synthesis will lead to more gains than previously achieved.

One strategy has been to upregulate genes at potential control points in the synthesis of fatty acids and TAGs (Weselake et al., 2009). As already noted, ACCase helps to control flux into fatty acids, and modest increases in oil yield have been achieved by expressing a eukaryotic form of the enzyme in the prokaryotic plastid compartment. Overexpression of DGAT, the committed step of the most direct route to TAG, has also given positive results in several cases (Liu et al., 2012). Nevertheless, yield improvements due to single enzyme modifications, while significant, have not been particularly dramatic; rather, flux into TAG may be affected more by overall expression of entire pathways than by individual checkpoint enzymes.

It has long been known that lipid biosynthetic enzymes are expressed in a synchronised fashion during seed development. In recent years, several 'master regulators' of seed development have been shown to affect lipid production via *Wrinkled1* (*WRI1*), a transcription factor originally named for an *Arabidopsis* mutant whose seeds were shrivelled and oil poor. *WRI1*, which is conserved between plants as divergent as oil palm and *Arabidopsis*, coordinates expression of genes for much of fatty acid and oil synthesis, as well as parts of the glyco-lytic pathway, and early tests suggest that it will be valuable tool in upregulation of seed oils (Marchive et al., 2014).

7.4 Control of Triacylglycerol Composition

As highlighted in Section 6.3, TAG and phospholipids frequently have radically different fatty acid compositions, despite the synthesis of many unusual fatty acids on phospholipid substrates. Even when only the usual membrane fatty acids are involved, TAG and membrane composition are not necessarily correlated, as might be expected given the functional constraints on membrane fluidity and other physical attributes. Control of partitioning into TAG has implications not only at the basic level, but in the oilseed industry. For example, efforts to develop high 18:1 soybean oil were delayed by 'yield drag' that may have been due to unintended alteration of membrane composition (Clemente and Cahoon, 2009).

Several mechanisms probably enhance partitioning of unusual fatty acids into seed oils. Plants that accumulate unusual fatty acids typically possess acyl-transferases able to accommodate these acyl units. Not only the DGAT and PDAT responsible for assembly of TAG itself, but isozymes for PA synthesis sometimes show selectivity for uncommon acyl-CoAs. Another intriguing possibility is that TAG is assembled in distinct subdomains of the ER. For example, different isoforms of DGAT, one of which preferentially incorporates conjugated fatty acids into tung oil in vivo, can be detected in nonoverlapping sections of ER in developing tung seeds (Shockey et al., 2006). Finally, the editing reactions discussed in Section 7.2 can selectively remove inappropriate acyl groups from phospholipids. Recently, several LPCATs that readily select 18:1-CoA over hydroxy-18:1-CoA for PC synthesis were shown to have the opposite selectivity in the reverse reaction (Lager et al., 2013). Thus, a single enzyme can simultaneously reduce the concentration of an unusual fatty acid in phospholipid and convert it to a CoA substrate suitable for TAG biosynthesis.

7.5 Triacylglycerols in Vegetative Tissues

Although most plant TAG is concentrated in oil-rich seeds, fruits and pollen grains, vegetative tissues contain TAG, either in the form of the hydrophobic droplets within plastids known as plastoglobuli, or in oleosin-poor lipid bodies resembling those of avocado and olive fruits. Except in a few specialised cells such as vanilla leaf epidermis, lipid bodies are not particularly prominent in healthy vegetative tissue. For example, in their classic paper on TAG storage in different plant parts, Yatsu et al. (1971) began their purification of leaf lipid bodies with a cream separator and 50 pounds of cabbage. However, vegetative TAG rises during senescence and in response to environmental stresses such as phosphate deficiency (Section 6), when membrane degradation products are scavenged. Recent evidence suggests that PDAT-generated TAG plays a critical role in diverting excess fatty acid to β-oxidation (Fan et al., 2014). Because far more plant biomass is associated with vegetative than reproductive tissues, the potential of leaf and stem TAG for biodiesel production is under investigation. For example, overexpressing both the DGAT and a seed transcription factor known to upregulate *WRI1* doubled the amount of fatty acid extractable from tobacco leaves (Andrianov et al., 2010).

7.6 Triacylglycerol Engineering: Some Case Studies

Much of the information presented in this chapter has direct application in the tailoring of vegetable oils for edible or industrial use. Oilseed rape, the source of canola oil, has typically been the first crop to benefit, given its close relationship to the model plant *Arabidopsis*, but there has been progress with other oilseeds. In the United States, soybean oil is by far the dominant oil crop, although it is now second to palm oil in the world market. Efforts to modify soybean oil have been delayed by the complexity of its genetics compared to Arabidopsis. For example, a major goal of breeding programmes has been to reduce the high proportions of polyunsaturated fatty acids in soybean oil that reduce its shelf life and make it unsuitable as a frying oil. These problems were formerly dealt with by partial hydrogenation of the oil, a process now known to produce damaging *trans* fats.

Because soybean seeds express three *FAD2* and four *FAD3* desaturase genes, finding appropriate combinations of alleles to decrease polyunsaturates in oil without affecting agronomic characteristics has been a long process (Clemente and Cahoon, 2009). Traditional breeding using mutagenised seed ultimately achieved reduction of 18:3 from 8% to about 3% by crossing plants now known to have mutations in *FAD3-1a* and *1b* (Figure 6). Individually, the

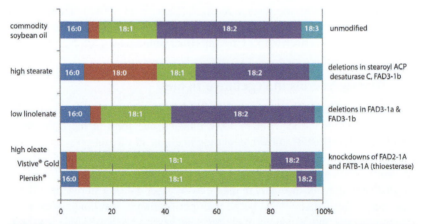

FIGURE 6 Fatty acid composition of soybean oils with reduced polyunsaturation. Mutations and genetic modifications used to obtain cultivars with low 18:3, high 18:0 and high 18:1 are listed. High 18:1 lines were obtained by RNAi of conventionally bred low 18:3 lines.

two mutations had only reduced 18:3 to about 5%. More recently, soybeans that also incorporate genetically modified traits have begun reaching consumers. Two current high oleate lines, Monsanto's Vista® Gold and Pioneer/Dupont's Plenish® (Figure 6), both employ RNAi to reduce expression of an 18:1 desaturase (FAD2) and the acyl-ACP thioesterase that releases 16:0 for export from the plastids. These oils may be used both as olive oil substitutes and for applications that would require hydrogenation of ordinary soybean oil.

Efforts are also underway to modify crops to produce marketable volumes of unusual fatty acids. Aside from a few specialty niches such as production of γ-linolenic acid and botanical substitutes for fish oil, most of the focus has been on vegetable oil components that could replace industrial feedstocks currently supplied by the petroleum industry. Proponents of the latter contemplate either modification of traditional oilseed crops or development of new crops such as *Camelina sativa* to be reserved for industrial uses. Many unusual fatty acids have been expressed in transgenic plants. In fact, identification of these compounds in genetically modified plants is now a routine step towards identifying the relevant genes, and has spawned many patents showing proof of concept for development of new oils. However, the resulting oil profiles and yields are frequently disappointing. Even if the enzyme in question is strongly expressed and active, accumulation of the unique fatty acid may be depressed by poor incorporation of the new fatty acids into TAG, preferential turnover of the unusual acyl groups, or even inhibition of fatty acid synthesis (Cahoon and Schmid, 2008; Bates et al., 2014). Approaches to overcome these problems have included stacking genes for fatty acid modification with acyltransferases from the source plant, adding constructs that increase substrate or downregulate competing reactions, and reducing β-oxidation during oil deposition. For example, although castor oil contains about 90% of ricinoleate, transformation of *Arabidopsis* with castor hydroxylase resulted in seed oils with no more than 17% ricinoleate and half their normal TAG. In this case, depressed yield was overcome and ricinoleate raised to 25% by cotransforming with a castor bean DGAT2 (Bates et al., 2014). Similarly, seed oils of *Arabidopsis* transformed with *E. coli* cyclopropane fatty acid synthase reached only 9% cyclopropane fatty acid despite the use of host plants modified for high levels of the 18:1 substrate. This level was ultimately raised to 35% by cotransformation with an LPAAT gene from a plant that normally makes cyclopropane fatty acids. However, further work will be needed to overcome severe agronomic problems in the high cyclopropane seeds (Yu et al., 2014).

8. PROTECTIVE LIPIDS: CUTIN, WAXES, SUBERIN AND SPOROPOLLENIN

All above-ground tissues of plants are protected against desiccation and pathogens by a thin layer of lipids called the cuticle. The cuticle, which is produced by epidermal cells and intimately associated with the cell wall, consists of two major types of lipids, cutin and waxes. Cutin is a complex, insoluble polyester of glycerol and fatty acid derivatives with a range of oxygen-containing functional groups. Sixteen- and 18-carbon fatty acids with one or more hydroxyl groups are particularly common (Figure 7). Many details of cutin's three-dimensional structure remain unresolved, although considerable progress has been made in identifying necessary hydroxylases, acyltransferases and ABC transporters, as well as transcription factors mediating responses of the cuticle to drought and other environmental factors. In addition, cutin localisation helps to determine which organs will fuse during plant development (Yeats and Rose, 2013).

The cutin mesh is both permeated by and overlaid with cuticular wax. The appearance and reflective characteristics of plant epidermis can vary depending on whether the outer layer of wax forms a smooth, glossy layer or is roughened by wax crystals. Chemically, cuticular waxes are complex mixtures including very-long-chain alkanes, aldehydes and ketones as well as wax esters and their building blocks. Visual screening of plant surfaces has allowed isolation of mutants blocked in many of the biosynthetic steps of wax synthesis, and a number of genes encoding these enzymes have now been isolated (Yeats and Rose, 2013; Li-Beisson et al., 2013).

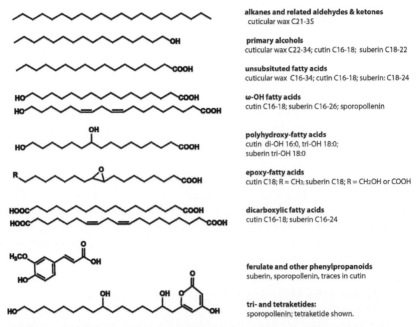

alkanes and related aldehydes & ketones
cuticular wax C21-35

primary alcohols
cuticular wax C22-34; cutin C16-18; suberin C18-22

unsubsituted fatty acids
cuticular wax C16-34; cutin C16-18; suberin: C18-24

ω-OH fatty acids
cutin C16-18; suberin C16-26; sporopollenin

polyhydroxy-fatty acids
cutin di-OH 16:0, tri-OH 18:0;
suberin tri-OH 18:0

epoxy-fatty acids
cutin C18; R = CH3; suberin C18; R = CH2OH or COOH

dicarboxylic fatty acids
cutin C16-18; suberin C16-24

ferulate and other phenylpropanoids
suberin, sporopollenin, traces in cutin

tri- and tetraketides:
sporopollenin; tetraketide shown.

FIGURE 7   Monomers of cuticular waxes and polyesters. Proportions differ among plant species and tissues. The multiple functional groups of hydroxy- and dicarboxylic fatty acids, as well as the presence of glycerol in both polyesters, permit complex patterns of cross-linking. Ferulate is esterified to ω-hydroxy-fatty acids in suberin, but may be able to cross-link further via its ring hydroxyl group.

A second polyester, suberin, is characteristic of bark, wound callus and specialised tissues such as the endodermis that controls entry into the root vascular system. Unlike cuticle, which coats the outer surface of epidermal cells, suberin is normally deposited between a plant cell's membrane and its primary wall. Patterns of suberin deposition vary between tissues and plant species. For example, cork cells accumulate a thick, even layer of suberin shortly before apopotosis, whereas endodermal cells develop distinctive suberised strips that prevent soil water from penetrating further into the root without crossing an endodermis membrane.

Suberin's resistance to chemical and enzymatic digestion has made it difficult to study. Like cutin, it is based on a polyester of glycerol and fatty acids enriched in carboxyl and hydroxyl groups, although suberin fatty acids are typically longer than those of cutin (Figure 7). A specific class of glycerol acyltransferases necessary for assembly of endodermis suberin has been identified in *Arabidopsis*. In addition to their aliphatic components, suberins also feature large proportions of phenylpropanoids, and knockdowns of these components also appear to affect overall permeability of suberin. A single feruloyl-CoA acyltransferase transferase expressed in *Arabidopsis* endodermis can esterify this phenylpropanoid group either to fatty alcohols or to ω-hydroxy-fatty acids (Li-Beisson et al., 2013).

Sporopollenin, a polymer with particularly deep evolutionary roots, protects pollen, air-borne plant spores and the zygotes of plants' closest algal relatives (Figure 7). Sporopollenins are notoriously resistant to chemical analysis, and their monomer composition remains controversial. However, genetic approaches have revealed participation of hydroxy-fatty acids, phenylpropanoids and oxygenated polyketides in the assembly of *Arabidopsis* sporopollenin (Grienenberger et al., 2010).

8.1 Fatty Acid Elongation and Wax Production

Not only protective polymers, but certain seed oils, the wax esters stored by jojoba, and sphingolipids (Section 9) contain fatty acids longer than the 16- and 18-carbon moieties produced by the plastidial fatty acid synthase. Fatty acids up to 30 carbons long are produced at the ER by sequential addition of two-carbon units from malonyl-CoA to fatty acids exported by the plastids. As in normal fatty acid synthesis, a series of four enzymatic reactions is required. Acyl-CoA substrates of differing length require different β-ketoacyl-CoA synthases (KCS) for the initial, rate-limiting condensation step, and specific members of a large KCS family have been implicated in elongation of fatty acids for cuticular wax, root suberin and TAG. In contrast, mutations in β-ketoacyl-CoA reductase, β-hydroxyacyl-CoA dehydratase and enoyl-CoA reductase typically affect elongation products used in multiple pathways (Li-Beisson et al., 2013). The same substrates that initiate fatty acid elongation, 16- and 18-carbon acyl-CoAs and malonyl-CoA, are likewise substrates for the polyketide synthases involved in sporopollenin production (Grienenberger et al., 2010).

As described in Section 6.3, the ER contains an array of fatty acid modification enzymes, including hydroxylases employed in the production of polyester monomers. In addition, it houses both a pathway that reduces fatty acids to the fatty alcohols required for wax ester synthesis, and a decarbonylation pathway that generates the odd-chain long-chain alkanes, aldehydes, ketones and secondary alcohols also associated with epicuticular wax.

Two distinct classes of wax synthases have been identified in plants. Enzymes such as those responsible for the storage wax of jojoba seeds are monofunctional, catalysing dehydration synthesis between acyl-CoA and fatty alcohols. In the case of jojoba, the most common substrates in each category are monounsaturated 18-, 22- or 24-carbon chains. However, the best studied example of cuticular wax formation involves a member of the second class, a bifunctional wax synthase/DGAT (Li-Beisson et al., 2013).

9. SPHINGOLIPID BIOSYNTHESIS

Sphingolipids are usually considered minor constituents of plant lipids, accounting for 5% or less of most lipid extracts. This fact, and the more complex methods needed for their identification and characterisation, have resulted in a comparative lack of information on plant sphingolipid biosynthesis and function. Nevertheless, sphingolipids can make up 40% of plant plasma membrane lipids and substantial amounts of the tonoplast, with glucosylceramides and inositolphosphorylceramides constituting major fractions. In addition to participating in lipid rafts and membrane trafficking, plant sphingolipids have been implicated in signal transduction events that help plants respond to cold, drought and pathogens (Markham et al., 2013).

As in other eukaryotes (Chapter 10), sphingolipid biosynthesis begins with the condensation of palmitoyl-CoA and serine to form 3-keto-sphinganine. The plant enzyme is a heteromer homologous to the fungal and animal proteins, and deletions of a single subunit are lethal in *Arabidopsis*. The 3-keto-sphinganine is typically reduced to sphinganine, and further modifications including desaturation or hydroxylation at Δ^4 and desaturation at Δ^8 result in a pool of long-chain bases more diverse than in mammals or fungi (Markham et al., 2013).

In *Arabidopsis*, there is strong evidence that one ceramide synthase isozyme pairs dihydroxy-long-chain bases with 16:0-CoA, while two other isozymes combine trihydroxy-long-chain bases with 24:1, 26:1 or other very-long-chain fatty acids. The dihydroxy-base/16:0 combinations are primarily associated with glucosylceramides, while trihydroxy-bases are channelled preferentially to the glycosyl-inositolphosphorylceramide pathway. Mutants lacking the ceramide synthase that consumes dihydroxy-bases are at a disadvantage during stress, while mutations lacking both trihydroxy-base consuming isoforms are lethal (Markham et al., 2013). It will be interesting to see to what extent the *Arabidopsis* model of ceramide synthesis holds in crop plants that add large proportions of Δ^4-unsaturated long-chain bases and ceramides with more complex head groups to the mix.

10. OXYLIPINS AS PLANT HORMONES

Like animals, plants produce a range of physiologically active oxylipins via the oxidation of polyunsaturated fatty acids released from membranes. For example, the action of lipoxygenases on linolenic acid released from membranes gives rise to 9- and 13-hydroperoxylinolenic acids. Figure 8 details some of the fates of 13-hydroperoxylinolenic acid, notably biosynthesis of

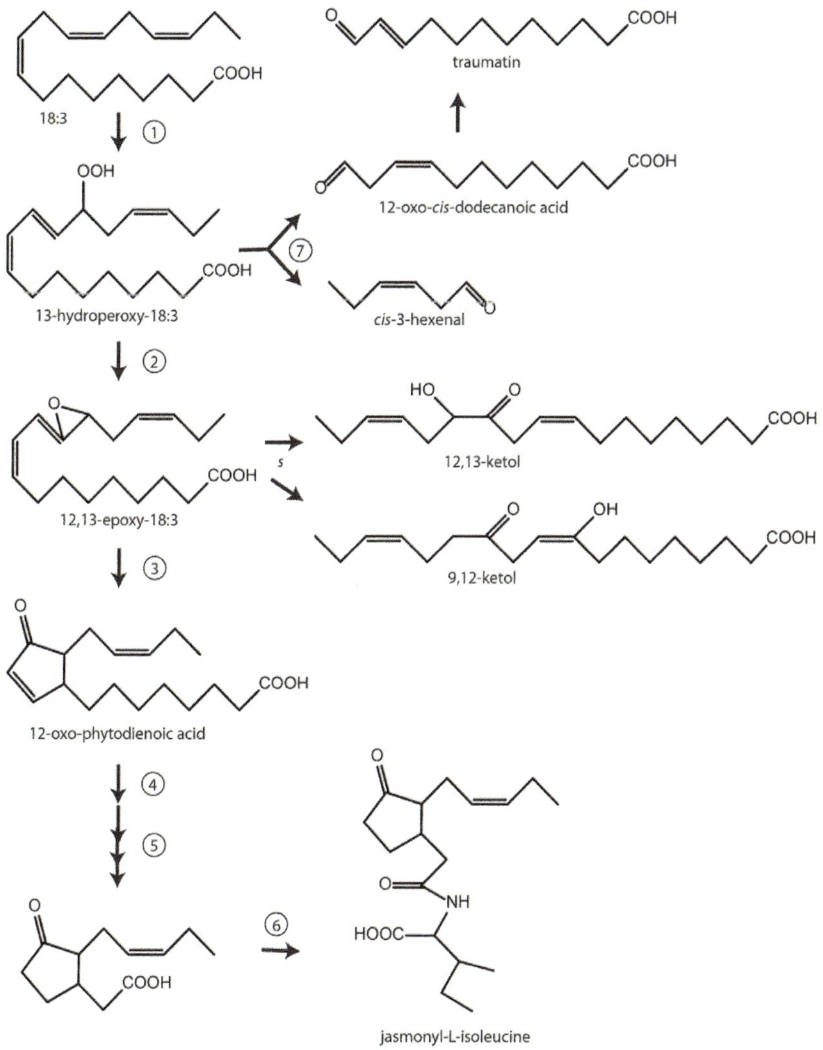

FIGURE 8 Metabolism of 18:3 to some oxylipin derivatives. 1, Lipoxygenase; 2, allene oxide synthase; 3, allene oxide cyclase; 4, 12-oxo-phytodienoic acid reductase; 5, β-oxidation (3 rounds); 6, jasmonyl-isoleucine synthetase (JAR1); 7, hydroperoxide lyase; s, spontaneous.

jasmonic acid, a key defensive hormone reminiscent of mammalian prostaglandins. Production of jasmonate requires cooperation between multiple cellular compartments. Conversion of 13-hydroperoxylinolenic acid to oxophytodienoic acid (OPDA), also a bioactive compound, occurs in the plastids. OPDA then travels to the peroxisomes, where reduction followed by three rounds of β-oxidation generates jasmonic acid itself. The best understood responses to jasmonate require its conjugation to isoleucine. Jasmonyl-L-isoleucine, which is assembled in the cytoplasm, ultimately helps to derepress genes associated with stress responses, wound healing and some aspects of development (Kombrink, 2012). A number of other jasmonate derivatives, including volatile methyl jasmonate that may mediate interplant communication, have been characterised.

Two other branch points from 13-hydroperoxides are noted in Figure 8. Allene oxide, in addition to undergoing enzymatic conversion towards jasmonate, may also spontaneously decompose to ketols. The C18 hydroperoxide itself may be cleaved at the *trans*-11,12-double bond by hydroperoxide lyase (reaction 7) to produce a C12 compound, 12-oxo-*cis*-9-dodecenoic acid C6 aldehyde, *cis*-3-hexenal. The acid is subsequently metabolised to 12-oxo-*trans*-10-dodecenoic acid, the wound hormone traumatin, and a volatile hexanal whose 'freshly cut grass' scent in at least some cases attracts predatory insects that help keep herbivores such as caterpillars under control.

11. STEROL AND ISOPRENOID BIOSYNTHESIS

In the plant kingdom, isoprenoids represent the most diverse range of natural products with over 25,000 lipophilic structures known, ranging from small, volatile compounds to rubber. Quantitatively, the photosynthetic apparatus is probably the primary consumer of isoprenoids, since carotenoids, plastoquinone and the phytol tail of chlorophyll all belong to this group (Hemmerlin et al., 2012; Vranová et al., 2013). Given that vital plant hormones such as gibberellin and abscisic acid, plus many defensive compounds, are isoprenoids, the early steps of this pathway have been studied intensely. However, surprisingly, it was not until the late 1990s that researchers realised that plants have two very different pathways for production of isopentenyl pyrophosphate (IPP), the five-carbon central precursor of all isoprenoids (Hemmerlin et al., 2012). For several decades it had been known that, as in mammals and yeast (Chapter 11), plants produce IPP via the mevalonate (MVA) pathway, in which hydroxymethylglutaryl-CoA formed from three molecules of acetyl-CoA is reduced to mevalonic acid by the highly regulated enzyme, hydroxymethylglutaryl-CoA reductase. Plants contain multiple well-studied hydroxymethylglutaryl-CoA reductase genes that are differentially expressed during development and in response to such stimuli as light, wounding and infection. It was incorrectly assumed that the MVA pathway was localised in both cytosol and plastids and produced all classes of isoprenoids. However, it is now clear that, while IPP for steroid synthesis is indeed generated by a cytosolic MVA pathway, IPP biosynthesis in plastids occurs via a seven enzyme nonmevalonate route known as the 2-C-methyl-D-erythritol 4-phosphate (MEP) pathway and shared by cyanobacteria and *E. coli*. Important control points in the plant MEP pathway include the initial step, condensation of pyruvate with glyceraldehyde 3-phosphate to produce 1-deoxy-D-xylulose-5-P, reduction of this product to MEP, and the production of IPP and its isomer, dimethylallylpyrophosphate, via hydroxy-3-methylbut-2-enyl diphosphate reductase (Vranová et al., 2013). The MEP pathway generates classic plant photosynthetic isoprenoids such as phytols and carotenoids and, with the MVA pathway, contributes to production of certain mono-, di-, sesqui- and tetraterpenoids. Typically the MEP pathway is upregulated in photosynthetic tissues, while the MVA pathway is especially active in non-green plant parts (Vranová et al., 2013).

Up until the cyclisation step, sterol synthesis in plants follows the common eukaryotic pathway (Chapter 11). Three C5 IPP units from the MVA pathway join head to tail to produce a C15 compound, farnesyl pyrophosphate. Two farnesyl pyrophosphates are then united head to head to form squalene, the progenitor of the C30 isoprenoids from which sterols are derived. The plant squalene synthetase, like its mammalian homologue, is found in the ER, and the reaction proceeds via a presqualene pyrophosphate intermediate. In the last step prior to cyclisation, squalene is converted to squalene 2,3-epoxide (Benveniste, 2004). Beyond this point, the pathway diverges, with a family of oxidosqualene synthases shunting the precursor into mixtures of triterpenoid secondary products that vary widely across the plant kingdom (Xue et al., 2014). The one oxidosqualene synthase encoded by all plants investigated is cycloartenol synthase. Unlike animals and fungi, which rely on lanosterol synthase for the initial cyclisation leading to sterol production, even plants with the ability to produce small quantities of lanosterol produce sterols primarily via cycloartenol (Ohyama et al., 2009) (Figure 9).

Plants lack a monolithic membrane sterol comparable to cholesterol. Rather, a complex series of reactions including opening of the cyloartenol cyclopropane ring, double bond formation and isomerisation, demethylation of ring carbons, and methylation of the side chain results in mixtures of membrane sterols whose proportions differ not only among plant species but from tissue to tissue (Benveniste, 2004). Common plant sterols – phytosterols in the medical literature – include sitosterol, campesterol, and stigmasterol (Figure 9). In addition, sterol esters, sterol glycosides, and acylated sterol glycosides are common plant constituents whose physiological significance is under scrutiny. Both cold adaptation and pathogenesis drastically alter free and derivatised sterol pools. Sterols are also critical to plants as precursors of the brassinosteroids, a class of steroid hormones required for light-induced development, fertility and stress responses (Vriet et al., 2013). Although brassinosteroid signal transduction pathways are quite different from their animal counterparts, steroid hormone biosynthesis

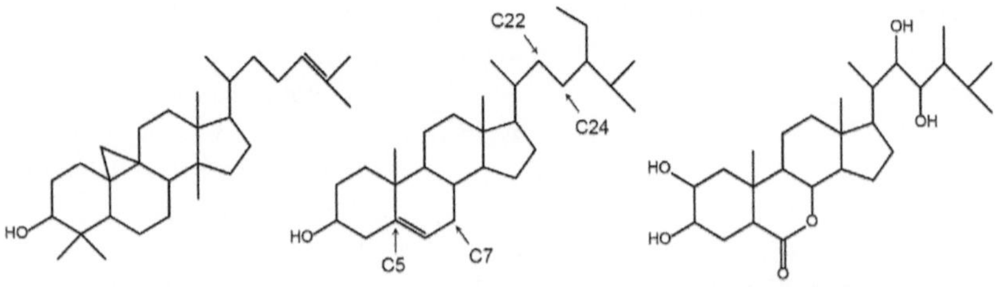

FIGURE 9 Plant steroids. Cycloartenol, rather than lanosterol, is the first cyclised precursor of sterols and plants. Sitosterol (24α-ethyl cholesterol), shown here, is the most common plant sterol, but plant membranes generally contain mixtures of sterols. Other prominent phytosterols differ from sitosterol as follows. Campesterol, 24α-methyl; stigmasterol, C22 double bond; dihydrospinasterol, move double bond from C5 to C7; spinasterol, move C5 double bond to C7, add C22 double bond; dihydrobrassicasterol, 24β-methyl; brassicasterol, 24β-methyl, add C22 double bond. Brassinolide, the most active known steroid hormone in plants, plays roles in morphogenesis, fertility and stress responses.

includes interesting parallels. For example, the gene for a 5α-reductase in the brassinosteroid pathway can complement the corresponding reductase in the testosterone pathway (Li et al., 1997).

12. FUTURE PROSPECTS

In recent years, the contributions of plastids and ER to both acyl lipid and isoprenoid metabolism have become increasingly clear, particularly in the model plant *Arabidopsis*. Superficially redundant components are now understood to allow a sedentary organism maximal flexibility in responding to resource availability, an inconsistent environment, and a throng of herbivores and pathogens. New tools for mapping metabolic fluxes and transcriptional profiles should improve our understanding of the control of this web of reactions under different stresses and developmental programmes. Similar approaches should inform breeding to improve yield and composition of lipid-based commodities by identification of control points, bottlenecks and situations in which results depend on incremental contributions by entire suites of enzymes.

Many questions also remain in fundamental lipid biochemistry. The mechanisms by which lipids flow between compartments are poorly understood, as is the degree to which metabolic channelling influences the distribution of molecular species. The assembly and architecture of cutin, suberin and other polyesters still require extensive investigation. Finally, the manifold roles of acyl lipids and isoprenoids in signal transduction are under intense scrutiny. In addition to brassinosteroids, jasmonates and other lipid-derived hormones, plant development and survival depend on a complex network of molecules including PI phosphates, DAGs, sphingolipids, *N*-acylphosphatidylethanolamine and prenyltransferases.

ACKNOWLEDGEMENTS

Thanks are due to John B. Ohlrogge for his contributions to earlier editions of this chapter.

REFERENCES

Andersson, M.X., 2013. Chloroplast contact to the endoplasmic reticulum and lipid trafficking. Adv. Photosynth. Respir. 36, 155–167.

Andersson, M.X., Dörmann, P., 2009. Chloroplast membrane lipid biosynthesis and transport. Plant Cell Monographs 13, 125–159.

Andre, C., Haslam, R.P., Shanklin, J., 2012. Feedback regulation of plastidic acetyl-CoA carboxylase by 18:1-acyl carrier protein in *Brassica napus*. Proc. Natl. Acad. Sci. U.S.A. 109, 10107–10112.

Andrianov, V., Borisjuk, N., Pogrebnyak, N., Brinker, A., Dixon, J., et al., 2010. Tobacco as a production platform for biofuel: overexpression of Arabidopsis DGAT and LEC2 genes increases accumulation and shifts the composition of lipids in green biomass. Plant Biotechnol. J. 8, 277–287.

Bates, P.D., Johnson, S.R., Cao, X., Li, J., Nam, J.-W., et al., 2014. Fatty acid synthesis is inhibited by inefficient utilization of unusual fatty acids for glycerolipid assembly. Proc. Natl. Acad. Sci. U.S.A. 111, 1204–1209.

Bates, P.D., Stymne, S., Ohlrogge, J., 2013. Biochemical pathways in seed oil synthesis. Curr. Opin. Plant Biol. 16, 358–364.

Benveniste, P., 2004. Biosynthesis and accumulation of sterols. Annu. Rev. Plant Biol. 55, 429–457.

Boudière, L., Michaud, M., Petroutsos, D., et al., 2014. Glycerolipids in photosynthesis: composition, synthesis and trafficking. Biochim. Biophys. Acta 1837, 470–480.

Cahoon, E.B., Schmid, K.M., 2008. Metabolic engineering of the content and fatty acid composition of vegetable oils. Adv. Plant Biochem. Mol. Biol. 1, 161–200.

Chapman, K.D., Dyer, J.M., Mullen, R.T., 2012. Biogenesis and functions of lipid droplets in plants. J. Lipid Res. 53, 215–226.

Chen, M., Thelen, J., 2013. *ACYL-LIPID DESATURASE2* is required for chilling and freezing tolerance in *Arabidopsis*. Plant Cell 25, 1430–1444.

Clemente, T.E., Cahoon, E.B., 2009. Soybean oil: genetic approaches for modification of functionality and total content. Plant Physiol. 151, 1030–1040.

Fan, J., Yan, C., Roston, R., Shanklin, J., Xu, C., 2014. Arabidopsis lipins, PDAT1 acyltransferase, and SDP1 triacylglycerol lipase synergistically direct fatty acids toward β-oxidation, thereby maintaining membrane lipid homeostasis. Plant Cell 26, 4119–4134.

Graham, I.A., 2008. Seed storage oil mobilization. Annu. Rev. Plant Biol. 59, 115–142.

Grienenberger, E., Kim, S.S., Lallemand, B., Geoffroy, P., Heintz, D., et al., 2010. Analysis of TETRAKETIDE alpha-PYRONE REDUCTASE function in *Arabidopsis thaliana* reveals a previously unknown, but conserved, biochemical pathway in sporopollenin monomer biosynthesis. Plant Cell 22, 4067–4083.

Guy, J.E., Whittle, E., Moche, M., Lengqvist, J., Lindqvist, Y., Shanklin, J., 2011. Remote control of regioselectivity in acyl-acyl carrier protein-desaturases. Proc. Natl. Acad. Sci. U.S.A. 108, 16594–16599.

Hemmerlin, A., Harwood, J.L., Bach, T.J., 2012. A raison d'être for two distinct pathways in the early steps of plant isoprenoid biosynthesis? Prog. Lipid Res. 51, 95–148.

Horn, P.J., James, C.N., Gidda, S.K., Kilaru, A., Dyer, J.M., et al., 2013. Identification of a new class of lipid droplet-associated proteins in plants. Plant Physiol. 162, 1926–1936.

Hurlock, A.K., Roston, R.L., Wang, K., Benning, C., 2014. Lipid trafficking in plant cells. Traffic 15, 915–932.

Kombrink, E., 2012. Chemical and genetic exploration of jasmonate biosynthesis and signaling paths. Planta 236, 1351–1366.

Kunst, L., Browse, J., Somerville, C., 1988. Altered regulation of lipid biosynthesis in a mutant of *Arabidopsis* deficient in chloroplast glycerol-3-phosphate acyltrasferase activity. Proc. Natl. Acad. Sci. U.S.A. 85, 4143–4147.

Lager, I., Yilmaz, J.L., Zhou, X.-R., Jasieniecka, K., Kazachkov, M., et al., 2013. Plant acyl-CoA:lysophosphatidylcholine acyltransferases (LPCATs) have different specificities in their forward and reverse reactions. J. Biol. Chem. 288, 36902–36914.

Li, J., Biswas, M.G., Chao, A., Russell, D.W., Chory, J., 1997. Conservation of function between mammalian and plant steroid 5α-reductases. Proc. Natl. Acad. Sci. U.S.A. 94, 3554–3559.

Li-Beisson, Y., Shorrosh, B., Beisson, F., Andersson, M.X., Arondel, V., et al., 2013. Acyl-lipid metabolism. Arabidopsis Book 11 tab.0161.

Liu, Q., Siloto, R., Lehner, R., Stone, S.J., Weselake, R.J., 2012. Acyl-CoA:diacylglycerol acyl-transferase: molecular biology, biochemistry and biotechnology. Prog. Lipid Res. 51, 350–377.

Lou, Y., Schwender, J., Shanklin, J., 2014. FAD2 and FAD3 desaturases form heterodimers that facilitate metabolic channeling *in vivo*. J. Biol. Chem. 289, 17996–18007.

Marchive, C., Nikovics, K., To, A., Lepiniec, L., Baud, S., 2014. Transcriptional regulation of fatty acid production in higher plants: molecular bases and biotechnological outcomes. Eur. J. Lipid Sci. Technol. 116, 1332–1343.

Markham, J.E., Lynch, D.V., Napier, J.A., Dunn, T.M., Cahoon, E.B., 2013. Plant sphingolipids: function follows form. Curr. Opin. Plant Biol. 16, 350–357.

McConn, M., Browse, J., 1998. Polyunsaturated membranes are required for photosynthetic compe-tence in a mutant of Arabidopsis. Plant J. 15, 521–530.

Moellering, E.R., Muthan, B., Benning, C., 2010. Freezing tolerance in plants requires lipid remodeling at the outer chloroplast membrane. Science 330, 226–228.

Moreno-Pérez, A.J., Venegas-Calerón, M., Vaistij, F.E., Salas, J.J., Larson, T.R., et al., 2014. Effect of a mutagenized acyl-ACP thioesterase FATA allele from sunflower with improved activity in tobacco leaves and Arabidopsis seeds. Planta 239, 667–677.

Ohyama, K., Suzuki, M., Kikuchi, J., Saito, K., Muranaka, T., 2009. Dual biosynthetic pathways to phytosterol via cycloartenol and lanosterol in *Arabidopsis*. Proc. Natl. Acad. Sci. U.S.A. 106, 725–730.

Overath, P., Stumpf, P.K., 1964. Fat metabolism in plants. XXIII. Properties of a soluble fatty acid synthetase from avocado mesocarp. J. Biol. Chem. 239, 4103–4110.

Parthibane, V., Iyappan, R., Vijayakumar, A., Venkateshwari, V., Rajasekharan, R., 2012. Serine/threonine/tyrosine protein kinase phosphorylates oleosin, a regulator of lipid metabolic func-tions. Plant Physiol. 159, 95–104.

Sasaki, Y., Nagano, Y., 2004. Plant acetyl-CoA carboxylase: structure, biosynthesis, regulation, and gene manipulation for plant breeding. Biosci. Biotechnol. Biochem. 68, 1175–1184.

Schmidt, M.A., Herman, E.M., 2008. Suppression of soybean oleosin produces micro-oil bodies that aggregate into oil body/ER complexes. Mol. Plant 1, 910–924.

Shanklin, J., Guy, J.E., Mishra, G., Lindqvist, Y., 2009. Desaturases: emerging models for understanding functional diversification of diiron-containing enzymes. J. Biol. Chem. 284, 18559–18563.

Shockey, J.M., Gidda, S.K., Chapital, D.C., Kuan, J.C., Dhanoa, P.K., et al., 2006. Tung tree DGAT1 and DGAT2 have nonredundant functions in triacylglycerol biosynthesis and are localized to different subdomains of the endoplasmic reticulum. Plant Cell 18, 2294–2313.

Sparace, S.A., Kleppinger-Sparace, K.F., 1993. Metabolism in nonphotosynthetic, nonoilseed tissues. In: Moore Jr., T.S. (Ed.), Lipid Metabolism in Plants. CRC Press, Boca Raton, FL, pp. 569–589.

Sutoh, K., Sanuki, N., Sakaki, T., Imai, R., 2010. Specific induction of TaAAPT1, an ER- and Golgi-localized ECPT-type aminoalcoholphosphotransferase, results in preferential accumulation of the phosphatidylethanolamine membrane phospholipid during cold acclimation in wheat. Plant Mol. Biol. 70, 519–531.

Tanoue, R., Kobayashi, M., Katayama, K., Nagata, N., Wada, H., 2014. Phosphatidylglycerol biosynthesis is required for development of embryos and normal membrane structures of chloroplasts and mitochondria in *Arabidopsis*. FEBS Lett. 588, 1680–1685.

Vranová, E., Coman, D., Gruissem, W., 2013. Network analysis of the MVA and MEP pathways for isoprenoid synthesis. Annu. Rev. Plant Biol. 64, 665–700.

Vriet, C., Russinova, E., Reuzeau, C., 2013. From squalene to brassinolide: the steroid metabolic and signaling pathways across the plant kingdom. Mol. Plant 6, 1738–1757.

Weselake, R.J., Taylor, D.C., Rahmana, M.H., Shah, S., Laroche, A., et al., 2009. Increasing the flow of carbon into seed oil. Biotechnology Adv. 27, 866–878.

Xue, Z., Duan, L., Liu, D., Guo, J., Ge, S., et al., 2014. Divergent evolution of oxidosqualene cyclases in plants. New Phytol. 193, 1022–1038.

Yang, W., Simpson, J.P., Li-Beisson, Y., Beisson, F., Pollard, M., Ohlrogge, J.B., 2012. A land-plant-specific glycerol-3-phosphate acyltransferase family in Arabidopsis: substrate specificity, *sn*-2 preference, and evolution. Plant Physiol. 160, 638–652.

Yatsu, L.Y., Jacks, T.J., Hensarling, T.P., 1971. Isolation of spherosomes (oleosomes) from onion, cabbage, and cottonseed tissues. Plant Physiol. 48, 675–682.

Yeats, T.H., Rose, J.K.C., 2013. The formation and function of plant cuticles. Plant Physiol. 163, 5–20.

Yu, X.-H., Prakash, R.R., Sweet, M., Shanklin, J., 2014. Coexpressing *Escherichia coli* cyclopropane synthase with *Sterculia foetida* lysophosphatidic acid acyltransferase enhances cyclopropane fatty acid accumulation. Plant Physiol. 164, 455–465.

Chapter 5

Fatty Acid Handling in Mammalian Cells

Richard Lehner,[1,2] Ariel D. Quiroga[1,3]

[1]Group on Molecular and Cell Biology of Lipids, University of Alberta, Edmonton, AB, Canada; [2]Departments of Pediatrics and Cell Biology, University of Alberta, Edmonton, AB, Canada; [3]Institute of Experimental Physiology (IFISE-CONICET), Faculty of Biochemical and Pharmacological Sciences, National University of Rosario, Suipacha, Rosario, Argentina

ABBREVIATIONS

AADAC Arylacetamide deacetylase
ACC Acetyl-CoA carboxylase
ACBP Acyl-CoA binding protein
ACSL Acyl-CoA synthetase/ligase
ACSVL Acyl-CoA synthetase very long chain
AMPK AMP-activated protein kinase
ATGL Adipose triglyceride lipase
BAT Brown adipose tissue
Ces or CES Carboxylesterase
ChREBP Carbohydrate responsive element binding protein
CoA Coenzyme A
CPT Carnitine palmitoyltransferase
DAG Diacylglycerol
DGAT Diacylglycerol acyltransferase
ER Endoplasmic reticulum
FABP Fatty acid binding protein
FAS Fatty acid synthase
FATP Fatty acid transport protein
GlyK Glycerol kinase
G0S2 G0/G1 switch gene 2
HDL High-density lipoprotein
HSL Hormone-sensitive lipase
IKK IκB kinase
JNK c-Jun N-terminal kinase
LAL Lysosomal acid lipase
LDs Lipid droplets
LXR Liver-X-receptor
MAPK, p38 mitogen-activated protein kinase
MAG Monoacylglycerol

Biochemistry of Lipids, Lipoproteins and Membranes. http://dx.doi.org/10.1016/B978-0-444-63438-2.00005-5

149

MGAT Monoacylglycerol acyltransferase
MGL Monoglyceride lipase
Mogat/MOGAT Genes encoding MGAT
PKA cAMP-dependent protein kinase
PKG Protein kinase G
PPAR Peroxisome proliferator-activated receptor
ROS Reactive oxygen species
PUFA Polyunsaturated fatty acid
RXR Retinoid-X-receptor
SLC27A Solute carrier 27 gene family A
SREBP Sterol regulatory element binding protein
TAG Triacylglycerol
TLR Toll-like receptor
VLDL Very-low-density lipoprotein
WAT White adipose tissue

1. INTRODUCTION

Long-chain fatty acids are crucial for a number of functions in mammals. They are (1) essential constituents of phospholipids that form biological membranes (Chapters 3 and 7); (2) substrates for adenosine triphosphate (ATP) production in the muscle, heart and liver are used for heat generation in muscle and brown adipose tissue (BAT); (3) covalent modifiers of diverse proteins (Chapter 13); (4) signalling molecules in metabolic processes; and (5) regulators of gene expression. This chapter focuses on the synthesis, transport and intracellular use of fatty acids, with the emphasis on the regulation of their storage as triacylglycerol (TAG) in lipid droplets (LDs) and their release from these storage depots.

2. FATTY ACID BIOSYNTHESIS

Most fatty acids in the body are obtained from diet, but palmitic acid (C16:0), and to a much lesser extent myristic acid (C14:0) and stearic acid (C18:0), can be synthesised de novo from excess carbohydrate in 'lipogenic tissues' such as liver, adipose and lactating mammary glands. Fatty acids can undergo desaturation and elongation in the cell (Chapter 6), yielding a variety of long-chain saturated, monounsaturated and polyunsaturated species. Importantly, linoleic acid (18:2n-6) and α-linolenic acid (18:3n-3) are essential fatty acids and must be obtained from diet. The metabolism of glucose to fatty acids proceeds first via the pentose phosphate pathway and glycolysis to yield acetyl-coenzyme A (CoA). The pentose phosphate pathway also produces nicotinamide adenine dinucleotide phosphate (NADPH), a hydrogen carrier used in fatty acid synthesis. David Rittenberg and Konrad Bloch were the first to demonstrate in early 1940s that acetate is the building block used for fatty acid synthesis. A decade later, Salih Wakil discovered a requirement for ATP and bicarbonate for the synthesis of malonyl-CoA from acetyl-CoA, a reaction catalysed by acetyl-CoA carboxylase (ACC). The synthesis of fatty acids, mainly palmitic acid, is catalysed by cytosolic fatty acid synthase (FAS):

$$\text{acetyl-CoA} + 7\text{malonyl-CoA} + 14\text{NADPH} + 14\text{H}^+$$
$$\rightarrow \text{palmitic acid} + 7\text{HCO}_3^- + 8\text{CoA} + 14\text{NADP}^+ + 6\text{H}_2\text{O}$$

The contribution of de novo synthesised fatty acids to overall lipid metabolism is dependent on the diet. When consuming diets low in fat (10% calories) and high in carbohydrates (75% calories), de novo lipogenesis can contribute up to 50% of fatty acids in plasma lipids (mainly TAG). However, when a Western-type diet containing 30–40% calories from fat and 45–55% calories from carbohydrates is consumed, the contribution of de novo lipogenesis to plasma lipids is nearly undetectable. Because humans consume diets high in fat content, de novo lipogenesis might not normally play an important role in the overall lipid metabolism. However, hepatic de novo lipogenesis is upregulated in obesity-associated insulin resistance (hyperinsulinenia) and type II diabetes mellitus (Chapter 19), despite the presence of high flux of exogenous fatty acids to the liver, and can contribute up to 25–30% of hepatic and circulating lipids.

2.1 Acetyl-CoA Carboxylase

ACC catalyses the formation of malonyl-CoA from acetyl-CoA (Tong, 2013). The multifunctional protein contains biotin carboxylase, biotin carboxyl carrier protein and carboxyltransferase. Mammals express ACC1 and ACC2 isoforms (also known as ACCα and ACCβ), which share 73% amino acid sequence identity. ACC1 is cytosolic, while the additional 140 amino acid residues in the N-terminal of ACC2 target this isoform to the outer mitochondrial membrane. ACC1 is the predominant form expressed in lipogenic tissues where fatty acid biosynthesis is robust, while ACC2 is mainly expressed in tissues with low lipogenic capacity but high fatty acid oxidation rates, such as heart and skeletal muscle. In these tissues, malonyl-CoA produced by ACC2 negatively regulates fatty acid β-oxidation by inhibiting carnitine palmitoyltransferase (CPT)-I, which catalyses the transport of long-chain fatty acyl-CoAs into the mitochondria. Ablation of *Acaca*[1] gene (encoding ACC1/ACCα) expression in mice is embryonically lethal. Liver- or adipose-specific ablation of *Acaca* expression in mice reduced lipid accumulation in these tissues. On the other hand, mice in which the expression of *Acacb* gene (encoding ACC2/ACCβ) was ablated have elevated fatty acid oxidation, increased energy expenditure, reduced body fat and body weight, improved insulin sensitivity, smaller heart size but with normal function, and normal life span and fertility. Reduction of both *Acaca* and *Acacb* expression by antisense oligonucleotides reverses hepatic steatosis and hepatic insulin

---

1. Nomenclature: Human genes are identified by capital letters in italic (e.g. *XXX*); rodent genes are identified by the first letter in capital followed by small case letters in italic (e.g. *Xxx*). Protein abbreviations are in capital letters regardless the species, unless specified.

resistance in rats fed high-fat diet. However, there is some controversy about the role of these key enzymes in regulating fatty acid oxidation. Liver-specific ACC1 and ACC2 double knockout mice have increased hepatic lipid levels and might be caused by reduced fatty acid oxidation. It has been suggested that ACC might be important in the regulation of acetyl-CoA levels. Acetyl-CoA is used as a substrate for posttranslational modification (lysine acetylation) and regulation of many metabolic genes. In fact, almost every enzyme in the glycolytic and glucose and fatty acid oxidation pathway is acetylated, which suggests that ACC inhibition might trigger multiple mechanisms to repress fatty acid oxidation.

2.1.1 Regulation of Acetyl-CoA Carboxylase by Allosteric Mechanisms and Phosphorylation

The activities of both ACC1 and ACC2 in mammals are allosterically activated by citrate through increasing the V_{max} without affecting the K_m for acetyl-CoA. A homodimer of a small cytosolic protein MIG12 promotes ACC polymerisation and activation. The activation by MIG12 is negatively regulated by Spot14 protein, which is related to MIG12 and can form MIG12/Spot14 heterodimers. ACC is inactivated by phosphorylation catalysed by AMP-activated protein kinase (AMPK) and cAMP-dependent protein kinase (PKA). PKA is activated by glucagon and epinephrine, mediated by an increase in cellular cAMP concentrations, importantly when glucagon signalling predominates over insulin in the liver (i.e. during fasting). AMPK is an energy state sensor and is activated by increased AMP/ATP ratio that occurs in the energy deficient state (i.e. starvation, exercise and hypoxia). Thus AMPK inhibits anabolic (ATP-consuming) pathways and stimulates catabolic (ATP-generating) pathways. However, the importance of AMPK phosphorylation of ACC2 as a regulatory step in cardiac fatty acid oxidation has come under scrutiny because cardiac fatty acid oxidation appears to proceed normally in double knock-in mice in which the AMPK phosphorylation sites of ACC2 were mutated (Zordoky et al., 2014). Lack of a correlation between ACC2 phosphorylation and fatty acid oxidation in skeletal muscle has been also reported (Alkhateeb et al., 2011).

2.1.2 Transcriptional Regulation

The levels of hepatic ACC are low during fasting/starvation and high during carbohydrate feeding. Induction of ACC production by carbohydrate has been attributed to insulin action. The importance of insulin-mediated control of ACC abundance is supported by low fatty acid synthesis in untreated diabetes mellitus (low insulin) and restoration of fatty acid biosynthesis after administration of insulin. Conversely, ACC levels are increased and fatty acid synthesis is augmented in obese models with elevated glucose and insulin levels. The rapid changes in ACC abundance due to feeding and fasting indicate coordinated transcriptional activation and repression. ACC1 and ACC2 are encoded by two separate genes. The *Acaca* gene is transcribed from at least three different promoters (PI-III). Transcription from a PIII promoter yields *Acaca* messenger

RNA (mRNA) that is translated into an ACC1 protein with N-terminal amino acid sequence that lacks one of the AMPK phosphorylation sites. At least two separate promoters are used for the transcription of *Acacb* mRNA. The complexity of *Acaca/b* gene transcription (different promoters used in different tissues and different mammalian species) has so far prevented a unified blueprint for the regulation of *Acaca/b* gene transcription. However, it has been shown that the *Acaca* promoter binds three important lipogenic transcription factors: sterol regulatory element binding protein-1c (SREBP1-c), liver-X-receptor (LXR) and carbohydrate responsive element binding protein (ChREBP).

Insulin activates SREBP-1c through the insulin receptor-PKB/Akt-mTORC1 pathway (Ferre and Foufelle, 2010). SREBP-1c is synthesised as a precursor transmembrane protein associated with the endoplasmic reticulum (ER). Its processing into an active nuclear transcription factor involves transport from the ER to the Golgi compartment, where proteolytic processing of the precursor SREBP-1c protein releases soluble N-terminal transcription factor fragment that is transported to the nucleus. The retention of the SREBP-1c precursor in the ER is regulated by the presence of the protein Insig-1, and the export of SREBP-1c from the ER is dependent on the escort protein SCAP (sterol regulatory element-binding protein cleavage-activating protein). While the expression of SREBP-1c is increased by carbohydrate feeding/insulin, it is decreased by unsaturated fatty acids. Unsaturated fatty acids stabilise Insig-1 (preventing Insig-1 degradation), which results in retention of SREBP-1c in the ER and prevents its proteolytic activation (Ye and DeBose-Boyd, 2011). Mice deficient in SREBP-1c in the liver have reduced levels of ACC and lower fatty acid synthesis, a finding that demonstrates a requirement for SREBP-1c in this process. SREBP-1c appears to be the dominant regulator of lipogenesis in the liver, but ablation of SREBP-1c did not affect expression of fatty acid synthesis genes in adipose tissue. ChREBP appears to be the dominant regulator of lipogenesis in this tissue because ChREBP-deficient mice have reduced adipose depots, downregulated expression of lipogenic genes, and diminished fatty acid synthesis. ChREBP binds to the PI promoter of the rat *Acaca* gene. ChREBP nuclear localisation is inhibited by PKA- and AMPK-mediated phosphorylation. Increased glucose concentration augments glucose catabolism and formation of xylulose-5-phosphate, which activates protein phosphatase 2A leading to dephosphorylation of ChREBP and its translocation to the nucleus.

2.2 Fatty Acid Synthase

Mammalian FAS is a soluble cytosolic protein. Similar to ACC1-deficient mice, global ablation of *Fasn* gene (encoding FAS) expression in mice results in embryonic lethality. On the other hand, liver or adipose-specific FAS-deficient mice are viable. Functional FAS is a homodimer of 273 kDa subunits. Each monomer contains seven catalytic elements that are required for the biosynthetic process. The acyltransferase component 'loads' acetyl-CoA and malonyl-CoA onto the FAS complex, resulting in the formation of thioester-enzyme intermediates; acyl carrier protein translocates the various thioester intermediates among the catalytic sites of β-ketoacyl reductase, β-hydroxylacyl dehydratase and enoyl reductase. Thioesterase is a chain-terminating enzyme that releases the product (mainly palmitic acid).

Similarly to *Acaca*, *Fasn* gene expression is transcriptionally regulated by SREBP-1c, LXR and ChREBP. In addition, binding of upstream stimulatory factors 1 and 2 to E-boxes in *Fasn* proximal promoter is required for insulin-dependent upregulation of *Fasn* expression. Fasting rapidly reduces *Fasn* expression; however, in mice fasted for 14 h, FAS activity remains similar to ad-lib fed mice, possibly because of the long half-life of the protein. Counterintuitively, FAS activity is initially inhibited by insulin for a period of up to 15 min after insulin administration before an insulin-stimulated increase in activity is observed. This suggests acute regulation of the protein, possibly by posttranslational modification. Phosphorylation and acetylation of FAS have been reported but the physiological consequence of these modifications remains to be elucidated.

Attenuation of hepatic FAS activity would be expected to be protective against hepatic lipid accumulation. However, liver-specific FAS-deficient mice instead developed severe hepatic steatosis when fed a zero-fat diet or on prolonged fasting. This phenotype can be corrected by activation of peroxisome proliferator-activated receptor (PPAR)-α with a synthetic agonist, which suggests that FAS provides a ligand for this important transcription factor that regulates fatty acid oxidation and mitochondrial biogenesis. Semenkovich's group has identified 16:0/18:1-glycerophosphocholine as an endogenous ligand for PPARα, and the production of this phospholipid molecular species was found to be dependent on FAS activity (Chakravarthy et al., 2009).

3. FATTY ACID UPTAKE, ACTIVATION AND TRAFFICKING

Dietary fatty acids are present as components of various lipids, mainly in TAG with a smaller contribution from glycerophospholipids and steryl esters. These lipids are hydrolysed in the lumen of the small intestine to release fatty acids. Fatty acids are also present in the blood compartment. These are either released from white adipose tissue (WAT) depots or from circulating lipoproteins (Chapters 16). The mechanism by which fatty acids enter cells from the intestinal lumen or from the blood compartment continues to be a matter of controversy. Some studies support diffusion of protonated (uncharged) fatty acids across the plasma membrane. However, because at neutral pH fatty acids are amphipathic molecules with nonpolar aliphatic chains and polar head groups, flip–flop of such charged molecules from the outer leaflet to the inner leaflet of the membranes is not favoured. Many studies have implicated protein-facilitated mechanisms, and a number of plasma membrane fatty acid transporters have been characterised. About 90% of fatty acid uptake is now believed to be protein mediated. Long-chain fatty acids have very low water solubility (1–10 nM). In the intestinal lumen, fatty acids partition into bile-salt-containing micelles, and in the blood, they are bound by albumin. Albumin in the plasma ($\approx$600 μM) can accommodate up to 2 mM fatty acid concentration; therefore only very little fatty acid exists 'free' in the aqueous phase. Once fatty acids are translocated to the cytosolic side of the plasma membrane, they are trapped inside the cells by either binding to small molecular mass fatty acid binding proteins (FABPs) that can accommodate up to 300 μM fatty acid or by activation to acyl-CoA by fatty acid transport proteins (FATPs)/very-long-chain acyl-CoA synthetases (ACSVLs) or by fatty acyl-CoA synthetases/ligases (ACSLs). Activation of a fatty acid to its CoA thioester is an energy-dependent process requiring ATP:

$$\text{Fatty acid} + \text{ATP} + \text{CoA} \rightarrow \text{Fatty acyl-CoA} + \text{AMP} + \text{PPi}$$

Acyl-CoA serves as a high-energy acyl donor for numerous esterification reactions and is also required for delivery of fatty acids for oxidation. CoA is derived from pantothenic acid, an essential vitamin that must be obtained from diet or from the intestinal flora. The transport of fatty acids/fatty acyl-CoAs to their various metabolic fates depends on many factors, including the type of FATP/ACSVL or ACSL that activated the fatty acid because the various ACSLs are associated with diverse metabolic functions.

3.1 CD36

CD36 was the first mammalian plasma membrane fatty acid transporter discovered (Abumrad et al., 1993). The 472 amino acid protein contains two membrane-spanning domains and several palmitoylation sites and presents with an apparent molecular mass of 88 kDa because of abundant glycosylation. CD36 belongs to the class B scavenger receptor family, and in addition to long-chain fatty acids, it also binds native and oxidised lipoproteins, thrombospondin-1, amyloid B and other ligands. CD36 is expressed in many cells and tissues including the intestine, adipose tissue, heart, skeletal muscle and macrophages. CD36 abundance in hepatocytes is normally very low, but is increased in fatty liver disease due to increased transcriptional activity of LXR and PPARγ, which upregulate *Cd36* expression. Mice lacking CD36 have impaired fatty acid uptake into the muscle, heart and adipose tissue, and excess fatty acids are delivered to the liver leading to steatosis. In humans, CD36 deficiency is associated with reduced insulin signalling and hyperlipidaemia. CD36 is also localised on the apical surface of taste bud cells in the tongue, where it contributes to fat taste perception/preference.

3.2 FATPs/ACSVL

Six FATPs/ACSVLs encoded by the solute carrier 27 gene family A (*SLC27A*) have been characterised (Anderson and Stahl, 2013). Overexpression of any of FATPs/ACSVLs increases fatty acid uptake. Each FATP/ACSVL is preferentially expressed in specific cell types, although there is some overlap in tissue-specific expression (Table 1). Although the catalytic mechanism is still not clear, all these proteins harbour intrinsic ACSL activity and are, in principle, bifunctional proteins (transporter and enzyme).

FATPs/ACSVLs share 20–40% sequence identity with ACSL1 and contain a 300 amino acid FATP signature motif present in the adenylate-generating super-family of enzymes, including ACSL, a 100 amino acid AMP-binding region, and a lipocalin motif. FATP/ACSVL membrane topology models propose a protein with a single membrane-spanning domain, a short extracellular/lumenal, and a long intracellular segment. The intracellular region contains both AMP-binding region and ACSL catalytic domain. Functional FATPs/ACSVLs operate as dimers. FATP1/ACSVL4, FATP2/ACSVL1 and FATP5/ACSVL6 function predominantly at the plasma membrane, but some FATPs/ACSVLs are also localised on internal membranes including the ER, mitochondria and peroxisomes.

TABLE 1 Tissue Distribution and Function of the FATP/ACSVL Protein Family

Protein	Main Tissues of Expression	Loss-of-Function Phenotype (Mice)
FATP1 ACSVL4 SLC27A1	WAT, BAT, skeletal muscle, heart	Reduced fatty acid uptake into muscle and adipose tissue; increased fatty acid uptake into liver, reduced thermogenesis
FATP2 ACSVL1 SLC27A2	Liver, kidney	Protection from high-fat diet-induced hepatic steatosis, decreased oxidation of very-long-chain fatty acids in peroxisomes
FATP3 ACSVL3 SLC27A3	Skin, endothelial cells, adrenal gland	Knockdown decreased fatty acid activation, but not uptake
FATP4 ACSVL5 SLC27A4	Small intestine, skin, heart, skeletal muscle, brain, WAT, BAT	Neonatal death, dehydration, skin abnormalities, no change in oleic acid activation, a significant decrease in most glycerophospholipids, cholesteryl esters and acylceramide suggests ER-localised activity
FATP5 ACSVL6 SLC27A5	Liver	Reduced fatty acid uptake, reduced triacylglycerol storage, increased fatty acid synthesis, defective conjugation of bile salts, reduced food intake and resistance to high-fat diet-induced weight gain
FATP6 ACSVL2 SCL27A6	Heart	N/A

BAT, brown adipose tissue; WAT, white adipose tissue; N/A, not available.

3.3 Acyl-CoA Synthetases

In addition to FATPs/ACSVLs, cells and tissues express several long-chain fatty ACSLs (Grevengoed et al., 2014). Five different ACSLs (ACSL1, 3, 4, 5 and 6) vary in their cell/tissue expression, intracellular localisation and function (Table 2). Consistent with their role in regulating fatty acid esterification and oxidation, *Acsl* expression is differentially regulated by nutritional status (fasting/refeeding) and diet (high fat/carbohydrate).

ACSL1 – The enzyme is localised on the plasma membrane, ER, nucleus, mitochondria and LDs. The association with mitochondria/LDs is consistent with the role of ACSL1 in fatty acid oxidation. In the liver, skeletal muscle and heart, *Acsl1* expression is upregulated by fasting, while in WAT, ACSL1 activity is reduced by fasting/exercise and increased by insulin. This highlights the different metabolic function of ACSL1 between catabolic (muscle/heart) and anabolic (white adipose) tissues. Therefore, in adipose tissue PPARγ regulates *Acsl1* expression, while in oxidative tissues such as liver, heart and skeletal muscle, PPARα increases transcription of the *Acsl1* gene.

ACSL3 – The enzyme is present on the ER and LDs. Knockdown of *Acsl3* expression in the liver downregulated reporter activities of anabolic transcription

TABLE 2 Tissue Distribution and Function of the ACSL Proteins

Protein	Main Tissues of Expression	Loss-of-Function Phenotype (ko or kd)
ACSL1	WAT, BAT, skeletal muscle, heart, liver	Reduced fatty acid incorporation into TAG in liver, reduced fatty acid oxidation in heart/muscle/BAT, impaired thermogenesis, decreased 20:4CoA in macrophages and PGE$_2$ formation
ACSL3	Most tissues	Kd decreases hepatic lipogenesis and function of specific transcription factors and co-activators (LXR, SREBP1c, ChREBP, PPARγ)
ACSL4	Adrenal gland, testis, ovary, brain	Defective eicosanoid metabolism (kd and human mutations) Mental retardation in humans (mutations)
ACSL5	High in the small intestine, lower in WAT, BAT, liver, lung, adrenal gland	Diminished glycerolipid and steryl ester synthesis in liver kd, but normal fat absorption and weight gain in global ko mice
ACSL6	Brain	Reduced proliferation and neurite outgrowth (kd in neuroblastoma cell line)

BAT, brown adipose tissue; WAT, white adipose tissue; ko, knockout; kd, knockdown; N/A, not available.

factors (SREBP1c, ChREBP, LXR, PPARγ), and it is suggested that ACSL3-generated acyl-CoA and/or metabolites of ACSL3-derived acyl-CoA are used to upregulate lipogenesis.

ACSL4 – The enzyme is proposed to play an important role in neural development because of its high expression in the brain, preference for arachidonic acid (20:4n-6) and eicosapentaenoic acid (20:5n-3), and cognitive delay associated with loss-of-function mutations. The precise function of ACSL4 has not yet been elucidated, but involvement in transport of synaptic vesicles and channelling of polyunsaturated fatty acids (PUFAs) to synthesis of specific phospholipid species have been postulated.

ACSL5 – Although the enzyme is highly expressed in intestinal mucosa, ACSL5-deficient mice showing more than 60% reduction of total ACSL activity do not exhibit fat malabsorption or resistance to high-fat diet-induced weight gain. On the other hand, attenuation of ACSL5 activity in hepatocytes reduced incorporation of exogenously supplied oleic acid into phospholipids, TAG and cholesteryl ester.

ACSL6 – Attenuation of expression of *Acsl6* in neural cells blocked proliferation and neurite outgrowth, suggesting an important role of this enzyme in provision of crucial substrates for neural cell processes. ACSL6 enhances uptake of docosahexaenoic acid (22:6n-3) and its incorporation into phospholipids.

3.4 Fatty Acid Binding Proteins

Fatty acids are powerful biological detergents, therefore they are bound by the abundantly expressed 14–15 kDa FABPs present in the cytosol (Storch and Thumser, 2010) (Table 3). FABPs are expressed in a tissue-specific manner, and although the seven characterised FABPs may not share high amino acid identity (20–70%), they fold into a similar tertiary structure consisting of a 10-stranded antiparallel β-barrel and an N-terminal helix-turn-helix motif. Most of FABPs bind one fatty acid molecule per molecule of FABP, with the exception of FABP1 (also called liver FABP or LFABP), which has a binding pocket that can accommodate two fatty acids. FABP1 is also more promiscuous and can bind other hydrophobic molecules. FABP6 (also called ileal bile acid binding protein or ILBP) binds primarily bile salts, but can also bind fatty acids. The precise function of FABPs in many cells/tissues has not yet been determined, although divergent and sometime tissue-specific roles have been proposed, including targeting of fatty acids to anabolic and catabolic processes, regulation of TAG storage, assisting lipoprotein assembly, and involvement in intracellular signalling.

FABP1 (liver FABP, LFABP) was initially characterised in the liver, but is also expressed in the small intestine (duodenum and jejunum). FABP1 appears to be the most promiscuous FABP, as it was reported to bind a variety of small molecules in addition to fatty acids including bile acids, lysophosphatidic acid, prostaglandins, fatty acyl-CoA, bilirubin and haem. Some studies suggested that FABP1 might be involved in the delivery of ligands for PPARα activation

TABLE 3 Tissue Distribution and Function of the FABP Protein Family

Protein	Main Tissues of Expression	Loss-of-Function Phenotype (Mice)
FABP1 LFABP	Liver, duodenum/ jejunum	Reduced fatty acid oxidation (liver and intestine), protection against steatosis on high-fat diet, decreased apoB lipoprotein assembly/secretion, mice develop hyperinsulinaemia on both carbohydrate or high-fat diet
FATP2 IFABP	Duodenum, jejunum	Reduced TAG synthesis, sensitive to developing fatty liver and metabolic syndrome on high-fat diet
FABP3 HFABP	Heart, skeletal muscle, brain	Decreased fatty acid oxidation, cold intolerance, increased glucose use
FABP4 AFABP aP2	WAT, BAT, macrophages	Increase intracellular FA, decreased HSL-mediated lipolysis, protection against diet-induced atherosclerosis (macrophage FABP4 knockout in apoE knockout mice)
FABP5 KFABP	WAT, BAT, skin, macrophages	Reduced PPAR signalling
FABP7 BFABP	Embryonic and perinatal brain	N/A

BAT, brown adipose tissue; HSL, hormone-sensitive lipase; WAT, white adipose tissue; N/A, not available.

in the nucleus; however, FABP1 deficient mice showed normal regulation of PPARα target genes. On the other hand, PPARα upregulates *Fabp1* expression. FABP1 enhances apoB-lipoprotein secretion, which suggests involvement of this protein in the delivery of fatty acids to the ER for esterification into lipo-protein-destined TAG. Hepatic fatty acid oxidation is also diminished in fasted FABP1 knockout mice, which implies deficient transport of fatty acids to the oxidative pathway. Interestingly, attenuation of fatty acid oxidation in FABP1 knockout mice was not accompanied by increased accumulation of intracellular TAG, even when the mice were challenged with a high-fat diet.

FABP2 (intestinal FABP, IFABP) is present in the duodenum and jejunum, where it directs fatty acids to TAG synthesis. Alanine 54 to threonine (A54T) polymorphism was found to be associated with insulin resistance in Pima Indians. This condition might be the result of increased TAG secretion due to augmented fatty acid binding affinity of FABP2.

FABP3 (heart/skeletal muscle FABP, HFABP) is most abundant in heart and skeletal muscle, but on exposure to cold, its expression is also induced in BAT. Mice lacking FABP3 have significantly reduced fatty acid oxidation, decreased tolerance to exercise and increased reliance on glucose for energy production, which strongly suggests a role of FABP3 in fatty acid targeting to mitochondria.

FABP4 (adipose tissue/macrophage FABP, AFABP, aP2) is highly upregulated during the differentiation of preadipocytes to adipocytes, a process stimulated by insulin, glucocorticoids, (PUFAs), and dependent on the activation of the master regulator PPARγ. FABP4 binds to the N-terminus of hormone-sensitive lipase (HSL) and stimulates its activity by preventing fatty acid (product)-mediated inhibition of the lipase by removing fatty acids from the lipid monolayer interface. Ablation of *Fabp4* expression in mice results in increased expression of *Fabp5* (encoding epithelial/keratinocyte FABP, KFABP) and only a mild metabolic phenotype. Conversely, combined ablation of both FABPs protects mice from developing insulin resistance and metabolic syndrome. Therefore, unlike the noncompensating roles of FABP1/2 in the small intestine, FABP4 and FABP5 exhibit similar metabolic functions. Interestingly, Hotamisligil's group has shown that FABP4/FABP5 double knockout mice have increased levels of palmitoleic acid (16:1n-9) in adipose tissue/plasma, and an inverse correlation between this lipid and insulin resistance has been suggested. The same group has also demonstrated that lipolytic stimulation results in secretion of FABP4 from adipose tissue via a nonclassical pathway and that circulating FABP4 increases hepatic glucose production, thereby acting as a gluco-regulatory hormone.

FABP5 (epithelial/keratinocyte FABP, KFABP) is present in WAT, BAT, macrophages, endothelial cells, neural tissues and breast tumours. The protein interacts with PPARδ, and a role of FABP5 in activation of PPARα has also been demonstrated. Therefore, in addition to having a similar function as FABP4, it might also have a cell-specific signalling role.

FABP6 (ileal lipid binding protein, ILBP) binds fatty acids, but binds primarily conjugated bile acids that have been absorbed in the ileum. Mice lacking FABP6 are protected from gallstone disease.

FABP7 (brain fatty acid binding protein, BFABP) has been detected in embryonic and perinatal stages in the brain, but very low levels are seen in the adult brain when FABP3 predominates.

Given the current information, it appears that FABP1, 2 and 3 play a role in buffering and trafficking fatty acids for various metabolic functions. It has emerged that FABP4 plays an active role in the regulation of adipose tissue lipolysis and, unlike the other FABPs, can also be secreted into circulation where it acts as a gluco-regulatory adipokine. The roles of FABP5 and 7 have remained poorly defined, while FABP6 is primarily involved in bile acid uptake and transport.

3.5 Acyl-CoA Binding Protein

A single Acyl-CoA binding protein (ACBP) has been identified and characterised in mammals (Bloksgaard et al., 2014). The 10kDa soluble protein is expressed in all tissues, where it binds medium- and long-chain fatty acyl-CoAs with high affinity (K_d ~1–10 nM), but does not bind fatty acids or any other hydrophobic molecules. High expression occurs in tissues most active in fatty acid metabolism, including liver and adipose tissue. The expression of *Acbp* is regulated by PPARs and SREBP-1c. As expected, ACBP regulates transport of acyl-CoA to various metabolic pathways, including TAG, phospholipid and cholesteryl ester synthesis. The protein can also extract acyl-CoA from membranes, and sequestration of acyl-CoA by ACBP prevents inhibition of FAS, ACC and ACSL. Inactivation of *Acbp* expression (global knockout) in mice results in the development of skin/fur phenotype during suckling/weaning transition (by day 16) that is characterised by reduced nonesterified very-long-chain fatty acids and decreased synthesis of epidermal protein-bound ceramides containing ultra-long fatty acyl chains resulting in increased *trans*-epidermal water loss because of impaired epidermal barrier. Skin levels of TAG that may provide substrates for acylceramide synthesis are also reduced in ACBP-deficient mice, as is decreased expression of lipogenic and cholesterol biosynthetic genes due to reduced SREBP1c and 2 levels in the nucleus. Interestingly, mice with liver-specific deletion of ACBP do not have any overt phenotype, while skin-specific ablation of *Acbp* expression yields similar phenotype to the global knockout mice. This suggests that the loss of keratinocyte ACBP, and not systemic ACBP, is solely responsible for the observed metabolic dysfunction. The phenotype of the global ACBP knockout mice is similar to that of mice deficient in stearoyl-CoA desaturase 2, very-long chain fatty acid elongase 4, FATP4/ACSVL5, and, to some extent, stearoyl-CoA desaturase 1 or very-long chain fatty acid elongase 3, all of which present with a milder skin phenotype (Chapter 6).

4. FATTY ACID STORAGE AS TRIACYLGLYCEROL IN LIPID DROPLETS

Most cells have the capacity to use fatty acids for the synthesis and storage of TAG, but adipose tissue, liver and intestine are the most adept, while skeletal muscle and heart also accumulate substantial TAG storage during fasting or exercise. Excessive TAG storage in LDs occurs when intracellular levels of fatty acids exceed the requirement for cellular processes such as membrane synthesis, mitochondrial β-oxidation and lipoprotein production (liver and intestine). TAG is synthesised by either the glycerol-3-phosphate (Kennedy) pathway or the monoacylglycerol (MAG) pathway (Figure 1). The glycerol-3-phosphate pathway is present in most cell types and requires glucose for the production of the glycerol-3-phosphate backbone to which fatty acids are esterified.

In the liver, glycerol-3-phosphate is also synthesised by glycerol kinase (GlyK) using glycerol that is obtained through complete hydrolysis of glycerolipids. The MAG pathway is primarily responsible for re-esterification of 2-MAG derived from dietary TAG. The MAG pathway might also contribute to TAG synthesis in other cells such as hepatocytes and adipocytes, where MAG might be generated intracellularly through lipolysis of pre-existing lipids. Both pathways require fatty

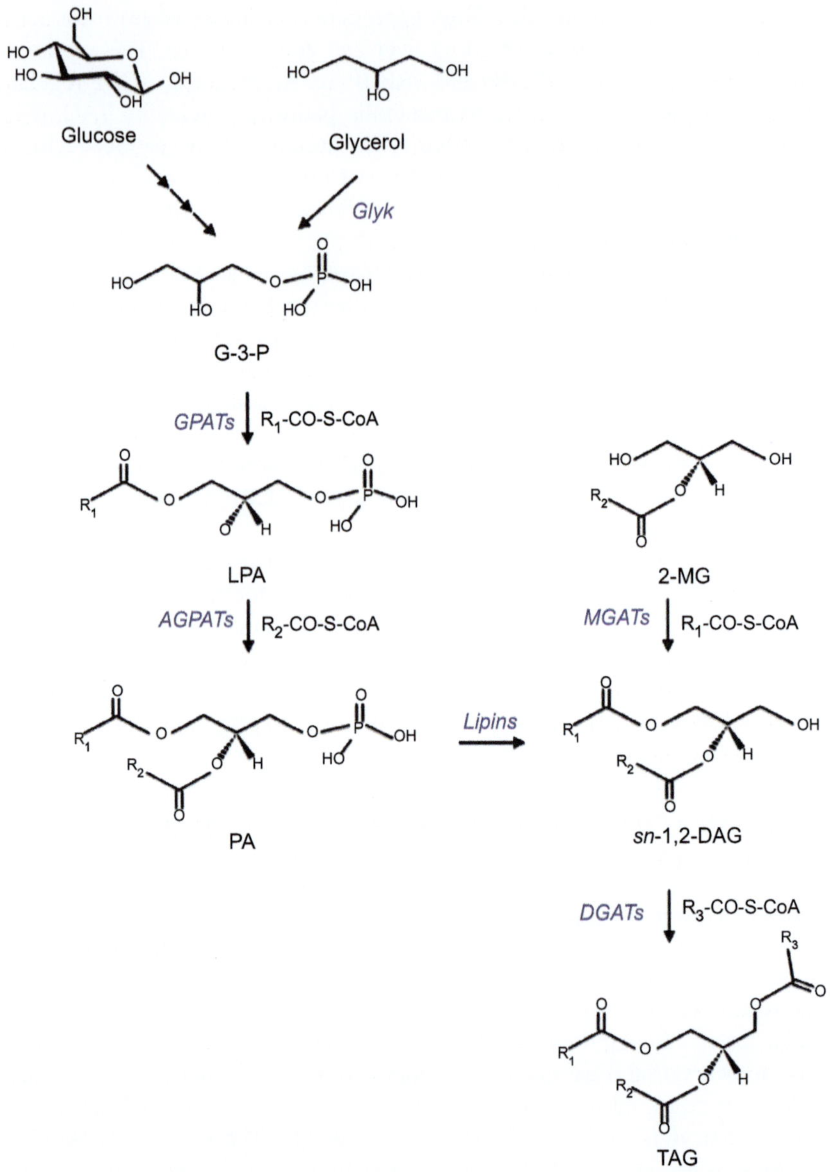

FIGURE 1 Triacylglycerol synthesis by glycerol-3-phosphate and 2-monoacylglycerol pathways. AGPATs, acylglycerolphosphate acyltransferases; DAG, diacylglycerol; DGATs, diacylglycerol acyltransferases; G-3-P, glycerol-3-phosphate; GPATs, glycerol-3-phosphate acyltransfeases; GlyK, glycerol kinase; LPA, lysophosphatidic acid; MAG, monoacylglycerol; MGATs, monoacylglycerol acyltransferases; PA, phosphatidic acid; TAG triacylglycerol.

acyl-CoAs as acyl donors. Enzymes catalysing the synthesis of TAG, phosphatidylcholine (PC) and phosphatidylethanolamine (PE) from glycerol-3-phosphate are shared until the formation of *sn*-1,2-diacylglycerol (DAG) (Chapter 7). Esterification of 2-MAG to DAG is catalysed by monoacylglycerol acyltransferases (MGATs) and that of DAG to TAG by diacylglycerol acyltransferases (DGATs).

4.1 Monoacylglycerol Acyltransferases

Three different human genes encoding enzymes with MGAT activity have been identified: *MOGAT1*, *MOGAT2* and *MOGAT3* (Hall et al., 2012). All three MGATs are ~38 kDa enzymes localised on the ER; however, they differ in tissue distribution and species-specific expression. *MOGAT1* gene encoding MGAT1 is most highly expressed in the adipose tissue, liver and gallbladder, and lower expression is observed in the small intestine and kidney. *MOGAT2* and *MOGAT3* are highly expressed in the small intestine and liver. MGAT3 is only present in higher mammals including humans and is absent in rodents. MGAT3 also possesses significant DGAT activity and therefore can be viewed as a TAG synthase (an enzyme able to convert 2-MAG to TAG).

In rodents, hepatic *Mogat1* expression is upregulated approximately 40-fold in diet-induced obese/steatotic mice in a PPARγ-dependent manner. Knockdown of hepatic *Mgat1* expression by antisense oligonucleotides reduced expression of lipogenic genes, but did not change hepatic steatosis or obesity and surprisingly led to increased hepatic DAG content. Despite DAG accumulation, hepatic insulin signalling and glucose tolerance in these mice were significantly improved (Hall et al., 2014). A functional role of MGAT1 in dietary fat absorption in the small intestine has not yet been determined.

Mouse MGAT2 is 81% identical to human MGAT2 and 53% identical to mouse MGAT1. Mice deficient in MGAT2 do not suffer from fat malabsorption (possibly because of MGAT1 compensation), but the rate of absorption is slower (Yen et al., 2009). In addition, global MGAT2 deficiency in mice confers resistance to diet-induced obesity and glucose intolerance, and the mice present with increased energy expenditure. Unlike *Mogat1* expression, *Mogat2* expression is relatively insensitive to a high-fat diet, although liver expression of *Mogat2* is increased about 10-fold in obese insulin resistant leptin-deficient (*ob/ob*) mice.

4.2 Diacylglycerol Acyltransferases

Two enzymes with DGAT activity (in addition to MGAT3) have been identified (DGAT1 and DGAT2) (Yen et al., 2008). The two DGATs share no homology in primary amino acid sequences and are encoded by distinct gene families. Both murine *Dgat* genes are ubiquitously expressed and are significantly upregulated during adipogenesis. *Dgat1* gene expression is upregulated in the liver during fasting, which is consistent with a role for this enzyme in re-esterification of exogenously supplied (adipose tissue-derived) fatty acid. On the other hand, hepatic *Dgat2* gene expression is upregulated by refeeding (insulin) and suppressed by fasting, suggesting that it might function in concert with the de novo lipogenic pathway. It is not yet clear whether TAG synthesised by the two DGATs are deposited in different LDs and used for different metabolic purposes. This scenario is a distinct possibility given that the two DGATs are differentially regulated, and studies in mice lacking either of the two DGAT have shown that the enzymes perform nonredundant physiological functions.

DGAT1 is ~55 kDa polytopic ER membrane protein with eight predicted transmembrane domains. Interestingly, membrane topology analysis placed the active site of DGAT1 on the lumenal side of the ER. The lumenal orientation of the active side would not pose any problem for the DGAT substrate DAG, because this molecule can flip–flop freely within the lipid bilayer. However, active transport of acyl-CoA across the membrane would be required, because this polar molecule is synthesised on the cytosolic side of the ER and cannot permeate the ER membrane. Presence of an ER-localised carnitine transport system similar to that catalysing shuttling of acyl-CoA in mitochondria has been suggested, but not characterised. Therefore, the topology of DAG esterification by DGAT1 requires further exploration. DGAT1 can also esterify retinol and long-chain (wax) alcohols in addition to DAG. The *Dgat1* gene belongs to the acyl-CoA, cholesterol acyltransferase gene family that is part of a membrane-bound *O*-acyltransferase (*MBOAT/Mboat*) gene superfamily. DGAT1-deficient mice are viable, possibly owing to only a moderate reduction in hepatic TAG, while plasma TAG concentration remains normal after a 4-h fast, suggesting a relatively minor role of DGAT1 in hepatic lipoprotein production. The mice are resistant to diet-induced hepatic TAG accumulation, adiposity is decreased by ~50%, they are more sensitive to insulin, and they exhibit higher energy expenditure. Importantly, the protection against high-fat diet-induced metabolic complications was not due to altered energy intake, because DGAT1 deficient mice eat just as much or more food as the wild-type mice and have no detectable malabsorption.

DGAT2 is a ~42 kDa protein that localises on the ER, mitochondrial-associated membranes (regions of the ER tethered to mitochondria) and also LDs. The location on LDs is interesting because LDs have a phospholipid monolayer rather than a bilayer, and this precludes intercalation of authentic transmembrane proteins. It has been proposed that the hydrophobic (membrane interacting) region of DGAT2 can form a 'hairpin-like' structure, which would allow stable partitioning of the protein into phospholipid monolayers. Because of its presence on LDs, it has been proposed that DGAT2 plays a role in LD expansion. *DGAT2/Dgat2* gene belongs to a seven-member family that contains the three *MOGAT/Mogat* genes and two genes encoding wax synthases. DGAT2 contains an amphipathic neutral lipid-binding domain with a consensus amino acid sequence FLXLXXXn, where n is a nonpolar amino acid. This sequence is also present in several other nonrelated neutral lipid metabolising enzymes, including acyltransferases and lipases. DGAT2 co-localises with stearoyl-CoA desaturase 1, which suggests that it might be linked with esterification of endogenously synthesised monounsaturated fatty acids. DGAT2 appears to be the dominant DGAT for regulation of TAG metabolism in vivo. Newborn DGAT2-deficient mice die within a few hours after birth due to a drastic reduction of TAG content in tissues and plasma and a severely impaired epidermal barrier protection. Diminished *Dgat2* expression in liver of adult mice by antisense oligonucleotide technology resulted in markedly decreased hepatic TAG content and improved steatosis in obese mice. Importantly, the perinatal lethality of DGAT2-deficient mice indicated that DGAT1 could not compensate for the loss of DGAT2, again suggesting that the two DGATs have nonoverlapping functions.

Because of the metabolic phenotypes of mice lacking DGAT1 or hepatic DGAT2 (antisense oligonucleotide experiments), there has been significant pharmacological interest in producing DGAT1 and DGAT2 inhibitors. However, while mice express both *Dgat1* and *Dgat2* in the intestine and therefore DGAT2 can to a certain degree compensate for the loss of DGAT1 and maintain fat absorption, humans do not express *DGAT2* in the intestine, and DGAT1 deficiency (loss-of-function mutations) leads to severe malabsorption that can result in death during infancy. Therefore, the use of DGAT1 inhibitors in humans needs re-evaluation. On the other hand, despite the early lethality of DGAT2-deficient mice, inhibition of hepatic DGAT2 in adults might reduce steatosis and plasma lipid concentrations and improve insulin sensitivity in obese, insulin-resistant patients.

4.3 LD Biogenesis

TAG synthesised by DGATs on the ER has only limited solubility in the ER bilayer and is thus efficiently deposited into LDs. LDs are heterogeneous in size, varying from 0.5 to 200 μm. LDs contain a monolayer of amphipathic lipid, primarily PC, around a neutral lipid core composed of TAG with some steryl and retinyl esters. It is well accepted that LDs begin their life within the ER, but it is not clear how enrichment of the monolayer with PC (or exclusion of other ER phospholipids) is regulated. One plausible explanation is the existence of domains within the ER specialised in LD biogenesis (Wang et al., 2010). PC serves an important surfactant role to stabilise LDs. The choice of PC is determined by its biophysical properties. This phospholipid is cylindrical in shape and therefore is more suited than other phospholipids to cover the LD surface area and lower surface tension. How is growth (expansion) of LDs achieved? Several mechanisms have been described, including coalescence, ripening and in situ core expansion (Thiam et al., 2013). Coalescence occurs when surface tension increases by introduction of lipids with negative curvature (cholesterol, DAG, phosphatidylethanolamine (PE), fatty acids) or by depletion of PC. For example, inactivation of PC synthesis by targeted deletion of CTP:phosphocholine cytidylyltransferase α or PE *N*-methyltransferase in mice (Chapter 7) results in the formation of large LDs. Ripening of LDs involves a transfer of core lipids between two LDs. The direction of transfer is determined by the pressure difference (Laplace pressure) between the inside and outside of the curved liquid surface. Transfer of TAG from smaller to larger LDs occurs because small LDs have a higher Laplace pressure compared to larger LDs. Fat-specific protein (FSP, known also as CideC) promotes ripening in adipose tissue and formation of a large unilocular LD from smaller LDs during adipogenesis. Enzymes catalysing TAG synthesis (glycerol-3-phosphate acyltransferase 4, acylglycerolphosphate acyltransferase 3, DGAT2 and some ACSLs) have all been localised to the LD surface (Wilfling et al., 2014) suggesting in situ core expansion mechanism of LD growth. Other proteins that have been implicated in LD growth are Berardinelli-Seip congenital lipodystrophy type 2 (known as seipin) and fat storage-inducing transmembrane (known as FIT); however, the precise role of these proteins in the process is not yet clear.

4.3.1 Perilipins

Besides surfactant lipids, LDs are stabilised by the presence of specific proteins. Direct interaction of proteins with the LD surface occurs either through the presence of amphipathic helices or by an intercalation into the phospholipid monolayer via a hairpin of two α-helices. The protein composition of LDs is tissue/cell specific and is determined by the energy status of the cell. This is in concert with the role of LDs in the regulation of energy storage, provision of signalling molecules and detoxification of the detergent properties of fatty acids. Several hundred proteins of diverse functions have been identified in various LD preparations from diverse cells/tissues. The bona fide protein components of LDs are the perilipin (from Greek, meaning surrounding lipid) protein family (Sztalryd and Kimmel, 2014), sometimes referred to as the PAT family of proteins based on the names of the founding members: **P**erilipin 1 (Plin1), **A**dipophilin (now annotated as perilipin 2, Plin2) and **T**ip47 (**t**ail-**i**nteracting **p**rotein of **47**kDa, now annotated as perilipin 3, Plin3). Two additional perilipins have been characterised: perilipin 4 (Plin4, known also as S3-12) and perilipin 5 (Plin5, known also as LSDP5, OXPAT, MLDP, PAT1).

Perilipin 1 is most highly expressed in adipose tissue and the first LD-associated protein described by Constantine Londos' group in the early 1990s. The protein plays a crucial role in modulating TAG storage/lipolysis. During the fed state (high insulin) perilipin 1 is dephosphorylated and promotes lipid storage by sequestering adipose triglyceride lipase (ATGL) activator CGI58 and by preventing binding of HSL to the LD surface. Stimulation of β-adrenergic signalling during fasting results in PKA-dependent phosphorylation of perilipin 1 and structural reorganisation of the protein such that a protein called comparative gene identification 58 (CGI-58) is released. CGI-58 can then activate ATGL, and HSL is no longer prevented from binding to LDs. Ablation of perilipin 1 expression in mice results in increased lipolysis during the fed state, reduced WAT and enhanced ectopic fat deposition, which is consistent with the antilipolytic barrier function of the protein. Humans with loss-of-function mutations in perilipin 1 present with partial lipodystrophy, dyslipidaemia and type 2 diabetes mellitus.

Perilipin 2 is ubiquitously expressed, but cannot compensate for the functional loss of perilipin 1 in adipose tissue. Nevertheless, increased perilipin 2 also attenuates lipolysis in tissues by decreasing ATGL-LD interaction. Perilipin 2 is most abundant in the liver, heart and skeletal muscle. Loss of perilipin 2 in mice through genetic ablation or antisense oligonucleotide administration activates hepatic lipolysis and fatty acid oxidation, thereby decreasing lipid accumulation. An α-helix disruptive polymorphism that replaces serine 251 with proline in human perilipin 2 results in a gain-of-function phenotype characterised by augmented lipid storage in small LDs and decreased production of apoB-containing lipoproteins. The mechanism by which perilipin 2 links lipid storage and secretion is not entirely understood, but might be related to its lipid sequestration capacity.

Perlipin 3 is also ubiquitously expressed, but is especially high in the small intestine, liver and macrophages. The protein exhibits apolipoprotein-like capacity to reorganise liposomes into discs. This suggests that perilipin 3 might be involved in the organisation of the LD phospholipid monolayer. Experimentally, perilipin 3 is found predominantly on small (primordial) LDs that have small amounts of core lipids. This is consistent with its inability to inhibit lipolysis, a function innate to the other perilipins. Depletion of hepatic perilipin 3 by antisense oligonucleotide treatment results in diminished lipid deposition, while increased perilipin 3 concentrations are observed in steatotic mice. Therefore, like perilipin 2, perilipin 3 regulates hepatic lipid levels, but the mechanism is currently unclear.

Perilipin 4 abundance in WAT, heart and skeletal muscle is regulated by a PPARγ-dependent mechanism. Loss-of-function studies did not provide insight into its role in adipose tissue metabolism, but in the heart, loss of perilipin 4 is associated with concomitant loss of perilipin 5 and diminished lipid storage.

Perilipin 5 is primarily present in oxidative tissues, including the heart, skeletal muscle and liver. As such, perilipin 5 is metabolically and physically connected to LDs and mitochondria. Overexpression of perilipin 5 inhibits lipolysis and therefore promotes steatosis, while decreased perilipin 5 abundance reduces lipid storage. Perilipin 5 expression and phosphorylation in tissues increases during fasting and endurance exercise through PPARα/δ-mediated transcriptional activation. Phosphorylation of perilipin 5 on serine 155 is required for the activation of PGC1α (a transcriptional co-activator regulating mitochondria biogenesis and expression of fatty acid oxidation genes). It seems counterintuitive to increase perilipin 5 expression and inhibit lipolysis in conditions when the need for fatty acid oxidation for energy production is at its highest. One explanation is that perilipin 5 regulates the amount of fatty acid release by dampening extensive lipolysis. In this way, perilipin 5 could have a protective function against the formation of reactive oxygen species (ROS) and hence lipotoxicity from excessive oxidation.

5. FATTY ACID USE FOR ENERGY

Long-chain fatty acids are an important energy source for the skeletal muscle and heart. In the liver, ketone bodies are produced by incomplete fatty acid oxidation and used as fuel by other tissues. Fatty acids can be catabolised through α-, β- and ω-oxidation. In mammals, β-oxidation of fatty acids takes place in mitochondria and peroxisomes and provides energy for oxidative phosphorylation, and in the liver generates acetyl-CoA for ketone body production. The process of β-oxidation involves sequential removal of 2-carbon units of fatty acids. Some dietary fatty acids are methylated (e.g. phytanic acid present in dairy products and tissues of ruminants) and cannot be oxidised by the conventional β-oxidation pathways. They are oxidised by the peroxisomal α-oxidation pathway requiring a specific α-hydroxylation at the α-carbon of the methylated fatty acid. Liver and kidney also oxidise fatty acids through ω-oxidation on the ER. This is normally a minor pathway for oxidation of long-chain fatty acids, but in conditions such as diabetes mellitus, chronic alcohol consumption and starvation it significantly contributes to elimination of potentially toxic levels of fatty acids. The pathway is catalysed by the cytochrome P450 family of enzymes, alcohol dehydrogenase and aldehyde dehydrogenase yielding dicarboxylic acids, which are subsequently catabolised via the β-oxidation pathway in peroxisomes. Fatty acids must be activated to their respective CoA derivative by ACSLs prior to subsequent oxidation. In addition to exogenous fatty acids, hydrolysis of intracellular TAG in LDs is an important source of fatty acids used not only for energy production, but also for regulation of the fatty acid oxidation pathway.

5.1 Lipolysis of Lipid Storage

Lipolysis is a highly regulated process involving a number of enzymes and co-factors. The location, substrate specificity and stereoselectivity of lipases determine the metabolic pathways to which the released fatty acids are channelled (Figure 2). Studies in patients with mutations in lipase genes and in animal models in which the expression of lipolytic enzymes has been ablated have shown that lipases play a crucial role in energy metabolism, including fatty acid oxidation, thermogenesis, lipoprotein production and intracellular signalling.

Although lipases vary greatly in their primary amino acid sequence, they are all serine esterases belonging to the α/β hydrolase superfamily. The α/β hydrolase classification alludes to the secondary structure of the proteins, in which α-helices follow β-sheets with the central β-sheet core containing at least five parallel strands. In most lipases, the nucleophilic serine residue is part of a Ser–Asp/Glu–His triad, but some lipase active sites contain a Ser–Asp dyad. The catalytic serine residues lie within the GXSXGA/G lipase/esterase signature motif. Another structural feature of many lipases is the presence of a loop or 'lid' domain that protects the hydrophobic active site from the aqueous environment. The lid also plays an important role in substrate selectivity of lipases.

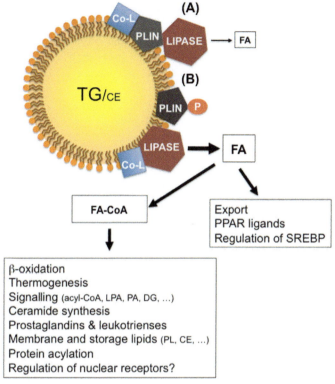

FIGURE 2 Lipolysis of triacylglycerol stored in LDs and the fate of released fatty acids. (A) Lipolysis is very low in energy-sufficient conditions (fed state when insulin signalling predominates). Perilipins (PLIN) either block access of lipases to the LDs or sequester lipase activators (colipase, Co-L). (B) When energy is required (fasting, stress), hormone and catecholamine signalling leads to phosphorylation/rearrangement of perilipins on LDs, lipases gain access to LDs and their respective activators, and robust hydrolysis of stored triacylglycerol ensues. Released fatty acids are used for various cell/tissue-dependent processes.

Lipases are generally soluble enzymes (only a few are transmembrane proteins), yet they act on insoluble lipid substrates at the water/lipid interface and follow interfacial activation kinetics rather than the Michaelis–Menten kinetics. The catalytic mechanism depends strongly on the organisation of lipids at the interface, as well as lipid composition to orchestrate the physical properties of the surface at which the lipases act: membrane bilayers (organelles), monolayers (LDs, lipoproteins), micelles (intestinal absorption), or oil-in-water emulsions. The adsorption of lipases to lipid surfaces necessitates significant structural rearrangement of the proteins. The opening of the lid domain creates a large hydrophobic surface and provides an entry for the substrate to the active site. The 'open lid' conformation of lipases at the lipid interface is stabilised by an interaction of the lipase with a co-factor (sometimes called a co-lipase or an activating protein), which enhances the catalytic activity of the enzyme.

Intracellular lipolysis is compartmentalised to the cytosol, ER and late endosomes/lysosomes. Much has been learnt about intracellular lipolysis from studies in WAT. However, care should be taken not to generalise the mechanism of WAT lipolysis to other tissues because different sets of lipases may be used in the hydrolysis of stored lipid. Lipolysis of stored TAG in WAT is regulated by nutritional needs. WAT lipolysis is suppressed during the postprandial state (insulin action predominates), while during fasting (when catecholamines and stress hormones predominate) lipolysis of stored TAG is stimulated. Lipolysis on LDs occurs by two main mechanisms. These involve the β-adrenergic stimulation of lipolysis by catecholamines and the lipolysis induced by natriuretic peptide. The first pathway involves the catecholamine-induced increment in cAMP levels, leading to PKA activation and phosphorylation of lipases and/or co-factors that permit access of the lipases to LDs. The natriuretic peptide pathway is interesting given its role in enhancing thermogenesis. After exposure to cold, p38 mitogen-activated protein kinase (MAPK) pathway is activated, leading to increased thermogenic activity in WAT. When natriuretic peptide binds to its receptor, increased cGMP levels and activation of protein kinase G (PKG) occurs. This kinase then phosphorylates the same targets as PKA independently of β-adrenergic stimulation. Fatty acids that are released from LDs in WAT are bound by FABP4, trafficked to plasma membrane where they are exported to albumin in the circulation, and delivered to tissues for use. The mobilisation of WAT TAG stores to fatty acids and glycerol is catalysed by a sequential action of ATGL HSL and monoglyceride lipase (MGL). A similar sequence of events appears to occur in the heart, muscle and pancreatic β-cells.

5.1.1 Adipose Triglyceride Lipase

ATGL is ~54 kDa lipase, also annotated as patatin-like phospholipase domain containing protein 2 (PNPLA2) or desnutrin, that catalyses the first step of TAG conversion to DAG (Lass et al., 2011). ATGL contains a Ser–Asp catalytic dyad, and its C-terminus encompasses a hydrophobic region that is required for binding to LDs. The stereospecificity of ATGL in vivo is not yet entirely clear, but in vitro ATGL shows preference for the sn-2 and sn-1 positions of TAG, yielding sn-1,3- and sn-2,3-DAG, respectively. ATGL is highly expressed in adipose tissue, but also to a lesser extent in other tissues including heart, muscle, intestine, liver and pancreatic β-cells, suggesting a wider role of the enzyme in energy homeostasis. The crucial role of ATGL in TAG lipolysis has been demonstrated in mice in which the gene encoding ATGL has been disrupted. ATGL-deficient mice present with increased TAG stores in most tissues, but the largest TAG deposits were found in the heart, which leads to premature death of mice from cardiac dysfunction. ATGL-deficient mice die within 6 h of exposure to 4 °C, implicating the ATGL in the provision of fatty acids for thermogenesis and as ligands for transcription factors and transcriptional co-activators that regulate fatty acid oxidation (PGC1α and PPARα). Specific deletion of ATGL in the liver results in 4-fold increased TAG storage without affecting plasma glucose, glucose tolerance, insulin sensitivity or plasma lipid levels, while hepatic fatty acid oxidation was decreased.

ATGL is a hormone-regulated enzyme. It is activated by the co-activator CGI-58 (also annotated as α/β-hydrolase domain-containing 5, ABHD5) and inhibited by G0/G1 switch gene 2 (G0S2). In adipose tissue, CGI-58 associates with perilipin 1 in the fed state (when insulin signalling predominates) and is not available for activation of ATGL. Lipolytic stimulus (fasting, stress) results in the production of cAMP, leading to activation of protein kinase A and phosphorylation of perilipin 1. CGI-58 is then released from perilipin 1 and is free to activate ATGL. Perilipin 5 is the likely functional replacement in tissues where perilipin 1 is not expressed. Loss-of-function mutations in CGI-58 results in neutral lipid storage disease with ichthyosis in humans (Chanarin-Dorfman syndrome), characterised by TAG accumulation in tissues, ichthyosis and muscle weakness. However, there are some significant phenotypic differences between CGI-58 and ATGL deficiencies, and this suggests that CGI-58 plays another role in addition to ATGL activation. G0S2 was initially identified as a protein expressed during cell cycle transition from G0 to G1 phase. G0S2 protein expression in WAT is increased by insulin (carbohydrate feeding) and PPARγ agonists. G0S2 inhibits ATGL by directly interacting with the enzyme's patatin-like domain even in the presence of CGI-58. Because G0S2 abundance is rapidly increased by insulin (an antilipolytic signal), it is plausible that G0S2 provides a mechanism for the acute attenuation of lipolysis.

5.1.2 Hormone-Sensitive Lipase

It was believed for a long time that HSL was the only enzyme regulating TAG hydrolysis in WAT and BAT. However, characterisation of HSL-deficient mice, with the discovery and characterisation of ATGL, showed that HSL is primarily a DAG lipase (with a preference for the *sn*-3 position), cholesteryl ester lipase and retinyl ester lipase, although it can also hydrolyse TAG. Despite demonstrated TAG lipase activity in vitro, HSL does not compensate for the loss of ATGL in vivo. HSL is mainly expressed in WAT, BAT and steroidogenic tissues and to a much lesser extent in other tissues. It is thought to be absent from the human liver, and low-level expression is detected in the mouse liver. The detected levels could be due to its presence in Kupffer cells rather than hepatocytes. HSL activity is increased during fasting by a mechanism that includes phosphorylation of cytosolic HSL by PKA and its translocation to LDs. The recruitment of HSL to the LDs is dependent on perilipin 1. Mice lacking HSL are not overweight; fatty acid release from adipose tissue is decreased by less than 40%, and WAT accumulates DAG (Haemmerle et al., 2002). A loss of functional HSL in mice (knockouts) and humans (mutations) results in reduced PPARγ signalling (Zechner and Langin, 2014), and consequently, adipogenesis, lipogenesis and lipid uptake/TAG synthesis are diminished, leading to lipodystrophy. Therefore, HSL plays an important role in providing ligands (fatty acids) to activate the essential adipogenesis transcription factors PPARγ and/or retinoid-X-receptor-α (PXPα). There are, however, significant phenotypic differences between the HSL-null mice and humans. Mice present with decreased hepatic TAG and consequently decreased very-low-density lipoprotein (VLDL) production, but increased high-density lipoproteins (HDL). In contrast, humans with loss-of-function HSL mutations present with increased hepatic TAG and have increased plasma TAG concentrations and decreased HDL concentrations and develop diabetes mellitus. These results perhaps point to important differences in lipid metabolic processes in mice and humans (mice carry cholesterol mainly on HDL, while humans carry cholesterol mainly on LDL), but also to possible age-dependent effects, because the studies in mice were done in young animals when WAT was still overtly normal and not lipodystrophic.

5.1.3 Monoglyceride Lipase

MGL catalyses the release of fatty acid from MAG and does not exhibit hydrolytic activity toward TAG or DAG. It is localised in the cytosol and on LDs, but the mechanism that regulates the distribution between the two cellular compartments is unknown. MGL-deficient mice have decreased release of fatty acids and glycerol from WAT and, consequently, diminished hepatic TAG accumulation and reduced VLDL production (Taschler et al., 2011).

While the biology and mechanisms of lipolysis by the ATGL–HSL–MGL pathway are well understood, the same lipolytic sequence is not applicable to all tissues. TAG lipolysis occurs in ATGL-deficient tissues after stimulation with PPARα agonist, and most tissues (such as the human liver) do not express HSL. In addition, lipolysis in the epidermis is not affected by the ablation of *Atgl* expression. Therefore, additional lipases exist, and some have been identified and characterised, but their function in energy metabolism is less clear. Below is a brief description of additional intracellular lipases that have been demonstrated to play a role in TAG metabolism.

5.1.4 PNPLA3

PNPLA3, also known as adiponutrin and Ca^{2+}-independent phospholipase A2 epsilon (iPLA2ε), shares homology with ATGL including the Ser–Asp catalytic dyad (Smagris et al., 2014). The protein is present in WAT, the liver and adrenal glands. PNPLA3 is found on LDs, but also in other cytoplasmic compartments. PNPLA3 is regulated by nutritional, hormonal and pharmacological factors, but in the opposite direction to ATGL. The authentic physiological substrate of PNPLA3 is not known; however, PNPLA3 catalyses the hydrolysis of TAG in vitro and may also function as a lipase in vivo. Interestingly, hepatic TAG was mostly unaffected when active PNPLA3 was overexpressed, while the expression of an inactive PNPLA3 (I148M) variant is associated with steatosis in humans. Ablation of PNPLA3 expression in mice did not affect body composition, energy homeostasis, hepatic lipid metabolism, glucose homeostasis or insulin sensitivity, suggesting that the PNPLA3 I148M is a gain-of-function mutant.

5.1.5 ER-Associated Lipolysis

It has been known since the early 1980s that the ER harbours lipid hydrolases (carboxylesterases); however, their characterisation has only recently begun. Even less is known about the lipid substrates for these enzymes. It is possible that these lipases access lipids in LDs that fuse with the ER or LDs that are present in the ER lumenal compartment of lipoprotein secreting cells (Chapter 16). Carboxylesterases (Ces for murine origin or CES for human origin) are 60–62 kDa glycoproteins localised to the lumen of the ER. Carboxylesterase-mediated lipolysis of MAG and DAG in vitro was demonstrated in the early 1980s, but the extent of their contribution to lipid hydrolysis in intact cells was not determined until 30 years later. Carboxylesterases contain Ser–Glu–His catalytic triad and other hallmarks of lipolytic enzymes, including a hydrophobic crevice entry into the active site and a lid domain.

Mouse carboxylesterase 1d (Ces1d, also annotated as carboxylesterase 3 or Ces3) and its human orthologue carboxylesterase 1 (CES1 or CES1D), also known as TAG hydrolase and cholesteryl ester hydrolase, are expressed in the liver, WAT, BAT, and, to a lesser extent, small intestine, heart and kidney. Genetic or chemical inactivation of Ces1d/CES1D significantly decreases TAG and apolipoprotein B secretion (Wei et al., 2010). Ces1d was shown to catalyse basal (nonstimulated) lipolysis in differentiated adipocytes. Interestingly, Ces1d deficiency was not accompanied by hepatic TAG accumulation because of increased hepatic fatty acid oxidation and decreased lipogenesis.

Mouse carboxylesterase Ces1g (also termed Es-x) shares about 76% protein sequence identity with mouse Ces1d, including the residues of the esterase/lipase catalytic triad, neutral lipid-binding domain, but less amino acid sequence similarity is present in the lid domain, which suggests different substrate specificity. Ces1g is primarily present in the liver and small intestine. Overexpression of Ces1g reduced TAG accumulation and augmented fatty acid β-oxidation, while ablation of *Ces1g* expression resulted in hepatic TAG accumulation, weight gain and increased apoB-lipoprotein production (Quiroga et al., 2012). Ces1g releases n-3 PUFAs from TAG, inhibits SREBP processing and consequently diminishes de novo lipogenesis.

Arylacetamide deacetylase (AADAC) shares amino acid sequence homology with HSL. AADAC is a type II membrane protein with its active site facing the lumen of the ER. In humans, AADAC protein is present in the liver and intestine; however, lower *Aadac* expression was also observed in the adrenal cortex/medulla and pancreas. Overexpression of AADAC decreased hepatocyte TAG storage and lipoprotein secretion and increased fatty acid β-oxidation (Lo et al., 2010). In vitro lipase assay suggested preference for DAG, which is in accordance with the homology of this protein with HSL. AADAC tissue distribution and activity suggests a role for this enzyme in modulating lipoprotein assembly and fatty acid oxidation.

5.1.6 Lysosomal Acid Lipase

Lysosomal acid lipase (LAL) is related to gastric lipase and is primarily involved in hydrolysis of cholesteryl esters and TAG derived from endocytosed plasma lipoproteins. LAL has optimum activity at pH 4–5, which is in agreement with its localisation in late endosomes/lysosomes. During prolonged starvation, LAL activity may also become important in hydrolysing cytosolic TAG stores to generate fatty acids for β-oxidation through a process termed lipophagy (Singh et al., 2009). In humans, LAL deficiency causes two related diseases, Wolman disease and cholesteryl ester storage disease. Both diseases result in excessive lipid accumulation in tissues and hepatomegaly.

5.2 Fatty Acid Oxidation

To be oxidised in mitochondria, fatty acids need to reach the mitochondrial matrix. Fatty acids with chain lengths of 12 carbons or fewer enter mitochondria without the help of membrane transporters. However, most fatty acids are 14 or more carbons in length and require membrane transporters to enter the mitochondrial matrix. This is fulfilled in a simple, yet efficient manner. The transport is initiated by one of the ACSLs located in the outer mitochondrial membrane. Formed acyl-CoAs can be either transported into the mitochondrial matrix and oxidised to produce ATP, or used by mitochondrial glycerol-3-phosphate acyltransferase to synthesise lysophoshatidic acid, the first step in the glycerol-3-phosphate pathway (Chapter 7). Acyl-CoA destined for mitochondrial oxidation becomes the substrate for a membrane-bound CPT-I, which catalyses the production of acyl-carnitine from acyl-CoA. Acyl-carnitine is transported across the outer mitochondrial membrane and reaches the inner mitochondrial membrane, where it is received by carnitine:acylcarnitine translocase, a protein that transports both carnitine and acyl-carnitine to the cytosol and to the mitochondrial matrix, respectively. Once the acyl-carnitine is in the mitochondrial matrix, CPT-II catalyses the reversible transfer between carnitine and CoA to regenerate acyl-CoA that then enters the β-oxidation pathway. It is important to note that the carnitine-mediated entry process is the rate-limiting step for oxidation of fatty acids in mitochondria. Mitochondrial fatty acid oxidation occurs in three main steps. The first step is β-oxidation, where fatty acids undergo oxidative removal of successive two-carbon units as acetyl-CoA, starting at carbon atom 3, the β-carbon of the fatty acyl chain. The process involves the action of acyl-CoA dehydrogenase, enoyl-CoA hydratase, hydroxyacyl-CoA dehydrogenase and ketoacyl-CoA thiolase. A 16-carbon fatty acid (palmitic acid) undergoes seven passes through this oxidative sequence, losing two carbons as acetyl-CoA on each cycle to yield total of eight molecules of acetyl-CoA. The second step, which also occurs in the mitochondrial matrix, involves the oxidation of acetyl-CoA to CO_2 in the citric acid cycle. In the third step, flavin adenine dinucleotide and nicotinamide adenine dinucleotide (NADH) produced during β-oxidation are used by the electron transport chain to generate ATP. Electrons derived from the first two steps pass to O_2 via the mitochondrial respiratory chain, providing the energy for ATP synthesis by oxidative phosphorylation. The net yield for the complete oxidation of a palmitic acid molecule is 129 ATP molecules. Some naturally occurring fatty acids contain an odd number of carbon atoms. These fatty acids are also oxidised by β-oxidation through removing two carbons as acetyl-CoA in each round of the oxidative process. However, the final products of the thiolytic cleavage of fatty acid with an odd number of carbon atoms are acetyl-CoA and propionyl-CoA (a three carbon molecule). Propionyl-CoA metabolism requires three more catabolic steps to be completely metabolised into succinyl-CoA that can be finally metabolised to oxaloacetate, a substrate for gluconeogenesis.

5.2.1 Regulation of Fatty Acid Oxidation

Fatty acid oxidation is tightly regulated at several points of the pathway to achieve a balance between energy production and expenditure. The rate-limiting step of fatty acid oxidation is the transport of fatty acyl-CoA to the mitochondria matrix through the carnitine system. This is important because acyl-CoA formed on the cytosolic side of mitochondria can either be directed to lipid biosynthesis or fatty acid oxidation. Acyl-CoA translocated to the mitochondrial matrix is committed to fatty acid oxidation. The first key regulatory checkpoint of fatty acid oxidation is performed by malonyl-CoA, the first intermediate in the cytosolic biosynthesis of long-chain fatty acids from acetyl CoA (Section 2). Malonyl-CoA inhibits CPT-I, thus ensuring dominance of anabolic (synthesis) over catabolic (oxidation) processes. Fatty acid oxidation is additionally regulated by the energetic balance of the cell with [NADH]/[NAD$^+$] ratio serving as the determinant of the cellular energetic status; excess NADH inhibits β-hydroxyacyl-CoA dehydrogenase, limiting the conversion of L-α-hydroxyacyl-CoA to α-ketoacyl-CoA, the third step in the β-oxidation loop. In conditions of low ATP levels (energy deficient state), the concentration of AMP increases and AMPK phosphorylates and inhibits several lipogenic target enzymes, including ACC1/2. A decrease in malonyl-CoA eliminates the inhibition of fatty acyl-carnitine transport into the mitochondria and allows β-oxidation to proceed to replenish ATP stores.

Other regulatory mechanisms involve the action of transcription factors over certain genes to obtain a response in a prolonged manner. The PPAR family of nuclear receptors affects many metabolic pathways in response to a variety of fatty acid-like ligands. PPARα acts in tissues with highly oxidative function such as muscle, BAT and liver to increase expression of a set of enzymes required for fatty acid oxidation, including ACSL1, CPT-I, CPT-II and fatty acyl-CoA dehydrogenases. Fasting or long-term starvation triggers a PPARα response to catabolise lipids for energy supply. Similarly, in response to reduced blood glucose concentration, glucagon-mediated signalling augments cAMP concentrations, which activates the transcription factor CREB responsible for turning on expression of a subset of genes catalysing lipid catabolism.

5.3 Fatty Acids and Thermogenesis

Uncoupling protein (UCP1) is a mitochondrial protein that uncouples mitochondrial respiration from ATP synthesis, resulting in heat production (Townsend and Tseng, 2014). UCP1 is highly expressed in mitochondria in BAT. Similar to WAT, BAT takes up glucose and fatty acids, synthesises TAG, and stores this lipid in LDs. In fact, activated BAT has such a large capacity for glucose and lipid uptake that it can normalise hyperglycaemia and hyperlipidaemia in mouse models of diabetes and dyslipidaemia. Fatty acids are taken up into BAT via CD36 and FATP1 and esterified into TAG that is stored in multilocular LDs, which is different from WAT storage. Fatty acids are released by lipolysis of TAG stores through the action of ATGL, HSL and MGL. ATGL-deficient mice are cold intolerant because ATGL-derived fatty acids are necessary for the activation of PPARα and in particular PGC1α, which is an important transcriptional regulator of UCP-1 expression. Ablation of PGC1α in mice results in significant impairment of cold-induced thermogenesis despite normal BAT formation. Conversely, expression of PGC1α in WAT results in the induction of expression of thermogenesis genes, including UCP1. Adipose tissue-specific ACSL1 or CPT-I ablation in mice results in cold intolerance, indicating their importance in regulating fatty acid activation and transport into mitochondria in BAT.

6. FATTY ACIDS AND SIGNALLING

Fatty acids in cells are converted to acyl-CoA and esterified to lysophosphatidic acid, DAG, ceramide, phospholipids, endocannabinoids (arachidonoyl ethanolamide, 2-arachidonoyl glycerol) and other lipids that are known activators or ligands of proteins mediating signal transduction processes. As an example, lysophosphatidic acid is mitogenic, DAG activates protein kinase C (PKC) and protein kinase D (PKD), while endocannabinoids serve as regulators of energy homeostasis. One of the major questions regarding lipids as regulators of signal transduction is their origin. For instance, is the 'signal producing' lysophosphatidic acid generated de novo, is it produced by combined activities of phospholipase D and phospholipase A1 or A2, or do both de novo and phospholipid degradation pathways produce this signal mediator? Do both fatty acids/DAG synthesised de novo and released by lipolysis of pre-existing stores activate PPARs and protein kinases? Some of these questions have been answered, but many unknowns still remain.

6.1 Fatty Acids as Ligands for Nuclear Receptors

Long-chain fatty acids bind and activate all PPARs and have been proposed to function as endogenous ligands for these receptors (Nakamura et al., 2014). In oxidative tissues where PPARα and PPARδ function predominate, ATGL-catalysed lipolysis has been demonstrated to regulate PPAR-mediated processes such as fatty acid oxidation in the skeletal muscle, heart and liver, and thermogenesis in BAT. In lipogenic tissues such as WAT, where PPARγ predominates, HSL has been shown to regulate PPARγ-mediated processes, such as adipogenesis, glucose and fatty acid uptake, and lipogenesis. ATGL-catalysed hydrolysis might also affect PPARγ function in WAT, but it is not clear whether this is a direct effect or occurs via DAG species that are then substrates for HSL. Depending on the method, the binding constants of fatty acids for PPARs are between 10 nM and 1–5 μM. Regardless of the experimental method used for determining fatty acid affinity for PPARs, fatty acid concentration in cells would be expected to be in the low nM range because of high affinity binding capacity of FABPs that are abundantly expressed in cells. Therefore, it is likely that intracellular concentrations of fatty acids are near or below the dissociation constants for PPARs and act as physiological regulators of their activity.

Unsaturated fatty acids activate RXRs, obligatory heterodimeric partners of other nuclear receptors including PPARs, LXR, farnesoid-X-receptor, retinoic acid receptor, vitamin D receptor and thyroid hormone receptor. Hepatocyte nuclear factor 4α, an important transcription factor that regulates hepatic functions such as detoxification, alcohol metabolism, bile acid metabolism, lipoprotein secretion, lipogenesis and glucose metabolism contains a fatty acid-binding pocket, and the crystal structure of the protein has revealed the presence of a fatty acid in the ligand-binding pocket. However, the role of the bound fatty acid in the transcriptional activity of this factor is not clear.

6.2 Plasma Membrane Fatty Acid Signalling Receptors

Long-chain fatty acids signal through two plasma membrane G-protein-coupled receptors, GPR40 (also known as FFA1 or FFAR1) and GPR120 (also known as FFA4 or FFAR4) to regulate a number of physiological processes, including sensing dietary fatty acids in the gastrointestinal tract and regulation of energy metabolism through mediating secretion of insulin and incretins.

The natural agonists for GPR40 are medium- to long-chain fatty acids (C12-22) and conjugated linoleic acid. GPR40 is expressed in pancreatic β-cells, intestine and tongue. Fatty acid-activated GPR40 amplifies glucose-stimulated insulin secretion from pancreatic β-cells (Mancini and Poitout, 2013). The importance of GPR40 in the potentiation of insulin secretion by fatty acids has been demonstrated in mice in which *Ffar1* expression was ablated by RNA interference or gene disruption and by pharmacological inactivation of GPR40. Ablation of *Ffar1* expression in mice diminished secretion of insulin, incretins, glucose-dependent insulinotropic peptide, GLP-1 and glucagon. Overexpression of GPR40 in β-cells prevents high-fat diet-induced hyperglycaemia and improves glucose tolerance and insulin secretion. GPR40 expression and signalling is reduced in pancreatic cells isolated from patients with type 2 diabetes mellitus, therefore pharmacological GPR40 agonists would be expected to improve glycaemic control and long-term clinical trials are underway to address agonist safety and efficacy.

Despite the positive effect of GPR40 activation by fatty acids little is known about the mechanism that links activated GPR40 to augmented insulin secretion. It was predicted that the signalling cascade involves the canonic phospholipase C hydrolysis of phosphatidylinositide yielding sn-1,2-DAG and inositol-3-phosphate, which activate PKC and release Ca^{2+} from the ER, respectively. However, that activation of GPR40 leads to influx of extracellular Ca^{2+}. In addition, PKD1 and not PKC was the activated kinase. Activation of PKD1 results in depolymerisation of actin and potentiation of insulin release from insulin-containing granules.

In contrast to GPR40, GPR120 has a broader expression profile and is present in intestine, WAT, macrophages, arcuate nucleus of hypothalamus, pancreas and other tissues (Ichimura et al., 2014). GPR120 is activated by medium- to long-chain unsaturated fatty acids. Stimulation of GPR120 in distal intestine (colon) by fatty acids increases plasma GLP-1, indicating that similar to GPR40 the receptor senses unabsorbed fatty acids and exerts feedback inhibition of gastrointestinal motility. Activation of GPR120 by n-3 PUFA in macrophages elicits anti-inflammatory responses including attenuation of lipopolysaccharide-induced phosphorylation and activation of IκB kinase (IKK) and c-Jun N-terminal kinase (JNK). The importance of GPR120 in mediating anti-inflammatory responses was demonstrated by the inability of GPR120 deficient mice to exert n-3 fatty acid-induced protection against inflammation. Knockdown of expression of *Ffar4* expression in 3T3-L1 preadipocytes rendered these cells unable to differentiate into adipocytes, suggesting a role of the receptor in adipocyte maturation. This is supported by studies in GPR120-deficient mice in which decreased adipocyte differentiation and lipogenesis were observed, accompanied with severe hepatic steatosis, hyperglycaemia and glucose intolerance. Tissue-specific ablation of *Ffar4* expression is necessary to discern tissue-specific effects of GPR120. Nevertheless, it is apparent that PUFA regulates energy metabolism via GPR120 and that activation of GPR120 might be beneficial as treatment of metabolic syndrome.

7. FATTY ACIDS AND DISEASE PATHOGENESIS

The role of fatty acids in the pathogenesis of several diseases has been extensively studied. In this section, we will briefly describe the role of fatty acids in three pathophysiological conditions: insulin resistance, ER stress and inflammation.

7.1 Insulin Resistance

Increased circulating fatty acids and TAG have been strongly implicated in and correlated with impaired insulin signalling and glucose intolerance in obesity and type 2 diabetes mellitus (Chapter 19). Ectopic TAG accumulation plays an important role in insulin resistance in obesity and type 2 diabetes mellitus. For example, muscle TAG levels are increased in patients with type 2 diabetes mellitus and correlate with insulin resistance in individuals without diabetes mellitus. Likewise, hepatic steatosis is associated with hepatic insulin resistance. When fatty acid and glucose plasma concentrations become elevated, cellular levels of NADH and ATP also increase, causing inhibition of Krebs cycle enzymes and exit of citrate from the mitochondria. Citrate activates ACC and production of malonyl-CoA, which inhibits CPT-I. This prevents fatty acid oxidation, resulting in redirection of fatty acids toward esterification and accumulation of lipid intermediates (DAG, ceramides), which have been implicated in a etiology of insulin resistance. On the other hand, in energy-deficient states preferential oxidation of TAG occurs. An important regulator of this energy metabolism switch is AMPK, which is activated in response to increased AMP/ATP ratio. AMPK inhibits ACC, thereby increasing fatty acid oxidation and decreasing TAG accumulation. Accordingly, AMPK activators, such as 5-aminoimidazole-4-carboxamide-1-ß-D-ribofurano-side, metformin, thiazolidinedione and adiponectin, decrease intrahepatic lipid content and improve hepatic insulin action. The molecular mechanism by which lipid accumulation causes hepatic insulin resistance possibly involves activation of one or more serine/threonine kinases, such as PKC, IKKβ and JNK1, which may directly disrupt hepatic insulin signalling via serine/threonine phosphoryla-tion of insulin receptor and/or insulin receptor substrate. Both long-chain fatty acid-CoA and DAG can stimulate various PKC isoforms that have been linked to insulin resistance. Excess of palmitoyl-CoA augments synthesis of ceramide, which in turn can activate JNK1. JNK1 has been implicated in fatty acid-induced insulin resistance in vitro and in muscle insulin resistance of obese individuals.

Lowering fatty acid concentrations in plasma by inducing adipogenesis in mice improves insulin sensitivity. Such beneficial effects were observed in trans-genic mice overexpressing DGAT1 in adipose tissue, in mice with enhanced glucose transport into adipose tissue, or in mice in which adipogenesis was increased by PPARγ agonism. On the other hand, PUFA suppresses the expres-sion and processing of SREBP-1c and therefore de novo lipogenesis and pre-vents insulin resistance induced by saturated fat diet.

7.2 ER Stress

Pathological events that perturb ER homeostasis affect folding of proteins, Ca^{2+} flux and lipid metabolism, which leads to accumulation and aggregation of mis-folded proteins in the ER lumen and ultimately to ER stress. Cells respond to this kind of stress by triggering a cytoprotective response called the unfolded protein response, which aims to restore normal ER function (Back and Kaufman, 2012). The unfolded protein response is sensed by three transmembrane ER proteins; protein kinase RNA-dependent-like ER kinase, inositol requiring ER-to-nucleus signal kinase 1α and activating transcription factor 6. Under normal physiological conditions, these proteins are maintained in an inactive state by interaction with the ER lumenal chaperone BiP. During an ER stress situation, BiP is released from these proteins to bind misfolded proteins, which leads to activation of the three stress sensors and transmission of the stress information to the cytosol and eventually to the nucleus, where expression of specific genes is activated to remedy the stress situation.

There is a significant connection between the ER homeostasis and lipids. Not only is the ER the main site of phospholipid synthesis, but phospholipids are critical for maintaining ER integrity and homeostasis. The role of fatty acids in eliciting ER stress and cell death has been clearly demonstrated. Palmitic acid, the main fatty acid in saturated fat diets, is mainly incorporated into the *sn*-1-position of glycerolipids, but cells do not deal well with excess of this fatty acid (or any other saturated fatty acid) and reach a limit for converting saturated fatty acids to less harmful lipids (e.g. TAG). TAG containing three palmitic acid molecules is solid at 37 °C and therefore not a favourable lipid species for storage. When cells cannot 'detoxify' excess palmitic acid, by desaturation or conversion to membrane or storage lipids or by oxidation, palmitic acid and its derivatives palmitoyl-CoA and ceramides disrupt the ER homeostatic mechanisms and elicit ER stress and cell dysfunction. In agreement with the concept that saturated fatty acids are harmful while unsaturated fatty acids are protective, overexpression of stearoyl-CoA desaturase-1 (Chapter 6) is sufficient to prevent lipotoxicity of palmitic acid. The importance of conversion of fatty acids to TAG as a protective mechanism against fatty acid-induced lipotoxicity was demonstrated by elegant studies in Jean Schaffer's laboratory. The mechanisms by which saturated fatty acids activate ER stress may involve PKC and Ca^{2+} signalling. Inhibition of PKCδ reduces palmitic acid-induced ER stress and apoptosis. Palmitic acid inhibits the ER-localised SERCA pump, which maintains cellular Ca^{2+} homeostasis. Palmitic acid-mediated reduction of ER Ca^{2+} content might be a prerequisite for activation of ER stress.

7.3 Inflammation

A rise in circulating fatty acids can activate toll-like receptors (TLRs), which in turn trigger the nuclear factor-κB (NF-κB) pathway and induce downstream expression of proinflammatory cytokines (Fessler et al., 2009). Fatty acids are active modulators of metabolism and inflammation in adipose tissue, liver, pancreas, skeletal muscle and the vessel wall. There is a close relationship between the molecular species of circulating fatty acid and inflammation. Replacing dietary saturated fatty acids with monounsaturated fatty acids and PUFA is associated with reduced insulin resistance and cardiovascular risk. Saturated fatty acids have the capacity to provoke an inflammatory response in target cells; one known mechanism is by activation of the cell-surface TLR2 and TLR4. TLRs are a family of pattern recognition receptors with a central function in innate immunity, and are expressed in both macrophages and adipocytes. In adipocytes, TLR2 and TLR4 stimulation causes NF-κB activation and subsequent proinflammatory cytokine release. Both TLR2 and TLR4 are upregulated in hypertrophic WAT, and inactivation of either TLR2 or TLR4 protects against WAT inflammation, macrophage infiltration, insulin resistance and fatty liver induced by high-fat diet or lipid infusions. In addition, some of the beneficial effects of PUFA consumption may be mediated by TLR4. PUFA blunts the proinflammatory response generated by saturated fatty acids. High intake of PUFA also prevents activation of TLR4 by its natural agonist lipopolysaccharide in human blood monocytes. In addition to proinflammatory responses, lipopolysaccharide-induced TLR4 activation induces lipolysis in WAT, which increases circulating fatty acids. Further support for a role of TLR4 in lipolysis was obtained from TLR4 mutant mice, which showed protection from high fat diet-induced elevation of plasma free fatty acids.

8. FUTURE DIRECTIONS

Despite major advances in our understanding of fatty acid synthesis and use for various cellular functions, detailed mechanistic information of enzymes and proteins that use them is lacking. The mechanism of action has not been determined for most enzymes catalysing TAG synthesis because the enzymes have not been purified and biochemically characterised. Many lipid biosynthetic and hydrolytic reactions are catalysed by two or more enzymes with the same activity, yet it is not clear how the specificity for one particular enzyme is determined. It is likely that the enzymes are compartmentalised and/or form complexes, so that intermediates are channelled through lipid synthetic or hydrolytic pathways, but the mechanisms of formation of such complexes in situ and compositions of such complexes are largely unknown. Information with respect to specificity of enzymes for particular fatty acid or position of fatty acid within a complex lipid is vastly incomplete. The localisation of acyltransferases, lipases and co-factors has been principally deduced from overexpression of tagged-proteins, which may result in erroneous conclusions when overexpression mislocalises proteins, particularly if they have a binding factor or form equimolar multimers, and/or a given overexpressed protein requires a cellular co-factor for activation/inhibition that might be present in limited quantities. Most of the enzymes discussed in this chapter undergo one or more posttranslational modification, such as phosphorylation and acetylation, yet the functional significance of such modifications is not well understood for many of the proteins discussed. The function of many enzymes/proteins has been deduced from genetic ablation of the encoding gene. However, in certain instances, animals may compensate for the long-term loss of the activity by eliciting global changes in metabolism that are not obvious in acute deletion of the protein. Therefore development and use of specific inhibitors would provide complementary information to studies using gene knockout or RNA interference. Many of the metabolic pathways have been delineated in animal models (mostly rodents); however, it is important to keep in mind that rodent fatty acid metabolism might be very different from that in humans. Finally, ongoing debate concerns the topic of whether TAG accumulation in tissues protects against lipotoxicity (ER stress) or promotes lipotoxicity (inflammation).

Thus, despite significant advances between 2000 and 2010 in our understanding of fatty acid metabolism, future studies are likely to identify additional regulators of fatty acid metabolic pathways and their relevance to health and disease.

REFERENCES

Alkhateeb, H., Holloway, G.P., Bonen, A., June 2011. Skeletal muscle fatty acid oxidation is not directly associated with AMPK or ACC2 phosphorylation. Appl. Physiol. Nutr. Metab. = Physiologie appliquee, nutrition et metabolisme 36 (3), 361–367. PubMed PMID: 21574785.

Abumrad, N.A., el-Maghrabi, M.R., Amri, E.Z., Lopez, E., Grimaldi, P.A., August 25, 1993. Cloning of a rat adipocyte membrane protein implicated in binding or transport of long-chain fatty acids that is induced during preadipocyte differentiation. Homology with human CD36. J. Biol. Chem. 268 (24), 17665–17668. PubMed PMID: 7688729.

Anderson, C.M., Stahl, A., April–June 2013. SLC27 fatty acid transport proteins. Mol. Aspects Med. 34 (2–3), 516–528. PubMed PMID: 23506886. Pubmed Central PMCID: 3602789.

Bloksgaard, M., Neess, D., Faergeman, N.J., Mandrup, S., March 2014. Acyl-CoA binding protein and epidermal barrier function. Biochim. Biophys. Acta 1841 (3), 369–376. PubMed PMID: 24080521.

Back, S.H., Kaufman, R.J., 2012. Endoplasmic reticulum stress and type 2 diabetes. Annu. Rev. Biochem. 81, 767–793. PubMed PMID: 22443930. Pubmed Central PMCID: 3684428.

Chakravarthy, M.V., Lodhi, I.J., Yin, L., Malapaka, R.R., Xu, H.E., Turk, J., et al., August 7, 2009. Identification of a physiologically relevant endogenous ligand for PPARα in liver. Cell 138 (3), 476–488. PubMed PMID: 19646743. Pubmed Central PMCID: 2725194.

Ferre, P., Foufelle, F., October 2010. Hepatic steatosis: a role for de novo lipogenesis and the transcription factor SREBP-1c. Diabetes Obes. Metab. 12 (Suppl. 2), 83–92. PubMed PMID: 21029304.

Fessler, M.B., Rudel, L.L., Brown, J.M., October 2009. Toll-like receptor signaling links dietary fatty acids to the metabolic syndrome. Curr. Opin. Lipidol. 20 (5), 379–385. PubMed PMID: 19625959. Pubmed Central PMCID: 3099529.

Grevengoed, T.J., Klett, E.L., Coleman, R.A., 2014. Acyl-CoA metabolism and partitioning. Annu. Rev. Nutr. 34, 1–30. PubMed PMID: 24819326.

Hall, A.M., Kou, K., Chen, Z., Pietka, T.A., Kumar, M., Korenblat, K.M., et al., May 2012. Evidence for regulated monoacylglycerol acyltransferase expression and activity in human liver. J. Lipid Res. 53 (5), 990–999. PubMed PMID: 22394502. Pubmed Central PMCID: 3329399.

Hall, A.M., Soufi, N., Chambers, K.T., Chen, Z., Schweitzer, G.G., McCommis, K.S., et al., July 2014. Abrogating monoacylglycerol acyltransferase activity in liver improves glucose tolerance and hepatic insulin signaling in obese mice. Diabetes 63 (7), 2284–2296. PubMed PMID: 24595352. Pubmed Central PMCID: 4066334.

Haemmerle, G., Zimmermann, R., Hayn, M., Theussl, C., Waeg, G., Wagner, E., et al., February 15, 2002. Hormone-sensitive lipase deficiency in mice causes diglyceride accumulation in adipose tissue, muscle, and testis. J. Biol. Chem. 277 (7), 4806–4815. PubMed PMID: 11717312.

Ichimura, A., Hara, T., Hirasawa, A., 2014. Regulation of energy homeostasis via GPR120. Front. Endocrinol. 5, 111. PubMed PMID: 25071726. Pubmed Central PMCID: 4093656.

Lass, A., Zimmermann, R., Oberer, M., Zechner, R., January 2011. Lipolysis - a highly regulated multi-enzyme complex mediates the catabolism of cellular fat stores. Prog. Lipid Res. 50 (1), 14–27 PubMed PMID: 21087632. Pubmed Central PMCID: 3031774. Epub 2010/11/23. eng.

Lo, V., Erickson, B., Thomason-Hughes, M., Ko, K.W., Dolinsky, V.W., Nelson, R., et al., February 2010. Arylacetamide deacetylase attenuates fatty-acid-induced triacylglycerol accumulation in rat hepatoma cells. J. Lipid Res. 51 (2), 368–377. PubMed PMID: 19654421. Pubmed Central PMCID: 2803239. Epub 2009/08/06. eng.

Mancini, A.D., Poitout, V., August 2013. The fatty acid receptor FFA1/GPR40 a decade later: how much do we know? Trends Endocrinol. Metab. 24 (8), 398–407. PubMed PMID: 23631851.

Nakamura, M.T., Yudell, B.E., Loor, J.J., January 2014. Regulation of energy metabolism by long-chain fatty acids. Prog. Lipid Res. 53, 124–144. PubMed PMID: 24362249.

Quiroga, A.D., Li, L., Trotzmuller, M., Nelson, R., Proctor, S.D., Kofeler, H., et al., December 2012. Deficiency of carboxylesterase 1/esterase-x results in obesity, hepatic steatosis, and hyperlipidemia. Hepatology 56 (6), 2188–2198. PubMed PMID: 22806626.

Storch, J., Thumser, A.E., October 22, 2010. Tissue-specific functions in the fatty acid-binding protein family. J. Biol. Chem. 285 (43), 32679–32683. PubMed PMID: 20716527. Pubmed Central PMCID: 2963392.

Sztalryd, C., Kimmel, A.R., January 2014. Perilipins: lipid droplet coat proteins adapted for tissue-specific energy storage and utilization, and lipid cytoprotection. Biochimie 96, 96–101. PubMed PMID: 24036367.

Smagris, E., BasuRay, S., Li, J., Huang, Y., Lai, K.M., Gromada, J., Cohen, J.C., Hobbs, H.H., January, 2015. Pnpla3I148M knockin mice accumulate PNPLA3 on lipid droplets and develop hepatic steatosis. Hepatology 61 (1), 108–118. http://dx.doi.org/10.1002/hep.27242. Epub 2014 Oct 1, PubMed PMID: 24917523.

Singh, R., Kaushik, S., Wang, Y., Xiang, Y., Novak, I., Komatsu, M., et al., April 30, 2009. Autophagy regulates lipid metabolism. Nature 458 (7242), 1131–1135. PubMed PMID: 19339967. Pubmed Central PMCID: 2676208. Epub 2009/04/03. eng.

Tong, L., March 2013. Structure and function of biotin-dependent carboxylases. Cell Mol. Life Sci. 70 (5), 863–891. PubMed PMID: 22869039. Pubmed Central PMCID: 3508090.

Thiam, A.R., Farese Jr., R.V., Walther, T.C., December 2013. The biophysics and cell biology of lipid droplets. Nat. Rev. Mol. Cell Biol. 14 (12), 775–786. PubMed PMID: 24220094.

Taschler, U., Radner, F.P., Heier, C., Schreiber, R., Schweiger, M., Schoiswohl, G., et al., May 20, 2011. Monoglyceride lipase deficiency in mice impairs lipolysis and attenuates diet-induced insulin resistance. J. Biol. Chem. 286 (20), 17467–17477. PubMed PMID: 21454566. Pubmed Central PMCID: 3093820.

Townsend, K.L., Tseng, Y.H., April 2014. Brown fat fuel utilization and thermogenesis. Trends Endocrinol. Metab. 25 (4), 168–177. PubMed PMID: 24389130. Pubmed Central PMCID: 3972344.

Wang, H., Wei, E., Quiroga, A.D., Sun, X., Touret, N., Lehner, R., June 15, 2010. Altered lipid droplet dynamics in hepatocytes lacking triacylglycerol hydrolase expression. Mol. Biol. Cell 21 (12), 1991–2000. PubMed PMID: 20410140. Pubmed Central PMCID: 2883943. Epub 2010/04/23. eng.

Wilfling, F., Thiam, A.R., Olarte, M.J., Wang, J., Beck, R., Gould, T.J., et al., 2014. Arf1/COPI machinery acts directly on lipid droplets and enables their connection to the ER for protein targeting. eLife 3, e01607. PubMed PMID: 24497546. Pubmed Central PMCID: 3913038.

Wei, E., Ben Ali, Y., Lyon, J., Wang, H., Nelson, R., Dolinsky, V.W., et al., March 3, 2010. Loss of TGH/Ces3 in mice decreases blood lipids, improves glucose tolerance, and increases energy expenditure. Cell Metab. 11 (3), 183–193. PubMed PMID: 20197051. Epub 2010/03/04. eng.

Ye, J., DeBose-Boyd, R.A., July 2011. Regulation of cholesterol and fatty acid synthesis. Cold Spring Harbor Perspect. Biol. 3 (7). PubMed PMID: 21504873. Pubmed Central PMCID: 3119913.

Yen, C.L., Cheong, M.L., Grueter, C., Zhou, P., Moriwaki, J., Wong, J.S., et al., April 2009. Deficiency of the intestinal enzyme acyl CoA:monoacylglycerol acyltransferase-2 protects mice from metabolic disorders induced by high-fat feeding. Nat. Med. 15 (4), 442–446. PubMed PMID: 19287392. Pubmed Central PMCID: 2786494.

Yen, C.L., Stone, S.J., Koliwad, S., Harris, C., Farese Jr., R.V., November 2008. Thematic review series: glycerolipids. DGAT enzymes and triacylglycerol biosynthesis. J. Lipid Res. 49 (11), 2283–2301. PubMed PMID: 18757836. Epub 2008/09/02. eng.

Zordoky, B.N., Nagendran, J., Pulinilkunnil, T., Kienesberger, P.C., Masson, G., Waller, T.J., et al., August 15, 2014. AMPK-dependent inhibitory phosphorylation of ACC is not essential for maintaining myocardial fatty acid oxidation. Circ. Res. 115 (5), 518–524. PubMed PMID: 25001074.

Zechner, R., Langin, D., August 5, 2014. Hormone-sensitive lipase deficiency in humans. Cell Metab. 20 (2), 199–201. PubMed PMID: 25100058.

Chapter 6

Fatty Acid Desaturation and Elongation in Mammals

Laura M. Bond,[1,*] Makoto Miyazaki,[3,*] Lucas M. O'Neill,[1,*]
Fang Ding,[4] James M. Ntambi[1,2]

[1]Department of Biochemistry, University of Wisconsin-Madison, Madison, WI, USA; [2]Department of Nutritional Science, University of Wisconsin-Madison, Madison, WI, USA; [3]Department of Medicine, Division of Renal Diseases and Hypertension, University of Colorado, Anschutz Medical Campus, Aurora, CO, USA; [4]Institute of Animal Genetics and Breeding, College of Animal Science and Technology, Sichuan Agricultural University, Wenjiang, Sichuan, PR, China

ABBREVIATIONS

BCFA Branched chain fatty acid
ChREBP Carbohydrate response element-binding protein
D5D $\Delta 5$ desaturase
D6D $\Delta 6$ desaturase
DHA Docosahexaenoic acid
ER Endoplasmic reticulum
HUFA Highly unsaturated fatty acids
KAR Ketoacyl-CoA reductase
LXR Liver X receptor
MUFA Monounsaturated fatty acids
PCD Palmitoyl-CoA desaturase
PL Phospholipids
PPAR Peroxisome proliferator activated receptor
PUFA Polyunsaturated fatty acids
SCD Stearoyl-CoA desaturase
SFA Saturated fatty acid
SREBP Sterol regulatory element-binding protein

1. INTRODUCTION

Fatty acids (FA) synthesised de novo and those obtained from the diet are modified through metabolic pathways that include desaturation and elongation. These reactions occur primarily in the endoplasmic reticulum (ER) and produce a variety of long-chain saturated, monounsaturated and polyunsaturated fatty acids (SFA,

---

* These authors contributed equally to this work.

Biochemistry of Lipids, Lipoproteins and Membranes. http://dx.doi.org/10.1016/B978-0-444-63438-2.00006-7

MUFA and PUFA) consisting of 16 carbons or more. These modified FA have a variety of fates and functions. Long-chain fatty acyl-coenzyme A (CoA) is esterified into triacylglycerols (TAG), which constitute the body's major and most efficient energy stores (Chapter 5). Phospholipids (PL) are essential for membrane structure and function. PL chain length and the number and position of double bonds can markedly influence the fluidity, permeability and stability of biological membranes. By influencing membrane physical properties, FA can alter the function of integral membrane proteins including receptors and those proteins involved in ion channelling, endocytosis and exocytosis. Additionally, the release of specific FA from membrane PL stores can lead to the generation of precursors for certain signalling molecules such as eicosanoids (Chapter 9), pheromones, growth regulators and hormones. Long-chain FA exert transcriptional and post-translational control over many biological pathways. Certain FA modulate the activity of specific nuclear receptors and other proteins involved in the regulation of gene expression. FA are also involved in the covalent modification of numerous proteins reported to be crucial for cell signalling cascades and for the function of certain viruses (Chapter 13). The advancement of biochemical, molecular and genetic techniques and the availability of radiolabelled substrates have greatly contributed to our understanding of fatty acid elongation and desaturation. This chapter will review our current knowledge regarding the function and regulation of mammalian fatty acid desaturation and elongation.

1.1 Nomenclature and Sources of Long-Chain Fatty Acids

Fatty acids are carboxylic acids containing hydrocarbon chains of variable lengths. The length can range from 4 to 36 carbons or more depending on the mammalian species. FA can be further defined by the degree of unsaturation (double and triple bonds), and the number and position as well as the stereochemistry (*cis* or *trans*) of the double bonds. There are four systems used for naming FA: common names, systematic nomenclature, Δ nomenclature and ω (or n) nomenclature. Δ nomenclature and ω (or n) nomenclature denote the position of the double bond counting from the carboxylic acid carbon and the terminal methyl carbon, respectively. For example, the systematic name of palmitoleic acid is 9-palmitoleic acid and the condensed formula is 16:1Δ9c, which denotes an 18-carbon fatty acid with one double bond at position 9 in the *cis* configuration. The ω (or n) nomenclature for palmitoleic acid is 16:1n7. The ω (or n) nomenclature is particularly useful for describing PUFA, which contain more than one double bond and have physiological properties of PUFA dependent on the position of the first unsaturation relative to the methyl end and not the carboxylic acid end. The number of carbons and the number of double bonds are also listed. For example, α-linolenic acid is denoted as 18:3(ω-3) or 18:3n3 and refers to an 18-carbon chain with three double bonds, with the first double bond in the third position from the methyl end. Double bonds are *cis* and separated by a single methylene (CH_2) group unless otherwise noted (Table 1).

TABLE 1 Nomenclature and Bond Positions of Major Long-Chain Fatty Acids

Common Name	Abbreviation	Double Bond Positions
Palmitic acid	16:0	
Palmitoleic acid	16:1n7	Δ9
Sapienic acid	16:1n10	Δ6
Stearic acid	18:0	
Oleic acid	18:1n9	Δ9
Vaccenic acid	18:1n7	Δ11
Linoleic acid	18:2n6	Δ9,12
α-Linolenic acid	18:3n3	Δ9,12,15
γ-Linolenic acid	18:3n6	Δ6,9,12
Stearidonic acid	18:4n3	Δ6,9,12,15
Arachidic acid	20:0	
Paullinic acid	20:1n7	Δ13
Gondoic acid	20:1n9	Δ11
Dihomo-γ-linolenic acid	20:3n6	Δ8,11,14
Meadic acid	20:3n9	Δ5,8,11
Arachidonic acid	20:4n6	Δ5,8,11,14
Eicosapentaenoic acid	20:5n3	Δ5,8,11,14,17
Behenic acid	22:0	
Erucic acid	22:1n9	Δ13
Adrenic acid	22:4n6	Δ7,10,13,16
n-6 Docosapentaenoic acid	22:5n6	Δ4,7,10,13,16
n-3 Docosapentaenoic acid	22:5n3	Δ7,10,13,16,19
Docosahexaenoic acid	22:6n3	Δ4,7,10,13,16,19
Lignoceric acid	24:0	
Nervonic acid	24:1n9	Δ15

All bonds are of a *cis* geometric configuration.

Branched chain fatty acids (BCFAs) are also present in mammals, albeit generally at very low levels. Synthesis of BCFAs occurs in the skin and much less in internal tissues. BCFAs compose 10–20% dry weight of vernix caseosa, a waxy substance coating the skin of newborn babies (Ran-Ressler et al., 2008). Branched chain hydroxy fatty acids were recently discovered, and certain BCFAs have been reported to improve insulin sensitivity and reduce adipose tissue inflammation through their activation of G-protein coupled receptor 120 (GPR120) (Yore et al., 2014).

Most mammalian cells have the capacity to synthesise FA from glucose in a pathway called de novo lipogenesis, which uses products from glycolysis and two key cytosolic enzymes, acetyl-CoA carboxylase and fatty acid synthase (fatty acid synthesis, transport and storage are described in Chapter 5). This pathway generates long-chain saturated fatty acid (SFA), mainly palmitate (16:0). Palmitate synthesised de novo or derived from the diet is transported to the ER. On the ER membrane, two major fatty acid enzymatic modifications – elongation and desaturation – occur to yield longer chain SFA and unsaturated FA.

The types of FA obtained through dietary intake and de novo synthesis are insufficient to meet the varied demands of cells; therefore, these FA undergo substantial metabolism and rearrangement to facilitate key biological processes. Our understanding of how the array of fatty acyl chains is derived and altered, and what regulates the modification of fatty acyl chains by elongation and desaturation, is described in the sections that follow.

2. ELONGATION REACTIONS OF LONG-CHAIN FATTY ACIDS

2.1 Microsomal Fatty Acid Elongation

The predominant pathway for fatty acid elongation occurs on the ER and uses malonyl-CoA and fatty acyl-CoAs as substrates for addition of two carbon atoms to the existing fatty acid substrates. First, fatty acyl-CoA synthetases activate FA through the addition of a CoA moiety. Unlike cytosolic fatty acid synthesis by the type I fatty acid synthase complex, separate enzymes catalyse the four sequential reactions of fatty acid elongation. Microsomal chain elongation utilises SFA, MUFA and PUFA, but γ-linolenate (18:3n6) is the most active substrate. The four consecutive and independent reactions are: (1) condensation of an acyl-CoA molecule and the 3-carbon malonyl-CoA substrate to form ketoacyl-CoA; (2) a reduction by the 3-ketoacyl-CoA reductase using NADPH to form 3-hydroxyacyl-CoA; (3) a dehydration of 3-hydroxyacyl-CoA to generate *trans*-2,3-enoyl-CoA; (4) a reduction of enoyl-CoA to generate a 2-carbon-extended acyl-CoA product (Figure 1).

2.1.1 Ketoacyl-CoA Synthase

Ketoacyl-CoA synthases perform the condensing step, which is the initial and rate-limiting step that determines fatty acyl specificity and results in the addition

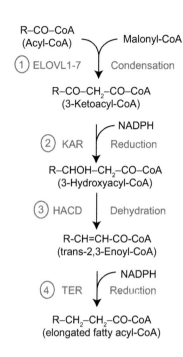

R–CO–CoA
(Acyl-CoA)

ⓘ ELOVL1-7

Malonyl-CoA

Condensation

R–CO–CH$_2$–CO–CoA
(3-Ketoacyl-CoA)

② KAR

NADPH

Reduction

R–CHOH–CH$_2$–CO–CoA
(3-Hydroxyacyl-CoA)

③ HACD

Dehydration

R-CH=CH-CO-CoA
(trans-2,3-Enoyl-CoA)

④ TER

NADPH

Reduction

R–CH$_2$–CH$_2$–CO-CoA
(elongated fatty acyl-CoA)

FIGURE 1 Reactions in the 2-carbon chain elongation of long-chain fatty acids in the ER. ELOVL, elongation of long-chain fatty acids; KAR, 3-ketoacyl-CoA reductase; HACD, hydroxyacyl-CoA dehydratase; TER, *trans*-2,3-enoyl-CoA reductase.

of the 2-carbon moiety. These enzymes, called ELOVLs, have been identified in both mice and humans. They share common evolutionarily conserved motifs despite low overall sequence similarity. None of the identified proteins have been crystallised or structurally determined, but structure predictions based on hydropathy plots indicate that they are polytopic proteins with 5–7 membrane spanning segments. The elongation activities are oriented toward the cytoplasmic surface of the ER membrane. Seven distinct fatty acid elongase subtypes (ELOVL1–7) are present in the mouse, rat and human genome. The ELOVLs are divided into three major classes: (1) isoforms that elongate SFA and MUFA (ELOVL1, 3, 6 and 7), (2) isoforms that are PUFA (ELOVL2 and 4) and (3) ELOVL5 isoform that uses a broader range of involved in substrates (16–22 carbons) (Figures 1 and 2).

ELOVL1 (previously named Ssc1) is involved in the biosynthesis of very-long-chain SFA. ELOVL1 is ubiquitously expressed in mammalian tissues with very high levels of expression found in the myelin of the central nervous system. Significant reductions of ELOVL1 activity and mRNA occur in the brain of Quaking and Jumpy mice, two models of myelination deficiency, suggesting that very-long-chain SFA and sphingolipid formation are essential in myelination (Guillou et al., 2010).

ELOVL2 (formerly denoted as Ssc2) is involved in the biosynthesis of the n-3 and n-6 series of highly unsaturated fatty acids (HUFA). This isoform elongates 20:4n6, 20:5n3 and 22:5n3 (Figure 4). Male *Elovl2*$^{-/-}$ mice are infertile, demonstrating that this elongase isoform is required for synthesis of 28:5n6

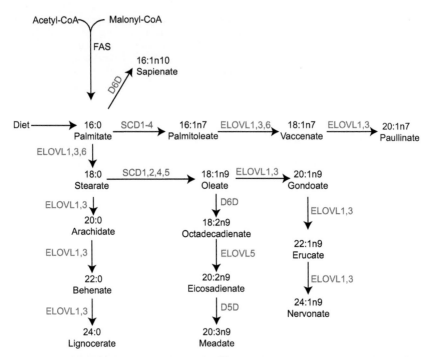

FIGURE 2   Synthesis of saturated and n-9 series of mono- and polyunsaturated fatty acids by desaturation and elongation from dietary and de novo synthesised palmitate. FAS, fatty acid synthase.

and 30:5n6 in testes, which are necessary for male fertility. ELOVL2 expression is highest in the testes and liver, but has also been found in white adipose tissue, kidney, brain and lung. *Elovl2−/−* mice also revealed that synthesis of docosahexaenoic acid (DHA) by ELOVL2 is an important regulator of hepatic lipid homeostasis. Interestingly, *Elovl2−/−* mice are protected from high-fat diet-induced hepatic steatosis; yet, with chow feeding, a simultaneous increase in hepatic fatty acid synthesis and oxidation is observed, yielding no net change in hepatic lipid content (Pauter et al., 2014).

ELOVL3 (originally cloned as Cig30) is a glycoprotein and was the first elongase identified in mammals. ELOVL3 expression is highly induced in brown adipose tissue on cold exposure, by administration of norepinephrine and by high calorie intake. ELOVL3 is also expressed in white adipose tissue, skin and liver. This enzyme elongates SFA and MUFA containing up to 24 carbons. The physiological role of ELOVL3 has been studied through the aid of mouse models with targeted disruption of *Elovl3* (Westerberg et al., 2006). ELOVL3-deficient mice show decreased activity in the elongation of C16–C22 saturated acyl-CoAs after short-term (3 days) cold exposure. The most prominent dysfunction in *Elovl3−/−* mice is evident in the skin. ELOVL3 is expressed in the sebocytes of the sebaceous glands, which produce the lubricant, sebum; consequently, *Elovl3−/−* mice have a sparse hair coat with hyperplastic sebaceous glands. In addition, defective water repulsion and structural abnormalities in the skin barrier are present in the ELOVL3-deficient mice, indicating the important role of long-chain FA in skin and hair function as well as skin barrier formation. $Elovl3^{-/-}$ mice accumulate gondoic acid (20:1n9) in the neutral lipids of hair, suggesting that gondonyl-CoA is a preferred substrate for ELOVL3.

Mutations in *Elovl4* were identified as being responsible for Stargardt's and autosomal dominant macular dystrophy. ELOVL4 is highly expressed in rod and cone photoreceptor cells in the retina, a tissue with high contents of DHA (22:6n3). ELOVL4 is involved in the biosynthesis of PUFA, but has also been reported to elongate very long-chain SFA (Figure 4). Homozygous $Elovl4^{-/-}$ mice die within a few hours of birth, highlighting the importance of ELOVL4 in growth and development. Lethality of $Elovl4^{-/-}$ mice is thought to be due to lack of skin PUFA that are essential for epidermal barrier function (Vasireddy et al., 2007). ELOVL4 dysfunction may also play a role in eye disease. Mice that overexpress a truncated version of human ELOVL4 exhibit photoreceptor degradation. This raises the possibility that either the lack of ELOVL4 function (PUFA synthesis) or the accumulation of the truncated form of the protein causes macular dystrophy (Karan et al., 2005).

ELOVL5 (formerly designated HELO1) was identified based on sequence homology with yeast Elo2p, and its function overlaps with that of Elovl2. ELOVL5 is capable of elongating C18–22 carbons of PUFA but not those with more than 22 carbons (Figure 4). Although expression of ELOVL5 is found in most human tissues, it is highly expressed in the testes and adrenal gland, where significant amounts of adrenic acid (22:4n6) and docosapentaenoic acid (22:5n3) are found. $Elovl5^{-/-}$ mice develop hepatic steatosis due to decreased PUFA abundance and increased nuclear sterol regulatory element-binding protein (SREBP) levels (Moon et al., 2009). Interestingly, fertility is altered in female, but not male, $Elovl5^{-/-}$ mice.

ELOVL6 (formerly designated LCE) was identified as an SREBP target gene and is highly expressed in lipogenic tissues, such as liver and adipose tissue. It catalyses the elongation of saturated and monounsaturated acyl-CoAs with 12–16 carbons. $Elovl6^{-/-}$ mice exhibit hepatic steatosis but remain insulin sensitive (Matsuzaka et al., 2007). This dissociation of obesity and insulin resistance is especially intriguing because $Elovl6^{-/-}$ mice have increased levels of 16:0, a saturated fatty acid proposed to promote insulin resistance and inflammation.

ELOVL7 is the most recently identified elongase and is expressed in prostate, kidney, pancreas and adrenal gland (Tamura et al., 2009). ELOVL7 expression is elevated in pancreatic cancer cells, and knockdown of ELOVL7 decreases cancer cell growth. The substrate specificity of ELOVL7 is not clear. A study using a pancreatic cancer cell line suggests that ELOVL7 elongates saturated fatty acid chains 18–22 carbons in length; another study investigated the activity of purified ELOVL7 and reported the highest activity with 18:3n3 and 18:3n6 (Naganuma et al., 2011).

2.1.2 Ketoacyl-CoA Reductase

Ketoacyl-CoA reductase (KAR) utilises NADPH (in preference to NADH) to form β-hydroxy acyl-CoA (Figure 1). The flow of reducing equivalents from NADPH or NADH to the β-keto reductase appears to involve two other ER membrane proteins: cytochrome b_5 and cytochrome p450 reductase. KAR was identified along with *trans*-2,3 enoyl-CoA reductase (described below) using a homology search with the yeast enzyme TSC13. Unlike the ELOVL family of enzymes, only a single KAR has been identified in mammals (Moon and Horton, 2003). The cDNA encodes a protein that is predicted to contain 312 amino acids with a conserved dilysine ER retention motif at the C-terminus and four putative transmembrane domains. The enzyme is present in all human and mouse tissues with the highest expression in tissues that are directly involved in fatty acid metabolism.

2.1.3 β-Hydroxyacyl-CoA Dehydratase

β-Hydroxyacyl-CoA dehydratase (HACD) catalyses the third reaction in chain elongation: the dehydration of 3-hydroxyacyl-CoA to enoyl-CoA (Figure 1). Four mammalian HACD proteins (HACD1–4) have been identified. These ER-resident enzymes have unique tissue expression patterns and interact with ELOVL enzymes (Ikeda et al., 2008). HACD1 is expressed in heart and skeletal muscle, HACD2 and 3 are expressed ubiquitously and HACD4 is expressed in the small intestine, placenta, lung and peripheral blood leukocytes.

2.1.4 trans-2,3-Enoyl CoA Reductase

trans-2,3-Enoyl-CoA reductase catalyses the final step of the chain elongation pathway and requires NADPH as electron donor. The reductase protein is presumed to be 308 amino acids in length with five transmembrane domains. The tissue distribution of *trans*-2,3-enoyl-CoA reductase mirrors that of KAR (Moon and Horton, 2003).

2.2 Mitochondrial Fatty Acid Elongation

Fatty acid elongation has also been reported to occur in mitochondria. Mitochondrial elongation entails the successive addition and reduction of acetyl units in a reversal of fatty acid oxidation. Although fatty acid β-oxidation and mitochondrial chain elongation occur in the same organelle, reversal of FAD-dependent acyl-CoA dehydrogenase of β-oxidation is not energetically favourable. This reaction is substituted by NADPH-dependent enoyl-CoA reductase, to produce overall negative free energy for the sequence (Seubert and Podack, 1973). pH optima studies suggest that enoyl-CoA reductase from liver mitochondria is distinct from the ER reductase. The function of the mitochondrial elongation system is unclear but has been proposed to participate in the biogenesis of mitochondrial membranes and in the transfer of reducing equivalents between carbohydrates and lipids.

2.3 Peroxisomal Fatty Acid Elongation

Peroxisomal fatty acid elongation has been proposed to function similar to the mitochondrial fatty acid elongation system, using acetyl-CoA as 2-carbon unit substrates and occurring as a reversal of β-oxidation. Peroxisomal *trans*-2-enoyl-CoA reductase replaces acyl-CoA dehydrogenase to enable a thermodynamically favourable reaction (Das et al., 2000). Specific roles for fatty acid elongation in peroxisomes are not well-defined.

3. DESATURATION OF LONG-CHAIN FATTY ACID IN MAMMALS

3.1 Δ9 Desaturase

The Stearoyl-CoA desaturase (SCD) proteins reside in the ER where they catalyse the biosynthesis of MUFA from SFA that are either synthesised de novo or derived from the diet (Figure 2). The preferred substrates for SCD are palmitoyl- and stearoyl-CoA, which are converted to palmitoleoyl- and oleoyl-CoA, respectively. The MUFA synthesised by SCD are then used as substrates for the synthesis of various lipid classes including PL, TAG and cholesteryl esters (CE).

The presence of a Δ9 desaturation system in mammals was first reported in 1960. Four genes encoding Δ9 desaturases (*Scd1*, *Scd2*, *Scd3* and *Scd4*) have been cloned in mice, and two isoforms (*SCD1* and *SCD5*) are present in humans. All mammalian Δ9 desaturases contain four transmembrane domains and three regions of conserved histidine motifs, which coordinate the diiron centre and are essential for catalytic activity. In contrast to their structural similarity, the substrate specificity, tissue distribution and physiological importance of the desaturases are different, as described below.

The mechanism of desaturation involves NADPH, the flavoprotein cytochrome b_5 reductase, the electron acceptor cytochrome b_5 and molecular oxygen. SCD introduce a single double bond into a spectrum of methylene-interrupted fatty acyl-CoA substrates (Figure 3). The desaturation of a fatty acid is an oxidation reaction that requires molecular oxygen and two electrons; however, oxygen itself is not incorporated into the fatty acid chain but is released in the form of water (Figure 3).

All mouse Δ9 desaturase genes are localised on chromosome 19 and form a cluster within the 200 kb region. Mouse *Scd1* was first identified in differentiated 3T3-L1 adipocytes and is ubiquitously expressed with the highest expression in lipogenic tissues, including liver, adipose tissue and sebaceous glands. SCD1 desaturates saturated fatty acyl-CoA molecules between 12 and 19 carbons. $Scd1^{-/-}$ mice are protected from diet- and genetic-induced obesity and have improved glucose sensitivity. SCD1 deficiency also yields dramatic sebocyte dysfunction. Mouse models with tissue-specific deletion of SCD1 demonstrate the role of SCD1 in liver and skin in regulating energy balance, insulin sensitivity and de novo lipogenesis. The function of MUFA is further discussed in Section 3.4.1.

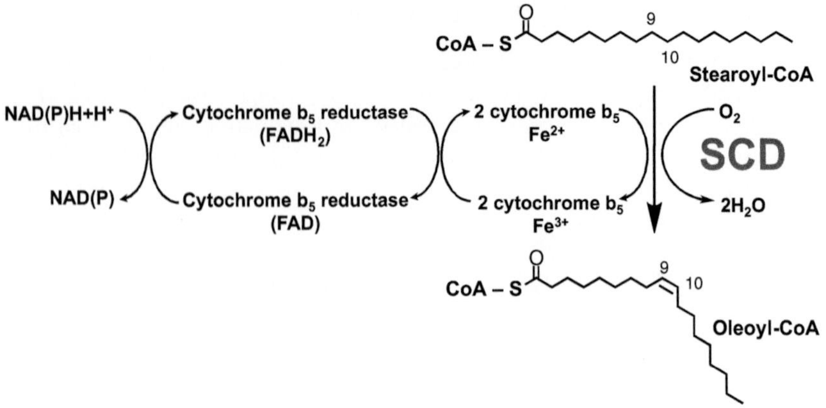

FIGURE 3   Desaturation of stearoyl-CoA by stearoyl-CoA desaturase (SCD). The reaction is dependent on O_2 and electron transfer between NAD(P)H, cytochrome b_5 reductase and cytochrome b_5.

Scd2 was the second stearoyl-CoA desaturase gene cloned from differentiated mouse 3T3-L1 preadipocytes (Cantley et al., 2013). Similar to SCD1, SCD2 mainly converts palmitoyl-CoA and stearoyl-CoA to palmitoleoyl-CoA and oleoyl-CoA, respectively (Figure 2). SCD2 is predominantly expressed in the brain and is developmentally induced during the neonatal myelination period. *Scd2* mRNA is expressed to a lesser extent in kidney, spleen and lung where it is induced in response to a high-carbohydrate diet. In addition, SCD2 mRNA is expressed in pancreatic β cells and is downregulated during lymphocyte development. *Scd2* expression in liver is high in the embryo and neonate but decreases in adult mice whereas *Scd1* expression is induced after weaning. Consistent with the expression of SCD2 occurring in the early stages of life, neonatal *Scd2*$^{-/-}$ mice have decreased MUFA and TAG content in the liver and skin. In addition, *Scd2*$^{-/-}$ neonatal mice have a skin permeability barrier defect with decreased levels of ceramides including acylceramides. It is now clear that SCD1 expression is crucial in adult mice, whereas SCD2 is important in the synthesis of MUFA that are required for maintaining normal epidermal barrier function and lipid biosynthesis during early liver and skin development.

SCD3 was discovered in 2001 and is expressed in mature sebocytes present in exocrine glands, such as sebaceous, meibomian, preputial and harderian glands (Bond and Ntambi, 2013). SCD3 expression has also been detected in white and brown adipose tissue. Compared to other mammalian Δ9 desaturases, SCD3 mainly utilises saturated fatty acyl-CoAs with 14–16 carbons (Figure 2). SCD3 is not able to desaturate fatty acyl-CoAs with more than 17 carbons including stearoyl-CoA. For this reason, the name stearoyl-CoA desaturase-3 is rather a misnomer, and this Δ9 desaturase has also been referred to as palmitoyl-CoA desaturase (PCD).

Three SCD isoforms – SCD1, SCD2 and SCD3 – are expressed in skin, underscoring the importance of MUFA production on skin function. However, their expression patterns are not redundant; in situ hybridisation demonstrates that SCD1 and SCD3 are expressed in undifferentiated and differentiated sebocytes, respectively, whereas SCD2 is expressed in keratinocytes (Bond and Ntambi, 2013). $Scd1^{-/-}$ mice exhibit severe sebocyte dysfunction, resulting in alopecia, atrophy of the sebaceous gland and a decrease in sebum production. Therefore, *Scd1* gene expression is required for proper differentiation of sebocytes and hair follicle development.

SCD4 was identified in 2004 and is expressed in the heart in mice. SCD4 is able to desaturate 16:0-CoA and 18:0-CoA to 16:1n7-CoA and 18:1n9-CoA, respectively (Figure 2); however, the desaturase activity is significantly lower than that of other SCD isoforms (Miyazaki et al., 2006).

Much less is known about the human SCD isoforms. Human *SCD1* and *SCD5* are located on chromosomes 10 and 4, respectively. Another *SCD* pseudogene, which is not transcribed, is located on chromosome 17. Phylogenetic analysis suggests that *SCD1* is the human orthologue to mouse *Scd1*. Human SCD1 is expressed ubiquitously and has been shown to positively correlate with skeletal muscle lipids and obesity (Hulver et al., 2005). Also, four single-nucleotide polymorphisms in SCD1 were found to be associated with decreased BMI and abdominal adiposity, as well as improved insulin sensitivity (Warensjo et al., 2007). Orthologues of *SCD5* (formerly called *ACOD4* and *SCD2*) have not been found in genomes of rodents but have been identified in other primates. However, despite distinct evolutionary origins, human SCD5 and mouse SCD2 share similar patterns of tissue expression, as both are highly expressed in the brain and pancreas. Mutations in human SCD5 are associated with cleft lip. SCD5 is highly expressed in the human brain and pancreas (Cantley et al., 2013).

3.2 Δ5 and Δ6 Desaturases

Δ5 Desaturase (D5D, or FADS1) and Δ6 desaturase (D6D, or FADS2) are membrane-bound desaturases catalysing the synthesis of n-3 and n-6 series of HUFA from dietary linoleic acid (18:2n6) and α-linolenic acid (18:3n3) (Figure 4) (Nakamura and Nara, 2004). Human *D5D* and *D6D* genes are localised on chromosome 11 and form a cluster within the 100 kb region of the chromosome. The mouse homologues are located on chromosome 19 and also occur as a cluster with a similar exon/intron organisation, suggesting that these desaturases arose evolutionarily by gene duplication. The *D5D* and *D6D* genes both consist of 12 exons and 11 introns spanning 17.2 and 39.1 kb regions, respectively. The proximity of their promoters suggests that the transcription of the *D5D* and *D6D* genes is coordinately controlled by common regulatory sequence within the 11 kb region (Nakamura and Nara, 2004).

The two enzymes share structural characteristics. Both enzymes have an N-terminal cytochrome b_5 domain-carrying heme-binding motifs, two membrane-spanning domains and three histidine boxes characteristic of membrane desaturases. However, the first His residue of the third His-box is replaced with glutamine (QXXHH instead of HXXHH). The glutamine residue is essential for desaturase activity because mutations of this amino acid residue to histidine

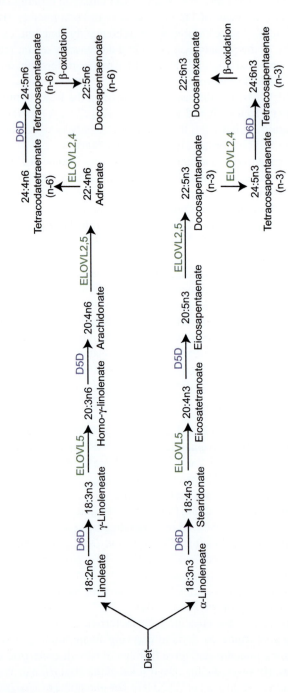

FIGURE 4 Synthesis of n-6 and n-3 series of polyunsaturated fatty acids from dietary essential fatty acids linoleic and α-linolenic acids, respectively, by elongation and desaturation.

or isoleucine abolish the activity of the enzyme (Nakamura and Nara, 2004). D6D and D5D are widely expressed in human tissues with the highest levels in liver. D6D catalyses the Δ6 desaturation of PUFA of both 18 carbons and 24 carbons. This desaturase converts linoleic acid (18:2n6) and α-linolenic acid (18:3n3) to γ-linolenic acid (18:3n6) and stearidonic acid (18:4n3), respectively. ELOVL5 then elongates these FA by two carbons to synthesise 20:3n6 and 20:4n3. After desaturation and elongation by D6D and the appropriate elongase, respectively, D5D introduces another double bond at the 5-position of 20-carbon FA 20:3n6 and 20:4n3 to yield 20:4n6 and 20:5n3 (Figure 4).

In addition, D6D is able to desaturate palmitate (16:0) to sapienic acid (16:1n10), a major fatty acid of human sebum. In fact, D6D is highly expressed in the sebaceous gland in human skin. Thus, mammalian D6D, unlike other desaturases, can desaturate a wide range of FA with 16–24 carbons.

The characterisation of $D6d^{-/-}$ mice reveals the critical role of HUFA in fertility, as D6D deficiency impairs sperminogenesis, resulting in infertility. Stroud et al. also reported that $D6d^{-/-}$ mice develop intestinal and dermal ulceration (Stroud et al., 2009).

3.3 Fatty Acid Desaturase 3

Fatty acid desaturase 3 (*FADS3*) was cloned in 2000. Like human *FADS1* and *FADS2*, human *FADS3* is located on chromosome 11 and has 12 exons and 11 introns. Human *FADS3* has an open reading frame of 1338 bp encoding 445 amino acids, whereas the mouse gene possesses a 1350 bp open reading frame encoding 449 amino acids (Nakamura and Nara, 2004). FADS3 mRNA has been shown to be expressed ubiquitously in human, rat and baboon. The predicted amino acid sequence contains all of the conserved structural features of D6D, including three histidine boxes that are found in carboxyl-end desaturases. FADS3 shares high sequence identity with D5D (52%) and D6D (62%). However, despite their apparent sequence similarity, FADS3, D5D and D6D appear to have unique functions and FADS3 does not catalyse Δ5, Δ6 or Δ9 desaturation. Recently, rat FADS3 was reported to be a *trans*-vaccenate Δ13 desaturase and synthesises *trans* 11, *cis*13-18:2 (Rioux et al., 2013).

3.4 Functions of Fatty Acids Synthesised by Δ9, Δ6 and Δ5 Desaturases

3.4.1 Monounsaturated Fatty Acids (n-9)

As shown in Figure 2, mammals have all enzymes required for the synthesis of MUFA from acetyl-CoA. The roles of MUFA are diverse and crucial in living organisms. Palmitoleic acid and oleic acid are the major MUFA of TAG, CE, wax esters (WE) and membrane PL. The ratio of SFA to MUFA contributes to membrane fluidity and changes in the acyl chain composition of CE and TAG affect lipoprotein metabolism. Apart from being the components of more complex lipids, MUFA are mediators of signal transduction, cellular differentiation and metabolic homeostasis. Regulation of MUFA levels has the potential to affect a variety of key physiological variables, which include insulin sensitivity, metabolic rate, adiposity, atherosclerosis, cancer and obesity.

Understanding the physiological role of MUFA has been advanced by studies in a mouse model with a targeted disruption of the *Scd1* gene. The phenotypes of *Scd1*$^{-/-}$ mice suggest that while MUFA are essential for skin function, decreased MUFA abundance is associated with beneficial metabolic phenotypes, such as reduced adiposity and increased insulin sensitivity. *Scd1*$^{-/-}$ mice exhibit cutaneous abnormalities with atrophic sebaceous glands and narrow eye fissure with atrophic meibomian glands, suggesting an important role of MUFA in skin and eyelid function (Sampath and Ntambi, 2011). The major purpose of these glands is to secrete complex lipid lubricants, termed sebum and mebum, respectively. These lubricants contain WE, TAG and CE and prevent the evaporation of moisture from the skin and the eyeball. *Scd1*$^{-/-}$ mice are deficient in TAG, CE and WE, which contain MUFA as acyl chains. In addition to a decrease in tissue neutral lipids, *Scd1* deficiency results in resistance to diet-induced obesity (Sampath and Ntambi, 2011).

Tissue-specific knockout models have elucidated the contribution of SCD in skin, liver and adipose tissue toward regulating whole body energy homeostasis. Skin-specific deletion of SCD1 (SKO) recapitulates the hypermetabolic phenotype of global *Scd1*$^{-/-}$ mice and protects mice from high-fat diet-induced obesity. SKO mice also have sebaceous gland atrophy. This mouse model illustrates the communication between skin and peripheral tissues and how this crosstalk regulates energy homeostasis. Peripheral tissues, such as liver and white adipose tissue, exhibit increased expression of genes involved in energy expenditure. It is not known if increased energy expenditure stems from disruption of skin insulation, resulting in heat loss, or from a circulating factor, secreted by skin that acts on peripheral tissues (Sampath and Ntambi, 2011).

Liver-specific *Scd1*$^{-/-}$ mice are protected from high-carbohydrate – but not high-fat – diet-induced obesity (Sampath and Ntambi, 2011). This phenotype is due to impaired hepatic de novo lipogenesis, driven by reductions in nuclear SREBP-1c and ChREBP (carbohydrate response element-binding protein). Liver-specific knockout of *Scd1* also yields hypoglycaemia with high-sucrose feeding because of impaired gluconeogenesis. Interestingly, oleate supplementation normalised these phenotypes. Surprisingly, the combined deletion of SCD1 in adipose tissue and liver does not protect mice from high-fat diet-induced obesity (Flowers et al., 2012). This lack of protection may stem from the presence of other SCD isoforms in adipose tissue. Another study utilised antisense oligonucleotide treatment to inhibit SCD1 expression in adipose tissue and liver and observed protection from diet-induced obesity (Jiang et al., 2005). The discrepancy between studies may stem from off-target effects of the ASO treatment. Namely, ASO may be inhibiting other *Scd* genes in adipose tissue or could also act on skin *Scd1*, which would facilitate obesity protection as described above.

An SCD product, palmitoleate, has recently gained significant attention for its putative function as a beneficial 'lipokine'. In 2009, this MUFA was reported to improve glucose sensitivity and repress hepatic lipogenesis (Cao et al., 2008). However, the effects of 16:1n7 are controversial and subsequent studies do not confirm the beneficial effects on this fatty acid. Guo et al. reported that palmitoleate induced hepatic steatosis but represses hepatic inflammation (Guo et al., 2012). The inconsistencies between these studies may stem from the nature in which palmitoleate levels were modulated. Cao et al. achieved increased palmitoleate through HFD feeding and deletion of FABP4 and FABP5, Guo et al. used dietary supplementation of palmitoleate, and Ntambi and colleagues utilised genetic manipulation of a palmitoleate-synthesising enzyme, SCD1. Other studies have demonstrated additional effects of palmitoleate in cellular signalling. Palmitoleate, produced by SCD, is covalently attached to Wnt proteins; this post-translational modification is mediated by porcupine and is required for Wnt protein secretion and signalling (Rios-Esteves and Resh, 2013) (see Chapter 13). Palmitoleate has also been shown to increase lipolysis in white adipocytes through induction of peroxisome proliferator activated receptors (PPAR) α transcriptional activity (Bolsoni-Lopes et al., 2013).

To determine the differential metabolic roles of palmitoleate and oleate, liver transgenic mouse models expressing human SCD5, which synthesises oleate, and mouse SCD3, which synthesises palmitoleate, were created. This study determined that hepatic-derived oleate represses both de novo lipogenesis and fatty acid oxidation in epididymal white adipose tissue (Burhans et al., 2015).

3.4.2 Polyunsaturated Fatty Acids (n-3 and n-6)

Requirements for PUFA cannot be met by de novo metabolic processes in mammalian tissues. Animals are dependent on plants (or insects) for providing double bonds in the Δ12 and Δ15 positions of the two major precursors of the n-6 and n-3 fatty acids, linoleic and α-linolenic acids (Figure 5). These two FA

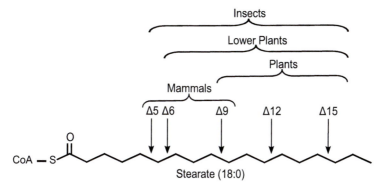

FIGURE 5 Positions of fatty acyl chain desaturation by enzymes of animals, plants, insects and lower plants.

therefore are called essential FA. Linoleic and α-linolenic acid are parents to the n-6 and n-3 PUFA families, respectively, which have important physiological and structural roles. PUFA have diverse functions critical for many biological processes in numerous tissues; this commentary only provides a brief summary of the physiological roles of PUFA.

The physiological functions of PUFA have been elucidated through rodent studies of essential fatty acid deficiency and through mouse models modulating activity of PUFA-synthesising enzymes. Essential fatty acid deficiency in rodents causes dry skin, dermatitis and massive water loss through the skin.

Arachidonic acid (20:4n6) is one PUFA synthesised by the D6D/D5D pathway that has important physiological functions (Figure 4). In many tissues and cell types, 20:4n-6 is esterified to the *sn*-2 position of membrane PL, and is used for the eicosanoid-mediated signalling to perform specialised cell functions. Arachidonic acid is hydrolysed from the PL by phospholipases prior to enzymatic conversion into eicosanoids. Eicosanoids are autocrine/paracrine hormones that mediate a variety of localised reactions, such as inflammation, homeostasis and protection of digestive tract epithelium (see Chapter 9). In a study of D6D-deficient human subjects, arachidonic acid supplementation reversed symptoms of food intolerance and growth retardation, which supports the essential role of eicosanoids in the protection of digestive tract mucosa in humans (Nwankwo et al., 2003).

DHA (22:6n3) is abundant in excitable membranes in the retina and brain, particularly in PL of the rod outer segment of retina and of synaptic vesicles. DHA is proposed to function in retina and brain by facilitating transport of 11-*cis*-retinal and by altering the stability of rhodopsin. Consistent with DHA's roles in retinal and brain molecular function, DHA supplementation has been reported to improve cognitive skills in the elderly and in infants, although this effect is controversial. DHA also appears to be essential for fertility, as DHA supplementation restores impaired spermatogenesis in male $D6d^{-/-}$ mice (Roqueta-Rivera et al., 2010).

Omega-3 FA are touted for their anti-inflammatory and insulin-sensitising properties. Oxygenated metabolites of DHA, called resolvins and protectins, are proposed to be anti-inflammatory and facilitate cell survival on oxidative stress (Serhan and Petasis, 2011). In adipose tissue, omega-3 FA promote insulin sensitivity and alleviate chronic tissue inflammation by binding to G-protein coupled receptors, such as GPR120 (Kalupahana et al., 2011). Studies suggest that DHA supplementation protects against many inflammatory diseases such as rheumatoid arthritis, cystic fibrosis, ulcerative colitis, asthma, cancer and cardiovascular disease (Calder, 2006); however, the efficacy of DHA in ameliorating these conditions is controversial.

PUFA are important regulators of signalling pathways. Additionally, PUFA released from membrane PL by agonist-induced stimulation of phospholipases are involved in signal transduction through modulation of activities of cellular proteins including isoforms of protein kinase C and in translocation of specific enzymes to membranes (see Chapter 9). Specific types of FA modulate many of these diverse functions. Both n-3 and n-6 PUFA regulate gene expression by affecting the activities of a number of transcription factors. As noted below, PUFA regulate the transcription of genes involved in desaturation and elongation.

4. TRANSCRIPTIONAL REGULATION OF DESATURASES AND ELONGASES

4.1 Sterol Regulatory Element Binding Proteins

SREBPs belong to the basic-helix-loop-helix-leucine zipper (bHLH-Zip) family of transcription factors (Horton et al., 2002). Three isoforms of SREBP have been identified in mammalian cells: SREBP-1a, SREBP-1c and SREBP-2. SREBP-1a activates all SREBP target genes involved in cholesterol and fatty acid metabolism, whereas SREBP-2 predominantly activates transcription of genes involved in cholesterol synthesis (see Chapter 11). SREBP-1c is the main isoform in liver and preferentially controls genes involved in de novo lipogenesis, including Δ5, Δ6 and Δ9 desaturase and ELOVL6 (Figure 6). SREBP-1c is activated by insulin and the presence of certain FA.

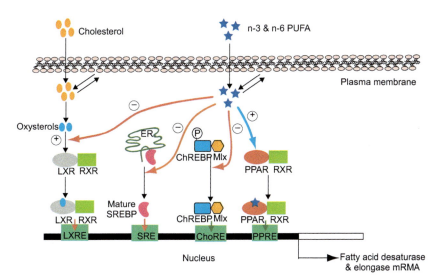

FIGURE 6 Coordinated transcriptional regulation of fatty acid desaturases and elongases in mammals. PUFA, polyunsaturated fatty acids; +, stimulation; −, inhibition; LXR, liver X receptor; RXR, retinoid X receptor; SREBP, sterol regulatory element-binding protein; ChREBP, carbohydrate response element-binding protein; Mlx, Max-like receptor; PPARα, peroxisome proliferator activated receptor α; LXRE, liver X receptor response element; SRE, sterol response element; ChoRE, carbohydrate response element; PPRE, peroxisome proliferator response element; P, phosphate group.

Dietary long-chain saturated FA regulate hepatic lipogenic effects through SREBP-1c and PGC-1β (Lin et al., 2005). SREBP-1c on its own is a very weak transcriptional activator and requires the assistance of coactivators. With regard to induction of SREBP-1c through 18:0, Ntambi and colleagues have shown that the de novo conversion of 18:0 to 18:1n9, via SCD, is required for the induction of PGC-1β (Sampath et al., 2007). However, the mechanism by which 18:1n9 induces PGC-1β is unknown. In contrast to long-chain FA, short-chain FA and PUFA are implicated as anti-lipogenic (Sun et al., 2013).

One or two sterol response elements are present in promoters of the *Scd1*, *Scd2*, *D6d*, *D5d* and *Elovl6* genes. A series of studies demonstrated that SREBP-1c is involved in the PUFA-mediated repression of transcription of the desaturases and *Elovl6* in the liver (Caputo et al., 2011). Several mechanisms underlying PUFA repression of SREBP-1c have been proposed: (1) blocking the proteolytic processing of SREBP-1c, (2) reducing the stability of SREBP-1c mRNA and (3) acting as antagonists of the liver X receptor (LXR), which activates SREBP-1c transcription. Specifically, n-3 PFA suppress insulin-induced SREBP-1c transcription via reduced transactivating capacity of LXRα (Howell III et al., 2009).

4.2 Liver X Receptors

The liver X receptors, LXRα (NR1H3) and LXRβ (NR1H2), belong to the nuclear hormone receptor superfamily. As ligand-activated transcription factors, LXRs regulate cholesterol homeostasis and lipoprotein remodelling at both the organismal and the cellular levels. A synthetic LXR agonist T0901317 was developed as an anti-atherogenic agent due to its ability to increase the level of high-density lipoprotein. LXRs regulate hepatic fatty acid biosynthesis in both an SREBP-1c-dependent and -independent manner. The accumulation of lipids and increased lipogenesis by T0901317 are primarily explained by increased expression of SREBP-1c, whose promoter contains two LXR response elements (LXRE). SREBP-1c downstream target genes include the fatty acid biosynthetic genes *Acc*, *Fasn* and *Scd1*, which contain functional LXREs and can be directly regulated by LXRs.

The expression of LXR in liver is activated by insulin and in turn increases the mRNA abundance of SREBP-1c and ChREBP (Faulds et al., 2010). Furthermore, lipogenic gene expression was abolished in insulin-injected LXRα/β double knockout mice. The mechanisms of insulin-activated LXR expression are unknown but may involve an endogenous ligand for LXRα/β or post-translational modifications. PUFA repress LXR transcriptional activity by inhibiting oxysterol binding (Figure 6).

An LXR response element is likely also present in the promoters of *D6d*, *D5d* and *Elovl6*. Reduced expressions of SREBP-1c and its target genes, including *Fasn* and *Scd1*, are detected in *Lxrα*[−/−] and *Lxrα/Lxrβ*[−/−] mice, but not in *Lxrβ*[−/−] mice. Meanwhile, the LXR agonist T0901317 promotes hepatic expression of *Srebp1c*, *Fasn* and *Scd1* in both wild-type and *Lxrβ*$^{-/-}$ mice, but not *Lxrα/Lxrβ*$^{-/-}$ or *Lxrα*$^{-/-}$ mice, which is consistent with the critical lipogenic role of LXRα (Faulds et al., 2010).

4.3 Peroxisome Proliferator Activated Receptors

PPARs are transcription factors belonging to the superfamily of nuclear receptors. Three isoforms (α, δ and γ) have been identified. PPARα, when activated, promotes fatty acid oxidation, ketone body synthesis and glucose sparing. PPARδ is ubiquitously expressed and may be involved in the regulation of cholesterol and lipid metabolism. PPARγ is mainly expressed in adipose tissue, where it is involved in the induction of adipogenesis and stimulates triglyceride accumulation. When activated, PPARs bind to the peroxisome proliferator response element (PPRE) of target genes, which in turn activates their transcription. PPREs are present in the promoters of SCD1 and D6D. However, it is important to note that although all PPARs can active SCD1 and D6D alone, they can have differing effects on the overall de novo lipogenesis program.

Induction of SCD1 by the synthetic PPARα agonist Wy14643 was masked in the liver of PPARα null mice. PUFA are considered to be one of the endogenous PPARα ligands (Figure 6). However, PUFA repression of lipogenic genes including *Scd1* still occurs in the liver of *Pparα*$^{-/-}$ mice, demonstrating that the repression of desaturases by PUFA is not mediated by PPARα. PPARα also strongly activates D6D gene expression and plays a crucial role in the feedback regulation of the synthesis of HUFA. D6D mRNA abundance is not increased in PPARα null mice fed essential fatty acid-deficient diets, although nuclear SREBP-1c is elevated in both *Pparα*$^{-/-}$ and wild-type mice. The role that PPARα plays in the regulation of elongases has not been well established.

PPARγ controls the expression of genes required to maintain the phenotype of mature adipocytes; therefore, a loss in the activity or expression of PPARγ leads to a loss in adipocyte function. The inhibition of adipogenesis caused by SCD2 depletion was associated with a decrease in PPARγ mRNA and protein; however, in the mature adipocytes loss of SCD2 only diminished PPARγ protein levels. Additionally, SCD2 does not seem to be regulating the production of a PPARγ ligand, since the inhibition of adipogenesis by SCD2 depletion could not be restored by the addition of the PPARγ-specific ligand, rosiglitazone (Christianson et al., 2008).

4.4 Carbohydrate Response Element Binding Protein

Like SREBP, ChREBP is a transcription factor of the bHLH family and is involved in glucose and lipid metabolism. There are two identified ChREBP isoforms, ChREBPα and ChREBPβ, and they are transcribed from a single gene using alternative promoters. Activation of ChREBPα induces expression of the more potent transcription factor ChREBPβ (Herman et al., 2012). ChREBP is regulated at the post-translational level; glucose metabolism stimulates the dephosphorylation and subsequent nuclear translocation of ChREBP. ChREBP is also regulated at the transcriptional level and is a direct target of LXR and ChREBP itself. However, glucose-mediated activation is still required for increased ChREBP gene expression to yield active, nuclear ChREBP (Denechaud et al., 2008).

Although a carbohydrate response element (ChoRE) has not yet been identified in the promoter of any desaturase or elongase, SCD1 is likely a target of ChREBP; in $Chrebp^{-/-}$ mice, $SCD1$ mRNA abundance is significantly lower than in wild-type mice. The decreased expression of SCD1 is hypothesised to be a direct effect of ChREBP rather than SREBP-1c, because no notable difference was detected in SREBP-1 or SREBP-2 expression or processing (Iizuka et al., 2004).

Recently it was shown that PUFA suppress the ability of ChREBP to activate its target genes, liver-type pyruvate kinase and fatty acid synthase, by decreasing ChREBP mRNA stability and inhibiting ChREBP nuclear translocation (Figure 6). Additionally, PUFA also decreased Max-like receptor expression. These data suggest that reduced nuclear abundance of the ChREBP/Max-like receptor complex likely accounts for PUFA-mediated repression of desaturases and elongases (Dentin et al., 2005).

In addition to glucose, insulin has been shown to activate ChREBP (Laplante and Sabatini, 2010). The ChREBP gene promoter contains a Pit-1, Oct-1/Oct-2 and Unc-86 (POU) homeodomain protein biding motif (ATGCTAAT) within the proximal promoter region of the ChREBP gene. Additionally, ChREBP promoter activity and endogenous ChREBP expression are negatively regulated by the POU homeodomain protein octamer transcription factor-1 (Oct-1). In contrast to this negative regulation, insulin was shown to stimulate ChREBP expression [68]. The stimulatory effect of insulin on ChREBP was dependent on the presence of the POU protein-binding site in the promoter region, and insulin treatment reduced Oct-1 expression level.

5. SUMMARY AND FUTURE DIRECTIONS

Lipids are no longer considered to be just a static store of energy. Rather, these molecules are widely recognised as modulators of whole body lipid and glucose metabolism. A common function of FA desaturases and elongases is the production of long-chain FA. These FA maintain the physical attributes of membrane PL and stored TAG. In addition, FA and their metabolites are used by cells for other functions, especially cell signalling. The regulation of the desaturases and elongases demonstrates that they have shared, as well as distinct, mechanisms. Recent progress in elucidating regulatory mechanisms of gene expression of desaturases and elongases presents exciting examples of the sophisticated control of mammalian gene transcription that is achieved by a combination of multiple transcription factors such as SREBP-1c, LXR, PPARγ, PPARα and ChREBP. With new knowledge on the regulation of transcription and the increasing prevalence of disorders of lipid metabolism, there is renewed interest in understanding the physiological roles of long-chain FA. Further research into this topic will be key to understanding the nuanced functions of long-chain FA and their contribution to disease etiology and prevention. Current studies explore the distinct signalling properties of specific FA – in their nonesterified form or as a constituent of a complex lipid – and how these FA-mediated molecular events facilitate tissue crosstalk and regulate whole body lipid metabolism and energy expenditure.

REFERENCES

Bolsoni-Lopes, A., Festuccia, W.T., Farias, T.S., Chimin, P., Torres-Leal, F.L., Derogis, P.B., De Andrade, P.B., Miyamoto, S., Lima, F.B., Curi, R., Alonso-Vale, M.I., 2013. Palmitoleic acid (n-7) increases white adipocyte lipolysis and lipase content in a PPARalpha-dependent manner. Am. J. Physiol. Endocrinol. Metab. 305, E1093–E1102.

Bond, L.M., Ntambi, J.M., 2013. Stearoyl-CoA desaturase isoforms 3 and 4: Avenues for tissue-specific Δ9 desaturase activity. In: Ntambi, J.M. (Ed.), Stearoyl-CoA Desaturase Genes in Lipid Metabolism. Springer, New York.

Burhans, M.S., Flowers, M.T., Harrington, K.R., Bond, L.M., Guo, C.A., Anderson, R.M., Ntambi, J.M., 2015. Hepatic oleate regulates adipose tissue lipogenesis and fatty acid oxidation. J. Lipid Res. 56, 304–318.

Calder, P.C., 2006. n-3 polyunsaturated fatty acids, inflammation, and inflammatory diseases. Am. J. Clin. Nutr. 83, 1505S–1519S.

Cantley, J.L., O'neill, L.M., Ntambi, J.M., Czech, M.P., 2013. The cellular function of stearoyl-CoA Desaturase-2 in development and differentiation. In: Ntambi, J.M. (Ed.), Stearoyl-CoA Desaturase Genes in Lipid Metabolism. Springer, New York.

Cao, H., Gerhold, K., Mayers, J.R., Wiest, M.M., Watkins, S.M., Hotamisligil, G.S., 2008. Identification of a lipokine, a lipid hormone linking adipose tissue to systemic metabolism. Cell 134, 933–944.

Caputo, M., Zirpoli, H., Torino, G., Tecce, M.F., 2011. Selective regulation of UGT1A1 and SREBP-1c mRNA expression by docosahexaenoic, eicosapentaenoic, and arachidonic acids. J. Cell. Physiol. 226, 187–193.

Christianson, J.L., Nicoloro, S., Straubhaar, J., Czech, M.P., 2008. Stearoyl-CoA desaturase 2 is required for peroxisome proliferator-activated receptor γ expression and adipogenesis in cultured 3T3-L1 cells. J. Biol. Chem. 283, 2906–2916.

Das, A.K., Uhler, M.D., Hajra, A.K., 2000. Molecular cloning and expression of mammalian per-oxisomal trans-2-enoyl-coenzyme A reductase cDNAs. J. Biol. Chem. 275, 24333–24340.

Denechaud, P.-D., Bossard, P., Lobaccaro, J.-M.A., Millatt, L., Staels, B., Girard, J., Postic, C., 2008. ChREBP, but not LXRs, is required for the induction of glucose-regulated genes in mouse liver. J. Clin. Invest. 118, 956–964.

Dentin, R., Benhamed, F., Pegorier, J.P., Foufelle, F., Viollet, B., Vaulont, S., Girard, J., Postic, C., 2005. Polyunsaturated fatty acids suppress glycolytic and lipogenic genes through the inhibition of ChREBP nuclear protein translocation. J. Clin. Invest. 115, 2843–2854.

Faulds, M.H., Zhao, C., Dahlman-Wright, K., 2010. Molecular biology and functional genomics of liver X receptors (LXR) in relationship to metabolic diseases. Curr. Opin. Pharmacol. 10, 692–697.

Flowers, M.T., Ade, L., Strable, M.S., Ntambi, J.M., 2012. Combined deletion of SCD1 from adipose tissue and liver does not protect mice from obesity. J. Lipid Res. 53, 1646–1653.

Guillou, H., Zadravec, D., Martin, P.G., Jacobsson, A., 2010. The key roles of elongases and desaturases in mammalian fatty acid metabolism: insights from transgenic mice. Prog. Lipid Res. 49, 186–199.

Guo, X., Li, H., Xu, H., Halim, V., Zhang, W., Wang, H., Ong, K.T., Woo, S.L., Walzem, R.L., Mashek, D.G., Dong, H., Lu, F., Wei, L., Huo, Y., Wu, C., 2012. Palmitoleate induces hepatic steatosis but suppresses liver inflammatory response in mice. PLoS One 7, e39286.

Herman, M.A., Peroni, O.D., Villoria, J., Schön, M.R., Abumrad, N.A., Blüher, M., Klein, S., Kahn, B.B., 2012. A novel ChREBP isoform in adipose tissue regulates systemic glucose metabolism. Nature 484, 333–338.

Horton, J.D., Goldstein, J.L., Brown, M.S., 2002. SREBPs: activators of the complete program of cholesterol and fatty acid synthesis in the liver. J. Clin. Invest. 109, 1125–1131.

Howell III, G., Deng, X., Yellaturu, C., Park, E.A., Wilcox, H.G., Raghow, R., Elam, M.B., 2009. N-3 polyunsaturated fatty acids suppress insulin-induced SREBP-1c transcription via reduced trans-activating capacity of LXRα. Biochim. Biophys. Acta (BBA)-Mol Cell Biol. Lipids 1791, 1190–1196.

Hulver, M.W., Berggren, J.R., Carper, M.J., Miyazaki, M., Ntambi, J.M., Hoffman, E.P., Thyfault, J.P., Stevens, R., Dohm, G.L., Houmard, J.A., Muoio, D.M., 2005. Elevated stearoyl-CoA desaturase-1 expression in skeletal muscle contributes to abnormal fatty acid partitioning in obese humans. Cell Metab. 2, 251–261.

Iizuka, K., Bruick, R.K., Liang, G., Horton, J.D., Uyeda, K., 2004. Deficiency of carbohydrate response element-binding protein (ChREBP) reduces lipogenesis as well as glycolysis. Proc. Natl. Acad. Sci. U. S. A. 101, 7281–7286.

Ikeda, M., Kanao, Y., Yamanaka, M., Sakuraba, H., Mizutani, Y., Igarashi, Y., Kihara, A., 2008. Characterization of four mammalian 3-hydroxyacyl-CoA dehydratases involved in very long-chain fatty acid synthesis. FEBS Lett. 582, 2435–2440.

Jiang, G., Li, Z., Liu, F., Ellsworth, K., Dallas-Yang, Q., Wu, M., Ronan, J., Esau, C., Murphy, C., Szalkowski, D., Bergeron, R., Doebber, T., Zhang, B.B., 2005. Prevention of obesity in mice by antisense oligonucleotide inhibitors of stearoyl-CoA desaturase-1. J. Clin. Invest. 115, 1030–1038.

Kalupahana, N.S., Claycombe, K.J., Moustaid-Moussa, N., 2011. (n-3) Fatty acids alleviate adipose tissue inflammation and insulin resistance: mechanistic insights. Adv. Nutr. 2, 304–316.

Karan, G., Lillo, C., Yang, Z., Cameron, D.J., Locke, K.G., Zhao, Y., Thirumalaichary, S., Li, C., Birch, D.G., Vollmer-Snarr, H.R., Williams, D.S., Zhang, K., 2005. Lipofuscin accumulation, abnormal electrophysiology, and photoreceptor degeneration in mutant ELOVL4 transgenic mice: a model for macular degeneration. Proc. Natl. Acad. Sci. U.S.A. 102, 4164–4169.

Laplante, M., Sabatini, D.M., 2010. mTORC1 activates SREBP-1c and uncouples lipogenesis from gluconeogenesis. Proc. Natl. Acad. Sci. U.S.A. 107, 3281–3282.

Lin, J., Yang, R., Tarr, P.T., Wu, P.H., Handschin, C., Li, S., Yang, W., Pei, L., Uldry, M., Tontonoz, P., Newgard, C.B., Spiegelman, B.M., 2005. Hyperlipidemic effects of dietary saturated fats mediated through PGC-1beta coactivation of SREBP. Cell 120, 261–273.

Matsuzaka, T., Shimano, H., Yahagi, N., Kato, T., Atsumi, A., Yamamoto, T., Inoue, N., Ishikawa, M., Okada, S., Ishigaki, N., Iwasaki, H., Iwasaki, Y., Karasawa, T., Kumadaki, S., Matsui, T., Sekiya, M., Ohashi, K., Hasty, A.H., Nakagawa, Y., Takahashi, A., Suzuki, H., Yatoh, S., Sone, H., Toyoshima, H., Osuga, J., Yamada, N., 2007. Crucial role of a long-chain fatty acid elongase, Elovl6, in obesity-induced insulin resistance. Nat. Med. 13, 1193–1202.

Miyazaki, M., Bruggink, S.M., Ntambi, J.M., 2006. Identification of mouse palmitoyl-coenzyme A Delta9-desaturase. J. Lipid Res. 47, 700–704.

Moon, Y.A., Hammer, R.E., Horton, J.D., 2009. Deletion of ELOVL5 leads to fatty liver through activation of SREBP-1c in mice. J. Lipid Res. 50, 412–423.

Moon, Y.A., Horton, J.D., 2003. Identification of two mammalian reductases involved in the two-carbon fatty acyl elongation cascade. J. Biol. Chem. 278, 7335–7343.

Naganuma, T., Sato, Y., Sassa, T., Ohno, Y., Kihara, A., 2011. Biochemical characterization of the very long-chain fatty acid elongase ELOVL7. FEBS Lett. 585, 3337–3341.

Nakamura, M.T., Nara, T.Y., 2004. Structure, function, and dietary regulation of delta6, delta5, and delta9 desaturases. Annu. Rev. Nutr. 24, 345–376.

Nwankwo, J.O., Spector, A.A., Domann, F.E., 2003. A nucleotide insertion in the transcriptional regulatory region of FADS2 gives rise to human fatty acid delta-6-desaturase deficiency. J. Lipid Res. 44, 2311–2319 Epub September 1, 2003.

Pauter, A.M., Olsson, P., Asadi, A., Herslof, B., Csikasz, R.I., Zadravec, D., Jacobsson, A., 2014. Elovl2 ablation demonstrates that systemic DHA is endogenously produced and is essential for lipid homeostasis in mice. J. Lipid Res. 55, 718–728.

Ran-Ressler, R.R., Devapatla, S., Lawrence, P., Brenna, J.T., 2008. Branched chain fatty acids are constituents of the normal healthy newborn gastrointestinal tract. Pediatr. Res. 64, 605–609.

Rios-Esteves, J., Resh, M.D., 2013. Stearoyl CoA desaturase is required to produce active, lipid-modified Wnt proteins. Cell Rep. 4, 1072–1081.

Rioux, V., Pedrono, F., Blanchard, H., Duby, C., Boulier-Monthean, N., Bernard, L., Beauchamp, E., Catheline, D., Legrand, P., 2013. Trans-vaccenate is Delta13-desaturated by FADS3 in rodents. J. Lipid Res. 54, 3438–3452.

Roqueta-Rivera, M., Stroud, C.K., Haschek, W.M., Akare, S.J., Segre, M., Brush, R.S., Agbaga, M.P., Anderson, R.E., Hess, R.A., Nakamura, M.T., 2010. Docosahexaenoic acid supplementation fully restores fertility and spermatogenesis in male delta-6 desaturase-null mice. J. Lipid Res. 51, 360–367.

Sampath, H., Miyazaki, M., Dobrzyn, A., Ntambi, J.M., 2007. Stearoyl-CoA desaturase-1 mediates the pro-lipogenic effects of dietary saturated fat. J. Biol. Chem. 282, 2483–2493.

Sampath, H., Ntambi, J.M., 2011. The role of stearoyl-CoA desaturase in obesity, insulin resistance, and inflammation. Ann. N. Y. Acad. Sci. 1243, 47–53.

Serhan, C.N., Petasis, N.A., 2011. Resolvins and protectins in inflammation resolution. Chem. Rev. 111, 5922–5943.

Seubert, W., Podack, E.R., 1973. Mechanisms and physiological roles of fatty acid chain elongation in microsomes and mitochondria. Mol. Cell Biochem. 1, 29–40.

Stroud, C.K., Nara, T.Y., Roqueta-Rivera, M., Radlowski, E.C., Lawrence, P., Zhang, Y., Cho, B.H., Segre, M., Hess, R.A., Brenna, J.T., Haschek, W.M., Nakamura, M.T., 2009. Disruption of FADS2 gene in mice impairs male reproduction and causes dermal and intestinal ulceration. J. Lipid Res. 50, 1870–1880.

Sun, H., Jiang, T., Wang, S.B., He, B., Zhang, Y.Y., Piao, D.X., Yu, C., Wu, N., Han, P., 2013. The effect of LXR alpha, ChREBP and Elovl6 in liver and white adipose tissue on medium- and long-chain fatty acid diet-induced insulin resistance. Diabetes Res. Clin. Pract. 102, 183–192.

Tamura, K., Makino, A., Hullin-Matsuda, F., Kobayashi, T., Furihata, M., Chung, S., Ashida, S., Miki, T., Fujioka, T., Shuin, T., Nakamura, Y., Nakagawa, H., 2009. Novel lipogenic enzyme ELOVL7 is involved in prostate cancer growth through saturated long-chain fatty acid metabolism. Cancer Res. 69, 8133–8140.

Vasireddy, V., Uchida, Y., Salem, N., Kim, S.Y., Mandal, M.N., Reddy, G.B., Bodepudi, R., Alderson, N.L., Brown, J.C., Hama, H., Dlugosz, A., Elias, P.M., Holleran, W.M., Ayyagari, R., 2007. Loss of functional ELOVL4 depletes very long-chain fatty acids (> or =C28) and the unique omega-O-acylceramides in skin leading to neonatal death. Hum. Mol. Genet. 16, 471–482.

Warensjo, E., Ingelsson, E., Lundmark, P., Lannfelt, L., Syvanen, A.C., Vessby, B., Riserus, U., 2007. Polymorphisms in the SCD1 gene: associations with body fat distribution and insulin sensitivity. Obesity (Silver Spring) 15, 1732–1740.

Westerberg, R., Mansson, J.E., Golozoubova, V., Shabalina, I.G., Backlund, E.C., Tvrdik, P., Retterstol, K., Capecchi, M.R., Jacobsson, A., 2006. ELOVL3 is an important component for early onset of lipid recruitment in brown adipose tissue. J. Biol. Chem. 281, 4958–4968.

Yore, M.M., Syed, I., Moraes-Vieira, P.M., Zhang, T., Herman, M.A., Homan, E.A., Patel, R.T., Lee, J., Chen, S., Peroni, O.D., Dhaneshwar, A.S., Hammarstedt, A., Smith, U., Mcgraw, T.E., Saghatelian, A., Kahn, B.B., 2014. Discovery of a class of endogenous mammalian lipids with anti-diabetic and anti-inflammatory effects. Cell 159, 318–332.

Chapter 7

Phospholipid Synthesis in Mammalian Cells

Neale D. Ridgway

Departments of Pediatrics and Biochemistry & Molecular Biology, Dalhousie University, Halifax, NS, Canada

ABBREVIATIONS

AGPAT 1-Acylglycerol-3-phosphate acyltransferases
CCT CTP:phosphocholine cytidylyltransferase
CEPT Choline/ethanolamine phosphotransferase
CHT1 Choline transporter 1
CK Choline kinase
CPT Choline phosphotransferase
CTL Choline transporter-like
DAG Diacylglycerol
DGK Diacylglycerol kinase
DHAP Dihydroxyacetone-phosphate
ER Endoplasmic reticulum
GPAT Glycerol-3-phosphate acyltransferase
LPLAT Lyso-phospholipid acyltransferases
MAM Mitochondrial-associated membrane
NLS Nuclear localisation signal
NR Nucleoplasmic reticulum
PA Phosphatidic acid
PPAR Peroxisome proliferator-activated receptors
PC Phosphatidylcholine
PE Phosphatidylethanolamine
PG Phosphatidylglycerol
PI Phosphatidylinositol
PIPs Phosphatidylinositol polyphosphates
PS Phosphatidylserine
PSD Phosphatidylserine decarboxylase
PSS Phosphatidylserine synthase
SREBP Sterol-regulatory element-binding protein
TAG Triacylglycerol
VLDL Very-low-density lipoprotein

Biochemistry of Lipids, Lipoproteins and Membranes. http://dx.doi.org/10.1016/B978-0-444-63438-2.00007-9

209

1. INTRODUCTION

Phospholipids consist of a glycerol backbone with hydrophobic fatty acids esterified at *sn*-1 and *sn*-2 hydroxyl groups, and a hydrophilic head group (choline, ethanolamine, inositol, serine or glycerol) attached by a phosphodiester linkage to the *sn*-3 hydroxyl group (Figure 1). This amphipathic property of phospholipids drives the formation of membrane bilayers that provide a selective barrier to the extracellular environment and define intracellular organelles in eukaryotic and prokaryotic cells. Another equally important function of phospholipids is as precursors for signalling molecules, such as diacylglycerol (DAG), phosphatidic acid (PA), eicosanoids (Chapter 9) and lyso-phospholipids. In addition, phospholipids are essential components of lipoproteins (Chapters 15 and 16), bile (Chapter 12) and lung surfactant, lipid–protein complexes that are secreted from specialised cells into the extracellular space. Phospholipids are represented in all three kingdoms; an overview of bacterial and plant phospholipid synthesis is presented in Chapters 3 and 4, respectively. This chapter is concerned with the metabolism and functions for the major phospholipid classes in mammalian cells. Selected

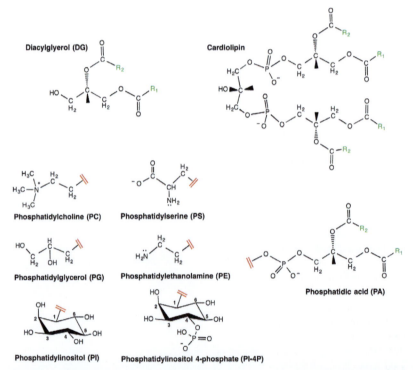

FIGURE 1   Structure of the phospholipid classes. The structures of diacylglycerol, cardiolipin and phosphatidic acid are shown (R_1 and R_2 indicate the position of *sn*-1 and *sn*-2 fatty acid chains). The site of attachment of polar head groups to phosphatidic acid to form the major phospholipids (PC, PS, PG, PE PI and PI-4P) is indicated by the hatched red lines.

aspects of phospholipid metabolism in yeast are included since many important insights have been gained through the use of this experimental organism.

2. BIOSYNTHESIS OF PHOSPHATIDIC ACID AND DIACYLGLYCEROL

PA and DAG (structures shown in Figure 1) are the precursors for discrete phospholipid classes and as such their relative proportions within the cell affect the synthesis of respective end products (Figure 2). The synthesis of PA begins with the activation of long-chain fatty acids to their CoA-esters by acyl-CoA synthetases. In mammals, 11 different long-chain acyl-CoA synthetases catalyse this reaction in a variety of cellular locations (Chapter 5). PA is synthesised by the successive addition of two fatty acyl-CoAs to glycerol-3-phosphate, a product of glycolysis. PA is the precursor for CDP-DAG that is used in the synthesis of phosphatidylinositol (PI) and cardiolipin, and is dephosphorylated to DAG for the synthesis of phosphatidylcholine (PC) and phosphatidylethanolamine (PE) as well as triacylglycerol (TAG). Thus, the dephosphorylation of PA to DAG is a key step that controls phospholipid biosynthesis and energy storage in the form of TAG.

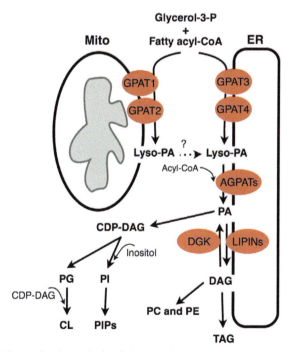

FIGURE 2 Pathways for the synthesis of phosphatidic acid (PA) and diacylglycerol (DAG) on the outer leaflet of the endoplasmic reticulum (ER) and mitochondria. GPAT, glycerol-3-phosphate acyltransferase; AGPAT, 1-acylglycerol-3-phosphate acyltransferase; Mito, mitochondria; PIPs, phosphatidylinositol polyphosphates.

2.1 Glycerol-3-Phosphate Acyltransferases

The initial step in the synthesis of PA is catalysed by a family of four glycerol-3-phosphate acyltransferases (GPAT) that esterify a fatty acyl-CoA to the *sn*-1 position of glycerol-3-phosphate (Figure 2). GPAT1 and GPAT2 reside on the outer mitochondrial membrane, whereas GPAT3 and GPAT4 are found on the endoplasmic reticulum (ER). GPAT4 moves from the ER to the surface of lipid droplets where it contributes to the synthesis of TAG and PC required for the expansion of this organelle. A related dihydroxyacetone-phosphate (DHAP) acyltransferase catalyses the synthesis of 1-acyl-DHAP in peroxisomes, which is subsequently converted to 1-alkyl-DHAP, a precursor for ether lipids and plasmalogens that contain a *sn*-1-alkyl or *sn*-1-alk-1′-enyl moiety (Braverman and Moser, 2012). GPAT1 in the liver and adipose tissue of mice is increased by insulin and reduced by fasting, indicating a role in channelling fatty acids into TAG for storage in lipid droplets or export in lipoproteins. Human GPAT deficiencies have not been identified but *Gpat1*$^{-/-}$ mice have reduced liver and adipose TAG and elevated ketone body production, indicating that the enzyme diverts fatty acids towards storage and away from β-oxidation. Unlike other GPATs, GPAT1 has a preference for palmitoyl-CoA, and *Gpat1*$^{-/-}$ mice have reduced palmitate incorporation into phospholipids. GPAT2 is also in the mitochondria but is not highly expressed in lipogenic tissues nor is it hormonally regulated. GPAT3 is expressed predominately in the ER of differentiated adipocytes where it also directs fatty acids into TAG for storage in lipid droplets. Experiments using *Gpat4*$^{-/-}$ mice indicate an important role for this enzyme in liver TAG synthesis. Manipulation of GPAT expression in mice or cultured cells causes dramatic changes in TAG synthesis and relatively minor changes in phospholipid synthesis. Thus GPATs and downstream enzymes can efficiently channel fatty acids into TAG for storage without affecting the availability of PA or DAG for phospholipid synthesis. On the other hand, blocking the CDP-choline or CDP-ethanolamine pathways in mice or cultured cells leads to diversion of DAG into TAG.

2.2 1-Acylglycerol-3-Phosphate Acyltransferases

Lyso-PA is acylated at the *sn*-2 position with fatty acyl-CoAs by a large gene family of 1-acylglycerol-3-phosphate acyltransferases (AGPAT) or lyso-phospholipid acyltransferases (LPLAT, see Section 8). Lyso-PA-specific AGPAT1, 2, 3 and 5 are integral membrane proteins of the ER that are expressed in tissues with increased capacity for TAG synthesis, such as adipocytes and liver. AGPAT1 and 2 are expressed in adipose tissue and liver where they primarily utilise 18:1 and 18:2 fatty acyl-CoAs as substrates for PA synthesis. An important role of AGPAT2 in PA synthesis is indicated by abnormal hepatic TAG accumulation and reduced white adipose tissue mass (termed lipodystrophy) in mice and humans in which the enzyme is deleted or mutated. Collectively, the AGPAT gene family remains poorly understood due to complex tissue expression patterns, functional redundancy and broad in vitro substrate specificity.

The presence of GPAT1 and GPAT2 in mitochondria and AGPATs in the ER suggests that lyso-PA must be transported between these organelles for efficient metabolic coupling of the pathways for TAG and phospholipid synthesis (Figure 2). While specific transfer proteins have not been identified, a discrete subdomain of the ER that contacts the outer mitochondrial membrane, termed the mitochondrial-associated membrane (MAM), could be a site for substrate transfer and communication between these organelles (Vance, 1990). GPAT1 and acyl-CoA synthetases are present at the MAM but AGPATs have not been identified in this compartment.

2.3 PA Phosphatases

PA is the immediate precursor for the synthesis of CDP-DAG, cardiolipin and phosphatidylinositol (PI), and is dephosphorylated to DAG for incorporation into PC and PE by the CDP-choline/ethanolamine (Kennedy) pathways or is further esterified at the *sn*-3 position by DAG acyltransferases to make TAG (Chapter 5). Thus the dephosphorylation of PA is a pivotal step that partitions substrate to different branches of glycerolipid synthesis. Although PA phosphatase (PAP) activity in the ER was identified in the 1970s, it proved resistant to purification and it was not until 2006 that yeast PAP (Pah1) was finally characterised (Han et al., 2006). The discovery of Pah1 led to the identification of a mammalian homologue called LIPIN1 that was previously identified as the causative mutation in *fld* mice that exhibit liver steatosis and severe lipodystrophy (Peterfy et al., 2001). In mammals, three *LIPIN* genes are expressed in most tissues but in variable proportions; *LIPIN1* is expressed at high levels in skeletal muscle and adipose tissue, while *LIPIN2* and *LIPIN3* are more highly expressed in the liver and brain. All three enzymes are soluble and magnesium dependent, and have a conserved C-terminal catalytic domain, a serine-rich phosphorylation site and a nuclear localisation signal (NLS). LIPINs are soluble enzymes but interact with membranes to access their substrate PA by a phosphorylation-dependent mechanism; phosphorylated LIPIN is inactive in the cytoplasm whereas the dephosphorylated form is associated with the ER and nuclear envelope. In the case of LIPIN1, the phosphorylated form is primarily in the cytoplasm but dephosphorylation also promotes import into the nucleus (Figure 3). LIPINs contain a transcriptional co-activator motif within the catalytic region that interacts with the transcriptional co-activator PGC-1α and peroxisome proliferator-activated receptors (PPAR) (Csaki et al., 2013). Import of LIPIN1 into the nucleus promotes (1) transcriptional activation of PPARγ and PPARα leading to adipocyte differentiation and hepatic fatty acid oxidation, respectively, and (2) suppression of the master lipogenic transcription factor sterol-regulatory element-binding protein (SREBP) (Figure 3). Thus, LIPIN catalytic activity increases the availability of DAG for lipid synthesis whereas its nuclear co-activator activity regulates tissue-specific fatty acid oxidation and storage.

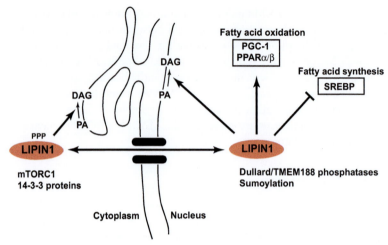

FIGURE 3 Regulation of LIPIN1 activity by nuclear–cytoplasmic shuttling. Phosphorylation of LIPIN1 by mTORC1 causes retention in the cytoplasm for synthesis of DAG on the ER and nuclear membranes. Dephosphorylation by LIPIN1 phosphatases or sumoylation promotes nuclear import, transcriptional co-activator activity, inhibition of SREBP and synthesis of nuclear DAG.

The lipodystropic phenotype of *lipin1*$^{-/-}$ mice suggests an important role in adipose tissue differentiation and hepatic lipid metabolism. However, in contrast to the mouse phenotype, rare *LIPIN1* mutations in humans cause rhabdomyolysis, breakdown of skeletal muscle. *Lipin2*$^{-/-}$ mice have normal lipid homeostasis but develop fatty livers if challenged with at high-fat diet. With increased age, these animals also develop tremors and ataxic gait related to elevated PA levels in cerebellar Purkinje cells. *LIPIN2* mutations in humans cause Majeed disease, which is characterised by anaemia, fever and joint pain. It has been speculated that many of the features of human LIPIN deficiency are caused by imbalance in signalling pathways that are activated by PA and DAG.

2.4 Diacylglycerol Kinases

DAG is also phosphorylated to PA in an ATP-dependent reaction catalysed by a large family of DAG kinases (DGK) (Cai et al., 2009). Ten separate genes encode mammalian DGKs that are divided into five functional subtypes based on domain organisation. However, all DGKs have a common catalytic domain and at least two C1 domains that could localise the enzymes to DAG-containing membranes and/or regulate enzyme activity. DGKs localise to a variety of intracellular membranes where they are primarily involved in the attenuation of DAG-dependent signalling and activation of PA signalling pathways. Although DGKs have been implicated in metabolic regulation, for example by PA-dependent activation of the mTOR nutrient-sensing pathway, there is no evidence that DGKs directly influence metabolic flux between glycerolipid species (Figure 2).

Yeasts have a single DGK that utilises CTP as a phosphate donor, is structurally unrelated to the mammalian enzymes, and converts TAG-derived DAG into PA for phospholipid synthesis.

3. PHOSPHATIDYLCHOLINE BIOSYNTHESIS AND REGULATION

3.1 Phosphatidylcholine Synthesis by the CDP-Choline/Kennedy Pathway

Phosphatidylcholine (PC) is a choline-containing, zwitterionic phospholipid (structure shown in Figure 1) that constitutes between 30 and 60% of the phospholipid mass of eukaryotic cell membranes. The quaternary amine choline is essential for PC function and cannot be replaced by primary or secondary amine analogues without adversely affecting cell viability. Given its abundance in membranes it is not surprising that depriving cells of PC by inhibiting synthesis or the availability of precursors results in growth arrest and apoptosis. In addition to an essential role in membrane structure, PC is also a component of secreted lipoproteins, bile and lung surfactant. If PC availability is limited, the assembly and secretion of these lipid–protein complexes is inhibited, causing the sequestration of lipids within cells and interfering with essential extracellular functions. PC is a reservoir for second messengers such as fatty acids, DAG, PA and lyso-PC that can be released on activation of lipases in response to primary signalling factors. Finally, PC is an important precursor for the synthesis of phosphatidylserine (PS) and sphingomyelin through donation of its PA and phosphocholine moiety, respectively.

In the 1950s, Kennedy and Weiss identified the de novo pathways for the synthesis of PC and PE that utilise the activated nucleotide intermediates CDP-choline and CDP-ethanolamine, respectively. The three-step CDP-choline/Kennedy pathway is the primary source of PC in all mammalian cells and utilises choline that is salvaged from PC degradation or transported from the extracellular space (Figure 4). PC can also be synthesised from pre-existing phospholipids, namely the methylation of PE and acylation of lyso-PC, but the contribution of these pathways to cellular PC synthesis is small and tissue specific. The CDP-choline and CDP-ethanolamine pathways involve similar reactions that are, in some cases, catalysed by enzymes that utilise both choline and ethanolamine intermediates as substrates.

3.2 Choline Transporters

The initial step in the CDP-choline/Kennedy pathway is uptake of choline into cells by three classes of transporters: the high-affinity (K_D 2 µM), sodium-coupled choline transporter 1 (CHT1 or SLC5A7); intermediate-affinity (K_D 10–100 µM), sodium-independent choline transporter-like (CTL) proteins of

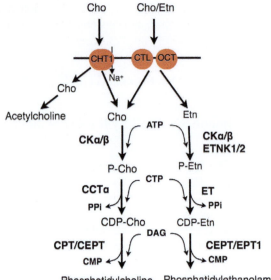

FIGURE 4 Synthesis of phosphatidylcholine and phosphatidylethanolamine by the CDP-base/ Kennedy pathways. Cho, choline; CK, choline kinase, CCT, CTP:phosphocholine cytidylyltransferase; CPT, choline phosphotransferase, CEPT, choline/ethanolamine phosphotransferase; CHT, choline transporter 1, CTL, choline transporter-like, ETNK, ethanolamine kinase, ET, CTP: phosphoethanolamine cytidylyltransferase; EPT, ethanolamine phosphotransferase, OCT, organic cation transporter.

the SLC44A family; and low-affinity (K_D >100 μM), organic cation transporters (OCT) of the SLC22 family (Traiffort et al., 2013) (Figure 4). CHT1 is expressed primarily in the presynaptic termini of cholinergic neurons where it mediates choline uptake for acetylcholine synthesis. Facilitated transport of choline by CTL and OCT occurs in other cell types. The CTL/SLC44A transporters, which consist of five related members with 10–12 transmembrane domains, contribute the bulk of plasma membrane sodium-independent choline transport activity that is required for PC synthesis. CTL1/SLC44A1 is also expressed in mitochondria where it mediates choline uptake for oxidation to betaine as well as export of choline following PC catabolism. CTL1 and CHT1 can be distinguished from low-affinity transporters by their sensitivity to inhibition by the choline analogue hemicholinium-3. While choline transport is not rate limiting for PC synthesis, the rapid phosphorylation of choline to phosphocholine by choline kinases (CK) drives the net import into the cell.

3.3 Choline Kinase

CKα and CKβ are cytosolic enzymes encoded by separate genes (*CHKA* and *CHKB*) that catalyse the ATP-dependent phosphorylation of choline and/or ethanolamine (Figure 4). CKα and CKβ form homo- (α/α, β/β) or hetero-dimers (α/β), with the α/β heterodimer accounting for the majority of CK activity (Aoyama et al., 2002; Sher et al., 2006). CKα is ubiquitously expressed and accounts for the majority of kinase activity, and deletion of its gene in mice results in embryonic death (Wu et al., 2010). CKβ is also widely expressed in tissues but gene deletion is nonlethal, instead causing a hind-limb muscular dystrophy-like phenotype. Human mutations in *CHKB* also cause a rare form of congenital muscular dystrophy (Mitsuhashi et al., 2011). Although CK does not catalyse the rate-limiting step for PC synthesis, there are instances in which increased activity results in elevated PC synthesis, notably in oncogenic Ras-transformed cancer cells that have increased total CK activity (2- to 3-fold) and increased phosphocholine mass. CKα activity and phosphocholine mass are also increased in a variety of cancer cells and tumours. Inhibition of CKα activity prevents proliferation and promotes apoptosis of cancer cells by a mechanism that involves disruption of the epidermal growth factor signalling pathway.

3.4 CTP:Phosphocholine Cytidylyltransferase

Phosphocholine is converted to the high-energy nucleotide derivative CDP-choline by CTP:phosphocholine cytidylyltransferase (CCT) (Figure 4). Under most experimental conditions, CCT catalyses the rate-limiting step in the CDP-choline pathway and is subject to regulation by multiple transcriptional and posttranscriptional mechanisms (Cornell and Northwood, 2000). The human *PCYT1a* and *PCYT1b* genes encode CCTα and two CCTβ splice variants (β1 and β2), respectively. The CCTα and β2 isoforms have three functional domains: a conserved catalytic domain immediately followed by a 50 amino acid membrane-binding amphipathic helix (domain M), and a C-terminal serine-rich phosphorylation (P) domain. Domain M allows the enzyme to interact reversibly with membranes. CCTβ1 is structurally similar but lacks domain P. CCTα differs from β isoforms by the presence of N-terminal NLS.

The mRNA for CCTα is expressed in most human and mouse tissues, particularly in the liver and lung, whereas expression of the β isoforms is lower and they have a tissue-restricted distribution. The relative contribution of the CCTα and β isoforms to PC synthesis has been determined by gene knockout studies in mice. *Pcyt1b*$^{-/-}$ mice are viable but have gonadal degeneration and reproductive abnormalities. In contrast, disruption of the *Pcyt1a* gene is embryonic lethal. However, tissue-specific knockout of CCTα in mice has revealed that the CDP-choline pathway has essential roles in specific cell types. *Pcyt1a*$^{-/-}$ macrophages are viable but sensitive to cholesterol-induced cell death, and the null mice are susceptible to diet-induced atherosclerosis. Similarly, mice with a disruption of the *Pcyt1a* gene in lung epithelial cells developed normally in utero but have reduced synthesis of dipalmitoyl-PC, an essential component of lung surfactant, leading to respiratory failure at birth. Liver-specific knockout of *Pcyt1a* in mice had no effect on viability but reduced PC secretion in very-low-density lipoprotein (VLDL), PC efflux to HDL and secretion of PC in bile. Collectively this suggests that CCTα is essential during early embryonic development; however, conditional knockouts revealed tissue-specific functions that could be partially compensated by synthesis of CDP-choline by CCTβ and/or PC synthesis by the PE methylation pathway. Recently, loss-of-function mutations in human *PCYT1A* were identified for the rare genetic disorder spondylometaphyseal dysplasia (SMD) with cone–rod dystrophy, which is characterised by short stature, thinning and bowing of the legs, and visual defects. Other rare human biallelic *PCYT1A* mutations result in fatty liver, type 2 diabetes and lipodystrophy, suggesting an important role in adipocyte function. Since these human CCTα mutations are nonlethal and phenotypically heterogeneous, other genetic and environmental modifiers may influence the consequences of restricted PC synthesis.

3.5 Posttranscriptional Regulation of CTP:Phosphocholine Cytidylyltransferase

The activity of CCTα is primarily regulated by reversible association with membranes, a property termed amphitropism (Figure 5). The cycling of CCTα

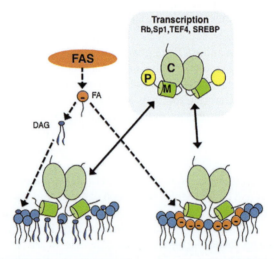

FIGURE 5 Regulation of CTP:phosphocholine cytidylyltransferase (CCT) by transcriptional and posttranscriptional mechanisms. Transcription of the *PCYT1A* gene is driven by the indicated transcription factors. CCTα is primarily regulated by reversible association with nuclear and ER membranes that are enriched in fatty acids and/or DAG produced directly or indirectly by fatty acid synthase (FAS) and lipid degradation. Association of CCTα with membranes results in dephosphorylation and increased catalytic activity, which increases PC synthesis, reduces the membrane content of the lipid activators and releases CCTα to its inactive state. C, catalytic domain; M, membrane-binding domain; P, phosphorylation domain.

between its active membrane-bound and inactive soluble forms is regulated by membrane composition and structure, and involves interplay among the catalytic domain, the amphipathic domain M and the C-terminal serine phosphorylation domain P. Domain M has a disordered structure in the soluble enzyme but forms an α-helix that penetrates into the interior of membranes that are enriched in anionic lipids (such as fatty acids) or have curvature stress due to the presence of DAG or PE. The interaction of domain M with membranes enriched in anionic lipids involves electrostatic interactions with basic arginine and lysine residues (Johnson et al., 2003). PE and DAG induce negative curvature of membranes by decreasing the packing between polar head groups. The resultant curvature stress induced by these lipids in a planar membrane is relieved by insertion of domain M into voids in the membrane (Figure 5). Similarly, domain M also inserts into membranes with positive curvature, suggesting that it senses voids between phospholipid head groups.

In the absence of membranes, CCTα activity is inhibited by association of an active site α-helix with the C-terminal segment of domain M, which disengages in the presence of membranes to allow full catalytic activity. This conformational change increases the k_{cat} by >80-fold and reduces the K_m for CTP by >10-fold (Wang and Kent, 1995). Thus CCTα senses the relative amounts of CDP-choline pathway substrates (fatty acids and DAG) by the influence these lipids have on membrane structure and regulates the activity of the CDP-choline pathway accordingly. With increased PC synthesis, membranes become more planar and less negatively charged, leading to dissociation and inhibition of CCTα. Thus CCTα activity and PC synthesis respond rapidly to the relative content of substrates and products of the CDP-choline pathway.

In its soluble form, human CCTα is phosphorylated on 16 serine residues and undergoes dephosphorylation after translocation to membranes. CCTα mutants that mimic the phosphorylated and dephosphorylated states have decreased and increased membrane localisation, respectively. However, constitutively dephosphorylated CCTα mutants are similarly translocated and activated by oleate and support growth of CCTα-deficient cells, suggesting a subtle negative regulatory effect of phosphorylation on membrane binding (Yang and Jackowski, 1995). Recently, domain P phosphorylation was implicated in the curvature-sensing activity of domain M by altering its affinity for curved versus planar membranes. In this model, the phosphorylated domain P directs domain M binding to positively curved membranes, such as the ER, whereas the dephosphorylated forms would bind indiscriminately to planar membranes, such as the nuclear envelope (Chong et al., 2014).

A defining feature of CCTα is its localisation to the nucleus and translocation to the nuclear envelope in response to lipid activators. CCTα is imported into the nucleus via the α-importin pathway using an N-terminal NLS. CCTα is cytoplasmic in cells that secrete PC or have rapidly proliferating membranes, such as differentiating B-cells, pulmonary epithelial cells, lipopolysaccharide-treated primary macrophages or during TAG synthesis and deposition into lipid droplets. This suggests that signals for increased production of PC cause the enzyme to move from the nucleus to the cytoplasm or ER. Increased PC and membrane synthesis during the cell cycle is also associated with changes in CCTα activity. Activation of quiescent macrophages with growth factors causes a burst of PC degradation, releasing lipid activators that stimulate CCTα activity and PC synthesis into the S phase followed by a decline during G2/M. CCTα activation during the cell cycle is accompanied by decreased phosphorylation and nuclear export to the ER.

Despite the presence of nuclear CCTα in many cell types, expression of a CCTα mutant lacking the NLS is sufficient to support the growth of CCTα-deficient cells, indicating a role in other nuclear-specific functions. Cells contain tubular membrane invaginations of the nuclear envelope that extend deep into the nucleoplasm termed the nucleoplasmic reticulum (NR). CCTα is involved in formation of the NR by a mechanism that involves increasing the positive membrane curvature of the NR to form membrane tubules (Lagace and Ridgway, 2005). NR formation by CCTα is independent of catalytic activity and requires Domain M, which is capable of deforming planar membranes enriched in anionic phospholipids into thin tubules. These effects on the nuclear envelope highlight an interesting feature of CCTα that is shared by other proteins that modify membrane structure with membrane-sensing amphipathic helices. CCTα binds to planar membranes that are enriched in lipid activators, such as the nuclear envelope, and induces a localised discontinuity between the inner and the outer membrane bilayers that leads to increased positive curvature. Increased positive curvature then attracts more CCTα, leading to a progressive extension of narrow lipid tubules. The significance of this in terms of CCTα activity and PC synthesis requires further investigation but suggests that CCTα translocates to different membranes depending on their lipid composition and curvature (Taneva et al., 2012).

3.6 CDP-Choline:1,2-DAG Choline Phosphotransferase

The final step in the CDP-choline pathway is the transfer of the phosphocholine moiety from CDP-choline to the *sn*-3 hydroxyl of DAG by CDP-choline:1,2-DAG choline phosphotransferase (CPT) and the related enzyme CDP-choline:1,2-DAG choline/ethanolamine phosphotransferase (CEPT) (Figure 4). CPT exclusively utilises CDP-choline, whereas either CDP-choline or CDP-ethanolamine is a substrate for CEPT. Both enzymes share 60% sequence identity, are embedded in ER or Golgi membranes by multiple transmembrane segments, and have a predicted amino alcohol catalytic motif oriented towards the cytoplasm (Henneberry and Mcmaster, 1999). A CPT activity specific for synthesis of platelet-activating factor from 1-alkyl-2-acetyl DAG has been identified based on sensitivity to reducing agents but the protein and gene have not yet been isolated. The relative contribution of CEPT to PC and PE synthesis is not well understood but CEPT can complement the loss of CPT in yeast, indicating a significant contribution to PC synthesis. CEPT is localised to the ER and nuclear envelope, the major sites of PC synthesis in mammalian cells. On the other hand, CPT is present in the Golgi apparatus and likely contributes to the regulation of PC and DAG that control secretion from this organelle. CPT and CEPT activities are primarily controlled by the availability of the substrates CDP-choline and DAG, and changes in expression are not associated with the altered rates of PC synthesis. However, CPT mRNA and activity is increased as part of a programmed expansion of ER membranes during the unfolded protein response (Fagone et al., 2007).

3.7 Phosphatidylethanolamine *N*-Methyltransferase

Although early studies recognised that methionine methyl groups were required for choline synthesis, it was not until 1960 that Bremer and Greenburg identified a microsomal PE *N*-methyltransferase (PEMT) activity in hepatocytes that catalysed the successive transfer of three methyl groups from *S*-adenosylmethionine to the amino head group of PE to make the choline head group of PC (Figure 6). Mono- or dimethylated PE are not released by PEMT, indicating a concerted reaction that involves tight binding of reaction intermediates. PEMT was purified as a low molecular mass (20 kDa) integral membrane protein that localised to the ER and MAM (Ridgway and Vance, 1987).

The enzyme is almost exclusively expressed in hepatocytes, where it contributes approximately 30% of total PC synthesis (the majority synthesised via the CDP-choline pathway). If either the hepatic CDP-choline or the PEMT pathways are disrupted by genetic or pharmacological inhibition, the other pathway

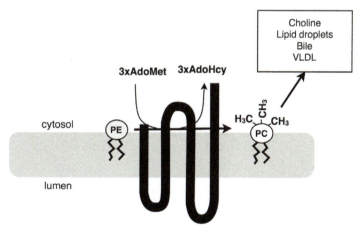

FIGURE 6 Methylation of phosphatidylethanolamine (PE) by PE *N*-methyltransferase (PEMT) in the ER membrane. The model of PEMT is based on ER membrane topology predicted by Shields et al. (2003). The hepatocyte- and adipocyte-specific functions of PEMT-derived PC are indicated. AdoMet, *S*-adenosylmethionine; AdoHcy, *S*-adenosylhomocysteine.

is upregulated to compensate for the loss of biosynthetic capacity. Recently, PEMT was also detected in differentiating adipocytes where it makes PC that is incorporated into the surface monolayer of TAG-rich lipid droplets. PC made by methylation of PE has specific functions that are essential for life. First and foremost, PEMT provides the body with a source of choline, an essential nutrient that can usually be obtained in adequate amounts from the diet. The importance of this endogenous pathway for choline synthesis is exemplified by studies of *Pemt* knockout mice. When fed a diet with sufficient choline, *Pemt*$^{-/-}$ mice have no overt signs of choline or PC deficiency. However, when placed on a choline-deficient diet, the animals die due to liver failure after 3 days. Liver failure is due to acute reduction in PC synthesis via the CDP-choline pathway that leads to accumulation of intercellular TAG and a reduced PC/PE ratio in the plasma membrane that compromises cellular integrity. The liver is susceptible to damage following the cessation of PC synthesis in *Pemt*$^{-/-}$ mice because of the large amount of PC that is exported from hepatocytes in VLDL and bile. When *Pemt*$^{-/-}$ mice were crossed with a strain that is deficient in ABCB4, which transports PC into the bile duct for secretion in bile, the double knockout mice survived for at least 90 days. Thus, hepatic function can be maintained under conditions of PC deficiency if the efflux of PC from the liver is inhibited.

The results from radiolabelling experiments in cultured hepatocytes and *Pemt* knockout mice also demonstrate a specific requirement for incorporation of PEMT-derived PC into hepatic VLDL. Hepatocytes from *Pemt*$^{-/-}$ mice secrete less TAG in VLDL, and circulating TAG, cholesterol and apoB are also reduced in *Pemt*$^{-/-}$ mice fed a high-fat/cholesterol-rich diet. Importantly, crossing the *Pemt*$^{-/-}$ mice into an LDL receptor null background, which is susceptible to diet-induced atherosclerosis, resulted in a significant reduction in atherosclerosis. *Pemt*$^{-/-}$ mice are also resistant to the weight gain and reduced insulin sensitivity caused by a high-fat diet (Jacobs et al., 2010). Resistance to the effects of a high-fat diet is related to choline deficiency since inclusion of choline in the diet caused the *Pemt*$^{-/-}$ mice to respond similarly to control animals. While PEMT is a potential target to ameliorate the effects of fat- and cholesterol-enriched diets, *Pemt*$^{-/-}$ mice have a potentially damaging increase in liver fat content or steatosis caused by a shift in TAG from peripheral organs to the liver.

4. PHOSPHATIDYLETHANOLAMINE BIOSYNTHESIS AND REGULATION

4.1 Functions of Phosphatidylethanolamine

Phosphatidylethanolamine (PE) adopts a conical shape due to the relative volume difference between its small ethanolamine head group and long, unsaturated acyl chains at the *sn*-1 and *sn*-2 positions (structure shown in Figure 1). Due to its shape and lack of a charged head group, PE induces negative curvature when incorporated into membranes, which is reported to enhance membrane fusion and molecular folding of some membrane proteins (Chapter 1). In addition, the primary amino group of PE is relatively reactive and forms protein and lipid conjugates in a number of important biological pathways. For example, the formation and expansion of autophagic vesicles require the conjugation of Atg8 with the amino group of PE by the ubiquitin-like ligases Atg7 and Atg3 (see Chapter 13). *N*-acyl-PE, a precursor for *N*-acylethanol-amine (such as anandamide) that has potent analgesic effects in the central nervous system by interaction with cannabinoid receptors, is formed by transfer of arachidonate from a phospholipid to the ethanolamine head group. PE reversibly interacts with retinal to form *N*-retinylidene-PE in photoreceptor cells of the eye. ABCA4 then transports this conjugate across disc membranes as part of the visual cycle in the retina. PE also supplies the phosphoethanol-amine moiety on glycosylphosphatidylinositol anchors that link many signalling proteins to specific lipid domains in the plasma membrane (Chapter 13).

PE is synthesised de novo by the CDP-ethanolamine/Kennedy pathway, which shares similarities and enzymes with the analogous pathway for PC synthesis (Figure 4). In addition, PE is synthesised from pre-existing phospholipids by decarboxylation of PS in the mitochondria and acylation of lyso-PE. The relative importance of the decarboxylation versus CDP-ethanolamine pathways for PE synthesis depends on the availability of PS and ethanolamine, and is cell specific.

4.2 The CDP-Ethanolamine Pathway

Ethanolamine is acquired from the diet, but is also produced in cells by the degradation of sphingosine 1-phosphate (Chapter 10) and PE. The transporters that import ethanolamine into cells are poorly characterised. The transport of choline into cells by CTL1 is inhibited by ethanolamine, suggesting that these precursors for the Kennedy pathways compete for the SLC44A transporters. Once in the cell, ethanolamine is phosphorylated by two ethanolamine-specific kinases (ETNK1 and ETNK2) that account for the majority of ethanolamine phosphorylation activity (Gustin et al., 2008) (Figure 4). Both kinases are highly expressed in the liver and $Etnk2^{-/-}$ mice have an 80% reduction in hepatic etha-nolamine kinase activity. However, $Etnk2^{-/-}$ mice develop normally and do not have defects in lipid metabolism, suggesting functional redundancy between the isoforms. Interestingly, $Etnk2^{-/-}$ female mice had reduced litter sizes and increased instances of intrauterine death and placental thrombosis. CKα and CKβ also phosphorylate ethanolamine in vitro, but to a lesser extent than choline. Phosphoethanolamine levels are normal in $Chkb^{-/-}$ mice but reduced in the livers of $Chka^{+/-}$ mice, suggesting a significant contribution to ethanolamine kinase activity by CKα isoforms.

The synthesis of CDP-ethanolamine from CTP and phosphoethanolamine is rate-limiting for PE synthesis and catalysed by CTP:phosphoethanolamine cytidylyltransferase (ET), a product of the human PCYT2 gene. Unlike other enzymes in the Kennedy pathways that have dual specificity, ET does not utilise phosphocholine. The enzyme contains two cytidylyltransferase catalytic motifs separated by a short linker region that contains an alternately spliced exon. Both catalytic motifs are required for ET activity. ET is primarily activated by phosphorylation by protein kinase C in the linker region and C-terminal catalytic domain. A complete knockout of *Pcyt2* in mice is embryonic lethal, indicating an essential role for de novo PE synthesis during development. Heterozygous (*Pcyt2*$^{+/-}$) and liver-specific *Pcyt2* knockout mice fed a normal diet display increased VLDL secretion, insulin resistance, hepatic steatosis and defective fatty acid oxidation. These are features of human metabolic syndrome and are the result of accumulation of hepatic TAG that is derived from DAG that is normally utilised for PE synthesis. Hepatic deletion of CCTα does not cause this phenotype, suggesting that DAG used for PE and TAG synthesis, but not PC synthesis, is from a shared pool. As noted in Section 3.6, the final step in the pathway is catalysed by the dual specificity CEPT that is localised to the ER and nuclear envelope. This enzyme is specific for 1-palmitoyl-2-docosahexaenoyl DAG species that are enriched in newly synthesised PE in cultured hepatocytes but rapidly remodelled to other molecular species. CEPT also utilises 1-alkyl-2-acyl DAG for the synthesis of plasmalogens. A CDP-ethanolamine-specific phosphotransferase EPT1 ubiquitously expressed in mouse tissues is manganese dependent and exclusively uses CDP-ethanolamine. The relative contribution of this enzyme to PE synthesis is unknown.

4.3 The Phosphatidylserine Decarboxylase Pathway for PE Synthesis

In contrast to the CDP-ethanolamine pathway, which produces PE in the ER, PE made by decarboxylation of PS resides on the outer surface of the inner mitochondrial membrane (IMM) (Figure 7). The PS substrate for phosphatidylserine decarboxylase (PSD) is synthesised in the ER by the base-exchange enzymes PSS1 and PSS2 (Section 5.2), and is imported into the mitochondria at contact sites with the ER termed the MAM. Both PSS1 and PSS2 are localised to the MAM, effectively coupling synthesis of PS with import to the site of decarboxylation. The import of PS into mitochondria is the rate-limiting step in PE synthesis, is ATP-dependent, requires contact between the inner and the outer mitochondrial membranes, but does not require cytosolic factors. A protein complex in yeast called ERMES (ER/mitochondrial encounter structure) tethers the ER and mitochondria at 10- to 30-μm-wide contact zones, but the identity of similar tethering factor(s) in mammalian cells remains elusive. Cultured cells can derive PE exclusively from the mitochondrial PS decarboxylation pathway by the efficient export of PE to the ER and other organelles by a poorly understood mechanism (Figure 7).

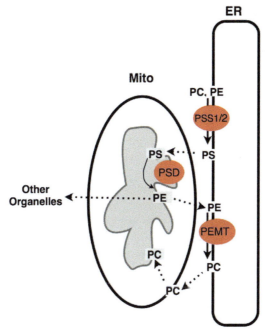

FIGURE 7 The synthesis of PS and PE at mitochondrial–ER contact zones. The figure shows a concerted pathway for PS synthesis by phosphatidylserine synthases (PSS1/2), transport to the mitochondria for decarboxylation to PE and export of PE to the ER and other organelles. PEMT is enriched at ER–mitochondrial contact sites and could make PC for import into the mitochondria. ER, endoplasmic reticulum; PSD, phosphatidylserine decarboxylase; Mito, mitochondria.

PSD has a pyruvoyl prosthetic group and undergoes autocatalytic cleavage to form the active mature enzyme. Disruption of *Psd1* in mice causes early embryonic lethality. Fibroblasts isolated from *Psd1* knockout embryos contain mitochondria that are fragmented and misshapen, indicating a defect in fusion activities due to insufficient PE. Similarly, a 50% reduction in PSD1 expression in cultured cells causes structural and respiratory defects that cannot be compensated for by the CDP-ethanolamine/Kennedy pathway. Yeasts have two genes that encode PSD; a mitochondrial PSD1 and Golgi-associated PSD2. Deletion of either PSD gene is not lethal but deletion of both renders the yeast auxotrophic for ethanolamine.

5. PHOSPHATIDYLSERINE BIOSYNTHESIS AND REGULATION

5.1 Functions of Phosphatidylserine

Eukaryotic phospholipids are composed of approximately 5–15% phosphatidylserine (PS) (structure shown in Figure 1), which at cellular pH is an anionic lipid that imparts significant electronegative charge to membranes. PS is not evenly distributed in cells but tends to concentrate in the plasma membrane and endosomal compartments. In addition, PS is asymmetrically distributed in the PM; 70–90% is found on the cytosolic leaflet as the result of ATP-dependent flippase activity. The negative charge of PS makes it an important recognition or docking site for soluble proteins. For example, PS recognition by the calcium-dependent C2 domains of conventional PKC isoforms facilitates membrane association and kinase activation. Calcium-dependent discoidin-type C2 and γ-carboxyglutamic acid motifs in factor V and pro-thrombin, respectively, mediate PS binding on the surface of activated plate-lets during clotting. The pleckstrin homology domain of Akt also interacts with PS to facilitate targeting to the PM and activation of pro-survival signal-ling pathways. Intracellular annexin V binds PS at the plasma membrane in a calcium-dependent manner and could be involved in negative regulation of coagulation. Annexin V is also used as a probe to monitor the exposure of PS on the extracellular leaflet of the PM during apoptosis, which is a signal for engulfment and digestion by phagocytic cells.

5.2 The Serine Base-Exchange Pathways: Phosphatidylserine Synthases 1 and 2

In mammalian cells, PS is synthesised by a calcium-dependent reaction in which the head group of PC or PE is replaced by serine. The reaction is cat-alysed by PS synthase 1 (PSS1), which utilises PC, whereas PS synthase-2 (PSS2) uses PE (Kuge and Nishijima, 1997) (Figure 7). The mechanism for PS synthesis in mammals differs from that of yeast and bacteria, which have PS synthases that utilise CDP-DAG and serine as substrates. PSS1-deficient mam-malian cells were isolated based on PS auxotrophy; cell growth was dependent on the presence of exogenous PS, PE or ethanolamine in the media. The growth requirement for exogenous PE indicates that PSS1 is a major contributor to PE synthesis by decarboxylation of its product in the mitochondria. In vitro assays of PSS1-deficient cells detected residual PSS2 activity that utilised PE as a sub-strate, and indicated that PC was the preferred substrate for PSS1. PSS1-defi-cient cells were further mutagenised to generate a clone with only 5% residual PS synthase activity that was dependent on exogenous PS for survival. Depen-dence on exogenous PS in the absence of functional PSS1 and PSS2 indicates that they are the principal PS biosynthetic enzymes in mammals.

The cDNA for PSS1 was cloned by complementation of the PSS1-deficient cells, whereas the PSS2 cDNA was identified by sequence alignment of PSS1 with DNA databases. The enzymes share approximately 30% sequence identity and are predicted to be integral membrane proteins localised to the ER. PSS1 and PSS2 are not found in the bulk ER but are instead localised to the MAM where newly synthesised PS is transported to the mitochondria for decarbox-ylation to PE. A reoccurring theme in phospholipid biosynthesis is the expres-sion of two or more enzymes catalysing the same step in a pathway. Similarly,

PS synthesis involves two apparently redundant enzymes that produce PS in the MAM but utilise different substrates. The functional overlap was confirmed when it was observed that $Pss1^{-/-}$ mice were indistinguishable from normal littermates, indicating that PSS2 can compensate for loss of PSS1 activity (Vance and Vance, 2009). $Pss2^{-/-}$ mice were also similar to wild-type littermates and PS levels were relatively unaffected in tissues due to increased PSS1 expression and reduced PS degradation. However, PSS2 is highly expressed in testis and $Pss2^{-/-}$ male mice were infertile as a result of lack of compensation by PSS1. Strikingly, crossing $Pss1^{-/-}$ and $Pss2^{-/-}$ mice yielded viable animals with only one functional Pss allele and 10% PSS activity but only a slight reduction in tissue PS mass. However, ablation of both Pss alleles is lethal, indicating an absolute requirement for PS during development.

PSS1 and PSS2 are inhibited in vitro by inclusion of PS in a base-exchange assay. Product inhibition of PSS1 and PSS2 is in accord with studies showing that supplementation of cultured cells with PS reduces the rate of de novo PS synthesis. Mutants of PSS1 and PSS2 have been identified that have increased activity when expressed in cells and are resistant to suppression by exogenous PS. In both instances a conserved arginine residue (position 95 and 97 in PSS1 and PSS2, respectively) is mutated to lysine, suggesting that a direct electrostatic interaction with PS could be involved in inhibition. Alanine-scanning mutagenesis of PSS1 identified five additional polar residues that mediated the inhibition of PSS1 activity by PS. Whether PSS is directly regulated by PS or indirectly by another PS-derived factor or by other transcriptional or posttranscriptional mechanisms is unknown.

Lenz–Majewski syndrome (LMS) is a rare spontaneous heterozygous-dominant disorder characterised by intellectual disability and craniofacial and distal limb abnormalities, notably increased bone density. Exome sequencing of DNA from affected individuals identified three separate mutations in $PSS1$ that did not affect PSS1 expression or in vitro enzyme activity but increased the rates of PS synthesis in patient fibroblasts and rendered the enzyme resistant to suppression by exogenous PS. The LMS mutations were in predicted cytoplasmic loops of PSS1 and are close to residues that are involved in product inhibition by PS. It is speculated that perturbed PS synthesis in LMS adversely affects calcium homeostasis and bone mineralisation during development. The phenotypes of the gain-of-function PSS1 allele and disorders of PC biosynthesis (Section 3) suggest that phospholipid metabolism is essential for normal skeletal development.

6. PHOSPHATIDYLINOSITOL AND POLYPHOSPHORYLATED PHOSPHATIDYLINOSITOL

6.1 Cellular Functions of PI and Its Phosphorylated Derivatives

The inositol head group of PI is a cyclohexane carbohydrate with hydroxyl groups at all six positions. Although there are nine possible stereoisomers of inositol, *myo*-inositol is most commonly found in nature. In PI, the 2-hydroxyl group of *myo*-inositol is attached by a phosphoester linkage to PA (structure shown in Figure 1). *Myo*-inositol is derived from the diet, salvaged from degradation pathways or synthesised de novo. The synthesis of *myo*-inositol occurs by a two-step pathway that involves the initial isomerisation of glucose 6-phosphate to *myo*-inositol-3-phosphate followed by dephosphorylation to *myo*-inositol. *Myo*-inositol is primarily synthesised in the kidney where it is an important osmolyte. PI comprises approximately 5–10% of the phospholipid mass of cells and is enriched in ER, its primary site of synthesis. The inositol head group of PI is uncharged and thus the lipid has an overall negative charge due to the phosphodiester moiety. PI is both a structural element in membranes and a precursor for bioactive derivatives that are phosphorylated at the 3, 4 and 5 positions of the inositol ring (Section 6.3).

6.2 Biosynthesis of CDP-DAG and PI

The activated nucleotide intermediate CDP-DAG is the substrate for synthesis of PI, phosphatidylglycerol (PG) and cardiolipin (diphosphatidylglycerol) (Figure 2). CDP-DAG is found at very low concentrations in mammalian cells (<0.01%) as a result of rapid conversion to phospholipid products. CDP-DAG is synthesised by CDP-DAG synthase (CDS) 1 and 2 in mammalian cells, and by a single Cds1 enzyme in yeast. CDS1 is an integral membrane protein that is uniformly localised to the ER and nuclear envelope. Tam41 is a related CDS that localises to the mitochondria and is specifically involved in production of CDP-DAG for PG and cardiolipin synthesis (Figure 9, Section 7.2). Overexpression studies in cultured cells indicate that phosphatidylinositol synthase (PIS), and not CDS1 and 2, is rate limiting for PI synthesis. PIS, which is encoded by a single ubiquitously expressed gene, is localised to the ER with CDS1 but is also present in vesicles that bud from the ER or at the tip of extending ER tubules. The membranes containing active PIS make contact with plasma membrane–ER contact sites, suggesting that this could be a mechanism for delivery of PI to the PM for synthesis of phosphatidylinositol polyphosphates (PIPs). Based on density gradient fractionation, CDS1 and 2 are excluded from the PIS-containing vesicles, indicating that CDP-DAG must be incorporated to maintain the synthesis and delivery of PI to other organelles.

6.3 Functions of PIPs

Phosphorylation of the inositol head group of PI at the 3, 4 and/or 5 positions results in seven mono-, di- and triphosphorylated derivatives that have a diverse signalling activities (Figure 8). In contrast to the relative abundance of PI, PIPs are found at lower concentrations (<0.1–0.001%) in cells and undergo rapid synthesis, turnover and interconversion. In 1984, the Berridge lab discovered that extracellular messengers activate phospholipase C-mediated hydrolysis

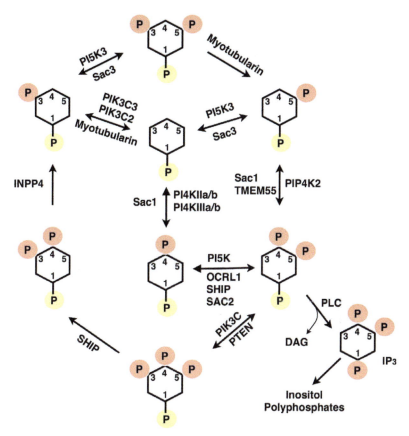

FIGURE 8 Kinases and phosphatases that interconvert phosphatidylinositol polyphosphate species. Shown are the inositol head groups of PIPs with phosphates attached at the 3, 4 and 5 positions, as well as the various kinases and phosphatases that metabolise each species. The degradation of PI-4,5P$_2$ by phospholipase C (PLC) and the formation of inositol polyphosphates from inositiol-1,4,5P$_3$ (IP$_3$) is illustrated.

of phosphatidylinositol-4,5-bisphosphate (PI-4,5P$_2$) to generate the second messengers DAG and inositol-3,4,5P$_3$ (IP$_3$). Transient accumulation of DAG activates PKC while IP$_3$ promotes the release of calcium from intracellular stores by interaction with ligand-gated channels. IP$_3$ is further phosphorylated to produce a complex array of soluble mono- and pyro-phosphorylated inositol polyphosphates that have diverse roles in regulation of cell signalling, vesicle transport, and cell death and development (Chakraborty et al., 2011) (Figure 8). At the same time that PI-4,5P$_2$ was identified as a precursor for second messengers, other researchers found that PI-4,5P$_2$ bound to proteins involved in cytoskeleton dynamics, which eventually led to identification of hundreds of PI-4,5P$_2$-binding proteins or 'effectors' that regulate secretion, ion channels, gene expression and many more essential cellular activities.

The position and number of phosphates decorating the inositol head group dictate the recruitment of specific effector proteins to membranes. Numerous PIP-binding protein modules have been identified such as pleckstrin homology (PH), FYVE, PX and ENTH domains that recognise the orientation of phosphate groups attached to the inositol ring of PIPs. For example, PI-3P in the endosomes is involved in recruitment of FYVE domain-containing proteins that modify membrane and cargo movement through endocytic and exocytic compartments. PH domains are less specific for individual PIPs and mediate recruitment of proteins to a variety of organelles, usually in association with other domains to allow coincident detection of specific metabolic signals. Due to their intrinsic cone shape and anionic head group, PIPs have the potential to induce positive membrane curvature. Curvature induction by PIPs is aided by the binding of ENTH domains, a protein module with intrinsic positive curvature that enhances curvature induction by PIPs. Since PIPs are rapidly metabolised by kinases and phosphatases at specific cell sites (Section 6.4), they temporally and spatially regulate the recruitment of these effector proteins to membranes.

6.4 PI Kinases and Phosphatases

PIPs are not uniformly distributed in cellular membranes. PI-3P is present in the endosomal pathway, PI-4P is enriched in the Golgi apparatus and PM, and PI-4,5P_2 is found in the PM and endocytic compartments. However, static measurement of lipids with such low abundance and rapid turnover does not provide a meaningful assessment of cellular function. Rather, the activities of PIPs are defined by temporal and spatial synthesis and degradation at membrane sites by kinases and phosphatases (Figure 8). PI-4P is synthesised from PI by four ATP-dependent mammalian PI 4-kinases (PI4K): PI4KIIα, PI4KIIβ, PI4KIIIα and PI4KIIIβ. PI4KIIα and PI4KIIIβ synthesise PI-4P in the Golgi and endosomes where the lipid is required for membrane trafficking. PI-4P in the Golgi apparatus and ER is dephosphorylated to PI by Sac1. PI4KIIβ and PI4KIIIα are in other endomembranes including the PM and could be involved in providing the precursor for PI-4,5P_2 synthesis by a family of three PI 5-kinases (PIP5K1). PI-4,5P_2 is also synthesised from PI-5P by a 4-kinase activity encoded by the *PIP4K2* gene family. Human genetic defects in the *PIP5K1C* gene cause congenital contractile deficiency 3, a lethal disorder that causes respiratory and muscle contraction defects related to insufficiency in PI-4,5P_2 synthesis in the brain. In contrast, defects in the PI-4,5P_2 5-phosphatase OCRL1 cause Lowe disease, which is characterised by cognitive deficits and kidney dysfunction, possibly related to delayed turnover of PI-4,5P_2 and defective receptor transport in the endocytic pathway (Nicot and Laporte, 2008).

PI-3P and PI-3,5P_2 are enriched in the endosomes and lysosomes where they are involved in recruitment of PH, FYVE and PX containing proteins that regulate membrane trafficking and autophagy. PI-3P is synthesised in the endosomal compartment by the class III kinase PIK3C3 (Vps34 in yeast) and in other organelles by class II PI 3-kinases PIK3C2. PI-3P is phosphorylated to PI-3,5P$_2$ by PIP5K3, which contains an FYVE required for PI-3P recognition. PI-3,5P$_2$ is dephosphorylated to PI-3P by Sac3, while myotubularins dephosphorylate PI-3P to PI. Several genetic disorders characterised by defects in endosomal trafficking are caused by mutations in Sac3 and myotubularin genes, such as Charcot–Marie–Tooth and X-linked myotubular myopathy.

A separate class of PI 3-kinases (referred to as class I enzymes or PIKC3A to D) in the plasma membrane is recruited to activated growth factor receptors by associated regulatory subunits and there phosphorylate PI-4,5P$_2$ to produce PI-3,4,5P$_3$, a potent activator of the mitogenic Akt/PKB signalling pathway. Production of PI-3,4,5P$_3$ leads to recruitment of the protein kinase Akt (via a PH domain) at the plasma membrane where it is subsequently phosphorylated and activated by PDK1 and mTORC1. Activated Akt then phosphorylates downstream targets that promote cell survival and proliferation. In opposition to PI-3,4,5P$_3$-mediated mitogenic signalling is the 3-phosphatase PTEN and the 5-phosphatase SHIP1 and 2, which remove the respective phosphates from PI-3,4,5P$_3$ and inhibit Akt signalling. Somatic mutations in *PIKC3* isoforms that result in increased kinase activity and activation of the Akt pathway are observed in many types of cancer. Similarly, cancer promoting loss-of-function mutations in PTEN result in sustained PI-3,4,5P$_3$ signalling, leading to the designation of this phosphatase as an anti-oncogene. The 4-phosphatases INPP4 and TMEM55 also degrade PI-3,4P$_2$ and PI-4,5P$_2$, respectively, to the PI-monophosphates.

7. BIOSYNTHESIS OF PHOSPHATIDYLGLYCEROL AND CARDIOLIPIN

7.1 Functions of PG and Cardiolipin

PG is a relatively minor lipid, comprising approximately 1% of mammalian membrane phospholipid mass, but has very important physiological activities (structure shown in Figure 1). PG is present at high concentrations in lung surfactant (5–17% of lipid mass) where it stabilises alveolar structure and contributes to the innate immune response. PG is also the precursor for the synthesis of cardiolipin, a diphosphatidylglycerol that makes up 10–15% of the phospholipid mass of the mitochondria. Cardiolipin is an unusual tetra-acylated phospholipid (structure shown in Figure 1) that is found exclusively in mitochondria (concentrated in the IMM) where it is essential for the activity of the respiratory chain by interaction with the ATP/ADP carrier, and complexes III and IV. Protein import into mitochondria, mitochondrial dynamics (fission and fusion) and release of apoptotic factors are also regulated by cardiolipin (Osman et al., 2011).

7.2 PG and Cardiolipin Biosynthetic Pathways

PG and cardiolipin are synthesised in the mitochondria via a concerted pathway that utilises CDP-DAG on the IMM (Figure 9). Mitochondrial CDP-DAG is synthesised by Tam41, which catalyses the rate-limiting step and utilises PA that is imported into the mitochondria by an unknown mechanism. There is conflicting evidence that CDS1/2-derived CDP-DAG is imported into mitochondria for cardiolipin synthesis. CDP-DAG is condensed with glycerol-3-phosphate by phosphatidylglycerol phosphate synthase (PGPS) to form phosphatidylglycerol phosphate (PGP). PGP is dephosphorylated by a mitochondrial phosphatase (PTPMT1) to form PG. The final step involves transfer of a PA group from CDP-DAG to the 3-hydroxyl group on the glycerol moiety of PG by cardiolipin synthase (CLS). The fatty acid composition of cardiolipin is different than its CDP-DAG substrate, suggesting that significant remodelling of acyl chains occurs after synthesis. Newly synthesised cardiolipin is deacylated by a mitochondrial PLA2 to form monolysocardiolipin (MLCL), which then has linoleate added by MLCL acyltransferases to form the mature functional lipid. This is an example of a fatty acid remodelling pathway for phospholipids, which are described in more detail in Section 8. The human X-linked disorder Barth syndrome is characterised by cardiomyopathy and heart failure, accumulation of MLCL and reduced amounts of cardiolipin with an abnormal fatty acid composition. The mutation responsible for Barth syndrome is in *TAZ1*, which encodes an

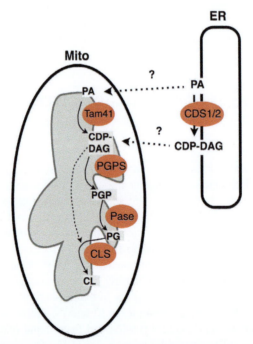

FIGURE 9 Synthesis of phosphatidylglycerol and cardiolipin in the mitochondria. CDS, CDP-diacylglycerol synthase; CLS, cardiolipin synthase; PGPS, phosphatidylglycerol phosphate synthase; Pase, PGP phosphatase.

acyltransferase that specifically adds linoleate to MLCL to make cardiolipin. The reduced level of cardiolipin and elevated MLCL in Barth syndrome are proposed to compromise the activity of energetically demanding tissues such as the heart.

8. FATTY ACID REMODELLING OF PHOSPHOLIPIDS

Just as the polar head groups of phospholipids are interconverted by exchange and methylation reactions, the fatty acid composition of individual phospholipids can be altered or 'remodelled' by fatty acid hydrolysis and reesterification reactions (Figure 10). While there are a limited number of polar head groups, combinations and permutations of the fatty acid attached to the *sn*-1 and *sn*-2 positions result in thousands of different phospholipid species (Chapter 2).

The enzymes that catalyse the final steps in phospholipid synthesis have limited selectively for fatty acyl-species of DAG or CDP-DAG and thus the fatty acid composition of newly synthesised phospholipids is similar to precursors. Once phospholipids are synthesised, however, they undergo extensive remodelling of fatty acids, primarily at the *sn*-2 position, by the concerted action of phospholipase A2 (PLA2) and LPLATs, which are composed of AGPAT and membrane-bound *O*-acyltransferase (MBOAT) gene families. William Lands identified this

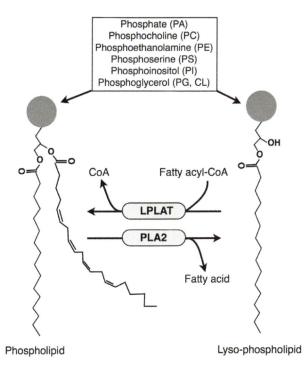

FIGURE 10 Remodelling of the *sn*-2 fatty acids of phospholipids. The concerted action of phospholipase A2 and lysophospholipid acyltransferases (LPLAT) exchange fatty acids at the *sn*-2 position of phospholipids.

remodelling cycle in 1958 as a route to increase the diversity and asymmetry of phospholipid fatty acids. Altering the fatty acid composition of phospholipids has far reaching impact on membrane and cellular function (Hishikawa et al., 2014). The introduction of polyunsaturated fatty acids (i.e. arachidonic and eicosapentanoic acid) produces phospholipid species that are involved in cell signalling, and increases the fluidity and permeability of membranes. As well, increasing the proportion of polyunsaturate-rich phospholipids at the expense of saturated species and lyso-lipids promotes negative curvature and head group packing defects in membranes. These changes in membrane structure also affect the structure and activity of integral and peripheral membrane proteins. Specific remodelling pathways in the lung are responsible for production of dipalmitoyl-PC that is essential for the activity of secreted lung surfactant.

The hydrolysis of fatty acids from the *sn*-2 position of phospholipids is catalysed by intracellular calcium-dependent and -independent PLA2 (Chapter 8). The specific PLA2s involved in remodelling are poorly understood but the activity is tightly coupled with reesterification to prevent accumulation of toxic lyso-phospholipids. Fatty acylation of lyso-phospholipids is potentially catalysed by 13 different members of the AGPAT and MBOAT families that are localised to the ER (Hishikawa et al., 2014). The LPLAT family utilises the major lyso-phospholipids as substrates in vitro but often have overlapping substrate specificities, causing uncertainly in assignment of specific in vivo functions to individual LPLAT enzymes. However, evidence is accruing that these enzymes maintain an optimal fatty acid composition in individual phospholipids that is essential for membrane function. For example, LPCAT2 (AGPAT11) and LPCAT3 (MBOAT5) use lyso-PC and polyunsaturated fatty acyl-CoAs as substrates for the synthesis of PC enriched in arachidonic and eicosapentanoic acids. In liver, the content of polyunsaturated fatty acids in PC is maintained by the liver X receptor-dependent transcription of *Lpcat3*, which prevents accumulation of saturated fatty acid-rich PC and induction of ER stress (Rong et al., 2013). AGPAT9 (LPCAT1) is highly expressed in type II alveolar cells of the lung and specifically incorporates saturated fatty acids such as palmitate fatty into lyso-PC for secretion in lung surfactant. Lyso-platelet activating factor is also acetylated by APGAT9, indicating a specialised role in pro-inflammatory pathways in other tissues. Further studies using genetic manipulation of LPLAT expression in animals and cells will provide insight into the role of these enzymes in phospholipid and membrane function.

9. FUTURE DIRECTIONS

- Although phospholipid biosynthetic pathways are well delineated, there is still a limited understanding of how phospholipids and their precursors are transported between organelles. A major challenge will be to identify the specificity of lipid transport proteins and site-to-site transport mechanisms (see Chapter 14).
- Membrane contact sites exist between most major organelles. How important are these contact zones for lipid synthesis and transport, and how are they formed and maintained?

- Rare human genetic disorders involving aberrant phospholipid metabolism have been identified. However, the advent of personalised genetics (i.e. exome sequencing) will lead to identification of more human genetic disorders that will challenge our conventional understanding of the role phospholipids play in cellular function.
- Establishing the specificity of LPLATs (AGPAT/MBOAT) involved in fatty acid remodelling of phospholipids will require a combination of in vitro enzymology and in vivo genetic approaches.
- Many phospholipid biosynthetic enzymes are potential therapeutic targets but remain poorly characterised due to membrane association and utilisation of lipophilic substrates. Large-scale expression, purification and membrane reconstitution of these proteins are required for in-depth kinetic and structure analysis as pharmaceutical targets.

REFERENCES

Aoyama, C., Ohtani, A., Ishidate, K., 2002. Expression and characterization of the active molecular forms of choline/ethanolamine kinase-alpha and beta in mouse tissues, including carbon tetrachloride-induced liver. Biochem. J. 363, 777–784.

Braverman, N.E., Moser, A.B., 2012. Functions of plasmalogen lipids in health and disease. Biochim. Biophys. Acta 1822, 1442–1452.

Cai, J., Abramovici, H., Gee, S.H., Topham, M.K., 2009. Diacylglycerol kinases as sources of phosphatidic acid. Biochim. Biophys. Acta 1791, 942–948.

Chakraborty, A., Kim, S., Snyder, S.H., 2011. Inositol pyrophosphates as mammalian cell signals. Sci. Signal 4 (188), re1. http://dx.doi.org/10.1126/scisignal.2001958.

Chong, S.S., Taneva, S.G., Lee, J.M., Cornell, R.B., 2014. The curvature sensitivity of a membrane-binding amphipathic helix can be modulated by the charge on a flanking region. Biochemistry 53, 450–461.

Cornell, R.B., Northwood, I.C., 2000. Regulation of CTP:phosphocholine cytidylyltransferase by amphitropism and relocalization. Trends Biochem. Sci. 25, 441–447.

Csaki, L.S., Dwyer, J.R., Fong, L.G., Tontonoz, P., Young, S.G., Reue, K., 2013. Lipins, lipinopathies, and the modulation of cellular lipid storage and signaling. Prog. Lipid Res. 52, 305–316.

Fagone, P., Sriburi, R., Ward-Chapman, C., Frank, M., Wang, J., Gunter, C., Brewer, J.W., Jackowski, S., 2007. Phospholipid biosynthesis program underlying membrane expansion during B-lymphocyte differentiation. J. Biol. Chem. 282, 7591–7605.

Gustin, S.E., Western, P.S., Mcclive, P.J., Harley, V.R., Koopman, P.A., Sinclair, A.H., 2008. Testis development, fertility, and survival in Ethanolamine kinase 2-deficient mice. Endocrinology 149, 6176–6186.

Han, G.S., Wu, W.I., Carman, G.M., 2006. The Saccharomyces cerevisiae Lipin homolog is a Mg2+-dependent phosphatidate phosphatase enzyme. J. Biol. Chem. 281, 9210–9218.

Henneberry, A.L., Mcmaster, C.R., 1999. Cloning and expression of a human choline/ethanolaminephosphotransferase: synthesis of phosphatidylcholine and phosphatidylethanolamine. Biochem. J. 339, 291–298.

Hishikawa, D., Hashidate, T., Shimizu, T., Shindou, H., 2014. Diversity and function of membrane glycerophospholipids generated by the remodeling pathway in mammalian cells. J. Lipid Res. 55, 799–807.

Jacobs, R.L., Zhao, Y., Koonen, D.P., Sletten, T., Su, B., Lingrell, S., Cao, G., Peake, D.A., Kuo, M.S., Proctor, S.D., Kennedy, B.P., Dyck, J.R., Vance, D.E., 2010. Impaired de novo choline synthesis explains why phosphatidylethanolamine N-methyltransferase-deficient mice are protected from diet-induced obesity. J. Biol. Chem. 285, 22403–22413.

Johnson, J.E., Xie, M., Singh, L.M., Edge, R., Cornell, R.B., 2003. Both acidic and basic amino acids in an amphitropic enzyme, CTP:phosphocholine cytidylyltransferase, dictate its selectivity for anionic membranes. J. Biol. Chem. 278, 514–522.

Kuge, O., Nishijima, M., 1997. Phosphatidylserine synthase I and II of mammalian cells. Biochim. Biophys. Acta 1348, 151–156.

Lagace, T.A., Ridgway, N.D., 2005. The rate-limiting enzyme in phosphatidylcholine synthesis regulates proliferation of the nucleoplasmic reticulum. Mol. Biol. Cell 16, 1120–1130.

Mitsuhashi, S., Ohkuma, A., Talim, B., Karahashi, M., Koumura, T., Aoyama, C., Kurihara, M., Quinlivan, R., Sewry, C., Mitsuhashi, H., Goto, K., Koksal, B., Kale, G., Ikeda, K., Taguchi, R., Noguchi, S., Hayashi, Y.K., Nonaka, I., Sher, R.B., Sugimoto, H., Nakagawa, Y., Cox, G.A., Topaloglu, H., Nishino, I., 2011. A congenital muscular dystrophy with mitochondrial structural abnormalities caused by defective de novo phosphatidylcholine biosynthesis. Am. J. Hum. Genet. 88, 845–851.

Nicot, A.S., Laporte, J., 2008. Endosomal phosphoinositides and human diseases. Traffic 9, 1240–1249.

Osman, C., Voelker, D.R., Langer, T., 2011. Making heads or tails of phospholipids in mitochondria. J. Cell Biol. 192, 7–16.

Peterfy, M., Phan, J., Xu, P., Reue, K., 2001. Lipodystrophy in the fld mouse results from mutation of a new gene encoding a nuclear protein, lipin. Nat. Genet. 27, 121–124.

Ridgway, N.D., Vance, D.E., 1987. Purification of phosphatidylethanolamine N-methyltransferase from rat liver. J. Biol. Chem. 262, 17231–17239.

Rong, X., Albert, C.J., Hong, C., Duerr, M.A., Chamberlain, B.T., Tarling, E.J., Ito, A., Gao, J., Wang, B., Edwards, P.A., Jung, M.E., Ford, D.A., Tontonoz, P., 2013. LXRs regulate ER stress and inflammation through dynamic modulation of membrane phospholipid composition. Cell Metab. 18, 685–697.

Sher, R.B., Aoyama, C., Huebsch, K.A., Ji, S., Kerner, J., Yang, Y., Frankel, W.N., Hoppel, C.L., Wood, P.A., Vance, D.E., Cox, G.A., 2006. A rostrocaudal muscular dystrophy caused by a defect in choline kinase beta, the first enzyme in phosphatidylcholine biosynthesis. J. Biol. Chem. 281, 4938–4948.

Shields, D.J., Lehner, R., Agellon, L.B., Vance, D.E., 2003. Membrane topography of human phosphatidylethanolamine N-methyltransferase. J. Biol. Chem. 278, 2956–2962.

Taneva, S.G., Lee, J.M., Cornell, R.B., 2012. The amphipathic helix of an enzyme that regulates phosphatidylcholine synthesis remodels membranes into highly curved nanotubules. Biochim. Biophys. Acta 1818, 1173–1186.

Traiffort, E., O'regan, S., Ruat, M., 2013. The choline transporter-like family SLC44: properties and roles in human diseases. Mol. Aspects Med. 34, 646–654.

Vance, D.E., Vance, J.E., 2009. Physiological consequences of disruption of mammalian phospholipid biosynthetic genes. J. Lipid Res. 50, 132–137.

Vance, J.E., 1990. Phospholipid synthesis in a membrane fraction associated with mitochondria. J. Biol. Chem. 265, 7248–7256.

Wang, Y., Kent, C., 1995. Identification of an inhibitory domain of CTP:phosphocholine cytidylyltransferase. J. Biol. Chem. 270, 18948–18952.

Wu, G., Sher, R.B., Cox, G.A., Vance, D.E., 2010. Differential expression of choline kinase isoforms in skeletal muscle explains the phenotypic variability in the rostrocaudal muscular dystrophy mouse. Biochim. Biophys. Acta 1801, 446–454.

Yang, W., Jackowski, S., 1995. Lipid activation of CTP:phosphocholine cytidylyltransferase is regulated by the phosphorylated carboxyl-terminal domain. J. Biol. Chem. 270, 16503–16506.

Chapter 8

Phospholipid Catabolism

Robert V. Stahelin[1,2]

[1]*Department of Biochemistry & Molecular Biology, Indiana University School of Medicine-South Bend, South Bend, IN, USA;* [2]*Department of Chemistry & Biochemistry, University of Notre Dame, Notre Dame, IN, USA*

ABBREVIATIONS

C1P Ceramide 1-phosphate
cPLA$_2$ Cytosolic phospholipase A$_2$
DAG Diacylglycerol
ER Endoplasmic reticulum
IP$_3$ Inositol 1,4,5-triphosphate
iPLA$_2$ Calcium-independent phospholipase A$_2$
PA Phosphatidic acid
PAF Platelet-activating factor
PH Pleckstrin homology
PIP$_2$ Phosphatidylinositol 4,5-bisphosphate
PLA$_1$ Phospholipase A$_1$
PLA$_2$ Phospholipase A$_2$
PLB Phospholipase B
PLC Phospholipase C
PLD Phospholipase D
PLase phospholipase
sPLA$_2$ Secreted phospholipase A$_2$

1. INTRODUCTION

1.1 The Enzymatic Activities of Phospholipases

Phospholipases (PLases) are a family of enzymes that hydrolyse phospholipid substrates. The major classes of PLases, which include A$_1$, A$_2$, C and D PLases, catalyse hydrolysis at different positions in a phospholipid molecule (Figure 1). Although these enzymes catalyse site-specific hydrolytic reactions, they share the property of increased catalytic efficiency for lipid substrates in a membrane when compared to a monomeric substrate. Specifically, PLases have higher activity when their substrate is aggregated such as in a micelle, mixed micelle or lipid bilayer. The aggregation of phospholipid substrate into micelles occurs when their concentration is above the critical micelle concentration.

Biochemistry of Lipids, Lipoproteins and Membranes. http://dx.doi.org/10.1016/B978-0-444-63438-2.00008-0

237

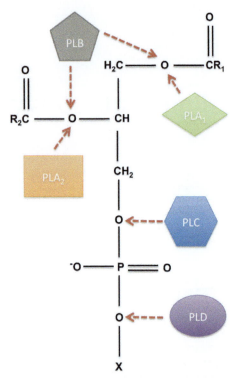

FIGURE 1 Hydrolysis of phospholipids by phospholipase. A basic phospholipid structure is shown with sites of hydrolysis indicated for PLA_1, PLA_2, PLB, PLC and PLD. R_1 and R_2, position of fatty acyl chains; X, phospholipid head group.

The different classes of PLases and their sites of attack on a phospholipid molecule are shown in Figure 1. Phospholipases A (PLA) hydrolyse either the *sn*-1 acyl ester (PLA_1) or the *sn*-2 acyl ester (PLA_2). In addition, a group of enzymes termed phospholipases B (PLB) hydrolyse both acyl chains. Phospholipase C (PLC) hydrolyses the glycerophosphate bond while phospholipase D (PLD) cleaves the phosphodiester bond to liberate the phospholipid head group. There are also lysophospholipases that remove the acyl chain from the *sn*-1 or *sn*-2 position of a lysophospholipid.

The different classes of PLases have important and diverse functions that are fundamental to life. The removal of acyl chains or phospholipid head group by PLases and the subsequent remodelling of these hydrolytic products to make new lipids have important roles in lipid signalling, membrane trafficking and nutrient digestion. The PLase activity associated with bee and snake venoms contribute to toxicity and inflammatory reactions. The PLases identified to date are soluble enzymes that transiently interact with lipid substrates. This suggests that recruitment of PLases to membranes is a regulated step that controls product formation. Some of these diverse regulatory mechanisms will be described throughout this chapter.

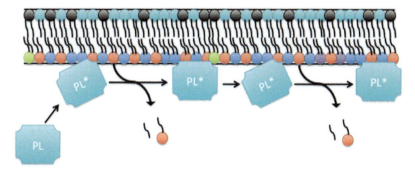

FIGURE 2 A model for interfacial catalysis by PLases on a membrane or micelle interface. A PLA$_2$ is shown in solution (PL) or interfacially bound on a membrane (PL*). PL associates with the membrane and undergoes a two-dimensional search for substrate. The interfacial binding surface locks into the membrane to provide access to substrate. The PL* binds a phospholipid substrate and hydrolyses a fatty acid from the *sn*-2 position, producing a 1-acyl lysophospholipid. PL* remains bound to the interface and scoots in two dimensions across the membrane to access more substrate. Multiple catalytic cycles continue through this scooting action. *Adapted from Bahnson (2005).*

1.2 Phospholipase Interactions at the Membrane Interface

PLases are most active in catalysis when their substrate is in an aggregated state. Thus, as noted earlier, membrane or micelle association (as in the gut during digestion) is an important determinant of their activity. The type of interfacial binding and the mode of regulation are instrumental in affecting substrate accessibility and catalysis. A general model of interfacial binding and subsequent catalysis is illustrated in Figure 2. Here, the enzyme binds to the membrane interface through a combination of electrostatic and hydrophobic interactions depending on the class of PLase. Often, association with the membrane interface induces conformational changes in the enzyme that increase substrate binding and activity. The active enzyme forms a complex with substrate, which is followed by substrate hydrolysis and formation of an enzyme–product complex that ultimately results in product release. At this point, the enzyme is still bound at the interface and can proceed through further rounds of catalysis by a scooting mechanism across the two-dimensional surface of the membrane (Bahnson, 2005).

1.3 Substrate Activation of PLases at Membrane Surfaces

Since PLases have much greater activity at the membrane interface compared to monomeric substrate in solution, the association of PLases with membranes influences the kinetics of hydrolysis. Thus, the rate of membrane association should be considered in enzyme assays, and the preferred method is to use micelles or lipid bilayers that allow PLase membrane recruitment and retention through the aforementioned scooting mechanism. These assays have been developed for secreted PLA$_2$ (sPLA$_2$), which has high affinity for anionic membranes (Gelb et al., 1995; Berg et al., 2001). If the enzyme concentration is low relative to the lipid concentration, sPLA$_2$ will remain membrane bound following release of product and scoot along the interface to promote further catalysis (Bahnson, 2005). This method is advantageous as enzyme kinetics can be measured without factoring in enzyme association and dissociation from the membrane interface. It is also an effective method of assessing active site inhibitors that do not alter enzyme conformation or membrane-binding kinetics (Gelb et al., 1995; Berg et al., 2001). However, allosteric inhibitors, which bind outside the active site and alter membrane binding, should be assessed with other methods to determine the mechanism by which they alter interfacial PLase binding.

The packing of lipid bilayers or micelles will also affect the enzyme kinetics of PLases. Thus, highly ordered lipid packing density will generally lower the rate of hydrolysis. In contrast, if the membrane is more fluid or has packing defects, catalysis will be increased due to the enhanced accessibility to substrate.

1.4 Conformation Activation of PLases at Membrane Surfaces

Membrane association of PLases often induces conformational changes that increase activity compared to the soluble enzyme (Figure 2). A well-known example of this type of interfacial association-induced conformational activation occurs for porcine pancreatic sPLA$_2$. Analysis of this enzyme by nuclear magnetic resonance and X-ray crystallography revealed a structural change in the N-terminus when bound to micelles containing anionic lipids. Lipid binding to a site distinct from the catalytic site may also be an important factor in the allosteric regulation of other PLases (Berg et al., 2001). Assessment of PLase membrane orientation and depth of penetration at the membrane interface has been examined by electron paramagnetic resonance (EPR). PLases harbouring a nitroxide spin label on an amino acid residue can be analysed by EPR, which measures the space between the spin label and a labelled lipid. By incorporating the nitroxide label at numerous sites in the PLase structure, a calculation of depth of penetration and protein orientation at the membrane interface can be performed. This method has been used to examine the orientation of membrane-bound group IIA sPLA$_2$ as well as the depth of penetration and orientation of the C2 domain from group IVA cytosolic phospholipase A$_2$ (cPLA$_2$) (Malmberg and Falke, 2005). Additionally, sPLA$_2$ from bee venom was shown to insert into membranes using aliphatic residues adjacent to the catalytic site. Thus, the composite of electrostatic, aromatic and aliphatic interactions at the membrane interface can define the membrane-bound orientation of PLases as well as their depth of penetration, which may enhance substrate accessibility to the active site. More recently, studies using molecular dynamics simulations and hydrogen–deuterium exchange mass spectrometry have been an effective means of understanding the structural transitions and docking mechanism of group IVA cPLA$_2$.

1.5 Enzymatic Assays for PLases

A number of assays are available to assess PL enzyme activity (Van den Bosch and Aarsman, 1979; Reynolds et al., 1991; Hendrickson, 1994; Lucas and Dennis, 2005). Lipid substrates with a radioactive tracer (tritium or carbon-14) at a specific position allow for sensitive detection of product. For instance, if cells are labelled with a radioactive lipid precursor, such as arachidonic acid, that is incorporated into phospholipids, cultured cells or isolated cell membranes can be used as substrates to probe for the PLase activity of choice. This method has also been used extensively to measure the in vitro activity of different classes of PLases using micelles and lipid bilayers that contain phospholipids that are isotopically labelled at different positions.

Assays using phospholipid substrates that have been synthetically modified with fluorescent or colourimetric groups have also been used to monitor PL activity (Reynolds et al., 1991; Hendrickson, 1994). Since these substrates have an additional chemical modification, they may not recapitulate the activity of PLases towards endogenous lipid substrates. However, they provide for robust in vitro analysis, especially when comparing a wild-type enzyme to mutants in order to understand the mechanisms of substrate recognition and hydrolysis. In some cases, fluorescent substrates have been used to visualise PLase activity in live cells. For a fluorescent assay to be successful, however, the substrate and product must be physically separated for analysis or have different fluorescent properties. One well-established approach for physical separation utilises a fluorescent displacement assay that allows for use of native phospholipid substrates (Wilton, 1990, 1991; Richieri and Kleinfeld, 1995). Here, the enzyme product (e.g. a long-chain fatty acid) will dislodge a bound fluorescent reporter from a protein that interacts with the enzyme product. The result is a change in fluorescence as the fluorescent reporter is displaced into solution.

2. THE PHOSPHOLIPASE A FAMILY

Since the mid-1990s, a myriad of structure–function, genetic and cell signalling studies have been performed to understand the fundamental physiology and pathophysiology of the A_1, A_2 and B families of PLases. Due to space limitations, we present several detailed examples of how basic PLase structure and function relate to their kinetics, signalling and physiological functions. We refer readers to many excellent reviews that will expand on information presented herein.

2.1 The Phospholipase A_1 Family

The phospholipase A_1 (PLA_1) family members are esterases that liberate the sn-1 acyl chain from phospholipids, resulting in the formation of a fatty acid and sn-2 acyl lysophospholipid. PLA_1s are found in mammals, snake and insect venoms and parasites. Some members of this family are thought to have diverse functions, including digestion of dietary fat, blood clotting and regulation of muscle activity. However, there is still a dearth of information on the PLA_1 family, including a lack of structural and functional data, and an overall understanding of their physiological roles. PLA_1s can hydrolyse anionic phospholipids at acidic pH or neutral lipids at physiological pH. Additionally, enzymes within this class that cleave both the *sn*-1 and the *sn*-2 acyl groups and act on lysophospholipids are referred to as phospholipase B (PLB). Further, enzymes that have limited selectivity for the *sn*-1 or *sn*-2 acyl position, releasing a fatty acid and lysophospholipid, are referred to as phospholipase A (PLA).

2.2 *Escherichia coli* Phospholipases A

E. coli contain an outer membrane phospholipase A that hydrolyses a number of lipid species. Specifically, this enzyme has PLA_1, PLA_2 and lysophospholipase activity for the *sn*-1 or *sn*-2 acyl chain. In addition, the enzyme can utilise monoacylglycerol and diacylglycerol (DAG) as substrates. The enzyme is unique among PLases as it is an integral membrane protein containing a β-barrel composed of 12 strands in an antiparallel formation that crosses the outer bacterial membrane (Snijder and Dijkstra, 2000). The β-barrels of two PLAs form a dimerisation interface that includes regions extending above the outer membrane as well as within the membrane interior (Snijder and Dijkstra, 2000). The active site, which resides above the outer membrane, contains two calcium ions as well as three key amino acids (serine, histidine and asparagine) that form the catalytic triad. It is important to note that the outer membrane of *E. coli* is rich in lipopolysaccharide and thus it is unlikely that the enzyme degrades *E. coli* phospholipids. In agreement with this conclusion, mutations of *E. coli* PLA have revealed little change in bacterial growth and lipid metabolism. Instead, the enzyme is proposed to be involved in release of proteinaceous toxins.

2.3 Phospholipase B

PLBs harbour catalytic activity for the *sn*-1 and *sn*-2 acyl positions of phospholipids. It is also referred to as a lysophospholipase because of its hydrolytic specificity for both acyl chains. Some of the better-characterised PLBs are from fungi, some of which have been proposed to serve as virulence factors in pathogenic fungi. PLBs identified include those from *Penicillum notatum* (Masuda et al., 1991), *Saccharomyces cerevisiae* and *Candida albicans*. *Mycobacteria* have also been shown to encode a copy of PLB.

2.4 Phospholipase A_2

The first phospholipase A_2 (PLA_2) activity was discovered in pancreatic fluid in the late 1800s. Later, a similar PLA_2 activity was identified in snake venom that had membrane lytic activity towards red blood cells. Subsequently, a large number of PLA_2s have been identified that are categorised into five basic groups: cPLA$_2$, sPLA$_2$, calcium-independent phospholipases A$_2$ (iPLA$_2$), platelet-activating factor (PAF) acetylhydrolases and lysosomal PLA$_2$s. The enzymes are further grouped into discrete enzyme classes based on structure, function and catalytic activity (Table 1).

sPLA$_2$s are small (14 kDa) extracellular enzymes that require high concentrations of calcium for full activity. These enzymes are divided into groups that are found in plants, snake and bee venoms and mammals (Table 1). A defining feature of the family is a catalytic site composed of key aspartate and histidine residues but lacking a catalytic serine (Figure 3). sPLA$_2$s are proposed to catalyse ester hydrolysis by two different catalytic mechanisms. The first mechanism proposed in 1981 is termed the proton-relay mechanism (Verheij et al., 1981). Because sPLA$_2$s lack a catalytic serine residue found in many other esterases, it was proposed that a molecule of water substitutes for serine in the catalytic triad to mediate hydrolysis of the *sn*-2 acyl chain. In the mid-1990s a second mechanism, based on structural analysis of the active site, proposed that two water molecules function in catalysis, one of which interacts with calcium (Rogers et al., 1996). sPLA$_2$s usually contain numerous disulphide bonds that provide stability when they are secreted from cells into the oxidising environment of the extracellular space.

2.4.1 Group I sPLA$_2$

Group I sPLA$_2$s are split into the group IA enzymes, from snake venoms, and the group IB enzymes, the pancreatic sPLA$_2$s. A number of structures of these enzymes have been solved, allowing for several proposed mechanisms of catalysis (Yuan and Tsai, 1999). The group IA enzymes utilise hydrophobic and aromatic interactions to interact with a membrane and gain access to lipid substrate, utilising a deep tunnel-like cavity in the catalytic site. These enzymes do not have a high degree of specificity for the *sn*-2 acyl chain length. The group IB enzymes, which are involved in breakdown of dietary phospholipids, are synthesised as pro-enzymes to prevent nonspecific hydrolysis of pancreatic lipids. Thus, these enzymes need to be cleaved by a protease to become active. Their activity is significantly increased on binding to bile salt mixed micelle substrates in the gut where they primarily hydrolyse dietary phospholipids to liberate *sn*-2 acyl chains and lysophospholipids that become part of a mixed micelle. Bile salts are anionic molecules and in consonance with their charge, group IB enzymes have been shown to be much more active in the presence of anionic membranes when compared to zwitterionic membranes.

2.4.2 Group II sPLA$_2$s

The group II enzymes include the subgroup IIA that are present in the venom of snakes. The IIA subgroup also includes a mammalian sPLA$_2$ that hydrolyses arachidonic acid from the *sn*-2 position of phospholipids. Interestingly, this enzyme was first discovered in patients with rheumatoid arthritis, and secreted

TABLE 1 Summary of PLA$_2$ Families

Group	Name/ Abbreviation	Source/Location	Size (kDa)	Calcium Requirement	Amino Acid Involved in Catalysis	Arachidonic Acid Selectivity at the sn-2 Position
IA	sPLA$_2$	Cobra, krait venom	13–15	μM	His	Not selective
IB	sPLA$_2$	Mammalian pancreas	13–15	μM	His	Not selective
IIA	sPLA$_2$	Human (synovial fluid), platelets, viper venom	13–15	μM	His	Not selective
IIB	sPLA$_2$	Gaboon viper venom	13–15	μM	His	Not selective
IIC	sPLA$_2$	Rat and mouse testis	15	μM	His	Not selective
IID	sPLA$_2$	Human/mouse pancreas and spleen	14–15	μM	His	Not selective
IIE	sPLA$_2$	Human/mouse brain, heart and uterus	14–15	μM	His	Not selective
IIF	sPLA$_2$	Mouse testis and embryo	16–17	μM	His	Not selective
III	sPLA$_2$	Human/bee/lizard/scorpion	15–18	μM	His	Not selective
IVA	cPLA$_2\alpha$	Human U937 cells/ platelets/RAW 264.7 cells/ rat kidney	85	nM for membrane-binding and subsequent activity	Ser	Selective

IVB	cPLA$_2$β	Human brain, heart, liver and pancreas	114	nM for cardiolipin-binding and subsequent activity	Ser	—
IVC	cPLA$_2$γ	Human heart and muscle	61	No	Ser	—
V	sPLA$_2$	Mammalian heart and lungs, macrophages	14	μM	His	Not selective
VI A-1	iPLA$_2$ or iPLA$_2$-A	P388D1 macrophages, CHO cells	84–85	No	Ser	Not selective
VI A-2	iPLA$_2$-B	Human lymphocytes and testis	88–90	No	Ser	Not selective
VIB	iPLA$_2$γ or iPLA$_2$-2	Human heart and muscle	88	No	Ser	Not selective
VIIA	PAF-AH	Human/bovine/mouse/porcine plasma	45	No	Ser	Not selective, selective for acetyl group
VIIB	PAF-AH (II)	Human/bovine liver and kidney	40	No	Ser	Not selective, selective for acetyl group
VIIIA	PAF-AH1b α1	Human brain	26	No	Ser	Not selective, selective for acetate
VIIIB	PAF-AH1b α2 (same as VIIIA but active as dimer)	Human brain	26	No	His	Not selective

Continued

TABLE 1 Summary of PLA$_2$ Families—cont'd

Group	Name/Abbreviation	Source/Location	Size (kDa)	Calcium Requirement	Amino Acid Involved in Catalysis	Arachidonic Acid Selectivity at the sn-2 Position
IX	sPLA$_2$	Snail venom	14	μM	His	—
X	sPLA$_2$	Human leucocytes, spleen and thymus	14	μM	His	Not selective
XIA	PLA$_2$-I	Green rice shoots	12.4	μM	His	—
XIB	PLA$_2$-II	Green rice shoots	12.9	μM	His	—
XII	sPLA$_2$	Mammalian heart, kidney, muscle and skin	18.7	nM	His	Not selective
XIII	Parvovirus	Parvovirus	<10, motif within larger protein	—	—	—
XIV	Bacteria and fungi	Symbiotic fungi and Streptomyces	13–19	—	—	—
XV	Lysosomal LPLA$_2$	Human macrophages	45	No	Ser	Not selective

—, poorly characterised or unknown.
Adapted from Six and Dennis (2000), Balsinde (2002), Cayman Chemical Company.

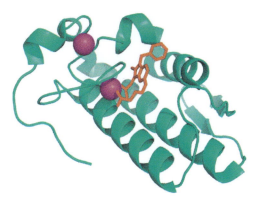

FIGURE 3 Structure of human type IIA sPLA$_2$ from synovial fluid. sPLA$_2$ binds two calcium ions (magenta) and in this example is cocrystallised with the inhibitor indole 6 (PDB ID: 1DB5). *Adapted from Schevitz (1995).*

enzyme levels dramatically increase during inflammation (e.g. sepsis or acute inflammation). It is currently unclear whether the arachidonic acid released by the enzyme is a precursor for inflammatory eicosanoids and leukotrienes (Chapter 9). In contrast to many of the sPLA$_2$s that hydrolyse phosphatidylcholine (PC), group IIA sPLA$_2$ tightly binds to membranes enriched in anionic phospholipids (e.g. phosphatidylglycerol). Despite having much stronger affinity for anionic membranes compared to zwitterionic membranes, sPLA$_2$ hydrolyses the *sn*-2 position of PC when membrane bound. The high affinity for anionic membranes as well as proteoglycans such as heparin is attributed to several cationic clusters that give this enzyme a large positive electrostatic potential.

The ability of group IIA sPLA$_2$ to migrate past the outer peptidoglycan layer of Gram-positive bacteria suggested that the enzyme has antibacterial properties (Buckland and Wilton, 2000). Further, its selectivity for binding anionic membranes is consistent with the enrichment of phosphatidylglycerol in the outer leaflet of Gram-positive bacteria. The low binding affinity for zwitterionic membranes is important as the exoplasmic leaflet of mammalian cell membranes is enriched with PC and sphingomyelin. This low affinity for mammalian cell membranes likely reduces the incidence of undesirable hydrolysis of host cell lipids. A number of other pieces of evidence support the antibacterial activity of group IIA sPLA$_2$. First, the enzyme is secreted from cells involved in host defence, such as macrophages. Second, mice deficient in the enzyme are more susceptible to Gram-positive bacterial infections. Other types of sPLA$_2$s include groups IIB, a snake venom enzyme, and groups IIC, D and F (Murakami et al., 2014).

2.4.3 Group III sPLA$_2$s

The group III sPLA$_2$ includes those isolated from bee venoms. This enzyme, which was discussed earlier, harbours an aliphatic patch surrounding the active site that can penetrate the membrane bilayer. The enzyme provokes a type 2 immune response in mice by hydrolysing host membrane phospholipids leading to interleukin-33 release, T helper cell response and activation of lymphoid cells. This suggests that the host mounts a protective immune response against the bee venom sPLA$_2$.

2.4.4 Group V sPLA$_2$s

Group V sPLA$_2$s were shown to be involved in eicosanoid synthesis and phagocytosis in studies using a knockout mouse (Balestrieri and Arm, 2006). The enzyme binds to heparan sulphate proteoglycans on the extracellular matrix, which facilitates internalisation of the enzyme. The enzyme hydrolyses phospholipids on the outer plasma membrane to liberate fatty acids. When the enzyme is added to cells it also induces activation of the group IV cPLA$_2$. The group V enzyme contains a key tryptophan residue on its interfacial-binding surface that facilitates binding to and hydrolysis of membranes rich in zwitterionic lipids such as PC.

2.4.5 Group X sPLA$_2$s

Group X sPLA$_2$ specifically binds zwitterionic membranes where it catalyses the release of fatty acids ultimately leading to eicosanoid production. This enzyme is not thought to act intracellularly like the group IIA and V enzymes as it is not internalised when added to cells in culture. The group X enzyme lacks the large cationic binding surface that mediates binding of group IIA and V enzymes to heparin sulphate proteoglycans. Like the group V enzymes, group X sPLA$_2$ contains a tryptophan residue at position 67 that is essential for the binding and hydrolysis of PC-enriched membranes. Group X also hydrolyses phospholipids on the surface of lipoproteins. The active site of group X sPLA$_2$ is similar to that of the group IA and IB enzymes, which includes bound Ca^{2+} and the presence of two water molecules. Group X sPLA$_2$ derived from epithelial tissues is important in eicosanoid generation in asthma. The neutrophil-derived enzyme is also involved in inflammation in a mouse model of abdominal aortic aneurysms.

2.4.6 Other sPLA$_2$s

Several other sPLA$_2$s have been identified that are grouped in IX, XI, XIII and XIV subclasses. The group IX PLA$_2$s are found in snail venom and group XI enzymes, which have low molecular mass (~12.5 kDa), have been identified in rice. Group XIV PLA$_2$s are found in bacteria and fungi and contain two disulphide bonds. Group XII enzymes are from type 2 helper T cells and are notably larger in mass (~20 kDa) than other sPLA$_2$s. Moreover, two splice forms were identified (XII-1 and XII-2) that exhibited a different pattern of subcellular localisation. Subsequently, a mammalian group XIIB protein was identified with a leucine substitution at the critical histidine residue of the catalytic site. Thus, the group XIIB enzyme does not hydrolyse the *sn*-2 acyl position of phospholipids and its biological role is not understood. A sPLA$_2$ motif (classified in group XIII) at the N-terminus of a parvovirus capsid protein was shown to facilitate release of the virus from the host endosomal compartment.

2.5 Cytosolic Phospholipases A₂

cPLA$_2$, also known as group IV cPLA$_2$, is a peripheral membrane protein that transiently interacts with the cytoplasmic surface of organelle membranes via a Ca^{2+}-binding N-terminal C2 domain. This ~85 kDa enzyme contains a C-terminal lipase domain. Unlike the sPLA$_2$ family members that require millimolar concentrations of Ca^{2+} for activity, cPLA$_2$ requires only nanomolar spikes in cytosolic Ca^{2+} to induce association with the cytoplasmic surface of the Golgi apparatus, endoplasmic reticulum (ER) and nuclear envelope. cPLA$_2$ primarily interacts with membranes enriched in zwitterionic lipids via a patch of aliphatic residues near the Ca^{2+}-binding site in the C2 domain. cPLA$_2$ then liberates arachidonic acid from the *sn*-2 position of membrane phospholipids, PC being the primary endogenous substrate. The primary function of cPLA$_2$ is the release of arachidonate for the production of pro-inflammatory prostaglandins and leukotrienes (Ghosh et al., 2006; Leslie et al., 2010).

2.5.1 cPLA₂ Isoforms

The cPLA$_2$ family includes the α isoform, which is implicated in a number of pathophysiological processes (Ghosh et al., 2006; Leslie et al., 2010; Bonventre, 1999; Bonventre et al., 1997). Additional β, γ, δ, ε and ζ isoforms encoded by separate genes have been discovered (Ghosh et al., 2006), but mechanistic information is often lacking and their role in cellular physiology and pathophysiological processes is poorly understood. Whereas the α, β, δ, ε and ζ isoforms contain an N-terminal C2 domain, the γ isoform lacks the C2 domain and instead is isoprenylated (Chapter 13), which provides a membrane retention signal. Unlike cPLA$_2$α, which has a high degree of specificity for phospholipids with arachidonic acid at the *sn*-2 position, cPLA$_2$γ is not selective (Ghosh et al., 2006). Additionally, cPLA$_2$β binds anionic lipids, including phosphoinositides and cardiolipin, which facilitate hydrolysis of its main substrate PC. Strikingly, the interaction of cPLA$_2$β with phosphoinositides does not require Ca^{2+} but binding to cardiolipin does. The interaction of cPLA$_2$β with cardiolipin is mediated by cationic residues adjacent to the calcium-binding sites in the N-terminal C2 domain. In contrast, phosphoinositide binding was nonselective and seemed to utilise nonspecific electrostatic association. Similar to the β isoform, cPLA$_2$δ interaction with cardiolipin is enhanced by calcium. In contrast to other cPLA$_2$s, cPLA$_2$δ has extensive *sn*-1 acyl hydrolytic activity.

2.5.2 The C2 Domain of cPLA₂

The C2 domain is an antiparallel β-sandwich that contains three Ca^{2+}-binding loops with variable amino acids but often cationic and polar amino acids that facilitate binding to anionic membranes. cPLA$_2$α harbours a unique C2 domain that contains a cluster of aliphatic and aromatic residues just above the Ca^{2+}-binding site (Perisic et al., 1998; Dessen et al., 1999) (Figure 4). In the absence of calcium, or at resting cytoplasmic concentrations of calcium, the C2 domain

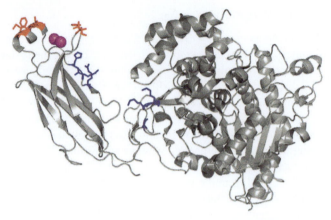

FIGURE 4 Structure of the group IVA cPLA$_2$. cPLA$_2$ has an N-terminal C2 domain (left side) that binds two calcium ions (magenta spheres) (PDB ID: 1CJY). The calcium-binding region is flanked by two loop regions containing aliphatic and aromatic residues (red) that selectively insert into the hydrocarbon region of membranes containing zwitterionic phospholipids. The C2 domain also contains a region enriched in basic residues (blue) that interacts with the anionic lipid ceramide 1-phosphate. The catalytic domain (right side of image) contains another region enriched in basic residues (blue) that interacts with anionic phospholipids including phosphoinositides. The catalytic domain accesses its should be its substrate when membrane bound, and selectively liberates arachidonic acid from the *sn*-2 position of phospholipids. *Adapted from Dessen et al. (1999).*

has low affinity for membranes containing zwitterionic phospholipids. When calcium levels increase and bind to the C2 domain, a large reduction in the desolvation penalty is observed, which facilitates insertion of the hydrophobic residues into the interior of the membrane. With small increases in cytoplasmic calcium, cPLA$_2\alpha$ has been observed to translocate to the Golgi apparatus, and as calcium continues to rise, targeting can be observed at the ER and nuclear envelope. Notably, the enzymes downstream of cPLA$_2\alpha$, 5-lipoxygnease and cyclooxygenase, which utilise arachidonic acid as substrate, are localised to these regions. Colocalisation of cPLA$_2\alpha$ with cyclooxygenase has also been observed.

Golgi membrane targeting at lower concentrations of calcium may be attributed to the ability of the C2 domain to bind the sphingolipid ceramide 1-phosphate (C1P) through basic residues adjacent to the hydrophobic/aromatic loop regions (Figure 4). Recently, C1P was shown to be synthesised at the Golgi membrane by the ceramide kinase. An increase in Golgi levels of C1P significantly contributes to eicosanoid production. Other residues in the C2 domain Ca^{2+}-binding loops can be mutated to alter intracellular targeting. For instance, if aliphatic/aromatic residues in the C2 loop are mutated to residues that facilitate binding to anionic membranes, cPLA$_2\alpha$ can be targeted to the plasma membrane when calcium levels rise. The molecular basis for targeting of other cPLA$_2$ isoforms to cellular membranes is not yet understood.

2.5.3 cPLA$_2$ Catalytic Domain

The catalytic domain of cPLA$_2\alpha$ contains Ser228, Asp549 and Arg200 residues that are required for hydrolytic activity. Similarly, other cPLA$_2$ isoforms also harbour these conserved Ser, Asp and Arg residues that, although structural information is not available, are likely required for catalysis. cPLA$_2\alpha$ is selective for arachidonate in the *sn*-2 position of membrane phospholipids, particularly PC, but also has activity towards lyso-PC with palmitate in the *sn*-1-position. Thus, despite mechanistic research into cPLA$_2\alpha$ function it may have cellular roles in lipid metabolism that are not yet understood. Phosphorylation of cPLA$_2\alpha$ at multiple sites also regulates its cellular and in vitro activity. Ser505 phosphorylation in cPLA$_2\alpha$ by mitogen-activated protein kinase promotes increased cPLA$_2\alpha$ localisation at membranes and increases activity and arachidonic acid release. Phosphorylation of cPLA$_2\alpha$ also occurs at Ser727. Cell-specific stimuli may be important determinants of site-specific phosphorylation of cPLA$_2\alpha$ and regulation of its membrane translocation and activation response. The catalytic domain also contains a cluster of basic residues that mediate phosphoinositide binding, increased activity and arachidonic acid release (Figure 4). Although this cationic cluster does not influence translocation to organelle membranes, it seems to be important for phagosomal membrane translocation in macrophages.

2.5.4 cPLA$_2$ and Health and Disease

Genetic manipulation of cPLA$_2\alpha$ expression in rodents and cell culture models has led to a better understanding of how the enzyme influences cellular homeostasis and disease (Bonventre, 1999; Bonventre et al., 1997; Kita et al., 2006). cPLA$_2\alpha$ has an important role in embryo implantation and fertility (Bonventre et al., 1997). cPLA$_2\alpha$ has also been implicated in the pathophysiology of allergic inflammation, asthma, lung cancer metastasis, spinal cord injury, Alzheimer's disease and Niemann Pick type C cholesterol storage disorder. Inhibition of a recently discovered C1P transport protein caused increased cPLA$_2\alpha$ activity and eicosanoid production. Thus, the correlation between C1P levels and cPLA$_2\alpha$ activity provides evidence of a C1P/eicosanoid axis that contributes to the aforementioned diseases. cPLA$_2\delta$ has been implicated in psoriasis but the role of other cPLA$_2$ isoforms in disease will require more investigative effort.

2.6 Group VI Intracellular PLA$_2$

Group VI calcium-independent PLA$_2$ (iPLA$_2$) is encoded by a single gene but has multiple splice variants of which iPLA$_2\beta$ and iPLA$_2\gamma$ are the most-well studied. Unlike cPLA$_2$, iPLA$_2$ does not require calcium to bind membranes but is regulated by protein–protein interactions, ATP, phosphorylation, proteolysis and calmodulin. Despite not directly interacting with calcium, iPLA$_2\beta$ is indirectly regulated by calcium levels in the cell. As calcium levels decrease, calmodulin binding to the C-terminus of iPLA$_2\beta$ increases, leading to increased enzyme activity, elevated lyso-PC and modulation of Ca^{2+} uptake into cells.

iPLA$_2$ hydrolyses phospholipids rather promiscuously with respect to head group and acyl chain composition, and unlike other PLA$_2$s, the enzyme has lysophospholipase and transacylase activity. It has also been proposed that the alternative splice variants of iPLA$_2$ lacking the catalytic domain may interact with and inhibit the full-length enzyme.

iPLA$_2$ has a protective role against bacterial and fungal infections, and is also a potential lead for targeted therapy in cancer. The enzyme is also implicated in diabetes and fatty liver disease, suggesting that pharmacotherapy aimed at iPLA$_2$ inhibition may be a useful treatment option. iPLA$_2$β is a substrate for caspase-3, which enhances enzyme activity and contributes to apoptotic cell death. The mechanism is through generation of lyso-PC, which is a secreted bioactive "eat me" signal for phagocytic cells that remove apoptotic cell corpses from tissues.

2.7 Group VII and VIII PAF Hydrolases

PAF acetylhydrolases are classified as group VII and VIII PLA$_2$s (Tjoelker and Stafforini, 2000). The group VIIA forms are secreted into the plasma while the VIIB form is intracellular. The VIIA enzyme hydrolyses the acetyl group from the sn-2 position of PAF and also hydrolyses lipids that contain oxidised fatty acids. Interestingly, this VIIA enzyme is associated with lipoproteins where it hydrolyses oxidised fatty acids from the sn-2 position of PC, primarily liberating oxidised derivatives of arachidonic acid. This activity is implicated in athero-sclerosis as it releases oxidised fatty acid and lyso-PC that have pro-inflammatory activity in the vessel wall. Thus the group VIIA enzyme is considered to be an attractive drug target and marker of cardiovascular disease. The group VIIB and group VIII enzymes have acetylhydrolase activity but low activity towards oxidised phospholipids (Tjoelker and Stafforini, 2000).

2.8 Group XV Lysosomal PLA$_2$

Lysosomal PLA$_2$, classified as a group XV PLA$_2$, has an acidic pH optima and broad substrate specificity towards phospholipids. The enzyme is expressed in many cells and tissues but is highly expressed in alveolar macrophages where it is required for hydrolysis and clearance of lung surfactant. Mice with a targeted deletion of the lysosomal PLA$_2$ gene accumulate lamellar lipid inclusions in alveolar macrophages, referred to as phospholipidosis. Mice that are deficient in lysosomal PLA$_2$ and apolipoprotein E also have increased susceptibility to atherosclerosis when fed a high cholesterol diet. Recently, the crystal structure of lysosomal PLA$_2$ was solved to reveal structural homology to lecithin:cholesterol acyltransferase, an enzyme that synthesises cholesterol esters in plasma lipoproteins. The structure also revealed an amphipathic helix in the N-terminal region that is required for interfacial activation of the lysosomal PLA$_2$ at membrane surfaces.

3. PHOSPHOLIPASE C

3.1 Bacterial PLC

PLC was first discovered in bacteria in the early 1940s when alpha-toxin from *Clostridium perfringens* was shown to hydrolyse the *sn*-3 glycerophosphodiester bond of phospholipids to release a phospho-base (head group) and DAG (Figure 1). Structurally, this toxin is similar to PLCs from other bacteria, such as *Bacillus cereus*, and contains a catalytic domain and PLAT (*p*olycystin-1, *l*ipoxygenase, *a*lpha-*t*oxin) domain, which is structurally similar to the calcium-binding C2 domains. PLAT domains have membrane-binding activity and, in the case of bacterial PLC, mediate the membrane binding and penetration of the toxin so the catalytic domain can bind and degrade PC to DAG. Ultimately, degradation of membrane phospholipids by the PLC alpha-toxin, and generation of DAG and other lipid signalling factors, contributes to the ability of bacteria to invade and destroy host cells. Notably, alpha-toxin is necessary for *C. perfringens* to escape phagocytosis and induce cytotoxicity in macrophages, leading to clostridial myonecrosis (gangrene) in infected mammals. Structural analysis of bacterial PLCs from *B. cereus* was instrumental in gaining a molecular understanding of enzyme catalysis by the PC hydrolysing form as well as a phosphatidylinositol-specific enzyme. This structure, which at the time revealed a novel all alpha-helical fold, demonstrated the presence of Zn^{2+} in the active site, which was proposed to be responsible for coordinating substrate.

3.2 Mammalian Phospholipases C

Mammalian PLCs (β, γ, δ, ϵ, ζ and η isoforms) are encoded by six separate genes and primarily involved in lipid signalling (Bunney and Katan, 2011). Like the bacterial enzymes, mammalian PLCs are phosphodiesterases but in this case the primary substrate is phosphatidylinositol 4,5-bisphosphate (PIP_2) that is degraded to a soluble signalling factor inositol triphosphate (IP_3) and membrane-associated DAG (see Chapter 7). The generation of IP_3 and DAG can cause myriad signalling events, the most well characterised being activation of the protein kinases C family of serine/threonine kinases through an IP_3-dependent increase in cytosolic calcium levels and a direct interaction of PKC C1 domains with DAG. IP_3 stimulates a rise in cytosolic calcium by binding to gated calcium channels in the ER. The DAG signal is attenuated by DAG kinase-dependent phosphorylation to phosphatidic acid (PA) (Chapter 7). The regulation and interaction of PLCs with cellular membranes are mediated by their lipid-binding pleckstrin homology (PH) and C2 domains that have isoform-specific lipid-binding properties. With the exception of the ζ isoform, all PLC isoforms harbour a PH domain that mediates membrane association by interaction with the head group of phosphoinositides. All PLC isoforms contain a C2 domain as well as four calcium-binding EF hand motifs. Additionally, protein–protein interactions motifs (SH2 and SH3 domains) are found in the γ

isoform, while the ε isoform harbours a Ras-associating domain and a guanine nucleotide exchange factor domain. The first structure of a mammalian PLC, solved in 1996 by L.-O. Essen, suggested a phosphoinositide-mediated recruitment of the PH domain to membranes that enhanced C2 membrane interactions and the accessibility of the catalytic domain for its substrate PIP_2. Recent structural analysis of PLCβ3 revealed that the enzyme's activity is inhibited in the basal state due to an interaction of a C-terminal helix with part of the catalytic site. Release of this intramolecular interaction seems to require major energetic input through multiple lipid–protein and protein–protein interactions to induce enzyme activation.

PLC catalysis proceeds via a two-step mechanism in which the first step is an intramolecular attack by the 2-hydroxyl on the inositol ring on the 1-phosphate group to generate a cyclic phosphodiester intermediate (IP_3) and DAG. The cyclic IP_3 is the product of bacterial PLC catalysis. However, in the case of mammalian PLCs, the cyclic IP_3 is retained in the active site and hydrolysed by water to produce the acyclic IP_3 product. In both cases, DAG that is generated is retained in the membrane bilayer.

4. PHOSPHOLIPASE D

The PLD family of enzymes catalyses the hydrolysis of the phosphodiester bond with the head group to generate PA and free head group, for example, choline in the case of PC hydrolysis (Figure 1). PA itself can activate a variety of protein and lipid kinases, or can be deacylated to lyso-PA and signal through members of the EDG family of G-protein coupled receptors. PA can be dephosphorylated to DAG, which activates the protein kinases C family. PA is emerging as an important regulator of cellular proliferation and membrane trafficking. As a consequence, altered PLD expression and PA metabolism have been implicated in the pathophysiological of cancer and Alzheimer and Parkinson diseases (Zhang and Frohman, 2014; Bruntz et al., 2014). PA can also change the physical properties of membranes, inducing negative membrane curvature that is required for vesicular transport. The first PLD was discovered in the late 1940s by D.J Hanahan and I.L. Chaikoff, and for many decades research was focused on this plant enzyme. A PLD gene from plants was cloned and sequenced in 1994, which subsequently led to identification of PLD genes in mammals. This was an important discovery that initiated several decades of research to decode the downstream signalling role of PA in different organisms.

4.1 PLD Structure and Enzymology

PLD catalyses a transphosphatidylase reaction that involves a covalent phosphatidyl intermediate with an active site histidine and release of the soluble phospholipid head group (i.e. choline). Water or an alcohol then act as the nucleophile on the phosphatidylhistidine to release PA or generate a phospholipid with a new head group, respectively. In vivo, the primary nucleophile is water, leading to release of PA. However, exposure of the enzyme to high concentrations of a primary or secondary alcohol in vitro favours the synthesis of phosphatidylalcohols. This later reaction is a useful assay for measuring the activity of PLD in intact cells or lysates. While PC is the major substrate of PLD, some PLD isoforms can utilise phosphatidylethanolamine or cardiolipin as substrates. The structure of mammalian PLD has not been solved but several structures of bacteria or spider venom enzymes have been elucidated. The first structure for the *Streptomyces* sp. PLD revealed two active site histidine residues that were proposed to be nucleophiles for the covalent enzyme intermediate. Later structural work clearly showed that His^{170} was the nucleophile that forms the phosphatidylhistidine intermediate. The catalytic site of PLDs has two histidine, lysine and aspartic acid (HKD) motifs that are necessary for catalytic activity. As noted above, a histidine residue in this motif serves as the nucleophile that forms the covalent intermediate that is quickly hydrolysed by water or alcohol.

4.2 PLD Isoforms

Two PLD isoforms, PLD1 and PLD2, have been identified in mammals. These two isoforms are ~50% identical and contain PH and PX regulatory domains. A motif positioned between the HDK catalytic domain interacts with PIP$_2$, which is an important regulator of PLD activity. One plant PLD isoform has a similar structure, but other plant isoforms lack PH and PX domains and contain a C2 domain that mediates calcium regulation of activity. Mammalian PLD1 and 2 are active on the cytoplasmic surface of the plasma membrane, Golgi apparatus and ER, where they are regulated by a complex set of interactions with PIP$_2$, protein kinases, small GTPases and adapter proteins (Gomez-Cambronero, 2014). Small GTPases both interact with and are regulated by PLDs. For instance, the region around the PX domain of PLD2 has guanine nucleotide exchange activity for rac2 and rhoA. Activation of rac2 by PLD2 controls actin cytoskeleton organisation and cell motility, important determinates for cancer cell invasiveness and metastasis.

5. FUTURE DIRECTIONS

From 2004 to 2014 there has been tremendous growth in understanding how PLases control physiological and pathophysiological processes. Despite this, there is still a dearth of information on what guides spatial and temporal cues for site-specific recruitment of different types of PLases. Given the complex cellular roles of the PLases and different cellular and tissue-specific patterns of expression, secretion and cellular localisation, this becomes a timely but difficult question to answer. How one enzyme such as cPLA$_2\alpha$ has critical roles in many different pathological conditions is also an intriguing avenue of research. Additionally, the emergence of PLD as an important determinant of cancer cell migration and invasion warrants future investigation. With advancements in structural analysis, cellular imaging, lipidomic and animal models of disease, many of these questions can be answered in the coming years.

REFERENCES

Bahnson, B.J., 2005. Structure, function, and interfacial allosterism in phospholipase A2: insight from the anion-assisted dimer. Arch. Biochem. Biophys. 433, 96–106.

Balestrieri, B., Arm, J.P., 2006. Group V sPLA(2): classical and novel functions. Biochim. Biophys. Acta Mol. Cell Biol. Lipids 1761, 1280–1288.

Balsinde, J., Winstead, M.V., Dennis, E.A., 2002. Phospholipase A(2) regulation of arachidonic acid mobilization. FEBS Lett. 531, 2–6.

Berg, O.G., Gelb, M.H., Tsai, M.D., Jain, M.K., 2001. Interfacial enzymology: the secreted phospholipase A(2)-paradigm. Chem. Rev. 101, 2613–2653.

Bonventre, J.V., 1999. The 85-kD cytosolic phospholipase A2 knockout mouse: a new tool for physiology and cell biology. J. Am. Soc. Nephrol. 10, 404–412.

Bonventre, J.V., Huang, Z., Taheri, M.R., O'leary, E., Li, E., Moskowitz, M.A., Sapirstein, A., 1997. Reduced fertility and postischaemic brain injury in mice deficient in cytosolic phospholipase A2. Nature 390, 622–625.

Van den Bosch, H., Aarsman, A.J., 1979. Review on methods of phospholipase-A determination. Agents Actions 9, 382–389.

Bruntz, R.C., Lindsley, C.W., Brown, H.A., 2014. Phospholipase D signaling pathways and phosphatidic acid as therapeutic targets in cancer. Pharmacol. Rev. 66, 1033–1079.

Buckland, A.G., Wilton, D.C., 2000. The antibacterial properties of secreted phospholipases A(2). Biochim. Biophys. Acta Mol. Cell Biol. Lipids 1488, 71–82.

Bunney, T.D., Katan, M., 2011. PLC regulation: emerging pictures for molecular mechanisms. Trends Biochem. Sci. 36, 88–96.

Dessen, A., Tang, J., Schmidt, H., Stahl, M., Clark, J.D., Seehra, J., Somers, W.S., 1999. Crystal structure of human cytosolic phospholipase A2 reveals a novel topology and catalytic mechanism. Cell 97, 349–360.

Gelb, M.H., Jain, M.K., Hanel, A.M., Berg, O.G., 1995. Interfacial enzymology of glycerolipid hydrolases – lessons from secreted phospholipases A(2). Annu. Rev. Biochem. 64, 653–688.

Ghosh, M., Tucker, D.E., Burchett, S.A., Leslie, C.C., 2006. Properties of the Group IV phospholipase A(2). Prog. Lipid Res. 45, 487–510.

Gomez-Cambronero, J., 2014. Phospholipase D in cell signaling: from a myriad of cell functions to cancer growth and metastasis. J. Biol. Chem. 289, 22557–22566.

Hendrickson, H.S., 1994. Fluorescence-based assays of lipases, phospholipases, and other lipolytic enzymes. Anal. Biochem. 219, 1–8.

Kita, Y., Ohto, T., Uozumi, N., Shimizu, T., 2006. Biochemical properties and pathophysiological roles of cytosolic phospholipase A(2)s. Biochim. Biophys. Acta Mol. Cell Biol. Lipids 1761, 1317–1322.

Leslie, C.C., Gangelhoff, T.A., Gelb, M.H., 2010. Localization and function of cytosolic phospholipase A2alpha at the Golgi. Biochimie 92, 620–626.

Lucas, K.K., Dennis, E.A., 2005. Distinguishing phospholipase A(2) types in biological samples by employing group-specific assays in the presence of inhibitors. Prostaglandins Other Lipid Mediat 77, 235–248.

Malmberg, N.J., Falke, J.J., 2005. Use of EPR power saturation to analyze membrane-docking geometries or peripheral proteins: applications to C2 domains. Annu. Rev. Biophys. Biomol. Struct. 34, 71–90.

Masuda, N., Kitamura, N., Saito, K., 1991. Primary structure of protein moiety of *Penicillium notatum* phospholipase B deduced from the cDNA. Eur. J. Biochem. 202, 783–787.

Murakami, M., Taketomi, Y., Miki, Y., Sato, H., Yamamoto, K., Lambeau, G., 2014. Emerging Roles of Secreted Phospholipase A2 Enzymes, third ed. 107, 105–113.

Perisic, O., Fong, S., Lynch, D.E., Bycroft, M., Williams, R.L., 1998. Crystal structure of a calcium-phospholipid binding domain from cytosolic phospholipase A2. J. Biol. Chem. 273, 1596–1604.

Reynolds, L.J., Washburn, W.N., Deems, R.A., Dennis, E.A., 1991. Assay strategies and methods for phospholipases. Methods Enzymol. 197, 3–23.

Richieri, G.V., Kleinfeld, A.M., 1995. Continuous measurement of phospholipase A2 activity using the fluorescent probe ADIFAB. Anal. Biochem. 229, 256–263.

Rogers, J., Yu, B.Z., Serves, S.V., Tsivgoulis, G.M., Sotiropoulos, D.N., Ioannou, P.V., Jain, M.K., 1996. Kinetic basis for the substrate specificity during hydrolysis of phospholipids by secreted phospholipase A2. Biochemistry 35, 9375–9384.

Schevitz, R.W., Bach, N.J., Carlson, D.G., Chirgadze, N.Y., Clawson, D.K., Dillard, R.D., Draheim, S.E., Hartley, L.W., Jones, N.D., Mihelich, E.D., Olkowski, J.L., Snyder, D.W., Sommers, C., Wery, J.-P., 1995. Structure-based design of the first potent and selective inhibitor of human non-pancreatic secretory phospholipase A2. Nat. Struct. Biol. 2, 458–465.

Six, D.A., Dennis, E.A., 2000. The expanding superfamily of phospholipase A(2) enzymes: classification and characterization. Biochim. Biophys. Acta 1488, 1–19.

Snijder, H.J., Dijkstra, B.W., 2000. Bacterial phospholipase A: structure and function of an integral membrane phospholipase. Biochim. Biophys. Acta Mol. Cell Biol. Lipids 1439, 1–16.

Tjoelker, L.W., Stafforini, D.M., 2000. Platelet-activating factor acetylhydrolases in health and disease. Biochim. Biophys. Acta Mol. Cell Biol. Lipids 1488, 102–123.

Verheij, H.M., Slotboom, A.J., Dehaas, G.H., 1981. Structure and function of phospholipase-A2. Rev. Physiol. Biochem. Pharmacol. 91, 91–203.

Wilton, D.C., 1990. A continuous fluorescence displacement assay for the measurement of phospholipase A2 and other lipases that release long-chain fatty acids. Biochem. J. 266, 435–439.

Wilton, D.C., 1991. A continuous fluorescence-displacement assay for triacylglycerol lipase and phospholipase C that also allows the measurement of acylglycerols. Biochem. J. 276, 129–133.

Yuan, C.H., Tsai, M.D., 1999. Pancreatic phospholipase A(2): new views on old issues. Biochim. Biophys. Acta Mol. Cell Biol. Lipids 1441, 215–222.

Zhang, Y., Frohman, M.A., 2014. Cellular and physiological roles for phospholipase D1 in cancer. J. Biol. Chem. 289, 22567–22574.

Chapter 9

The Eicosanoids: Cyclooxygenase, Lipoxygenase and Epoxygenase Pathways

William L. Smith,[1] Robert C. Murphy[2]
[1]*Department of Biological Chemistry, University of Michigan Medical School, Ann Arbor, MI, USA;* [2]*Department of Pharmacology, University of Colorado-Denver, Aurora, CO, USA*

ABBREVIATIONS

2-AG 2-Arachidonoylglycerol
AA Arachidonic acid
BLT (1 or 2) Leukotriene B4 receptor (subclass 1 or 2)
COX Cyclooxygenase
cPLA$_2$ Cytosolic phospholipase A$_2$
CRE cAMP response element
cysLT (1 or 2) Cysteinyl leukotriene receptor (subclass 1 or 2) for which leukotriene C4, D4 and E4 are agonists
CYP4F(x) Cytochrome P450 isozyme that carries out ω-oxidation of leukotriene B4 (subclass x)
EET Epoxyeicosatetraenoic acid
EP Prostaglandin E receptor
ER Endoplasmic reticulum
FLAP 5-Lipoxygenase-activating protein
HETE Hydroxyeicosatetraenoic acid
HpETE Hydroperoxyeicosatetraenoic acid
LO Lipoxygenase
LT Leukotriene (followed by letter to designate structural type)
NSAID Nonsteroidal anti-inflammatory drug
PG Prostaglandin (followed by letter to designate structural type)
PGHS Prostaglandin endoperoxide H synthase
PLA$_2$ Phospholipase A$_2$
sPLA$_2$ Secretory phospholipase A$_2$
TXAS TxA synthase

Biochemistry of Lipids, Lipoproteins and Membranes. http://dx.doi.org/10.1016/B978-0-444-63438-2.00009-2

1. INTRODUCTION

1.1 Background, Terminology, Structures and Nomenclature

The term 'eicosanoids' is widely used to denote a group of bioactive, oxygenated, long chain polyunsaturated fatty acids containing 18–22 carbons (Smith, 2008). The major precursor of these compounds is the 20 carbon essential fatty acid arachidonic acid (AA; all *cis* 5,8,11,14-eicosatetraenoic acid). The pathways leading to oxygenated AA derivatives are known collectively as the 'arachidonate cascade' (Figure 1). There are three major pathways within this cascade: the cyclooxygenase (COX), lipoxygenase (LO) and epoxygenase (EPOX) pathways. In each case, these pathways are named after the enzyme(s) that catalyses the committed step. The prostanoids, which include the prostaglandins and thromboxanes, are formed via the COX pathway. The first part of our discussion will focus on the prostanoids. Later in this chapter, we describe the LO and EPOX pathways. Because of space limitations, we cite relevant reviews and not primary references in most of the chapter. Additional information about many aspects of the topics covered here can be found in a collection of detailed reviews published in 2011 in a special thematic issue of *Chemical Reviews*.

The major eicosanoids are products of nutritionally essential ω-6 essential fatty acids. The essentiality of this group of fatty acids relates in large part to their functions as eicosanoids (Smith, 2008; Zheng et al., 2011). Prostanoids and leukotrienes were discovered as vasoactive substances active on reproductive and pulmonary smooth muscle, respectively. The structures of the more stable prostanoids were determined in the late 1950s and early 1960s by workers at the Karolinksa Institute in Sweden and Unilever Laboratories in the Netherlands. The leukotrienes were initially known as slow-reacting substances of anaphylaxis. Their structures were determined in the early 1980s. P450-dependent oxidations and epoxidations of polyunsaturated fatty acids were discovered and related to certain vaso-activities in the early 1980s. More recently, soluble epoxide hydrolases have been recognised as pharmacologically important agents converting epoxy acids to inactive products. Details of the pathways and the enzymes and receptors for these various pathways have been characterised since 1965. Because eicosanoid overproduction is associated with a number of pathologies, potent enzyme inhibitors and receptor antagonists have been developed that are widely used therapeutically.

2. PROSTANOIDS

2.1 Prostanoid Structures and Pathways

The structures and biosynthetic interrelationships of the most important prostanoids are shown in Figure 2 (Smith, 2008). PG is the abbreviation for prostaglandin, and Tx is the abbreviation for thromboxane. Naturally occurring

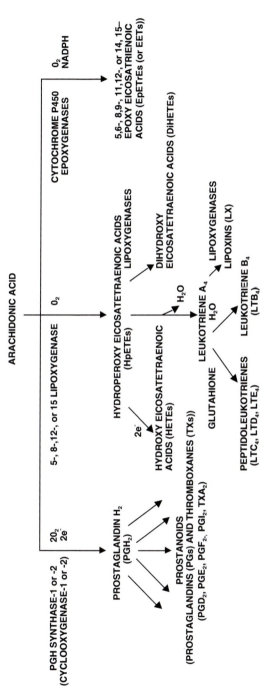

FIGURE 1 Cyclooxygenase, lipoxygenase and epoxygenase pathways leading to the formation of eicosanoids from arachidonic acid.

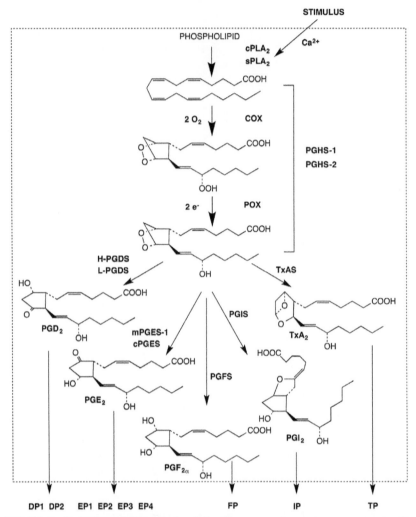

FIGURE 2   Structures and biosynthetic relationships among the most common prostanoids. cPLA$_2$, cytosolic phospholipase A$_2$; sPLA$_2$, nonpancreatic, secretory phospholipase A$_2$; PG, prostaglandin; PGHS, prostaglandin endoperoxide H synthase; COX, cyclooxygenase; POX, peroxidase; H-PGDS, hematopoietic PGD synthase; L-PGDS, lipocalin-type PGD synthase; cPGES, cytosolic PGE synthase; mPGES-1, microsomal PGE synthase-1; PGFS, PGF synthase; PGIS, PGI (prostacyclin) synthase; TXAS, TxA synthase. DP1, DP2, EP1, EP2, EP3, EP4, FP, IP and TP are designations for the G protein linked PG receptors.

prostaglandins contain a cyclopentane ring, a *trans* double bond between C-13 and C-14, and a hydroxyl group at C-15. The letters following the abbreviation PG indicate the nature and location of the oxygen-containing substituents present in the cyclopentane ring. Letters are also used to label thromboxane derivatives (e.g. TxA and TxB). The numerical subscripts indicate the number of carbon–carbon double bonds in the side chains emanating from the cyclopentane ring (e.g. PGE_1 vs. PGE_2). In general, those prostanoids with the '2' subscript are derived from AA; the '1' series prostanoids are formed from 8,11,14-eicosatrienoic acid (dihomo-γ-linolenic acid), and the '3' series compounds are derived from 5,8,11,14,17-eicosapentaenoic acid (EPA). Greek subscripts are used to denote the orientation of ring hydroxyl groups (e.g. $PGF_{2\alpha}$).

Prostanoids formed by the action of COXs have their aliphatic side chains emanating from C-8 and C-12 of the cyclopentane ring in the orientations shown in Figure 2. Prostanoids formed via the COX pathway have their side chains *cis* to the cyclopentane ring. Another group of prostanoids, known as isoprostanes, are formed from AA and other fatty acids by nonenzymatic autooxidation. The side chains of isoprostanes are primarily in the *trans* orientation. Somewhat surprisingly, isoprostanes and their metabolites are often found in greater quantities in urine than metabolites of prostanoids formed enzymatically via the COX. Particularly under pathological conditions that support autooxidation (e.g. CCl_4 toxicity), isoprostanes are produced in abundance and they may be general markers for oxidant stress (Milne et al., 2011).

2.2 Prostanoid Chemistry

Prostaglandins are soluble in lipid solvents below pH 3.0 and are typically extracted from acidified aqueous solutions with ether, chloroform/methanol or ethyl acetate. PGE, PGF and PGD derivatives are relatively stable in aqueous solution at pH 4–9 for short times; above pH 10, both PGE and PGD are subject to dehydration. PGI_2, which is also known as prostacyclin, contains a vinyl ether group that is very sensitive to acid-catalysed hydrolysis; PGI_2 is unstable below pH 8.0. The stable hydrolysis product of PGI_2 is 6-keto-$PGF_{1\alpha}$. PGI_2 formation is usually monitored by measuring 6-keto-$PGF_{1\alpha}$ formation. TxA_2, which contains an oxane–oxetane grouping in place of the cyclopentane ring, is hydrolysed rapidly ($t_{1/2} \sim 30$ s at 37 °C in neutral aqueous solution) to TxB_2; TxA_2 formation is assayed by quantifying TxB_2. Prostaglandin derivatives are often quantified with immunoassays or, increasingly more commonly, by mass spectrometry using deuterium-labelled internal standards.

3. PROSTANOID BIOSYNTHESIS

Prostanoids are not stored by cells, but rather are synthesised and then released rapidly (5–60 s) in response to extracellular hormonal stimuli. The pathway for stimulus-induced prostanoid formation as it might occur in a model cell is illustrated in Figure 2. Prostanoid formation occurs in three stages: (1) mobilisation of free AA (or 2-arachidonoylglycerol (2-AG); see below) from membrane phospholipids primarily via cytosolic phospholipase $A_2\alpha$ ($cPLA_2\alpha$) but involving in some cases nonpancreatic, secretory PLA_2s ($sPLA_2$s); (2) transformation of AA (or 2-AG) to the prostaglandin endoperoxide PGH_2 (or 2-PGH_2-glycerol) by a prostaglandin endoperoxide H synthase (PGHS; also known as COX); and (3) cell-specific conversion of PGH_2 (or 2-PGH_2-glycerol) to one of the major biologically active prostanoids by one of a series of different synthases.

3.1 Mobilisation of AA

Prostanoid synthesis is initiated by the interaction of various stimuli (e.g. bradykinin, angiotensin II, thrombin, PDGF, IL-1β) with their cognate cell surface receptors (Figure 2), which leads to increases in the concentration of cytosolic Ca^{2+} and activation of one or more cellular lipases. Although in principle, there are a variety of lipases that could participate in this AA-mobilisation phase, the most relevant are the high molecular weight, Ca^{2+}-dependent cPLA$_2\alpha$ and the Group IIA and Group V sPLA$_2$s (Dennis et al., 2011; Murakami et al., 2011) (see Chapter 8). An evolving consensus regarding the roles of PLA$_2$s in prostanoid synthesis is that cPLA$_2\alpha$ is involved directly in mobilizing AA for the constitutive PGHS-1 whereas sPLA$_2$s function coordinately with cPLA$_2\alpha$ in mobilizing AA for the inducible PGHS-2 (Smith, 2008; Murakami et al., 2011). It should be noted that cPLA$_2$s and sPLA$_2$s as well as the Ca^{2+}-independent iPLA$_2$s play many physiologic roles not directly related to eicosanoid formation (Dennis et al., 2011; Murakami et al., 2011).

3.2 Cytosolic and Secreted Phospholipases A$_2$

The predominant cPLA$_2$ isoform is cPLA$_2\alpha$, which is found in the cytosol of resting cells. Hormone-induced mobilisation of intracellular Ca^{2+} leads to the translocation of cPLA$_2\alpha$ to intracellular membranes. The translocation of cPLA$_2\alpha$ involves binding of Ca^{2+} to its N-terminal CalB (also called a C2) domain and then binding of the Ca^{2+}/CalB domain to intracellular membranes (Evans et al., 2004). At lower concentrations of intracellular Ca^{2+}, cPLA$_2\alpha$ translocates to the Golgi apparatus. As Ca^{2+} concentrations rise, cPLA$_2\alpha$ also translocates to the endoplasmic reticulum (ER) and contiguous membranes of the nuclear envelope (Evans et al., 2004). Thus, Ca^{2+} is involved in translocation of cPLA$_2\alpha$ but not directly in its catalytic mechanism (Dennis et al., 2011). cPLA$_2\alpha$ preferentially cleaves AA and EPA as opposed to other fatty acids from the *sn2* position of glycerophospholipids located on the cytosolic surfaces of intracellular membranes. Presumably, newly mobilised AA then can partially traverse the membrane bilayer where it can enter the COX active site of PGHSs embedded in the lumenal surface of intracellular membranes (Smith et al., 2011). The activity of cPLA$_2\alpha$ is regulated by phosphorylation by a variety of kinases and is augmented by interactions with ceramide-1-phosphophate and phosphatidylinositol 4,5-bisphosphate (Dennis et al., 2011).

There are 11 different nonpancreatic sPLA$_2$s (Murakami et al., 2011). Two of these enzymes, Group IIA and Group V sPLA$_2$s, have been shown to be involved in releasing AA for prostanoid synthesis in isolated cell systems (Smith, 2008). Stimulus-dependent AA mobilisation by these sPLA$_2$s appears to occur independent of the ability of these enzymes to be released from cells. Expression of Group IIA sPLA$_2$ requires the presence of cPLA$_2\alpha$. In general, it is difficult to determine if the effects of sPLA$_2$s on eicosanoid formation are direct or indirect. sPLA$_2$s do not exhibit specificity toward AA in hydrolysing fatty acids from phospholipids, and sPLA$_2$s can generate lysophosphatidylcholine, itself a bioactive signalling molecule.

sPLA$_2$s are compact, stable, relatively low molecular weight proteins. Several crystal structures have been determined including that of Group IIA sPLA$_2$ from human synovial fluid. PLA$_2$s require high concentrations of Ca^{2+} ($\sim$1 mM) for maximal activity. Ca^{2+} is directly involved in phospholipid substrate binding and catalysis by sPLA$_2$s. As noted above, unlike cPLA$_2\alpha$, sPLA$_2$ shows no specificity toward the acyl group at the *sn*2 position. The levels of sPLA$_2$s are regulated transcriptionally in response to cell activation.

cPLA$_2\alpha$ is directly involved in the immediate AA release that occurs when a cell is stimulated with a circulating hormone or protease and a large increase in intracellular Ca^{2+} concentration occurs. For example, thrombin acting through its cell surface receptor activates cPLA$_2\alpha$ in platelet cells to cause AA release that results in TxA$_2$ formation involving constitutive PGHS-1. This entire process occurs in seconds. cPLA$_2\alpha$ also plays a prominent role in 'late-phase' prostaglandin formation mediated by inducible PGHS-2, which typically occurs 1–4 h after cells have been exposed to an inflammatory mediator (e.g. endotoxin or interleukin-1β) or a growth factor (e.g. platelet-derived growth factor) to induce PGHS-2 and Ca^{2+} concentrations are moderately elevated (Smith, 2008).

3.3 Mobilisation of 2-AG and AA via Phospholipase C

2-AG is an efficient in vitro substrate for PGHS-2. Originally, it was found that PGHS-2 but not PGHS-1 converts 2-AG to 2-PGH$_2$-glycerol, and that this intermediate can be converted to various 2-prostanyl-glycerol derivatives although not to 2-thromboxane-glycerol (Rouzer and Marnett, 2011). It is unclear in which biological settings or to what degree 2-AG is utilised in vivo to form 2-prostanyl-glycerol derivatives. However, some of these products do have unique actions that distinguish them from the classical prostanoid homologs (Rouzer and Marnett, 2011). Although the pathway leading to the formation of 2-AG itself has not been defined for prostanoid metabolism, 2-AG could be formed from phosphatidylcholine through the sequential actions of phospholipase C and a diacylglycerol lipase. There is evidence that monoacylglycerol lipase in brain can cleave 2-AG to mobilise AA for PG biosynthesis. This provides for an alternate route for AA mobilisation not involving PLA$_2$s (Nomura et al., 2011).

3.4 Physico-Chemical Properties of PGHS That Catalyse Prostaglandin Endoperoxide H₂ Formation

Once AA is mobilised from phospholipids, it can be acted on by PGHSs to form PGH_2 (Tsai and Kulmacz, 2010; Smith et al., 2011; Dong et al., 2013) (Figure 2). There are two PGHS isozymes – PGHS-1 and PGHS-2. Figure 3(A) compares the domain structures of PGHS-1 and PGHS-2. A crystal structure of PGHS-1 with a bound inhibitor is presented in Figure 3(B). Crystal structures of PGHS-1 and PGHS-2 are essentially superimposable.

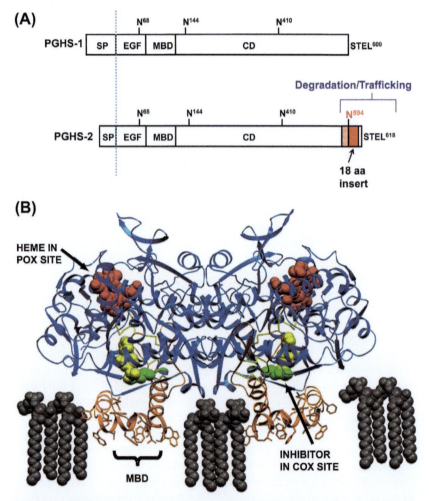

FIGURE 3 PGHS structures. (A) Domain structures of PGHS-1 and PGHS-2. (B) Ribbon diagram of the structure of ovine PGH synthase-1 homodimer interdigitated via its membrane-binding domain (MBD) into the luminal surface of the endoplasmic reticulum.

PGHS-1 was first purified from membrane fractions of bovine and ovine vesicular glands in the mid-1970s, and many biochemical studies have been performed using this PGHS isoform (Tsai and Kulmacz, 2010; Smith et al., 2011). The sequences of cDNA clones for PGHS-1 from many mammals indicate that the protein has a signal peptide of 24–26 amino acids that is cleaved to yield a mature protein of 574 amino acids. PGHS-2 was discovered in 1991 in phorbol ester-activated murine 3T3 cells as an immediate early gene product that was subsequently shown to encode a second catalytically active PGHS isoform. The mature, processed form of PGHS-2 from mammals has 587 amino acids. PGHS-1 and PGHS-2 from the same species have amino acid sequences that are 60% identical (Smith et al., 2011). The major sequence differences between the isoforms are in the signal peptides and the membrane-binding domains (residues 70–120 of PGHS-1) (Figure 3(A)). In addition, PGHS-2 contains a unique 18 amino acid insert near its carboxyl terminus.

Detergent-solubilised PGHSs are both homodimers with subunit molecular masses of about 72 kDa. The biological homodimers are seen in all crystal structures. Both isoforms are cotranslationally N-glycosylated at the same three positions (Figure 3(A)) and are hemoproteins containing a protoporphyrin IX prosthetic group (Figure 3(B)). PGHS monomers contain three major folding domains – an N-terminal epidermal growth factor-like domain of about 50 amino acids, an adjoining region containing about 70 amino acids that serves as the membrane-binding domain, and a C-terminal globular catalytic domain (Figure 3(A) and (B)). The C-terminal 18 amino acid extension of PGHS-2 and its associated posttranslational N-glycosylation site at Asn-594 are involved in two related processes – trafficking PGHS-2 to the Golgi apparatus and ER-associated degradation of PGHS-2 (Smith et al., 2011; Kang et al., 2007).

PGHSs interact with the lumenal surfaces of the ER and the inner and outer membranes of the nuclear envelope through their membrane-binding domains as depicted in the model shown in Figure 3(B). Although PGHS-1 and PGHS-2 are integral membrane proteins, their interactions with membranes do not involve typical transmembrane helices. Instead, analysis of the crystal structures and membrane domain labelling studies have established that PGHSs interact *monotopically* with only one surface of the membrane bilayer (Smith et al., 2011). The interaction involves four short amphipathic α-helices present in the membrane-binding domain (Figure 3(B)). The side chains of hydrophobic residues located on one surface of these helices interdigitate into and anchor PGHSs to the membrane.

3.5 Prostaglandin Endoperoxide H$_2$ Formation

PGHS-1 uses only nonesterified polyunsaturated fatty acids such as AA as substrates. In contrast, PGHS-2 can oxygenate AA and 2-AG about equally well, at least in vitro (Rouzer and Marnett, 2011). The PGHSs exhibit two different but complementary enzymatic activities (Figures 2 and 4): (a) a COX (*bis*-oxygenase), which catalyses the formation of PGG_2 (or 2-PGG_2-glycerol) from AA (or 2-AG) and two molecules of O_2, and (b) a peroxidase, which facilitates the two-electron reduction of the 15-hydroperoxyl group of PGG_2 (or 2-PGG_2-glycerol) to PGH_2 (or 2-PGH_2-glycerol) (Figures 2 and 4). The COX and peroxidase activities occur at distinct but interactive sites within the protein.

The initial step in the COX reaction is the stereospecific removal of the 13-pro-S hydrogen from AA. As depicted in Figure 4, an AA molecule becomes oriented in the COX active site with a kink in the carbon chain at C-9 (Tsai and Kulmacz, 2010; Smith et al., 2011). Abstraction of the 13-pro-S hydrogen and subsequent isomerisation lead to a carbon-centered

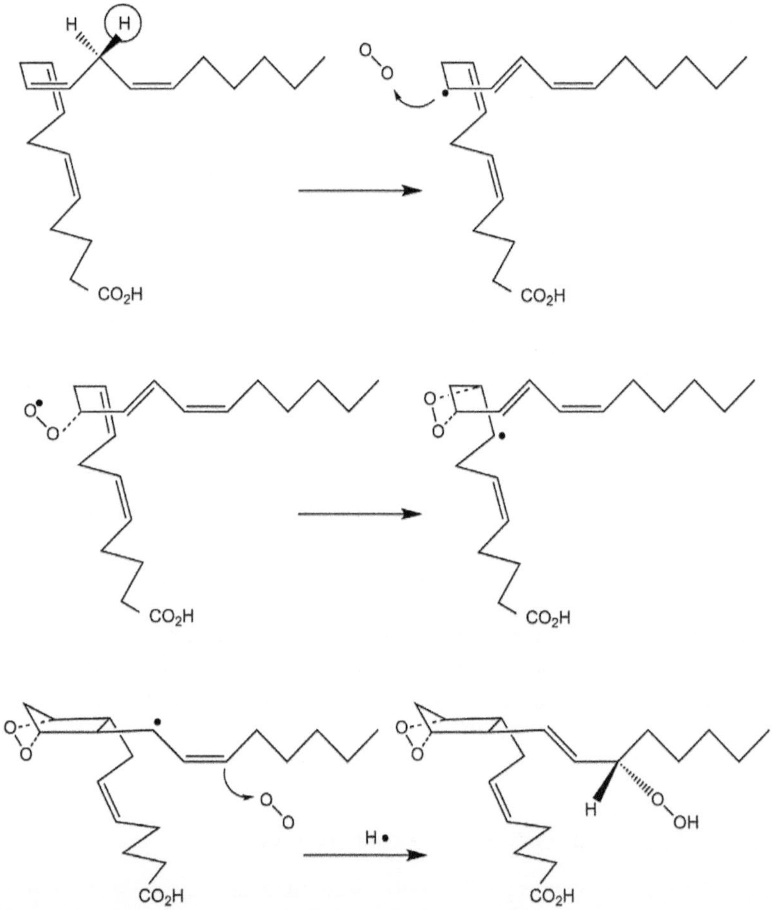

FIGURE 4 Mechanism for the cyclooxygenase reaction showing the conversion of arachidonic acid and two molecules of oxygen to PGG_2.

radical at C-11 and attack of molecular oxygen at C-11 from the side opposite that of hydrogen abstraction. The resulting 11-hydroperoxyl radical adds to the double bond at C-9, leading to intramolecular rearrangement to a bicyclic endoperoxide structure and formation of another carbon-centered radical at C-15. This radical then reacts with another molecule of oxygen and is reduced and protonated to produce PGG_2. The 15-hydroperoxyl group of PGG_2 undergoes a two-electron reduction to an alcohol yielding PGH_2 in a reaction catalysed by the peroxidase activity of PGHSs or other available cellular peroxidases (Figure 2).

3.6 PGHS Active Site

Depicted in Figure 5 is a model of the COX and peroxidase active sites of PGHS (Smith, 2008). The COX is an unusual activity that exhibits a requirement for hydroperoxide and undergoes a suicide inactivation (Tsai and Kulmacz, 2010; Smith et al., 2011). The reason for the hydroperoxide-activating requirement is that in order for the COX to function, a hydroperoxide must oxidise the haeme prosthetic group located at the peroxidase active site to an oxo-ferryl haeme radical cation. This oxidised haeme intermediate abstracts an electron from Tyr-385 located in the COX active site. The resulting Tyr-385 tyrosyl radical abstracts the 13-pro-S hydrogen from AA, initiating the COX reaction. Once the COX reaction begins, newly formed PGG_2 can serve as the source of the activating hydroperoxide. Prior to PGG_2 formation, ambient cellular hydroperoxides and peroxynitrite generated during inflammation serve to initiate haeme oxidation and COX catalysis. Ser-530, the site of acetylation of PGHS-1 by aspirin (Section 3.7), is shown within the COX active site in Figure 5. Also shown is the guanidino group of Arg-120 that serves as the counterion or hydrogen bonding partner for the carboxylate group of AA.

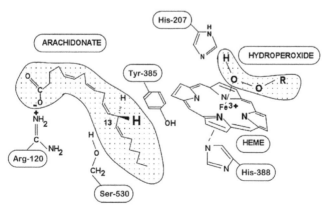

FIGURE 5 Model of the cyclooxygenase and peroxidase active sites of the ovine PGH synthase-1 with arachidonic acid bound to the cyclooxygenase site an alkyl hydroperoxide (ROOH) bound to the haeme group at the peroxidase site.

3.7 PGHS and Nonsteroidal Anti-Inflammatory Drugs

Prostaglandin synthesis is inhibited by both nonsteroidal anti-inflammatory drugs (NSAIDs) and anti-inflammatory steroids. Both PGHS isozymes are pharmacological targets of common NSAIDs (e.g. aspirin, ibuprofen, naproxen). However, only prostaglandin synthesis mediated by PGHS-2 is inhibited by anti-inflammatory steroids, which block the synthesis of PGHS-2, at least in part, at the level of transcription (Kang et al., 2007).

PGHS-2 is also inhibited by 'COX-2 inhibitors' including celecoxib. These drugs belong to a special class of NSAIDs somewhat more specific for this isoform and are often referred to as coxibs. The selectivity of these latter compounds depends on subtle structural differences between the COX active sites of PGHS-1 and PGHS-2 (Uddin et al., 2010; Patrono and Baigent, 2014). These same differences account, at least in part, for the ability of PGHS-2 but not PGHS-1 to bind and oxygenate 2-AG (Rouzer and Marnett, 2011). Rofecoxib, sold as Vioxx, is a coxib that was withdrawn from the market in the United States because of adverse cardiovascular effects. These effects are due in part to an increase in the ratio of the prothrombotic compound TxA_2 formed via platelet PGHS-1 and the anti-thrombotic and anti-hypertensive compounds PGE_2 and PGI_2 formed via vascular and cardiac PGHS-2 that occurs on preferential COX-2 inhibition (Yu et al., 2012; Patrono and Baigent, 2014).

The best-known NSAID is aspirin, acetylsalicylic acid. Aspirin binds to the COX active site, and, once bound, can acetylate Ser-530 (Figure 5) (Smith et al., 2011). Acetylation of this active site serine causes irreversible COX inactivation. Curiously, the hydroxyl group of Ser-530 is not essential for catalysis, but aspirin-acetylated Ser-530 protrudes into the COX site and interferes with the binding of AA.

Acetylation of PGHSs by aspirin has important pharmacological consequences. Besides the analgesic, anti-pyretic and anti-inflammatory actions of aspirin, low dose aspirin treatment – either one 'baby' aspirin (81 mg) daily or one regular aspirin (324 mg) every three days – is a useful anti-platelet cardiovascular therapy (Patrono and Baigent, 2014). This low-dosage regimen leads to pharmacokinetically selective inhibition of platelet TxA_2 formation (and platelet aggregation) without appreciably affecting the synthesis of other prostanoids in other cells. Circulating blood platelets lack nuclei and are unable to synthesise new protein at appreciable rates. Exposure of the PGHS-1 of platelets to circulating aspirin causes irreversible inactivation of the platelet enzyme. Of course, PGHS-1 (and PGHS-2) inactivation also occurs to a limited extent in other cell types, but cell types other than platelets can resynthesise PGHSs relatively quickly. For new PGHS-1 activity to appear in platelets, new platelets must be formed. Because the replacement time for platelets is about 12 days, it takes time for the circulating platelet pool to regain its original complement of active PGHS-1.

There are many NSAIDs other than aspirin (Patrono and Baigent, 2014). In fact, this is one of the largest niches in the pharmaceutical market – 30 million people in the United States take an NSAID daily. Like aspirin, other common non-prescription NSAIDs (e.g. ibuprofen, diclofenac) act by inhibiting the COX activity of PGHSs (Smith et al., 2011; Patrono and Baigent, 2014). However, unlike aspirin, most of these drugs cause reversible enzyme inhibition simply by competing with AA for binding. A well-known example of a reversible NSAID is ibuprofen. All currently available NSAIDs inhibit both PGHS-1 and PGHS-2. However, inhibition of PGHS-2 appears to be primarily responsible for both the anti-inflammatory and analgesic actions of NSAIDs. Dual inhibition of PGHS-1 and PGHS-2 with common NSAIDs causes unwanted ulcerogenic side effects (Patrono and Baigent, 2014). Indeed, even COX-2 inhibitors, which were developed to circumvent this problem, interfere with wound healing.

From 2004 to 2014 considerable attention has been focused on COX-2 inhibitors as prophylactic agents in the prevention of cancer, particularly colon cancer (Uddin et al., 2010). About 85% of tumors of the colon express elevated levels of PGHS-2 and both classical NSAIDs and COX-2 inhibitors reduce mortality due to colon cancer.

3.8 Allosteric Regulation of PGHS by Fatty Acids and COX Inhibitors

PGHSs are sequence homodimers, and crystal structures indicate that the two monomers composing a dimer are structurally identical. However, in solution PGHSs function as conformational heterodimers and exhibit half-of-sites catalytic activity with only one monomer functioning catalytically at any given time (Smith et al., 2011; Dong et al., 2013). One monomer acts as a regulatory, allosteric monomer (E_{allo}) and the other as a catalytic monomer (E_{cat}) (Figure 6). E_{cat} has a bound haeme, whereas E_{allo} does not. The COX activities of both PGHS-1 and PGHS-2 are allosterically modulated by fatty acids including common fatty acids that are not COX substrates. The binding of saturated and monounsaturated FAs to E_{allo} of PGHS-1 causes enzyme inhibition whereas binding of these same FAs to the E_{allo} of PGHS-2 increases enzyme activity to varying degrees. Thus, PGHSs are able to sample and respond to

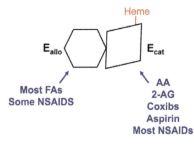

FIGURE 6 Depiction of PGHS conformational heterodimer showing interactions with common fatty acids, NSAIDs and coxibs.

the concentration and composition of the fatty milieu – the fatty acid "tone" – to which they are exposed. This property underlies the ability of PGHS-2 to function at low AA concentrations while PGHS-1 is effectively latent.

Fatty acids also affect the interactions of COX inhibitors with PGHSs. In general, binding of fatty acids to E_{allo} enhances the binding of NSAIDs and cox-ibs to E_{cat}, thereby augmenting inhibition of COX activity. In a few instances, COX inhibitors bind to E_{allo} and function allosterically to inhibit COX activity. The common NSAID naproxen functions in this way.

3.9 Regulation of PGHS-1 and PGHS-2 Gene Expression

PGHS-1 and PGHS-2 are encoded by separate genes (Kang et al., 2007). Apart from the first two exons, the intron/exon arrangements are similar. However, the *PGHS-2* gene (~8 kb) is considerably smaller than the *PGHS-1* gene (~22 kb). The *PGHS-1* gene is on human chromosome 9, while the *PGHS-2* gene is located on human chromosome 1.

Expression of the *PGHS-1* and *PGHS-2* genes is regulated in quite different ways (Kang et al., 2007; Smith, 2008). PGHS-1 is expressed more or less constitutively in almost all tissues. Cells use PGHS-1 to produce prostaglandins needed to regulate 'housekeeping activities' typically involving immediate responses to circulating hormones (Figure 2). PGHS-1 is expressed during cell development (e.g. during megakaryocyte maturation). The *PGHS-1* gene has a TATA-less promoter, a common feature of housekeeping genes. Induction of *PGHS-1* gene expression is regulated by SP1 elements in the PGHS-1 promoter and by downstream intronic elements (Kang et al., 2007). PGHS-1 protein is quite stable in cells (Kang et al., 2007).

PGHS-2 is absent from cells unless induced in response to cytokines, tumor promoters or growth factors. PGHS-2 produces prostanoids that function during early stages of cell differentiation or replication; in many cases the newly formed prostanoids elicit cAMP formation, which, in turn, promotes expression of specific genes (Hirata and Narumiya, 2011). There is some indirect evidence suggesting that at least some of the products formed via PGHS-2 operate at the level of the nucleus through peroxisomal proliferator-activated receptors to modulate transcription of specific genes (Smith, 2008).

Much of what is known about PGHS-2 comes from studies with cultured fibroblasts, endothelial cells and macrophages (Kang et al., 2007). Typically, PGHS-2 is induced rapidly (1–3 h) and dramatically (20- to 80-fold). Platelet-derived growth factor, phorbol ester and interleukin-1β induce PGHS-2 expression in fibroblasts and endothelial cells. Bacterial lipolysaccharide, interleukin-1β and tumor necrosis factor α stimulate PGHS-2 in monocytes and macrophages. Although only a limited number of tissues and cell types have been examined, it is likely that PGHS-2 can be induced in almost any cell or tissue with the appropriate stimuli. Importantly, PGHS-2 expression, but not PGHS-1 expression, can be completely inhibited by anti-inflammatory glucocorticoids such as dexamethasone (Kang et al., 2007).

Transcriptional activation of the *PGHS-2* gene is one important mechanism for increasing PGHS-2 protein expression. The *PGHS-2* promoter contains a TATA box and at least six functional regulatory elements including an overlapping E-box and cAMP response element (CRE) close to the TATA box, an AP1 site, a CAAT enhancer binding protein (C/EBP) sequence, a nuclear factor κB (NF-κB)-binding site and a downstream CRE (Kang et al., 2007). *PGHS-2* gene transcription is controlled by multiple signalling pathways including cAMP, protein kinase C (phorbol esters), viral transformation (*src*) and other pleiotropic pathways, such as those activated by growth factors, endotoxin and inflammatory cytokines. These latter agents (e.g. platelet-derived growth factor, lipopolysaccharide, interleukin-1β, tumor necrosis factor α) likely share convergent pathways involving the proximal CRE, the NF-κB site and the C/EBP elements, two transcription factors common to inflammatory responses, and one or more of the established mitogen-activated protein kinase cascades (ERK1/2, JNK/SAPK and p38/RK/Mpk2). In macrophages, bacterial endotoxin stimulates PGHS-2 expression through cooperative activation involving all of the known response elements in the *PGHS-2* promoter (Kang et al., 2007; Smith, 2008).

There are recent reports indicating that *PGHS-2* gene expression is also affected by methylation of the *PGHS-2* gene and by microRNAs. As noted above PGHS-1 is a stable protein whereas PGHS-2 is degraded relatively rapidly by the ER-associated degradation system involving the C-terminal insert of PGHS-2 (Smith, 2008; Kang et al., 2007) (Figure 3(A)).

3.10 Formation of Biologically Active Prostanoids from Prostaglandin Endoperoxide H_2

Although all the major prostanoids are depicted in Figure 2 as being formed by a single cell, prostanoid synthesis is somewhat cell specific (Smith, 2008). For example, platelets form mainly TxA_2, endothelial cells form PGI_2 as their major prostanoid, and PGE_2 is the major prostanoid produced by renal collecting tubule cells. The syntheses of PGD_2, PGE_2, $PGF_{2\alpha}$, PGI_2 and TxA_2 from PGH_2 are catalysed by PGD synthases, PGE synthases, $PGF_{2\alpha}$ synthase, PGI synthase and TxA synthase, respectively. The properties of these enzymes have been reviewed in detail recently (Smith et al., 2011). The impetus for many of these studies has been a search for therapeutically useful inhibitors that would block the formation of specific bioactive prostanoids.

Formation of $PGF_{2\alpha}$ involves a two-electron reduction of PGH_2 by aldo-ketoreductase 1B1 with NADPH as the reductant (Smith et al., 2011). All other prostanoids are formed from PGH_2 via isomerisation reactions involving no net change in oxidation state.

Both reduced glutathione-dependent and -independent PGD synthases have been isolated (Smith et al., 2011). The glutathione-dependent, hematopoietic PGD synthase also exhibits glutathione-*S*-transferase activity. A glutathione-independent, lipocalin PGD synthase has been purified from brain and may be involved in producing PGD_2 that is involved in sleep regulation (Smith et al., 2011). Structures of both of these proteins have been reported and inhibitors have been synthesised (Smith et al., 2011).

Several different proteins that have PGE synthase activity in vitro have been described. However, to date microsomal PGE synthase-1 (mPGES-1) is the only member of this group that has been established to be involved in PGE_2 synthesis in vivo. Knockout mice that lack a functional mPGES-1 exhibit an approximately 50% reduction in PGE_2-derived urinary metabolites. mPGES-1 is a member of the family of membrane-associated proteins of eicosanoid and glutathione metabolism (MAPEG) (Smith et al., 2011). Reduced glutathione is not consumed during PGE_2 formation by mPGES-1 but rather functions to facilitate cleavage of the endoperoxide group and formation of the 9-keto group. mPGES-1 expression is tightly coupled to PGHS-2 and, like that of PGHS-2, is inhibited by anti-inflammatory steroids such as dexamethasone. However, mPGES-1 is also present in renal collecting tubules cells that express high levels of constitutive PGHS-1. A constitutively expressed cytosolic PGE synthase (cPGES) appears to be coupled to PGHS-1 in cultured cells, but its role in PGE_2 synthesis in vivo remains to be established. Genetic knockout of cPGES does not lead to significant decreases in whole body PGE_2 formation (Smith et al., 2011).

PGI synthase and TxA synthase are hemoproteins with molecular weights of 50–55 kDa. Both of these proteins are cytochrome P450s (Smith et al., 2011) and, like PGHSs, undergo suicide inactivation during catalysis. TxA synthase is found in abundance in platelets and lung. PGI synthase is localised to endothelial cells, as well as both vascular and nonvascular smooth muscle. Both TxA synthase and PGI synthases are believed to be on the cytosolic face of the ER. PGH_2 formed in the lumen of the ER via PGHSs presumably diffuses across the membrane and is converted to a prostanoid end product on the cytosolic side of the membrane.

4. PROSTANOID CATABOLISM AND MECHANISMS OF ACTION

4.1 Prostanoid Catabolism

Once a prostanoid is formed on the cytosolic surface of the ER, it diffuses to the cell membrane and exits the cell probably via carrier-mediated transport. Prostanoids are local hormones that act very near their sites of synthesis. Unlike typical circulating hormones that are released from one major endocrine site, prostanoids are synthesised and released by virtually all organs. In addition, all prostanoids are rapidly inactivated. The initial step of inactivation of PGE_2 is oxidation to a 15-keto compound in a reaction catalysed by a family of 15-hydroxyprostaglandin dehydrogenases (Tai et al., 2002). Further catabolism involves reduction of the double bond between C-13 and C-14, ω-oxidation, and β-oxidation. 15-Hydroxyprostaglandin dehydrogenases are reported to be tumor suppressors acting by reducing PGE_2 levels (Myung et al., 2006).

4.2 Physiological Actions of Prostanoids

Prostanoids can act both in an autocrine fashion on the parent cell and in a paracrine fashion on neighbouring cells (Smith, 2008; Hirata and Narumiya, 2011). Typically, the role of a prostanoid is to coordinate the responses of the parent cells and neighbouring cells to a biosynthetic stimulus – a circulating hormone. The actions of prostanoids are mediated largely by G protein-linked prostanoid receptors of the seven-transmembrane domain receptor superfamily (Hirata and Narumiya, 2011).

One process that has been examined extensively is the platelet–vessel wall interaction involving PGI_2 and TxA_2 (Patrono and Baigent, 2014). TxA_2 is synthesised by platelets when they bind to subendothelial collagen that is exposed by microinjury to the vascular endothelium. Newly synthesised TxA_2 promotes subsequent adherence and aggregation of circulating platelets to the subendothelium. In addition, TxA_2 produced by platelets causes constriction of vascular smooth muscle. The net effect is to coordinate the actions of platelets and the vasculature in response to de-endothelialisation of arterial vessels. Thus, prostanoids can be viewed as local hormones, which coordinate the effects of circulating hormones and other agents (e.g. collagen) that activate their synthesis.

4.3 Prostanoid Receptors

Molecular characterisation of prostanoid receptors has occurred during the past 20 years driven largely by the work of Narumiya and coworkers (Hirata and Narumiya, 2011). These results, coupled with studies of PGHS-1 and PGHS-2 and prostanoid receptor knockout mice, have been critical in rationalizing the results of earlier studies on the physiological and pharmacological actions of prostanoids that were difficult to interpret because prostaglandins caused such a variety of seemingly paradoxical effects.

The seminal step in understanding the structures of prostanoid receptors and their coupling to second messenger systems resulted from the cloning of receptors for each of the prostanoids. Prostanoid receptor cloning began with the TxA/PGH receptor known as the TP receptor. A cDNA encoding this receptor was cloned using oligonucleotide probes designed from protein sequence data obtained from the TP receptor purified from platelets. The results confirmed biochemical predictions that the TP receptor was a seven-membrane spanning domain receptor of the rhodopsin family. Subsequent cloning of other receptors was performed by homology screening using receptor cDNA fragments as cross-hybridisation probes. All of these prostanoid receptors are of the G protein-linked receptor family.

It is now clear that there are nine PG distinct receptors each encoded by a different gene (Hirata and Narumiya, 2011). For example, in the case of PGE_2, four different prostaglandin E (EP) receptors have been identified and designated as EP1, EP2, EP3 and EP4 receptors. Based on studies with selective agonists for each of the EP receptors and their effects on second messenger production, it appears that EP1 is coupled to Ca^{2+} mobilisation, EP2 and EP4 are coupled via G_s to the stimulation of adenylate cyclase and EP3 receptors are coupled via G_i to the inhibition of adenylate cyclase. In addition, recent studies have shown that a number of prostanoid receptors are coupled to cell proliferation.

In order to determine the physiological roles of various prostanoid receptors a number of knockout mice have been developed (Hirata and Narumiya, 2011). These studies indicate that prostacyclin receptors are involved in at least some types of pain responses, EP3 receptors are involved in the development of fever, EP2 and EP4 function in allergy and bone resorption and EP1 receptors are involved in chemically induced colon cancer. The availability of cloned prostanoid receptors has led to searches for receptor agonists and antagonists that might provide some specificity beyond the currently available COX inhibitors, which prevent broadly the synthesis of all prostaglandins.

The realisation that *PGHS-2* is an immediate early gene associated with cell replication and differentiation suggests that prostanoids synthesised via PGHS-2 may have nuclear effects. As noted above there have been several reports indicating that prostanoid derivatives can activate some isoforms of peroxisomal proliferator-activated receptors (PPARs).

5. LEUKOTRIENES AND LIPOXYGENASE PRODUCTS

5.1 Introduction and Overview

Leukotrienes (LTs) are produced by the action of 5-lipoxygenase (5-LO) which inserts a diatomic oxygen at carbon atom-5 of AA yielding 5(*S*)-hydroperoxye-icosatetraenoic acid (5-HpETE) (Haeggstrom and Funk, 2011) (Figure 7). A subsequent dehydration reaction catalysed by the same enzyme, 5-LO, results in formation of leukotriene A_4 (LTA_4), the chemically reactive precursor of biologically active leukotrienes. As was the case with prostanoid biosynthesis, LT biosynthesis depends on the availability of AA as the 5-LO substrate, which typically requires the action of $cPLA_2\alpha$ to release AA from membrane phospholipids. Also, LTs are not stored in cells, but are rapidly synthesised and released following cellular activation. Interest in the LT family of AA metabolites arises from the potent biological activities of two products derived from LTA_4 – leukotriene B_4 (LTB_4) and leukotriene C_4 (LTC_4) (Figure 7). LTB_4 is a very potent chemotactic and chemokinetic agent for the human polymorphonuclear leukocyte, while LTC_4 powerfully constricts specific smooth muscle such as bronchial smooth muscle, and mediates leakage of vascular fluid in the process of edema (Nakamura and Shimizu, 2011). The name leukotriene was

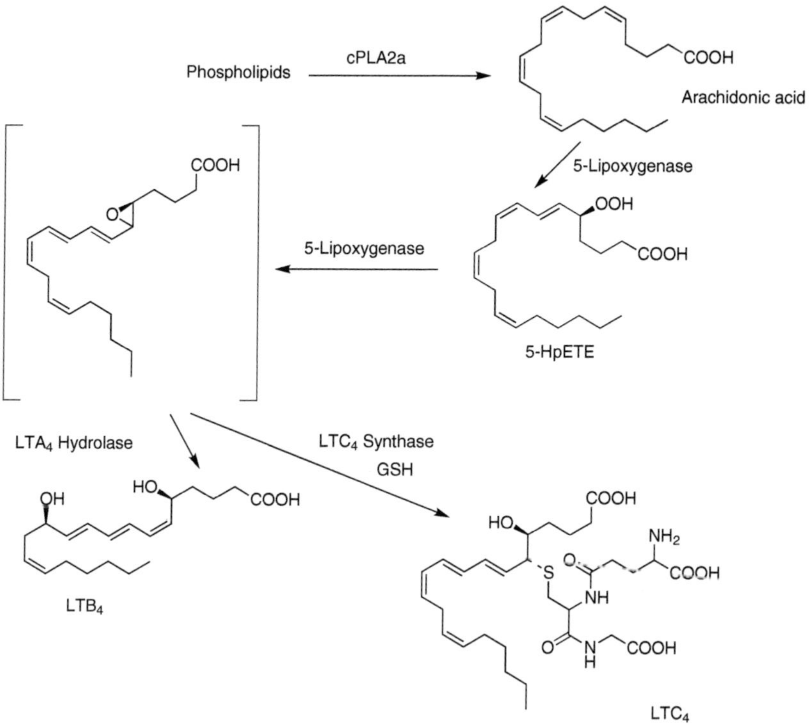

FIGURE 7 Biochemical pathway of the metabolism of arachidonic acid into the biologically active leukotrienes. Arachidonic acid released from phospholipids by cytosolic phospholipase $A_2\alpha$ is metabolised by 5-lipoxygenase to 5-hydroperoxyeicosatetraenoic acid (5-HpETE) and leukotriene A_4 (LTA$_4$), which is then enzymatically converted into leukotriene B_4 (LTB$_4$) or conjugated by glutathione to yield leukotriene C_4 (LTC$_4$).

conceived to capture two unique attributes of these molecules. The first part of the name relates to those white blood cells derived from the bone marrow that have the capacity to synthesise this class of eicosanoid, for example, the poly-morphonuclear *leukocyte*. The last part of the name refers to the unique chemical structure, a conjugated *triene*.

There are numerous other biochemical products of AA metabolism formed by LO enzymes other than 5-LO. Other monooxygenases expressed in mammalian cells include 12(S)-LO (platelet), 12(R)-LO (epidermis), 15-LO-1 and 15-LO-2. These enzymes are named in accordance with the carbon atom position of AA initially oxygenated even though other polyun-saturated fatty acids can be substrates. In addition, AA can be oxidised by specific isozymes of cytochrome P450, leading to a family of epoxyeico-satrienoic acids (EETs). Methyl terminus (ω-oxidised) AA as well as LO-like monohydroxyeicosatetraenoic acid (HETE) products are also formed by various P450s. In general, much more is known about the biochemical role of prostaglandins and LTs as mediators of biochemical events, but the other LO products or cytochrome P450 products are important metabolites; for example, the P450 metabolites play an important role in kidney function. In accord with the body of information available, various pharmacological tools are available to inhibit 5-LO as well as specific LT receptors (Nakamura and Shimizu, 2011).

5.2 Leukotriene Biosynthesis

The 5-LO (EC 1.13.11.34) is a nonhaeme metalloenzyme with iron coordinated in the active site by three histidine residues, an asparagine and an isoleucine carboxyl group. The human 77 kDa protein 5-LO catalyses the addition of molecular oxygen to the 1,4-*cis*-pentadienyl structural moiety closest to the carboxyl group of AA to yield a conjugated diene hydroperoxide, typical of all LO reactions. The mechanism of 5-LO is thought to be similar to that of 15-LO, which has been studied in great detail (Segraves et al., 2006; Haeggstrom and Funk, 2011) and begins with iron in the ferric state. In the reactive site, the pro-S hydrogen atom from carbon-7 of AA is removed as a proton and electron, which enters the molecular orbitals of Fe(III) to form Fe(II). Molecular oxygen then adds to carbon-5 of the resulting radical to yield the hydroperoxy radical. The hydroperoxy radical abstracts an electron from Fe^{2+} to yield a hydroperoxide anion that can readily remove a proton from a water ligand of the catalytic iron to yield 5(*S*)-HpETE [5(*S*)-hydroperoxy-6,8,11,14-(*E,Z,Z,Z*)-eicosatetraenoic acid]. A second enzymatic activity of 5-LO catalyses the stereospecific removal of the pro-R hydrogen atom at carbon-10 of the 5(*S*)-HpETE through a second cycle very similar in electron movement detail to the first 5-LO mechanism (Figure 8). In this case, Fe^{3+} with a hydroxide anion ligand removes an electron from the carbon (10)–hydrogen bond by a tunnelling mechanism and the nascent proton is picked up by the hydroxide anion ligand. The radical site at carbon-10 then delocalises over seven carbon atoms and the emerging high density of electron character at C-6 causes weakening of the peroxide (oxygen–oxygen bond) at C-5. The iron atom, now Fe^{2+}, can donate an electron to the forming hydroxyl radical to yield hydroxide anion, driving formation of the new carbon–oxygen bond of the epoxide. The hydroxide anion abstracts the proton from the water molecule coordinated with Fe^{3+} and thus regenerates Fe^{3+} with an hydroxyl anion as the sixth ligand. One unique feature of 5-LO is that the active enzyme with coordinated Fe^{3+} containing hydroxide anion as the sixth ligand is regenerated during each catalytic cycle by either the monooxygenase or the LTA_4 synthase mechanism. This is different from the COX mechanism where a second peroxidase mechanism is required to regenerate active enzyme.

The product of this reaction, 5(*S*),6(*S*)-oxido-7,9,11,14-(*E,E,Z,Z*)-eicosatetraenoic acid, is LTA_4, a conjugated triene epoxide that is highly unstable. LTA_4 undergoes rapid hydrolysis in water with a very short half-life. Nonetheless

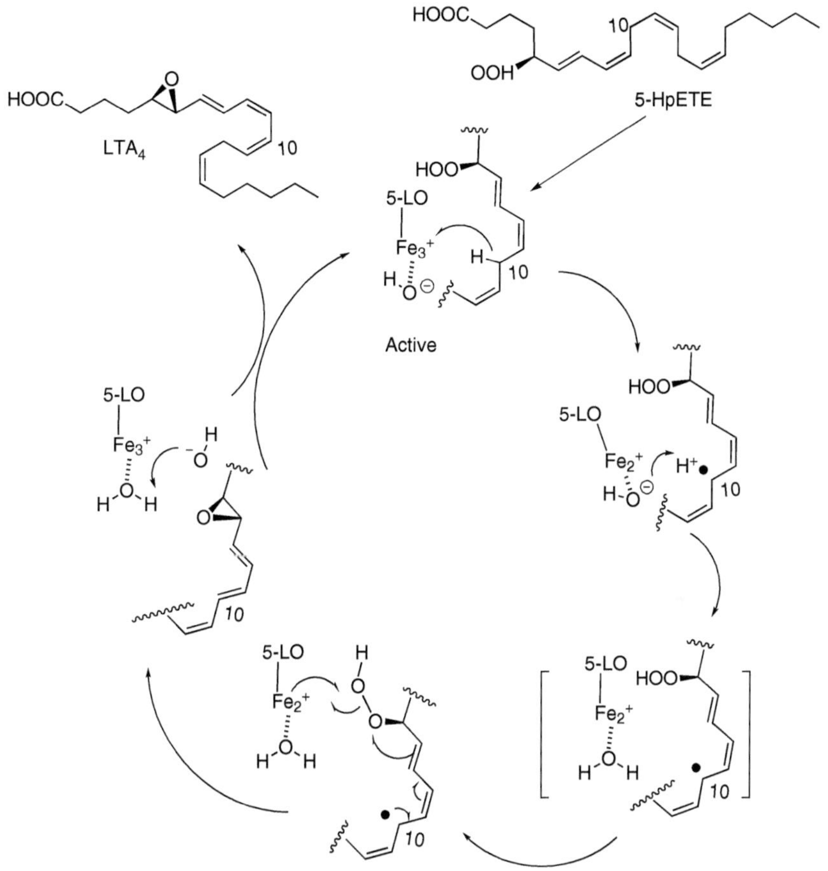

FIGURE 8 Suggested mechanism for 5-lipoxygense converting 5-HpETE to LTA₄. The oxidation state of the iron ion and the nature of the sixth ligand to the coordination sphere of iron as either water or hydroxide ion are indicated as well as the transformation of 5-HpETE into the triene epoxide, LTA₄.

within cells, LTA₄ is stabilised by binding to proteins that remove water from the immediate environment of the epoxide structure. The nonenzymatic hydrolysis products of LTA₄ include several biologically inactive and enantiomeric 5,12- and 5,6-diHETEs. However, the hydrolysis of LTA₄ catalysed by LTA₄ hydrolase (Haeggstrom and Funk, 2011) produces the biologically active LTB₄, 5(S),12(R)-dihydroxy-6,8,10,14-(Z,E,E,Z)-eicosatetraenoic acid (Figure 7). A second pathway for LTA₄ metabolism is prominent in cells expressing the enzyme LTC₄ synthase (Haeggstrom and Funk, 2011), which catalyses the addition of glutathione to carbon-6 of the triene epoxide yielding 5(S),6(R)-S-glutathionyl-7,9,11,14-(E,E,Z,Z)-eicosatetraenoic acid (LTC₄) (Figure 7). LTC₄ synthase is a unique glutathione (S) transferase localised on the nuclear membrane.

The formation of either LTC_4 or LTB_4 is controlled by the expression of either LTA_4 hydrolase or LTC_4 synthase in specific cell types. The human neutrophil, for example, expresses LTA_4 hydrolase and produces LTB_4 while the mast cell and eosinophil produce LTC_4, since they express LTC_4 synthase. Interestingly, cells have been found that do not have 5-LO, but express either LTA_4 hydrolase (e.g. erythrocytes and lymphocytes) or LTC_4 synthase (e.g. platelets and endothelial cells). Cells cooperate in the production of biologically active LTs through a process termed transcellular biosynthesis (Folco and Murphy, 2006) where a cell such as the neutrophil or mast cell generates LTA_4, which is released and then taken up by either a platelet to make LTC_4 or red blood cell to make LTB_4. Despite the chemical reactivity of LTA_4, this process is known to be highly efficient and approximately 60–70% of the LTA_4 produced in vivo can be presented to another cell for transcellular biosynthesis of cysteinyl leukotrienes.

5.3 Enzymes and Proteins Involved in Leukotriene Biosynthesis

5.3.1 5-Lipoxygenase

The human *5-LO* gene on chromosome 10 covers more than 80 kb of DNA and has 14 exons that encode a 673 amino acid protein (Haeggstrom and Funk, 2011). 5-LO purified from human, pig, rat and guinea pig leukocytes have close to 90% homology. It is interesting to note that the purified or recombinant enzyme requires several cofactors for activity, including Ca^{2+}, ATP, fatty acid hydroperoxides and phosphatidylcholine, in addition to the substrates AA and molecular oxygen. Purified 5-LO was found to catalyse the initial oxidation of AA to yield 5-HpETE as well as the second enzymatic reaction to convert 5-HpETE into LTA_4.

Low concentrations of Ca^{2+} (1–2 μM) are required for maximal activity of purified 5-LO by a mechanism that increases the translocation of 5-LO to the nuclear envelope through interaction with its C-2-like domain. Calcium ion increases the association of 5-LO with phosphatidylcholine vesicles, recapitulating events within the cell where 5-LO becomes associated with the nuclear envelope, which is rich in phosphatidylcholine (Brock, 2005). At the nuclear membrane AA is thought to be presented by a protein called 5-LO-activating protein (FLAP) (Haeggstrom and Funk, 2011). In cells, such as the neutrophil and mast cells where 5-LO is found in the cytosol, 5-LO is catalytically active only when bound to membranes, typically the nuclear envelope. ATP has a stimulatory effect on 5-LO at 20 nM and lipid hydroperoxides are important to initiate the 5-LO catalytic cycle by forming Fe(III) from the inactive Fe(II) form of 5-LO by the pseudoperoxidase mechanism.

Despite successful X-ray crystallographic analysis of several LOs, including 8(*R*)-LO and rabbit reticulocyte 15-LO, the crystal structure of 5-LO was very difficult to obtain due to protein instability in solution (auto-inactivation). Based on the structure of the two stable LOs, a lysine-rich region near the C-terminus was identified that confers instability to

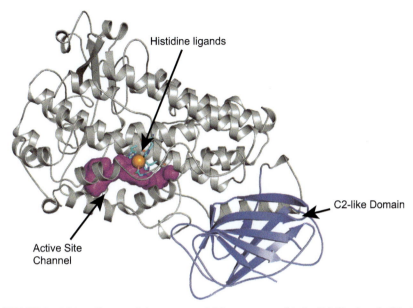

Histidine ligands

C2-like Domain

Active Site
Channel

FIGURE 9 Ribbon diagram of the structure of 5-lipoxygenase with the C-2-like domain (blue) and catalytic iron (orange). The coordinating histidine and C-terminus Ile are green. The putative binding site for AA is indicated in red.

5-LO. After mutation of KKK to ENL, a stable 5-LO was made and a crystal structure obtained at 2.4 Å resolution (Gilbert et al., 2011). The structure of stable 5-LO is very similar to other LOs with a C2-like domain responsible for the calcium ion-dependent binding to membranes and an α-helix-dominated catalytic domain that contains a nonhaeme iron (Figure 9). The active site of stable 5-LO appears to be a closed hydrophobic channel, into which AA can fit in a manner similar to that found for other LOs. A major difference is that the active site of 5-LO is shielded at one end by two amino acids, phenylalanine and tyrosine, that restrict substrate entry. This phenylalanine/tyrosine structure forces AA to enter the active site in the direction opposite to that suggested for 15(S)- and 8(R)-LOs.

5.3.2 5-LO-Activating Protein

FLAP was found during the course of development of the drug MK-886 by workers at Merck-Frosst in Montreal (Haeggstrom and Funk, 2011). They found that this drug bound to a novel protein that was essential for the production of LTs in stimulated intact cells and hence was named the 'five lipoxygenase-activating protein'. FLAP is a unique 161 amino acid-containing protein (18 kDa) that is suggested to act as a substrate transfer protein to stimulate 5-LO-catalysed formation of leukotrienes. LTC_4 synthase has 31% amino acid identity to FLAP in a highly conserved region possibly involved in lipid binding

(Haeggstrom and Funk, 2011). The three-dimensional structure of FLAP was determined by X-ray crystallography with a bound inhibitor (Ferguson et al., 2007). The inhibitor bounds to the protein where it is presumed to project into the nuclear phospholipid membrane, consistent with a site that would have access to free AA released by $cPLA_2\alpha$ following translocation and activation.

5.3.3 Leukotriene A_4 Hydrolase

LTA_4 hydrolase catalyses the stereochemical addition of water to form the neutrophil chemotactic factor LTB_4. LTA_4 hydrolase (EC 3.3.2.6, 69 kDa) contains one zinc atom per enzyme molecule that is essential for catalytic activity (Haeggstrom and Funk, 2011). LTA_4 hydrolase is also a member of a family of zinc metalloproteases, which led to the discovery of several drugs, such as bestatin and captopril, that are inhibitors of this enzyme. The aminopeptidase activity of LTA_4 hydrolase was found to have physiological relevance since it was demonstrated to be specific for cleaving a Pro–Gly–Pro (PGP) tripeptide derived from the degradation of extracellular matrix proteins (Snelgrove et al., 2010). PGP has chemotactic activity for neutrophils and is therefore a pro-inflammatory peptide. Specific inhibitors have been generated based on the crystal structure of LTA_4 hydrolase that specifically inhibit the LTA_4 hydrolase activity and not the aminopeptidase activity (Stsiapanava et al., 2014).

LTA_4 hydrolase is found in many cells including those that do not contain 5-LO, where it could play an important role in transcellular biosynthesis of LTB_4 through cell–cell cooperation. LTA_4 is thought to be localised to the cell cytosol, and is the only protein in the LT biosynthetic cascade that is not membrane bound or translocated to the nuclear membrane following cellular activation. The three-dimensional structure of LTA_4 hydrolase has been determined by X-ray crystallography and a detailed explanation of the enzymatic mechanism has been proposed (Haeggstrom and Funk, 2011).

5.3.4. Leukotriene A_4 Synthase

The conjugation of the tripeptide glutathione (γ-glutamyl-cysteinyl glycine) to the triene epoxide LTA_4 is carried out by LTC_4 synthase (EC 4.4.1.20). This enzyme is localised on the nuclear envelope and has little homology to soluble glutathione (S) transferases. LTC_4 synthase has some primary amino acid sequence homology to the recently described microsomal glutathione (S) transferases and FLAP (Haeggstrom and Funk, 2011). Several microsomal glutathione (S) transferases have been found that catalyse formation of LTC_4 from LTA_4 and glutathione. LTC_4 synthase has a restricted distribution and is found predominantly in mast cells, macrophages, eosinophils and monocytes. However, human platelets and endothelial cells also contain LTC_4 synthase. The drug MK-886 inhibits LTC_4 synthase with an IC_{50} of approximately 2–3 μM (Haeggstrom and Funk, 2011). The gene for LTC_4 synthase is located on human chromosome 5, distal to that of cytokine, growth factor and receptor genes relevant to the T-helper lymphocyte (TH2) response. The structure of human LTC$_4$ synthase was determined by X-ray crystallography (Ago et al., 2007; Niegowski et al., 2014). Using modelling programmes and the structure of LTC$_4$ synthase with GSH bound to the active site it appears that LTA$_4$ is held in a hydrophobic channel that places the C6 atom next to the bound thiol group of GSH. It has been proposed that removal of a proton from the GSH sulfhydryl group by Arg-104 facilitates attack at C-6 of LTA$_4$, and that Arg-31 provides the proton for the emerging carbinol at C5 and the resulting LTC$_4$.

5.4 Regulation of Leukotriene Biosynthesis

The biosynthesis of LTs within cells is highly regulated and depends not only on the availability of AA and molecular oxygen but also on the serine phosphorylation and subcellular location of 5-LO (Rådmark and Samuelsson, 2009). Resting neutrophils synthesise little if any LTs. However, following the elevation of intracellular Ca^{2+} concentrations, either by a physiological event such as phagocytosis or pharmacological manipulation with the calcium ionophore A23187, the neutrophil produces a substantial amount of 5-LO products, including LTs from either endogenous or exogenous AA. The need for increased intracellular ion concentrations is a distinguishing feature of 5-LO that differentiates this enzyme from other LOs and COX-1 and -2.

A major determinant of 5-LO activity is its translocation to the nuclear envelope (Figure 10). The site of localisation of active 5-LO is thought to be on the outer nuclear envelope. Calcium ion-dependent translocation brings 5-LO to the same region where FLAP and cPLA$_2\alpha$ translocate. Regulation of LT

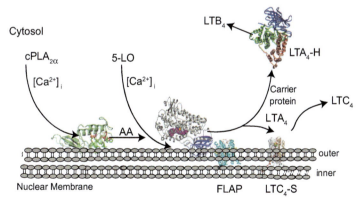

FIGURE 10 Proposed model for the location of biosynthetic events occurring during leukotriene (LT) biosynthesis at the nuclear envelope of cells. Arachidonic acid is released from membrane glycerophospholipids by nuclear membrane-associated cPLA$_2\alpha$ and is then presented to 5-lipoxygenase via 5-lipoxygenase-activating protein (FLAP). LTA$_4$ is either converted by nuclear envelope-associated LTC$_4$ synthase (LTC$_4$-S) into LTC$_4$ or carried to LTA$_4$ hydrolase (LTA$_4$-H). These events can occur from the cytosolic or nucleoplasmic side of the nuclear envelope.

biosynthesis thus involves the assembly of an LT biosynthetic machine at a nuclear envelope. It is this site where AA is released from nuclear membrane phospholipids, then converted to LTA_4 and ultimately conjugated with glutathione by nuclear membrane LTC_4 synthase. The mechanism by which 5-LO is trafficked to the nuclear envelope is controlled by its C2-like domain, which binds two Ca^{2+} ions and has a high affinity for phosphatidylcholine in the cytosolic face of the nuclear envelope (Brock, 2005). 5-LO phosphorylation at Ser-271 (MAPKAP kinase dependent) and Ser-663 (ERK-Z dependent) increases enzymatic activity and translocation to the nucleoplasm. However, phosphorylation of Ser-523 by protein kinase A inhibits 5-LO activity (see below).

The activity of 5-LO is also regulated by a suicide inactivation mechanism by LTA_4, likely by a covalent modification mechanism. Continued synthesis of LTs then requires synthesis of new 5-LO. The production of 5-LO is known to be regulated at the level of gene transcription as well as mRNA translation in addition to the translocation mechanism (Brock, 2005; Haeggstrom and Funk, 2011). The *5-LO* gene promoter contains several consensus-binding sites for known transcription factors including Sp1 and EGR-l. In a region located 212 to 88bp upstream from the transcription start site of the human *5-LO* gene *(ALOX5)*, a highly rich G+C region is found that contains the consensus sequence for Spl and EGR-I transcription factors (Haeggstrom and Funk, 2011). Furthermore, common variations have been found in human populations in which deletions of one or two of the Spl-binding motifs were observed (Haeggstrom and Funk, 2011). These genetic polymorphisms could have substantial effects on the induction of 5-LO transcription.

Various biochemical mechanisms can also alter LT biosynthesis within cells. Because of the complexity of 5-LO activation and the requirement of the calcium ion-dependent translocation, modification of signal transduction pathways are known to alter LT biosynthesis. For example, elevation of cellular cyclic AMP inhibits LT biosynthesis even when synthesis is stimulated by the powerful calcium ionophore A23187. Adenosine and A2A receptor agonists inhibit production of LTs in human neutrophils, likely through enhanced production of cyclic AMP and phosphorylation of Ser-523. Pharmacological agents have been developed to inhibit LT production through direct action on 5-LO. The drug zileuton likely reduces activated 5-LO Fe(III) to the inactive 5-LO Fe(II) or prevents the oxidation of Fe(II) by lipid hydroperoxides, the pseudoperoxidase step, by serving as a competitive substrate (Haeggstrom and Funk, 2011). Inhibitors of FLAP (Section 4.3.2) include MK-886 and the related agent BAY x1005. Because of its close similarity to the FLAP protein, LTC_4 synthase is inhibited by MK-886 albeit at higher concentrations (Haeggstrom and Funk, 2011).

5.5 Metabolism of Leukotrienes

The conversion of LTs into alternative structural entities is an important feature of inactivation of these potent biologically active eicosanoids. Metabolism of LTs is rapid and the exact pathway depends on whether the substrate is LTB_4 or LTC_4.

LTB$_4$ is rapidly metabolised through both oxidative and reductive pathways (Figure 10) (Murphy and Wheelan, 1998). The most prominent pathway present in human neutrophil (CYP4F3) and hepatocytes (CYP4F2) involves specific and unique cytochrome P450s of the CYP4F family. cDNAs encoding 16 different proteins have now been cloned and expressed in several animal species (Cui et al., 2001) and each of these enzymes efficiently converts LTB$_4$ into 20-hydroxy-LTB$_4$. 20-Hydroxy-LTB$_4$ has some biological activity since it is a competitive agonist for the LTB$_4$ receptor. In the human neutrophil, 20-hydroxy-LTB$_4$ is further metabolised into biologically inactive 20-carboxy-LTB$_4$ by CYP4F3. In the hepatocyte and other cell types, 20-hydroxy-LTB$_4$ is further metabolised by alcohol dehydrogenase to form 20-oxo-LTB$_4$, and then by fatty aldehyde dehydrogenase (AldDH) to form 20-carboxy-LTB$_4$ both of which reactions require NAD$^+$ (Figure 11).

A unique reductive pathway is highly expressed in cells such as keratinocytes, endothelial cells and kidney cells. This pathway consists of an initial oxidation of the 12-hydroxyl group of LTB$_4$ to a 12-oxo moiety followed by reduction of the conjugated dienone and double bond $\Delta^{10,11}$ (Figure 11). The products of the 12-hydroxyeicosanoid dehydrogenase pathway are devoid of biological activity and in certain cells represent the major pathway of inactivation. The 12-hydroxy dehydrogenase/15-oxo-prostaglandin-13-reductase, which is the committed enzyme of this pathway, has been crystallised and a three-dimensional structure reported.

A secondary metabolic pathway of LTB$_4$ in hepatocytes is β-oxidation that requires initial ω-oxidation of the C-20 methyl terminus to 20-carboxy-LTB$_4$ followed by CoA ester formation (Murphy and Wheelan, 1998). β-Oxidation takes place within both the peroxisome and the mitochondria of the hepatocyte. The importance of each oxidation processes in LTB$_4$ metabolism is highlighted in human subjects with various genetic abnormalities in peroxisomal metabolism. For example, Zellweger disease leads to a reduction in β-oxidation, and LTB$_4$ and 20-carboxy-LTB$_4$ are urinary excretion products. Individuals with a deficiency in fatty AldDH, termed Sjögren-Larsson syndrome, excrete measurable levels of LTB$_4$ and 20-hydroxy-LTB$_4$.

The metabolism of LTC$_4$ results in a major alteration of its biological activity since the initial metabolites, LTD$_4$ and LTE$_4$ (Figure 12), interact with different affinities with the cysteinyl LT (cysLT) receptors. The initial peptide cleavage reaction involves γ-glutamyl transpeptidase, an enzyme typically located on the plasma membrane, to yield LTD$_4$, which has a very high affinity for the cysLT receptors. Conversion of LTD$_4$ to LTE$_4$ likely involves a specific membrane-bound dipeptidase whose three-dimensional structure was recently described. Sulfidopeptide LTs can also be metabolised specifically at the sulfur atom through oxidation reactions initiated by reactive oxygen species. More specific metabolic processing of the sulfidopeptide LTs include ω-oxidation by cytochrome P450 followed by β-oxidation from the ω-terminus, resulting in a series of chain-shortened products (Murphy and Wheelan, 1998). Formation of 20-carboxy-LTE$_4$ results in complete inactivation of the biological activities

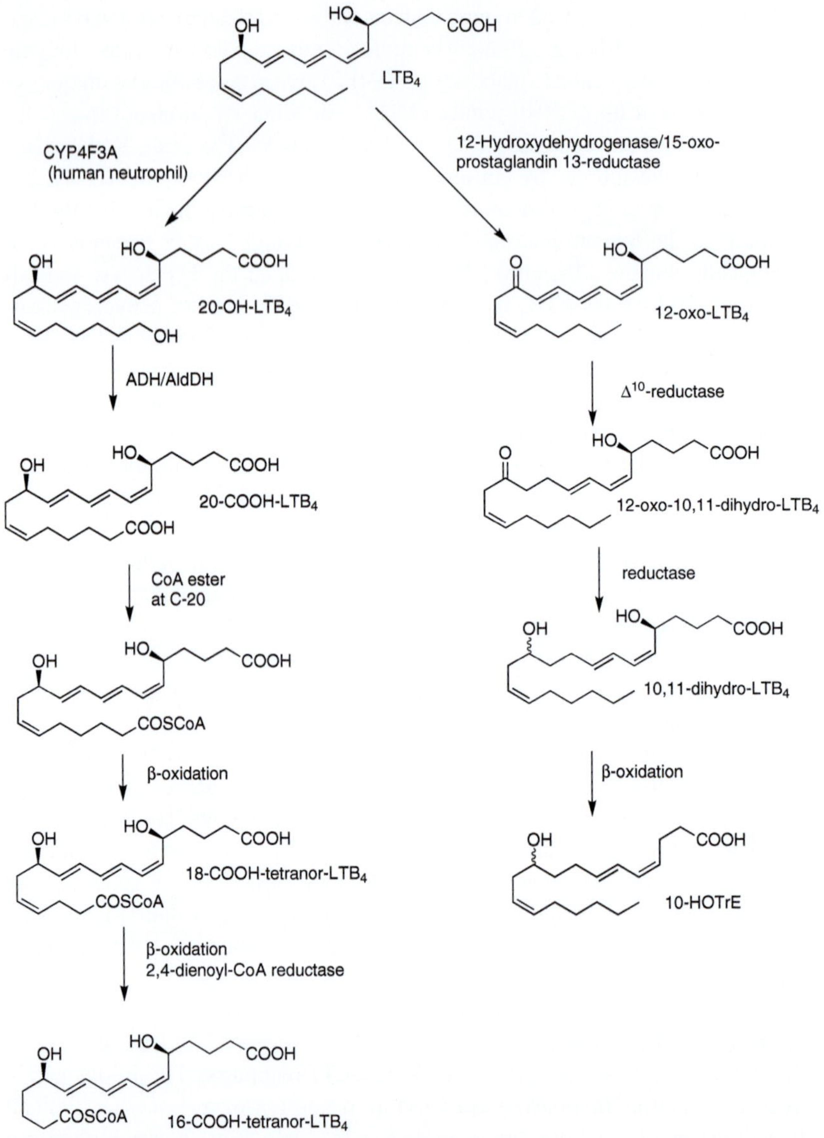

FIGURE 11 Common metabolic transformations of LTB_4 either by the cytochrome P450 (CYP4F-family members) and ω-oxidation followed by β-oxidation or by the 12-hydroxyeicosanoid dehydrogenase pathway, which leads to reduction of the $\Delta^{10,11}$ double bond. The oxidative enzymes alcohol dehydrogenase (ADH) and aldehyde dehydrogenase (AldDH) are also involved. HOTrE is hydroxyoctadecatrienoic acid.

of this molecule due to lack of cysLT receptor recognition. Acetylation of the terminal amino group in LTE_4 generates N-acetyl-LTE_4, an abundant metabolite in rodent tissue. In humans, some LTE_4 is excreted in urine and has been used to reflect whole body production of sulfidopeptide LT in vivo.

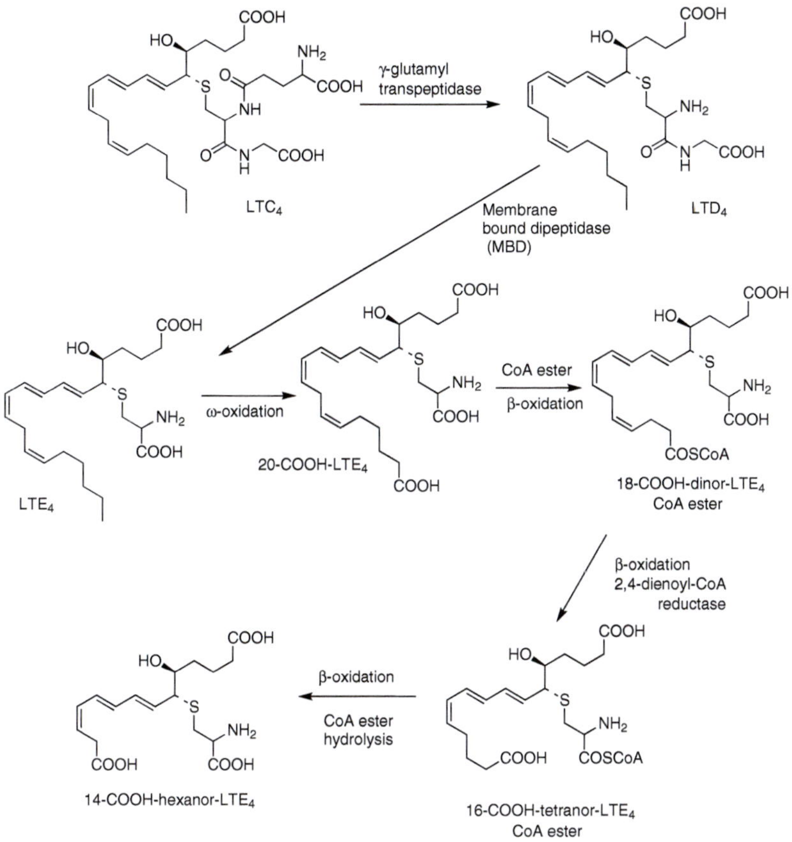

FIGURE 12 Common metabolic transformations of leukotriene C_4 (LTC_4) to the biologically active sulfidopeptide leukotrienes, LTD_4 and LTE_4. Subsequent ω-oxidation of LTE_4 by cytochrome P450 leads to the formation of 20-carboxy-LTE_4 that can undergo β-oxidation, after formation of the CoA ester, into a series of chain-shortened cysteinyl leukotriene metabolites.

5.6 Biological Activities of Leukotrienes

LTB_4 is thought to play an important role in the inflammatory process by way of its chemotactic and chemokinetic effects on the human polymorpho-nuclear leukocyte (Nakamura and Shimizu, 2011). LTB_4 induces the adher-ence of neutrophils to vascular endothelial cells and enhances the migration of neutrophils (diapedesis) into extravascular tissues. The chemotactic activity of LTB_4 is mediated through two specific G-protein-coupled receptors termed BLT1 and BLT2 (Nakamura and Shimizu, 2011). Human BLT1 (37 kDa) is expressed almost exclusively in polymorphonuclear leukocytes and to a much lesser extent in macrophages and the thymus. The human BLT gene *(LTB4R)* is located on chromosome 14. Several specific agents have been developed by pharmaceutical companies to inhibit the LTB_4 receptor; however, none has been fully developed for use in humans.

LTC$_4$ and the peptide cleavage products LTD$_4$ and LTE$_4$ cause bronchial smooth muscle contraction in asthma. These sulfidopeptide LTs also increase vascular leakage leading to edema. The discovery of LTs was a result of a search for the biologically active principle called 'slow reacting substance of ana-phylaxis' (Haeggstrom and Funk, 2011; Nakamura and Shimizu, 2011). Two G-protein-coupled receptors for the cysteinyl LT have been characterised and termed cysLT1 and cysLT2, while a third receptor that specifically binds to LTE$_4$ has been described as GPR99 (Kanaoka et al., 2013). The cysLT1 recep-tor has restricted tissue distribution and is most abundant in smooth muscle of the lung and small intestine; LTD$_4$ and LTC$_4$ both activate this receptor. Several drugs are now available that inhibit the cysLT1 receptor in human subjects; montelukast (IC$_{50}$ 2–5 nM), pranlukast (IC$_{50}$ 4–7 nM) and zafirlukast (IC$_{50}$ 2–3 nM) (Nakamura and Shimizu, 2011).

5.7 Other LO Pathways

Numerous LOs occur within the plant and animal kingdoms that have several aspects in common. First, nonhaeme iron is an essential component of the catalytic activity of these enzymes. Furthermore, these enzymes catalyse the insertion of molecular oxygen into polyunsaturated fatty acids, predominantly linoleic and AA, with the initial formation of lipid hydroperoxides. In general, the overall biological activities of the LO products are incompletely understood and the significance of 12- and 15-LO in humans is not fully defined.

5.7.1 12-Lipoxygenase

Two different 12-lipoxygenases (12-LO) catalyse the conversion of AA to 12-HpETE acid (Haeggstrom and Funk, 2011), which is subsequently reduced to 12-12-HETE (Figure 13). The human platelet expresses a 12-LO [12(*S*)-lipoxygenase, EC 1.13.11.31], the cDNA for which has been cloned and found to encode a 662 amino acid protein with a molecular weight of 75 kDa. A second 12-LO is present in other mammalian systems, including the mouse and rat, and is termed the 12(*R*)-LO. The 12(*R*)-LO is less well studied, but has been described as one of two epidermal LOs that oxy-genate AA as well as linoleate esterified to a ω-hydroxy ceramide pres-ent in the epidermal barrier (Zheng et al., 2011). A second epidermal LOs termed eLO-3 converts the 12(*R*)-LO hydroperoxide product into an epoxy alcohol. These reactions catalysed by the epidermal LOs are essential for formation of a water-tight epidermal barrier of the skin. In addition, sev-eral lines of evidence suggest that the 12-LO pathway of AA metabolism plays an important role in regulating cell survival, apoptosis and cancer (Haeggstrom and Funk, 2011).

The 12(*S*)-LO is similar in many respects to 15-lipoxygenase (15-LO) in terms of substrate specificity and capability of forming both 12-HpETE and 15-HpETE from AA. The 12(*S*)-LO has approximately 65% identity in primary

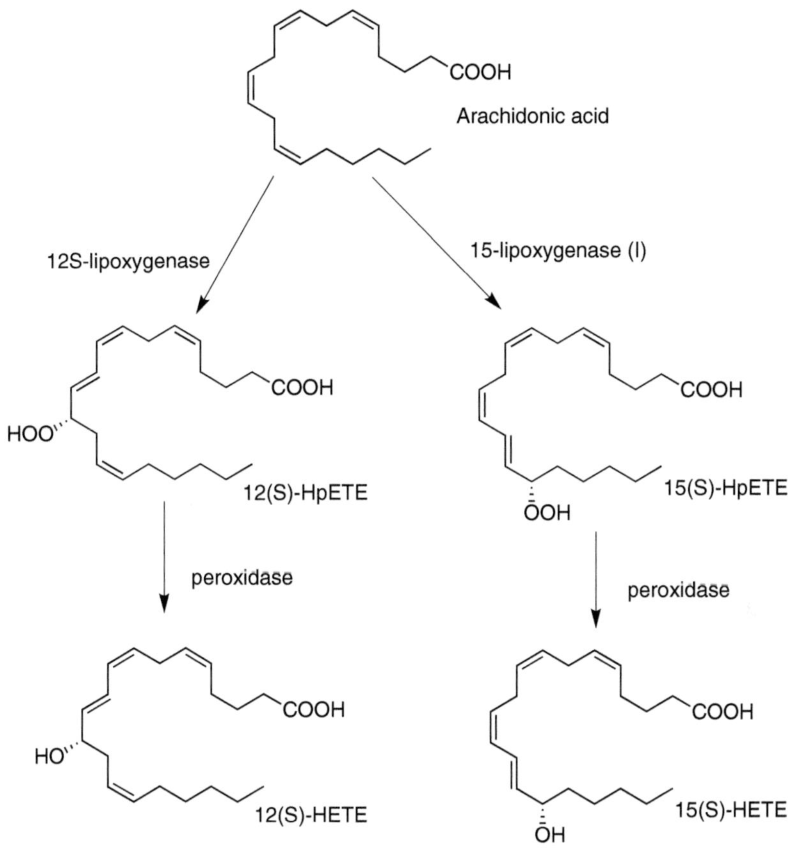

FIGURE 13 Metabolism of arachidonic acid by 12- and 15-lipoxygenase pathways with corresponding stereospecific formation of hydroperoxyeicosatetraenoic acids (HpETEs). Subsequent reduction of these hydroperoxides leads to the corresponding hydroxyeicosatetraenoic acids (HETEs) at either carbon-12 or carbon-15, which are thought to mediate biological activities of these enzymatic pathways.

structure to that of 15-LO from human reticulocytes. The 12(R)-LO is less well studied but can also oxygenate AA at C-15 and C-8. The three enzymes, 5/12/15-LO, are suicide inactivated.

5.7.2 15-Lipoxygenase

The oxidation of AA at C-15 is catalysed in humans by 15-LO-1 and 15-LO-2 (Figure 13), both of which are soluble proteins with a molecular weight of approximately 75 kDa. Many cells express these enzymes, which can also efficiently oxidise linoleic acid to 13-hydroperoxyoctadecadienoic acid and, to a lesser extent, 9-hydroperoxyoctadecadienoic acid. 15-LO can also oxygenate AA to 12-HpETE and 15-HpETE (Haeggstrom and Funk, 2011). The best substrate for mammalian 15-LO is cholesterol linoleate and many oxidised cholesterol esters are found in human atherosclerotic plaques. One distinguishing feature of 15-LO-1 is that it oxidises AA esterified to membrane phospholipids, thus forming esterified 15-HpETE. Expression of 15-LO-2 in the macrophage suggests a role for this enzyme in atherosclerosis (Magnusson et al., 2012).

The X-ray crystal structure of mammalian 15-LO-1 has revealed a β-barrel domain that is similar to C2 domains found on other proteins and directs the binding of this enzyme to phospholipid membranes (Haeggstrom and Funk, 2011), the source of either AA or phospholipids for the oxidation process. The catalytic domain, which contains the histidine-coordinated Fe^{+3}, holds the AA assisted by an ionic bond between Arg-403 and the ionised carboxyl group of AA. The methyl terminus of AA is thus placed deep within a hydrophobic-binding cleft in an arrangement that is similar to that of other LOs.

The structures of 15-LO-2 and 15-LO-1 are very similar (Kobe et al., 2014), providing insight into the enzymatic mechanism of substrate insertion and confirmation of the position of the polyunsaturated fatty acid tail in the active site. The oxygenase channel and catalytic iron atom are placed adjacent to the C-11–C-15 pentadienyl structure of the substrates so that a hydrogen atom from C-13 (of AA) can be abstracted and the triplet radical of diatomic oxygen attacks at C-15 to form the 15(S)-hydroperoxy radical product (Kobe et al., 2014). Mammalian 15-LO is involved in the production of more complicated eicosanoids including the biologically active lipoxins (Serhan, 2014). Lipoxins are formed by the sequential reaction of AA by 15-LO and 5-LO. For example, LTA_4 (the 5-LO product) can be converted to a lipoxin by 15-LO.

Recently, derivatives of DHA and EPA have been identified that can augment the resolution of inflammation (Serhan, 2014). These various pharmacologically active agents, which are called maresins, protectins and resolvins could, like lipoxins, be formed by the combined actions of several of the LOs described above (5-, 12-, and 15-LO). Studies of these new agents have contributed to our understanding of inflammatory regulation.

6. CYTOCHROME P450s AND EPOXYGENASE PATHWAYS

AA can be metabolised to a series of products characterised by the introduction of a single oxygen atom to form three different types of initial products catalysed by various cytochrome P450 mixed function oxidases (Figure 14) (Imig, 2012). The three classes of products include a series of HETEs formed by an allylic oxidation mechanism resulting in a family of conjugated dienes isomeric to the reduced products of a LO reaction. P450 metabolites formed by this mechanism have been characterised as 5-, 8-, 9-, 11-, 12-, and 15-HETE (Figure 14, middle reaction scheme), some of which are epimeric to the LO catalysed products. A second class of reactions involves oxidation of the terminal alkyl chain region of AA with placement of a hydroxyl group between the terminal carbon atom (ω) through the ω-4 position with formation of a family of ω-oxidised mono-HETEs such as

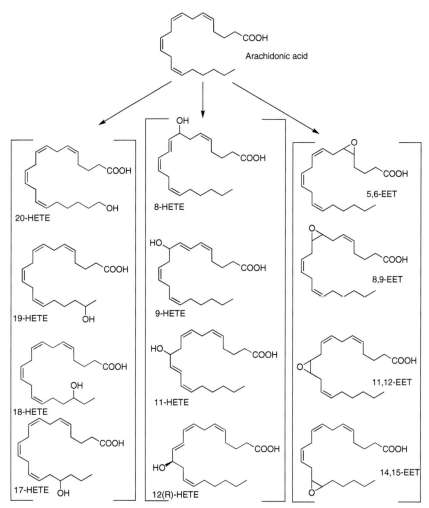

FIGURE 14 Metabolism of arachidonic acid by cytochrome P450 enzymes and the formation of three structurally distinct metabolite families. ω-Oxidation leads to a family of ω to ω-4 products including 20-HETE (ω-oxidation) and 19-HETE (ω-1 oxidation). The lipoxygenase-like mechanism of cytochrome P450 metabolism leads to the formation of six different conjugated dienols, for which the structures of four are indicated. One unique biologically active lipoxygenase-like P450 metabolite is 12(*R*)-HETE. The epoxygenase pathway leads to the formation of four regio-isomeric epoxyeicosatrienoic acid (EETs) all of which are biologically active.

20-20-HETE (Figure 14, left reaction scheme). Insertion of oxygen into the carbon–carbon bond results in the formation of a family of regio-isomeric *cis*-epoxyeicosatrienoic acids (EETs) from which the general pathway, the epoxygenase pathway, has been named (Figure 14, right reaction scheme). These regio-isomers include 14,15-, 11,12-, 8,9-, and 5,6-EETs, which can be formed either as an *R,S* or the *S,R* enantiomer.

6.1 Epoxygenase P450 Isozymes

With the availability of recombinant P450 isozymes, it is possible to identify those that metabolise AA. EET biosynthesis can be accomplished by CYPIA, CYP2B, CYP2C, CYP2D, CYP2G, CYP2J, CYP2N and CYP4A subfamilies (Spector and Kim, 2015). For each of these, unique EET regio-isomers are formed. For example, CYP2C8 produces $14(R),15(S)$-EET and $11(R),12(S)$-EET with optical purities of 86 and 81%, respectively. However, it is likely that more than a single P450 contributes to EET biosynthesis within a specific cell or tissue. Thus, individual AA epoxygenase metabolites may depend on expression of specific P450 isoforms. It is thought that the majority of EET biosynthesis in the kidneys of humans and rats is the result of CYP2C expression in these tissues. However, the induction of specific P450s can greatly alter the production of specific epoxygenase products.

6.2 Occurrence of Epoxyeicosatrienoic Acids

Various EETs have been measured in tissues as well as physiological fluids such as urine (Imig, 2013). Biologically active lipids originally defined as an endothelium-derived hyperpolarizing factor and an inhibitor of Na+/K+ATPase found in the thick ascending loop of Henle were structurally characterised as $11(R),12(S)$-EET and 20-HETE, both derived from cytochrome P-450-mediated metabolism of AA (Imig, 2012). Interestingly, EETs can readily form CoA esters and are substrates for reacylation of lysophospholipids and incorporation of these oxidised metabolites of AA into phospholipid membranes, a biochemical reaction not observed for prostaglandins, thromboxanes or leukotrienes. For example, human platelets contain 14,15-EET esterified within membrane phospholipids.

6.3 Metabolism of Epoxyeicosatrienoic Acids

A number of metabolic pathways operate on the primary epoxygenase metabolites of AA. Some of the more abundant pathways include CoA-dependent reesterification (noted above) as well as β-oxidation chain shortening. A unique pathway involves epoxide hydrolase, a cytosolic enzyme that hydrates EETs to the corresponding vicinal dihydroxyeicosatrienoic acids (Imig, 2012). A microsomal epoxide hydrolase also metabolises EETs, but at a somewhat lower rate. The soluble epoxide hydrolase has substrate specificity, both in terms of the stereochemistry of the EET and in terms of its position in the AA chain. As expected, nonenzymatic hydration of these epoxides occurs, especially under acidic conditions, and can be accelerated during the isolation of these AA metabolites. Therefore, it is sometimes difficult to distinguish between nonenzymatic and enzymatic hydration of EETs. The 5,6-EET is a poor substrate for cytosolic and microsomal epoxide hydrolase; however, 5,6-EET is a substrate for PGH synthase, leading to the formation of 5,6-epoxy-PGH$_1$. This reactive intermediate can subsequently be transformed into corresponding 5,6-epoxy-prostaglandins of the E, F and I series or into an epoxy thromboxane analog. All of the EETs can also be substrates for LOs that introduce molecular oxygen at any 1,4-*cis*-pentadienyl position not interrupted by the epoxide ring. The EETs can also be conjugated with reduced glutathione catalysed by glutathione (*S*) transferases. Studies of the metabolism of EETs by rat or mouse liver microsomal P450 revealed the formation of a series of diepoxyeicosadienoic acids as well as monohydroxyepoxyeicosatrienoic acids.

6.4 Biological Actions of Epoxyeicosatrienoic Acids

Metabolites of AA derived from the epoxygenase P450 pathway have been studied extensively in terms of their pharmacological properties. Potent effects have been observed on various ion channels, membrane-bound transport proteins, mitogenesis, PPARα agonists and activators of tyrosine kinase cascades (Imig, 2012). EETs likely play an important role in mediating Na^+/K^+-ATPase activity and inhibiting the hydroosmotic effect of arginine vasopressin in the kidney. A picture has emerged for an important role of EETs and 20-HETE in regulating renal vascular tone and fluid/electrolyte transport in the pathogenesis of hypertension. 20-HETE causes constriction of the renal arterials by inhibiting vascular smooth muscle K_{Ca} channels, leading to membrane depolarisation by calcium ion influx (Imig, 2013). The EETs are thought to dilate renal arterials by an opposite mechanism and, together with 20-HETE, contribute to autoregulation of renal blood flow whereby the glomerular filtration rate is held constant despite changes in blood pressure.

7. FUTURE DIRECTIONS

There is currently a reasonable understanding of the structures of PGHSs, but many of the relationships between structure and function remain to be identified. For example, the peroxidase activities of PGH synthases preferentially utilise alkyl hydroperoxides such as PGG_2 rather than hydrogen peroxide, yet the molecular basis for this specificity is not evident from simple observation of the structures. It will also be important to characterise further the membrane-binding domain of PGHSs in the context of the interaction with specific membrane lipids, and whether this domain plays a role in governing substrate entry into the COX site. The recent discovery of the allosteric regulation of PGHSs by fatty acids that are not substrates needs further exploration in the context of understanding how dietary fatty acids affect both the PG synthesis and the responses of PGHSs to NSAIDs and coxibs.

Our understanding of the different biological roles of PGHS-I and PGHS-2 is only beginning to emerge. The functions of these two isozymes in apoptosis, carcinogenesis, angiogenesis, respiration, inflammation and pain have not been well delineated. Of key importance is understanding the reason for the existence of the two PGHS isozymes, and how coupling occurs among PGHSs, upstream lipases, and downstream synthases and receptors.

From a pharmacological perspective, it will be important to understand the molecular basis for the adverse cardiovascular side effects of COX-2 inhibitors. The next few years should also see the development and testing of receptor antagonists that have the promise of being more specific than NSAIDs and COX-2 inhibitors. For example, it is thought that EP receptors play roles in colon cancer and that inhibition of one or more of these receptors may be useful in preventing colon carcinogenesis.

Considerable challenges remain in understanding the detailed biochemistry involved in the synthesis and release of biologically active LTs. Little is known about how these highly lipophilic molecules are released from cells; but even more curious is how the chemically reactive intermediate LTA$_4$, made on the perinuclear envelope, finds its way into a neighbouring cell in the process of transcellular biosynthesis.

The mechanism of 5-LO is still poorly understood; however, the detailed X-ray structure of 5-LO, FLAP, LTA$_4$ hydrolase and LTC$_4$ synthase have all opened new avenues of understanding the mechanisms of these enzymes that are responsible for leukotriene biosynthesis. Structure-driven drug design is now possible and has been realised for several of these enzymes in terms of specific inhibitors. Our understanding of the biological actions of leukotrienes in health and disease continues to expand. While initial discovery of this pathway was driven by the potential role of these molecules in pulmonary diseases such as asthma, research has revealed a critical role for metabolites of LO in heart disease, cancer and neuropathological disturbances, including traumatic brain injury, multiple sclerosis and perhaps even Alzheimer disease where components of inflammation play an important role. The P450 pathway of arachidonate metabolism has substantially advanced along with the potential for development of novel therapeutics such as soluble epoxide hydrolase inhibitors to address cardiovascular diseases.

REFERENCES

Ago, H., Kanaoka, Y., Irikura, D., Lam, B.K., Shimamura, T., Austen, K.F., Miyano, M., 2007. Crystal structure of a human membrane protein involved in cysteinyl leukotriene biosynthesis. Nature 448, 609–612.

Brock, T.G., 2005. Regulating leukotriene synthesis: the role of nuclear 5-lipoxygenase. J. Cell Biochem. 96, 1203–1211.

Cui, X., Kawashima, H., Barclay, T.B., Peters, J.M., Gonzalez, F.J., Morgan, E.T., Strobel, H.W., 2001. Molecular cloning and regulation of expression of two novel mouse CYP4F genes: expression in peroxisome proliferator-activated receptor alpha-deficient mice upon lipopolysaccharide and clofibrate challenges. J. Pharmacol. Exp. Ther. 296, 542–550.

Dennis, E., Cao, J., Hsu, Y., Magrioti, V., Kokotos, G., 2011. Phospholipase A2 enzymes: physical structure, biological function, disease implication, chemical inhibition, and therapeutic intervention. Chem. Rev. 111, 6130–6185.

Dong, L., Sharma, N.P., Jurban, B.J., Smith, W.L., 2013. Pre-existent asymmetry in the human cyclooxygenase-2 sequence homodimer. J. Biol. Chem. 288, 28641–28655.

Evans, J.H., Gerber, S.H., Murray, D., Leslie, C.C., 2004. The calcium binding loops of the cytosolic phospholipase A2 C2 domain specify targeting to Golgi and ER in live cells. Mol. Biol. Cell 15, 371–383.

Ferguson, A.D., McKeever, B.M., Xu, S., Wisniewski, D., Miller, D.K., Yamin, T.T., Spencer, R.H., Chu, L., Ujjainwalla, F., Cunningham, B.R., Evans, J.F., Becker, J.W., 2007. Crystal structure of inhibitor-bound human 5-lipoxygenase-activating protein. Science 317, 510–512.

Folco, G., Murphy, R.C., 2006. Eicosanoid transcellular biosynthesis: from cell–cell interactions to in vivo tissue responses. Pharmacol. Rev. 58, 375–388.

Gilbert, N.C., Bartlett, S.G., Waight, M.T., Neau, D.B., Boeglin, W.E., Brash, A.R., Newcomer, M.E., 2011. The structure of human 5-lipoxygenase. Science 331, 217–219.

Haeggstrom, J.Z., Funk, C.D., 2011. Lipoxygenase and leukotriene pathways: biochemistry, biology, and roles in disease. Chem. Rev. 111, 5866–5898.

Hirata, T., Narumiya, S., 2011. Prostanoid receptors. Chem. Rev. 111, 6209–6230.

Imig, J.D., 2012. Epoxides and soluble epoxide hydrolase in cardiovascular physiology. Physiol. Rev. 92, 101–130.

Imig, J.D., 2013. Epoxyeicosatrienoic acids, 20-hydroxyeicosatetraenoic acid, and renal microvascular function. Prostaglandins Other Lipid Mediators 104–105, 2–7.

Kanaoka, Y., Maekawa, A., Austen, K.F., 2013. Identification of GPR99 protein as a potential third cysteinyl leukotriene receptor with a preference for leukotriene E4 ligand. J. Biol. Chem. 288, 10967–10972.

Kang, Y.J., Mbonye, U.R., Delong, C.J., Wada, M., Smith, W.L., 2007. Regulation of intracellular cyclooxygenase levels by gene transcription and protein degradation. Prog. Lipid Res. 46, 108–125.

Kobe, M.J., Neau, D.B., Mitchell, C.E., Bartlett, S.G., Newcomer, M.E., 2014. The structure of human 15-lipoxygenase-2 with a substrate mimic. J. Biol. Chem. 289, 8562–8569.

Magnusson, L.U., Lundqvist, A., Karlsson, M.N., Skalen, K., Levin, M., Wiklund, O., Boren, J., Hulten, L.M., 2012. Arachidonate 15-lipoxygenase type B knockdown leads to reduced lipid accumulation and inflammation in atherosclerosis. PLoS One 7, e43142.

Milne, G., Yin, H., Hardy, K., Davies, S., Roberts, L., 2011. Isoprostane generation and function. Chem. Rev. 111, 5973–5996.

Murakami, M., Taketomi, Y., Sato, H., Yamamoto, K., 2011. Secreted phospholipase A2 revisited. J. Biochem. 150, 233–255.

Murphy, R.C., Wheelan, P., 1998. Pathways of leukotriene metabolism in isolated cell models and human subjects. In: Drazen, J.M. (Ed.), Five-lipoxygenase Products in Asthma. Marcel-Dekker, New York.

Myung, S.-J., Rerko, R.M., Yan, M., Platzer, P., Guda, K., Dotson, A., Lawrence, E., Dannenberg, A.J., Lovgren, A.K., Luo, G., Pretlow, T.P., Newman, R.A., Willis, J., Dawson, D., Markowitz, S.D., 2006. 15-Hydroxyprostaglandin dehydrogenase is an in vivo suppressor of colon tumorigenesis. Proc. Natl. Acad. Sci. U.S.A. 103, 12098–12102.

Nakamura, M., Shimizu, T., 2011. Leukotriene receptors. Chem. Rev. 111, 6231–6298.

Niegowski, D., Kleinschmidt, T., Ahmad, S., Qureshi, A.A., Marback, M., Rinaldo-Matthis, A., Haeggstrom, J.Z., 2014. Structure and inhibition of mouse leukotriene C4 synthase. PLoS One 9, e96763.

Nomura, D.K., Morrison, B.E., Blankman, J.L., Long, J.Z., Kinsey, S.G., Marcondes, M.C.G., Ward, A.M., Hahn, Y.K., Lichtman, A.H., Conti, B., Cravatt, B.F., 2011. Endocannabinoid hydrolysis generates brain prostaglandins that promote neuroinflammation. Science 334, 809–813.

Patrono, C., Baigent, C., 2014. Nonsteroidal anti-inflammatory drugs and the heart. Circulation 129, 907–916.

Rådmark, O., Samuelsson, B., 2009. 5-Lipoxygenase: mechanisms of regulation. J. Lipid Res. 50, S40–S45.

Rouzer, C.A., Marnett, L.J., 2011. Endocannabinoid oxygenation by cyclooxygenases, lipoxygenases, and cytochromes P450: cross-talk between the eicosanoid and endocannabinoid signaling pathways. Chem. Rev. 111, 5899–5921.

Segraves, E.N., Chruszcz, M., Neidig, M.L., Ruddat, V., Zhou, J., Wecksler, A.T., Minor, W., Solomon, E.I., Holman, T.R., 2006. Kinetic, spectroscopic, and structural investigations of the soybean lipoxygenase-1 first-coordination sphere mutant, Asn694Gly. Biochemistry 45, 10233–10242.

Serhan, C.N., 2014. Pro-resolving lipid mediators are leads for resolution physiology. Nature 510, 92–101.

Smith, W., Urade, Y., Jakobsson, P., 2011. Enzymes of the cyclooxygenase pathways of prostanoid biosynthesis. Chem. Rev. 111, 5821–5865.

Smith, W.L., 2008. Nutritionally essential fatty acids and biologically indispensable cyclooxygenases. Trends Biochem. Sci. 33, 27–37.

Snelgrove, R.J., Jackson, P.L., Hardison, M.T., Noerager, B.D., Kinloch, A., Gaggar, A., Shastry, S., Rowe, S.M., Shim, Y.M., Hussell, T., Blalock, J.E., 2010. A critical role for LTA4H in limiting chronic pulmonary neutrophilic inflammation. Science 330, 90–94.

Spector, A.A., Kim, H.Y., 2015. Cytochrome P epoxygenase pathway of polyunsaturated fatty acid metabolism. Biochim. Biophys. Acta 1851, 356–365.

Stsiapanava, A., Olsson, U., Wan, M., Kleinschmidt, T., Rutishauser, D., Zubarev, R.A., Samuelsson, B., Rinaldo-Matthis, A., Haeggstrom, J.Z., 2014. Binding of Pro-Gly-Pro at the active site of leukotriene A4 hydrolase/aminopeptidase and development of an epoxide hydrolase selective inhibitor. Proc. Natl. Acad. Sci. U.S.A. 111, 4227–4232.

Tai, H., Ensor, C., Tong, M., Zhou, H., Yan, F., 2002. Prostaglandin catabolizing enzymes. Prostaglandins Other Lipid Mediators 68-69, 483–493.

Tsai, A., Kulmacz, R., 2010. Prostaglandin H synthase: resolved and unresolved mechanistic issues. Arch. Biochem. Biophys. 493, 103–124.

Uddin, M.J., Crews, B.C., Blobaum, A.L., Kingsley, P.J., Gorden, D.L., McIntyre, J.O., Matrisian, L.M., Subbaramaiah, K., Dannenberg, A.J., Piston, D.W., Marnett, L.J., 2010. Selective visualization of cyclooxygenase-2 in inflammation and cancer by targeted fluorescent imaging agents. Cancer Res. 70, 3618–3627.

Yu, Y., Ricciotti, E., Scalia, R., Tang, S.Y., Grant, G., Yu, Z., Landesberg, G., Crichton, I., Wu, W., Pure, E., Funk, C.D., FitzGerald, G.A., 2012. Vascular COX-2 modulates blood pressure and thrombosis in mice. Sci. Transl. Med. 4, 132–154.

Zheng, Y., Yin, H., Boeglin, W.E., Elias, P.M., Crumrine, D., Beier, D.R., Brash, A.R., 2011. Lipoxygenases mediate the effect of essential fatty acid in skin barrier formation: a proposed role in releasing omega-hydroxyceramide for construction of the corneocyte lipid envelope. J. Biol. Chem. 286, 24046–24056.

Chapter 10

Sphingolipids

Anthony H. Futerman

Department of Biological Chemistry, Weizmann Institute of Science, Rehovot, Israel

ABBREVIATIONS

ASM Acid sphingomyelinase
CerS Ceramide synthases
CERT Ceramide transport protein
CNS Central nervous system
DAG Diacylglycerol
ER Endoplasmic reticulum
FAPP2 Four-phosphate adaptor protein 2
FFAT Two-phenylalanines in an acidic tract
GalCer Galactosylceramide
GlcCer Glucosylceramide
GSL Glycosphingolipid
HSAN1 Hereditary sensory neuropathy type I
3KSR 3-Ketosphinganine reductase
LCB Long chain base
LSD Lysosomal storage disease
MLD Metachromatic leukodystrophy
PC Phosphatidylcholine
PH Pleckstrin homology
PI Phosphatidylinositol
SAP Sphingolipid activator protein
S1P Sphingosine 1-phosphate
SL Sphingolipid
SM Sphingomyelin
SMase Sphingomyelinase
SphK Sphingosine kinase
SPT Serine palmitoyl transferase
START Steroidogenic acute regulatory protein-related lipid transfer

1. INTRODUCTION

Sphingolipids (SLs) are ubiquitous components of eukaryotic cell membranes and are found in species as diverse as fungi to mammals, and even in some bacteria and viruses (Merrill, 2008). Interest in SLs has blossomed over the past couple of decades due to two major discoveries, namely that in addition

Biochemistry of Lipids, Lipoproteins and Membranes. http://dx.doi.org/10.1016/B978-0-444-63438-2.00010-9

to their well-established roles as structural components of cell membranes, SLs also 'turnover' in a number of cellular signalling pathways (Hannun and Obeid, 2008; Maceyka and Spiegel, 2014), and are essential components of the so-called 'membrane-rafts' (Simons and Gerl, 2010). These two findings have revolutionised SL biology and stimulated research directions that could not have been foreseen 20 years ago.

In this chapter, I will first give an outline of SL structure, with particular emphasis on the unexpected diversity of structures that have come to light due to advances in the analytical techniques of lipidomics (see Chapter 2). This diversity, although perhaps suspected in days gone by, is nevertheless much larger than anticipated (Merrill, 2011), and has broad implications for understanding the role of SLs in cell biology. By way of example, the first committed step in SL biosynthesis was previously thought to occur via the condensation of palmitoyl-CoA with serine via the enzyme serine palmitoyl transferase (SPT), but it is now known that other acyl-CoAs can also be used by SPT, and of no less interest, other amino acids, such as alanine and glycine, can also be used by SPT. This finding is of significance for understanding the role of SLs in human diseases; for instance, hereditary sensory neuropathy type I (HSAN1) is an autosomal dominant genetic disorder caused by mutations in the SPT gene that alters the specificity of SPT such that it uses alanine, rather than serine. This finding exemplifies advances in SL biology that have impacted understanding of human disease.

After outlining SL structure, I will discuss SL metabolism, addressing SL synthesis, which unlike the synthesis of other membrane lipids is compartmentalised between the endoplasmic reticulum (ER) and the Golgi apparatus, and SL degradation. The latter has been extensively studied for decades since a number of human genetic diseases are caused by defects in the lysosomal enzymes that degrade SLs; these diseases are known collectively as lysosomal storage diseases (LSDs), and result in accumulation of undegraded SLs and glycosphingolipids (GSLs).

As noted above, the role of SLs in cellular signalling pathways is now well established, and will be discussed in detail, with emphasis not only on the two best-characterised SL signalling molecules, ceramide and sphingosine 1-phosphate (S1P), but also additional SLs such as ceramide-1-phosphate, for which less information is available. This section will emphasise that SL signalling is enormously complex since individual SLs in the same metabolic pathway can play opposite roles in signalling, necessitating exquisite control of levels of specific SLs in specific cell compartments. An additional layer of complexity in lipid signalling is the ability of certain SLs to segregate within the plane of the lipid bilayer into lipid microdomains or rafts, which can act as signalling platforms. Finally, the chapter will conclude with a discussion of the role of SLs in human disease. Again, this is an area in which SLs have been implicated to play vital roles, as a direct cause of human disease, such as in the LSDs, or downstream to the causative disease agent or mechanism. Nevertheless, even in the latter case, intervention in the SL pathway has been shown to have beneficial effects on the pathology of diseases as wide ranging as cancer, diabetes and psychiatric illness. Thus, although the chapter will end with a brief discussion on future directions, past experience has taught us that the field of SL biology is evolving so fast that predicting the future is little more than intelligent guesswork.

2. NOMENCLATURE AND STRUCTURE

As is the case for all membrane lipids, SLs are amphipathic molecules containing hydro*phobic* ('water-hating') and hydro*philic* ('water-loving') regions. SLs (which compose ~30% of cell membrane lipids) differ from glycerolipids (which compose ~50–60% of cell membrane lipids) inasmuch as the building block of the latter is glycerol (Chapter 7), to which two fatty acids are esterified, whereas the building block of SLs is a long chain base (LCB) (usually sphinganine or sphingosine in mammals) to which one fatty acid is attached via an amide bond to form ceramide. The difference between glycerolipids and SLs was first noted by Thudicum (1874), who pointed out in his remarkable treatise 'The Chemical Constituents of the Brain' that sphingomyelin (SM) (a major brain lipid) contained 'no glycerophosphoric acid on chemolyses'. Thudicum suggested that this finding was 'of the utmost interest and importance', a statement which has certainly been vindicated during the years since the publication of his work in 1874.

The structure of ceramide, the basic building block of SLs, is given in Figure 1, in which the blue portion of the molecule is the sphingoid LCB and the red portion corresponds to the fatty acid that is *N*-acylated to the LCB. As noted above, the structural complexity of SLs is huge, with thousands of possible structures varying in the three major structural regions of the SL molecule, the sphingoid LCB, the fatty acid and the head group.

First, is the sphingoid LCB, which can differ in length, stereochemistry, degree of saturation, branching, hydroxylation and structure of the head group moiety (Figure 2(A)). The most common sphingoid LCB is D-*erythro*-sphingosine (known by the nomenclature d18:1), which is the sphingoid LCB shown in Figure 1, and contains 18 carbon atoms, 16 of which are donated from palmitoyl-CoA and two from serine. However, the length of the LCB can be as short as 14 carbons, in d14:0 and d14:1-sphinganine/sphingosine, or more commonly in mammalian cells, C16- or C20-sphingosine. In terms of stereochemistry, the sphingoid LCB contains two chiral carbon atoms, at carbons 2 and 3. Natural SLs occur in the D-*erythro* (2S, 3R) configuration, but three additional stereoisomers exist, L-*erythro* (2R, 3S) (the enantiomer of D-*erythro*-), D-*threo* (2R, 3R) and L-*threo* (2S, 3S) (Figure 2(B)); this is of more than theoretical importance since the lack of metabolism of some of these stereoisomers has proved useful in generation of SL metabolites that could potentially be of therapeutic benefit in some human diseases. Mammalian sphingoid LCBs often contain a double bond in the 4,5 position (d18:1), but a double bond in the 8,9 position (d18:2)

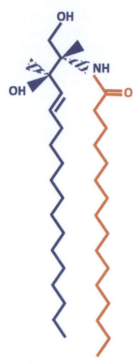

FIGURE 1   **Ceramide, the basic building block of SLs.** The blue portion of the molecule is the sphingoid LCB and the red portion the fatty acid, which is attached to the LCB via an amide bond. *Adapted from Tidhar and Futerman (2013).*

is also common in plants. In some lower organisms, such as *Caenorhabditis elegans*, the sphingoid LCB is branched at the 19 position. Hydroxylation of the LCB, in, for instance, the C6 position (in skin) or at C4 (in yeast), is also common. Finally, the use of amino acids other than serine can produce sphingoid LCBs known as deoxysphingosines, desoxysphingosines (Figure 3) and others. The existence of such a wide range of LCB structures has resulted in unexpected insight into some human diseases. Thus, although the generic structure of the sphingoid LCB has been known for decades, the many subtle yet important variations in LCB structure have proved to be of unexpected biological relevance.

The fatty acid attached to the sphingoid LCB via an amide bond can also vary significantly. While synthetic SLs containing fatty acids as short as hexanoic acid (C6) have been widely used experimentally (Figure 4) (Lipsky and Pagano, 1985), it is generally considered that C14-fatty acid (myristic acid) is among the shortest naturally occurring fatty acids found in SLs. Mammalian SLs contain a wide variety of fatty acids, ranging in length from C14 to C32 that are predominantly saturated, and can contain α- (Figure 4) or ω-hydroxyl groups.

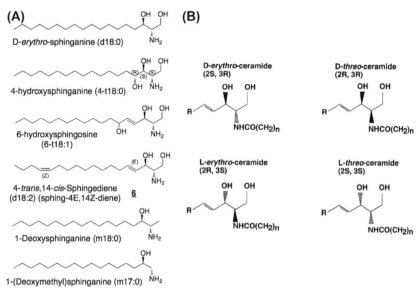

FIGURE 2   The variety of LCB structures found in SLs. Panel (A) *(taken from Merrill (2008))* shows a selection of LCBs and panel (B) the stereochemistry of the LCB *(adapted from Venkataraman and Futerman (2001))*. The four stereoisomers of ceramide are designated using the *R/S* nomenclature. D-*erythro*-ceramide and L-*erythro*-ceramide are enantiomers (mirror images), as are D-*threo*- and L-*threo*-ceramide; the *threo*- and *erythro*-ceramides are diastereoisomers. The *threo* forms differ from the *erythro* forms inasmuch as C2 and C3 are in the *cis* configuration in the *threo* forms but *trans* in the *erythro* forms.

The *N*-acylation of a sphingoid LCB results in the formation of ceramide, which is extremely hydrophobic, and the longer the acyl chain, the more hydrophobic it becomes. Thus, C32-ceramides (i.e. ceramide containing a fatty acid with 32 carbon units, i.e. dotriacontanoic acid) are found at high levels in skin, where their hydrophobic nature makes them an essential part of the skin permeability barrier. The hydrophobic nature of ceramide, along with other biophysical properties dependent on its structure, will be relevant later when discussing the ability of ceramide to segregate into domains ('rafts') in the plane of the lipid bilayer.

The final region of the SL molecule that shows considerable structural variety is the head group. Complex SLs are formed by the attachment of different head groups at C-1. Attachment of phosphorylcholine forms SM, and attachment of glucose or galactose is the first step in the formation of GSLs (Figure 5). The GSLs are the most structurally diverse class of complex SLs, and are normally classified as acidic or neutral. More than 500 different carbohydrate structures have been described in GSLs, with the main sugars being glucose, galactose, fucose, *N*-acetylglucosamine (GlcNAc), *N*-acetylgalactosamine (GalNAc) and sialic acid (*N*-acetylneuraminic acid).

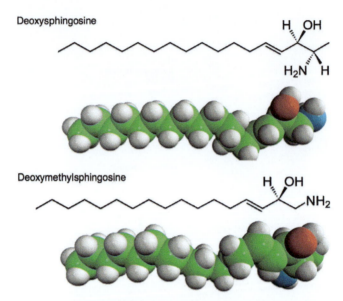

FIGURE 3   Structures of deoxysphingosine and deoxymethylsphingosine. The structure and 3D structure of deoxysphingosine (upper panel) and deoxymethylsphingosine (lower panel) are shown. Note that these LCBs cannot be further modified at the C-1 position. *Taken with permission from Avanti Polar Lipids* (http://www.avantilipids.com).

GSLs are classified as acidic (the gangliosides) or neutral, depending on whether or not they contain sialic acid. Figure 5(A) shows the structure of a typical GSL, namely ganglioside GM1 and Figure 5(B) shows the remarkable structural diversity in GSL structures along with their pathways of biosynthesis.

3. SPHINGOLIPIDS BIOSYNTHESIS

Unlike the synthesis of other major membrane lipids, namely sterols and glycerolipids, which takes place almost uniquely in the ER, the initial steps of SL synthesis take place in the ER, but latter steps take place for the most part in the Golgi apparatus; moreover, specific steps of SL synthesis occur on opposite surfaces of these organelles (lumenal vs cytosolic) (Futerman and Riezman, 2005). The reasons for these complex topological relationships are not entirely clear, although in some cases, such as SM and GSLs, synthesis in the lumen of the Golgi apparatus can be rationalised by the topological equivalence of the lumenal leaflet of the Golgi apparatus with the outer (extracellular) leaflet of the plasma membrane, where SM and GSLs mainly reside.

3.1 SPT and 3-Ketosphinganine Reductase

SL synthesis begins with the conversion of serine and fatty acyl-CoA into 3-ketosphinganine via SPT (Figure 6). SPT, the rate-limiting enzyme in SL biosynthesis, belongs to the α-oxoamine synthase family and is pyridoxal

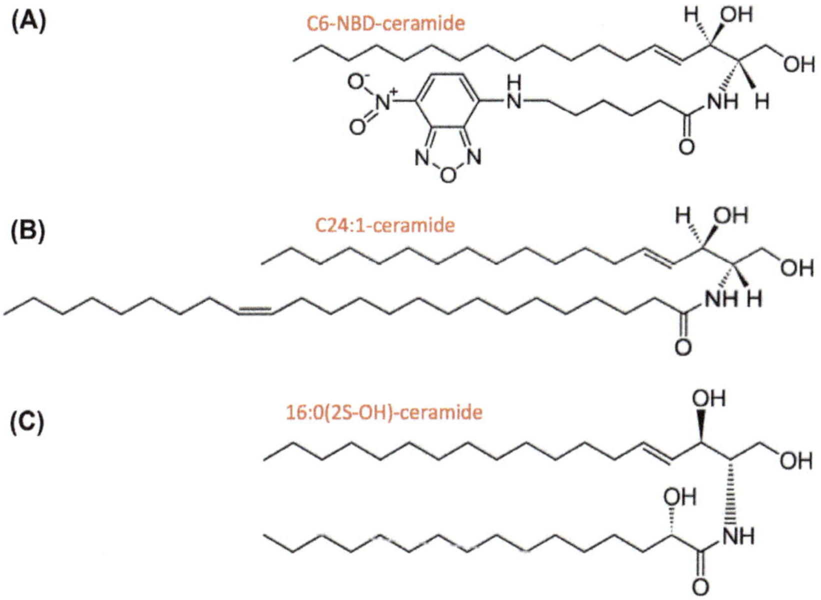

(A)

C6-NBD-ceramide

(B)

C24:1-ceramide

(C)

16:0(2S-OH)-ceramide

FIGURE 4   **A selection of ceramide structures.** (A) C6-NBD-ceramide (*N*-[6-[(7-nitro-2-1,3-benzoxadiazol-4-yl)amino]hexanoyl]-D-*erythro*-sphingosine). This synthetic ceramide derivative has been used extensively to study the intracellular transport and distribution of SLs. (B) C24:1-ceramide (d18:1/24:1) (*N*-nervonoyl-D-*erythro*-sphingosine). This ceramide occurs at high levels in a number of mammalian tissues. (C) 16:0(2S–OH)-ceramide (*N*-(2′-(*S*)-hydroxypalmitoyl)-D-*erythro*-sphingosine). This is an example in which ceramide contains a hydroxy-fatty acid. *Taken with permission from Avanti Polar Lipids (*http://www.avantilipids.com*).*

5′-phosphate dependent (Ikushiro and Hayashi, 2011). SPT is a heteromeric protein located in the ER of eukaryotic cells, but located in the cytoplasm of the gram-negative bacterium, *Sphingomonas paucimobilis*. The two mammalian SPT subunits are known as long chain base 1 and 2 (LCB1/LCB2) or as SPT1 and SPT2, with SPT2 found as two isoforms (LCB2a and LCB2b (also known as SPT3)). The SPT heterodimer can also associate with other small subunits; in yeast, a subunit known as Tsc3p is required for maximal SPT activity and in mammals, two subunits, ssSPTa and ssSPTb (*small subunits of SPT), substantially enhance SPT activity. Combinations of different subunits confer distinct specificities toward acyl-CoAs. Thus, the complex of LCB1/LCB2a/ssSPTa shows a strong preference for C16-CoA (palmitoyl-CoA), whereas LCB1/LCB2b/ssSPTa can also use C14-CoA (myristoyl-CoA), and ssSPTb confers specificity toward C18-CoA (stearoyl-CoA). LCB2b may be responsible for the generation of C14- and C16-sphingoid bases, making LCB2b functionally distinct from the LCB2a subunit. Thus, SPT can use at least three different acyl-CoAs for formation of 3-ketosphinganine, depending on the combination of the subunits (Lowther et al., 2012). Moreover, as noted above, serine is not the only amino acid to be used by SPT.

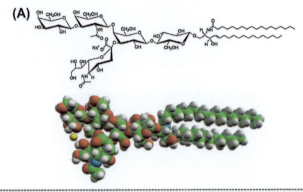

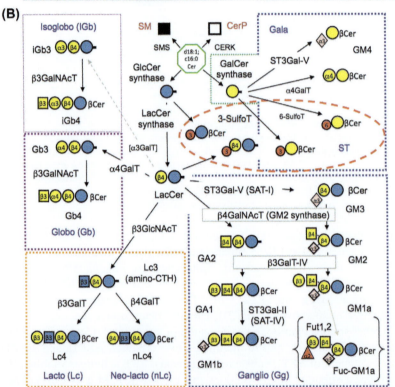

FIGURE 5 The structure of GSLs. (A) The chemical structure and 3D structure of ganglioside GM1 (*taken with permission from Avanti Polar Lipids* (http://www.avantilipids.com)). (B) A selection of GSL structures organised according to the GSL biosynthetic pathway. *Taken with permission from Merrill (2008).*

The next enzyme in the biosynthetic pathway is 3KSR, which reduces 3-ketosphinganine to dihydrosphingosine (sphinganine) in an NADPH-dependent manner. The active site of 3KSR appears to face the cytosol (Figure 7), and of all the enzymes in the SL biosynthetic pathway, 3KSR is perhaps the least

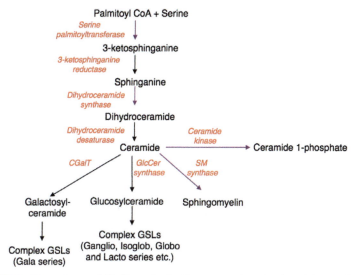

FIGURE 6 **The pathway of SL biosynthesis.** Enzymes are in red. *Adapted from Tidhar and Futerman (2013).*

well studied. However, the product of 3KSR, sphinganine, is a key intermediate in the SL biosynthetic pathway, as it is the substrate of the (dihydro)ceramide synthases (CerSs). CerSs can also *N*-acylate sphingosine, which is formed during SL degradation (see below).

3.2 Ceramide Synthases and Dihydroceramide Desaturase

Six CerS isoforms (CerS1–6) exist in mammals, which differ in the use of fatty acyl-CoAs for *N*-acylation of the sphingoid LCB (Table 1). The CerS are less specific toward the LCB than toward acyl-CoAs, since they are able to *N*-acylate a variety of LCBs, including natural LCBs such as sphinganine, sphingosine and phytosphingosine, and LCB analogues such as fumonisin B1. All six CerS contain a domain called the *T*ram-*L*ag-*C*LN8 (TLC) domain which is required for enzymatic activity, although the exact residues that determine specificity have not been delineated. For CerS activity, the Lag1p motif, a conserved stretch of 52 amino acids within the TLC domain, is essential.

The three-dimensional structure of the CerS, and their precise membrane topology (Figure 7), remains to be elucidated. Mammalian CerS also exhibit a distinct tissue distribution pattern (Table 1). However, ceramide species with distinct acyl chain lengths do not always correlate with CerS expression, suggesting that other factors apart from CerS expression and activity (perhaps posttranslational modification of CerS activity) determine the acyl chain length composition of ceramides and SLs in a particular tissue.

The next enzyme in the pathway (Figure 6) is dihydroceramide desaturase, the enzyme responsible for the formation of the 4,5-double bond on the sphingoid

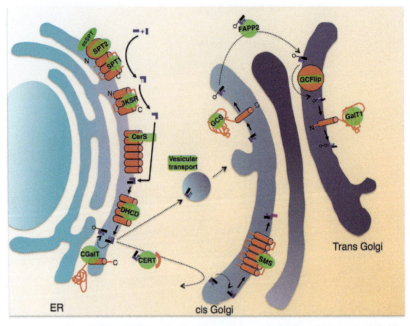

FIGURE 7 **The topology of SL synthesis.** The compartmentalisation of SL synthesis is illustrated. Enzymes are as follows: SPT, serine palmitoyl transferase; 3KSR, 3-ketosphinganine reductase; CerS, ceramide synthase; DHCD, dihydroceramide desaturase; CGalT, UDP-galactose:ceramide galactosyltransferase; CERT, ceramide transfer protein; SMS, SM synthase; GCS, GlcCer synthase; GCflip, GlcCer flippase; GalT1, LacCer synthase; FAPP2, four-phosphate adaptor protein 2. A schematic of SL structure is indicated as follows: Blue, sphingoid long chain base; thin blue line, 3-OH (as in 3-ketosphinganine); black, fatty acid; white line, 4,5-double bond in sphingoid base; black circle and line, sugar head group (i.e. glucose, galactose or GlcGal in the case of LacCer); purple rectangle, phosphorylcholine (as in SM). *Adapted from Tidhar and Futerman (2013).*

LCB, thus generating ceramide from dihydroceramide. Current data are consistent with the idea that the active site of this enzyme faces the cytosol, which is consistent with topology predictions implying three transmembrane domains.

3.3 Ceramide Transport between the ER and the Golgi Apparatus

Subsequent to its generation on the cytosolic surface of the ER, ceramide is transported to at least three other subcellular locations, where it is further metabolised (Figure 7). The first of these is the lumenal surface of the ER, where galactosylceramide (GalCer) is generated by the action of UDP-galactose:ceramide galactosyltransferase (CGalT), a type 1 membrane protein whose *N*-terminus and active site face the lumen. The mechanism by which ceramide is translocated to the lumenal leaflet of the ER (or of the Golgi apparatus, see below) is not known, and current data support two possibilities, namely

TABLE 1 Fatty Acid Specificity and Distribution of CerS in Mammalian Tissues

CerS	Predominant Fatty Acid Specificity of CerS	Tissue Distribution
CerS1	C18	Brain (neurons), cerebellum (Purkinje cells), skeletal muscle
CerS2	C22–C24	Brain (oligodendrocytes), lung, liver, intestine, adrenal gland, kidney, white adipose tissues
CerS3	C26	Skin, testis
CerS4	C18–C22	Heart, skin
CerS5	C16	Prostate gland and skeletal muscle
CerS6	C14, C16	Brain (hippocampus), kidney (glomeruli), small and large intestine, thymus

Tissue distribution is taken from Levy and Futerman (2010).
For Many Years, CerS were Known as Lass Genes (*L*ongevity *A*ssurance Gene Homologues).

spontaneous intrabilayer transport or facilitated (i.e. protein-mediated) transport. Most of the other enzymes of ceramide synthesis are located in the Golgi apparatus. One of the most interesting and unexpected discoveries in the area of SL biosynthesis over the past decade was the discovery that ceramide can be transported from the ER to the Golgi apparatus via a nonvesicular mechanism (Hanada et al., 2003).

The conversion of ceramide to SM occurs at the inner leaflet of the Golgi apparatus (Futerman and Riezman, 2005), and uniquely utilises ceramide that is transported via the *cer*amide *t*ransport protein, CERT (Hanada et al., 2009). CERT is a 68 kDa cytosolic protein that transports ceramide in an ATP-dependent and nonvesicular manner. Significant structural information is available about CERT (Figure 8). The *C*-terminus contains a sequence homologous to the steroidogenic acute regulatory protein-related lipid transfer (START) domain; the structure of the START domain has been determined, revealing a long amphiphilic cavity in which one ceramide molecule is buried.

At the far end of the cavity, the amide and hydroxyl groups of ceramide form a hydrogen bond network with specific amino acid residues that play key roles in stereospecific ceramide recognition. Ceramide is surrounded by the hydrophobic wall of the cavity, with size and shape dictating a length limit for ceramides. The *N*-terminus of CERT contains a pleckstrin homology (PH) domain

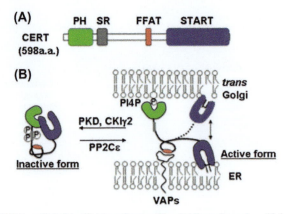

FIGURE 8   CERT-mediated trafficking of ceramide. (A) Domains and motifs in CERT. The PH domain, serine repeat (SR) motif, FFAT motif, and START domain. (B) Regulation of CERT activity by phosphorylation. Phosphorylation by PKD and CKI inactivates CERT and the dephosphorylation by PP2Cε activates CERT. Dephosphorylated CERT binds to PI-4P at the Golgi through the PH domain and also to VAPs at the ER through the FFAT motif to transfer ceramide from the ER to *trans*-Golgi. *Taken from Hanada et al. (2009).*

that specifically binds to phosphatidylinositol 4-monophosphate (PI-4P). PI-4P is the most abundant and preferentially distributed phosphoinositide in *trans* Golgi membranes, and the PH domain of CERT is required for its efficient association with the Golgi via PI-4P. Finally, a coiled–coil motif is found in the middle region of CERT containing the two-phenylalanines in an acid tract (FFAT) motif, which interacts with two ER-resident type II membrane proteins, vesicle-associated membrane protein-associated proteins (VAP-A and VAP-B). VAPs recruit FFAT-motif-containing proteins to the cytosolic surface of ER membranes, and it is this interaction that is crucial for the association of CERT with the ER membrane. On its delivery to the Golgi apparatus, ceramide disso-ciates from CERT, translocates to the luminal surface and becomes available to SM synthase 1 (SMS1), which transfers the phosphorylcholine head group from phosphatidylcholine (PC) to ceramide, to yield SM and diacylglycerol (DAG). The active site of SMS1 faces the lumen of the Golgi apparatus. An additional SM synthase, SMS2, resides in the plasma membrane, and may be responsible for synthesis of the SM in the plasma membrane that is involved in signalling pathways.

3.4 Formation of S1P

S1P is formed by a combination of the biosynthetic and degradative pathways. Sphingosine, which is generated by the hydrolysis of ceramide by ceramidases, can be recycled back to ceramide or can be phosphorylated by one of two sphingosine kinases, SphK1 and SphK2, forming S1P, a bioactive metabolite (see Section 5). S1P can then be either dephosphorylated by S1P-specfic phosphatases (SPP1 and

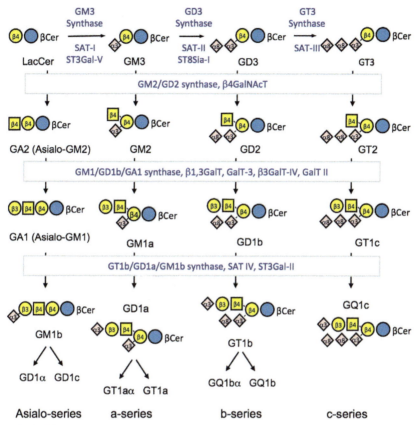

FIGURE 9 **The pathway and enzymes of GSL synthesis.** *Adapted from Merrill (2008).*

SPP2) or irreversibly degraded by S1P lyase (SPL) to phosphoethanolamine and hexadecenal, and subsequent entry into the glycerolipid pathway.

3.5 GSL Synthesis and Transport

Whereas SM synthesis in the lumen of the Golgi apparatus depends entirely on the delivery of ceramide via CERT, glucosylceramide (GlcCer) synthesis via GlcCer synthase (GCS) uses ceramide delivered via vesicular transport (Figure 7). GCS is a type III membrane protein whose active site faces the cytosol. GlcCer is the precursor for complex GSLs, which are synthesised by the sequential addition of sugar moieties (Figure 5). GlcCer is first converted to lactosylceramide (LacCer), which is subsequently sialylated to form gangliosides GM3, GD3 and GT3 (Figure 9), which are further converted by glycosyltransferases to more complex gangliosides (Sandhoff and Harzer, 2013). Of great interest is the recent discovery that distinct pathways of GSL synthesis also require facilitated nonvesicular transport of GlcCer, which is operated by the cytosolic GlcCer-transfer protein, four-phosphate adaptor protein 2 (FAPP2). Interestingly, GlcCer is channelled by vesicular and nonvesicular transport to two topologically distinct glycosylation pathways, the former for the synthesis of the ganglio-series GSLs and the latter for globo-GSL synthesis (D'Angelo et al., 2013).

In summary, SL synthesis is unexpectedly complex, with synthesis compartmentalised between the ER and the Golgi apparatus, utilising both vesicular and facilitated mechanisms to transport ceramide and GSLs. The molecular rationale for this complexity is not known, but presumably exists to provide multiple sites of regulation of the pathway.

4. SPHINGOLIPID DEGRADATION

Compared to SL biosynthesis, SL and GSL degradation is relatively straightforward, at least with respect to the cellular compartment where it occurs, with degradation taking place largely in the acid milieu of the lysosomes. Study of the lysosomal hydrolysis of SLs and GSLs has been motivated by their involvement in a number of inherited human diseases, the LSDs. Indeed, a disease is associated with essentially every enzyme in the pathway of GSL and SL degradation (Zigdon et al., 2014) (Figure 10). Nonlysosomal hydrolysis of SLs also occurs by hydrolases that work at neutral or alkaline pH. For instance,

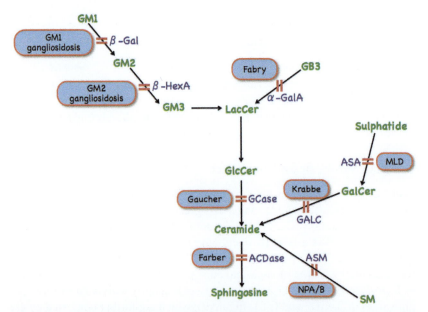

FIGURE 10 **The pathway of GSL and SL degradation.** Enzymes are in blue, lipids in green and the names of the diseases in enclosed boxes. Gal, galactosidase; Hex, hexosaminidase; ASA, arylsulphatase A; GALC, galactosylceramidase; ASM, acid sphingomyelinase; GCase, acid-β-glucosidase; ACDase, acid ceramidase. *Adapted from Zigdon et al. (2014).*

degradation of SM can be mediated via a neutral SMase located at the PM and an alkaline SMase located in the Golgi apparatus and in endosomes; a neutral β-glucosidase has recently been reported whose deficiency leads to impaired male fertility (Sandhoff and Harzer, 2013).

The lysosomal degradation of GSLs occurs sequentially with the stepwise release of monosaccharides from the reducing end of the oligosaccharide chain. Thus, β-galactosidase removes the terminal β-galactose residue from ganglioside GM1 to generate GM2, which is subsequently acted on by β-hexosaminidase. The function of β-hexosaminidase has been particularly well studied due to its role in the human genetic diseases, the GM2 gangliosidoses, a group of inherited metabolic disorders caused by mutations in any of three genes, the *HEXA* gene, resulting in Tay–Sachs disease, the *HEXB* gene, resulting in Sandhoff disease, and the *GM2A* gene, resulting in GM2 activator deficiency (Gravel et al., 2001). The *HEXA* and *HEXB* genes code for β-hexosaminidase α- and β-subunits, respectively, which dimerise to produce two forms of the enzyme, A (αβ) and B (ββ), and a minor form, S (αα). In both Tay–Sachs and Sandhoff diseases (α- and β-subunit deficiency, respectively), there is a deficit of hexosaminidase A (αβ), and as a consequence, massive accumulation of ganglioside GM2 occurs. Another well-studied enzyme is acid β-glucosidase, which cleaves GlcCer to ceramide (Figure 10); this enzyme is the cause of the most common LSD, Gaucher disease. The pathology of SL storage diseases will be discussed in more detail in Section 7.

Several of the lysosomal hydrolases also require the assistance of small glycoprotein cofactors, the SL activator proteins (SAPs), whose function appears to be somehow related to rendering the SLs more accessible to the hydrolases. Two genes encode the SAPs: one encodes the GM2-activator protein, and the other encodes the SAP-precursor protein, prosaposin; this protein is posttranslationally processed to four homologous mature proteins, saposins A–D. The hydrolases also need to bind to negatively charged membranes (Kolter and Sandhoff, 2005). Defects in the saposins can also lead to GSL/SL LSDs. Thus, although the pathways of GSL/SL degradation may appear somewhat less complex than those of SL biosynthesis, the role of this pathway in human disease has led to considerable insight into the roles that SL degradation plays in cell physiology and pathophysiology.

5. SPHINGOLIPID SIGNALLING AND ROLES IN CELL REGULATION

The basic paradigm of SL signalling is similar to that of all other lipid signalling pathways inasmuch as a cell receives an input, normally in the form of a soluble extracellular ligand which binds to a membrane receptor, which subsequently activates a membrane-associated enzyme that hydrolyses a membrane lipid to produce a lipid product that activates a downstream signalling pathway (Hannun et al., 2001). The classic example of lipid signalling is the turnover of phosphatidylinositol (PI), in which phosphorylated forms of PI are hydro-lysed by membrane-bound phospholipases to generate a membrane-bound lipid signalling molecule (DAG) and a soluble product, inositol 1,4,5-triphosphate. This paradigm dominated thinking in the field of lipid signalling, although study of SL signalling has led to the formulation of additional paradigms due to the relatively small number of targets to which SL signalling molecules (particu-larly ceramide) directly bind. This issue will be discussed in more detail below. Moreover, the large variety of specific SLs that function in intracellular signal-ling pathways, and in particular the fact that many SL signalling molecules are part of the same metabolic pathway, has led to novel ways of thinking about the connection between SL metabolism and SL signalling.

Two pieces of evidence first suggested that SLs act as signalling molecules. The first was obtained by Yusuf Hannun, who showed that sphingosine could directly inhibit the action of protein kinase C (Hannun et al., 1986), implying that sphingosine might antagonise the effects of DAG, a known activator of PKC, which is generated during PI turnover. The second came from Richard Kolesnick, who showed that SM could be hydrolysed to ceramide via the action of SMase (Kolesnick and Paley, 1987), suggesting that SMase acted in an analo-gous manner to the action of phospholipase C in PI signalling pathways. A short time later, Sarah Spiegel suggested that S1P was also a signalling molecule that could act as a second messenger (Zhang et al., 1991). Thus, even in the early days of studying the roles of SLs in signalling pathways, at least three SLs were implicated as second messengers, namely sphingosine, ceramide and S1P, leading to considerable confusion about the importance of these molecules and their relative contribution to SL signalling. Of these three, the latter two have received particular attention, and there is now no doubt that both ceramide and S1P act as genuine signalling molecules, although their modes of action are quite different.

Much work on SLs in signal transduction pathways has focused on the role of ceramide in apoptosis (Tirodkar and Voelkel-Johnson, 2012). A large num-ber of extracellular signals or stimuli elevate intracellular ceramide levels and induce apoptosis, including, but not exclusively, heat-shock, ionising radiation, oxidative stress, progesterone, vitamin D3, daunorubicin, TNFα, IL-1α, IL-1β, interferon-γ, Fas ligand, fenretinide, oxidised-LDL and nitric oxide. Intracel-lular ceramide levels can be elevated as a result of either de novo synthesis or SM hydrolysis by acid (acid sphingomyelinase (ASM)) or neutral SMase. Sig-nificant evidence for a role of ASM has come from the ASM knockout mouse, a mouse that was originally generated to study the LSD, Niemann-Pick A/B disease (Horinouchi et al., 1995), but which has proved hugely useful in delin-eating the roles of ASM, and the role of ceramide generated by ASM, in signal-ling pathways. Thus, apoptosis induced by ionising radiation is blocked in the ASM knockout mouse (Paris et al., 2001) and various other pathways have been shown to depend on ceramide generated via the action of ASM. However, it should be stressed that a variety of enzymes can generate ceramide, and almost every one of these enzymes has been implicated in one or other signalling pathway. Moreover, some of these enzymes are located in intracellular membranes, leading to the possibility that ceramide signalling plays an intracellular role in addition to its more established role at the plasma membrane.

Two major possibilities have been proposed concerning the mechanism by which ceramide induces apoptosis and plays roles in other signalling pathways. In the first, ceramide would act as a typical second messenger, inasmuch as on its production, it binds to proteins whose activity it regulates. Among the proteins regulated by ceramide are a ceramide-activated protein phosphatase, protein kinase Cζ, kinase suppressor of Ras, phospholipase A_2, cathepsin D, Jun-*N*-terminal kinases, c-Raf-1, the small G-proteins Ras and Rac and Src-like tyrosine kinases. In the second possibility, the unique biophysical properties of ceramide are responsible for its ability to act as a signalling lipid. Ceramide can self-associate in the plane of the membrane lipid bilayer (see Section 6), and by so doing provides the driving force that results in the fusion of GSL- and cholesterol-containing rafts into large signalling macrodomains (signalling platforms). For instance, many of the above stimuli activate ASM at the PM (Grassme et al., 1997), resulting in formation of ceramide-enriched membrane platforms that trap and cluster signalling proteins, thus allowing for signal initiation and amplification via concentration and oligomerisation of proteins associated with apoptotic signalling mechanisms (Figure 11). Such clustering has been shown for Fas receptors, which become concentrated in these macrodomains.

S1P is perhaps the most intriguing SL messenger inasmuch as it acts as both a first and a second messenger, and indeed the evidence for its role as a first messenger (Rosen et al., 2009) is somewhat stronger than for its role as a second messenger (Spiegel and Milstien, 2003) (Figure 12). As a first messenger (or paracrine agent), S1P regulates processes such as cytoskeletal rearrangement,

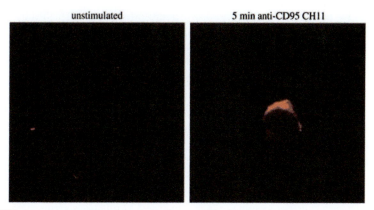

FIGURE 11 **An example of formation of ceramide-enriched membrane domains induced by CD95.** Stimulation of Jurkat T-lymphocytes via CD95 results in the formation of ceramide-enriched membrane platforms on the cell surface. Domains were detected using an anti-ceramide antibody. *Taken from Zhang et al. (2009).*

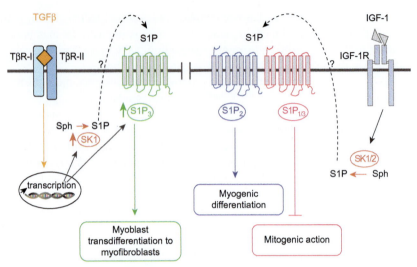

FIGURE 12 **S1P acts as a first and second messenger.** The figure shows an example of S1P signalling, modulated in this case by TGFβ and IGF-1 on myoblasts. S1P is generated intracellularly by SphK, and then secreted by the cell prior to its binding to an S1PR. *Taken from Donati et al. (2013).*

cell migration, angiogenesis, vascular maturation, embryonic development of the heart, and immunity and lymphocyte trafficking. As an intracellular second messenger, S1P, on its generation by SphK1 or 2, mediates calcium homeostasis, cell growth, tumorogenesis and suppression of apoptosis.

S1P can trigger signal transduction pathways by acting on the same cell from which it is secreted, or by acting in an autocrine manner. Inducers of cell proliferation and differentiation, including growth factors, GPCR agonists, cytokines, phorbol esters, vitamin D3 and antigens, increase intracellular S1P levels by activation of SK. The downstream effectors of S1P in intracellular pathways include IP$_3$-independent calcium mobilisation and DNA synthesis.

S1P is also secreted and subsequently binds to a family of G-protein-coupled receptors, the S1P receptors (S1PRs). Five cell-surface G-protein-coupled S1PRs have been identified and are expressed in a wide variety of tissues (Rosen et al., 2009). Significant amounts of information about the S1PRs have been obtained. For instance, of all these receptors, deletion of S1PR1 alone is embryonically lethal whereas deletion of S1PR2 and 3 results in viable offspring with variable phenotypes, suggesting compensation in receptor expression and signalling. A key finding in the field of S1PR signalling was the discovery that lymphocyte egress is reversibly modulated by S1PRs (Matloubian et al., 2004), and with the development of Fingolimod (FTY720), a pro-drug which is a nonselective S1PR agonist, and is in therapeutic use in the treatment of relapsing, remitting multiple sclerosis (see Section 8). The latter is a remarkable example of how studies on the basic science of SLs initiated in the early 1990s have led to development of drugs that interfere with SL signalling and lead to therapeutic benefit.

Upon the discovery of a role for more simple SLs in cell signalling, less attention was paid to complex GSLs. However, GSLs clearly also play vital roles in cell physiology. Early studies on complex GSLs were mainly devoted to attempting to understand their roles in the nervous system where GSLs, particularly gangliosides, are expressed at different levels in different regions of the brain during development, suggesting functional roles for gangliosides in brain development. The essential nature of GSLs for sustaining life was first shown in knockout mice lacking GlcCer synthase, the first enzyme in the pathway of GSL biosynthesis; these mice showed embryonic lethality (Ichikawa and Hirabayashi, 1998). A number of other knockout mice have been produced in the ganglioside biosynthetic pathway, and the overall consensus is that there might be considerable functional redundancy between different gangliosides; however, this view is gradually changing with the discovery that specific gangliosides can play specific roles in defined cellular events.

Gangliosides act as toxin receptors. In order for toxins to reach their intracellular targets, they bind to cell surface receptors, which are then internalised together with the toxins. Two toxins, namely, cholera toxins from *Vibrio cholera* and *Escherichia coli* heat-labile enterotoxin bind to ganglioside GM1, and Shiga toxin from *Shigella dysenteriae* binds to the neutral GSL, globotriaosylceramide (Gb3). GSLs at the cell surface are also involved in recognition events that are beneficial, rather than the toxic effects caused by toxin binding. Thus, a large body of data has shown roles for GSLs as antigens, as mediators of cell adhesion and as modulators of signal transduction. The list of receptor functions of GSLs is extensive. For instance, GSLs mediate E-selectin-dependent rolling and tethering, α-GalCer acts as a ligand recognised by a special group of immune T cells, known as invariant NKT cells, and 9-*O*-acetyl GD3, expressed in regions of cell migration and neurite outgrowth in the developing and adult nervous system, plays a role in neuronal motility. Together, these varied examples illustrate that the complex glycan structures of GSLs are involved in vital recognition events at the cell surface. Along with the roles of more simple SLs in cell signalling, it is apparent that both SLs and GSLs are critically involved in regulating cell physiology.

6. SPHINGOLIPID BIOPHYSICS

As is the case for all membrane lipid classes, the physiological functions of membrane SLs are governed by their biophysical properties. Due to their hydrophobic nature, lipids cannot undergo spontaneous interbilayer movement (and hence the need, for instance, of ceramide to undergo either vesicle-mediated transport or facilitated transport mediated by CERT between the ER and the Golgi apparatus; see Section 3), and SLs with bulky or charged head groups (such as GSLs or SM) do not undergo transbilayer movement (often referred to as 'flip–flop'). However, SLs do undergo lateral diffusion, and it is the discovery of their restricted lateral diffusion, due to their interactions with other membrane components, that led to the concept of microdomains ('lipid rafts'), which impinges on many aspects of SL physiology. Thus, protein–protein, lipid–lipid and lipid–protein interactions dictate the spatial organisation of membrane components and restrictions on lateral mobility, and SLs play a key role in this process.

Lipid domains have the capacity to segregate or co-localise membrane proteins, resulting in membrane regions with composition and physical properties that differ from the average properties of the bulk membrane (see Figure 11 for an example), allowing for the possibility that these regions might play specialised roles in biological processes. Lipid domains are often referred to as 'rafts', which are formed as a result of the preferential interactions between cholesterol and SLs, in particular SM. Rafts exist in a liquid-ordered phase (l_o), and are surrounded by a bulk fluid membrane in the liquid-disordered phase (l_d), which is composed of unsaturated PC. Rafts, believed to occur mainly on the outer leaflet of the plasma membrane where SLs are enriched, were first proposed by Simons and Ikonen (1997) based on studies on the preferential delivery of newly synthesised GSLs to the apical membrane of epithelial cells. To explain this phenomenon, it was proposed that clusters of GSLs serve as sorting platforms in the cell.

An important distinction needs to be made between rafts and membrane domains isolated due to their resistance to cold solubilisation with non-ionic detergents such as Triton X-100 (Brown and Rose, 1992). Some proteins, such as GPI-anchored proteins, cholesterol-linked and palmitoylated proteins, Src-family and α-subunits of heterotrimeric G proteins, can be recovered in these insoluble membrane fractions together with lipids. However, the relationship between detergent-insoluble domains and rafts is unclear, since lipid reorganisation occurs during detergent solubilisation, and the structure of detergent-resistant membranes differs from that of the membrane prior to detergent extraction. Moreover, the extraction procedure itself can induce domain formation.

Rafts are dynamic entities in equilibrium with the remaining membrane lipids, meaning that lipids move in and out of the domains on varying time scales. SM and ceramide are important players in raft formation. Ceramide is highly hydrophobic and undergoes complex polymorphic phase behaviour (Goñi and Alonso, 2009). When mixed with other lipids, ceramide strongly affects bilayer properties since ceramide is able to increase the order of both unsaturated and saturated phospholipid acyl chains. Ceramide also drives lateral phase separation in fluid phospholipid bilayers to form regions of ceramide-enriched gel domains and ceramide-poor fluid domains. Ceramide-gel domains are characterised by an atypically high degree of order perhaps allowing proteins, for example, receptors and ligands involved in signalling cascades, to partition between ceramide- and SM-gel domains, with changes in their partitioning driving signalling. When ceramide is generated in model membranes from SMase, ceramide also segregates into distinct domains. Besides its ability to drive phase separation, ceramide has other effects on membrane properties. For instance, ceramide destabilises lamellar phases, facilitating the lamellar-to-inverted hexagonal phase transition of certain lipids such as phosphatidylethanolamine (See Chapter 1). Finally, ceramide is able to undergo transbilayer movement (flip–flop) across the membrane, with some studies suggesting that ceramide flip–flop takes place in seconds, while others argue that flip–flop occurs in a few to tens of minutes or even hours.

Thus, much evidence points to an important biological role for ceramide via structural and physical changes in membranes. A number of biological processes are driven by biophysical changes induced by ceramide, and some of these are induced by activation of ASM at the cell surface. Although ASM is a lysosomal enzyme, it nevertheless appears to have sufficient activity at the pH found at the cell surface to generate enough ceramide to form signalling platforms. The ceramide-enriched membrane platforms trap and cluster receptors such as the CD95/CD95 ligand and a significant number of other ligands (Zhang et al., 2009), and this process appears essential for a number of signal transduction pathways that are mediated via ceramide. One example of this concerns the insulin receptor. On changing the SL acyl chain length, and thus altering the biophysical properties of the membrane, insulin receptor signalling is abrogated due to the inability of the receptor to translocate into SL-enriched microdomains (Figure 13).

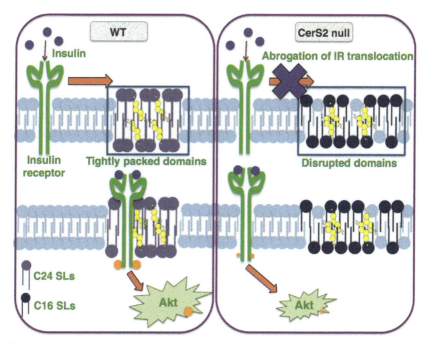

FIGURE 13 **Insulin receptor translocation upon altering the SL acyl chain length.** Membrane domains in wild-type (WT) liver are tightly packed, permitting translocation of the insulin receptor (IR). However, membrane domains in CerS2 null mouse liver are disrupted and insulin receptor translocation and phosphorylation are abrogated. *Taken from Park et al. (2014).*

7. SPHINGOLIPIDS IN DISEASE PATHOLOGY

One of the more significant advances in SL biology over the past decade has been the discovery that SLs are involved in a large variety of human diseases, ranging from the well-studied SL storage diseases to complex diseases in which changes in SL metabolism are not the causative agent of the disease, but nevertheless modulation of the SL pathway can reverse disease pathology in many cases. Such diseases include cancer, inflammation and infection, and neurological and psychiatric diseases. This section documents the various enzymes and proteins of SL metabolism that have been shown to be directly causative for disease. It should be stressed that much of the information discussed in this section has been generated relatively recently, and therefore this area is a growing focus of research.

7.1 Serine Palmitoyl Transferase

Serine is not the only amino acid to be used by SPT as recent studies have suggested that mutant forms of SPT are able to utilise alanine and glycine. This was discovered after the unravelling of the molecular basis of HSAN1 (Bejaoui et al., 2001), an autosomal dominant genetic disorder caused by mutations in LCB1 that causes structural perturbations and alters the specificity of SPT such that it uses alanine, rather than serine, to generate 1-deoxysphinganine. Remarkably, oral administration of L-serine prevented accumulation of deoxysphinganines and alleviated HSAN1 symptoms.

7.2 Ceramide Synthases

Ceramides with defined acyl chain lengths are involved in a number of human diseases, and although there is less information about the specific roles of CerS in human disease, it is very likely that CerS will also be implicated in a number of diseases, directly via mutations in their coding sequences or indirectly as a downstream response to other metabolic alterations. In terms of diseases caused by mutations in human CerS genes, a number of examples have come to light (Park et al., 2014). Thus, in Rhegmatogenous, a missense coding single-nucleotide polymorphism is found in the CerS2 gene. Mutations in the CerS3 gene lead to skin ichthyosis, and as a result of this mutation, levels of C26:0-ceramides, and of ceramides containing ultra-long acyl chains, are reduced in keratinocytes, attesting to the important roles that C26:0-ceramide generated by CerS3 plays in epidermal differentiation, barrier integrity and the formation of the cornified lipid envelope in humans. Moreover, the human phenotype is similar to that observed in the mouse CerS3 knockout. In terms of diseases that appear to be a downstream response to other metabolic alterations, changes in CerS levels have been reported in a number of diseases, with cancer providing the best example. Ceramide levels are elevated in head and neck squamous cell carcinomas (HNSCC), which additionally suggested roles for ceramide with particular acyl chain lengths in cancer pathology (Saddoughi and Ogretmen, 2013). In terms of therapies, gemcitabine (GEM) and doxorubicin (DOX) treatment decreased HNSCC xenograft tumour growth and progression in SCID mice, which was associated with increased C18-ceramide and CerS1 expression. Based on these data, a phase II clinical trial was designed to test the hypothesis that treatment with a GEM/DOX could be efficacious as a means to reconstitution of C18-ceramide signalling in HNSCC patients for whom first-line platinum-based therapy failed. CerS have also been implicated in the development of colon cancer, inasmuch as CerS6 expression is involved in tumour necrosis factor-related apoptosis-inducing ligand sensitivity in colon cells. CerS levels are also altered in breast cancer. Thus, there appears to be a correlation between the extent of proliferation and cell death in some cancers and the levels of expression or activity of specific CerS, which potentially might pave the way for therapies that regulate cancer cell growth via modulation of CerS activity. CerS have been shown to be involved in the etiology of a number of other diseases, including cardiomyopathy, multiple sclerosis and diabetes.

7.3 GSL Synthesis

Two disorders of GSL biosynthesis are known, resulting from deficiencies in GM3 and GM2 synthases (Platt, 2014). GM3 synthase deficiency, which was discovered in the Old Order Amish community in the United States, presents as a severe epilepsy syndrome whereas GM2 synthase deficiency results in spastic paraplegia. Since both of these enzymes also act as precursors for more complex GSLs, the disease-causing lipid has not been formally identified and the disease could in principle be caused by loss of the specific ganglioside species or by elevation in precursor GSLs; this is of course true for all diseases in which metabolism of a precursor in a biosynthetic pathway is defective. For the GSLs, it is highly probable that other diseases will be discovered, although they are likely to be relatively rare.

7.4 GSL Degradation

Rare diseases are not limited to the SL/GSL biosynthetic pathway, since a family of diseases is caused by defects in the degradation of GSLs. These are the sphingolipidoses, LSDs in which SLs accumulate due to the defective activity of one or other of the enzymes involved in their degradation. Indeed, compared to diseases in the biosynthetic pathway, this family of diseases has been relatively well studied, and in some cases, therapies are available. Among the diseases are the following (Platt, 2014): (1) Gaucher disease, the most common LSD, which is caused by a deficiency of acid β-glucosidase (GCase) (Figure 10). More than 350 different mutations have been identified in the gene encoding GCase, *GBA1* (Figure 14).

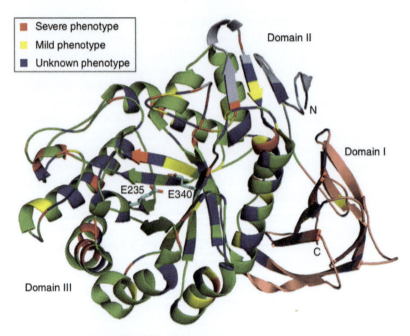

FIGURE 14 **Mutations in the GCase protein.** The 3D structure of the protein is shown, along with the different domains and the location of mutations that cause different forms of Gaucher disease.

In relatively rare cases, deficiency in Saposin C results in a variant form of Gaucher disease. Gaucher disease can be divided into three clinical subtypes, with types 2 and 3 displaying central nervous system (CNS) involvement. Genotype–phenotype correlations are often poor, with patients with the same mutation sometimes displaying widely different phenotypes, although some mutations do predispose to a particular disease type. Interestingly, mutations in GCase (the *GBA1* gene) are also associated with Parkinson's disease. (2) The second most common sphingolipidosis is Fabry disease, which is caused by mutations in the GLA gene, which codes for α-galactosidase A. Affected individuals accumulate Gb3, leading to severe multisystemic disease. Stroke, seizures, and heart and kidney disorders develop in the third or fourth decade and can lead to premature death. The age of onset is between 3 and 10 years in males and 6 and 15 years in females. (3) Metachromatic leukodystrophy (MLD, also known as arylsulphatase A deficiency) is caused by mutations in the *ARSA* gene, which codes for arylsulphatase A (ASA), the protein that catalyses the first step in 3-*O*-sulphogalactosylceramide (sulphatide) degradation. In rare cases, MLD results from mutations in saposin-B. Sulphatide accumulates in oligodendrocyte in the CNS and also in visceral tissues such as liver and kidney; however, functionally, storage only affects oligodendrocytes. The pathological hallmark is progressive widespread demyelination and reduced nerve conduction velocity

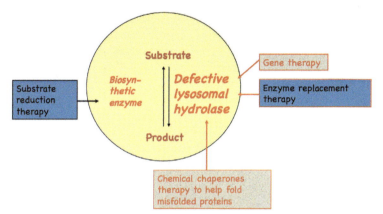

FIGURE 15 **Putative therapies for the sphingolipidoses.** Therapeutic options include enzyme replacement therapy, chaperone therapy to help misfold mutant enyzmes and substrate reduction therapy in which inhibition of GSL biosynthesis prevents accumulation of the offending GSL.

in the CNS and peripheral nervous system, resulting in various neurological symptoms, and finally death. (4) GM1 gangliosidosis is caused by mutations in the GLB1 gene, causing a deficiency in β-galactosidase activity, resulting in massive GM1 storage, mainly in the brain. The adult form (type III), with onset between 3 and 30 years, is characterised by cerebellar dysfunction, dystonia, slurred speech, short stature and mild vertebral deformities. (v) The GM2 gangliosidoses are disorders caused by β-hexosaminidase (β-Hex) deficiency, resulting in massive accumulation of GM2 ganglioside and related glycolipids, particularly in neurons. β-Hex consists of two major isoenzymes, hexosaminidase A (HEXA) and hexosaminidase B (HEXB). HEXA itself consists of two subunits, the α-subunit, encoded by the *HEXA* gene, and the β-subunit, encoded by the *HEXB* gene. Tay–Sachs disease is caused by mutations in *HEXA* and Sandhoff disease by mutations in *HEXB*. In rare cases, GM2 gangliosidosis results from mutations in the GM2A activator protein. All three GM2 gangliosidoses display a similar phenotype and neuropathology.

Since the sphingolipidoses are monogenic diseases, they are ripe for intervention by gene therapy. However, at present gene therapy is not a viable therapeutic option, and the main way that sphingolipidoses are treated is enzyme replacement therapy, in which a recombinant form of the enzyme is given to patients. This treatment does not work for CNS forms of the diseases due to the inability of the enzymes to cross the blood–brain barrier. Other treatment options are either in clinical use or in clinical trials (Figure 15).

7.5 Acid Sphingomyelinase and Ceramidase

Diseases caused by ASM or ceramidase deficiency were previously thought to be limited to the two sphingolipidoses, Niemann-Pick diseases A and B (NPA/B) and

Farber's diseases. The former is caused by mutations in the *SMPD1* gene, which codes for ASM, and has many of the same types of characteristics discussed above for the other sphingolipidoses. Likewise, Farber disease is caused by mutations in the *ASAH1* gene, which codes for acid ceramidase. However, ASM in particular has recently been implicated in a large number of other human diseases, which appear to be caused by changes in ceramide levels due to the defective degradation of SM to ceramide. Farber disease leads to changes in ceramide levels due to the defective degradation of ceramide. Thus, ASM and ceramidase activities are altered in a number of human diseases, but in contrast to NPA/B and Farber disease, these changes are not caused by genetic mutations. Three examples are as follows. (1) ASM and bacterial infections (Beckmann et al., 2014): ASM plays a crucial role in infection, with a number of pathogens directly connected to the requirement of ceramide-enriched membrane domains (see above) for the internalisation of bacteria, such as *Pseudomonas aeruginosa*. Thus, ASM-deficient mice are highly susceptible to pulmonary bacterial infections and infected mice die from sepsis because they are unable to clear the infection. Moreover, bacterial infections also play an important role in the progression of cystic fibrosis, which can be partially reversed by chemical inhibition of ASM. (2) ASM in inflammatory diseases: ASM is of importance in mast cell function, and inhibitors of ASM may be potential new anti-allergic agents. This is also the case for inflammatory bowel disease, in which mice with chronic colitis have increased levels of ceramide. Inhibition of ASM activity to treat the disease might involve small molecule inhibitors such as the anti-depressants imipramine or amitriptyline, which inhibit ASM activity. (3) ASM and psychiatric diseases (Kornhuber et al., 2014): Increased ASM activity has been found in peripheral mononuclear cells of patients suffering from depressive disorders and enhanced plasma ceramide levels are associated with depressive symptoms. It has been suggested that ceramide generated by ASM might be a missing link between immune inflammatory/oxidative stress dysregulation and disorders resulting from depressive symptoms. Indeed, the link between depressive disease and inflammation and/or metabolic syndrome may possibly be explained by changes in levels of ceramides.

The three diseases above exemplify the vital role that the ASM/ceramide pathway plays not only in normal cell physiology but also in the pathophysiology of human disease. It is legitimate to ask how one pathway can play vital roles in so many diverse diseases, and the answer to this question probably lies in the central role that the ASM pathway plays in cellular responses; thus, disruption of this pathway, directly or indirectly, has many consequences that lead to the vast repertoire of human diseases in which the ASM/ceramide axis is involved.

7.6 S1P, S1PR and SphK

The diverse roles of S1P in innate and adaptive immunity, including immunosurveillance, immune cell trafficking and differentiation, immune responses and endothelial barrier integrity, are mediated by its binding to one of five

G-protein-coupled receptors, named S1PR1–S1PR5 (Figure 12). Downstream signalling of these receptors is complex as they are differentially expressed and coupled to a varied set of heterotrimeric G proteins. Thus, activation of S1PR1 promotes cell migration and the egress of T and B lymphocytes from lymphoid tissues, whereas S1PR2 inhibits motility to promote retention of B cells. S1P produced inside cells by activation of SphK1 can be secreted and signals through its receptors in a paracrine and/or autocrine manner. Much less is known about the intracellular targets of S1P, although it is now known that S1P binds directly to the TNF receptor-associated factor 2 and cellular inhibitor of apoptosis 2.

S1P has been implicated in a large number of human diseases (Maceyka and Spiegel, 2014), such as autoimmune diseases, rheumatoid arthritis, inflammatory bowel disease and asthma, to name but a few. However, S1P is probably best known for the role that its analogue, fingolomid, plays in modulating multiple sclerosis (Brinkmann, 2009). Fingolimod is phosphorylated in vivo by SphK 2 to form the active moiety FTY720-phosphate, which binds to four of the five G-protein-coupled S1PR subtypes. Importantly, egress of lymphocytes from lymph nodes requires signalling of lymphocytic S1PR1 by endogenous S1P and the S1P mimetic, FTY720-phosphate, causes internalisation and degradation of cell membrane-expressed S1PR1, thereby antagonising S1P action at the receptor (Figure 16). Moreover, in models of human multiple sclerosis and demyelinating polyneuropathies, functional antagonism of lymphocytic S1PR1

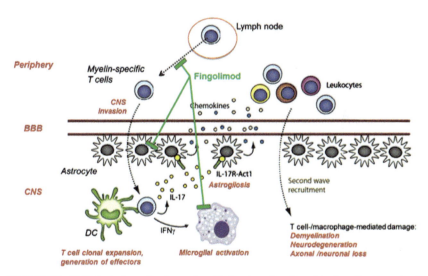

FIGURE 16 **Pathological mechanisms of multiple sclerosis relevant to fingolimod mode of action.** In multiple sclerosis, pathogenic T cells committed to T helper 1 (TH1) and TH17 lineages cross the blood–brain barrier (BBB) to invade the CNS. Treatment with fingolimod leads to a down-modulation of S1P1 receptors on lymphocytes, thereby preventing their egress from lymph nodes and the invasion into the CNS. *Taken from Bigaud et al. (2014).*

slows S1P-driven egress of lymphocytes from lymph nodes. Thus, FTY720 appears to act through immune-based and possible central mechanisms to reduce inflammation.

The examples given above suggest that intervention in the SL pathway is highly likely to provide novel sites for therapeutic intervention in the relatively large numbers of human disease in which SLs have been implicated. The novel roles that SLs play in human diseases were quite unexpected even a decade ago, and have provided huge impetus for study of this fascinating class of lipids.

8. PERSPECTIVES

Predicting future scientific directions is often an exercise in futility. However, in the case of SLs, exciting times lie ahead. The discovery that SLs are involved in vital intracellular signalling pathways has revolutionised the field and led to the discovery of the involvement of these pathways in many physiological and pathophysiological events. As a consequence, SLs have been implicated in an unexpectedly large number of human diseases. Thus it is anticipated that the lipids discovered nearly one and a half centuries ago by Thudicum will continue to be the focus of intense research studies, not only by S: and lipid aficionados but also by researchers interested in understanding basic regulatory mechanisms involved in cell physiology.

ACKNOWLEDGMENTS

I would like to particularly thank Dr. Al Merrill, the author of the chapter on sphingolipids in the previous edition of this book series, for his friendship over many years and for allowing me to use some figures from the previous edition.

REFERENCES

Beckmann, N., et al., 2014. Inhibition of acid sphingomyelinase by tricyclic antidepressants and analogons. Front. Physiol. 5, 331.

Bejaoui, K., et al., 2001. SPTLC1 is mutated in hereditary sensory neuropathy, type 1. Nat. Genet. 27 (3), 261–262.

Bigaud, M., et al., 2014. Second generation S1P pathway modulators: research strategies and clinical developments. Biochim. Biophys. Acta 1841 (5), 745–758.

Brinkmann, V., 2009. FTY720 (fingolimod) in multiple sclerosis: therapeutic effects in the immune and the central nervous system. Br. J. Pharmacol. 158 (5), 1173–1182.

Brown, D.A., Rose, J.K., 1992. Sorting of GPI-anchored proteins to glycolipid-enriched membrane subdomains during transport to the apical cell surface. Cell 68 (3), 533–544.

D'Angelo, G., et al., 2013. Vesicular and non-vesicular transport feed distinct glycosylation pathways in the Golgi. Nature 501 (7465), 116–120.

Donati, C., Cencetti, F., Bruni, P., 2013. New insights into the role of sphingosine 1-phosphate and lysophosphatidic acid in the regulation of skeletal muscle cell biology. Biochim. Biophys. Acta 1831 (1), 176–184.

Futerman, A.H., Riezman, H., 2005. The ins and outs of sphingolipid synthesis. Trends Cell Biol. 15 (6), 312–318.

Goñi, F.M., Alonso, A., 2009. Effects of ceramide and other simple sphingolipids on membrane lateral structure. Biochim. Biophys. Acta 1788 (1), 169–177.

Grassme, H., et al., 1997. Acidic sphingomyelinase mediates entry of *N*-gonorrhoeae into nonphagocytic cells. Cell 91 (5), 605–615.

Gravel, R.A., et al., 2001. The GM2 gangliosidosis. In: Scriver, C.R., et al. (Eds.), The Metabolic and Molecular Bases of Inherited Disease. McGraw-Hill Inc, New York, pp. 3827–3876.

Hanada, K., et al., 2009. CERT-mediated trafficking of ceramide. Biochim. Biophys. Acta 1791 (7), 684–691.

Hanada, K., et al., 2003. Molecular machinery for non-vesicular trafficking of ceramide. Nature 426 (6968), 803–809.

Hannun, Y.A., Obeid, L.M., 2008. Principles of bioactive lipid signalling: lessons from sphingolipids. Nat. Rev. Mol. Cell Biol. 9 (2), 139–150.

Hannun, Y.A., et al., 1986. Sphingosine inhibition of protein kinase C activity and of phorbol dibutyrate binding in vitro and in human platelets. J. Biol. Chem. 261 (27), 12604–12609.

Hannun, Y.A., Luberto, C., Argraves, K.M., 2001. Enzymes of sphingolipid metabolism: from modular to integrative signaling. Biochemistry 40 (16), 4893–4903.

Horinouchi, K., et al., 1995. Acid sphingomyelinase deficient mice: a model of types A and B Niemann-Pick disease. Nat. Genet. 10 (3), 288–293.

Ichikawa, S., Hirabayashi, Y., 1998. Glucosylceramide synthase and glycosphingolipid synthesis. Trends Cell. Biol. 8, 198–202.

Ikushiro, H., Hayashi, H., 2011. Mechanistic enzymology of serine palmitoyltransferase. Biochim. Biophys. Acta 1814 (11), 1474–1480.

Kolesnick, R.N., Paley, A.E., 1987. 1,2-diacylglycerols and phorbol esters stimulate phosphatidylcholine metabolism in GH3 pituitary cells. Evidence for separate mechanisms of action. J. Biol. Chem. 262 (19), 9204–9210.

Kolter, T., Sandhoff, K., 2005. Principles of lysosomal membrane digestion: stimulation of sphingolipid degradation by sphingolipid activator proteins and anionic lysosomal lipids. Annu. Rev. Cell Dev. Biol. 21, 81–103.

Kornhuber, J., et al., 2014. The ceramide system as a novel antidepressant target. Trends Pharmacol. Sci. 1–12.

Levy, M., Futerman, A.H., 2010. Mammalian ceramide synthases. IUBMB Life 62 (5), 347–356.

Lipsky, N.G., Pagano, R.E., 1985. A vital stain for the Golgi. Science 228, 745–747.

Lowther, J., et al., 2012. Structural, mechanistic and regulatory studies of serine palmitoyltransferase. Biochem. Soc. Trans. 40 (3), 547–554.

Maceyka, M., Spiegel, S., 2014. Sphingolipid metabolites in inflammatory disease. Nature 510 (7503), 58–67.

Matloubian, M., et al., 2004. Lymphocyte egress from thymus and peripheral lymphoid organs is dependent on S1P receptor 1. Nature 427 (6972), 355–360.

Merrill, A.H.J., 2008. Sphingolipids. In: Vance, D.E., Vance, J.E., (Eds.), Biochemistry of Lipids, Lipoproteins and Membranes. Elsevier, Amsterdam, pp. 364–397.

Merrill Jr., A.H., 2011. Sphingolipid and glycosphingolipid metabolic pathways in the era of sphingolipidomics. Chem. Rev. 111 (10), 6387–6422.

Paris, F., et al., 2001. Endothelial apoptosis as the primary lesion initiating intestinal radiation damage in mice. Science 293 (5528), 293–297.

Park, J.-W., Park, W.-J., Futerman, A.H., 2014. Ceramide synthases as potential targets for therapeutic intervention in human diseases. Biochim. Biophys. Acta 1841, 671–681.

Platt, F.M., 2014. Sphingolipid lysosomal storage disorders. Nature 510 (7503), 68–75.

Rosen, H., et al., 2009. Sphingosine 1-phosphate receptor signaling. Annu. Rev. Biochem. 78, 743–768.

Saddoughi, S.A., Ogretmen, B., 2013. Diverse functions of ceramide in cancer cell death and proliferation. Adv. Cancer Res. 117, 37–58.

Sandhoff, K., Harzer, K., 2013. Gangliosides and gangliosidoses: principles of molecular and metabolic pathogenesis. J. Neurosci. 33 (25), 10195–10208.

Simons, K., Gerl, M.J., 2010. Revitalizing membrane rafts: new tools and insights. Nat. Rev. Mol. Cell Biol. 11 (10), 688–699.

Simons, K., Ikonen, E., 1997. Functional rafts in cell membranes. Nature 387 (6633), 569–572.

Spiegel, S., Milstien, S., 2003. Sphingosine-1-phosphate: an enigmatic signalling lipid. Nat. Rev. Mol. Cell Biol. 4 (5), 397–407.

Thudicum, J.L.W., 1874. A Treatise on the Chemical Composition of the Brain, Hamden. Republished in 1962 by Archon Books.

Tidhar, R., Futerman, A.H., 2013. The complexity of sphingolipid biosynthesis in the endoplasmic reticulum. Biochim. Biophys. Acta 1833 (11), 2511–2518.

Tirodkar, T.S., Voelkel-Johnson, C., 2012. Sphingolipids in apoptosis. Exp. Oncol. 34 (3), 231–242.

Venkataraman, K., Futerman, A.H., 2001. Comparison of the metabolism of L-*erythro*- and L-*threo*-sphinganines and ceramides in cultured cells and in subcellular fractions. Biochim. Biophys. Acta 1530 (2–3), 219–226.

Zhang, H., et al., 1991. Sphingosine-1-phosphate, a novel lipid, involved in cellular proliferation. J. Cell Biol. 114, 155–167.

Zhang, Y., et al., 2009. Ceramide-enriched membrane domains–structure and function. Biochim. Biophys. Acta 1788 (1), 178–183.

Zigdon, H., Meshcheriakova, A., Futerman, A.H., 2014. From sheep to mice to cells: tools for the study of the sphingolipidoses. Biochim. Biophys. Acta 1841 (8), 1189–1199.

Chapter 11

Cholesterol Synthesis

Andrew J. Brown, Laura J. Sharpe

School of Biotechnology and Biomolecular Sciences, The University of New South Wales (UNSW Australia), Sydney, NSW, Australia

ABBREVIATIONS

24,25EC 24,25-Epoxycholesterol
ABCA1 ATP-binding cassette subfamily A member 1
ABCG1 ATP-binding cassette subfamily G member 1
ACAT Acyl-coenzyme A:cholesterol acyltransferase
ACAT2 Acetyl-CoA acetyltransferase, cytosolic
AMPK Adenosine monophosphate-activated protein kinase
ATP Adenosine triphosphate
Cdc2 kinase Cyclin-dependent kinase
COPII Coatomer protein II
CYP Cytochrome P450
DHCR14 $\Delta14$-Sterol reductase
DHCR24 $\Delta24$-Sterol reductase
DHCR7 7-Dehydrocholesterol reductase
EBP 3β-Hydroxysteroid-$\Delta8,\Delta7$-isomerase
ER Endoplasmic reticulum
FAD Flavin adenine dinucleotide
FDFT1 Farnesyl-diphosphate farnesyltransferase 1
FF-MAS Follicular fluid meiosis-activating sterol
FPPS Farnesyl pyrophosphate synthase
GGPPS Geranylgeranyl pyrophosphate synthase
gp78 Glycoprotein 78
HDL High density lipoprotein
HMG-CoA 3-hydroxy-3-methylglutaryl coenzyme A
HMGCR HMG-CoA reductase
HMGCS HMG-CoA synthase
HRD1 HMGCR degradation protein 1
HSD17B7 3-keto-steroid reductase
IDI1/IDI2 Isopentenyl-diphosphate Δ-isomerase 1 or 2
INSIG Insulin-induced gene
LBR Lamin B receptor
LDL Low density lipoprotein
LDLR LDL receptor

Biochemistry of Lipids, Lipoproteins and Membranes. http://dx.doi.org/10.1016/B978-0-444-63438-2.00011-0

LDM Lanosterol 14α-demethylase
LIPID MAPS Lipid metabolites and pathways strategy
LS Lanosterol synthase
LXR Liver X receptor
MARCH6 Membrane-associated ring finger (C3CH4) 6
MK Mevalonate kinase
MVD Diphosphomevalonate decarboxylase
NADPH Nicotinamide adenine dinucleotide phosphate
NSDHL Sterol-4-α-carboxylate 3-dehydrogenase
PMK Phosphomevalonate kinase
RNA Ribonucleic acid
SC4MOL Methylsterol monooxygenase 1
SC5D Lathosterol oxidase
Scap SREBP-cleavage activating protein
SLOS Smith–Lemli–Opitz Syndrome
SM Squalene monooxygenase
SQS Squalene synthase
SRE Sterol-regulatory element
SREBP Sterol-regulatory element binding protein
T-MAS Testis-meiosis-activating sterol
TRC8 Translocation in renal cancer from chromosome 8
UV Ultraviolet

1. INTRODUCTION

1.1 What Is Cholesterol?

For more than half a century, cholesterol has been a byword for heart disease risk. Although cholesterol undoubtedly contributes to disease, sterols are in fact an essential hallmark of eukaryotes, and may even have triggered the evolution of multicellular organisms. Sterols come in a variety of flavors. Fungi (like yeast) make ergosterol; plants synthesise a range of phytosterols; and most animals (like humans) make cholesterol itself (Galea and Brown, 2009). All of these sterols have a waxy consistency, sharing the familiar four-fused ring structure and 3β-hydroxyl group, but differing in the number and placement of double bonds and additional carbons on the side chain. Figure 1 displays the structures of some of the key molecules to be discussed in this chapter.

1.2 The Cholesterol Synthesis Pathway

The pathway for cholesterol synthesis involves a large number of intermediates, beginning with acetyl-coenzyme A (CoA), and a number of side branches along the way (Figure 2). Isoprenoids are formed from intermediates between mevalonate and squalene, while 24,25-epoxycholesterol (24,25EC) is formed after squalene through the 24,25EC shunt pathway. Lanosterol can be diverted

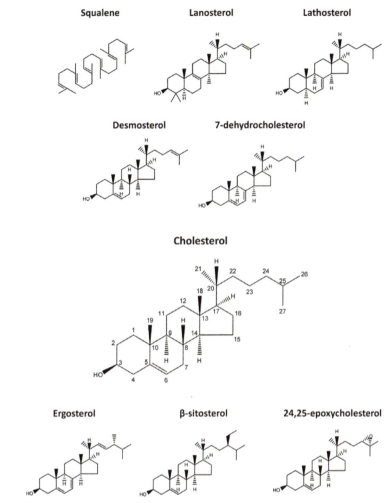

FIGURE 1 **Structures of key sterols and intermediates.** Squalene is an important precursor to cholesterol, which does not yet contain the fused rings. Squalene is cyclised into lanosterol. Lathosterol and desmosterol are intermediates that are commonly measured in blood samples. 7-dehydrocholesterol is an immediate precursor to cholesterol that can be converted to vitamin D. Ergosterol is the cholesterol equivalent found in yeast. β-sitosterol is a major phytosterol, that is a sterol found in plants. 24,25-epoxycholesterol is an oxysterol produced in a shunt of the cholesterol synthesis pathway.

into either the Kandutsch–Russell[1] or the Bloch[2] pathway, which are identical pathways except for the first and final steps. From 7-dehydrocholesterol, an intermediate in the Kandutsch–Russell pathway, vitamin D is formed by the

---

1. Named after Kandutsch and Russell, who deduced this route to cholesterol (Kandutsch and Russell, 1960).
2. Named in recognition of Konrad Bloch, whose elegant work toward elucidating the cholesterol synthesis pathway was recognised by a Nobel Prize in 1964. Johnston and Bloch realised that all the naturally occurring C27 sterols could not be fitted neatly into a single reaction sequence, anticipating what has become known as the Bloch pathway.

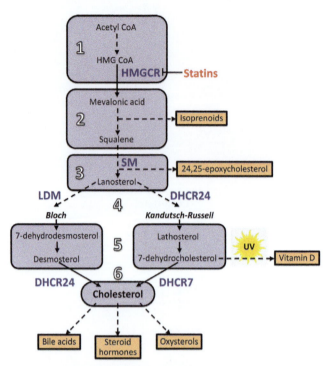

FIGURE 2 **Simplified cholesterol synthesis pathway.** An overview of the cholesterol synthesis pathway, showing key intermediates, enzymes, side branches and later products. Modules 1 to 6 are described in more detail in the section on enzymes (Section 5). Solid lines indicate a direct conversion, while dashed lines indicate multiple steps are required.

action of ultraviolet (UV) light. Once formed, cholesterol is converted to various steroid hormones, oxysterols and bile acids (Figure 2), and is incorporated into lipoproteins for secretion into the bloodstream (Chapter 16).

1.3 The Functions of Cholesterol

Cholesterol is essential for mammalian membranes, where it can enhance membrane fluidity and reinforce membranes. Akin to a 'molecular polyfiller', cholesterol plugs gaps in the phospholipid bilayer to reduce permeability. Cholesterol is also a key ingredient of the 'lipid raft' concept. These fluctuating nanoscale assemblies remain somewhat contentious but are proposed to organise membranes by providing platforms that function in membrane signalling and trafficking (Lingwood and Simons, 2010). Furthermore, cholesterol interacts with and regulates the activity of many integral membrane proteins, as indicated by the growing list of high-resolution protein structures containing cholesterol (Song et al., 2014). Moreover, cholesterol is the precursor for steroid hormones that maintain salt and water balance and reproductive function. Additionally, cholesterol in the liver is converted to bile acids, which aid fat digestion and absorption (Chapter 12).

1.4 Cholesterol in the Body

Although omnivorous humans may derive a substantial proportion of their cholesterol from their diet, the remainder must be synthesised. A 70 kg human synthesises approximately 700 mg of cholesterol per day (Figure 3).

On an equal weight basis, humans exhibit lower rates of cholesterol synthesis than many other animals (particularly rodents), and the liver is comparatively less important in contributing to total body cholesterol synthesis (Figure 4(A) and (B)) (Dietschy and Turley, 2002). Other tissues that also contribute significantly to total cholesterol synthesis are the intestines and skin. However, nearly all organs require a continuous supply of cholesterol to meet their needs for the synthesis of new membranes and as precursors of steroids in specialised tissues (Dietschy, 1984). Receptor-mediated uptake of lipoproteins such as low-density lipoprotein (LDL, so-called 'bad cholesterol'[3]) provide most cell types with some cholesterol that is supplemented by de novo cholesterol synthesis. Yet, the brain is an example where de novo synthesis provides all of the cholesterol in this lipid-rich organ, since circulating lipoproteins cannot cross the blood–brain barrier. It is also noteworthy that cholesterol synthesis in humans exhibits a diurnal cycle, with peak synthesis occurring in the early hours of the morning and the nadir occurring during the day (Figure 4(C)) (Galman et al., 2005).

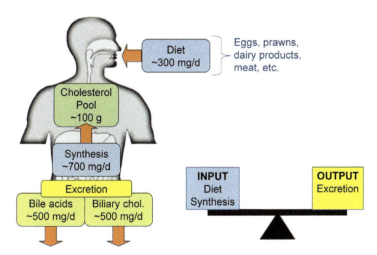

FIGURE 3 **Whole body cholesterol homeostasis.** Cholesterol homeostasis is achieved when input from diet and synthesis is matched by output through excretion in the bile (both as bile acids and biliary cholesterol).

---

3. LDL is known as 'bad cholesterol' because in excess it can accumulate in artery walls, a process termed atherosclerosis. So-called 'good cholesterol', high-density lipoprotein, HDL, is involved in the removal of cholesterol from cells and transporting it to the liver for excretion.

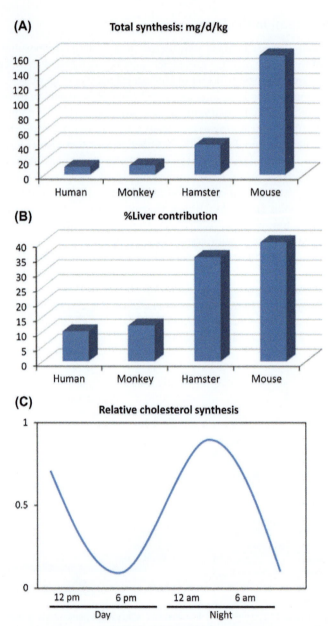

FIGURE 4 **Rates of cholesterol synthesis.** (A) Total cholesterol synthesis rates (mg per day per kg body weight) and (B) percentage contribution from the liver by species *(data from Dietschy and Turley, 2002)*. (C) Diurnal variation in cholesterol synthesis in humans *(after Galman et al., 2005)*.

1.5 Where Is Cholesterol Made within Cells?

Within cells, cholesterol synthesis occurs mostly in the membranes of the endoplasmic reticulum (ER). This membranous milieu makes sense considering that most enzymes acting after 3-hydroxy-3-methylglutaryl coenzyme A (HMG-CoA) reductase (HMGCR) convert hydrophobic substrates in membranes into hydrophobic products that also associate with membranes (Gaylor, 2002). There is also evidence for the peroxisomal localisation of the early steps in the biosynthetic pathway to the point of isoprenoid synthesis [i.e. acetyl-CoA acetyltransferase, cytosolic (ACAT2) to farnesyl pyrophosphate synthase (FPPS); see Section 5]. Although HMGCR is considered an ER enzyme, it may also be targeted to peroxisomes by an as yet unknown mechanism (Faust and Kovacs, 2014). Other cholesterogenic enzymes also redistribute from the ER to other organelles in response to certain stimuli, although the physiological significance of this redistribution is currently unclear (Sharpe and Brown, 2013). For example, $\Delta 24$-sterol reductase (DHCR24) may translocate to the nucleus during times of cellular stress, but it is not yet known how or why this occurs.

1.6 Cholesterol Homeostasis

Of course, cholesterol synthesis is not the only factor in contributing to cellular cholesterol homeostasis. As well as synthesis, cells can also acquire cholesterol by taking up LDL through the LDL receptor (LDLR, Chapter 17), and export cholesterol through efflux pumps like ATP-binding cassette transporters A1 and G1 (ABCA1 and ABCG1, Chapter 14). The transcription factors governing these processes are critical. Sterol regulatory element-binding proteins (SREBPs; see Section 7.1) increase cholesterol levels by upregulating expression of cholesterol synthesis genes and the LDLR. The nuclear receptor, liver X receptor (LXR), senses sterol excess and eliminates surplus cholesterol by upregulating ABCA1 and ABCG1 expression (Figure 5).

2. CHOLESTEROL SYNTHESIS – AN HISTORICAL OVERVIEW

Initially named 'cholesterine' by the French chemist Michel Chevreul in the same year that the French Revolution was ignited (1789), cholesterol has continued to intrigue and in the past century has been the most decorated small molecule in terms of Nobel Prizes awarded for its study. Elucidation of the structure of cholesterol in the 1920s and 1930s was a triumph of organic chemistry (Diels, 1951), but presented a new puzzle to solve: how was this strange four-fused ring structure synthesised biochemically?

This question engaged some of the best biochemical minds of the twentieth century. Eventually, teams led by Konrad Bloch and others were able to construct a workable biosynthetic pathway by tracing the origin of the single oxygen atom and each carbon atom (Bloch, 1965). In recognition of this heroic task, Bloch shared the 1964 Nobel Prize with Feodor Lynen whose

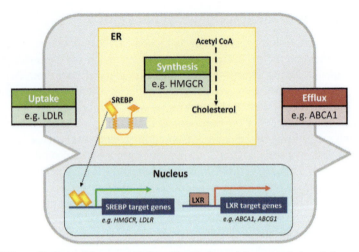

FIGURE 5 **Cholesterol homeostasis in cells.** Cholesterol homeostasis is a balance among synthesis, uptake and efflux. Green indicates an increase in cholesterol status, and red indicates a decrease in cholesterol status. See text for further details. *Figure adapted from Brown and Jessup (2009).*

discovery of the precursor acetyl-CoA in 1951 paved the way for elucidating early steps in cholesterol biosynthesis. Curiously, these early steps were also influenced by lessons from the rubber industry because acetate is also the precursor for this isoprene polymer. Isoprenoids compose the largest group of natural products, which can be found across all three domains of life (archea, bacteria and eukaryotes). Five-carbon isoprene units are assembled and modified in a myriad of ways, giving rise to tens of thousands of compounds, including many multicyclic structures like sterols. For the initial stages of cholesterol synthesis, six isoprene units formed from acetyl-CoA are built into the isoprenoid hydrocarbon squalene[4] (Figure 1), which then undergoes cyclisation to create the sterol backbone. Lanosterol[5] (Figure 1) is the resulting 30 carbon sterol, which then loses three methyl groups to be trimmed to the 27 carbons observed in cholesterol.

The oxygen atom in the signature hydroxyl group of sterols is introduced prior to cyclisation through an epoxidation of squalene involving molecular oxygen. Oxygen is also needed for other steps in cholesterol synthesis, with three molecules of O_2 required for the removal of each methyl group. In total, 11 molecules of O_2 are required for the synthesis of one molecule of cholesterol, making this a particularly oxygen-intensive process. Indeed, Konrad Bloch argued that cholesterol only evolved when sufficient atmospheric O_2 was available. Conversely, cholesterol was proposed to have evolved, at least in part,

---

4. Originally isolated from shark liver oil; *Squalus* being a genus of shark.
5. Originally isolated from wool fat; Lana being Latin for wool.

as an adaptation to the hazards of an O_2-rich environment by reducing membrane permeability to O_2 and providing a sink for cellular O_2 (Galea and Brown, 2009). Cholesterol synthesis is also energetically expensive, requiring ~100 ATP per molecule produced, compared to half that number of ATPs needed for fatty acid synthesis, and a tenth that number for glucose synthesis.

Identification and characterisation of the many enzymes required for cholesterol synthesis lagged behind the general ordering of the pathway (Figure 2), but have been greatly assisted by the molecular cloning of enzyme-encoding genes over the past couple of decades. A full list of the more than 20 enzymes involved in the conversion of acetyl-CoA to cholesterol is presented in Table 1.

3. TARGETING CHOLESTEROL SYNTHESIS THERAPEUTICALLY

First isolated from a mold, statins have become one of the most prescribed medicines in the world. Statins were discovered and developed due to a worrisome coronary epidemic in industrialised countries and mounting evidence that cholesterol was the culprit, or at least a nefarious accomplice. In 1913, Nikolai Anitschkov was reportedly the first to make a connection between cholesterol and heart disease, inducing atherosclerosis in rabbits by feeding them cholesterol (Li, 2009). In the 1950s, key epidemiological studies[6] firmly established raised blood cholesterol levels as a major risk factor for heart disease (Frantz and Moore, 1969). Thus, there was a need for the development of drug therapies to lower blood cholesterol levels, and hence combat the developing coronary epidemic. Existing compounds such as niacin, bile acid resins and fibrates (all still in clinical use today) were somewhat successful, but their ability to lower cholesterol levels was only modest, and not without side effects. The first drug to directly target the cholesterol biosynthetic pathway, triparanol[7], was approved in 1960. Triparanol inhibits DHCR24, one of the penultimate enzymes in cholesterol synthesis, resulting in the accumulation of desmosterol. However, it was withdrawn just 2 years later, due to many cases of cataract formation, hair loss and skin reactions (Gelissen and Brown, 2011). These adverse reactions were attributed to the accumulation of hydrophobic cholesterol precursors, particularly desmosterol. The cautionary tale of triparanol steered research efforts away from the distal end of the cholesterol biosynthesis pathway to an earlier step catalyzed by HMGCR, a key rate-limiting enzyme (Bucher et al., 1960) with a water-soluble precursor that should not accumulate to toxic levels.

In 1973, after a year of painstaking work assaying more than 3800 fungal strains, Akira Endo's team discovered a potent HMGCR inhibitor, mevastatin (also called compactin) (Endo, 2010). Under the trademark Mevacor, the

---

6. Decades later, a meta-analysis published on 61 prospective studies comprising data from ~900,000 individuals showed a clear positive association between blood cholesterol levels and cardiovascular disease risk at all age ranges examined (Lewington et al., 2007).
7. Trade name: MER/29.

TABLE 1 Cholesterol Synthesis Enzymes

Official Gene Symbol (HGNC)	Official Gene Name (HGNC)	Chromosome Location	Protein Symbol	Recommended Protein Name (UniProt)	Other Names	EC Number
ACAT2	Acetyl-CoA acetyltransferase 2	6q25.3	Thiolase 2	Acetyl-CoA acetyltransferase, cytosolic	ACAT2	2.3.1.9
HMGCS1	3-Hydroxy-3-methylglutaryl-CoA synthase 1 (soluble)	5p14–p13	HMGCS	Hydroxymethylglutaryl-CoA synthase, cytoplasmic	HMG-CoA synthase	2.3.3.10
HMGCR	3-Hydroxy-3-methylglutaryl-CoA reductase	5q13.3-q14	HMGCR	3-Hydroxy-3-methylglutaryl-coenzyme A reductase	HMG-CoA reductase	1.1.1.34
MVK	Mevalonate kinase	12q24	MK	Mevalonate kinase		2.7.1.36
PMVK	Phosphomevalonate kinase	1q21.3	PMK	Phosphomevalonate kinase	PMKase	2.7.4.2
MVD	Mevalonate (diphospho) decarboxylase	16q24.3	MVD	Diphosphomevalonate decarboxylase	MDDase	4.1.1.33
IDI1/IDI2	Isopentenyl-diphosphate Δ isomerase 1/ isopentenyl-diphosphate Δ isomerase 2	10p15.3	IDI1/IDI2	Isopentenyl-diphosphate Δ-isomerase 1/2	IPP isomerase 1/2	5.3.3.2

Gene	Full name	Location	Abbreviation	Protein name	Alternative name	EC number
FDPS	Farnesyl diphosphate synthase	1q22	FPPS	Farnesyl pyrophosphate synthase	FPP synthase, FPS	2.5.1.1, 2.5.1.10
GGPS1	Geranylgeranyl diphosphate synthase 1	1q43	GGPPS	Geranylgeranyl pyrophosphate synthase	GGPP synthase	2.5.1.1, 2.5.1.10, 2.5.1.29
FDFT1	Farnesyl-diphosphate farnesyltransferase 1	8p23.1–p22	SQS	Squalene synthase		2.5.1.21
SQLE	Squalene epoxidase	8q24.1	SM	Squalene mcnooxygenase		1.14.13.132
LSS	Lanosterol synthase (2,3-oxidosqualene-lanosterol cyclase)	21q22.3	LS	Lanosterol synthase	OSC, Oxidosqualene cyclase	5.4.99.7
CYP51A1	Cytochrome P450, family 51, subfamily A, polypeptide 1	7q21.2	LDM	Lanosterol 14-α demethylase	CYP51	1.14.13.70
TM7SF2	Transmembrane 7 superfamily member 2	11q13.1	DHCR14	$\Delta(14)$-sterol reductase		1.3.1.70
LBR	Lamin B receptor	1q42.1	LBR	Lamin-B receptor		1.3.1.70
MSMO1	Methylsterol monooxygenase 1	4q32–q34	SC4MOL	Methylsterol rronooxygenase 1	MSMO1	1.14.13.72
NSDHL	NAD(P)-dependent steroid dehydrogenase-like	Xq28	NSDHL	Sterol-4-α-carboxylate 3-dehydrogenase, decarboxylating		1.1.1.170

Continued

TABLE 1 Cholesterol Synthesis Enzymes—cont'd

HSD17B7	Hydroxysteroid (17-β) dehydrogenase 7	1q23	HSD17B7	3-keto-Steroid reductase		1.1.1.270
EBP	Emopamil binding protein (sterol isomerase)	Xp11.23–p11.22	EBP	3β-Hydroxysteroid-Δ(8),Δ(7)-isomerase	Sterol isomerase	5.3.3.5
SC5D	Sterol-C5-desaturase	11q23.3	SC5D	Lathosterol oxidase	Sterol-C5-desaturase	1.14.21.6
DHCR7	7-Dehydrocholesterol reductase	11q13.4	DHCR7	7-Dehydrocholesterol reductase		1.3.1.21
DHCR24	24-Dehydrocholesterol reductase	1p32.3	DHCR24	Δ(24)-Sterol reductase	Seladin-1	1.3.1.72

pharmaceutical company Merck developed a chemically related compound lovastatin in 1987, which became the first statin to reach patients (Li, 2009). Over subsequent years, a number of statins have become available, each tending to be more potent than their predecessors. From their humble fungal beginnings, statins are now considered the cornerstone in treatment of cardiovascular disease, providing benefit for the great majority of patients, but not tolerated by some. Moreover, statin therapy is expanding beyond the traditional cardiovascular area, with statins currently being tested for a remarkable diversity of diseases including Alzheimer's disease, asthma, lupus, sepsis, renal diseases and various cancers.

Other enzymes have also been targeted for the treatment of cardiovascular disease (Gelissen and Brown, 2011). For example, squalene synthase inhibitors have been developed, but from a marketing perspective they are not significantly distinct from statins in terms of efficacy or safety profile and have not been pursued to market.

3.1 How Statins Work to Decrease Blood Cholesterol Levels

Statins are reversible, competitive inhibitors of HMGCR that exploit a key feature of cholesterol homeostasis in hepatocytes. When cholesterol synthesis in hepatocytes is inhibited, the response is to maintain cholesterol status by upregulating the number of LDLRs on the surface of the cell via a transcriptional pathway mediated by SREBPs (Section 7.1). Since LDLR is responsible for clearance of LDL from the bloodstream, its upregulation decreases blood cholesterol levels and cardiovascular risk. Besides inhibiting cholesterol synthesis, statins also inhibit the formation of important nonsterol products downstream of HMGCR (Figure 2). These intermediates are involved in a variety of cell processes, including cell growth and differentiation (prenylated proteins; see Chapter 13), glycosylation (dolichol) and electron transport (ubiquinol). Inhibition of these products probably accounts for the non-cholesterol-related benefits often ascribed to statins. These so-called 'pleiotropic effects' include reduction of the inflammatory risk marker C-reactive protein. However, the idea that statins reduce cardiovascular risk by mechanisms other than LDL cholesterol lowering remains somewhat contentious. Inhibition of these nonsterol products of the pathway could contribute to some of the adverse effects reported for statins, such as muscle pain.

4. STEROL PATHWAY INTERMEDIATES

4.1 Overview

Although there are numerous intermediates in cholesterol synthesis, the scientific literature is unclear on their exact number and order, with some groups including additional sterols not mentioned by others. Because of this confusion, and the difficulties of systematically naming complex ring structures, the

LIPID MAPS (Metabolites And Pathways Strategy, lipidmaps.org) consortium established a nomenclature system in 2005. The intermediates in cholesterol synthesis indicated by LIPID MAPS are listed in Table 2(A–C). It is interesting to note that many of the intermediates were identified long before the enzymes were discovered.

4.2 Functions of the Intermediates

Cholesterol biosynthetic intermediates have potent biological effects that are not shared by cholesterol. For example, 24,25-dihydrolanosterol mediates proteolytic degradation of HMGCR, desmosterol is an activator of LXR, and FF-MAS and T-MAS may play a role in fertilisation. Like cholesterol, sterol intermediates can also be esterified to a fatty acid and/or undergo detoxification reactions in the liver. They also serve as substrates for bile acids and, in appropriate tissues, steroidogenesis. In addition, sterol intermediates are exported from tissues, and hence can be detected in the circulation. Circulating levels of sterol pathway intermediates are used as surrogate markers of cholesterol biosynthesis. The most commonly measured markers are squalene, lathosterol and desmosterol, which generally reflect hepatic cholesterol metabolism. In addition, circulating sterol intermediates may serve as markers of disease (Brown et al., 2014). Desmosterol and lathosterol are frequently perturbed under a variety of disease states[8]. For instance, desmosterol is increased in nonalcoholic steatohepatitis, and lathosterol is increased during insulin resistance. Elevated squalene is associated with risk of cardiovascular disease in women. Thus circulating levels of these sterol and nonsterol intermediates may be useful biomarkers in the diagnosis and treatment of human disease.

4.3 Diseases Resulting from Defective Cholesterol Synthesis

Genetic diseases that are caused by defects in a specific cholesterol synthesis gene lead to an accumulation of the intermediate that is acted on by the defective enzyme. The most common of these genetic disorders is deficiency of DHCR7, termed Smith–Lemli–Opitz Syndrome (SLOS), a developmental abnormality with mental retardation caused by defective signalling of the essential developmental protein Sonic Hedgehog. Since Sonic Hedgehog needs to be modified by cholesterol for its full signalling activity and its receptor Patched has a sterol-sensing domain spanning the membrane, the lack of cholesterol prevents Hedgehog signalling, thereby leading to SLOS. SLOS has an incidence of approximately 1 in 40,000, with a carrier rate of 1%. Over 150 different mutations in DHCR7 cause SLOS (Waterham and Hennekam, 2012). Table 3 provides an overview of genetic diseases of cholesterol synthesis and intermediates

---

8. As a side note, desmosterol levels (Bloch pathway) are generally higher in men than in women, but lathosterol levels (Kandutsch–Russell pathway) are comparable.

TABLE 2 Intermediates in Cholesterol Synthesis

Systematic Name (LIPIDMAPS)	Common Name	LIPIDMAPS ID	Enzyme	
			Product of	Substrate of
A: Early Steps				
Acetyl-CoA	Acetyl-CoA	LMFA07050029		Thiolase 2
Acetoacetyl-CoA	Acetoacetyl-CoA	LMFA07050030	Thiolase 2	HMGCS1
3S-Hydroxy-3-methyl-glutaryl-CoA	HMG-CoA, 3-hydroxy-3-methyl-glutaryl-CoA	LMFA07050028	HMGCS1	HMGCR
3R-Methyl-3,5-dihydroxy-pentanoic acid	Mevalonic acid	LMFA01050352	HMGCR	MK
3R-Methyl-3-hydroxypentanoic acid 5-phosphate	Mevalonate-P	LMFA01050415	MK	PMK
3R-Methyl-3-hydroxypentanoic acid 5-diphosphate	Mevalonate-PP	LMFA01050416	PMK	MVD
3-Methylbut-3-enyl pyrophosphate	Isopentenyl pyrophosphate, isopentenyl-diphosphate	LMPR01010008	MVD	IDI1/IDI2
3-Methylbut-2-enyl pyrophosphate	Dimethylallyl pyrophosphate	LMPR01010001	IDI1/IDI2	FPPS/GGPPS
Geranyl pyrophosphate	Geranyl diphosphate	LMPR0102010001	FPPS/GGPPS	FPPS/GGPPS
Farnesyl pyrophosphate	Farnesyl diphosphate	LMPR0103010002	FPPS/GGPPS	SQS

Continued

TABLE 2 Intermediates in Cholesterol Synthesis—cont'd

Systematic Name (LIPIDMAPS)	Common Name	LIPIDMAPS ID	Product of	Enzyme	
					Substrate of
Presqualene diphosphate	Presqualene diphosphate	LMPR0106010003	SQS		SQS
Squalene	Squalene	LMPR0106010002	SQS		SM
2,3S-epoxy-2,6,10,15,19,23-hexamethyltetracosa-6E,10E,14E,18E,22-pentaene	3S-Squalene-2,3-epoxide, monooxidosqualene (MOS)	LMPR0106010010	SM		LS
Lanosta-8,24-dien-3β-ol	Lanosterol	LMST01010017	LS		LDM/DHCR24
B: Bloch Pathway Intermediates					
4,4-Dimethyl-14α-hydroxymethyl-5α-cholesta-8,24-dien-3β-ol	Lanosta-8,24-dien-3β,30-diol	LMST01010124	LDM		LDM
4,4-Dimethyl-14α-formyl-5α-cholesta-8,24-dien-3β-ol	—	LMST01010222	LDM		LDM
4,4-Dimethylcholesta-8,14,24-trien-3β-ol	4,4-Dimethylcholesta-8,11,24-trienol, follicular fluid meiosis-activating sterol (FF-MAS)	LMST01010149	LDM		DHCR14/LBR
4,4-Dimethyl-5α-cholesta-8,24-dien-3β-ol	14-Demethyl-lanosterol, testis meiosis-activating sterol (T-MAS)	LMST01010176	DHCR14/LBR		SC4MOL

Continued

Name	Common name	LMST ID		
4α-Hydroxymethyl-4β-methyl-5α-cholesta-8,24-dien-3β-ol	—	LMST01010232	SC4MOL	SC4MOL
4α-Formyl-4β-methyl-5α-cholesta-8,24-dien-3β-ol	—	LMST01010229	SC4MOL	SC4MOL
4α-Carboxy-4β-methyl-cholesta-8,24-dien-3β-ol	4α-Carboxy-4β-methyl-zymosterol	LMST01010150	SC4MOL	NSDHL
4α-Methyl-5α-cholesta-8,24-dien-3-one	3-keto-4α-methyl-zymosterol	LMST01010237	NSDHL	HSD17B7
4α-Methyl-5α-cholesta-8,24-dien-3β-ol	4α-Methyl-zymosterol	LMST01010151	HSD17B7	HSD17B7
4α-Hydroxymethyl-5α-cholesta-8,24-dien-3β-ol	—	LMST01010234	SC4MOL	SC4MOL
4α-Formyl-5α-cholesta-8,24-dien-3β-ol	—	LMST01010226	SC4MOL	SC4MOL
4α-Carboxy-5α-cholesta-8,24-dien-3β-ol	4α-Carboxy-zymosterol	LMST01010152	SC4MOL	SC4MOL
5α-Cholesta-8,24-dien-3-one	Zymosterone	LMST01010168	NSDHL	NSDHL
5α-Cholesta-8,24-dien-3β-ol	Zymosterol	LMST01010066	HSD17B7	HSD17B7
5α-Cholesta-7,24-dien-3β-ol	5α-Cholesta-7,24-dien-3β-ol	LMST01010206	EBP	EBP
Cholest-5,7,24-trien-3β-ol	7-Dehydrodesmosterol	LMST01010121	SC5D	SC5D

TABLE 2 Intermediates in Cholesterol Synthesis — cont'd

Systematic Name (LIPIDMAPS)	Common Name	LIPIDMAPS ID	Enzyme	
			Product of	Substrate of
Cholest-5,24-dien-3β-ol	Desmosterol	LMST01010016	DHCR7	DHCR24
Cholest-5-en-3β-ol	Cholesterol	LMST01010001	DHCR24	
C: Kandutsch–Russell Pathway Intermediates				
5α-Lanost-8-en-3β-ol	24,25-Dihydrolanosterol, 24-dihydrolanosterol	LMST01010087	DHCR24	LDM
4,4-Dimethyl-14α-hydroxymethyl-5α-cholest-8-en-3β-ol	Lanost-8-en-3β,30-diol	LMST01010224	LDM	LDM
4,4-Dimethyl-14α-formyl-5α-cholest-8-en-3β-ol	-	LMST01010223	LDM	LDM
4,4-Dimethyl-5α-cholesta-8,14-dien-3β-ol	Dihydro-FF-MAS	LMST01010277	LDM	DHCR14/LBR
4,4-Dimethyl-5α-cholesta-8-en-3β-ol	Dihydro-T-MAS	LMST01010225	DHCR14/LBR	SC4MOL
4α-Hydroxymethyl-4β-methyl-5α-cholesta-8-en-3β-ol	–	LMST01010233	SC4MOL	SC4MOL
4α-Formyl-4β-methyl-5α-cholesta-8-en-3β-ol	–	LMST01010230	SC4MOL	SC4MOL

Systematic name	Common name	LMST ID		
4α-Carboxy-4β-methyl-5α-cholesta-8-en-3β-ol	–	LMST01010227	SC4MOL	NSDHL
4α-Methyl-5α-cholesta-8-en-3-one	–	LMST01010236	NSDHL	HSD17B7
4α-Methyl-5α-cholest-8-en-3β-ol	4α-Methylcholest-8-en-3β-ol	LMST01010197	HSD17B7	SC4MOL
4α-Hydroxymethyl-5α-cholesta-8-en-3β-ol	–	LMST01010235	SC4MOL	SC4MOL
4α-Formyl-5α-cholesta-8-en-3β-ol	–	LMST01010231	SC4MOL	SC4MOL
4α-Carboxy-5α-cholesta-8-en-3β-ol	–	LMST01010228	SC4MOL	NSDHL
5α-Cholest-8-en-3-one	–	LMST01010239	NSDHL	HSD17B7
5α-Cholest-8-en-3β-ol	Zymostenol	LMST01010096	HSD17B7	EBP
Cholest-7-en-3β-ol	Lathosterol	LMST01010089	EBP	SC5D
Cholesta-5,7-dien-3β-ol	7-Dehydrocholesterol, 7DHC	LMST01010069	SC5D	DHCR7
Cholest-5-en-3β-ol	Cholesterol	LMST01010001	DHCR7	

Notes: Common names are derived from lipidmaps.org and Ačimovič and Rozman (2013).

TABLE 3 Genetic Deficiencies of Cholesterol Synthesis Enzymes

Gene	Disease[a] (OMIM ref)	Brief Description	Accumulated Intermediate(s)
CYP51A1	Antley–Bixler syndrome (#201750)	An exceptionally rare craniosynostosis syndrome with genital anomalies and impaired steroidogenesis.	Lanosterol, 24,25-dihydrolanosterol
NSDHL	CHILD (#308050) and CK (#300831) syndromes	CHILD: An X-linked dominant disorder characterised by congenital hemidysplasia with ichythyosiform erythrodema and limb defects. Lethal in hemizygous males. CK: An X-linked recessive disorder characterised by cognitive impairment, seizures, and malformations.	C4-Methylsterols (e.g. FF-MAS and T-MAS)
EBP	CDPX2 (#302960)	X-linked dominant disorder. Clinically and genetically heterogeneous disorder characterised by punctiform calcification of the bones.	C4-Methylsterols (as above)
SC5D	Lathosterolosis (#607330)	Multiple congenital anomalies, mental retardation, and liver disease.	Lathosterol
DHCR7	SLOS (#270400)	Relatively common; an autosomal recessive multiple congenital malformation and mental retardation syndrome.	7-Dehydrocholesterol
DHCR24	Desmosterolosis (#602398)	A rare autosomal recessive disorder characterised by multiple congenital anomalies.	Desmosterol

[a]Abbreviations used: CHILD, congenital hemidysplasia with ichythyosiform erythrodema and limb defects; CDPX2, X-linked dominant chondrodysplasia punctata-2; SLOS, Smith–Lemli–Opitz Syndrome.

that subsequently accumulate. These diseases are likely the consequence of both an intermediate accumulation and the lack of cholesterol. The following sections describe the enzymes of the cholesterol biosynthetic pathway in more detail.

5. ENZYMES OF CHOLESTEROL BIOSYNTHESIS

5.1 Acetyl-Coenzyme A to Mevalonic Acid

Thiolase 2 is the first enzyme in the cholesterol synthesis pathway (see Figure 6), catalyzing the conversion of two molecules of acetyl-CoA to acetoacetyl-CoA. Hydroxymethylglutaryl-CoA synthase (HMGCS) acts next, requiring water to condense acetyl-CoA with acetoacetyl-CoA to form HMG-CoA. Cytosolic HMGCS1 synthesises mevalonic acid for cholesterol synthesis, whereas the related mitochondrial HMGCS2 produces HMG-CoA for ketogenesis and stimulates fatty acid oxidation.

Reduction of HMG-CoA to mevalonic acid by HMGCR requires two molecules of NADPH. HMGCR is the most widely studied of the cholesterol synthesis enzymes, largely since it is the target of the statin class of drugs, and was the first known rate-controlling step discovered in cholesterol synthesis. The degradation of this membrane-bound enzyme in response to sterol excess is discussed further in Section 7.2.1.

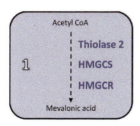

FIGURE 6 First steps to cholesterol synthesis.

5.2 Mevalonic Acid to Squalene

Mevalonate kinase (MK) uses ATP to phosphorylate mevalonic acid to produce mevalonate-P (see Figure 7), which is further phosphorylated to mevalonate-PP by phosphomevalonate kinase (PMK) by an ATP-dependent reaction. MK deficiency is a rare autosomal recessive disorder characterised by recurring episodes of fever, likely caused by the inability to produce isoprenoids. Diphosphomevalonate decarboxylase (MVD) catalyzes the ATP-dependent decarboxylation of mevalonate-PP to isopentenyl-PP, which is isomerised to dimethylallyl-PP by isopentenyl-diphosphate Δ-isomerase (IDI1/IDI2). These five carbon isoprenoid-pyrophosphates are condensed by either FPPS or geranylgeranyl pyrophosphate synthase (GGPPS) to produce the 10 and 15 carbon isoprenoids geranyl-PP and farnesyl-PP, respectively. Two molecules of farnesyl-PP are then converted to squalene by squalene synthase (SQS) in a two-step

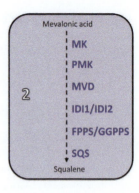

process requiring NADPH. Alternatively, these isoprenoid intermediates can be used for other purposes (i.e. protein prenylation). How cells maintain sufficient levels of these essential isoprenoids while also strictly regulating cholesterol synthesis is not yet fully understood.

5.3 Squalene Cyclisation to Lanosterol

Squalene monooxygenase (SM), the second rate-controlling step in cholesterol synthesis, introduces an epoxide group onto squalene using one molecule of O_2, FAD and NADPH (see Figure 8). This oxygen becomes the β-hydroxyl group on cholesterol. Addition of another epoxide group by SM diverts intermediates into the 24,25EC shunt pathway. SM has gained recent attention since it is also a key flux-controlling step that is regulated by cholesterol-sensitive proteolysis (see Section 7.2.1).

Lanosterol synthase (LS) catalyzes the cyclisation of squalene epoxide, argued to be the most complicated molecular rearrangement performed by a single enzyme, folding squalene into the four-fused ring structure of lanosterol (Figure 1).

FIGURE 8 **Squalene cyclisation to lanosterol.**

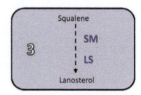

5.4 Lanosterol Metabolism in the Bloch and Kandutsch–Russell Pathways

After lanosterol, the pathway diverges into either the Bloch or Kandutsch–Russell pathway (see Figure 9). Note that these two routes are not exclusive and enzymes acting in each branch of the pathway are the same (Figure 10). Moreover, Δ(24)-sterol reductase (DHCR24, which uses NADPH and FAD) can utilise any of the

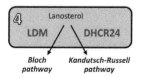

FIGURE 9 **Lanosterol diverges to the Bloch and Kandutsch–Russell pathways.**

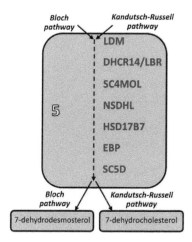

FIGURE 10 **The Bloch and Kandutsch–Russell pathways share the same enzymes but with different entry and exit enzymes.**

Bloch pathway intermediates and shuttle them into the Kandutsch–Russell pathway, thus intertwining these two paths to cholesterol. Lanosterol is a substrate for either lanosterol 14-α demethylase (LDM) or DHCR24, directing it into the Bloch or Kandutsch–Russell pathway, respectively. LDM is a cytochrome P450 that catalyzes three demethylation reactions, consuming three O_2 and NADPH molecules in the process. LDM acts either immediately on lanosterol, or after it is reduced to 24,25-dihydrolanosterol by DHCR24. Of note, one pathway is quantitatively more important than the other in several physiological and pathophysiological settings, although the reasons for this are not well understood. For example, in the brain, cholesterol synthesis occurs preferentially through the Bloch pathway in young mice, but the Kandutsch–Russell pathway in aged mice.

5.5 Lanosterol Conversion to Penultimate Intermediates

The product of LDM is a substrate for either Δ(14)-sterol reductase (DHCR14) or lamin-B receptor (LBR), which are both Δ14-sterol reductases that require NADPH. Subsequently, methylsterol monooxygenase 1 (SC4MOL), sterol-4-α-carboxylate 3-dehydrogenase, decarboxylating (NSDHL) and 3-keto-steroid reductase (HSD17B7) each act multiple times on the various intermediates to produce zymosterol (Bloch) or zymostenol (Kandutsch–Russell). See Table 2(B) and (C) for full details of the ordering and substrates for these enzymes. SC4MOL requires six molecules each of NAD(P)H and O_2, and NSDHL requires NAD^+ whereas HSD17B7 uses NADPH. Zymosterol and zymostenol are then substrates for 3β-hydroxysteroid-Δ(8),Δ(7)-isomerase (EBP) and lathosterol oxidase [SC5D, which requires O_2 and NAD(P)H] sequentially to make 7-dehydrodesmosterol (Bloch) or 7-dehydrocholesterol (Kandutsch–Russell). 7-Dehydrocholesterol is spontaneously converted to vitamin D by the action of UVB light in skin cells (keratinocytes). The UVB radiation opens the B-ring of 7-dehydrocholesterol, which then isomerises to form cholecalciferol (vitamin D_3).

5.6 Ultimate Enzymes in Cholesterol Synthesis

The final enzyme in the Kandutsch–Russell pathway is DHCR7, which produces cholesterol. In the Bloch pathway, 7-dehydrodesmosterol is converted to desmosterol by DHCR7, and then desmosterol is reduced to cholesterol by DHCR24 in the final reaction (Figure 11). Genetic deficiency of either DHCR24 or DHCR7 results in severe developmental abnormalities, as indicated in Table 3.

FIGURE 11 **The final steps in cholesterol synthesis.**

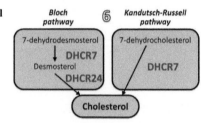

6. OXYSTEROLS

Cholesterol is converted to oxysterols by the addition of an hydroxyl, keto or epoxy group to the fused ring structure or side chain. Some oxysterols derived nonenzymically by free radical-mediated oxidation of cholesterol are toxic to cells, at least in vitro, for example 7-ketocholesterol. Other oxysterols are derived enzymically from cholesterol, mostly by cytochrome P450 enzymes in the initial stages of bile acid synthesis (CYP7A1, CYP27A1), for cholesterol elimination from specific organs like the brain (CYP24A1) and for innate immune modulation in macrophages (25-hydroxylase). Oxysterols can be potent regulators of cholesterol homeostasis that promote feedback inhibition of cholesterol synthesis. Indeed, in 1978 Andrew Kandutsch proposed that oxysterols are the principal regulators in what has become known as the Oxysterol Hypothesis of cholesterol homeostasis (Kandutsch et al., 1978). Subsequently, this hypothesis has been revised, attributing a more subordinate role to oxysterols and a greater role to cholesterol and its synthetic intermediates as key regulatory molecules (Gill et al., 2008). The main targets of oxysterols in the regulation of cholesterol are suppression of SREBP activation, through binding to Insig (discussed in Section 7.1.1), and the activation of ACAT – an enzyme that esterifies cholesterol.

24,25EC is an exceptional oxysterol in that it is formed not from cholesterol but in a shunt pathway that parallels cholesterol synthesis (Figure 2)

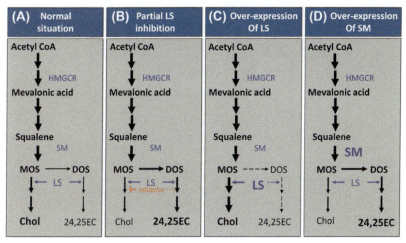

FIGURE 12   Manipulating 24,25EC levels. (A) In normal circumstances, 24,25EC levels are about 1:100–1000 of cholesterol levels. (B) If LS is partially inhibited, more dioxidosqualene (DOS) is formed by SM, and subsequently more 24,25EC. (C) If LS is overexpressed, monooxido-squalene (MOS) is acted on more quickly and thus more cholesterol is formed. (D) If SM is over-expressed, MOS is more readily converted to DOS, and thus more 24,25EC forms *Figure adapted from Wong et al. (2008).*

(Spencer et al., 1985). SM introduces a second epoxy group on the other end of squalene from the initial epoxidation, producing a diepoxy-squalene intermediate which then proceeds via the Kandutsch–Russell pathway to eventually produce 24,25EC. Under most conditions, 24,25EC mass is very low relative to cholesterol (~1 to 100–1000) (Figure 12(A)). Partial pharmacological inhibition of LS provides more monooxidosqualene substrate for SM and increases flux into the shunt pathway to augment production of 24,25EC (Figure 12(B)). Conversely, overexpressing LS ablates 24,25EC synthesis, while overexpressing SM increases it (Figures 12(C)and (D)). Using these approaches the role of 24,25EC in fine-tuning cholesterol regulation has been elucidated, and 24,25EC is a potent natural agonist for LXR, a nuclear receptor that regulates the expression of genes involved in cellular cholesterol export (Figure 4), among other processes. Because the small quantities of 24,25EC produced in the shunt pathway are sufficient for potent effects on cholesterol homeostasis, 24,25EC is able to fine-tune cholesterol levels to ensure smooth regulation instead of drastic fluctuation.

7. REGULATION OF CHOLESTEROL SYNTHESIS

Regulation of cholesterol synthesis can be explained in simple economic terms. As with every efficient economy, the supply of cholesterol is geared toward the cellular demand for the molecule. Making cholesterol de novo is energetically expensive, hence the cheapest option for the cell is to derive premade cholesterol by taking up circulating lipoproteins. Cholesterol synthesis is therefore aimed at supplementing that exogenous supply based on demand; production is decreased

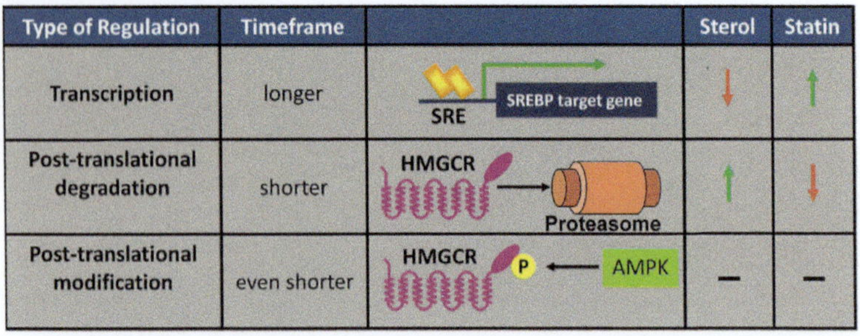

Type of Regulation	Timeframe			Sterol	Statin
Transcription	longer	SRE	SREBP target gene	↓	↑
Post-translational degradation	shorter	HMGCR	Proteasome	↑	↓
Post-translational modification	even shorter	HMGCR	P ← AMPK	–	–

FIGURE 13 **Levels of HMGCR regulation.** Transcription controls synthesis of HMGCR through SREBP. SREBP is downregulated by sterol, and upregulated by statin. HMGCR is post-translationally degraded by the proteasome, which is increased by sterol and decreased by statin. HMGCR is phosphorylated by AMP kinase in an example of a posttranslational modification unaffected by sterol levels.

when supply of exogenous cholesterol is adequate and ramped up in response to increased cholesterol demand. Because too much cholesterol is harmful to the cell, elaborate mechanisms have evolved to tightly regulate its levels by a negative feedback control mechanism. Indeed, cholesterol synthesis was one of the first examples of feedback control of a biosynthetic pathway. More than 80 years ago, Rudolph Schoenheimer[9] found that feeding mice cholesterol reduced its synthesis (Schoenheimer and Breusch, 1933). The following section outlines the major modes for regulation of cholesterol synthesis. Much of this work has focused on the rate-limiting step catalyzed by HMGCR, but it is important to note that other enzymes play critical roles, and will be areas of future investigation. Furthermore, different levels of regulation take effect over time (see Figure 13 for an example with HMGCR). The slowest regulation occurs through transcriptional downregulation, where the active enzyme remains but will not be replenished. Next is posttranslational degradation, which marks enzymes for destruction, and then posttranslational modification which is the fastest mode of regulation, as the modification can immediately affect enzyme activity.

7.1 Transcription

7.1.1 The Sterol-Regulatory Element Binding Protein Pathway

Most of the genes that encode cholesterol biosynthetic enzymes are controlled by SREBPs[10] (Sharpe and Brown, 2013). An unusual family of transcription factors, SREBP isoforms 1a, 1c and 2, are translated as inactive precursors anchored in

---

9. Incidentally, Schoenheimer introduced a young Konrad Bloch to cholesterol metabolism, when he became part of Schoenheimer's research team at Columbia.

10. The elegant cell biology of the SREBP pathway has been elucidated over the past couple of decades by the Dallas laboratory of Joseph Goldstein and Michael Brown (Goldstein et al., 2006), who incidentally also discovered LDLR, for which they were awarded the Nobel Prize in 1985 (Brown and Goldstein, 1985).

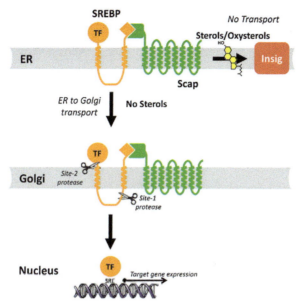

FIGURE 14   SREBP processing pathway. When sterol levels are sufficient, SREBP and Scap bind to Insig and remain inactive in the ER. When sterol levels are low, SREBP and Scap travel to the Golgi for processing by Site-1 protease and Site-2 protease. This releases the active transcription factor (TF), which travels to the nucleus and upregulates target genes. *Figure adapted from Du et al. (2004).*

ER membranes by two transmembrane segments. To become activated, precursor SREBPs are transported by COPII vesicles to the Golgi for proteolytic activation that releases a soluble transcription factor that enters the nucleus and activates genes involved in cholesterol synthesis and uptake. The SREBP cholesterol homeostatic machinery resides in the membranes of the ER, which are particularly low in cholesterol, thus allowing finely tuned responses to small fluctuations in cell cholesterol status. The cholesterol homeostatic machinery consists of Scap (SREBP-cleavage activating protein) and Insig (insulin-induced gene). Scap is an SREBP chaperone that senses cholesterol in the ER membrane, both through cholesterol directly binding to Scap and also by altering membrane properties. When there is too much cholesterol, the Scap–SREBP complex binds to the tethering protein Insig and is retained in the ER. Similarly, oxysterols can bind to Insig and also result in retention of the Scap–SREBP complex in the ER. Conversely, when cholesterol levels are low, Scap adopts a conformation that exposes a COPII recognition site and the Scap–SREBP complex is transported from the ER in COPII transport vesicles. When SREBP reaches the Golgi apparatus it is cleaved sequentially by site-1 protease and site-2 protease (Figure 14), releasing the active N-terminal basic helix–loop–helix domain (Figure 14). The transcription factor portion is imported into the nucleus where it upregulates expression of target genes. There are three isoforms of SREBP, with -1c involved in fatty acid metabolism, -2 involved in cholesterol metabolism and -1a in both.

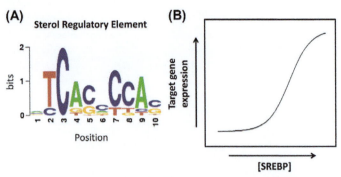

FIGURE 15 **SRE composition and cooperativity.** (A) This logo was created using known SRE sequences. The overall height at each base position indicates the sequence conservation at that position, while the height of each base individually indicates the frequency of the base at that position. (B) Two SREs function cooperatively to produce a sigmoidal response. The response of a target gene with one SRE (e.g. LDLR) would be linear.

7.1.2 Sterol-Response Element Regulated Transcription

SREBP activates its target genes by binding to sterol regulatory elements (SREs) in their promoters. An SRE is typically defined as a 10 bp element in the gene promoter that is compositionally similar to the canonical SRE in the LDLR (ATCACCCCAC). However, there can be considerable variation among these bases, as depicted by Figure 15(A). SREs have been identified in many cholesterol synthesis genes in rodents, but only some of these have been located in the human promoters. It has been suggested that the exact sequence and location of the SRE may not be conserved between species, but that the general region may be the same, suggesting some shifting during evolution, where the precise location and sequence of SREs are not necessary, but the sterol-responsive region is required (Prabhu et al., 2014).

Four cholesterol synthesis genes (HMGCS, FDFT1, DHCR24 and DHCR7) have two closely spaced SREs in their human promoters. Furthermore, in at least three cases the dual SREs function cooperatively to upregulate gene expression. This means that a higher threshold of SREBP is required to activate these genes (Figure 15(B)) so that the expensive process of cholesterol synthesis is not inappropriately induced. This is in contrast to the LDLR that has only one SRE and a linear response to SREBP levels, meaning that it is more quickly upregulated, and cholesterol can be acquired through the energetically 'cheaper' uptake of LDL.

7.2 Posttranslational Regulation

7.2.1 Enzyme Degradation

The proteolytic degradation of the key control enzymes HMGCR and SM is a mechanism that leads to a rapid reduction of cholesterol synthesis (Figure 13). Both of these membrane-bound enzymes are degraded by the proteasome in response to sterol excess; however, the factors that target the enzyme for degradation are different (Sharpe et al., 2014). First, sterols signal for the enzymes to be ubiquitylated by an E3 ligase. In this process, the 76-amino acid protein ubiquitin is attached to the target protein, which is subsequently poly-ubiquitylated, marking it for degradation by the proteasome. In the case of SM, the E3 ligase is MARCH6, but for HMGCR, several E3 ligases have been proposed, including gp78, TRC8, HRD1 and also MARCH6. An intriguing difference between sterol-stimulated degradation of HMGCR and SM is that degradation of the former is stimulated by oxysterols and metabolic intermediates of the mevalonate pathway, whereas SM degradation is accelerated by cholesterol. This may allow differential production of cholesterol and nonsterol products, since HMGCR acts prior to the isoprenoid (nonsterol) branch point, whereas SM acts at a later stage from which mostly sterols (including cholesterol) are formed. In contrast to SM and HMGCR, the final enzyme in the Bloch pathway, DHCR24, is remarkably stable under changing sterol conditions, although its activity is affected by other posttranslational mechanisms as described in the following section (Luu et al., 2014).

7.2.2 Phosphorylation

The addition of a phosphate group to serine, threonine or tyrosine can have significant effects on enzyme activity. While it is known from large-scale mass spectrometry studies that many cholesterol synthesis enzymes are phosphorylated, very little has been done to determine the functional consequences of this modification. Only three enzymes have more detailed information.

HMGCR was one of the first substrates identified for AMP kinase[11], and this phosphorylation inhibits its activity. AMP kinase is activated by AMP levels and in turn phosphorylates serine residue 872 on HMGCR, thereby inhibiting its activity. It can also be dephosphorylated by protein phosphatase 2A (PP2A), which fully restores enzyme activity.

On the other hand, phosphorylation of DHCR24 by an as yet unknown kinase enhances its activity (Luu et al., 2014). Mutation of any of three known phosphorylation sites (T110, Y299 or Y507) resulted in decreased DHCR24 activity. Additionally, the inhibition of protein kinase C had a similar effect, although the specific target residues remain unknown.

LBR is phosphorylated by Cdc2 kinase at S71, which regulates its major role in chromatin binding. However, it is not known whether this also affects its additional role in cholesterol synthesis (Sharpe and Brown, 2013).

7.2.3 Competitive Inhibition

Enzyme activity can be modulated by competitive inhibitors. Statins are a prime example of how this has been exploited to control hypercholesterolemia. Statins

---

11. Originally named HMGCR kinase (Beg et al., 1973).

are structural analogs of HMG-CoA and competitively inhibit the reduction of HMG-CoA to mevalonic acid. In a more recent example, 24,25EC competitively inhibits the activity of DHCR24. Inhibition was also observed with other sterols with side-chain modifications (e.g. 25-hydroxycholesterol and 27-hydroxycholesterol), which makes sense since this is where DHCR24 acts on the sterol. It is likely that similar examples exist for other enzymes, and there are a number of drugs that have been developed to target specific enzymes (e.g. AY-9944 for DHCR7 and LEK-935 for LDM), although these are not used clinically.

7.3 Other Modes of Regulation

There is a growing realisation that the 95% of the genome that is nonprotein coding harbours a growing list of noncoding RNAs that modulate a variety of biological processes including cholesterol homeostasis (Sharpe and Brown, 2013). Although so far none appear to directly regulate cholesterol synthesis, it is likely just a matter of time before one or more such noncoding RNAs are discovered. However, micro RNA 33a/b (miR-33), which is intronic to SREBP, inhibits the expression of genes such as ABCA1 and ABCG1, thereby decreasing cholesterol efflux. This means that when SREBP is activated to increase cholesterol synthesis, miR-33 is also activated to decrease cholesterol efflux, leading to more efficient stabilisation of cholesterol levels.

As with most proteins, multiple isoforms exist for many cholesterol synthesis enzymes as a result of alternative splicing. These may have different functions, as has been demonstrated for HMGCR. A shorter isoform of HMGCR, in which exon 13 is skipped, is inactive, with proportionally more of this inactive isoform expressed in response to higher levels of sterols, as yet another mode of feedback control (Medina and Krauss, 2013).

8. SUMMARY

Sterols are vital hallmarks of eukaryotic life. Fungicides target sterol synthesis, as do statins, the drugs used to lower cholesterol levels in humans. Largely due to the popularity of the statins, the control of sterol synthesis has been synonymous with regulation at only HMGCR out of more than 20 enzymes involved. Yet, there is growing evidence that other enzymes are also regulated and rate limiting, and relatedly that intermediates in cholesterol synthesis accumulate under various conditions and have important regulatory functions distinct from those of cholesterol. In light of these discoveries, important questions remain to be addressed. How are other enzymes in cholesterol synthesis regulated to control production of cholesterol and important bio-active intermediates? Are other enzymes appropriate therapeutic targets? What is the functional relevance of having two pathways (Bloch versus Kandutsch–Russell) converging to produce cholesterol? These and other questions remind us that there is still much to learn about this fascinating and vital pathway.

ACKNOWLEDGEMENTS

We thank members of the Brown Lab and students in the UNSW Human Biochemistry Course for their constructive criticism of this chapter. We are also incredibly grateful to all the researchers past and present who contributed to the knowledge that underpins our current understanding of cholesterol synthesis, but who could not be cited due to reference constraints.

REFERENCES

Ačimovič, J., Rozman, D., 2013. Steroidal triterpenes of cholesterol synthesis. Molecules 18, 4002–4017.

Beg, Z.H., Allmann, D.W., Gibson, D.M., 1973. Modulation of 3-hydroxy-3-methylglutaryl coenzyme A reductase activity with cAMP and wth protein fractions of rat liver cytosol. Biochem. Biophys. Res. Commun. 54, 1362–1369.

Bloch, K., 1965. The biological synthesis of cholesterol. Science 150, 19–28.

Brown, A.J., Ikonen, E., Olkkonen, V.M., 2014. Cholesterol precursors: more than mere markers of biosynthesis. Curr. Opin. Lipidol. 25, 133–139.

Brown, A.J., Jessup, W., 2009. Oxysterols: sources, cellular storage and metabolism, and new insights into their roles in cholesterol homeostasis. Mol. Aspects Med. 30, 111–122.

Brown, M.S., Goldstein, J.L., 1985. A receptor-mediated pathway for cholesterol homeostasis. Nobel Lect. 284–324.

Bucher, N.L., Overath, P., Lynen, F., 1960. Beta-hydroxy-beta-methyl-glutaryl coenzyme A reductase, cleavage and condensing enzymes in relation to cholesterol formation in rat liver. Biochim. Biophys. Acta 40, 491–501.

Diels, O., 1951. Description and importance of the aromatic basic skeleton of the steroids. Nobel Lect. 259–264.

Dietschy, J.M., 1984. Regulation of cholesterol metabolism in man and in other species. Klin. Wochenschr. 62, 338–345.

Dietschy, J.M., Turley, S.D., 2002. Control of cholesterol turnover in the mouse. J. Biol. Chem. 277, 3801–3804.

Du, X., Pham, Y.H., Brown, A.J., 2004. Effects of 25-hydroxycholesterol on cholesterol esterification and SREBP processing are dissociable: implications for cholesterol movement to the regulatory pool in the endoplasmic reticulum. J. Biol. Chem. 279, 47010–47016.

Endo, A., 2010. A historical perspective on the discovery of statins. Proc. Jpn. Acad. Ser. B Phys. Biol. Sci. 86, 484–493.

Faust, P.L., Kovacs, W.J., 2014. Cholesterol biosynthesis and ER stress in peroxisome deficiency. Biochimie 98, 75–85.

Frantz Jr., I.D., Moore, R.B., 1969. The sterol hypothesis in atherogenesis. Am. J. Med. 46, 684–690.

Galea, A.M., Brown, A.J., 2009. Special relationship between sterols and oxygen: were sterols an adaptation to aerobic life? Free Radic. Biol. Med. 47, 880–889.

Galman, C., Angelin, B., Rudling, M., 2005. Bile acid synthesis in humans has a rapid diurnal variation that is asynchronous with cholesterol synthesis. Gastroenterology 129, 1445–1453.

Gaylor, J.L., 2002. Membrane-bound enzymes of cholesterol synthesis from lanosterol. Biochem. Biophys. Res. Commun. 292, 1139–1146.

Gelissen, I.C., Brown, A.J., 2011. Drug targets beyond HMG-CoA reductase: why venture beyond the statins? Front. Biol. 6, 197–205.

Gill, S., Chow, R., Brown, A.J., 2008. Sterol regulators of cholesterol homeostasis and beyond: the oxysterol hypothesis revisited and revised. Prog. Lipid Res. 47, 391–404.

Goldstein, J.L., Debose-Boyd, R.A., Brown, M.S., 2006. Protein sensors for membrane sterols. Cell 124, 35–46.

Kandutsch, A.A., Russell, A.E., 1960. Preputial gland tumor sterols. 3. A metabolic pathway from lanosterol to cholesterol. J. Biol. Chem. 235, 2256–2261.

Kandutsch, A., Chen, H., Heiniger, H., 1978. Biological activity of some oxygenated sterols. Science 201, 498–501.

Lewington, S., Whitlock, G., Clarke, R., Sherliker, P., Emberson, J., Halsey, J., Qizilbash, N., Peto, R., Collins, R., 2007. Blood cholesterol and vascular mortality by age, sex, and blood pressure: a meta-analysis of individual data from 61 prospective studies with 55,000 vascular deaths. Lancet 370, 1829–1839.

Li, J.J., 2009. Triumph of the Heart: The Story of the Statins. Oxford University Press, Oxford.

Lingwood, D., Simons, K., 2010. Lipid rafts as a membrane-organizing principle. Science 327, 46–50.

Luu, W., Zerenturk, E.J., Kristiana, I., Bucknall, M.P., Sharpe, L.J., Brown, A.J., 2014. Signaling regulates activity of DHCR24, the final enzyme in cholesterol synthesis. J. Lipid Res. 55, 410–420.

Medina, M.W., Krauss, R.M., 2013. Alternative splicing in the regulation of cholesterol homeostasis. Curr. Opin. Lipidol. 24, 147–152.

Prabhu, A.V., Sharpe, L.J., Brown, A.J., 2014. The sterol-based transcriptional control of human 7-dehydrocholesterol reductase (DHCR7): evidence of a cooperative regulatory program in cholesterol synthesis. Biochim. Biophys. Acta 1841, 1431–1439.

Schoenheimer, R., Breusch, F., 1933. Synthesis and destruction of cholesterol in the organism. J. Biol. Chem. 103, 439–448.

Sharpe, L.J., Brown, A.J., 2013. Controlling cholesterol synthesis beyond 3-hydroxy-3-methylglutaryl-CoA reductase (HMGCR). J. Biol. Chem. 288, 18707–18715.

Sharpe, L.J., Cook, E.C.L., Zelcer, N., Brown, A.J., 2014. The UPS and downs of cholesterol homeostasis. Trends Biochem. Sci. 39, 527–535.

Song, Y., Kenworthy, A.K., Sanders, C.R., 2014. Cholesterol as a co-solvent and a ligand for membrane proteins. Protein Sci. 23, 1–22.

Spencer, T.A., Gayen, A.K., Phirwa, S., Nelson, J.A., Taylor, F.R., Kandutsch, A.A., Erickson, S.K., 1985. 24(S),25-Epoxycholesterol. Evidence consistent with a role in the regulation of hepatic cholesterogenesis. J. Biol. Chem. 260, 13391–13394.

Waterham, H.R., Hennekam, R.C., 2012. Mutational spectrum of Smith-Lemli-Opitz syndrome. Am. J. Med. Genet. Part C Semin. Med. Genet. 160C, 263–284.

Wong, J., Quinn, C.M., Gelissen, I.C., Brown, A.J., 2008. Endogenous 24(S),25-epoxycholesterol fine-tunes acute control of cellular cholesterol homeostasis. J. Biol. Chem. 283, 700–707.

Chapter 12

Bile Acid Metabolism

Paul A. Dawson
Division of Pediatric Gastroenterology, Hepatology and Nutrition, Department of Pediatrics,
School of Medicine, Emory University, Atlanta, GA, USA

ABBREVIATIONS
ASBT Apical sodium-dependent bile acid transporter
ABC ATP binding cassette
AKR1C4 Aldo-keto reductase 1C4 (AKR1C4; 3α-hydroxysteroid dehydrogenase)
AKR1D1 Aldo-keto reductase 1D1 (ΛKR1D1; Δ4-3-oxosteroid 5β-reductase)
BAAT Bile acid: amino acid *N*-acyltransferase
BSEP Bile salt export pump
CMC Critical micellar concentration
CREB cAMP response element-binding protein
CTX Cerebrotendinous xanthomatosis
CYP27A1 Sterol 27-hydroxylase
CYP46A1 Cholesterol 24-hydroxylase
CYP7A1 Cholesterol 7α-hydroxylase
CYP7B1 Oxysterol 7α-hydroxylase
CYP8B1 Sterol 12α-hydroxylase
FATP5 Fatty acid transport protein 5 (SLC27A5)
FXR Farnesoid X receptor
GLP-1 Glucagon-like peptide-1
HSD3B7 3β-Hydroxy-Δ5-C27-steroid oxidoreductase
IBABP Ileal bile acid binding protein
MRP Multidrug resistance protein
NTCP Na^+-taurocholate cotransporting polypeptide
OATP Organic anion transporting polypeptide
OST Organic solute transporter
PXR Pregnane X receptor
VDR Vitamin D receptor

1. INTRODUCTION

Bile acids have been a source of fascination to chemists and biochemists since investigations of biliary constituents began in the first decade of the nineteenth century and possibly earlier. Over the past 200 years, much has been learnt regarding the function, structure, physical properties and metabolism of these

Biochemistry of Lipids, Lipoproteins and Membranes. http://dx.doi.org/10.1016/B978-0-444-63438-2.00012-2

essential detergents and signalling molecules. Bile acids are stable and indigestible molecules that perform a variety of indispensable functions in the liver and gastrointestinal tract. Although best known for their ability to form micelles and facilitate absorption of lipids in the gut, the physiological functions of bile acids extend well beyond their role as simple detergents. The major functions of bile acids include: (1) inducing bile flow and hepatic secretion of biliary lipids (phospholipid and cholesterol); (2) digestion and absorption of dietary fats, particularly cholesterol and fat-soluble vitamins; (3) regulation of cholesterol homoeostasis, by which bile acids promote cholesterol intake by facilitating intestinal cholesterol absorption while also promoting cholesterol elimination as water-soluble end products of cholesterol catabolism; and (4) gut antimicrobial defence through direct bacteriostatic actions of bile acid–fatty acid mixed micelles in the proximal intestine and by inducing expression of antimicrobial genes in the distal small intestine.

Bile acids are planar amphipathic compounds possessing a characteristic four-ring perhydrocyclopentanophenanthrene nucleus and a carbon side chain attached to the D ring (Figure 1). In all vertebrates examined, bile acids are synthesised from cholesterol, thereby converting poorly soluble membrane lipids into water-soluble derivatives that are excreted in bile. As a group, these molecules are called 'bile salts' or 'bile acids' and are included in the ST04 category (sterol lipids: bile acids and derivatives) in the LIPID MAPS Lipid comprehensive classification system (http://www.lipidmaps.org/). In nature, bile salts consist mainly of sulphate conjugates of bile alcohols and of taurine

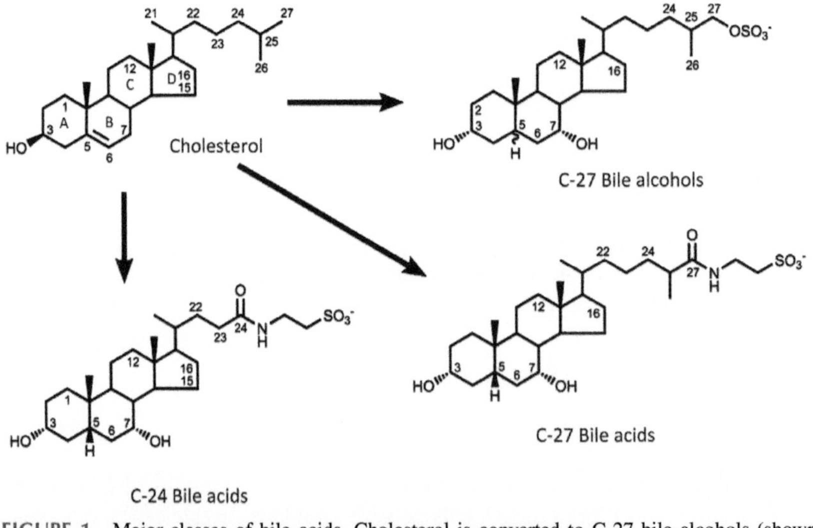

FIGURE 1   Major classes of bile acids. Cholesterol is converted to C-27 bile alcohols (shown as the sulphate conjugate), C-27 bile acids (shown as the taurine conjugate) and C-24 bile acids (shown as the taurine conjugate). The default structure is shown with 3α- and 7α-hydroxyl groups for all three classes. *From Hofmann and Hagey (2008). Adapted with permission from the publisher.*

(or glycine) amino-acyl amidated conjugates of bile acids. Conjugation with these hydrophilic moieties results in formation of molecules that are membrane-impermeable, an essential property that allows bile salts to be concentrated in bile and maintained at high concentration in the lumen of the small intestine. Most bile salts can be assigned to one of three general classes, based on the type of terminal polar group (alcohol vs acid) and the length of their side chain (27-carbon vs 24-carbon). The three major classes are 27-carbon (C-27) bile alcohols, C-27 bile acids and 24-carbon (C-24) bile acids; the general structure of their steroid nucleus, side chain and the position of their hydroxyl groups are shown in Figure 1. This chapter focuses primarily on C-24 bile acids (with a five carbon side chain and carboxyl group at the C-24 position), which are the predominant form of bile salts in most mammals, including humans (Hofmann et al., 2010).

2. BILE ACID STRUCTURE AND PHYSICAL PROPERTIES

Cholesterol is a C-27 sterol with a hydroxyl group at the C-3 position, a double bond at the C 5/6 position, and a steroid nucleus consisting of four fused carbon rings (designated A, B, C and D). During its conversion to bile acids, cholesterol undergoes changes that include hydroxylation at the C-7 position, shortening of the 8-carbon isooctane side chain to a 5-carbon isopentanoic acid side chain, epimerization of the C-3 hydroxyl group to a 3α conformation, and reduction of the double bond to yield a structure in which the A and B rings are oriented in a cis configuration. The resultant products, typically called 5β bile acids, are the major bile acid species in mammals, although bile acids with their A and B rings in a trans configuration (called 5α or allo bile acids) are present in some lower vertebrates. A comparison of the structure of cholesterol and a common bile acid, cholic acid, is shown in Figure 2. For bile acids, the carboxyl group at the C-24 position and the various hydroxyl groups often reside on the same side of the molecule, thereby forming a hydrophilic surface that opposes the hydrophobic face. As a result, bile acids are amphipathic molecules and may have potent detergent properties.

The names, abbreviations and structures of bile acids commonly found in mammals are shown in Figure 3(A). The relative hydrophilicity/hydrophobicity is also indicated for some of the most abundant bile acid species (Figure 3(B)). The molecular structure of an individual bile acid species determines its physical properties, metabolism, signalling potential and function. For example, the aqueous solubility of a bile acid is influenced by the position, orientation, and number of hydroxyl groups such that hydrophilicity decreases (hydrophobicity increases) in the progression from trihydroxy bile acids such as muricholic acid to the monohydroxy bile acid lithocholic acid (Figure 3(B)). The side chain of bile acids is typically linked by a covalent bond to an amino acid, taurine or glycine.

This amino acyl conjugation also affects the aqueous solubility of bile acids, with taurine conjugates being more soluble than glycine conjugates, which in

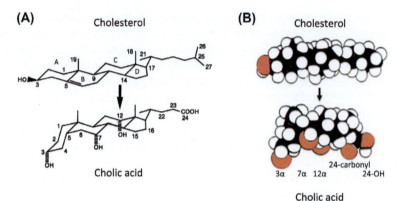

FIGURE 2 Conversion of cholesterol to bile acid. (A) In the pathway for bile acid synthesis, the stereo-configuration of the cholesterol steroid ring structure is altered, and the side chain is shortened (for C-24 bile acids). Saturation of the cholesterol double bond induces a cis-configuration at the junction of the A/B rings, resulting in a kink in the planar steroid nucleus. (B) Space-filling models of cholesterol and a bile acid (cholic acid) are shown, with carbon atoms indicated in black, hydrogen atoms in white and oxygen atoms in red. The positions of the 3α-hydroxyl, 7α-hydroxyl, 12α-hydroxyl, 24-carbonyl and 24-hydroxyl groups are indicated. Bile acids are amphipathic molecules with the hydroxyl groups aligning to form a hydrophilic surface opposing the hydrophobic face of the carbon steroid nucleus. *From Li and Chiang (2014). Adapted with permission from the publisher.*

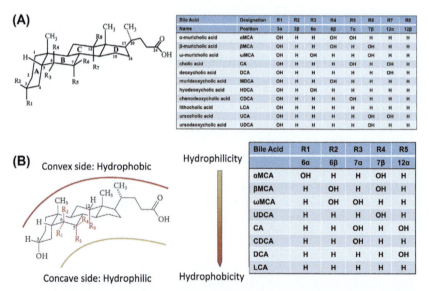

FIGURE 3 Structure and relative hydrophobicity of the common bile acids. (A) Structure of the most common bile acids in humans and rodents. (B) Structure and hydrophobicity of bile acids. Hydroxyl groups that are oriented in the α-orientation are located below the steroid nucleus and are axial to the plane of the steroid nucleus. Hydroxyl groups that are in the β-orientation are located above the steroid nucleus and are equatorial to the plane of the steroid nucleus. *From Thomas et al. (2008). Adapted with permission from the publisher.*

turn are more soluble than unconjugated bile acids. The solubility of bile acids will depend on their ionisation state, which is a function of their pKa (acid dissociation constant; degree of ionisation), and their aggregation state. The pKa values for bile acids range from ~5 for unconjugated bile acids to ~3.8 for glycine conjugates, and less than 2 for taurine conjugates. As a result, a significant fraction of unconjugated bile acids is protonated and insoluble under the pH conditions present in the small intestine during digestion (pH values of approximately 6–7), whereas most conjugated bile acids will be ionised and present in aqueous solution as sodium salts. Above a certain threshold concentration in aqueous solution called the critical micellar concentration (CMC), bile acid molecules will stack to form macromolecular aggregates or micelles. In vivo, bile acids preferentially complex with other polar lipids to form so-called mixed micelles. As with their aqueous solubility, the structure of the bile acid also affects its CMC, the size of the micelle that forms and the number of bile acid molecules per micelle. In fact, the capacity of the bile acid-containing micelle to incorporate other lipids varies directly with bile acid hydrophobicity, such that micelles containing hydrophobic bile acids such as glycodeoxycholate are able to incorporate more cholesterol than micelles containing hydrophilic bile acids such as tauro-α-muricholate, which is a relatively poor detergent.

3. BIOSYNTHESIS OF BILE ACIDS

Hepatic bile acid synthesis is the major pathway for cholesterol catabolism. In species such as humans and hamsters, only a minor fraction of newly synthesised cholesterol is used directly for hepatic bile acid synthesis. Most is drawn from the large hepatic pool that includes lipoprotein-derived cholesterol, thereby linking the pathways for hepatic metabolism of lipoproteins and cholesterol with bile acid biosynthesis. In a 70-kg human, approximately 0.5 g of cholesterol is converted to bile acids each day (~7 mg per day per kg body weight) to balance the loss of bile acids in the faeces. This accounts for more than 90% of the breakdown of cholesterol (vs approximately 10% for steroid biosynthesis), and faecal loss of bile acids accounts for almost half of all the cholesterol eliminated from the body each day. In mice, approximately 1.25 mg of cholesterol is converted to bile acids each day (~50 mg per day per kg body weight), accounting for a similar fraction of the daily whole body loss of cholesterol.

3.1 Biosynthetic Pathway

Bile acids synthesised by the hepatocyte are called primary bile acids to distinguish them from their metabolic products, called secondary bile acids, which are formed through a variety of reactions carried out by the gut microbiota. Bile acid synthesis is a complex process requiring 17 distinct enzymatic reactions and movement of intermediates between multiple compartments of the cell, including the cytosol, endoplasmic reticulum (ER), mitochondria and peroxisomes. It is likely that cytosolic carrier proteins and membrane transporters mediate the movement of biosynthetic intermediates into and out of these cell compartments, but those mechanisms still remain largely uncharacterised. It was originally thought that bile acids were synthesised by one major pathway, called the 'classical' or neutral pathway that favours biosynthesis of the primary bile acid cholic acid (Russell, 2003). However, subsequent studies culminated in the discovery of a second pathway, the 'alternative' or acidic pathway that favours biosynthesis of the primary bile acid chenodeoxycholic acid in humans and 6-hydroxylated bile acid species such as muricholic acid and hyocholic acid in mice and rats. The major steps and chemical modifications in the classical and alternative pathways for bile acid biosynthesis in humans are summarised in Figure 4. The biosynthetic enzymes that carry out those reactions are listed in Table 1.

In the classical pathway, modification of the steroid rings precedes side chain cleavage and is initiated by the hepatic enzyme cholesterol 7α-hydroxylase (CYP7A1). In the alternative pathway, the side chain of cholesterol is modified before the steroid ring structure is altered. The first step in the alternative pathway involves hydroxylation of the cholesterol side chain and is carried out by one of several different enzymes that are present in liver and extra-hepatic tissues such as brain and lung (Figure 4). After hydroxylation of the cholesterol side chain by C-24, C-25 or C-27 sterol hydroxylases, the products then undergo C-7 hydroxylation by distinct oxysterol 7α-hydroxylases, including CYP7B1 and to a lesser extent CYP39A1. Of these alternative hydroxylation pathways, the C-27 hydroxylation carried out by mitochondrial sterol 27-hydroxylase (CYP27A1) is quantitatively most important for bile acid synthesis. CYP27A1 is expressed in many tissues and cell types in addition to hepatocytes, and 27-hydroxylation may be a general mechanism for removal of cholesterol from cells such as macrophages. Although the 24-hydroxylation pathway is only a minor contributor to bile acid synthesis, conversion of cholesterol to 24S-hydroxycholesterol by the brain-specific cholesterol 24-hydroxylase (CYP46A1) is the major mechanism for cholesterol elimination from the brain. In that pathway, cholesterol is converted to an oxysterol product (24S-hydroxycholesterol) capable of crossing the blood–brain barrier and entering the systemic circulation, where it is carried on lipoprotein particles to the liver for catabolism to bile acids. It should also be noted that the substrate specificity of CYP46A1 is not limited to cholesterol, and the enzyme may also function in metabolism of neurosteroids or even drugs in the central nervous system (Russell et al., 2009).

In the classical pathway, CYP7A1, an ER-bound microsomal cytochrome P450 enzyme, is rate limiting for bile acid synthesis. Although cholesterol is the major in vivo substrate, in vitro studies have shown that CYP7A1 can bind and hydroxylate cholesterol, cholest-4-en-3-one and oxysterols such as 25-hydroxycholesterol and 27-hydroxycholesterol. Insight to the catalytic mechanism and substrate specificity of CYP7A1 and other cytochrome P450 sterol hydroxylases comes from resolution of the structure of human CYP7A1 by X-ray crystallography

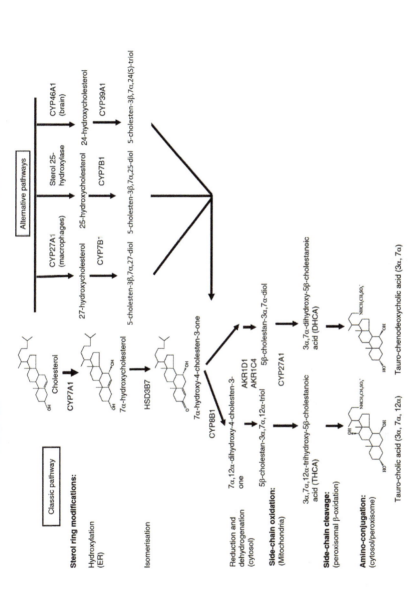

FIGURE 4 Major pathways for bile acid synthesis in humans. The major classical (also called neutral) pathway for bile acid synthesis begins with cholesterol 7α-hydroxylase (CYP7A1). Bile acid intermediates synthesised via this pathway are substrates for the sterol 12α-hydroxylase (CYP8B1), the rate-determining step in production of cholic acid. In the minor alternative (also called acidic) pathway for bile acid synthesis, cholesterol is first hydroxylated on its side chain by sterol 27-hydroxylase (CYP27A1), sterol 25-hydroxylase or sterol 24-hydroxylase (CYP46A1). Subsequent hydroxylation of the steroid nucleus is catalysed by oxysterol 7α-hydroxylase (CYP7B1) or to a lesser extent by the distinct oxysterol 7α-hydroxylase, CYP39A1. The classical and alternative pathways converge at the enzymatic step for isomerisation of the steroid ring. The alternative pathway preferentially produces chenodeoxycholic acid. After side chain oxidation and cleavage, bile acids are amino acyl-amidated to taurine or glycine. *From Chiang (2013). Adapted with permission from the publisher.*

TABLE 1 Enzymes of the Classical (Neutral) and Alternative (Acidic) Pathways for Bile Acid Synthesis

	Enzyme	Symbol	MW	Tissue	Organelle	Human Disease	Mouse KO
1	Cholesterol 7α-hydroxylase	CYP7A1	57,660	Liver	ER	Yes	Yes
2	Sterol 27-hydroxylase	CYP27A1	56,900	Many	Mito	Yes	Yes
3	Cholesterol 24-hydroxylase	CYP46A1	56,821	Brain-enriched	ER	No	Yes
4	Cholesterol 25-hydroxylase	CH25H	31,700	Many	ER	No	Yes
5	Oxysterol 7α-hydroxylase	CYP7B1	58,255	Liver-enriched	ER	Yes	Yes
6	Oxysterol 7α-hydroxylase	CYP39A1	54,129	Many	ER	No	No
7	3β-Hydroxy-Δ5-C27 steroid oxidoreductase	HSD3B7	40,929	Many	ER	Yes	Yes
8	Sterol 12α-hydroxylase	CYP8B1	58,078	Liver	ER	No	Yes
9	Δ4-3-oxosteroid 5β-reductase	AKR1D1	37,377	Many	Cytosol	Yes	No
10	3α-hydroxysteroid dehydrogenase	AKR1C4	37,095	Many	Cytosol	No	No
11	Very long chain acyl CoA synthetase; VLCS	SLC27A2	70,312	Liver, gut	Perox	No	Yes
12	Bile acid CoA synthetase; FATP5; BACS	SLC27A5	70,312	Liver	PM/ER/Perox	Yes	Yes

13	α-Methylacyl-CoA racemase; peroxisomal multifunctional enzyme 2	*AMACR*	42,359	Liver, kidney	Perox/Mito	Yes	Yes
14	Enoyl-CoA hydratase/3-hydroxy CoA dehydrogenase	*EHHADH;*	79,494	Many	Perox/Mito	Yes	Yes
15	Branched-chain acyl CoA oxidase	*ACOX2*	76,826	Many	Perox	No	No
16	D-bifunctional protein; 17β-hydroxysteroid dehydrogenase IV	*HSD17B4*	78,686	Many	Perox	Yes	Yes
17	Thiolase sterol carrier protein 2	*SCP2*	58,993	Liver-enriched	Perox	Yes	Yes
18	Bile acid CoA: Amino acid N-acyltransferase	*BAAT*	46,296	Liver	Perox	Yes	No

KO, knockout; ER, endoplasmic reticulum; mito, mitochondria; PM, plasma membrane; perox, peroxisome.

(Tempel et al., 2014). The structure revealed a typical mammalian P450 fold with the sequence organised into four β-pleated sheets and 12 α-helices, which form the core around the haem prosthetic group. With regard to mechanism, analysis of the ligand bound structures indicated that the hydrophobic side chain of the sterol substrate enters the active site first. Based on these analyses, the authors propose a model by which CYP7A1 is embedded in the outer leaflet of the ER membrane with its active site facing toward the inner leaflet. The cholesterol substrate sits in the inner leaflet of the membrane with its 3β-hydroxy group facing the ER lumen and its side chain extending toward the outer leaflet. Following changes in the membrane lipid composition or protein conformation, cholesterol enters the substrate access channel of CYP7A1 and moves into close proximity of the haem group in the active site. In the cytochrome P450 structures solved to date, this cholesterol orientation (side chain near the haem) is a general feature of the mammalian P450 enzymes that hydroxylate sterol substrates (CYP11A1, CYP46A1, CYP7A1).

Following 7α-hydroxylation, the next step for bile acid biosynthesis in the classical pathway is catalysed by the enzyme 3β-hydroxy-Δ5-C27-steroid oxidoreductase (HSD3B7). This reaction converts 7α-hydroxycholesterol to 7α-hydroxy-4-cholestene-3-one (often abbreviated as C4), which is a common precursor for cholic acid and chenodeoxycholic acid synthesis and is measured as a marker of hepatic bile acid synthesis. In the subsequent step, 7α-hydroxy-4-cholestene-3-one can be hydroxylated by the microsomal cytochrome P450 enzyme sterol 12α-hydroxylase (CYP8B1) or proceed directly to the steroid ring isomerisation and saturation reactions that are catalysed by the Δ4-3-oxosteroid 5β-reductase (aldo-keto reductase 1D1; AKR1D1) and the 3α-hydroxysteroid dehydrogenase (aldo-keto reductase 1C4; AKR1C4). If 7α-hydroxy-4-cholestene-3-one is first hydroxylated by CYP8B1, the 7α,12α-hydroxy-4-cholestene-3-one product is converted in subsequent steps to a trihydroxy bile acid (cholic acid). In contrast, bypassing the step catalysed by CYP8B1 leads to formation of a dihydroxy bile acid. Indeed, CYP8B1 is a major determinant of the ratio of cholic acid to chenodeoxycholic acid in the bile acid pool of humans. In mice, CYP8B1 controls the ratio of cholic acid to muricholic acid, because chenodeoxycholic acid is further metabolised to muricholic acid by distinct 6-hydroxylation reactions in that species. In this fashion, CYP8B1 plays a critical role in determining the chemical and functional properties of the bile acid pool.

The classical and alternative biosynthetic pathways converge at the isomerisation step catalysed by HSD3B7, converting the 7α-hydroxylated intermediates of cholesterol or oxysterols to their 3-oxo, Δ4 forms. Although oxysterol products of the alternative pathway's early hydroxylation reactions are substrates for CYP8B1, these precursors preferentially bypass sterol 12α-hydroxylation and proceed directly to the steroid ring isomerisation and saturation reactions in vivo. As such, the alternative pathway favours synthesis of chenodeoxycholic acid or its 6-hydroxylated products over cholic acid. Following the steroid ring modifications, β-oxidation and shortening of the side chain by three carbon atoms occurs in peroxisomes (Figure 4). The di- and trihydroxycoprostanoic acid derivatives formed from cholesterol are activated to their corresponding CoA-thioesters by very-long-chain acyl-CoA synthetase (VLCS; SLC27A2; FATP2). After transport into peroxisomes, these intermediates are converted to their respective (25S)-isomers by α-methylacyl-CoA racemase (AMACR) or enoyl-CoA hydratase/3-hydroxy CoA dehydrogenase (EHHADH; L-Bifunctional protein; peroxisomal multifunctional enzyme-1). This is an indispensable step for bile acid synthesis because the subsequent enzyme, branched-chain acyl-CoA oxidase (ACOX2), accepts only (S)-isomers as substrates. D-bifunctional protein (HSD17B4) then catalyses the subsequent hydration and dehydrogenation reactions prior to side chain cleavage by thiolase 2/sterol carrier protein 2. Finally, bile acids are conjugated to glycine or taurine in a reaction catalysed by bile acid-CoA:amino acid N-acyltransferase (BAAT). Interestingly, there are multiple enzymes with bile acid-CoA ligase activity capable of producing bile acid CoA-thioesters (Hubbard et al., 2006). During bile acid biosynthesis, the bile acid-CoA thioester intermediate is synthesised by VLCS (SLC27A2; FATP2). In contrast, a distinct enzyme bile acid CoA synthetase (BACS; SLC27A5; FATP5) is used to generate CoA-thioesters for the previously synthesised unconjugated bile acids that have returned to the liver in the enterohepatic circulation. Regardless of the source of the bile acid-CoA thioester, this step is considered rate limiting in bile acid amidation, and only a single enzyme (BAAT) catalyses the next step. The conjugation step mediated by BAAT is remarkably efficient, as more than 98% of bile acids secreted by the liver into bile are amidated to taurine or glycine. In most nonmammalian species, the amino acid primarily used for conjugation is taurine. However, bile acids may be amino acylated with taurine or glycine in mammals (Moschetta et al., 2005). In humans, the ratio of glycine to taurine-conjugated bile acid in the bile acid pool is approximately three to one, whereas in mice, bile acids are almost exclusively conjugated to taurine. Indeed, the taurine/glycine conjugation pattern varies considerably between different mammalian species, ranging from almost exclusively taurine in the rat, mouse, sheep and dog, to glycine greater than taurine in the pig, hamster, guinea pig, and human, to exclusively glycine in the rabbit. This specificity is controlled by the BAAT enzyme and, to a lesser degree, availability of the taurine precursor As noted above, taurine conjugated bile acids have a lower pKa than their respective glycine conjugates and are more likely to remain ionised and membrane-impermeant. However, both the glycine and taurine amide linkages are relatively resistant to cleavage by the pancreatic carboxypeptidases that would be encountered by bile acids in the digestive process, as compared to other amino acids. As such, the evolutionary and functional significance underlying selection of the particular amino acids used for conjugation is unknown. Regardless of whether the bile acids are conjugated to taurine or glycine, the physiological consequence of conjugation is decreased passive diffusion across cell membranes, increased water-solubility at acidic pH, and increased resistance to precipitation in the presence of high concentrations of calcium as compared to unconjugated bile acid species. These properties are critical for maintaining the high concentration of bile acids required to facilitate lipid digestion and absorption.

Although humans typically synthesise approximately 0.5 g of bile acids per day to maintain the bile acid pool, the maximal rate of bile acid synthesis can be as high as 4–6 g per day. With regard to the source of these bile acids, the classical (CYP7A1) pathway is quantitatively more important than the alternative pathway in rodents and humans and accounts for the bulk of the bile acids made by the liver. These estimates are derived from in vivo biochemical tracer studies using labelled precursors selective for the individual biosynthetic pathways and are strongly supported by evidence from human subjects with inborn errors in bile acid biosynthesis and from genetically modified mouse models. In mice, the classical pathway accounts for ~70% of total bile acid synthesis in adults and is the predominant pathway in neonates. In humans, biochemical tracer studies suggested that the classical pathway accounts for more than 90% of bile acid synthesis. That estimate is in good agreement with the finding that fecal levels of bile acids were reduced more than 90% in an adult human subject with an inherited *CYP7A1* deficiency. In contrast to mice, the alternative pathway appears to be the predominant biosynthetic pathway in human neonates (Schwarz et al., 1997). Indeed, *CYP7A1* activity is low or absent in newborns. As such, infants with inherited *CYP7B1* defects are unable to synthesise mature bile acids and present with severe life-threatening liver disease.

3.2 Inherited Defects in Bile Acid Synthesis

Although inherited defects in bile acid biosynthesis are rare, the study of these disorders has played an important role in advancing our understanding of bile acid metabolism. The findings have also been key for determining the human relevance of experimental observations from mice engineered to knock out important genes in bile acid biosynthesis. The severe phenotype that is often associated with these inborn errors, which includes liver failure and death in early childhood, illustrates the importance of bile acids for hepatic and intestinal function. Indeed, the pathophysiological effects include reduction of bile acid pool size, loss of bile acid-dependent bile flow, decreased biliary excretion of cholesterol and xenobiotics, reduced intestinal absorption of cholesterol and fat-soluble vitamins, and accumulation of cytotoxic bile acid biosynthetic intermediates. Disease-associated mutations in humans have been identified in nine of the 18 enzymes involved in bile acid biosynthesis (Table 1). Many of these genes have also been knocked out in mice. The list of genes with inborn errors includes cholesterol 7α-hydroxylase (*CYP7A1*), sterol 27-hydroxylase (*CYP27A1*), oxysterol 7α-hydroxylase (*CYP7B1*), 3β-hydroxy-Δ5-C27-steroid oxidoreductase (*HSD3B7*), Δ4-3-oxosteroid 5β-reductase (*AKR1D1*), α-methylacyl-coenzyme A racemase (*AMACR*),

D-bifunctional protein (*HSD17B4*), bile acid-CoA ligase (*SLC27A5*), and *BAAT* (Clayton, 2011). In addition, disorders that disrupt peroxisome biogenesis such as Zellweger syndrome, neonatal adrenoleucodystropy and infantile Refsum disease also affect bile acid synthesis. These defects in genes required for peroxisome formation indirectly affect the expression and activity of peroxisomal enzymes, including those involved in the final enzymatic side chain cleavage steps of bile acid biosynthesis. As such, patients with peroxisome biogenesis disorders accumulate atypical mono-, di- and tri-hydroxy C-27 bile acids.

A single enzyme defect is not generally sufficient to eliminate production of all bile acids, because multiple biosynthetic pathways exist. However, the result is often a severe reduction in the levels of normal primary bile acids and the appearance of bile acid biosynthetic intermediates and abnormal bile acid species, which can be detected in serum or urine by methods such as fast atom bombardment ionisation-mass spectrometry and gas chromatography-mass spectrometry. Depending on the step in the biosynthetic pathway and the nature of the mutation (i.e. missense vs nonsense mutation or deletion), the consequences of these enzyme defects are varied, with the most severe producing neonatal cholestatic liver disease or neurological disease later in life. The effects of inherited defects in *CYP7A1* and *CYP7B1* on bile acid synthesis were introduced in Section 3.2. Only a few human subjects homozygous for a *CYP7A1* deficiency have been identified and the natural course of that disorder has not been described. Those adult *CYP7A1*-deficient patients presented with hyperlipidaemia that was resistant to treatment with an HMG-CoA reductase inhibitor (statin), but were otherwise asymptomatic. Despite the absence of detectable *CYP7A1* protein, a low level of cholesterol 7α-hydroxylase activity could still be detected in the liver biopsy from an affected subject, suggesting an additional enzyme may be partially compensating in humans. This is not the case in mice, in which genetic inactivation of *CYP7A1* results in complete loss of hepatic cholesterol 7α-hydroxylase activity. *CYP7B1* deficiency is associated with a serious and often fatal neonatal liver disease. It should be noted that mutations in *CYP7B1* are also a rare cause of spastic paraplegia (spastic paraplegia-5A, SPG5A), a recessive neurological disorder. However, the pathogenesis of this progressive motor-neuron degenerative disease is likely due to loss of *CYP7B1*-dependent metabolism of cholesterol and neurosteroids in the central nervous system rather than alterations in bile acid synthesis.

The most commonly reported inherited defect in bile acid synthesis is *HSD3B7*-deficiency, which affects both the classical and alternative pathways. *HSD3B7*-deficiency is characterised by progressive intrahepatic cholestasis, and the clinical manifestations include unconjugated bilirubinemia, jaundice, elevated serum aminotransferase levels, steatorrhoea, fat-soluble vitamin deficiency, pruritus, and poor growth. Neonatal cholestasis has also been reported for inherited defects in *AKR1D1*, and these patients present with low or absent primary bile acids and elevated levels of bile acid biosynthetic intermediates.

Cerebrotendinous xanthomatosis (CTX) is a rare, inherited disease caused by mutations in the mitochondrial enzyme *CYP27A1*. In CTX, bile acid synthesis via the alternative pathway is greatly reduced, and synthesis via the classical pathway is rerouted such that primary bile acid production is diminished but not eliminated. Although CTX is associated with a period of neonatal cholestasis, its pathophysiology is mainly related to deposition of cholesterol and oxysterols such as cholestanol in tissues, especially the tendons and brain. As a result the major clinical manifestations present later, typically in the second or third decade of life, and are characterised by a slow progressive neurological dysfunction, cerebellar ataxia, premature atherosclerosis, cataracts and tendinous xanthomas. Finally, defects in the enzymes responsible for bile acid conjugation, bile acid CoA synthetase (BACS, also called BACL; SLC27A5) and bile acid CoA: amino acid *N*-acyltransferase (BAAT), have been identified and characterised. Patients with these enzyme defects present with high serum levels of unconjugated bile acids and malabsorption of dietary fat and fat-soluble vitamins, illustrating the importance of conjugation to maintain the high intraluminal concentrations of bile acids required for efficient lipid digestion and absorption. Early bile acid replacement therapy with bile acids such as cholic acid and chenodeoxycholic acid (or their conjugates in the case of bile acid conjugation defects) can suppress or even reverse the biochemical abnormalities associated with many of the inherited bile acid biosynthetic defects and may be life-saving. The basis for this therapeutic approach is two-fold. First, administration of exogenous primary bile acids partially restores the endogenous bile acid pool to drive hepatic bile flow and facilitate intestinal lipid absorption. Second, administration of exogenous cholic acid or chenodeoxycholic acid inhibits endogenous bile acid synthesis through feedback mechanisms described in Section 3.3. This blocks production and accumulation of the cytotoxic biosynthetic intermediates and abnormal bile acid species that are involved in the disease pathogenesis or progression.

3.3 Regulation of Bile Acid Biosynthesis

The biosynthesis of bile acids is tightly controlled and regulated by bile acids, oxysterols, cytokines, nutrients, growth factors, diurnal rhythm and hormones such as thyroid hormone, glucocorticoids and insulin (Chiang, 2013). These various factors operate primarily by controlling expression of *CYP7A1*, the rate-limiting step in bile acid biosynthesis. This section will focus on the negative feedback regulation of *CYP7A1* by bile acids, a major physiological pathway for maintaining bile acid homoeostasis. Studies dating from the 1960s showed that *CYP7A1* activity and hepatic bile acid biosynthesis increased when bile acid return to the liver was blocked by administration of bile acid sequestrants or by biliary diversion. Conversely, feeding bile acids would suppress *CYP7A1* activity and endogenous bile acid biosynthesis. This negative feedback regulation of *CYP7A1* was later shown to be transcriptional and involve the bile acid-activated nuclear receptor, farnesoid X-receptor (FXR, gene symbol: *NR1H4*). Initially, it was thought that the regulation of *CYP7A1* expression by bile acids involved a pathway by which activation of FXR in hepatocytes increased transcription of the atypical orphan nuclear receptor, small heterodimer partner (SHP, gene symbol: *NR0B2*). SHP then antagonised the actions of positive-acting transcription factors important for *CYP7A1* gene expression, such as liver receptor homologue-1 (LRH-1, gene symbol: *NR5A2*) and hepatocyte nuclear factor-4α (HNF4α, gene symbol: *NR2A1*) (Parks et al., 1999; Makishima et al., 1999). However, this model could not explain a series of puzzling experimental findings, including the observations that intravenous infusion of bile acids into bile-fistula rats was ineffective at downregulating hepatic *CYP7A1* expression compared to intraduodenal infusion of bile acids and that bile acids only weakly suppressed CYP7A1 expression when added directly to isolated hepatocytes in culture (Pandak et al., 1995).

Recent investigation culminated in the finding that the major pathway for feedback regulation of bile acid synthesis involves intestinal FXR and gut-liver signalling via the polypeptide hormone fibroblast growth factor-19 (FGF19; mouse ortholog: FGF15) (Inagaki et al., 2005). In that pathway, which is summarised in Figure 5, bile acids activate FXR in ileal enterocytes to induce synthesis of FGF15/19. After its release by the enterocyte, FGF15/19 is carried in the portal circulation to the hepatocyte, where it signals through its cell surface receptor, a complex of the βKlotho protein and fibroblast growth factor receptor-4 (FGFR4), to repress *CYP7A1* expression and bile acid synthesis. The dominant role of this pathway as the major physiological mechanism responsible for feedback repression of *CYP7A1* expression is strongly supported by results obtained using knockout mouse models, including FGFR4, βKlotho, FGF15, and tissue-specific FXR and βklotho knockout mice. There is also compelling evidence that this pathway is operative in humans, where circulating FGF19 levels inversely correlate with markers of hepatic bile acid synthesis. After binding FGF15/19, FGFR4/βKlotho signals through the docking protein fibroblast growth factor substrate 2 (FRS2α) and the tyrosine phosphatase known as tyrosine-protein phosphatase non-receptor type 11 (Shp2) to stimulate extracellular-signal-regulated kinase (ERK1/2) activity and block the activation of *CYP7A1* gene expression by HNF4α and LRH-1. The transcriptional repression likely involves SHP, because FGF15/19-mediated regulation of *CYP7A1* is blunted in SHP null mice. However, details of exactly how these factors and phosphorylation pathways interact to repress CYP7A1 transcription are still being clarified.

While *CYP7A1* is the rate-limiting enzyme for bile acid synthesis, *CYP8B1* activity determines the amount of cholic acid versus chenodeoxycholic acid (or muricholic acid) that is synthesised. Thus it is not surprising that hepatic *CYP8B1* expression is also regulated by bile acids, cholesterol, diurnal rhythm and hormones such as insulin and thyroid hormone. With regard to bile acids and in contrast to *CYP7A1*, genetic evidence in mice suggests that both the hepatic FXR-SHP pathway and the gut-liver FXR-FGF15/19-FGFR4/βKlotho

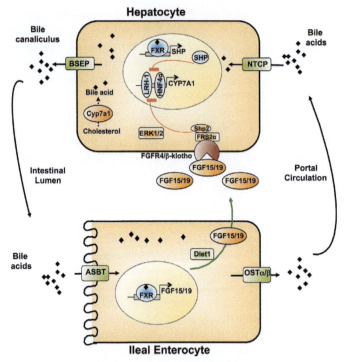

FIGURE 5 Mechanisms for feedback negative regulation of hepatic bile acid synthesis. In the major physiological pathway, intestinal bile acids are taken up by the ASBT and activate the nuclear receptor FXR to induce FGF15/19 expression in ileal enterocytes. The basolateral secretion of FGF15/19 protein may be facilitated by the endosomal membrane glycoprotein, Diet1. FGF15/19 is then carried in the portal circulation to the liver, where it binds to its cell-surface receptor, a complex of the receptor tyrosine kinase, FGFR4, and the associated protein βklotho. FGFR4/βKlotho then signals through the docking protein FRS2α and the tyrosine phosphatase Shp2 to stimulate ERK1/2 phosphorylation and block activation of CYP7A1 gene expression by the nuclear factors HNF4α and LRH1. In a direct pathway that may be more significant under pathophysiological conditions, bile acids can activate FXR in hepatocytes to induce expression of SHP, an atypical orphan nuclear receptor. SHP interacts with LRH-1 and HNF4α to block their activation of CYP7A1 transcription.

signalling pathways contribute to regulation of *CYP8B1* expression by bile acids. In this fashion, the machinery controlling bile acid synthesis and bile acid pool composition responds to changes in intestinal and hepatic bile acid levels.

In contrast to the classical (CYP7A1) pathway, bile acids do not directly regulate *CYP27A1* expression or the alternative pathway for bile acid synthesis. Instead, the alternative pathway appears to be primarily regulated by cholesterol delivery to the mitochondrial inner membrane, the site of sterol 27-hydroxylation. The mechanism may be similar to that controlling adrenal steroid hormone biosynthesis, in which transfer of cholesterol to the mitochondria inner membrane by steroidogenic acute regulatory protein (StAR; STARD1) is rate limiting. The liver expresses several members of the StAR-related lipid transfer (START) domain family of proteins, and one or more of these proteins could be involved in cholesterol delivery for bile acid synthesis (see Chapter 14). However, the molecular details of cholesterol trafficking and intramitochondrial cholesterol delivery in hepatocytes are not well understood. As such, the mechanisms regulating cholesterol flux through the alternative pathway for bile acid biosynthesis under physiological and pathophysiological states remain undefined.

3.4 Secondary Metabolism of Bile Acids

After their synthesis in the liver, the primary bile acids are secreted into bile and pass into the small intestine, where they begin to encounter the gut microbiota. In the small intestine, most of the conjugated bile acids are efficiently absorbed intact. However a fraction of the luminal bile acids in the distal small intestine undergo bacterial deconjugation (cleavage of the amide bond linking the bile acid to glycine or taurine), a process that continues to near completion in the colon. After deconjugation, the bile acids can be reabsorbed or undergo additional bacterial transformations for conversion to secondary bile acids. If reabsorbed, unconjugated bile acids are carried back to the liver in the enterohepatic circulation and efficiently reconjugation to taurine or glycine, a cycle originally termed 'damage and repair' by Hofmann (Hofmann and Hagey, 2014). The newly 'repaired' (reconjugated) bile acids are then secreted into bile along with any conjugated bile acids that have returned from the intestine in the enterohepatic circulation and any newly synthesised conjugated bile acids.

In the gut, primary bile acids are biotransformed to secondary bile acids by the microbiota through reactions that include oxidation and epimerization of the 3-, 6-, 7- or 12-hydroxyl groups to form oxo or iso epimers (e.g. epimerization of the 7α-hydroxy group of chenodeoxycholic acid to form the 3α,7β-dihydroxy bile acid, ursodeoxycholic acid), and dehydroxylation at the C-7 position (Dawson and Karpen, 2015). The common bacterial modifications of bile acids that occur in humans and mice are summarised in Figure 6. The most significant of these changes is 7-dehydroxylation, which generates the major secondary bile acid species. In humans, the process of 7-dehydroxylation converts cholic acid to deoxycholic acid, a dihydroxy bile acid with hydroxyl groups at the C-3 and C-12 positions, and converts chenodeoxycholic acid or ursodeoxycholic acid to lithocholic acid, a mono-hydroxy bile acid with a hydroxyl group at the C-3 position. In rats and mice, the 3,6,7-trihydroxy bile acids (α, β, ω-muricholic acid; hyocholic acid) are converted to the 3,6-dihydroxy bile acids, hyodeoxycholic acid (3α,6α-dihydroxy-5β-cholanoic acid) and murideoxycholic acid (3α,6β-dihydroxy-5β-cholanoic acid). There is an absolute requirement for bile acids to be deconjugated prior to 7-dehydroxylation. For that reason, deconjugation-resistant synthetic bile acids such as cholylsarcosine, where cholic acid is conjugated to sarcosine (N-methylglycine) instead of glycine or taurine, are also resistant to dehydroxylation. The secondary bile acids generated by the gut microbiota can also be absorbed and undergo

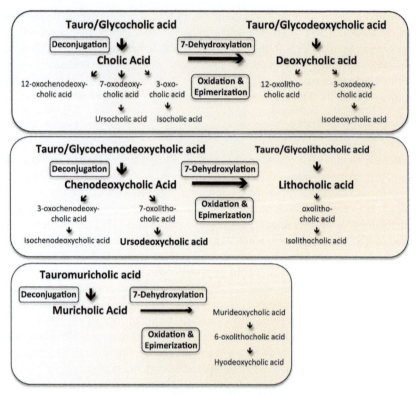

FIGURE 6   Scheme for bacterial metabolism of bile acids. Bile acids are biotransformed by the gut microbiota by reactions that deconjugation, 7-dehydroxylation, oxidation and epimerization. The major pathways and products are indicated in bold. Oxidation yields the respective oxo-bile acid species, which can be epimerised, converting the 7α-hydroxy bile acids to their 7β-hydroxy epimers (ursocholic acid, ursodeoxycholic acid), and 3α-hydroxy bile acids to their 3β-hydroxy epimers (isocholic acid, isodeoxycholic acid, isochenodeoxycholic acid and isolithocholic acid). In the human caecum, a significant fraction of the bile acids present have been converted to their respective 3β-epimers. In rats and mice, muricholic acid species (α, β, ω-muricholic acid) are converted to the 3,6-dihydroxy bile acids, murideoxycholic acid (3α,6β-dihydroxy-5β-cholanoic acid) and hyodeoxycholic acid (3α,6α-dihydroxy-5β-cholanoic acid).

'repair' by the hepatocyte. In addition to reconjugation with taurine/glycine, the hepatic repair reactions include re-epimerization of iso (3β-hydroxy) bile acids to their 3α-hydroxy form, reduction of oxo groups to hydroxyl groups, and rehydroxylation at the C-7 position to generate the original primary bile acid. Hepatic bile acid 7α-rehydroxylation is carried out to varying degrees in many species, but does not occur in humans. These interspecies differences in hepatic bile acid 7α-rehydroxylation activity are reflected in the amount of conjugated deoxycholic acid present in bile, which can range from 0% to 10% of the bile acids in rats, mice, guinea pigs, prairie dogs and hamsters, to 15–30% in dogs and humans, and greater than 90% in rabbits. The bacterial bile acid 7-dehydroxylation activity in the gut is also an important determinant of the amount of deoxycholic acid in the bile acid pool, because the activity is restricted to only a limited number of bacterial species, the abundance of which can vary many-fold between individuals and under different dietary or pathophysiological conditions.

The bacterial biotransformation of bile acids carried out in the gut is important for several reasons. First, these modifications decrease the aqueous solubility of bile acids and increase their hydrophobicity, resulting in a marked lowering of their monomeric concentration in the aqueous phase of the luminal contents. This in turn reduces intestinal bile acid absorption and increases bile acid loss in the faeces. Second, the input of secondary bile acids from the intestine influences the composition of the circulating pool of bile acids. This in turn affects the physiological and pathophysiological actions of the bile acid pool because secondary bile acids have distinct detergent properties, signalling activities and toxicities compared to their primary bile acid precursors. Third, bile acids affect the composition of the gut microbiota, with complex downstream effects on whole body metabolism and disease pathogenesis.

4. ENTEROHEPATIC CIRCULATION OF BILE ACIDS

The concept of an enterohepatic circulation by which constituents secreted by the liver into the intestine are returned via the portal circulation was postulated more than 300 years ago by Mauritius van Reverhorst and the seventeenth century Neapolitan mathematician Giovanni Alfonso Borelli (Hofmann and Hagey, 2014). Anatomically, the gut–liver circulation can be subdivided into a portal and extraportal pathway. The extraportal pathway consists primarily of the lymphatic drainage from the intestine into the superior vena cava. Although this process is important for the chylomicron lipoprotein particle-mediated transport of cholesterol, triglycerides, fat-soluble vitamins and phospholipids, it plays little role in bile acid absorption. In contrast, bile acids undergo a portal enterohepatic circulation. In that process, bile acids are secreted by the liver with other biliary constituents into the biliary tract and stored in the gallbladder (in most, but not all species). In response to a meal, the gallbladder contracts and empties its contents into the lumen of the small intestine, where bile acids function to facilitate lipid digestion and absorption. Unlike dietary lipids, there is only limited absorption of bile acids in the proximal small intestine. Most bile acids travel to the distal small intestine (ileum), where they are almost quantitatively reabsorbed and exported into the portal circulation. In the bloodstream, bile acids are bound to albumin and carried back to the liver for uptake and resecretion into bile. In adult humans, the enterohepatic circulation maintains a whole body bile acid pool size of approximately 2–4 g. The bile acid pool cycles several times per meal or about six cycles per day. Almost 95% of the bile acids that pass into the intestine are reabsorbed, such that only about

0.2–0.6 g of bile acids escape reabsorption and are eliminated in the faeces each day. Viewed as its individual components, the bile acid enterohepatic circulation consists of a series of storage chambers (the gallbladder and small intestine), valves (the sphincter of Oddi and ileocaecal valve), mechanical pumps (the hepatic canaliculi, the biliary tract, and the small intestine) and chemical pumps (primarily the hepatocyte and ileal enterocyte, with their respective transporters). The major transporter proteins responsible for the enterohepatic of bile acids are listed in Table 2 and are shown schematically in Figure 7. Their properties and roles in bile acid metabolism are discussed in more detail in the Sections 4.1 and 4.2.

4.1 Hepatic Bile Acid Transport

Bile acids are taken up from the bloodstream across the hepatocyte sinusoidal (basolateral) membrane by both sodium-dependent and sodium-independent mechanisms. The Na^+-taurocholate cotransporting polypeptide (NTCP; gene symbol: *SLC10A1*) is primarily responsible for the sodium-dependent uptake of conjugated bile acids, which constitute most of the whole body bile acid pool. In contrast, members of the organic anion transporting polypeptide (OATP) family are primarily responsible for hepatic uptake of unconjugated bile acids. NTCP is a 349-amino acid (approximately 45-kDa) polytopic membrane glycoprotein and founding member of the SLC10 family of solute carriers (SLC, solute carrier; http://slc.bioparadigms.org/), which also includes the ileal apical sodium-dependent bile acid transporter (ASBT, gene symbol: *SLC10A2*). NTCP is expressed almost exclusively by hepatocytes and functions as an electrogenic sodium-solute co-transporter. Its transport properties have been characterised extensively in vitro, and NTCP efficiently transports all the major glycine and taurine-conjugated bile acids, but only weakly transports unconjugated bile acids. A primary role for NTCP in the clearance of conjugated bile acids is strongly supported by recent genetic evidence in humans and mice. Plasma levels of conjugated bile acids were found to be dramatically elevated (25-–100-fold) in a patient with an inherited NTCP deficiency. Alternative transport mechanisms were able to maintain only a limited level of hepatic uptake and enterohepatic cycling of conjugated bile acids in the absence of NTCP. However, in contrast to the inborn errors of bile acid synthesis, the patient exhibited no jaundice, itching (pruritus), fat-malabsorption (steatorrhoea), or obvious liver disease (Vaz et al., 2015). Recent results reported for an NTCP knockout mouse model also support a primary role for NTCP in the hepatic clearance of conjugated bile acids.

Members of the OATP family mediate sodium-independent bile acid transport across the hepatocyte sinusoidal membrane and are primarily responsible for uptake of unconjugated bile acids. The bile acid-transporting OATPs are polytopic membrane glycoproteins that share no sequence identity with the

TABLE 2 Major Transport Proteins of the Enterohepatic Circulation of Bile Acids

Transport Protein (Gene)	Location	Function	Human Disease	Mouse KO
Hepatocyte				
NTCP (*SLC10A1*)	BLM	Na$^+$-dependent uptake of conjugated BAs	Hypercholanemia	Yes
OATP1B1, 1B3 (*SLCO1B1, 1B3*) Oatp1b2 (rodents)	BLM	Na$^+$-independent uptake of OAs and unconjugated BAs	Rotor syndrome	Yes
MRP3 (*ABCC3*)	BLM	ATP-dependent export of OAs and modified BAs	None	Yes
MRP4 (*ABCC4*)	BLM	ATP-dependent export of OAs and modified BAs	None	Yes
OSTα-OSTβ (*SLC51A; SLC51B*)	BLM	Na$^+$-independent export of BAs	None	Yes
FIC1 (*ATP8B1*)	CM	ATP-dependent PS flipping	Progressive familial cholestasis type 1	Yes
BSEP (*ABCB11*)	CM	ATP-dependent export of BAs	Progressive familial cholestasis type 2	Yes
MDR3 (*ABCB4*)	CM	ATP-dependent export of PC	Progressive familial cholestasis type 3	Yes

Continued

TABLE 2 Major Transport Proteins of the Enterohepatic Circulation of Bile Acids—cont'd

Transport Protein (Gene)	Location	Function	Human Disease	Mouse KO
MRP2 (*ABCC2*)	CM	ATP-dependent export of OAs and modified BAs	Dubin–Johnson syndrome	Yes
Cholangiocyte, Ileal Enterocyte, Kidney Proximal Tubule Cell				
ASBT (*SLC10A2*)	ApM	Na$^+$-dependent uptake of BAs	Primary bile acid malabsorption	Yes
IBABP (*FABP6*)	Cytosol	Cytosolic BA transport	None	Yes
OSTα-OSTβ (*SLC51A; SLC51B*)	BLM	Na$^+$-independent export of BAs	None	Yes
MRP3 (*ABCC3*)	BLM	ATP-dependent export of OAs and modified BAs	None	Yes

ApM, apical membrane; BA, bile acid; BLM, basolateral membrane; CM canalicular membrane; KO, knockout; OA, organic anion.

sodium-dependent bile acid transporters. The OATPs were originally assigned to the SLC21 transporter gene family. However, a new species-independent nomenclature system was adopted in 2004, in which OATP refers to protein isoforms (OATP1B1, OATP1B3, etc.) and SLCO ('O', referring to members of the OATP family; *SLCO1B1*, *SLCO1B3*, etc.) to the individual genes. The OATP/SLCO-type transporters constitute a large family with 11 human genes and 16 rat/mouse genes that fall within six subgroups. In humans, OATP1B1 (gene symbol: *SLCO1B1*; original protein name: OATP-C) and OATP1B3 (gene symbol: *SLCO1B3*; original protein name: OATP8) account for most hepatic sodium-independent bile acid clearance. The mouse genome encodes only a single transporter gene in the OATP1B subfamily (Oatp1b2; gene symbol: *Slco1b2*), because SLCO1B1 and SLCO1B3 arose in primates by gene duplication after divergence from rodents. In mice, Oatp1b2 accounts for most hepatic sodium-independent bile acid clearance. A primary role for OATP1B1/OATP1B3 and Oatp1b2 in the hepatic clearance of unconjugated bile acids is

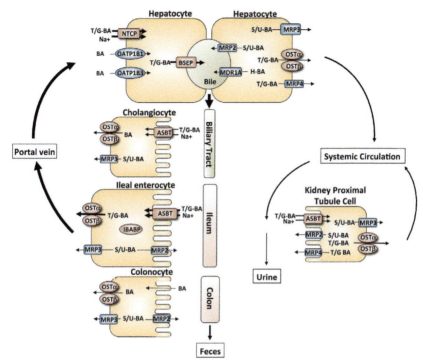

FIGURE 7   Enterohepatic circulation of bile acids showing the major tissues and transport proteins. After their synthesis or reconjugation, taurine and glycine-conjugated bile acids (T/G-BA) are secreted into bile by the canalicular bile salt export pump (BSEP). Bile acids modified by sulfation (S-BA) or glucuronidation (U-BA) are secreted into bile by the multidrug resistance-associated protein-2 (MRP2) or multidrug resistance protein-1 (MDR1A). A fraction of the bile acids secreted into bile acids undergo cholehepatic shunting. In this pathway, conjugated bile acids are taken up by cholangiocytes via the apical sodium-dependent bile acid transporter (ASBT), and exported across the basolateral membrane via the heteromeric transporter OSTα-OSTβ, and possibly MRP3, for return to the hepatocyte via the periductular capillary plexus. Ultimately, bile acids pass through the biliary tree and empty into the intestinal lumen. Bile acids are poorly absorbed in the proximal small intestine, but efficiently taken up by the ileal enterocytes via the ASBT. The bile acids bind to the ileal bile acid binding protein (IBABP) in the cytosol, and are efficiently exported across the basolateral membrane into the portal circulation by OSTα-OSTβ. MRP3 is a minor contributor to basolateral export of native bile acids from the enterocyte, but may have a more significant role in export of modified bile acids. Although most bile acids are absorbed in the small intestine, colonocytes express appreciable levels of MRP3 and OSTα-OSTβ. These carriers may serve to export unconjugated bile acids (BA) that were taken up by passive diffusion from the lumen of the colon. After their absorption from the intestine, bile acids travel back to the liver, where they are cleared by the Na$^+$-taurocholate cotransporting polypeptide (NTCP) and members of the organic anion transport protein family, OATP1B1 and OATP1B3. Under cholestatic conditions, unconjugated, conjugated, or modified bile acids can be effluxed across the basolateral (sinusoidal) membrane of the hepatocyte by OSTα-OSTβ, MRP3 or MRP4 into the systemic circulation. Under normal physiological conditions, a fraction of the bile acid escapes first pass hepatic clearance and enters the systemic circulation. The free bile acids are filtered by the renal glomerulus, efficiently reclaimed by the ASBT in the proximal tubules, and exported back into the systemic circulation, thereby minimising their excretion in the urine. A fraction of the glucuronidated or sulphated bile acids can also be exported across the apical membrane by MRP2. *From Dawson et al. (2010). Adapted with permission from the publisher.*

strongly supported by recent genetic evidence. In humans, the combined loss of *SLCO1B1* and *SLCO1B3* on chromosome 12 causes Rotor syndrome, a rare benign disorder characterised by elevated plasma levels of conjugated bilirubin and also unconjugated bile acids (van de Steeg et al., 2012). In mice, genetic deletion of the entire orthologous *Slco1a/1b* locus or *Slco1b2* alone results in defective hepatic clearance of unconjugated bile acids but not conjugated bile acids.

After their uptake, bile acids pass through the interior of the hepatocyte to the canalicular membrane for secretion into bile. Because bile acids are potent detergents, their uptake and export must be carefully balanced to avoid excess intracellular accumulation. It is also likely that bile acids are bound or sequestered in some fashion during their transit, and the evidence to date argues against a role for vesicular transport. In plasma, bile acids are transported bound to albumin, and cytosolic proteins such as members of the aldo-keto reductase (AKR) family may serve a similar role in hepatocyte intracellular transport. However, the underlying mechanisms remain poorly defined, and it is unclear whether bile acids returning in the enterohepatic circulation and newly synthesised bile acids share common transport pathways through the hepatocyte. Regardless of the mechanisms, the conjugated bile acids are shuttled to the canalicular membrane for secretion into bile by the ATP-dependent bile salt export pump (BSEP; gene symbol: *ABCB11*). This 160-kDa polytopic membrane glycoprotein is a member of the ATP binding cassette (ABC) transporter family and closely related to the multidrug resistance protein-1 (MDR1)/P-glycoprotein. When analysed in transfected mammalian cells, BSEP efficiently transports conjugated and unconjugated bile acids. However, in vivo, bile acids are first conjugated to glycine or taurine prior to secretion, as evidenced by the low proportions of unconjugated bile acids typically found in bile (less than 5%). A primary role for BSEP in the canalicular secretion of bile acids is strongly supported by genetic evidence in humans and mice. In humans, mutations in *ABCB11* are the cause of progressive familial intrahepatic cholestasis type 2 (PFIC-2), which is characterised by biliary bile acid concentrations less than 1% of normal. Mutations in *ABCB11* that impair synthesis, cellular trafficking, or stability of the BSEP protein lead to severe disease, including neonatal progressive cholestasis, jaundice, and hepatobiliary cancers. *Abcb11* deficiency also leads to progressive liver disease in mice that have been backcrossed onto a C57BL/6J background, although only a mild phenotype is observed in other background strains. In hepatocytes, a small amount of the bile acid is modified by the addition of sulphate or glucuronide and secreted into bile by other ABC transporters, primarily the multidrug resistance protein-2 (MRP2; gene symbol *ABCC2*). In mice and rats and to a lesser extent in humans, bile acids can also be modified by hydroxylation of additional sites on the steroid nucleus. These atypical bile acid species are secreted into bile by MRP2 and possibly the breast cancer-related protein (BCRP; gene symbol: *ABCG2*) or MDR1/P-glycoprotein (gene symbol: *ABCB1*).

Sinusoidal membrane bile acid transport proceeds in the direction of uptake under normal physiological conditions. However, bile acids can also exit the hepatocyte by efflux back into blood to protect the cell from bile acid overload under pathophysiological conditions in which canalicular secretion is impaired. This sinusoidal membrane efflux may also play a role under physiological conditions in the liver, analogous to the 'hepatocyte-hopping' that was recently demonstrated for conjugated bilirubin and drugs (van de Steeg et al., 2010). Under conditions of an increased bile acid load returning in the portal circulation, such as after a meal, a fraction of the bile acids absorbed by hepatocytes closest to the portal vein can be rerouted to blood by the sinusoidal membrane efflux transporters and then taken up again by downstream hepatocytes closer to the central vein for secretion into bile. This mechanism recruits additional hepatocytes within the liver lobule for bile acid clearance and could serve as an additional safeguard to protect the periportal hepatocytes from bile acid overload. The major transporters involved in hepatocyte sinusoidal membrane bile acid efflux are the ABC transporter, multidrug resistance protein-4 (MRP4; gene symbol *ABCC4*), and the heteromeric organic solute transporter (OST) OSTα-OSTβ (gene symbols: *SLC51A*, *SLC51B*). The ABC transporter multidrug resistance protein-3 (MRP3; gene symbol: *ABCC3*) may also be involved, but more so for glucuronidated or sulphated bile acids.

4.2 Intestinal Absorption of Bile Acids

Bile acids are taken up by passive mechanisms in the proximal small intestine and colon and active transport in distal ileum (Figure 7). As a result of bacterial metabolism and species differences in bile acid conjugation, the bile acid pool may include glycine conjugates, unconjugated bile acids, and more hydrophobic bile acid species. A portion of these bile acid species would be protonated in the intestinal lumen and will passively diffuse across the apical membrane of enterocytes or colonocytes. However, most bile acids in the intestinal lumen are conjugated and ionised. As such, their uptake requires a specific transporter. This active transport across the enterocyte apical brush border membrane is present only in the distal part of the small intestine (terminal ileum). Physiologically, this is important because it ensures that luminal bile acid concentrations will remain sufficiently high down the length of the small intestine to aid in fat absorption and reduce bacterial growth. In the terminal ileum, bile acids are almost quantitatively reabsorbed by the ASBT, a 348-amino acid (approximately 45-kDa) polytopic membrane glycoprotein that shares sequence identity with NTCP. ASBT also functions as an electrogenic sodium-solute co-transporter and is remarkably specific for bile acids. The ASBT transports all the major species of bile acids, including unconjugated species, but is primarily responsible for uptake of conjugated bile acids and unconjugated trihydroxy bile acids that are membrane impermeant. A primary role for the ASBT in intestinal absorption of bile acids is strongly supported by genetic evidence in humans and mice. In humans, inherited mutations in ASBT causes primary bile acid malabsorption, an idiopathic intestinal disorder associated with congenital diarrhoea, fat-malabsorption and interruption of the enterohepatic circulation of bile acids. In mice, inactivation of the ASBT largely abolished intestinal bile acid absorption (Dawson et al., 2003).

As in hepatocytes, bile acids in the ileal enterocyte are likely transported bound to cytosolic proteins to prevent their interaction with membranes and reduce cytotoxicity. The best-characterised enterocyte bile acid carrier protein is the ileal bile acid binding protein (IBABP; also called the ileal lipid binding protein; gene symbol: *FABP6*), a member of the fatty acid binding protein family (see Chapter 5). IBABP is an abundant small 14-kDa protein that constitutes almost two percent of cytosolic protein in ileal enterocytes. IBABP expression parallels that of the ASBT and is induced by bile acids acting through FXR. Although capable of binding fatty acids, IBABP preferentially binds bile acids, with a stoichiometry of two or three bile acids per molecule of IBABP. In terms of binding specificity, IBABP has higher affinity for taurine-conjugated bile acids versus glycine-conjugated and unconjugated species. A role for IBABP in intestinal bile acid transport is supported by genetic evidence in mice (Praslickova et al., 2012). In the IBABP null mouse, apical to baso-lateral transport of taurocholate is reduced when analysed using the inverted gut sac method of Wilson and Wiseman. Bile acid metabolism is also altered, suggesting that IBABP may also be involved in enterocyte bile acid sensing. However, no inherited defects or polymorphisms affecting bile acid metabolism have yet been described for the human IBABP.

After being internalised by the ileal enterocyte, bile acids are shuttled to the basolateral membrane and exported into the portal circulation by a hetero-meric transporter, OSTα-OSTβ (gene symbols: *SLC51A* and *SLC51B*) (Dawson et al., 2010). This unusual facilitative carrier shares no sequence identity with the other bile acid transport proteins, and requires expression of two distinct protein subunits encoded on separate chromosomes. The larger OSTα subunit is a 340-amino acid membrane protein with seven predicted transmembrane domains, whereas the smaller OSTβ subunit is a 128-amino acid predicted type I membrane protein. OSTα-OSTβ has been characterised using in vitro cell-based models. Co-expression of both subunits and their association is required for trafficking of the protein complex to the plasma membrane and solute trans-port. OSTα-OSTβ operates as a facilitative bidirectional carrier with a broad substrate specificity that includes all the major species of bile acids, as well as solutes such as prostaglandins, steroids and steroid sulphates. In agreement with their role in bile acid export, gene expression for both subunits is strongly induced by bile acids acting through FXR. A primary role for OSTα-OSTβ in ileal basolateral membrane bile acid export is strongly supported by genetic evidence in mice. Inactivation of the OSTα gene significantly impairs ileal bile acid transport and alters bile acid signalling through the FXR-FGF15/19-FGFR4 signalling pathway.

Although best characterised in the ileal enterocyte, the ASBT and OSTα-OSTβ are also expressed in other epithelia, including the cholangiocytes lining the biliary tract, the gallbladder epithelium and the kidney proximal tubules cells. As in the ileum, ASBT and OSTα-OSTβ function in the kidney to reabsorb most of the bile acids in the adjacent lumen of the proximal tubules for transport in plasma back to the liver. However, in the biliary tract and gallbladder, ASBT and OSTα-OSTβ are thought to function in cholehepatic or cholecystohepatic shunting of bile acids. In those shunt pathways, a fraction of the biliary or gallbladder bile acids are reabsorbed and emptied into the portal circulation for reuptake by hepatocytes liver. The physiological significance of this pathway is still unclear, but may be acting to stimulate hepatic bile flow.

5. BILE ACIDS AS SIGNALLING MOLECULES

Some of the recent and perhaps most exciting discoveries in the field concern the ability of bile acids to function as signalling molecules. Bile acids act as endogenous ligands to activate nuclear and G-protein-coupled receptors (Thomas et al., 2008; Li and Chiang, 2014; de Aguiar Vallim et al., 2013). Among the best-characterised bile acid receptors are the nuclear receptor FXR and the G-protein-coupled receptor TGR5. However, bile acids can also act through other nuclear receptors (pregnane X receptor, PXR; vitamin D receptor, VDR), other G protein coupled receptors (muscarinic receptors; sphingosine-1-phosphate receptor-2), and through signal transduction pathways such as those mediated by protein kinase C, Jun N-terminal kinase, or MAPK/ERK (Hylemon et al., 2009).

Although bile acids were noted to activate isoforms of protein kinase C and modulate cell growth, the concept of bile acids as hormones or signalling molecules was not firmly established until the discovery in 1999 that bile acids activate FXR (Parks et al., 1999; Makishima et al., 1999). Both conjugated and unconjugated bile acids are natural agonists for FXR, with the following rank order of potency: chenodeoxycholic acid > lithocholic acid ≈ deoxycholic acid > cholic acid. However, it should be noted that some natural bile acids such as ursodeoxycholic acid and muricholic acid are unable to activate FXR and may even be competitive inhibitors or antagonists. As such, FXR activity is affected by changes in the composition of the bile acid pool. FXR is mainly expressed in liver, intestine, kidney and adrenal, with only low levels of expression reported in tissues such as adipose, heart, pancreas, artery and circulating macrophage. Consistent with its gastrointestinal expression and bile acid ligand specificity, FXR controls a network of genes to regulate hepatic bile acid synthesis, the enterohepatic cycling of bile acids, and protective mechanisms to prevent bile acid overload and cytotoxicity. In the liver, FXR suppresses expression of NTCP, increases bile acid conjugation and induces expression of efflux transporters such as BSEP and OSTα-OSTβ. Thus, activation of FXR in hepatocytes reduces bile acid uptake and synthesis, and increases export across the canalicular membrane into bile or across the sinusoidal membrane back into blood. In the small intestine, FXR reduces expression of the ASBT, and induces expression of IBABP, OSTα-OSTβ and the polypeptide hormone FGF15/19 to protect the ileal enterocyte from bile acid accumulation. FXR also plays important protective roles in the gastrointestinal tract beyond its effects on bile acid homoeostasis. These functions of FXR in the gut include inducing expression of genes important for intestinal barrier function, antimicrobial defence, and inhibiting inflammation and cell proliferation (Inagaki et al., 2006; Gadaleta et al., 2011).

In 2002, two groups independently demonstrated that the membrane-bound G-protein coupled receptors TGR5 (also called membrane-type bile acid receptor, M-BAR; GPR131; G protein-coupled bile acid receptor 1, GPBAR1; gene symbol: *GPBAR1*) was activated by low concentrations of bile acids. TGR5 is a Gα$_s$-coupled receptor that stimulates adenylate cyclase, increases intracellular cAMP levels, and signals through downstream pathways such as those mediated by protein kinase A and the transcription factor CREB (cAMP response element-binding protein). However, unlike classical G-protein coupled receptors such as the β-adrenergic receptor, TGR5 does not appear to undergo desensitisation when exposed to high concentrations of bile acids for extended periods. Both conjugated and unconjugated bile acids bind and activate TGR5, with hydrophobic bile acids being the most potent activators in the following rank order of potency: lithocholic acid > deoxycholic acid > chenodeoxycholic acid > cholic acid. As a result of differences in their bile acid specificity for activation and their cellular localisation (intracellular vs cell surface), changes in the composition of the bile acid pool will differentially affect signalling via FXR and TGR5. Furthermore, these ligand specificity differences have been exploited to identify synthetic agonists that selectively activate the two receptors, greatly facilitating the study of their in vivo roles. TGR5 is widely expressed, with highest levels in gallbladder and low to moderate levels in liver and the intestinal tract. In liver, TGR5 is not expressed by hepatocytes, but is moderately expressed by the nonparenchymal cells, including cholangiocytes, sinusoidal endothelial cells, Kupffer cells and resident immune cells. In these cells, TGR5 plays a role in modulating the immune system and hepatic microcirculation. In the biliary tract, TGR5 functions to stimulate gallbladder filling and to couple biliary bile acid concentrations to bile acid reabsorption and fluid secretion. In the small intestine and colon, TGR5 is not expressed by enterocytes or colonocytes, but rather is highly expressed by enteroendocrine cells, where it signals to stimulate incretin hormone glucagon-like peptide-1 (GLP-1) release in response to bile acids. TGR5 is also expressed in the gut enteric nervous system, where it mediates the effects of bile acids on intestinal motility. Beyond the liver and the gastrointestinal tract, TGR5 functions in other tissues and cell types, including brown adipose, muscle, brain and macrophages. Recent evidence has implicated TGR5 in modulation of the immune system and the inflammatory response, in energy and glucose metabolism, and in the central nervous system pathways for pain and itch. Thus it is increasingly evident that beyond their classical role as detergents in the digestive process and end products of cholesterol catabolism, bile acids are important regulators of liver and gastrointestinal function, lipid and glucose metabolism, and energy homoeostasis.

6. FUTURE DIRECTIONS

Investigation since the last edition of 'Biochemistry of Lipids, Lipoproteins, and Membranes' has uncovered additional details regarding the pathways for bile acid synthesis, transport, signalling and the regulation of their metabolism. With regard to synthesis and transport, identity of the bile acid biosynthetic enzymes and plasma membrane transporters is well established and supported by biochemical and genetic evidence. However, many questions remain to be answered regarding the intracellular movement and compartmentalisation of bile acids. Resolution of the structures of the biosynthetic enzymes and transporters by X-ray crystallography will continue to provide insights to their substrate specificity and catalytic mechanism of action. Even more important, these studies will be part of a concerted effort, along with cell biology and genetic approaches, to extend our understanding of the molecular mechanisms responsible for intracellular movement of cholesterol into the bile acid biosynthetic pathway and the intracellular trafficking of bile acids, their biosynthetic intermediates, and derivatives. An exciting byproduct of this line of investigation is predicted to be further insights into the molecular mechanisms by which bile acids are delivered to and engage intracellular targets. These targets include nuclear receptors such as FXR, PXR, and VDR, transcription factors such as yes-associated protein 1 (YAP), and as yet unidentified mediators of bile acid signalling or cytotoxicity.

With regard to the transcriptional regulation of bile acid homoeostasis, details of the pathways continue to emerge. A future key area of research will be the role of protein modifications such as methylation, acetylation and sumoylation, and the interaction of transcription factors with chromatin. However, much of the future investigation will remain focus on the bile acid signalling, the interaction of bile acids with the microbiome, and the complex role of bile acids in the regulation of lipid, carbohydrate, and energy metabolism. These pathways provide promising targets for pharmacological intervention in a variety of liver, gastrointestinal and metabolic diseases, and have garnered considerable attention in and outside the bile acid field. Clearly, bile acids are much more than simple detergents and much remains to be discovered.

REFERENCES

de Aguiar Vallim, T.Q., Tarling, E.J., Edwards, P.A., 2013. Pleiotropic roles of bile acids in metabolism. Cell Metab. 17, 657–669.

Chiang, J.Y., 2013. Bile acid metabolism and signaling. Compr. Physiol. 3, 1191–1212.

Clayton, P.T., 2011. Disorders of bile acid synthesis. J. Inherited Metab. Dis. 34, 593–604.

Dawson, P.A., Haywood, J., Craddock, A.L., Wilson, M., Tietjen, M., Kluckman, K., Maeda, N., Parks, J.S., 2003. Targeted deletion of the ileal bile acid transporter eliminates enterohepatic cycling of bile acids in mice. J. Biol. Chem. 278, 33920–33927.

Dawson, P.A., Hubbert, M.L., Rao, A., 2010. Getting the mOST from OST: role of organic solute transporter, OSTalpha-OSTbeta, in bile acid and steroid metabolism. Biochim. Biophys. Acta 1801, 994–1004.

Dawson, P.A., Karpen, S.J., 2015. Intestinal transport and metabolism of bile acids. J. Lipid Res. 56, 1085–1099.

Gadaleta, R.M., Van Erpecum, K.J., Oldenburg, B., Willemsen, E.C., Renooij, W., Murzilli, S., Klomp, L.W., Siersema, P.D., Schipper, M.E., Danese, S., Penna, G., Laverny, G., Adorini, L., Moschetta, A., Van Mil, S.W., 2011. Farnesoid X receptor activation inhibits inflammation and preserves the intestinal barrier in inflammatory bowel disease. Gut 60, 463–472.

Hofmann, A.F., Hagey, L.R., 2008. Bile acids: chemistry, pathochemistry, biology, pathobiology, and therapeutics. Cell Mol. Life Sci. 65, 2461–2483.

Hofmann, A.F., Hagey, L.R., 2014. Key discoveries in bile acid chemistry and biology and their clinical applications: history of the last eight decades. J. Lipid Res. 55, 1553–1595.

Hofmann, A.F., Hagey, L.R., Krasowski, M.D., 2010. Bile salts of vertebrates: structural variation and possible evolutionary significance. J. Lipid Res. 51, 226–246.

Hubbard, B., Doege, H., Punreddy, S., Wu, H., Huang, X., Kaushik, V.K., Mozell, R.L., Byrnes, J.J., Stricker-Krongrad, A., Chou, C.J., Tartaglia, L.A., Lodish, H.F., Stahl, A., Gimeno, R.E., 2006. Mice deleted for fatty acid transport protein 5 have defective bile acid conjugation and are protected from obesity. Gastroenterology 130, 1259–1269.

Hylemon, P.B., Zhou, H., Pandak, W.M., Ren, S., Gil, G., Dent, P., 2009. Bile acids as regulatory molecules. J. Lipid Res. 50, 1509–1520.

Inagaki, T., Choi, M., Moschetta, A., Peng, L., Cummins, C.L., Mcdonald, J.G., Luo, G., Jones, S.A., Goodwin, B., Richardson, J.A., Gerard, R.D., Repa, J.J., Mangelsdorf, D.J., Kliewer, S.A., 2005. Fibroblast growth factor 15 functions as an enterohepatic signal to regulate bile acid homeostasis. Cell Metab. 2, 217–225.

Inagaki, T., Moschetta, A., Lee, Y.K., Peng, L., Zhao, G., Downes, M., Yu, R.T., Shelton, J.M., Richardson, J.A., Repa, J.J., Mangelsdorf, D.J., Kliewer, S.A., 2006. Regulation of antibacterial defense in the small intestine by the nuclear bile acid receptor. Proc. Natl. Acad. Sci. U.S.A. 103, 3920–3925.

Li, T., Chiang, J.Y., 2014. Bile acid signaling in metabolic disease and drug therapy. Pharmacol. Rev. 66, 948–983.

Makishima, M., Okamoto, A.Y., Repa, J.J., Tu, H., Learned, R.M., Luk, A., Hull, M.V., Lustig, K.D., Mangelsdorf, D.J., Shan, B., 1999. Identification of a nuclear receptor for bile acids. Science 284, 1362–1365.

Moschetta, A., Xu, F., Hagey, L.R., Van Berge Henegouwen, G.P., Van Erpecum, K.J., Brouwers, J.F., Cohen, J.C., Bierman, M., Hobbs, H.H., Steinbach, J.H., Hofmann, A.F., 2005. A phylogenetic survey of biliary lipids in vertebrates. J. Lipid Res. 46, 2221–2232.

Pandak, W.M., Heuman, D.M., Hylemon, P.B., Chiang, J.Y., Vlahcevic, Z.R., 1995. Failure of intravenous infusion of taurocholate to down-regulate cholesterol 7 alpha-hydroxylase in rats with biliary fistulas. Gastroenterology 108, 533–544.

Parks, D.J., Blanchard, S.G., Bledsoe, R.K., Chandra, G., Consler, T.G., Kliewer, S.A., Stimmel, J.B., Willson, T.M., Zavacki, A.M., Moore, D.D., Lehmann, J.M., 1999. Bile acids: natural ligands for an orphan nuclear receptor. Science 284, 1365–1368.

Praslickova, D., Torchia, E.C., Sugiyama, M.G., Magrane, E.J., Zwicker, B.L., Kolodzieyski, L., Agellon, L.B., 2012. The ileal lipid binding protein is required for efficient absorption and transport of bile acids in the distal portion of the murine small intestine. PLoS One 7, e50810.

Russell, D.W., 2003. The enzymes, regulation, and genetics of bile acid synthesis. Annu. Rev. Biochem. 72, 137–174.

Russell, D.W., Halford, R.W., Ramirez, D.M., Shah, R., Kotti, T., 2009. Cholesterol 24-hydroxylase: an enzyme of cholesterol turnover in the brain. Annu. Rev. Biochem. 78, 1017–1040.

Schwarz, M., Lund, E.G., Lathe, R., Bjorkhem, I., Russell, D.W., 1997. Identification and characterization of a mouse oxysterol 7alpha-hydroxylase cDNA. J. Biol. Chem. 272, 23995–24001.

van de Steeg, E., Stranecky, V., Hartmannova, H., Noskova, L., Hrebicek, M., Wagenaar, E., Van Esch, A., De Waart, D.R., Oude Elferink, R.P., Kenworthy, K.E., Sticova, E., Al-Edreesi, M., Knisely, A.S., Kmoch, S., Jirsa, M., Schinkel, A.H., 2012. Complete OATP1B1 and OATP1B3 deficiency causes human Rotor syndrome by interrupting conjugated bilirubin reuptake into the liver. J. Clin. Invest. 122, 519–528.

van de Steeg, E., Wagenaar, E., Van Der Kruijssen, C.M., Burggraaff, J.E., De Waart, D.R., Elferink, R.P., Kenworthy, K.E., Schinkel, A.H., 2010. Organic anion transporting polypeptide 1a/1b-knockout mice provide insights into hepatic handling of bilirubin, bile acids, and drugs. J. Clin. Invest. 120, 2942–2952.

Tempel, W., Grabovec, I., Mackenzie, F., Dichenko, Y.V., Usanov, S.A., Gilep, A.A., Park, H.W., Strushkevich, N., 2014. Structural characterization of human cholesterol 7alpha-hydroxylase. J. Lipid Res. 55, 1925–1932.

Thomas, C., Pellicciari, R., Pruzanski, M., Auwerx, J., Schoonjans, K., 2008. Targeting bile-acid signalling for metabolic diseases. Nat. Rev. Drug Discov. 7, 678–693.

Vaz, F.M., Paulusma, C.C., Huidekoper, H., De Ru, M., Lim, C., Koster, J., Ho-Mok, K., Bootsma, A.H., Groen, A.K., Schaap, F.G., Oude Elferink, R.P., Waterham, H.R., Wanders, R.J., 2015. Sodium taurocholate cotransporting polypeptide (SLC10A1) deficiency: conjugated hypercholanemia without a clear clinical phenotype. Hepatology 61, 260–267.

Chapter 13

Lipid Modification of Proteins

Marilyn D. Resh

Cell Biology Program, Memorial Sloan Kettering Cancer Center, New York, NY, USA

ABBREVIATIONS

ACAT Acyl-CoA: cholesterol acyltransferase
APT Acylprotein thioesterase
2BP 2-Bromopalmitate
DGAT Diacylglycerol acyltransferase
DHHC Asp-His-His-Cys
FTase Farnesyl transferase
FTI Farnesyl transferase inhibitor
GGTase Geranylgeranyl transferase
GOAT Ghrelin O-acyltransferase
GPI Glycosylphosphatidylinositol
Hhat Hedgehog acyltransferase
LPLAT Lysophospholipid acyltransferase
MARCKS Myristoylated alanine rich C-kinase substrate
MBOAT Membrane bound O-acyltransferase
NMT N-myristoyltransferase
PAT Palmitoyl acyltransferase
PPT Palmitoyl protein thioesterase
SCD Stearoyl-CoA desaturase

1. INTRODUCTION

More than 1000 proteins are known to contain covalently linked lipophilic groups, including fatty acids, cholesterol, isoprenoids, phospholipids and diacylglyceryl lipids (Resh, 2013, 2012; Linder and Deschenes, 2007; Bijlmakers, 2009; Berndt et al., 2011; Paulick and Bertozzi, 2008). Biochemical studies have revealed that each of these hydrophobic moieties is attached by a different enzyme that recognises distinct characteristics of the lipid and protein substrates. Most lipid modification reactions involve formation of a thioester or thioether link with the reactive thiol group of cysteine within the target protein, while other reactions occur via amide or oxyester bonds to glycine or serine residues, respectively. Although lipid modified proteins are mainly membrane bound, the functional consequences of having different lipids attached to a protein are varied. By studying the biochemical

Biochemistry of Lipids, Lipoproteins and Membranes. http://dx.doi.org/10.1016/B978-0-444-63438-2.00013-4

391

TABLE 1 Lipid Modifying Enzymes and Their Mammalian Protein Substrates

Lipidation Reaction	Enzyme(s)	Examples of Lipidated Substrates
N-Myristoylation	NMT1,2	c-Src, MARCKS, Arf1, Src family kinases, Gα subunits, AKAPs, recoverin, eNOS, GCAP1
S-Palmitoylation	DHHC family (>20)	H-Ras, N-Ras, GPCRs, PSD95, SNAP25, eNOS, iNOS, GAP43, Src family kinases, Gα subunits (Gαq, 11,12,13,14,15,16), AKAPs, transferrin receptor, GABA$_A$ receptor γ2 subunit, AMPA receptor subunits
De-(S)-Palmitoylation	APT1,2	Gα subunits, H-Ras, N-Ras
N-Palmitoylation	Hhat	Hedgehog family proteins (Shh, Ihh, Dhh)
Palmitoleoylation	Porcupine	Wnt family proteins
Octanoylation	GOAT	Ghrelin
Farnesylation	FTase	Ras proteins, LaminA, LaminB, CENP-E, CENP-F, Transducin γ subunit
Geranylgeranylation	GGTase I	Rho proteins, Rac proteins, Ral proteins, Rap proteins, Gγ subunit
Geranylgeranylation	GGTase II/REP	Rab proteins
GPI	~20 PIG and related enzymes	Alkaline phosphatase, 5′-nucleotidase, CD55, Thy1, NCAM, Acetylcholinesterase
PE conjugation	Atg 4, 7, 3	LC3

characteristics of lipidated proteins, we now have a greater understanding of how lipid modification can serve as a mechanism to regulate protein structure, localisation and function (Table 1).

2. ATTACHMENT OF FATTY ACIDS TO PROTEINS

Most fatty acylated proteins contain either myristate or palmitate or both. Despite the fact that these two fatty acids differ only by two methyl groups, the mechanisms of covalent attachment to proteins are completely different. Distinct fatty acyltransferases catalyse myristoylation and palmitoylation reactions, and the effects of myristate or palmitate attachment on protein localisation and function are different.

2.1 N-Myristoylation

N-myristoylation refers to covalent attachment of the 14-carbon fatty acid myristate to the N-terminus of a protein via an amide bond (Farazi et al., 2001). Although myristate is a relatively rare fatty acid, comprising less than 1% of the total fatty acids in animal tissues, this fatty acid, rather than the more abundant palmitate, is attached specifically to the N-terminal glycine of distinct target proteins. Approximately 0.5–0.8% of all eukaryotic proteins are estimated to be N-myristoylated, based on the consensus sequence Met-Gly-X-X-X-Ser/Thr. Algorithms to predict whether a protein is likely to be N-myristoylated (Maurer-Stroh et al., 2002) can be found at http://mendel.imp.ac.at/myristate/SUPLpredictor.htm.

2.1.1 Enzymology of N-Myristoylation

During translation, the N-terminal initiating methionine of a protein destined to be N-myristoylated is removed by the action of methionine aminopeptidase, exposing glycine as the N-terminal residue. Myristoyl-CoA:protein N-myristoyltransferase (NMT) is the enzyme that catalyses the transfer of myristate from myristoyl-CoA to the N-terminal glycine (Figure 1(A)). Mammals express two isozymes, NMT1 and NMT2, whereas only one NMT

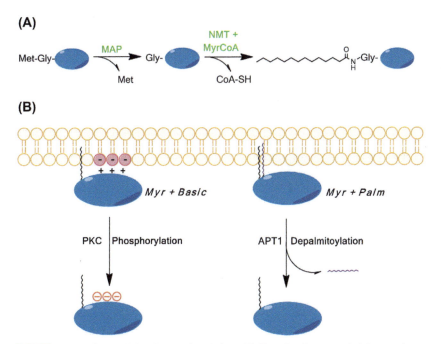

FIGURE 1 (A) Co-translational N-myristoylation; (B) Two signals are needed for membrane binding of N-myristoylated proteins. N-myristoylated proteins can be released from the membrane by phosphorylation or depalmitoylation.

gene is encoded in fungi. Extensive studies of NMT enzymology and three-dimensional crystal structures of yeast NMT bound to myristoyl-CoA and a peptide substrate have revealed that NMT catalyses N-myristoylation via an ordered Bi Bi kinetic mechanism (Bhatnagar et al., 1998). First, myristoyl-CoA binds to the enzyme in a conformation with four bends that resembles a question mark. The tail of the fatty acid chain is embedded within a deep, narrow binding pocket, limiting the chain length that can fit within the enzyme to 14 carbons and explaining the specificity of NMT for myristate rather than longer fatty acids. Next, the N-terminal region of the peptide or protein substrate binds to NMT to form a ternary complex. The presence of glycine is absolutely essential for recognition by NMT, and no other amino acid, not even alanine, at the N-terminus of the substrate can serve as an acceptor for N-myristoylation. Myristate is transferred to the N-terminal glycine, and CoA is released, followed by a release of the N-myristoylated protein.

N-myristoylation reactions can be recapitulated with short peptides in vitro, but in cells the reaction nearly always occurs co-translationally, typically before the first 100 amino acids of the newly synthesised polypeptide chain are polymerised on the ribosome. NMT is primarily a cytosolic enzyme, but it can also bind to ribosomes, and this likely facilitates the co-translational timing of the reaction. The situation is a bit different during apoptosis, programmed cell death, in which several examples of post-translational myristoylation have been described (Martin et al., 2011). Cleavage of proteins in apoptotic cells by caspases exposes an N-terminal glycine that is located within a cryptic myristoylation consensus sequence. Proteins such as Bid, actin, gelsolin and the Huntingtin protein Htt are post-translationally N-myristoylated during apoptosis.

2.1.2 Membrane Binding of N-myristoylated Proteins

Although attachment of myristate increases the hydrophobicity of the modified protein, not all myristoylated proteins are membrane bound (Resh, 2006a). Analyses by Peitzsch and McLaughlin first established the biophysical basis for interaction of myristoylated proteins with membranes (Peitzsch and McLaughlin, 1993). Myristate inserts hydrophobically into the bilayer interior. Each of the carbons in the fatty acyl chain provides a binding energy of 0.8 kcal/mol, and 10 of the 14 methylene carbons are embedded within the bilayer. The resultant binding energy of 8 kcal/mol corresponds to an effective dissociation constant for binding of an N-myristoylated peptide to a phospholipid bilayer of approximately 10^{-4} M, a binding energy that is not strong enough to stably anchor the modified protein to a biological membrane in the cell. To achieve stable membrane binding, N-myristoylated proteins require a second signal, typically a polybasic domain or a second fatty acid, usually palmitate (Figure 1(B)). Proteins with a 'myristate + basic' motif contain a cluster of positively charged residues that can be located near the N-terminus (e.g. c-Src, HIV-1 Gag) or >100 amino acids away from the N-terminus (e.g. MARCKS (myristoylated alanine rich C-kinase substrate)). The basic residues form electrostatic interactions with negatively charged phospholipids, such as phosphatidylserine and phosphatidylinositol phosphates, that are concentrated on the cytoplasmic leaflets of the plasma membrane and intracellular membranes. Neither myristate nor the basic amino acid cluster alone are sufficient for membrane binding, but when located within the same protein, the hydrophobic and electrostatic forces exhibit apparent cooperativity and act synergistically to increase membrane binding by several orders of magnitude. Alternatively, the second signal can be provided by attachment of palmitate (see Section 2.2), which provides the additional hydrophobicity needed to anchor the N-myristoylated protein to a membrane. Examples of proteins that are dually fatty acylated with myristate and palmitate include Src family kinases, Gα subunits, eNOS and A-kinase anchoring proteins (AKAPs).

2.1.3 Myristoyl Switches as Mechanisms for Reversible Membrane Binding

In a subset of N-myristoylated proteins, the myristate moiety can exist in two configurations: exposed or hidden. When myristate is exposed, it juts out from the protein surface and assists in directing membrane binding. In the hidden conformation, myristate is bound in a hydrophobic pocket within the modified protein, and the fatty acid is essentially sequestered from the aqueous milieu. Binding of calcium (recoverin), GDP (Arf), changes in pH (hisactophilin) or protein multimerisation (HIV-1 Gag) can trigger the myristoyl switch: myristate flips into the hydrophobic cleft, and the protein is released from the membrane into the cytosol (Ames et al., 1996; Tang et al., 2004).

An alternative mechanism to achieve regulated membrane binding is the myristoyl electrostatic switch (Murray et al., 1997). Some proteins have one or more phosphorylation sites within the polybasic motif. Phosphorylation reduces the electrostatic force between the positively charged amino acids and negatively charged membranes. For example, as a result of phosphorylation by protein kinase C within the polybasic domain, the MARCKS protein is released from the membrane.

Unlike the thioester bond that links palmitate to most S-palmitoylated proteins (Section 2.2), the amide bond between myristate and the protein is stable. Myristate remains attached to the modified protein throughout its lifetime in the cell, making N-myristoylation an essentially irreversible reaction. No natural mechanism for demyristoylation by specific cleavage of the amide bond between myristate and the N-terminal glycine has been identified. However, the *Shigella* pathogen expresses a cysteine protease that removes the N-terminal myristoylated glycine (Burnaevskiy et al., 2013). Using this cleavage mechanism for demyristoylation, Shigella can inactivate host N-myristoylated proteins involved in cell growth and signalling.

2.2 S-Palmitoylation

Palmitate is typically attached via thioester linkage to one or several cysteine residues within the polypeptide chain (Linder and Deschenes, 2007; Resh, 2006b) (Figure 2(A)). There is no distinct consensus sequence for S-palmitoylation, but analyses of hundreds of known S-palmitoylated proteins have led to the development of Web-based algorithms to predict potential palmitoylation sites (e.g. http://bioinfo.ncu.edu.cn/WAP-Palm.aspx; http://csspalm.biocuckoo.org/; http://14.139.227.92/mkumar/palmpred/). The fatty acid donor is palmitoyl-CoA, and the reaction occurs post-translationally. Fatty acid modification is

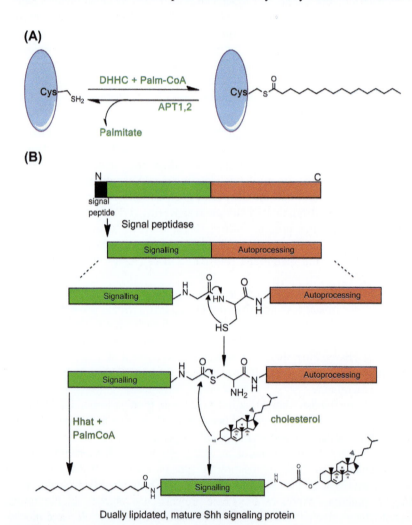

FIGURE 2 (A) S-Palmitoylation and depalmitoylation of proteins; (B) Processing and N-palmitoylation of Hedgehog proteins.

not limited to palmitate, because other medium and long chain fatty acids are attached to the S-palmitoylation site(s). Proteins that are S-palmitoylated may also be modified by N-myristoylation or prenylation. For these dually lipidated proteins, attachment of either myristate or a prenyl group is a prerequisite for subsequent S-palmitoylation. The S-palmitoylation reaction can be inhibited by treatment with 2-bromopalmitate (2BP), a nonmetabolisable palmitate analogue that has been shown to block palmitate incorporation into numerous proteins (Webb et al., 2000).

2.2.1 Enzymology of S-Palmitoylation and Depalmitoylation

The Asp-His-His-Cys (DHHC) family of palmitoyl acyltransferases (PATs) is primarily responsible for S-palmitoylation of proteins in cells (Linder and Jennings, 2013). The number of DHHC enzymes ranges from 7 in the yeast *Saccharomyces cerevisciae* to 23 in mammals. This family of membrane bound enzymes contains a highly conserved Asp-His-His-Cys (i.e. DHHC) sequence that is essential for catalysis, along with a cysteine-rich domain. The DHHC family of membrane proteins contain 4–6 transmembrane domains, and are localised to multiple intracellular locations, including the Golgi apparatus, endoplasmic reticulum (ER) and plasma membrane. Protein palmitoylation is catalysed on the cytosolic face of the membrane. The reaction proceeds via a two-step ping pong mechanism (Jennings and Linder, 2012). First, DHHC proteins undergo autoacylation in the presence of palmitoyl-CoA, resulting in release of CoA and formation of a palmitoylated acyl-enzyme intermediate. It has been suggested that the cysteine within the DHHC motif is the site of autopalmitoylation, but this has not yet been experimentally verified. Next, palmitate is transferred from the enzyme to the recipient protein. The mechanisms responsible for fatty acid and protein substrate selectivity for the different DHHC family members are not clear. Some of the family members can bind and transfer other fatty acids, such as myristate and stearate, in addition to palmitate. Each of the DHHC proteins appears to be responsible for palmitoylation of multiple cellular proteins, and many palmitoylated proteins can apparently be S-palmitoylated by more than one DHHC enzyme. Thus, a great deal of redundancy is built into the cellular S-palmitoylation machinery.

A key feature of S-palmitoylation is that the thioester bond between palmitate and the cysteine thiol is relatively labile. At least four protein palmitoyl thioesterases that cleave palmitate from S-palmitoylated proteins have been identified: acylprotein thioesterase 1 (APT1) and acylprotein thioesterase 2 (APT2), which are mostly localised in the cytoplasm, and the lysosomal enzymes palmitoyl protein thioesterase 1 (PPT1) and palmitoyl protein thioesterase 2 (PPT2) (Figure 2(A)). The mechanism by which APT1 and APT2 access their membrane-bound substrates (H-Ras, GAP-43) involves cycles of palmitoylation and depalmitoylation (Kong et al., 2013). Both APT1 and APT2 are palmitoylated on cysteine-2, which directs them to the membrane.

APT1 can then catalyse depalmitoylation of itself as well as of APT2, releasing these enzymes from the membrane. Studies of protein depalmitoylation have been enhanced by the identification of Palmostatin B, a thioesterase inhibitor (Dekker et al., 2010).

S-palmitoylated Ras proteins undergo a dynamic cycle of palmitoylation/depalmitoylation that mediates reversible membrane binding and dissociation. For example, H-Ras is palmitoylated by the DHHC enzyme complex DHHC9-GCP16 in the Golgi apparatus, transported to the plasma membrane, and then depalmitoylated. Once released from the plasma membrane, H-Ras relocalises to the Golgi apparatus, where it is re-palmitoylated (Rocks et al., 2005). Thus, at steady-state, populations of H-Ras are present at and signal from both the plasma membrane and the Golgi apparatus. This continuous cycle of palmitoylation/depalmitoylation is required for proper H-Ras localisation and signalling. Inhibition of depalmitoylation, with either an APT1 inhibitor or by attachment of a nonhydrolysable palmitate analogue to H-Ras, results in a random binding of H-Ras to all intracellular membranes and inhibition of Ras-mediated signalling.

2.2.2 Binding of S-palmitoylated Proteins to Membranes and Lipid Rafts

Most S-palmitoylated proteins are membrane bound and can be categorised in two types. One class is represented by transmembrane proteins that contain one or more membrane-spanning domains and are palmitoylated at a cysteine residue located within a cytoplasmic loop. Palmitoylation of transmembrane proteins, such as G protein coupled receptors or the transferrin receptor, is obviously not needed for membrane binding. Rather, palmitoylation regulates protein stability, subcellular targeting and intracellular trafficking. For example, inhibition of S-palmitoylation can alter the rates of delivery of transmembrane proteins to the plasma membrane, the Golgi apparatus and/or endosomes. The second class of S-palmitoylated proteins contains myristate, a prenyl group or no other lipid modification. When unmodified, these proteins are soluble and rely on palmitoylation to achieve stable membrane binding.

Palmitoylation plays a key role in directing many proteins to lipid rafts, membrane subdomains that are enriched in glycosphingolipids, cholesterol and phospholipids containing saturated fatty acids (Levental et al., 2010). Concentrating raft lipids induces the formation of liquid ordered domains that segregate away from the bulk lipid domains in the plasma membrane (Chapter 1). Proteins that contain covalently bound, saturated fatty acids preferentially partition into membrane rafts, while those with unsaturated, branched or bulky hydrocarbon groups are excluded. For many proteins, raft association increases the local concentration and interactions between and among signalling proteins in the plane of the membrane, thereby resulting in enhanced signal transduction.

2.3 N-Palmitoylation

A small subset of proteins contains palmitate attached via amide linkage to an N-terminal cysteine. These include hedgehog proteins (Hedgehog, Sonic Hedgehog, Indian Hedgehog, Desert Hedgehog) (Pepinsky et al., 1998) and the *Drosophila melanogaster* EGF-like ligands Spitz, Keren and Gurken (Miura et al., 2006). N-palmitoylation occurs as part of a series of post-translational protein modifications (Figure 2(B)). These reactions have been extensively characterised for Sonic hedgehog (Shh), but all other hedgehog family proteins likely undergo the same set of modifications (Buglino and Resh, 2012). Shh is synthesised as a 45 kDa precursor protein that contains an N-terminal signal sequence. Upon entry into the ER, the signal sequence is removed by signal peptidase. Shh undergoes an autocleavage reaction, mediated by the C-terminal half of the protein that generates an N-terminal 19 kDa fragment (denoted ShhN) and a C-terminal 26 kDa fragment that is degraded in the ER. ShhN is then modified by two separate lipid modifications. Palmitate is attached to the N-terminal cysteine, and cholesterol is linked to the C-terminal glycine (see Section 3). This dually lipidated molecule represents the mature, active Shh signalling protein. Mutation of the N-terminal cysteine or depletion of Hhat inactivates Shh, indicating that palmitoylation of Shh is crucial for effective signalling.

2.3.1 Cell Biology and Enzymology of N-Palmitoylation

Attachment of palmitate to Shh is dependent on passage of Shh through the secretory pathway, as Shh constructs in which the signal sequence has been removed are not N-palmitoylated. This suggests that N-palmitoylation occurs in the lumen of the ER. Following signal sequence cleavage, cysteine becomes the N-terminal amino acid of mature Shh. Unlike S-palmitoylation of internal cysteines, palmitate is linked via amide bond to the amino group of the N-terminal Shh cysteine. The molecular mechanism of N-palmitoylation is not yet defined. It is possible that palmitate is initially attached to the sulfhydryl group via thioester linkage and then rearranges through an intramolecular S-to-N shift to form an amide bond. Alternatively, the fatty acid could be directly attached to the N-terminus, similar to the reaction for N-myristoylation. The finding that Shh with a blocked N-terminus is not palmitoylated supports the latter mechanism. N-palmitoylation can also occur on the 45 kDa precursor protein, and mutations that prevent autocleavage and cholesterol attachment do not prevent Shh N-palmitoylation, indicating that attachment of palmitate and attachment of cholesterol are two separate, independent reactions.

The enzyme responsible for N-palmitoylation of mammalian hedgehog proteins is Hhat (Hedgehog acyltransferase); the *D. melanogaster* ortholog is known as Rasp (Buglino and Resh, 2008; Lee and Treisman, 2001). Genetic experiments in flies and mice revealed that mutation or loss of Hhat/Rasp results in production of nonpalmitoylated Shh/Hh and defective Shh signalling. Hhat, a 57 kDa protein, has been purified to homogeneity and shown to catalyse covalent linkage of palmitate to the N-terminal cysteine of Shh via amide bond. Hhat exhibits selectivity for this reaction, as it does not mediate attachment of palmitate to proteins such as H-Ras and Fyn, which are acylated by DHHC PATs, or Wnt proteins, which are modified by a different acyltransferase (see Section 2.4). The Hhat recognition motif encompasses the first 6 residues of mature Shh (CGPGRG), which are sufficient for Hhat-mediated N-palmitoylation in cells. Rasp also recognises the N-terminal sequence of Hh, as well as three other EGF-like ligands in flies: Spitz, Keren and Gurken (Miura et al., 2006). For both Hhat and Rasp, the presence of a positively charged residue within the N-terminal sequence of the protein substrate is required for N-palmitoylation.

Hhat is a member of the membrane bound O-acyltransferase (MBOAT) family, a group of 16 related enzymes that catalyse transfer of lipids (primarily fatty acids) to either proteins or other lipids (Hofmann, 2000). MBOAT proteins are membrane proteins predicted to contain 8–12 transmembrane domains. This family is involved in transfer of fatty acids to cholesterol (ACAT), diacylglycerol (DGAT), lysophospholipids (LPLAT) and secreted proteins (Hhat, Porcupine, GOAT). An MBOAT homology domain has been identified within the family that contains an invariant histidine implicated in catalysis. Mutation of this histidine within Hhat (H379) inhibits palmitoylation, primarily due to an increased K_m and decreased V_{max} of the enzyme for Shh. A second region of homology has been identified within MBOAT proteins that catalyse acylation of proteins, and mutations within this region have been shown to reduce Hhat activity and stability.

2.4 Acylation with Other Fatty Acids

2.4.1 Wnt Proteins Are Modified with Palmitoleic Acid

Wnt proteins (~20 in vertebrates) are secreted, lipid-modified proteins that regulate embryonic development and tissue renewal in adults. All Wnt proteins contain 22 cysteine residues, positioned with highly conserved spacing, and an N-terminal signal sequence. On entry into the ER, the signal sequence is cleaved, and Wnts are glycosylated and fatty acylated. The lipid modification, which is unique to Wnt proteins, was identified as palmitoleate (C16:1Δ9) (Takada et al., 2006). This *cis*-Δ9-monounsaturated fatty acid is linked to a conserved serine within Wnt via an oxyester bond. Palmitoleoylation is essential for the intracellular trafficking and secretion of Wnt proteins. Moreover, the three-dimensional crystal structure of *Xenopus* Wnt8 revealed the presence of an extension from the N-terminal domain that places the palmitoleic acid moiety within a hydrophobic groove of the Wnt receptor Frizzled (Janda et al., 2012). This suggests that fatty acylation is also required for binding of Wnt to its cell surface receptor and Wnt signalling activity.

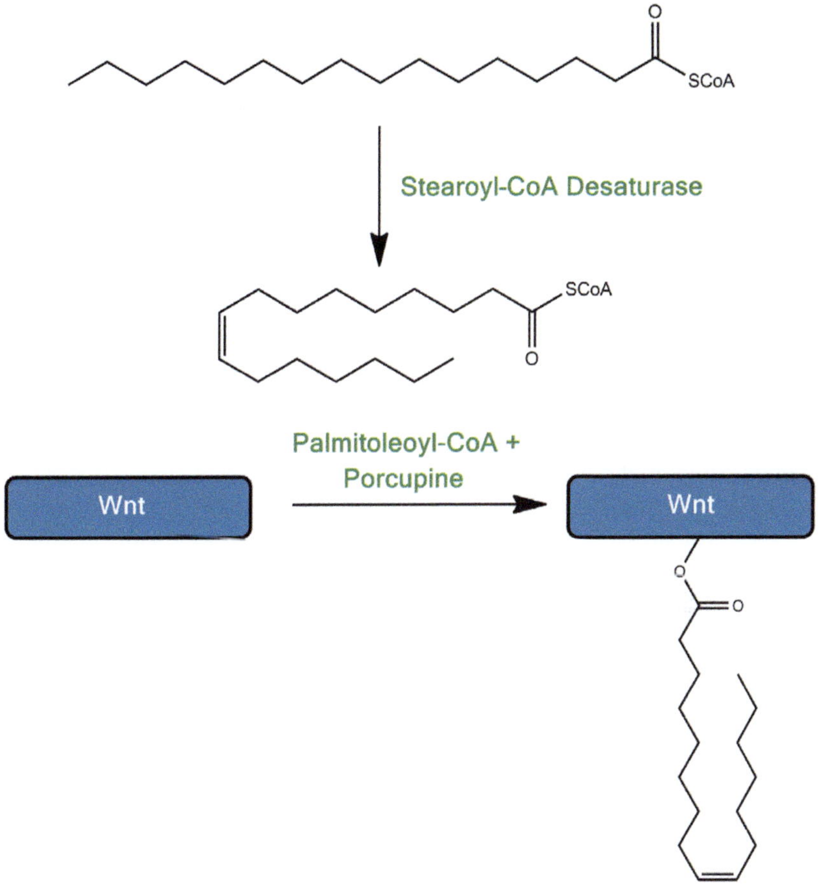

FIGURE 3   Desaturation of palmitate to palmitoleate precedes attachment to Wnt proteins.

Wnt fatty acylation is catalysed by Porcupine, an MBOAT family member localised to the ER. The reaction occurs in two steps (Figure 3). First, the monounsaturated fatty acid substrate for Porcupine is generated by another ER resident protein, stearoyl-CoA desaturase (SCD). SCD forms monounsaturated fatty acyl-CoAs from saturated precursors by placing a *cis*-double bond at position 9 in palmitoyl-CoA and stearoyl-CoA. Following conversion of palmitoyl-CoA to palmitoleoyl-CoA, Porcupine transfers the palmitoleoyl moiety to Wnt proteins. Porcupine exhibits acyl chain length preference: fatty acids shorter than, but not longer than, 16 carbons can be acylated to Wnt proteins (Rios-Esteves and Resh, 2013).

Porcupine recognises a conserved amino acid sequence surrounding the palmitoleoylated serine residue (CXCHGXSXXCXXKXC) in Wnt proteins. Threonine in place of serine can also serve as a palmitoleoylate acceptor, consistent with the designation of Porcupine as an O-acyl transferase. The requirement for the cysteines in this sequence, which are disulfide bonded, suggests that secondary structure within the Wnt protein may play a role in substrate recognition by Porcupine and is supported by studies showing that the integrity of the Wnt disulfide bonds is essential for its signalling function.

2.4.2 Octanoic Acid Is Attached to Ghrelin

Ghrelin is a small (28 amino acid) peptide hormone generated in the stomach by proteolytic cleavage from a 117 amino acid precursor. Binding of ghrelin to its receptor, growth hormone secretogogue receptor, results in appetite stimulation in mice and humans. To function as an appetite-stimulating hormone, ghrelin must be modified by covalent attachment of octanoate on serine-3. Ghrelin is the only protein known to be octanoylated. Acylation is catalysed by the MBOAT family member GOAT (ghrelin O-acyltransferase), which is expressed mainly in the stomach and intestine (Yang et al., 2008). GOAT recognises the first five amino acids of ghrelin (GSSFL), with glycine-1, serine-3 (the acylation site) and phenylalanine-4 being critical. As with the other MBOAT enzymes that acylate proteins (Hhat, Porcupine), the reaction occurs in the lumen of the ER after signal sequence cleavage. Given the important role of ghrelin in modulating weight gain and glucose metabolism, GOAT is an attractive therapeutic target for potential treatment of obesity and diabetes.

3. ATTACHMENT OF CHOLESTEROL TO HEDGEHOG PROTEINS

To date, hedgehog proteins are the only proteins known to contain covalently linked cholesterol (Mann and Beachy, 2004). Cholesterol attachment is tightly coupled to the hedgehog autocleavage reaction that occurs between glycine and cysteine and is mediated by the C-terminal half of the hedgehog protein precursor (Figure 2). The reaction is initiated by the cysteine sulfhydryl, which attacks the glycine carbonyl to form a thioester intermediate. The 3β-hydroxyl of cholesterol serves as a nucleophile to break the thioester bond and release the C-terminal 26 kDa domain of the hedgehog precursor. Cholesterol remains attached to the C-terminus of the 19 kDa ShhN protein via an ester link. Other sterols can replace cholesterol when the autoprocessing reaction is carried out in vitro, but there is an absolute requirement for the C3 hydroxyl group. Although autoprocessing and cholesterol modification contribute to long-range hedgehog signalling potential, hedgehog mutants that lack cholesterol are still active in many signalling contexts.

The mature hedgehog signalling protein contains two lipid modifications, palmitate and cholesterol (Figure 2). The question then arises as to how such a highly hydrophobic protein can be released from the cell and act as a morphogen to signal across multiple cell distances. At least two proteins that participate in hedgehog protein release from the cell have been identified (Tukachinsky et al., 2012).

Dispatched is a transmembrane protein that contains a sterol-sensing domain (SSD), a five membrane spanning domain motif found in proteins involved in cholesterol metabolism. The presence of the cholesterol moiety on hedgehog is required for binding of Dispatched to hedgehog and for its ability to promote release of lipidated hedgehog from the cell. However, this interaction alone is not sufficient for hedgehog release. A secreted protein, Scube2, is also required (Tukachinsky et al., 2012; Creanga et al., 2012). Scube2 binds to hedgehog in a cholesterol-dependent manner, although different structural features of cholesterol are recognised by Scube2 and Dispatched. Together, the two proteins act synergistically to release lipidated hedgehog from the producing cell.

4. ATTACHMENT OF ISOPRENOIDS TO PROTEINS

Prenyl groups that are attached to proteins are derived from 5-carbon intermediates in the cholesterol biosynthetic pathway (Chapter 11) to form farnesyl (15 carbons) and geranylgeranyl (20 carbons) (Figure 4(A)). The lipid donors in these two reactions are farnesyl diphosphate (FPP) and geranylgeranyl diphosphate (GGPP), respectively. Most proteins destined to be modified by prenylation contain a sequence known as the 'CAAX' box (cysteine-aliphatic-aliphatic-X) at the C-terminus (Nguyen et al., 2010). Three modification steps then occur (Figure 4(B)). First, the prenyl group is linked via a stable, thioether bond to the cysteine in the CAAX box. This reaction is catalysed by a protein prenyltransferase in the cytosol. The 'X' amino acid dictates the type of prenylation: farnesyl or geranylgeranyl. Next, the 'AAX' sequence is cleaved by RceI (Ras converting enzyme I) protease. Third, the newly generated cysteine at the C-terminus is carboxymethylated by Icmt (Isoprenylcysteine carboxymethyltransferase). The latter two modification reactions occur in the ER. Although the prenyl group is a permanent modification that is not removed, prenylated proteins can traffic between different membrane compartments or between membranes and the cytosol.

4.1 Farnesylation

When 'X' in the CAAX box is methionine, alanine, serine or glutamine, farnesyl is attached to the protein. The reaction is catalysed by farnesyl transferase (FTase), an $\alpha\beta$ heterodimer with a regulatory (α) and catalytic (β) subunit. FTase is a metalloenzyme that uses zinc to coordinate farnesyl binding to the thiol group of the cysteine in the CAAX box. FTase binds FPP first, then the protein substrate. The farnesyl moiety is transferred to the protein, followed by release of the farnesylated protein product. Protein farnesylation is an essential eukaryotic cell protein modification, and targeted deletion of FTase in mice leads to embryonic lethality.

(A)

(B)

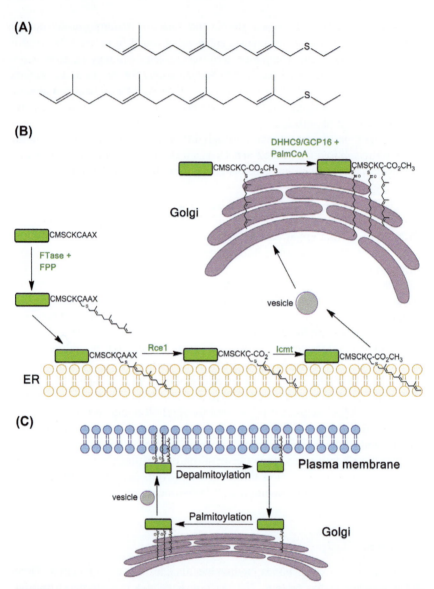

(C)

FIGURE 4 (A) Structures of thioether-linked farnesyl and geranylgeranyl groups; (B) Sequential processing and modification of H-Ras; (C) Cycling of H-Ras between the Golgi and the plasma membrane.

The role of farnesylation has been extensively studied for Ras proteins, which require this modification to bind to membranes and to mediate signal transduction in normal as well as malignant cells (Prior and Hancock, 2012). The hydrophobicity of farnesyl is not sufficient to stably anchor a protein to the membrane, and farnesylated proteins, like their myristoylated counterparts, require a second signal for membrane binding. The second signal can be a polybasic motif (K-Ras4B) or one or two palmitates attached to cysteine(s) just upstream of the farnesylated cysteine residue (H-Ras, N-Ras) (Figure 4(B)). A constitutive palmitoylation cycle between the plasma membrane and Golgi apparatus is required for the activity of H-Ras and N-Ras (Figure 4(C)).

Farnesyl electrostatic switches are used to regulate the movement of farnesylated K-Ras4B within the cell (Bivona et al., 2006). Protein kinase C-mediated phosphorylation of a serine within the polybasic cluster induces dissociation of phosphorylated K-Ras4B from the plasma membrane and translocation into the mitochondria. K-Ras4B can also be released from the plasma membrane and translocate to the Golgi apparatus when calcium/calmodulin binds to the polybasic cluster.

4.2 Geranylgeranylation

Two different enzymes catalyse geranylgeranylation of proteins in cells, geranylgeranyl transferase I and II (GGTase I and II) (Nguyen et al., 2010). Most proteins destined to be modified by GGTase I contain a CAAX box with leucine or isoleucine as the 'X' amino acid. These include small G protein members of the Rho, Rac and Rap families. GGTase I is a heterodimer that has the same catalytic α subunit as FTase, but a different β subunit. Given the identical nature of the catalytic subunits for FTase and GGTase I, it is not surprising that overlap in prenylated substrates exists. For example, some proteins that terminate in CAAL are farnesylated by FTase, proteins with CAAF can be modified by either FTase or GGTase I, and the RhoB protein (CKVL) contains both farnesyl and geranylgeranyl at its C-terminus. GGTase II is primarily dedicated to modification of Rab family proteins (60 members), which typically contain CC, CXC or CCX at their C-terminus. Rab escort protein (REP) binds Rab proteins in their GDP-bound state and presents them to GGTase II in a ternary complex. Two geranylgeranyl groups are generally transferred, although some Rab proteins only have one.

The geranylgeranyl group is sufficiently hydrophobic to confer stable membrane binding to the modified proteins. Although it is not removed from the modified protein during its lifetime in the cell, a mechanism exists to extract geranylgeranylated Rab and Rho proteins from the membrane using protein:protein interactions. Guanine nucleotide dissociation inhibitor (GDI) proteins are soluble proteins that bind to the GTP-bound forms of Rab and Rho proteins. At least one isoprenoid unit binds within a deep hydrophobic cavity of the GDI (Rak et al., 2003). As a result of isoprenoid sequestration, activated Rab and Rho proteins are removed from the membrane and then inactivated in the cytosol.

5. ATTACHMENT OF PHOSPHOLIPIDS AND DIACYLGLYCEROL LIPIDS TO PROTEINS

5.1 GPI Anchors

More than 150 eukaryotic proteins are anchored to the outer surface of the cell by a GPI (glycosylphosphatidylinositol) anchor. This lipid tail is attached to the C-terminus of the modified protein in a complex process that involves nearly one dozen steps (Paulick and Bertozzi, 2008; Orlean and Menon, 2007). All GPI-anchors contain a common core structure, consisting of $H_2N-(CH_2)_2-O-PO_3-$ mannose-mannose-mannose-glucosamine-phosphatidylinositol (Figure 5(A)). The core is first assembled in the ER. The reaction starts on the cytoplasmic side of the ER with the transfer of GlcNAc to diacyl PI. Following N-deacetylation, the GlcN-PI flips into the ER lumen, where palmitate is attached to the C2 position of inositol. Mannose residues are added from dolichol-phosphate-mannose donors followed by attachment of one or more phosphoethanolamines. Before the completed anchor can be attached, the recipient protein undergoes a C-terminal cleavage reaction. The GPI signal anchor sequence typically contains a stretch of 8–12 polar residues, the 'ω' amino acid (usually Gly, Ala, Cys, Ser, Asp or Asn) that will become the anchor attachment site, and 10–20 hydrophobic residues that are cleaved from the C-terminus. The GPI anchor is attached to the ω amino acid via an amide linkage. Additional alterations occur on the newly attached GPI anchor within the ER, including removal of one of the phosphoethanolamines and replacement of palmitate at the sn-2 position with stearate (in mammals) or 26:0 fatty acid (in yeast). In yeast, this latter re-acylation reaction is catalysed by GUP1, an MBOAT family member. The newly modified proteins are transported on COPII coated vesicles to the Golgi apparatus and then to the plasma membrane, where they localise on the outer leaflet of the plasma membrane.

GPI-anchored proteins require their anchors for membrane association and proper functioning, and mutations or loss of enzymes involved in anchor biosynthesis result in a lethal phenotype in a wide variety of species. However, it is not clear why such complicated structures are needed and/or how anchor formation is regulated. The presence of saturated fatty acids in the anchor promotes association of GPI-anchored proteins into lipid rafts, where they tend to accumulate in nanoclusters of 10–100 nm (Leventhal et al., 2010). It is well established that raft localisation is important for the signalling functions of many GPI-anchored proteins. GPI anchors also direct the modified protein to the apical surface of the plasma membrane in polarised mammalian cells. The GPI-anchor can be artificially removed from modified proteins by cleavage with PI-specific phospholipase C, resulting in protein release from the lipid bilayer, but it is not known whether this type of shedding occurs in vivo. It is likely that the GPI anchor confers unique properties to the modified protein that regulate the dynamics of plasma membrane association in a manner different from a transmembrane domain.

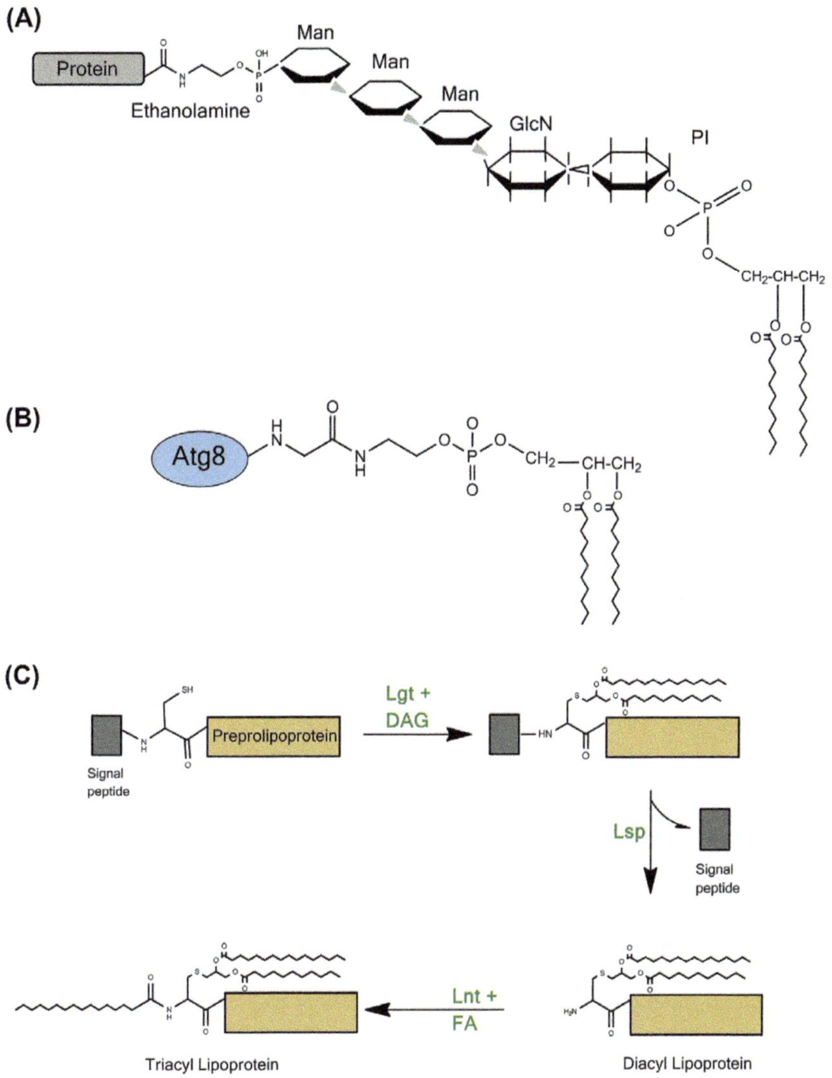

FIGURE 5 (A) Generalised structure of a GPI anchor; (B) Structure of PE-conjugated Atg8; (C) Bacterial lipoprotein processing.

5.2 PE Attachment to the Atg8 and LC3 Autophagy Proteins

Another mode of lipid attachment involves linkage of a phospholipid, PE, directly to a protein. An example of this type of modification is Atg8, an ubiquitin-like protein that regulates autophagy in yeast (Geng and Klionsky, 2008) (Figure 5(B)). Processing occurs via a series of steps mediated by autophagy-related (Atg) proteins. First, the C-terminal arginine in Atg8 is cleaved by a cysteine protease, Atg4. Next, the newly exposed C-terminal glycine is linked to a cysteine within Atg7 via a thioester bond. Third, Atg8 is transferred from Atg7 to Atg3, an E2-like enzyme that binds to Atg8 via a thioester linkage. Finally, Atg3 is released, and Atg8 is conjugated to PE via an amide bond. Attachment of PE to Atg8 promotes tight association of Atg8 with the autophagosome membrane and is essential for elongation and sealing of this double membrane structure. When the autophagosome matures, PE-conjugated Atg8 is cleaved by Atg4 to release free Atg8. The mammalian homologues of Atg8, LC3 and GABARAP, are also lipidated with PE through a process similar to that described in yeast. The lipidation process can be reconstituted in vitro and occurs much more efficiently on highly curved membrane surfaces, conditions that mimic the structure of the curved isolation membranes that constitute the initial autophagosome. This is mediated by the membrane curvature sensing ability of Atg3.

5.3 Bacterial Lipoproteins

In bacteria, an unusual but common lipid modification occurs at the N-terminus of more than 2000 proteins, resulting in the formation of a lipoprotein containing three fatty acids (Figure 5(C)) (Nakayama et al., 2012). Bacterial proteins destined to become lipoproteins contain an N-terminal signal peptide with a conserved sequence, termed the lipobox motif, just upstream of the signal peptidase cleavage site and a cysteine immediately following the cleavage site (Cys + 1). The biosynthetic pathway for generation of these lipoproteins comprises three steps. First, preprolipoprotein diacylglyceryl transferase (Lgt) catalyses attachment of a diacylglyceryl group from phosphatidylglycerol to the thiol of Cys + 1 via thioether linkage. Next, signal peptidase (Lsp) removes the signal peptide, exposing the modified Cys + 1 as the new N-terminus. Third, the α-amino group of the diacylglyceryl cysteine is acylated by Lnt (apoprotein N-acyltransferase) with a fatty acid (usually 16 or 18 carbons), to form the mature, triacylated lipoprotein. Gram-negative bacteria carry out all three steps to form a triacylated lipoprotein, while most Gram-positive bacteria lack Lnt and instead form diacylated lipoproteins with an acetylated N-terminus. Lgt and Lsp are essential for survival of all bacteria, and lipoproteins have been shown to play key roles in bacterial metabolism and host–pathogen interactions.

6. SPOTLIGHT ON INHIBITORS OF LIPID-MODIFYING ENZYMES AND THEIR ROLES IN DISEASE

6.1 Inhibitors of NMT

N-Myristoylation is required for the function of multiple proteins that are involved in both normal cell homeostasis as well as disease states. Inhibitors that block human NMT should therefore exhibit significant toxicity in normal human cells. Instead, approaches to target NMT in disease have focused on selectively inhibiting the enzymes from infectious parasites. The primary amino acid sequences of human and parasitic NMTs as well as the sequences comprising the myristoyl-CoA binding site are highly conserved, but the binding sites for protein substrates are not. This divergence has formed the basis for development of parasite-specific NMT inhibitors that interact with the peptide binding pocket of the enzyme. NMT is essential for viability of the parasites *Trypanosoma brucei* and *Leishmania major*, and a newly developed NMT inhibitor has been shown to kill trypanosomes and be curative in a mouse model of *T. brucei*-induced African sleeping sickness (Frearson et al., 2010). Inhibitors that target NMT from the *Plasmodium* parasite that causes malaria are also in preclinical development.

6.2 Inhibitors of S-Palmitoylation

One of the most commonly used inhibitors of S-palmitoylation is 2BP, a nonmetabolisable analogue of palmitate (Webb et al., 2000). Treatment of cells with 2BP blocks the S-palmitoylation of several dozen proteins, including Src family kinases, H-Ras, PSD95 and transmembrane receptors. 2BP enters cells and is converted to 2-bromopalmitoyl-CoA, but the presence of the 2-bromo group prevents breakdown of this fatty acyl-CoA by β-oxidation. 2BP has been shown to form an irreversible covalent adduct with DHHC PATs, as well as several known palmitoylated proteins. However, 2BP is not selective for protein palmitoylation, as it also inhibits a number of enzymes involved in lipid metabolism, including fatty acid-CoA ligase, glycerol-3-phosphate acyltransferase and triacylglycerol transferases. Thus, the mechanism of inhibition of protein palmitoylation may involve formation of an inhibitor: DHHC complex, formation of a covalent 2BP-modified substrate protein, or alterations in the levels or availability of intracellular palmitoyl-CoA.

A different approach has been taken to interfere with the function of S-palmitoylated proteins such as H- and N-Ras that rely on constitutive cycles of palmitoylation/depalmitoylation for activation at the plasma membrane (Figure 4(C)). Palmostatin B is an APT1 inhibitor that blocks depalmitoylation of H- and N-Ras. Treatment of cells with this small molecule inhibitor results in random redistribution of Ras proteins into all intracellular membranes and partial reversion of the oncogenic H-Ras transformed phenotype (Dekker et al., 2010).

6.3 Inhibitors of N-Palmitoylation and Palmitoleoylation

Given the importance of hedgehog and Wnt proteins in human diseases, the MBOAT enzymes Hhat and Porcupine are attractive targets for drug development. Aberrant expression of Shh in adults plays a key role in the initiation and maintenance of a number of human cancers, including pancreatic, lung and liver tumors. Hhat depletion or treatment with small molecule inhibitors of Hhat blocks Shh palmitoylation and signalling and reduces the growth of pancreatic cancer cells in vitro and in animal models of tumorigenesis (Petrova et al., 2013; Petrova and Resh, 2014). Likewise, Wnt signalling is one of the major pathways activated in human cancers. Small molecule inhibitors of Porcupine that block Wnt acylation, secretion and signalling have been developed. One of these compounds, LGK974, has been shown to block Wnt-driven breast cancer in animal models in vivo (Liu et al., 2013).

A cautionary note must be raised regarding inhibitors that target developmentally important signalling pathways, as these compounds would be teratogenic if used during pregnancy. Both Hedgehog and Wnt proteins, as well as Hhat and Porcupine, are required for normal embryonic development. For example, ingestion of the hedgehog pathway inhibitor cyclopamine, a natural product produced by the corn lily, results in developmental malformations such as holoprosencephaly and cyclopia. Point mutations in the Porcupine gene that occur in the X-linked disorder known as Goltz syndrome, or focal dermal hypoplasia (Wang et al., 2007), cause skin and digit abnormalities in affected individuals.

6.4 FTase and GGTase1 Inhibitors

Oncogenic mutations in Ras proteins occur in nearly one-third of all human cancers, making activated Ras a critical target in cancer therapeutics. Because farnesylation is required for the tumorigenic properties of the three major Ras isoforms, H-Ras, N-Ras and K-Ras4B, initial approaches focused on developing inhibitors of FTase. Approximately two dozen farnesyl transferase inhibitors (FTIs) have been synthesised and tested, including small molecules, peptidomimetics, FPP analogues and bisubstrate analogues (Berndt et al., 2011; Sebti, 2005). FTIs block H-Ras farnesylation, membrane binding and signalling in vitro and in cells. However, little or no effect of FTIs has been observed in nearly 75 different clinical trials when these compounds were tested for their ability to reduce the growth of solid tumors or hematopoietic cancers. Two key molecular mechanisms explain the failure of FTIs in the clinic. First, K-Ras is the most frequently mutated oncogene in human cancer, not H-Ras. Second, K-Ras can escape FTase inhibition because, on addition of FTIs, K-Ras instead is modified by geranylgeranylation and is fully active. GGTase1 inhibitors have been developed, but toxicity appears to be a major limiting factor for their use in vivo.

Although FTIs have not been successful as chemotherapeutics, these inhibitors show great promise in the treatment of a rare genetic disorder termed Hutchinson–Gilford progeria syndrome (HGPS). Patients with HGPS exhibit premature ageing and early death, primarily due to accelerated cardiovascular disease. HGPS is caused by a point mutation in the gene encoding laminA, a protein involved in maintaining the integrity of the nuclear lamina. The normal biosynthetic pathway for production of lamin A starts with the prelamin A precursor protein that contains a CAAX box at its C-terminus. Farnesylation triggers a zinc metalloprotease, ZMPSTE24, to cleave the 15 amino acid farnesylated tail, including the farnesylated cysteine to generate mature lamin A; this step is required for proper insertion into the nuclear lamina. In HGPS patients, a 50 amino acid region that contains the protease cleavage site is deleted.

The mutant protein, progerin, remains permanently farnesylated, accumulates in the nuclear lamina and disrupts integrity of the nuclear envelope. The progeria phenotype can be recapitulated in mice in which ZMPSTE24 is deleted or the HGPS mutation has been introduced. Clinical trials are currently being conducted using the FTI lonafarnib. Although not yet curative, the drug has significantly extended the lifespan of the treated patients (Gordon et al., 2014).

6.5 Defects in GPI Anchor Biosynthesis

Paroxysmal nocturnal haemoglobinuria (PNH) is a rare X-linked disorder in hematopoietic stem cells that causes severe haemolytic anaemia and bone marrow failure. More than 100 mutations have been identified in PNH patients within the PIG-A gene that encodes the protein required for transfer of N-acetylglucosamine from UDP-N-acetylglucosamine to phosphatidylinositol, the first step in the synthesis of GPI anchors (Luzzatto, 2006). The failure to correctly modify two GPI-anchored proteins, CD55 and CD59, leads to uncontrolled activation of complement, haemolysis and release of free haemoglobin, which likely explain the pathophysiology of the disease. Bone marrow transplantation and monoclonal antibodies that block the complement cascade have been used to treat PNH.

7. FUTURE DIRECTIONS AND CHALLENGES

The field of protein lipidation has expanded dramatically over the past decade, with the identification of new lipid modification reactions, and new lipid modifying enzymes and their protein substrates. The path ahead should now focus on delving deeper into the molecular mechanisms of the individual lipidation reactions. Three-dimensional structures of lipid modifying enzymes alone, in complex with a substrate, and in complex with an inhibitor will be required for a more complete molecular definition of reaction mechanisms. This is a daunting task, given the fact that many of these enzymes are multipass membrane proteins, but recent advances in in silico analysis of transmembrane proteins should accelerate structure determination. Crystal structures of lipid modified proteins will also be informative, as this information can reveal a role for lipidation in mediating protein:protein interactions. Our understanding of how protein and lipid substrates are selectively recognised by lipid transferases is also incomplete, particularly with regard to the large DHHC family of PATs. We need more comprehensive databases of lipidated proteins and better algorithms to predict the different types of protein lipidation. Another area of interest lies in the regulatory mechanisms that modulate protein S-acylation and deacylation states. Are all S-palmitoylated proteins continuously undergoing cycles of palmitoylation/depalmitoylation, and if so, how are lipidation dynamics regulated on a temporal and spatial level? Finally, inhibiting lipidation of proteins involved in human diseases is of great interest, and will require the development of drugs that can selectively, effectively and safely target the enzyme of interest.

REFERENCES

Ames, J.B., Tanaka, T., Stryer, L., Ikura, M., 1996. Portrait of a myristoyl switch protein. Curr. Opin. Struct. Biol. 6, 432–438.

Berndt, N., Hamilton, A.D., Sebti, S.M., 2011. Targeting protein prenylation for cancer therapy. Nat. Rev. Cancer 11, 775–791.

Bhatnagar, R.S., Futterer, K., Farazi, T.A., et al., 1998. Structure of N-myristoyltransferase with bound myristoylCoA and peptide substrate analogs. Nat. Struct. Biol. 5, 1091–1097.

Bijlmakers, M.J., 2009. Protein acylation and localization in T cell signaling. Mol. Membr. Biol. 26, 93–103.

Bivona, T.G., Quatela, S.E., Bodemann, B.O., et al., 2006. PKC regulates a farnesyl-electrostatic switch on K-Ras that promotes its association with Bcl-XL on mitochondria and induces apoptosis. Mol. Cell 21, 481–493.

Buglino, J.A., Resh, M.D., 2008. Hhat is a palmitoyl acyltransferase with specificity for N-palmitoylation of sonic hedgehog. J. Biol. Chem. 283, 22076–22088.

Buglino, J.A., Resh, M.D., 2012. Palmitoylation of Hedgehog proteins. Vitam. Horm. 88, 229–252.

Burnaevskiy, N., Fox, T.G., Plymire, D.A., et al., 2013. Proteolytic elimination of N-myristoyl modifications by the Shigella virulence factor IpaJ. Nature 496, 106–109.

Creanga, A., Glenn, T.D., Mann, R.K., Saunders, A.M., Talbot, W.S., Beachy, P.A., 2012. Scube/ You activity mediates release of dually lipid-modified Hedgehog signal in soluble form. Genes Dev. 26, 1312–1325.

Dekker, F.J., Rocks, O., Vartak, N., et al., 2010. Small-molecule inhibition of APT1 affects Ras localization and signaling. Nat. Chem. Biol. 6, 449–456.

Farazi, T.A., Waksman, G., Gordon, J.I., 2001. The biology and enzymology of protein N-myristoylation. J. Biol. Chem. 276, 39501–39504.

Frearson, J.A., Brand, S., McElroy, S.P., et al., 2010. N-myristoyltransferase inhibitors as new leads to treat sleeping sickness. Nature 464, 728–732.

Geng, J., Klionsky, D.J., 2008. The Atg8 and Atg12 ubiquitin-like conjugation systems in macro-autophagy. 'Protein modifications: beyond the usual suspects' review series. EMBO Rep. 9, 859–864.

Gordon, L.B., Massaro, J., D'Agostino Sr., R.B., et al., 2014. Impact of farnesylation inhibitors on survival in hutchinson-gilford progeria syndrome. Circulation 130, 27–34.

Hofmann, K., 2000. A superfamily of membrane-bound O-acyltransferases with implications for wnt signaling. Trends Biochem. Sci. 25, 111–112.

Janda, C.Y., Waghray, D., Levin, A.M., Thomas, C., Garcia, K.C., 2012. Structural basis of Wnt recognition by Frizzled. Science 337, 59–64.

Jennings, B.C., Linder, M.E., 2012. DHHC protein S-acyltransferases use similar ping-pong kinetic mechanisms but display different acyl-CoA specificities. J. Biol. Chem. 287, 7236–7245.

Kong, E., Peng, S., Chandra, G., et al., 2013. Dynamic palmitoylation links cytosol-membrane shuttling of acyl-protein thioesterase-1 and acyl-protein thioesterase-2 with that of proto-oncogene H-ras product and growth-associated protein-43. J. Biol. Chem. 288, 9112–9125.

Lee, J.D., Treisman, J.E., 2001. Sightless has homology to transmembrane acyltransferases and is required to generate active Hedgehog protein. Curr. Biol. 11, 1147–1152.

Levental, I., Grzybek, M., Simons, K., 2010. Greasing their way: lipid modifications determine protein association with membrane rafts. Biochemistry 49, 6305–6316.

Linder, M.E., Deschenes, R.J., 2007. Palmitoylation: policing protein stability and traffic. Nat. Rev. Mol. Cell Biol. 8, 74–84.

Linder, M.E., Jennings, B.C., 2013. Mechanism and function of DHHC S-acyltransferases. Biochem. Soc. Trans. 41, 29–34.

Liu, J., Pan, S., Hsieh, M.H., et al., 2013. Targeting Wnt-driven cancer through the inhibition of Porcupine by LGK974. Proc. Natl. Acad. Sci. U. S. A. 110, 20224–20229.

Luzzatto, L., 2006. Paroxysmal nocturnal hemoglobinuria: an acquired X-linked genetic disease with somatic-cell mosaicism. Curr. Opin. Genet. Dev. 16, 317–322.

Mann, R.K., Beachy, P.A., 2004. Novel lipid modifications of secreted protein signals. Annu. Rev. Biochem. 73, 891–923.

Martin, D.D., Beauchamp, E., Berthiaume, L.G., 2011. Post-translational myristoylation: Fat matters in cellular life and death. Biochimie 93, 18–31.

Maurer-Stroh, S., Eisenhaber, B., Eisenhaber, F., 2002. N-terminal N-myristoylation of proteins: prediction of substrate proteins from amino acid sequence. J. Mol. Biol. 317, 541–557.

Miura, G.I., Buglino, J., Alvarado, D., Lemmon, M.A., Resh, M.D., Treisman, J.E., 2006. Palmitoylation of the EGFR ligand Spitz by Rasp increases Spitz activity by restricting its diffusion. Dev. Cell 10, 167–176.

Murray, D., Ben-Tal, N., Honig, B., McLaughlin, S., 1997. Electrostatic interaction of myristoylated proteins with membranes: simple physics, complicated biology. Structure 5, 985–989.

Nakayama, H., Kurokawa, K., Lee, B.L., 2012. Lipoproteins in bacteria: structures and biosynthetic pathways. FEBS J. 279, 4247–4268.

Nguyen, U.T., Goody, R.S., Alexandrov, K., 2010. Understanding and exploiting protein prenyltransferases. Chembiochem: Eur. J. Chem. Biol. 11, 1194–1201.

Orlean, P., Menon, A.K., 2007. Thematic review series: lipid posttranslational modifications. GPI anchoring of protein in yeast and mammalian cells, or: how we learned to stop worrying and love glycophospholipids. J. Lipid Res. 48, 993–1011.

Paulick, M.G., Bertozzi, C.R., 2008. The glycosylphosphatidylinositol anchor: a complex membrane-anchoring structure for proteins. Biochemistry 47, 6991–7000.

Peitzsch, R.M., McLaughlin, S., 1993. Binding of acylated peptides and fatty acids to phospholipid vesicles: pertinence to myristoylated proteins. Biochemistry 32, 10436–10443.

Pepinsky, R.B., Zeng, C., Wen, D., et al., 1998. Identification of a palmitic acid-modified form of human Sonic hedgehog. J. Biol. Chem. 273, 14037–14045.

Petrova, E., Rios-Esteves, J., Ouerfelli, O., Glickman, J.F., Resh, M.D., 2013. Inhibitors of Hedgehog acyltransferase block Sonic Hedgehog signaling. Nat. Chem. Biol. 9, 247–249.

Petrova, E., Resh, M.D., 2014. Hedgehog acyltransferase as a target in pancreatic ductal adenocarcinoma. Oncogene. http://dx.doi.org/10.1038/onc.2013.575, advance online publication.

Prior, I.A., Hancock, J.F., 2012. Ras trafficking, localization and compartmentalized signalling. Semin. Cell Dev. Biol. 23, 145–153.

Rak, A., Pylypenko, O., Durek, T., et al., 2003. Structure of Rab GDP-dissociation inhibitor in complex with prenylated YPT1 GTPase. Science 302, 646–650.

Resh, M.D., 2006. Trafficking and signaling by fatty-acylated and prenylated proteins. Nat. Chem. Biol. 2, 584–590.

Resh, M.D., 2006. Palmitoylation of ligands, receptors, and intracellular signaling molecules. Sci. STKE 2006, re14.

Resh, M.D., 2012. Targeting protein lipidation in disease. Trends Mol. Med. 18, 206–214.

Resh, M.D., 2013. Covalent lipid modifications of proteins. Curr. Biol. 23, R431–R435.

Rios-Esteves, J., Resh, M.D., 2013. Stearoyl CoA desaturase is required to produce active, lipid-modified Wnt proteins. Cell Rep. 4, 1072–1081.

Rocks, O., Peyker, A., Kahms, M., et al., 2005. An acylation cycle regulates localization and activity of palmitoylated Ras isoforms. Science 307, 1746–1752.

Sebti, S.M., 2005. Protein farnesylation: implications for normal physiology, malignant transformation, and cancer therapy. Cancer Cell 7, 297–300.

Takada, R., Satomi, Y., Kurata, T., et al., 2006. Monounsaturated fatty acid modification of Wnt protein: its role in Wnt secretion. Dev. Cell 11, 791–801.

Tang, C., Loeliger, E., Luncsford, P., Kinde, I., Beckett, D., Summers, M.F., 2004. Entropic switch regulates myristate exposure in the HIV-1 matrix protein. Proc. Natl. Acad. Sci. U. S. A. 101, 517–522.

Tukachinsky, H., Kuzmickas, R.P., Jao, C.Y., Liu, J., Salic, A., 2012. Dispatched and scube mediate the efficient secretion of the cholesterol-modified hedgehog ligand. Cell Rep. 2, 308–320.

Wang, X., Reid Sutton, V., Omar Peraza-Llanes, J., et al., 2007. Mutations in X-linked PORCN, a putative regulator of Wnt signaling, cause focal dermal hypoplasia. Nat. Genet. 39, 836–838.

Webb, Y., Hermida-Matsumoto, L., Resh, M.D., 2000. Inhibition of protein palmitoylation, raft localization, and T cell signaling by 2-bromopalmitate and polyunsaturated fatty acids. J. Biol. Chem. 275, 261–270.

Yang, J., Brown, M.S., Liang, G., Grishin, N.V., Goldstein, J.L., 2008. Identification of the acyltransferase that octanoylates ghrelin, an appetite-stimulating peptide hormone. Cell 132, 387–396.

Chapter 14

Intramembrane and Intermembrane Lipid Transport

Frederick R. Maxfield, Anant K. Menon

Department of Biochemistry, Weill Cornell Medical College, New York, NY, USA

ABBREVIATIONS

ABC ATP-binding cassette
CERT Ceramide transporter
DHE Dehydroergosterol
ER Endoplasmic reticulum
ERC Endocytic recycling compartment
IM Mitochondrial inner membrane
LTP Lipid transfer protein
MAM Mitochondria-associated membrane
MCS Membrane contact site
MPD Mannose phosphate dolichol
NBD Nucleotide binding domain
OM Mitochondrial outer membrane
PI Phosphatidylinositol
PI4P Phosphatidylinositol 4-phosphate
PM Plasma membrane
SM Sphingomyelin
StAR *STeroidogenic Acute Response*
TGN *trans* Golgi network
TULIP Tubular lipid binding

1. INTRODUCTION

Every organelle in a eukaryotic cell has a distinct membrane protein composition, which is essential for the organelles to carry out their specialised functions. The mechanisms for sorting proteins among organelles are mainly based on specific protein–protein interactions that are now understood in great detail (Schmid et al., 2014). In general, certain membrane proteins are selected from the parent organelle by binding to cytoplasmic scaffolding protein complexes that also are involved in forming a bud or tubule that will form a transport vesicle. These transport vesicles then fuse with an acceptor organelle or the plasma membrane (PM). Selective removal and

Biochemistry of Lipids, Lipoproteins and Membranes. http://dx.doi.org/10.1016/B978-0-444-63438-2.00014-6

delivery of membrane vesicles generates the unique protein composition of each organelle.

The lipid composition also varies greatly among organelles (van Meer et al., 2008; Holthuis and Menon, 2014). The mechanisms for sorting lipids are not as well understood as the mechanisms for sorting proteins. Several general mechanisms for maintaining distinct lipid compositions in organelles will be introduced here and described in some detail in later sections. One mechanism is sorting during the formation of the vesicles and tubules required for vesicular trafficking processes. Certain lipids can be relatively enriched or depleted in the forming buds, and this will alter the lipid composition of the donor and acceptor organelles. A second mechanism for transporting lipids to selective organelles involves the use of cytoplasmic lipid transport proteins (LTPs) that can extract a lipid from one organelle and deliver it to another. Finally, lipids can be enzymatically modified in select organelles, leading to distinct lipid compositions. All these mechanisms to sort lipids depend on the proteins that are associated with the various organelles, so lipid distribution and sorting is inextricably linked with protein trafficking. In addition to movement between organelles, lipids are flipped between leaflets of a membrane, and the transbilayer distribution of lipids can have functional consequences.

Some important aspects of lipid distribution are illustrated in Figure 1. In mammalian cells, the PM has a high concentration of cholesterol (~35% of the lipid molecules), and its exofacial leaflet is enriched in sphingomyelin (SM) (van Meer et al., 2008).

The PM is a comparatively thick, planar bilayer with few voids between lipid headgroups (i.e. packing defects) (Holthuis and Menon, 2014). By comparison, the endoplasmic reticulum (ER), which is the site of synthesis of many lipids, including cholesterol, has only about 5% cholesterol (Radhakrishnan et al., 2008) and virtually no SM. The ER is a relatively thin, highly curved membrane with major packing defects between lipid groups (Holthuis and Menon, 2014). Phosphatidylserine (PS) is found in many organelles. At the PM, PS is nearly exclusively on the cytoplasmic leaflet, and it is present at high levels compared to other organelles, leading to a highly negative surface charge on the PM. This high negative charge can be used to recruit selective cytoplasmic proteins to the PM (Leventis and Grinstein, 2010). While PS is normally not detectable at the cell surface, there are specific circumstances when it becomes surface exposed. Thus, recognition and clearance of apoptotic cells by macrophages requires the dying cells to transfer PS from the cytoplasmic leaflet of the PM to the exofacial leaflet (Leventis and Grinstein, 2010). Activated blood platelets expose PS as a necessary step in promoting blood coagulation (Bevers and Williamson, 2010).

Many lipids are synthesised in the ER and then transported to other organelles. Compared to other organelles, the ER has a much higher level of unsaturated lipids, and it does not contain significant transbilayer lipid asymmetry, indicating that there are mechanisms for flipping lipids across this membrane relatively rapidly. An important question is how organelles retain their distinct

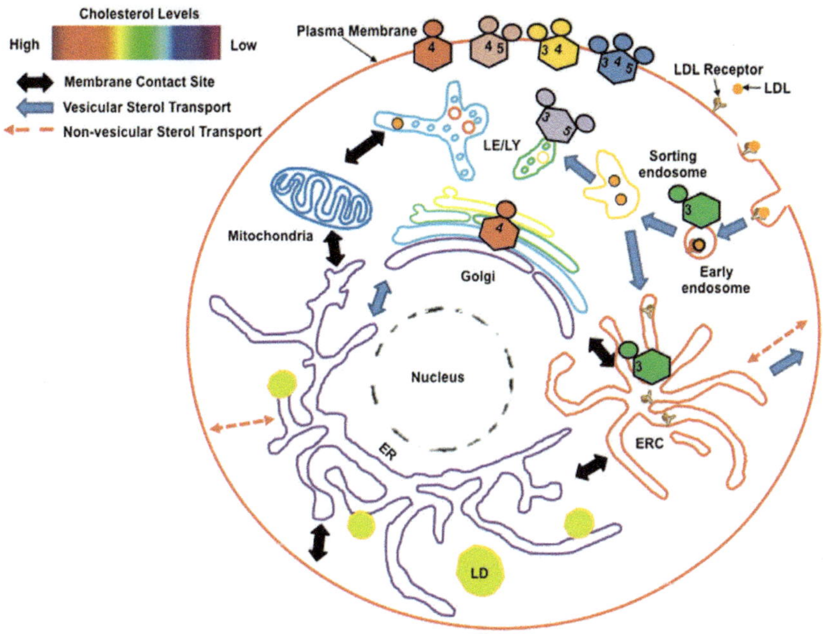

FIGURE 1 **Lipid distribution and transport in a mammalian cell.** Most phospholipids are synthesised in the ER and exit by vesicular or nonvesicular transport. Sphingomyelin and glycosphingolipids are synthesised in the Golgi and leave by vesicular transport. Organelles are colour-coded to reflect the relative level of cholesterol. (The cholesterol levels for the limiting membrane of the late endosomes and lysosomes are uncertain because of the cholesterol associated with lipoproteins.) As an example, pathways of intracellular cholesterol transport, both vesicular and nonvesicular are shown. Low density lipoprotein (LDL) particles are internalised by LDL receptors and transported to sorting endosomes. LDL-derived cholesterol, which is released on cholesterol ester hydrolysis in late endosomes and lysosomes, requires proteins NPC1 and NPC2 to leave these organelles. Excess cholesterol is esterified and packed into lipid droplets (LD). Phosphoinositides serve as organelle identity molecules. The structures and locations of phosphatidylinositol 3-phosphate, phosphatidylinositol 4-phosphate, phosphatidylinositol-(3,5)-bisphosphate, phosphatidylinositol (4,5)-bisphosphate and phosphatidylinositol (3,4,5)-trisphosphate are indicated by coloured inositol rings with numbered phosphates. ERC, endocytic recycling compartment.

lipid compositions in the presence of high levels of membrane vesicular transport in the secretory and endocytic pathways (Holthuis and Menon, 2014).

In the secretory pathway, the abundance of sterols and sphingolipids generally increases on passage from the ER through the Golgi to the PM (Figure 1). In mammalian cells, the cholesterol levels in the Golgi are generally intermediate between the levels in the ER and the PM. Secretory vesicles in yeast contain 2.3 times more ergosterol than the *trans* Golgi network (TGN) from which they are derived (Klemm et al., 2009). Similarly, in pancreatic beta cells the secretory vesicles containing insulin have high levels of cholesterol (Bogan et al., 2012).

In the endocytic pathway, the composition of early sorting endosomes is generally similar to the PM from which they are derived. In the endocytic recycling compartment, cholesterol is highly enriched (Mesmin and Maxfield, 2009), and SM (van Meer et al., 2008) and PS (Leventis and Grinstein, 2010) are also abundant. In late endosomes, the cholesterol ester and triglyceride cores of lipoproteins are hydrolysed to produce lipids for transfer to other organelles. Late endosomes contain inward invaginations and internal vesicles (sometimes termed multivesicular bodies) that are enriched in bis-monoacylglycerophosphate (BMP), which is not found at significant concentrations in other organelles, but constitutes about 15% of the lipids in late endosomes (Kobayashi et al., 1998; Gruenberg, 2001). These internal membranes are sites of digestion of glycerolipids and sphingolipids, thus avoiding damage to the limiting membrane of the endosomes (Kolter and Sandhoff, 2010).

Differences in lipid composition help to identify organelles. The phosphoinositides are some of the most important lipids for establishing organelle identity (Balla, 2013) (Figure 1). Phosphatidylinositol (PI), the precursor of phosphoinositides, is synthesised primarily in the ER, and typically represents about 15% of the total phospholipids found in eukaryotic cells (See Chapter 7). Phosphatidylinositol-(4,5)-bisphosphate ($PI(4,5)P_2$) is the most abundant phosphoinositide in the PM, where it represents about 1% of the phospholipids (Di Paolo and De Camilli, 2006). The Golgi apparatus is enriched in PI4P; early sorting endosomes are enriched in PI3P; and late endosomes and yeast vacuoles have high levels of $PI(3,5)P_2$ (Balla, 2013). These different phosphoinositides are specifically recognised by cytosolic proteins, for example through pleckstrin homology (PH) domains, leading to selective recruitment to the surface of specific organelles. Often the phosphoinositide-binding proteins also interact with other proteins (e.g. Rabs and other small GTPases) that carry out specific functions on these organelles. The mechanisms used to establish and maintain the phosphoinositide identity of organelles will be discussed later in this chapter.

2. VESICULAR TRAFFICKING OF LIPIDS

Large quantities of lipid are moved among organelles by vesicular trafficking in the endocytic and secretory pathways. It has been estimated, for example, that the $t_{1/2}$ for internalisation of lipids from the PM is about 15 min (Hao and Maxfield, 2000). Most of these lipids are returned to the PM by efficient recycling processes that allow the PM to retain its lipid composition even with such rapid endocytic flux (Maxfield and McGraw, 2004). Nearly all vesicle trafficking routes are associated with efficient recycling pathways, which can similarly help to maintain the lipid composition of the donor organelles.

Although most endocytosed lipids return to the PM, some lipids are selectively excluded from this efficient recycling. As one example, fluorescent lipid analogues with different degrees of unsaturation and hydrocarbon tail length follow different endocytic routes after internalisation from the PM (Mukherjee et al., 1999). Certain bacterial toxins (e.g. shiga or cholera toxins) bind to

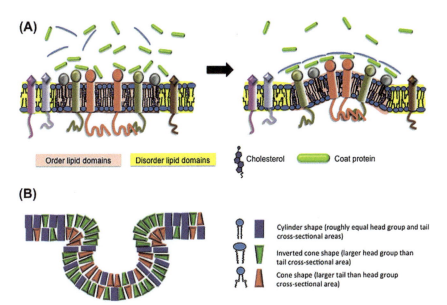

FIGURE 2 **Effect of lipid shapes on sorting.** Two putative mechanisms for sorting lipids during vesicle budding are illustrated. (A) Proteins associated with a particular type of microdomain (e.g. cholesterol-enriched ordered lipid domains) may preferentially associate with coat proteins that form a budding vesicle. (B) The shape of lipids can affect their ability to be accommodated in regions of high curvature present at sites of vesicle budding. Cone-shaped lipids (red) are preferred in regions of high negative curvature in the neck region of a vesicle bud, whereas inverted cone lipids (green) are better accommodated regions of positive curvature in the lumenal leaflet of the neck region. Different curvature preferences might also be found in the body of the budding vesicle or tubule.

glycolipids on the PM and follow a pathway from endosomes to the Golgi and finally the ER, where they use the protein translocation machinery to cross into the cytoplasm (Chinnapen et al., 2012).

The molecular basis for sorting of lipids is not well understood. Two types of mechanisms can be considered either based on formation of lipid nanodomains (Figure 2(A)) or on preferences of certain lipids to be included or excluded from highly curved buds and tubules that form transport vesicles (Figure 2(B)). Lipids can form nanodomains by association with proteins or by lipid–lipid sorting. In many cases, the sorting of lipids is altered by reducing the cholesterol content of cells, which would be consistent with membrane biophysical properties playing an important role in the sorting of lipids.

One possibility is that there could be coexisting nanodomains with different lipid compositions in a parent organelle and that different sets of proteins would be associated with these nanodomains. If one set of proteins contained motifs that would be recognised for incorporation into budding vesicles or tubules (e.g. by binding to coat proteins), then their associated lipids would also be preferentially included in the budding vesicles. As a specific example, some proteins preferentially associate with cholesterol-enriched ordered membrane domains, so sorting of these proteins into a transport vesicle would also enrich the vesicle in choles-terol and lipids with highly saturated acyl chains (Simons and Ikonen, 1997).

The second mechanism for lipid sorting is based on the shape and curvature preferences of various lipids (van Meer et al., 2008; Holthuis and Menon, 2014) (Figure 2(B)). Lipids are described as being cylindrical if the lateral areas of the headgroup and the acyl chain are nearly the same. Cone-shaped lipids have headgroup areas smaller than the area of the acyl chain, and inverted cone-shaped lipids have headgroups larger than the acyl chain area. Sites of vesicle and tubule budding often have high and complex curvature, including 'saddle' curvature at the bud neck. Lipids with different shapes can be preferentially included or excluded from regions of high curvature. For example, in a highly curved membrane vesicle, inverted cone-shaped lipids have a preference for the cytoplasmic leaflet, while cone-shaped lipids would fit better on the luminal side. These are generally not strong preferences because of the dynamic nature of the acyl chain motions, but these slight preferences can have significant effects during multiple cycles of vesicle formation.

An interesting example of the interplay of protein:lipizd interactions and cur-vature effects has been demonstrated for the ATP-dependent Drs2 phospholipid flippase in yeast (Xu et al., 2013; Hankins et al., 2015). By flipping PS from the lumenal to the cytoplasmic leaflet of the membrane, Drs2 imparts curvature to the bilayer and also increases the local negative charge. Both the curvature and the charge help to recruit an Arf GTPase activating protein (Gcs1) to the TGN and early endosome membranes. The activated Arf then plays a key role in recruiting AP-1 and clathrin to the bud site, and this leads to the formation of transport vesicles.

An important mechanism for maintaining distinct lipid compositions in various organelles is the precise localisation and activation of various lipid-modifying enzymes. The modifications of phosphoinositides during endocy-tosis and vesicle transport are examples of the importance of these processes. Many of the details of PI-kinase and phosphatase function in membrane transport pathways remain to be worked out, but a general outline is begin-ning to emerge (Balla, 2013). $PI(4,5)P_2$ is relatively abundant in the cytoplas-mic leaflet of the PM, where it plays an essential role in the binding of the clathrin adaptor protein, AP-2, which helps to recruit clathrin in early stages of coated pit formation. $PI(4,5)P_2$ 5-phosphatases are recruited to maturing coated pits and are activated in the coated pits or shortly after the formation of sealed vesicles. While at the PM, PI4P 5-kinases can maintain $PI(4,5)P_2$ levels, but after vesicle formation the 5-phosphatases convert this to PI4P. A PI3-kinase is also recruited to the forming coated pits, leading to production of $PI(3,4)P_2$ on the newly formed vesicles. At some point afterwards, the $PI(3,4)P_2$ is converted to PI3P, which is the main phosphoinositide species on the early endosomes. It is possible that the 4-phosphatase responsible for this step is IPNN4A/B, which is activated by the small GTPase Rab5 on early endosomes. While the precise orchestration of these enzymatic con-versions is still uncertain, a model in which budding vesicles recruit the enzymes necessary to carry out the next steps of phosphoinositide conversions may provide a general framework for maintaining distinct PI identities for various organelles.

3. NONVESICULAR TRANSPORT OF LIPIDS

Many lipids are synthesised in the ER and must be transported to other cellular destinations. While vesicle-mediated transport provides an obvious means of ferrying and sorting these lipids between stations of the secretory pathway, nonvesicular mechanisms come into play for lipid transfer involving organelles such as mitochondria that are not connected to the endomembrane system by vesicular transport (Lev, 2010; Holthuis and Menon, 2014). Interestingly, nonvesicular mechanisms are also important in moving lipids between organelles that are linked by the secretory pathway. Thus, cholesterol movement between the ER and PM occurs principally by nonvesicular means (Holthuis and Menon, 2014).

As presently understood, nonvesicular transport requires LTPs that exchange lipids between the cytoplasmic face of cellular organelles. These proteins may diffuse freely in the cytoplasm or be membrane bound. As discussed below, some LTPs may operate at membrane contact sites (MCSs) – regions of the cell where specific pairs of membranes are in close proximity.

3.1 Lipid Transport Proteins

Cytoplasmic LTPs were originally identified because of their ability to exchange lipids between microsomes and mitochondria (Wirtz and Zilversmit, 1968) or populations of vesicles in the test tube. LTP-mediated exchange has been extensively studied using such in vitro systems, and crystal structures are available for several LTPs, including members of the START domain family (Tsujishita and Hurley, 2000; Kudo et al., 2008) and Osh4/Kes1 (Im et al., 2005).

These studies reveal that the role of LTPs is at least two-fold: to accomplish the energetically difficult task of extracting a lipid from the membrane, and to provide an environment, a binding pocket, that shields the lipid from the aqueous medium of the cytoplasm. A flexible lid over the lipid-binding pocket allows LTPs to extract, present and deposit lipids ('lid open'), while also protecting them ('lid closed') during transit through the cytosol.

LTPs belonging to the START domain family are ubiquitous in plants, bacteria and animals (Clark, 2012). There are 15 mammalian START family members, which contain a domain of about 210 amino acids with an α-helix/β-sheet grip structure that has an interior pocket that can accommodate lipids (STARD3 complex with cholesterol is shown in Figure 3). The founding member of the family, STARD1 or StAR, plays an essential role in delivering cholesterol to the mitochondrial enzymes that convert sterol into steroid hormones. Several other START proteins bind cholesterol and/or oxysterols, but members of the STARD2 subfamily bind phospholipids (STARD2, 7 and 10) and STARD11 (more often

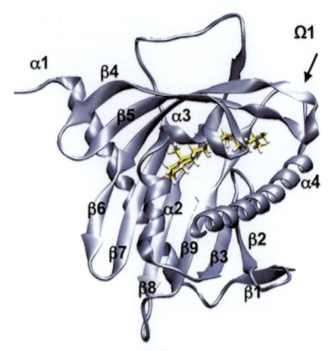

FIGURE 3 **Model of cholesterol in an START domain.** A cholesterol molecule was modelled in the interior binding pocket of STARD3 (MLN64) using a docking algorithm. This figure was originally included in Murcia et al. (2006) © the American Society for Biochemistry and Molecular Biology. The structure of the STARD3 START domain was obtained by X-ray crystallography (Tsujishita and Hurley, 2000).

named CERT (ceramide transporter)) (Kudo et al., 2008) binds ceramide. Conformational changes are required for lipids to enter or leave the internal cavity, but the precise nature of the conformational changes is still under investigation. One suggested mechanism, based on molecular dynamics simulations (Figure 3) (Murcia et al., 2006), involves movement of a loop (Ω1) that connects two antiparallel β strands. Movement of the loop away from the core of the protein opens the lid of the binding pocket in a way that would allow cholesterol to slide in or out. Using NMR measurements, it has been found that the Ω1 loop and a nearby α-helix are conformationally dynamic on the picosecond/nanosecond time scale, but it remains to be seen whether such motions can be demonstrated experimentally on interaction with membranes. An alternate model has been proposed in which the C-terminal α-helix (α 4) unfolds slightly to allow entry of exit of cholesterol (Roostaee et al., 2009). There is experimental evidence that this dynamic helix is stabilised by cholesterol binding to STARD1.

3.2 Membrane Contact Sites

Many LTPs are composed of only a lipid binding pocket. However, many other LTPs possess motifs that allow them to interact with specific membranes,

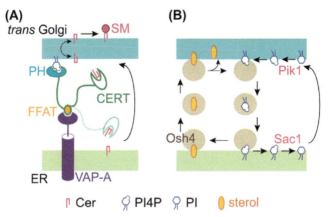

FIGURE 4 **Models of lipid exchange by LTPs.** (A) CERT-mediated ceramide exchange between the ER and *trans*-Golgi network in mammalian cells. CERT engages PI4P at the *trans*-Golgi via its PH domain, and VAP-A at the ER via its FFAT motif, catalysing transfer of ceramide between the two organelles. The illustration depicts CERT simultaneously engaging both membranes, creating a membrane contact site. On deposition at the *trans*-Golgi, ceramide flips spontaneously to the luminal side, where it is converted to sphingomyelin (SM). Although CERT facilitates bidirectional exchange of ceramide, its conversion to SM results in unidirectional flux. (B) Osh4, an oxysterol binding protein homologue in yeast, binds sterol or PI4P in a mutually exclusive manner. The figure shows Osh4 acquiring sterol at the ER and exchanging it for PI4P at the *trans*-Golgi network. PI4P is subsequently off-loaded at the ER where it is converted to PI by the ER-localised PI4P phosphatase Sac1. Osh4 then picks up another sterol molecule at the ER and the cycle continues. The PI released at the ER is transferred to the *trans*-Golgi network in exchange for PC; this requires a PI/PC exchanger (not depicted). At the *trans*-Golgi network it is converted to PI4P by the kinase Pik1. The hydrolysis of PI4P by Sac1 at the ER, and its synthesis by Pik1 at the *trans*-Golgi network, confers directionality to this system.

ensuring that lipid exchange is largely restricted to pairs of membrane compartments (Holthuis and Menon, 2014; Prinz, 2014). For example, the ceramide transport protein CERT has two membrane targeting motifs: a 'two phenylalanines in an acidic tract' (FFAT) motif through which it interacts with VAMP-associated proteins (VAP-A and VAP-B) on the ER, and a PH domain that promotes binding to PI4P in the Golgi apparatus. These interactions ensure that ceramide is transferred from the ER where it is synthesised to the Golgi apparatus where it is consumed in the synthesis of sphingolipids, without significant transfer to other membranes (see Chapter 10). To accomplish this, CERT could engage the ER and Golgi compartments alternately, or simultaneously. In the latter scenario, CERT would act as a physical bridge between the ER and a Golgi compartment, creating an MCS that would allow continuous exchange of ceramide (Figure 4(A)).

Like CERT, members of the mammalian oxysterol binding protein (OSBP) and yeast OSBP homology (Osh) families have FFAT and PH domain organelle targeting motifs. In contrast, yeast Osh4 is composed primarily of a lipid binding domain for sterols and PI-4P and does not have PH or FFAT motif. Nevertheless, Osh4 targets the ER and late Golgi compartments, and mutually exclusive binding of sterols and PI4P could therefore transfer both lipids between these membranes. It is proposed that Osh4 catalyses heterotypic lipid exchange (de Saint-Jean et al., 2011), driving sterol from the ER to late Golgi compartments as follows (Figure 4(B)). Osh4 binds sterol at the ER (where PI-4P levels are low) and deposits it in a late Golgi compartment in exchange for PI-4P. On a subsequent encounter with the ER, Osh4 releases PI-4P, which is promptly metabolised to PI by Sac1, an ER-localised PI-4P phosphatase. This allows Osh4 to pick up sterol again, and the cycle continues. The hydrolysis of PI-4-P ensures the unidirectional transfer of sterol between the ER and late Golgi complex. A similar mechanism has been described for cholesterol transport to the Golgi by mammalian OSBP.

While it is unclear whether CERT and Osh4/OSBP play a role in creating or maintaining an MCS, there is little doubt that MCSs exist in cells (Levine, 2004; Prinz, 2014). They are evident in electron micrographs that reveal regions in the cell in which two membranes persist within 30 nm of each other. They are commonly found between the ER and other cellular structures, notably the PM and mitochondria. Indeed, in yeast cells, approximately half the surface area of the PM is associated with the ER (Manford et al., 2012). Physical connections between the ER and mitochondria, now appreciated to be due to ER-mitochondria MCSs, were also recognised in early subcellular fractionation studies that revealed that a subfraction of the ER termed the 'mitochondria-associated membrane' (MAM) could be recovered with the rapidly sedimenting mitochondrial fraction during differential centrifugation of a cell homogenate (Vance, 2014). The MAM was subsequently shown to be enriched in phospholipid biosynthesis enzymes and to play a role in promoting phospholipid exchange between the ER and mitochondria, specifically the transport of PS from the ER to the mitochondrial inner membrane (IM) for decarboxylation to PE and the subsequent export of PE to the ER for methylation to PC (see Section 5.2).

MCSs are maintained by protein bridges (tethers) that are anchored in one membrane but interact directly or indirectly with lipids or proteins in another membrane. For example, the ER-localised extended synaptotagmins (E-Syt proteins, known as tricalbins in yeast) have been proposed to interact with the PM via their Ca^{2+} and PS-binding C2 domains (Manford et al., 2012; Prinz, 2014). These proteins represent some of the many tethers that maintain large regions of the ER in close proximity to the PM. Elimination of multiple tethers results in the quantitative loss of membrane contacts – thus, abrogation of six tethering proteins in yeast, including yeast counterparts of E-Syt and VAP-A, results in a 90% loss of contacts between the ER and PM (Manford et al., 2012). Although tethering proteins may help to create an MCS within which LTPs could operate, E-Syts have a structural domain – the SMP domain, related to the 'tubular lipid binding (TULIP)' superfamily of lipid-binding proteins (Kopec et al., 2010) – that appears to be able to facilitate lipid transport. Thus the tethers may represent a novel class of membrane-anchored LTPs. The crystal structure of an SMP domain from a mammalian

E-Syt (Schauder et al., 2014) reveals a hydrophobic groove that could accommodate the acyl chains of phospholipids while allowing the lipid headgroup to interact with water. Thus, a phospholipid in one membrane could be 'swiped' through the SMP domain to enter the apposing membrane leaflet at an MCS.

4. TRANSBILAYER MOVEMENT OF LIPIDS

Lipids move between membranes as discussed above, but also across them (Sanyal and Menon, 2009; Coleman et al., 2013; Hankins et al., 2015). The flipping of lipids across membrane bilayers is necessary to generate and maintain lipid asymmetry, support membrane growth and vesicle formation, and enable the biosynthesis of a variety of glycoconjugates such as N-glycoproteins and, in bacteria, cell wall peptidoglycan. For most polar lipids, the spontaneous rate of flip–flop is slow (time-scale ~ tens of hours), because it is energetically costly (>20 kcal/mol for common phospholipids) to transfer the lipid headgroup through the hydrophobic interior of the bilayer. However, for lipids such as ceramide, diacylglycerol and cholesterol that have a small polar headgroup, spontaneous flip–flop is rapid and occurs on a time-scale of seconds.

Cells deploy a variety of transport proteins – termed flippases, floppases and scramblases – to accelerate the slow flip–flop of polar lipids to a physiologically appropriate rate (Figure 5). Flippases and floppases are ATPases, whereas scramblases operate in an ATP-independent manner.

4.1 Flippases

Flippases move phospholipids inwards, from the exoplasmic to the cytoplasmic leaflet of membranes. They belong to the family of P-type ATPases (see below for flippase activity of some ABC transporters), the members of which include the ion transporting Na^+/K^+-ATPase and the Ca^{2+}-ATPase. These proteins couple ATP hydrolysis to the transport of substrates against their concentration gradient (Figure 5(A)). In so doing, they cycle between a number of conformational states, auto-phosphorylating and dephosphorylating a conserved aspartate residue within a conserved signature sequence (hence the designation 'P-type'). The flippase subfamily of P-type ATPases is classified as type IV, so flippases are referred to as P4-ATPases; several P4-ATPases have been linked to severe human diseases (e.g. Angelman syndrome) (Coleman et al., 2013). Although the activity of flippases was measured in the 1980s in pioneering work done with red blood cells, their molecular identity was only revealed more recently in yeast, where they constitute a family of five proteins: Dnf1, Dnf2, Dnf3, Drs2 and Neo1 (there are 14 P4-ATPases in humans) (Coleman et al., 2013; Hankins et al., 2015). Dnf1 and Dnf2 are localised to the PM, whereas Dnf3 and Drs2 are found in late Golgi compartments, and Neo1 localises to the early secretory pathway. Most, if not all, P4-ATPases associate with an accessory subunit (Cdc50) that is necessary for their correct subcellular localisation and may also

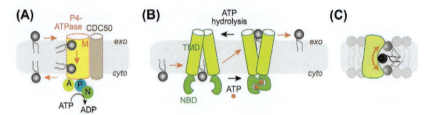

FIGURE 5 **Proteins mediating transbilayer lipid movement.** (A) Flippase activity is due to P4-ATPases and their associated CDC50 subunit. The P4-ATPase contains a transmembrane domain (M) with 10 predicted membrane spanning segments that form the lipid translocation pathway, and three domains (A, actuator; P, phosphorylation; N, nucleotide binding) that are embodied within cytosolically oriented loops. The N-domain binds ATP and serves as a built-in protein kinase, which phosphorylates the P-domain. The A-domain is an intrinsic protein phosphatase, which dephosphorylates the P-domain once during each catalytic cycle. Conformational changes that occur during this cycle lead ultimately to the translocation of a phospholipid from the exofacial (*exo*) to the cytoplasmic (*cyto*) leaflet. (B) The floppase ABC family of transporters contain two TMDs, and two NBDs. The transporter is open towards the cytoplasmic (*cyto*) side, but upon ATP binding, the NBDs dimerise prompting a conformational change such that the transporter now opens to the exofacial (*exo*) side. The system resets once ATP is hydrolysed and ADP is discharged. Phospholipids enter the transporter cavity from the cytoplasmic side and are expelled on the opposite side upon ATP binding. (C) Scramblases move phospholipids bidirectionally across the membrane, using a low energy pathway.

contribute to their activity. In the absence of one or more of these proteins, lipid asymmetry at the PM is perturbed – aminophospholipids are exposed at the cell surface – and vesicular transport can be affected.

4.2 Floppases

The transfer of lipids from the cytoplasmic to the exoplasmic leaflet of membranes is termed 'flop'. Flopping is catalysed by members of the ATP binding cassette (ABC) transporter family (Coleman et al., 2013), although some members of this family can also act as flippases. Eukaryotic ABC transporters are polytopic membrane proteins. They are synthesised as half transporters with six transmembrane spans denoted as the TMD (transmembrane domain) and a nucleotide-binding domain (NBD) that dimerise with other half transporters to form a functional unit. Alternatively, they are synthesised as full transporters with a modular organisation of TMD–NBD–TMD–NBD. In some cases the modular arrangement can be different. NBDs have the Walker A and Walker B motifs that are common among ATPases, as well as signature motifs of the ABC transporter family.

ABC transporters may use an alternating access mechanism for transporting lipid: depending on whether ATP is bound to the NBDs, the structure is open to the cytoplasmic side or luminal side of the membrane (Figure 5(B)). Lipid would enter the transporter on the cytoplasmic side in the absence of ATP and be extruded to the opposing leaflet as the transporter binds ATP and undergoes a conformational change. ATP hydrolysis within the NBD would then reset the protein to accept another lipid on the cytoplasmic side.

Humans have almost 50 different ABC transporters that are grouped into seven subfamilies (A–G) (Coleman et al., 2013). Members of many of these subfamilies transport lipids. For example, ABCA1 exports cholesterol and phospholipids to form HDL in the process known as 'reverse cholesterol transport' (See Chapter 15). Defects in ABCA1 result in Tangier disease. ABCA4 transports N-retinylidene PE and PE to the cytoplasmic face of disc membranes of photoreceptor cells in the retina. This activity minimises the build-up of toxic retinoids that are associated with retinal degeneration. Mutations in ABCA4 cause Stargardt macular degeneration. ABCB4 and ABCB11 export PC and bile salts, critical for canalicular bile formation (see Chapter 12). The multidrug efflux pump, P-glycoprotein/ABCB1, is an ABC transporter of the B-subfamily. ABCB1 is an unspecific phospholipid exporter. Whereas ABCA1 and ABCA4 are full transporters, the ABCG subfamily consists of half-transporters that homo- or hetero-dimerise to form functional complexes. Most ABCG complexes function in sterol transport.

4.3 Scramblases

Scramblases equilibrate lipids across the bilayer (Sanyal and Menon, 2009) (Figure 5(C)). They transport lipids at rates that are orders of magnitude greater than the rate of flip and flop catalysed by P4-ATPases and ABC transporters, where the rate of ATP hydrolysis (turnover <100 per second) controls the rate of coupled lipid transport. Phospholipid scramblases are required in the ER to promote uniform expansion of both leaflets of the bilayer after synthesis of phospholipids on the cytoplasmic side. Specific glycolipid scramblases are also required in the ER, where they play key roles in the assembly of lipid precursors needed for the biosynthesis of glycoproteins. None of the ER scramblases have been identified, although their activity has been recapitulated in vesicles reconstituted with ER membrane proteins (see below).

Interestingly, rhodopsin and other G protein-coupled receptors (GPCRs) are constitutively active as phospholipid scramblases (Goren et al., 2014). GPCRs are located at the PM of cells, and here their scramblase activity must be silent to preserve the transbilayer lipid asymmetry of the PM. However, during their biogenesis and transit to the PM, GPCRs may contribute to the overall phospholipid scramblase activity found in the ER. The specific case of rhodopsin deserves mention as rhodopsin is found in disc membranes of photoreceptor cells of the retina that also contain two flippases: Atp8a2, a P4-ATPase, and ABCA4 (see above). How these three different lipid transporters generate a homoeostatic balance that preserves disc membrane architecture is not clear.

PM-localised scramblases are important in exposing PS at the cell surface in activated platelets to promote blood coagulation, as well as in apoptotic cells, where PS acts as an 'eat-me' signal (Bevers and Williamson, 2010; Leventis and Grinstein, 2010). Members of the TMEM16 family of Ca^{2+}-activated ion channels have scramblase activity (Suzuki et al., 2010). This was explicitly verified for fungal homologues of the family that were shown to be Ca^{2+}-activated scramblases when purified and reconstituted into synthetic liposomes (Malvezzi et al., 2013; Brunner et al., 2014). Defects in TMEM16F cause Scott syndrome, a bleeding disorder associated with the lack of PS exposure on activated platelets. The mechanism of PS exposure in apoptotic cells is not known but requires members of the Xk-related (Xkr) protein family (Suzuki et al., 2013).

The mechanism of the TMEM16 Ca^{2+}-activated phospholipid scramblase is unknown, but it seems that a hydrophilic, membrane-exposed groove revealed in an X-ray structure of the protein (Brunner et al., 2014) would accommodate the headgroup of the phospholipid, allowing the lipid tails to engage the membrane. This arrangement would allow continuous passage of lipids from one leaflet of the bilayer to the other (Figure 5(C)). Other scramblases, such as the GPCR rhodopsin, may use a different mechanism; here, some aspect of protein architecture may create a lesion in the bilayer that promotes water penetration. Solvation of lipid headgroups by penetrant water would promote scrambling.

5. SPECIFIC EXAMPLES OF INTRACELLULAR LIPID TRANSPORT

5.1 Intracellular Trafficking of Cholesterol

Because sterols have a lower free energy of desorption from the bilayer than other lipids, they are good candidates for nonvesicular transport, and such transport has been documented. In yeast and in mammalian cells, inhibition of vesicular transport by genetic or pharmacological means does not make a large difference in transport of sterol from the ER to the PM. Thus *sec18-1* yeast cells, in which vesicular transport is blocked at 37 °C, or Brefeldin A-treated mammalian cells that are unable to move vesicles through the Golgi apparatus display no quantitative defect in sterol transport from the ER to the PM. Additionally, the endocytic recycling compartment (ERC) in cultured cells contains about 35% of the unesterified cholesterol in the cell. Using the fluorescent sterol dehydroergosterol (DHE, see below), it was shown that sterol in the ERC exchanges with other cellular pools within a few minutes (Mesmin and Maxfield, 2009) and that the rate and extent of exchange are only slightly affected by ATP-depletion. This indicates that most sterol transport into and out of the ERC is by an energy-independent process, which rules out all vesicular transport processes.

Rapid equilibration among organelles might be expected to dissipate differences in the sterol content of different organelle membranes, but we know that the cholesterol content of the PM is about seven-fold higher than the concentration in the ER (van Meer et al., 2008). Maintenance of PM cholesterol relies on the nature of the other lipid components in the bilayer. For example, sphingolipids and phospholipids with saturated acyl chains stabilise cholesterol in the membrane, and this type of lipid is enriched in the PM (high cholesterol), while unsaturated lipids predominate in the ER (low cholesterol) (van Meer et al., 2008; Holthuis and Menon, 2014).

The methods used to study sterol transport provide a good example of those used to study lipid transport in general. One type of methodology involves the use of isotope-labelled precursors that are metabolically incorporated into lipids. In the past, these were mainly radioisotopes such as [^{14}C]acetate, but with advances in analytical mass spectrometry nonradioactive isotopes can now also be used effectively. Cholesterol is mainly produced in the ER, so isolation of various organelles or the PM after a brief labelling period will allow a measurement of the rate of equilibration of sterol from the ER to other membranes. Conversely, isotope-labelled cholesterol can be incorporated into lipoproteins (usually as cholesterol esters) that are subsequently taken into cells by endocytosis. The redistribution of the internalised cholesterol can then be measured by isolating organelles or by examining re-esterification by acyl-CoA: cholesterol acyltransferase (ACAT), which occurs in the ER.

Fluorescence methods can also be used to observe and measure sterol transport in cells. A general issue with using fluorescence to study lipid transport is that attaching fluorophores to lipid molecules can greatly perturb the properties of the lipid. This is a reflection of the fact that lipid properties in a bilayer are based on many weak noncovalent interactions, and the fluorophores are often not much smaller than the parent lipid molecules. Thus, the behaviour of the fluorescent lipid derivatives is a combination of interactions of the fluorophore and the lipid with other membrane components. In the case of sterols, there are some naturally fluorescent sterols, such as DHE that appear to mimic the behaviour of cholesterol faithfully (Maxfield and Wustner, 2012) (Figure 6). In yeast, DHE is a naturally occurring sterol and an excellent mimic of the principal yeast sterol, ergosterol. The fluorescence of DHE comes from a weak fluorophore in the ring system of the sterol. DHE also has an additional double bond and a methyl group in the tail as compared to cholesterol. The synthetic fluorescent sterol cholestatrienol (CTL) more closely mimics cholesterol.

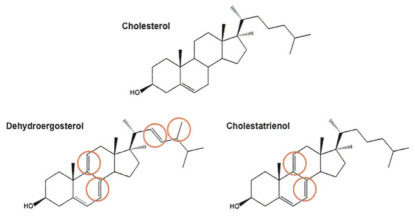

FIGURE 6 **Fluorescent sterols.** Chemical structures of cholesterol, dehydroergosterol and cholestatrienol. The circles highlight differences in structures as compared to cholesterol.

DHE (or CTL) can be delivered to cells by preparing complexes with methyl-beta cyclodextrin, which solubilises the cholesterol in aqueous buffers (Maxfield and Wustner, 2012). DHE can then be delivered to the PM with a brief (≈1 min) incubation with the cyclodextrin:DHE complexes (Maxfield and Wustner, 2012). The redistribution from the PM to other organelles can then be observed by fluorescence imaging with a microscope capable of near-ultraviolet imaging. DHE is easily photobleached, and this can be used to measure kinetics of transport. After selectively bleaching DHE in a small region, fluorescence recovery of DHE into that region provides a measurement of the rate of sterol transport into the bleached organelle. To quantify fluorescence recovery, DHE intensity in the bleached region is normalised to total cell intensity.

5.2 Transport of Phospholipids between the ER and Mitochondria

Mitochondria are rich in phospholipids, including the diglycerophospholipid cardiolipin (CL) that is characteristic of this organelle and important for its function (Claypool and Koehler, 2012). Most of these phospholipids are supplied by the ER (Osman et al., 2011; Tamura et al., 2014), which also provides precursors that are used to synthesise PE and CL in the mitochondrial inner membrane. The exchange of lipids between the ER and mitochondria occurs by nonvesicular mechanisms that are also relevant to lipid movement within mitochondria, because lipids deposited in the mitochondrial outer membrane (OM) must be further delivered to the inner membrane (IM) (Figure 7). Insights into how mitochondria acquire and export phospholipids have recently emerged, with many studies exploiting the readily detectable conversion of PS to PE as it traffics from the ER to the mitochondrial IM. Yeast mutants have been identified that are defective in PS transport, and PS trafficking has been used to quantify biochemically the contribution of proposed elements of the transport machinery.

Phospholipid exchange appears not to require cytosolic factors and likely occurs at sites where the ER is in close proximity to the OM. As described in Section 3.2, the ER makes contacts with a number of organelles, with ER-mitochondria contacts being among the first to have been biochemically isolated in the form of MAMs. These contacts require tethering proteins that bridge the two organelles. One such tethering complex, the ER-mitochondria encounter structure (ERMES), was identified in yeast. ERMES is made up of an ER membrane protein Mmm1, the OM proteins Mdm10 and Mdm34, and a soluble component Mdm12. Presumably Mmm1 interacts with Mdm10/34, to link the ER and OM, with Mdm12 contributing to the interaction. The integrity of ERMES may be regulated by the OM GTPase Gem1 and other factors. Mmm1, Mdm12 and Mdm34 possess TULIP domains, suggesting that PS and PE exchange may be directly mediated by the ERMES tethering complex. While this is possible, it cannot be the only pathway for lipid exchange as ERMES is found in yeast and not in higher eukaryotes, and ERMES mutants are viable and display only minor

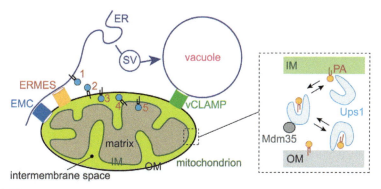

FIGURE 7 **Mitochondrial phospholipid transport.** Three contacts that mitochondria make with other organelles are depicted according to the protein complexes that mediate them: EMC and ERMES mediate ER-mitochondria contacts, and vCLAMP links mitochondria to the vacuole in yeast. Lipid transfer between the ER and mitochondria involves at least 5 stages. PS or PA is generated on the cytoplasmic face of the ER membrane (1); it is then transported to the cytoplasmic face of the mitochondrial OM (2), either directly via EMC or ERMES or indirectly via secretory vesicles (SV) to the vacuole followed by transfer mediated by vCLAMP; the lipid then flips across the OM (3) and is transferred across the intermembrane space to the IM (4). For PS, this is the site of decarboxylation to PE, which can then return to the ER by retracing the steps that have just been described. For conversion of PA to CL, PA must flip across the IM to the matrix side (5). Transfer of lipids between the IM and OM is depicted in the enlargement to the right. Transport of PA by the lipid transfer protein Ups1 and its cofactor Mdm35 is shown. Transfer may occur at IM–OM contact sites that are generated by protein complexes that span the intermembrane space (not shown).

changes in mitochondrial lipid uptake. Additional molecular contacts that could mediate lipid exchange into and out of mitochondria include the ER-membrane protein complex (EMC) that links ER with mitochondria (Lahiri et al., 2015), and vCLAMP, which tethers mitochondria to the vacuole (the yeast lysosome) (Klecker and Westermann, 2014) (Figure 7). It seems likely that redundant tethers and pathways may be involved in mitochondrial lipid homoeostasis.

For ER-derived PS to reach the IM, where it is decarboxylated to PS, it must be translocated across the OM and delivered to the IM (Figure 7). Similarly, phosphatidic acid must be transported to the IM to be converted to CL. The OM translocation step presumably requires scramblase proteins that have yet to be identified. Phospholipid exchange across the intermembrane space between the OM and IM could occur by processes similar to those discussed above for non-vesicular lipid movement between organelles. Thus, OM–IM contact sites exist, and protein complexes have been identified that may play a role in creating and maintaining these contacts. LTPs have also been identified in the intermembrane space. These proteins belong to the Ups family in yeast (PRELI family in mammals), and their stability and activity requires a protein co-factor (Mdm35 in yeast, TRIAP1/p53csv in mammals). Although many details remain to be worked out, it is likely that the combination of OM–IM contacts and LTPs in the intermembrane space is responsible for lipid transport within mitochondria (Osman et al., 2011; Tamura et al., 2014).

5.3 Glycolipid Scramblases Required for Protein *N*-Glycosylation

Assembly of the oligosaccharide donor for protein *N*-glycosylation requires flipping of three glycolipids from the cytoplasmic to the luminal side of the ER membrane (Sanyal and Menon, 2009). These glycolipids have exceptionally long hydrocarbon tails and very polar headgroups. How they are flipped across the ER membrane is a mystery, because the scramblases in question have not been identified. Nevertheless, activity assays provide compelling evidence that the glycolipid scramblases are ER membrane proteins with exquisite substrate specificity. $Glc_3Man_9GlcNAc_2$-PP-dolichol (G3M9-DLO), the oligosaccharide donor for *N*-glycosylation in yeast and humans, is built by sequentially adding sugars to dolichol phosphate.

Sugar addition occurs in two stages and on different sides of the ER. The first stage produces the key intermediate $Man_5GlcNAc_2$-PP-dolichol (M5-DLO) on the cytoplasmic face of the ER (Figure 8). M5-DLO is then flipped to the ER lumen, where G3M9-DLO synthesis is completed. The mannose and glucose residues needed for these latter reactions are provided by mannose phosphate dolichol (MPD) and glucose phosphate dolichol (GPD), both of which are synthesised on the cytoplasmic face of the ER and flipped to the luminal leaflet. Although the M5-DLO and MPD flippases are ATP-independent facilitators that move their substrates bidirectionally, consumption of M5-DLO and MPD in the ER lumen ensures directionality of transport. At least 25 human genetic defects are associated with the pathway of G3M9-DLO assembly.

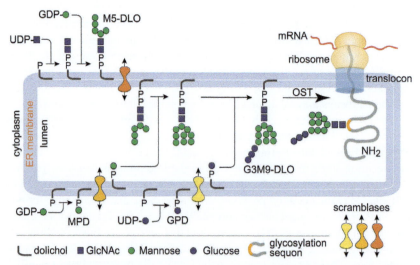

FIGURE 8 Glycolipid scrambling is necessary for protein *N*-glycosylation in the ER. Protein *N*-glycosylation occurs when oligosaccharyltransferase (OST) transfers the oligosaccharide moiety from $Glc_3Man_9GlcNAc_2$-diphosphate dolichol (G3M9-DLO) to Asn residues within glycosylation sequons in translocating proteins as they emerge from the translocon into the ER lumen. G3M9-DLO is synthesised in the ER in a multi-step, topologically split pathway that requires transbilayer movement (scrambling) of three glycolipids: M5-DLO, mannose phosphate dolichol (MPD) and glucosyl phosphoryl dolichol (GPD). See text for further details.

Glycolipid flipping across the ER membrane can be recapitulated in vesicles reconstituted with ER membrane proteins (Sanyal and Menon, 2009). For this assay, large unilamellar vesicles are reconstituted from phospholipids, detergent-solubilised ER membrane proteins and a trace amount of the lipid of interest, for example, [³H]M5-DLO or [³H]MPD. The vesicles are symmetric, with ~50% of the [³H]lipid in each leaflet. To assay flipping, the vesicles are incubated with a topological probe that captures or reacts specifically with only those [³H]lipids that are located in the outer leaflet of the vesicles. The mannose-binding lectin Concanavalin A is used to probe [³H]M5-DLO, and a membrane-impermeant oxidant is used to detect [³H]MPD. Capture of [³H] M5-DLO by Con A is assessed by solvent extraction, and [³H]MPD oxidation is quantified by thin layer chromatography. For vesicles that lack scramblase activity, only 50% of the total [³H]lipid pool will be detected corresponding to lipids in the outer leaflet. However, if [³H]lipids from the inner leaflet are transported across the bilayer, then the entire pool of [³H]lipids will be detected. Using such assays it was demonstrated that the ER glycolipid scramblases are highly specific. For example, Manβ-P-dolichol, the natural isomer of MPD is flipped >100-fold more rapidly than non-natural Manα-P-dolichol.

6. FUTURE DIRECTIONS

While there has been a resurgence of interest and remarkable recent progress in delineating the fine lipid composition of cellular membranes and the mechanisms by which lipids are transferred across and between membranes in the cell, there are many gaps in our knowledge. For example, none of the scramblases of the ER have been identified. Also, while the molecular details of MCSs both between and within organelles are beginning to emerge, their architecture and precise role in promoting lipid exchange is a subject for future investigation. The mechanisms for sorting lipids by vesicular transport are similarly only partially understood. Several LTPs have been identified, but we are still far from having molecular understanding of the mechanisms of selective transport among organelles. Given the importance of lipids for so many aspects of biology, all these areas are rich targets for further investigation.

ACKNOWLEDGEMENTS

We thank David Iaea and Mingming Hao for help with figures, and Thomas Günther-Pomorski for comments on Section 4. This work was supported by NIH grants R37 DK27083 (to F.R.M.) and R01 GM106717 and R21 EY024207 (to A.K.M.).

REFERENCES

Balla, T., 2013. Phosphoinositides: tiny lipids with giant impact on cell regulation. Physiol. Rev. 93, 1019–1137.

Bevers, E.M., Williamson, P.L., 2010. Phospholipid scramblase: an update. FEBS Lett. 584, 2724–2730.

Bogan, J.S., Xu, Y., Hao, M., 2012. Cholesterol accumulation increases insulin granule size and impairs membrane trafficking. Traffic 13, 1466–1480.

Brunner, J.D., Lim, N.K., Schenck, S., Duerst, A., Dutzler, R., 2014. X-ray structure of a calcium-activated TMEM16 lipid scramblase. Nature 516, 207–212.

Chinnapen, D.J., Hsieh, W.T., Te Welscher, Y.M., Saslowsky, D.E., Kaoutzani, L., Brandsma, E., D'auria, L., Park, H., Wagner, J.S., Drake, K.R., Kang, M., Benjamin, T., Ullman, M.D., Costello, C.E., Kenworthy, A.K., Baumgart, T., Massol, R.H., Lencer, W.I., 2012. Lipid sorting by ceramide structure from plasma membrane to ER for the cholera toxin receptor ganglioside GM1. Dev. Cell 23, 573–586.

Clark, B.J., 2012. The mammalian START domain protein family in lipid transport in health and disease. J. Endocrinol. 212, 257–275.

Claypool, S.M., Koehler, C.M., 2012. The complexity of cardiolipin in health and disease. Trends Biochem. Sci. 37, 32–41.

Coleman, J.A., Quazi, F., Molday, R.S., 2013. Mammalian P4-ATPases and ABC transporters and their role in phospholipid transport. Biochim. Biophys. Acta 1831, 555–574.

Di Paolo, G., De Camilli, P., 2006. Phosphoinositides in cell regulation and membrane dynamics. Nature 443, 651–657.

Goren, M.A., Morizumi, T., Menon, I., Joseph, J.S., Dittman, J.S., Cherezov, V., Stevens, R.C., Ernst, O.P., Menon, A.K., 2014. Constitutive phospholipid scramblase activity of a G protein-coupled receptor. Nat. Commun. 5, 5115.

Gruenberg, J., 2001. The endocytic pathway: a mosaic of domains. Nat. Rev. Mol. Cell Biol. 2, 721–730.

Hankins, H.M., Baldridge, R.D., Xu, P., Graham, T.R., 2015. Role of flippases, scramblases and transfer proteins in phosphatidylserine subcellular distribution. Traffic 16, 35–47.

Hao, M., Maxfield, F.R., 2000. Characterization of rapid membrane internalization and recycling. J. Biol. Chem. 275, 15279–15286.

Holthuis, J.C., Menon, A.K., 2014. Lipid landscapes and pipelines in membrane homeostasis. Nature 510, 48–57.

Im, Y.J., Raychaudhuri, S., Prinz, W.A., Hurley, J.H., 2005. Structural mechanism for sterol sensing and transport by OSBP-related proteins. Nature 437, 154–158.

Klecker, T., Westermann, B., 2014. Mitochondria are clamped to vacuoles for lipid transport. Dev. Cell 30, 1–2.

Klemm, R.W., Ejsing, C.S., Surma, M.A., Kaiser, H.J., Gerl, M.J., Sampaio, J.L., De Robillard, Q., Ferguson, C., Proszynski, T.J., Shevchenko, A., Simons, K., 2009. Segregation of sphingolipids and sterols during formation of secretory vesicles at the trans-Golgi network. J. Cell Biol. 185, 601–612.

Kobayashi, T., Gu, F., Gruenberg, J., 1998. Lipids, lipid domains and lipid-protein interactions in endocytic membrane traffic. Semin. Cell Dev. Biol. 9, 517–526.

Kolter, T., Sandhoff, K., 2010. Lysosomal degradation of membrane lipids. FEBS Lett. 584, 1700–1712.

Kopec, K.O., Alva, V., Lupas, A.N., 2010. Homology of SMP domains to the TULIP superfamily of lipid-binding proteins provides a structural basis for lipid exchange between ER and mitochondria. Bioinformatics 26, 1927–1931.

Kudo, N., Kumagai, K., Tomishige, N., Yamaji, T., Wakatsuki, S., Nishijima, M., Hanada, K., Kato, R., 2008. Structural basis for specific lipid recognition by CERT responsible for nonvesicular trafficking of ceramide. Proc. Natl. Acad. Sci. U. S. A. 105, 488–493.

Lahiri, S., Toulmay, A., Prinz, W.A., 2015. Membrane contact sites, gateways for lipid homeostasis. Curr. Opin. Cell Biol. 33C, 82–87.

Lev, S., 2010. Non-vesicular lipid transport by lipid-transfer proteins and beyond. Nat. Rev. Mol. Cell Biol. 11, 739–750.

Leventis, P.A., Grinstein, S., 2010. The distribution and function of phosphatidylserine in cellular membranes. Annu. Rev. Biophys. 39, 407–427.

Levine, T., 2004. Short-range intracellular trafficking of small molecules across endoplasmic reticulum junctions. Trends Cell Biol. 14, 483–490.

van Meer, G., Voelker, D.R., Feigenson, G.W., 2008. Membrane lipids: where they are and how they behave. Nat. Rev. Mol. Cell Biol. 9, 112–124.

Malvezzi, M., Chalat, M., Janjusevic, R., Picollo, A., Terashima, H., Menon, A.K., Accardi, A., 2013. Ca2+-dependent phospholipid scrambling by a reconstituted TMEM16 ion channel. Nat. Commun. 4, 2367.

Manford, A.G., Stefan, C.J., Yuan, H.L., Macgurn, J.A., Emr, S.D., 2012. ER-to-plasma membrane tethering proteins regulate cell signaling and ER morphology. Dev. Cell 23, 1129–1140.

Maxfield, F.R., McGraw, T.E., 2004. Endocytic recycling. Nat. Rev. Mol. Cell Biol. 5, 121–132.

Maxfield, F.R., Wustner, D., 2012. Analysis of cholesterol trafficking with fluorescent probes. Methods Cell Biol. 108, 367–393.

Mesmin, B., Maxfield, F.R., 2009. Intracellular sterol dynamics. Biochim. Biophys. Acta 1791, 636–645.

Mukherjee, S., Soe, T.T., Maxfield, F.R., 1999. Endocytic sorting of lipid analogues differing solely in the chemistry of their hydrophobic tails. J. Cell Biol. 144, 1271–1284.

Murcia, M., Faraldo-Gomez, J.D., Maxfield, F.R., Roux, B., 2006. Modeling the structure of the StART domains of MLN64 and StAR proteins in complex with cholesterol. J. Lipid Res. 47, 2614–2630.

Osman, C., Voelker, D.R., Langer, T., 2011. Making heads or tails of phospholipids in mitochondria. J. Cell Biol. 192, 7–16.

Prinz, W.A., 2014. Bridging the gap: membrane contact sites in signaling, metabolism, and organelle dynamics. J. Cell Biol. 205, 759–769.

Radhakrishnan, A., Goldstein, J.L., Mcdonald, J.G., Brown, M.S., 2008. Switch-like control of SREBP-2 transport triggered by small changes in ER cholesterol: a delicate balance. Cell Metab. 8, 512–521.

Roostaee, A., Barbar, E., Lavigne, P., Lehoux, J.G., 2009. The mechanism of specific binding of free cholesterol by the steroidogenic acute regulatory protein: evidence for a role of the C-terminal alpha-helix in the gating of the binding site. Biosci. Rep. 29, 89–101.

de Saint-Jean, M., Delfosse, V., Douguet, D., Chicanne, G., Payrastre, B., Bourguet, W., Antonny, B., Drin, G., 2011. Osh4p exchanges sterols for phosphatidylinositol 4-phosphate between lipid bilayers. J. Cell Biol. 195, 965–978.

Sanyal, S., Menon, A.K., 2009. Flipping lipids: why an' what's the reason for? ACS Chem. Biol. 4, 895–909.

Schauder, C.M., Wu, X., Saheki, Y., Narayanaswamy, P., Torta, F., Wenk, M.R., De Camilli, P., Reinisch, K.M., 2014. Structure of a lipid-bound extended synaptotagmin indicates a role in lipid transfer. Nature 510, 552–555.

Schmid, S.L., Sorkin, A., Zerial, M., 2014. Endocytosis: past, present, and future. Cold Spring Harb Perspect. Biol. 6, a022509.

Simons, K., Ikonen, E., 1997. Functional rafts in cell membranes. Nature 387, 569–572.

Suzuki, J., Denning, D.P., Imanishi, E., Horvitz, H.R., Nagata, S., 2013. Xk-related protein 8 and CED-8 promote phosphatidylserine exposure in apoptotic cells. Science 341, 403–406.

Suzuki, J., Umeda, M., Sims, P.J., Nagata, S., 2010. Calcium-dependent phospholipid scrambling by TMEM16F. Nature 468, 834–838.

Tamura, Y., Sesaki, H., Endo, T., 2014. Phospholipid transport via mitochondria. Traffic 15, 933–945.

Tsujishita, Y., Hurley, J.H., 2000. Structure and lipid transport mechanism of a StAR-related domain. Nat. Struct. Biol. 7, 408–414.

Vance, J.E., 2014. MAM (mitochondria-associated membranes) in mammalian cells: lipids and beyond. Biochim. Biophys. Acta 1841, 595–609.

Wirtz, K.W., Zilversmit, D.B., 1968. Exchange of phospholipids between liver mitochondria and microsomes in vitro. J. Biol. Chem. 243, 3596–3602.

Xu, P., Baldridge, R.D., Chi, R.J., Burd, C.G., Graham, T.R., 2013. Phosphatidylserine flipping enhances membrane curvature and negative charge required for vesicular transport. J. Cell Biol. 202, 875–886.

Chapter 15

High-Density Lipoproteins: Metabolism and Protective Roles Against Atherosclerosis

Gordon A. Francis
Department of Medicine, Centre for Heart Lung Innovation, University of British Columbia, Vancouver, BC, Canada

ABBREVIATIONS

apo Apolipoprotein
ABCA1 ATP-binding cassette transporter A1
ABCG1 ATP-binding cassette transporter G1
CE Cholesteryl ester
CETP Cholesteryl ester transfer protein
EL Endothelial lipase
HDL High-density lipoprotein
HL Hepatic lipase
LCAT Lecithin cholesterol acyltransferase
LDL Low-density lipoprotein
PC Phosphatidylcholine
PL Phospholipid
PLTP Phospholipid transfer protein
RCT Reverse cholesterol transport
rHDL Reconstituted HDL
SR-BI Scavenger receptor BI
TAG Triacylglycerol
VLDL Very-low-density lipoprotein

1. INTRODUCTION

Lipoproteins are complexes of lipids and proteins that solubilise cholesterol, triacylglycerols (TAG) and other lipophilic molecules in blood plasma and transport them between tissues. The lipoproteins in human plasma were initially named and characterised based on their density as determined by gradient ultracentrifugation, with high-density lipoproteins (HDLs) representing the smallest and most protein-rich of these particles. In contrast to lower density lipoproteins, the concentration of cholesterol in the plasma HDL fraction was

Biochemistry of Lipids, Lipoproteins and Membranes. http://dx.doi.org/10.1016/B978-0-444-63438-2.00015-8
437

found to correlate with protection against coronary heart disease (Barr et al., 1951). Subsequent epidemiologic studies in multiple countries confirmed these relationships, which suggested HDLs have protective activities distinct from the atherogenic properties of other lipoproteins, including low-density lipoproteins (LDLs). These observations have led to a huge body of research to understand the origin and metabolism of HDLs, their protective actions, and whether increasing HDL particle production or increasing HDL cholesterol levels can be used as means to reduce ischaemic cardiovascular disease (CVD) due to atherosclerosis (Francis, 2010).

HDLs originate as poorly lipidated apolipoproteins in the liver and intestine and undergo extensive modifications in the interstitial space and circulation through interactions with cell membranes, enzymes, transfer proteins and other lipoproteins that modify HDL structure and add additional proteins and lipids to the particles. After delivery of their lipid cargo to the liver, poorly lipidated HDL apolipoproteins can recycle multiple times in this pathway prior to eventual removal, mainly by the kidneys.

A major function of HDLs is the delivery of excess cholesterol, which is not catabolised in nonhepatic cells, from peripheral sites such as arterial wall macrophages and smooth muscle cells to the liver for excretion from the body in bile. This pathway is referred to as reverse cholesterol transport (RCT). HDLs also provide cholesterol to steroidogenic tissues for hormone synthesis and are thought to have additional beneficial effects, including reducing inflammation caused by cholesterol overload of cells, repairing damaged endothelial cells, reducing the endothelial expression of proatherogenic adhesion molecules, and reducing platelet aggregation and thrombus formation. In addition to the classic apolipoproteins, HDLs carry a host of other proteins in plasma, and are a major carrier of microRNAs for delivery from plasma to cells. The beneficial actions of HDLs appear to be reduced or lost on exposure to oxidants, systemic inflammation, and in diabetes mellitus, rendering the level of cholesterol carried in the HDL fraction a poor indicator of whether an individual's plasma HDLs are protective or not. In addition, genetic variants leading to extremely low or high HDL cholesterol (HDL-C) do not necessarily result in increased or decreased risk of ischaemic CVD. This chapter reviews the steps involved in the formation and metabolism of HDLs, the nature of known genetic variants leading to extremes of HDL-C levels, and proposed protective actions of HDLs that might be important for the prevention of atherosclerotic vascular disease.

2. HIGH-DENSITY LIPOPROTEIN FORMATION

2.1 HDL Composition and Subclasses

HDLs are the smallest but most dense of all plasma lipoproteins, based on their high level of protein per particle, representing approximately 50% of total mass. Like all lipoproteins, HDLs consist of a monolayer of surface phospholipids

(PLs) and apolipoproteins (apo) surrounding a core of hydrophobic cholesteryl esters (CEs), TAG and other dissolved lipids (e.g. lipid-soluble vitamins). Small amounts of unesterified cholesterol partition between the lipoprotein surface and hydrophobic core. The lipid content and composition of HDLs and other lipoprotein classes are indicated in Table 1 (Skipski, 1972), which shows that the total lipid content is inversely correlated with the density of a lipoprotein. The major lipid constituents of HDLs are PLs, most of which are phosphatidylcholines (PC), followed by sphingomyelins. Lyso-PC and other PLs (phosphatidylethanolamine, phosphatidylserine (PE) and phosphatidylinositol) constitute 5–18% of HDL PL mass. Fatty acid composition of HDL PLs in descending order is 16:0 (palmitic, 31.9 wt% total fatty acids), 18:2 (linoleic, 20.6%), 20:4 (arachidonic, 15.7%), 18:0 (stearic, 14.5%) and 18:1 (oleic, 12.3%) (Skipski, 1972). CEs are the next most abundant lipid of HDLs, formed in the circulation by the action of lecithin cholesterol acyltransferase (LCAT), as described in Section 3.1. The CEs are enriched in

TABLE 1 Lipid Composition of Lipoprotein Classes

	CM[a]	VLDL[b]	LDL[b]	HDL[b]
Density (g/ml)	<0.94	0.94–1.006	1.006–1.063	1.063–1.210
Total lipid (% wt)	98–99	90–92	75–80	40–48
Glycerolipids (% wt lipid)[c]	81–89	50–58	7–11	6–7
Cholesteryl esters (% wt lipid)	2–4	15–23	47–51	24–45
Unesterified cholesterol (% wt lipid)	1–3	4–9	10–12	6–8
Phospholipids (% wt lipid)[d]	7–9	19–21	28–30	42–51
PC (% wt PL)	57–80	60–74	64–69	70–81
SM (% wt PL)	12–26	15–23	25–26	12–14
Lyso PC (% wt PL)	4–10	~5	3–4	~3
Other (% wt PL)	6–7	6–10	2–10	5–10

[a]Chylomicrons (CM) were isolated during absorption of fat meals from plasma or lymph.
[b]VLDL, LDL and HDL were isolated from fasting plasma or serum.
[c]Most glycerolipids are triacylglycerols; only about 4% are diacylglycerols or monoacylglycerols.
[d]Phospholipid (PL), phosphatidylcholine (PC), sphingomyelin (SM).
Source: From Jonas and Phillips (2008); Adapted from Skipski (1972).

18:2 (linoleic) and 18:1 (oleic) fatty acids based on the substrate preference of LCAT for PC species with these fatty acids in the *sn*-2 position (Jonas and Phillips, 2008). The composition of fatty acids in PL and CE in fasting plasma is similar across lipoprotein classes, due to transfer of lipids among all the lipoproteins by cholesteryl ester transfer protein (CETP) and phospholipid transfer protein (PLTP), as indicated in Section 3.

HDLs are highly heterogeneous in terms of their size, lipid and apolipoprotein content, and their functional properties and can be separated into subclasses using various techniques (Jonas and Phillips, 2008). Based on ultracentrifugation and gel-filtration methods, HDLs are subdivided into HDL_1, HDL_2 and HDL_3 subclasses. HDL_1 are the largest and least dense and are enriched in apoE; this subclass of HDLs is the least abundant. HDL_3 are the smallest and densest. HDL_2 and HDL_3 are further divided based on nondenaturing gel electrophoresis into HDL_{2a}, HDL_{2b}, HDL_{3a}, HDL_{3b} and HDL_{3c} species, covering a density range from 1.085 to 1.171 g/ml and diameter range from 106 to 76 Å (Jonas and Phillips, 2008). HDL can also be separated by immunoaffinity chromatography into subclasses based on major apolipoprotein content: those containing apoAI but no apoAII (LpAI) and those containing both apoAI and apoAII (LpAI:AII). Minor proteins (apoE, apoCs) may or may not be present in significant amounts in these subclasses. Two-dimensional gel electrophoresis, using agarose in one dimension followed by nondenaturing polyacrylamide gel in the second, separates HDLs into α-migrating, pre-β-migrating and γ-migrating HDL subclasses of various sizes and compositions. Pre-β HDLs are present in low concentrations in plasma and somewhat higher levels in interstitial fluid. Despite their low abundance, pre-β HDLs are metabolically important because they represent nascent forms of HDL that are especially active in lipid removal from cells and subsequent cholesterol esterification by LCAT. This fraction contains discoidal particles containing two or three molecules of apoAI. A smaller component of pre-β HDLs comprises lipid-free or lipid-poor apoAI molecules perhaps in the monomeric state.

2.2 HDL Apolipoproteins

The assembly, structure, metabolism and receptor interactions of lipoproteins are determined by their apolipoprotein components. Apolipoproteins on HDLs include members of the apoA, C and E protein families. These proteins exchange on and off the surface of lipoproteins and interact with receptors and transfer proteins to facilitate the movement of lipids between cells and the liver. The exchangeable apolipoproteins contain amphipathic α-helical segments that allow them to bind to lipoproteins at the water lipid interface.

Apolipoprotein A-I (apoAI) is the most abundant protein of HDLs, comprising approximately 70% of HDL protein content. ApoAI is expressed mainly in the liver and intestine. Most apoAI is secreted as a pro-protein containing an N-terminal six-amino acid extension. Pro-apo-AI is rapidly cleaved by a circulating metalloproteinase to generate the mature 28.1 kDa, 243-amino acid protein (Fielding and Fielding, 2008). Like most plasma apolipoproteins, apoAI is organised into a series of 11- or 22-amino acid α-helical repeats linked together by helix-breaking glycine or proline residues (Fielding and Fielding, 2008). ApoAI secreted by hepatocytes is poorly lipidated and assumes a globular helical bundle formation that can self-associate and unfold as it is lipidated into discoidal and subsequently spherical particles with HDL maturation (Lund-Katz and Phillips, 2010). A small amount of apoAI is lipidated intracellularly prior to secretion; however, most lipidation of apoAI takes place after secretion.

The second most abundant apolipoprotein in HDLs is apoAII, which comprises about 20% of total HDL protein and is present on HDLs as a 17.4 kDa disulphide-linked dimer of two identical 77-amino acid monomers. ApoAII is secreted by hepatocytes as a dimer in a small discoidal particle that rapidly fuses with spherical HDL particles containing apoAI (LpAI) to generate apoAI- and apoAII-containing particles (LpAI:AII). Approximately 25% of HDLs contain apoAI as the sole apoA protein, while 75% contain apoAI and apoAII with other proteins. ApoAII is more hydrophobic than apoAI and can displace apoAI or make it more readily exchangeable from the HDL surface. LpAI:AII are more resistant to remodelling than LpAI (Gao et al., 2012). HDL also serves as a reservoir for the apoC proteins CI, CII and CIII for delivery to TAG-rich lipoproteins. ApoCI is an activator of LCAT and inhibitor of CETP on HDL, and on TAG-rich lipoproteins it can inhibit lipoprotein lipase (LPL) and lipoprotein uptake. ApoCII is an essential activator of LPL on TAG-rich lipoproteins. ApoCIII is an inhibitor of both LPL and TAG-rich lipoprotein remnant uptake.

ApoE is a 34.2 kDa glycoprotein protein expressed by many tissues, including the liver, brain, adipose tissue and arterial wall macrophages. HDL also serves as a reservoir of apoE for transfer to TAG-rich lipoproteins to facilitate the clearance of their remnants. On HDLs, apoE promotes the uptake of whole HDL particles by the liver, a minor pathway of HDL particle clearance. ApoD is an atypical apolipoprotein in the lipocalin family that is carried on HDLs and functions as a multiligand protein that transports hydrophobic molecules for processes including nerve regeneration. ApoM is another lipocalin carried mainly in HDL that binds a variety of hydrophobic molecules, including sphingosine-1-phosphate, involved in maintaining endothelial cell integrity and other antiatherogenic processes.

All the apoA, C and E apolipoproteins can mediate cholesterol efflux from cells via ATP-binding cassette transporter A1 (ABCA1), with apoAI being the most physiogically relevant due to its much higher abundance; apoE is the main mediator of ABCA1-dependent cholesterol efflux for HDL particle formation in the brain.

2.3 ABCA1 and Initial Formation of HDLs

Lipidation of apoAI and other exchangeable apolipoproteins to produce nascent, discoidal HDLs is the rate-limiting step in HDL particle formation. Fibroblasts from patients with the low HDL syndrome Tangier disease have severe defects in efflux of PLs and cholesterol to lipid-free apoAI (Francis et al., 1995). Tangier disease is caused by mutations in ABCA1 (Oram and Heinecke, 2005), a member of the ABC transporter superfamily of proteins that mediate trafficking of substrates across membranes. This family includes the classic ABC transporter *p*-glycoprotein, as well as several other ABC transporters involved in cholesterol trafficking, including ABCG1, G5 and G8. ABCA1 is a 226-amino acid integral membrane protein and a full ABC transporter, consisting of two half transporters with similar structure (Lund-Katz and Phillips, 2010). ABCA1 is expressed in most tissues, with highest expression in the liver, small intestine, adipose tissue, lung and cholesterol-loaded macrophages.

ABCA1 is upregulated in response to cholesterol loading of cells, via oxysterol dependent activation of the nuclear liver X receptors (LXRα and/or LXRβ) on the promoter region of the ABCA1 gene. The LXR partner retinoid X receptor (RXR) is activated by retinoic acid, and binding of either LXR or RXR ligands to their receptors can activate ABCA1 transcription. In addition, the interaction of lipid-poor HDL apolipoproteins with cells increases cellular ABCA1 post transcriptionally, by interfering with calpain-dependent proteolysis of the ABCA1 protein. This provides an additional feedback mechanism to increase ABCA1 levels when acceptors for lipid transport are available (Oram and Heinecke, 2005). On cholesterol removal from cells, ABCA1 is degraded rapidly with a half-life of 1–2 h.

ApoAI and other HDL apolipoproteins are lipidated by ABCA1 in a two-step process (Figure 1(A–C)). Lipid-free or lipid-poor HDL apolipoproteins bind directly and with high affinity (but low capacity) to ABCA1 in the plasma membrane of cells. This direct apolipoprotein-ABCA1 interaction is essential, but not sufficient for removal of lipids from cells. Most apolipoprotein lipidation is believed to occur indirectly, following binding of apolipoproteins to membrane lipid domains created by the actions of ABCA1. Binding and hydrolysis of ATP by the two cytoplasmic, nucleotide-binding domains of ABCA1 controls the conformation of the transmembrane domains of the transporter. This conformational change makes the lipid extrusion pocket of ABCA1 available to translocate PL substrates from the cytoplasmic to the exofacial (outer) leaflet of the membrane bilayer (Phillips, 2014). ABCA1 thus actively transports PLs across the membrane, with a preference for PC, although both PE and sphingomyelin can also be transferred. Movement of cholesterol to the outer leaflet of the membrane appears to occur simultaneously with PL transfer, although ABCA1 can promote PL efflux alone when membranes are depleted of cholesterol. This is consistent with PLs being the primary ABCA1 substrate (Oram and Heinecke, 2005).

While the exact mechanism of ABCA1-mediated PL translocation and nascent HDL formation is not fully known, the following model has been proposed (Phillips, 2014). Transport of PLs to the exofacial leaflet of the plasma membrane by ABCA1 causes the membrane to exovesiculate to relieve the membrane strain caused by the asymmetric packing of PL molecules (Figure 1(A) and (B)) ApoAI,

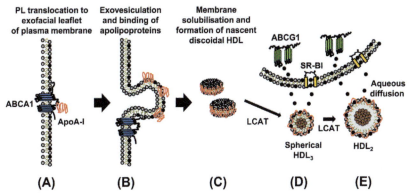

(A)   **(B)**   **(C)**   **(D)**   **(E)**

FIGURE 1   Model of HDL formation by ABCA1 and subsequent mechanisms of cholesterol efflux to HDL (Phillips, 2014). (A) ABCA1 translocates phospholipids from the inner (cytoplasmic) to outer (exofacial) leaflet of the plasma membrane. ApoAI binds to ABCA1 and increases ABCA1 activity by preventing calpain-dependent proteolysis. (B) Asymmetrical packing of phospholipids causes exovesiculation of the plasma membrane to relieve membrane strain. ApoAI and other HDL apolipoproteins bind to lattice defects in the exofacial leaflet of the plasma membrane. (C) Detergent-like activity of the apolipoproteins induces bilayer fragmentation and formation of nascent discoidal HDLs of varying sizes, with two to three molecules of apoAI or other apolipoproteins. (D, E) LCAT esterifies cholesterol on nascent HDL, creating hydrophobic cholesteryl esters that move to the core of discs, converting them to spherical HDL_3. ABCG1 promotes transport of intracellular cholesterol to the plasma membrane. SR-BI makes cholesterol available for efflux from the plasma membrane. Both mechanisms promote passive, aqueous diffusion of cholesterol to HDL particles. At low HDL concentrations, HDLs can also bind directly to and receive cholesterol from SR-BI (not shown). LCAT esterifies cholesterol on HDL_3, converting them to larger HDL_2 particles. See text for further details. *Figure courtesy of Joshua Dubland.*

apoE, and other HDL apolipoproteins have detergent-like properties that allow them to penetrate lattice defects in the membrane and destabilise these exovesicular regions. This results in solubilisation and release of regions of the membrane bilayer as discoidal HDL particles.

ApoAI in discoidal HDLs is thought to assume a belt conformation consisting of two to three apoAI molecules arranged in an antiparallel fashion around the edge of the PL disc (Segrest et al., 1999) (Figure 1(C)). The discoidal HDLs are formed in various sizes, with evidence suggesting that increased ABCA1 activity in cells results in formation of larger discoidal particles that mature into larger spherical HDL particles (Lyssenko et al., 2013). Additional support for this concept is provided in individuals heterozygous for ABCA1 mutations (Asztalos et al., 2001) or with lysosomal cholesterol storage diseases leading to impaired ABCA1 expression (Bowden et al., 2011). These subjects show reduced levels of larger α-1 HDL particles in plasma. These results suggest that the level of ABCA1 expression in cells is a key predictor of the HDL particle sizes they produce, and that none of the steps after ABCA1-mediated efflux in the HDL remodelling cascade can overcome an initial reduction in ABCA1 activity to produce the larger HDL particles.

Evidence from tissue-specific ABCA1 knockout studies suggests that a large percentage of apoAI lipidation by ABCA1 takes place immediately following secretion from the hepatocyte. Some lipid-poor apoAI may also circulate to the periphery, where it interacts with ABCA1 on cholesterol-loaded cells, particularly arterial wall macrophages, to form nascent HDLs (Oram and Heinecke, 2005). It is likely, however, that most lipid-poor apoAI in peripheral tissues is derived from the surface of mature HDL particles during their metabolism in the circulation. In this way, apoAI on HDL, initially formed by the liver, can repeat the apoAI-ABCA1 interaction for further HDL formation to promote RCT from peripheral cells.

The cellular pool of cholesterol removed by ABCA1 is the same pool that is esterified in the endoplasmic reticulum by acyl-CoA:cholesterol acyltransferase (ACAT) and stored in the cytosol as CE droplets. ABCA1 depletes this pool of cholesterol, and cholesterol available for re-esterification by ACAT, presumably by first shifting it to the outer leaflet of the plasma membrane for removal by apolipoproteins. ABCA1 also appears to mobilise cholesterol produced by autophagic hydrolysis of cytosolic CE droplets in lysosomes (Ouimet et al., 2011). It is estimated that approximately 2/3 of cholesterol efflux from cholesterol-loaded cells, such as arterial wall macrophages, is mediated by active processes, with ABCA1 being the predominant pathway (Adorni et al., 2007).

2.4 Additional Mechanisms of Cholesterol Efflux to HDLs

ABCG1 is another ABC transporter that is upregulated via LXR in response to cholesterol loading of cells, which translocates cholesterol and oxysterols across membranes (Phillips, 2014). ABCG1 functions as a homodimer in endosomes, where it actively transports cholesterol from the endoplasmic reticulum to the plasma membrane. In contrast to ABCA1, ABCG1 facilitates cholesterol efflux to preformed HDL, but not to lipid-free apoAI. ABCG1 increases the cholesterol content of the plasma membrane for release by passive diffusion through the aqueous layer surrounding cells to HDLs, in a process that does not require HDL binding to the cell surface (Figure 1(D) and (E)). Deficiency of ABCG1 in mice, however, does not result in a change in plasma HDL-C and has variable effects on atherogenesis and arterial wall inflammation. The overall importance of ABCG1 in HDL metabolism and RCT, therefore, remains to be determined.

Scavenger receptor class B, type 1 (SR-BI) plays a role in cholesterol uptake from HDLs in the liver (Section 3.6) and is also a mediator of cell cholesterol efflux to preformed HDLs. At lower concentrations of HDL, SR-BI facilitates the diffusion of cholesterol from the plasma membrane to HDLs via a hydrophobic channel created by SR-BI (Lund-Katz and Phillips, 2010). At saturating concentrations of HDL, SR-BI facilitates aqueous diffusion of cholesterol to HDLs by increasing cholesterol content of the plasma membrane (Figure 1(D) and (E)). SR-BI-dependent cholesterol efflux is more efficient to larger HDL particles due to their higher PL content and therefore absorptive capacity.

Despite an apparently significant role in mediating cholesteryl ester uptake by tissues from HDLs, the importance of SR-BI as a mediator of cholesterol efflux to HDLs from macrophages remains controversial.

Diffusion of cholesterol from cells to HDLs relies on the partial solubility of cholesterol in aqueous solution, and can be facilitated by serum albumin. Esterification by LCAT (Section 3.1) of cholesterol in nascent HDL and following aqueous diffusion onto spherical HDLs (Figure 1(C–E)) helps to maintain the predominantly unidirectional flux of cholesterol from cells to HDLs. Aqueous diffusion of cholesterol from cholesterol-loaded macrophages is estimated to account for ~30% of total cholesterol efflux (Phillips, 2014).

3. HIGH-DENSITY LIPOPROTEIN REMODELLING AND LIPID TRANSFER

Within the plasma, nascent HDLs are further metabolised by enzymes and lipid transfer proteins that ultimately promote delivery of cholesterol from peripheral cells to the liver for excretion. In addition, these pathways regenerate lipid-free or lipid-poor apoAI that can participate in further rounds of RCT (Figure 2).

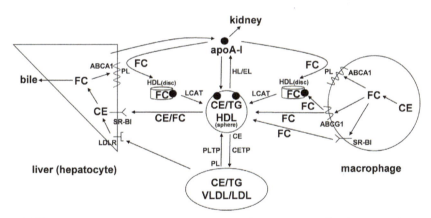

FIGURE 2   Overview of pathways involved in HDL-mediated cholesterol efflux from macrophages and reverse cholesterol transport to the liver. ApoAI produced by the liver and intestine acquires cholesterol (here indicated as FC, or free cholesterol) and phospholipids (PLs) via the ABCA1 transporter to form nascent discoidal HDL particles. Nonlipidated or poorly lipidated apoAI is cleared by the kidney. Cholesterol efflux from macrophages to HDL particles is also promoted by the ABCG1 transporter and SR-BI. Cholesterol in discoidal HDL is converted to cholesteryl esters (CE) by LCAT, leading to the formation of spherical HDL particles. PLTP mediates the transfer of PL from VLDL into HDLs, thereby contributing to HDL maturation. Mature HDL particles can be remodelled to smaller particles with the release of apoAI by hepatic lipase (HL), and through phospholipid hydrolysis by endothelial lipase (EL). In humans, HDL-CE can be transferred to the VLDL/LDL pool by CETP and take up by endocytosis into hepatocytes via the LDL receptor (LDLR). HDL CE and FC are also transferred directly to hepatocytes via SR-BI-mediated selective uptake. Cholesterol taken up by the liver can be recycled back into the ABCA1 pathway, secreted into bile as either bile acids or FC, or assembled into lipoprotein particles and secreted back into the circulation (not shown). See text for further details. *From Lund-Katz and Phillips (2010) with permission.*

LCAT esterifies tissue-derived cholesterol in HDLs. The lipid transfer proteins, CETP and PLTP, transfer CE, TAG and PL between HDLs and other lipoproteins. Endothelial lipase (EL) and hepatic lipase (HL) hydrolyse fatty acids in PL and TAG, further remodelling HDLs. SR-BI in the liver mediates the final selective uptake of HDL CE and cholesterol for excretion in bile, while allowing lipid-poor apoAI to recirculate again in the RCT pathway. The protein cargo of HDLs is also in a constant state of transfer, with proteins on other lipoproteins and circulating in plasma (Vaisar et al., 2007).

3.1 Lecithin Cholesterol Acyltransferase

Maturation of nascent discoidal to spherical HDL particles takes place in plasma through the actions of LCAT (Lund-Katz and Phillips, 2010; Kunnen and Van Eck, 2012). LCAT is expressed and secreted primarily from the liver and in lesser amounts from the brain and testes. LCAT expression is relatively insensitive to dietary or pharmacologic manipulation. It is synthesised as a 416-amino acid N- and O-glycosylated protein with a molecular mass of ~63 kDa. In plasma, most LCAT activity is associated with HDL (α-activity), with the remainder bound to and esterifying cholesterol on LDL and other apoB-containing lipoproteins (β-activity). LCAT is a phospholipase A_2, with a preference for PC (lecithin) containing oleic (18:1) or linoleic (18:2) acid in the sn-2 position. The hydrolysed fatty acid is initially bound to Ser-181 of LCAT, which is then the acyl donor for transesterification to the 3-β-hydroxyl group on the A ring of cholesterol to form CE. In plasma, apoA-I is the main activator of LCAT, with discoidal HDLs being the preferred substrates. LCAT binds with high affinity to amphipathic helices between amino acids 143 and 187 of apoA-I. Other apolipoproteins can also activate LCAT in vitro, in particular apoE, which may play a physiologic role in LCAT activity in the brain and on circulating apoB-containing lipoproteins. CEs produced by LCAT are hydrophobic and move to the core of the discoidal HDLs, converting the discoidal particles to spheres. At this point, the HDLs (HDL_3) have a surface monolayer composed of PLs and apolipoproteins, and a hydrophobic core containing predominantly CE.

Further esterification of cholesterol by LCAT on the HDL_3 surface allows HDL to accommodate additional cholesterol onto its surface by aqueous diffusion. Esterification of this additional cholesterol transforms smaller HDL_3 to larger HDL_2. Esterification of cholesterol on the HDL surface maintains a concentration gradient of cholesterol between cell plasma membranes and the HDL surface, which may generate a further flow of cholesterol to HDLs and prevent transfer back to the cell. LCAT activity on HDLs is reduced as HDL particle size increases, which might be due to CE product inhibition of the enzyme. Models of apoAI organisation on spherical HDLs propose three molecules of apoAI arranged in either a trefoil cage (Huang et al., 2011) or a helical dimer and hairpin conformation (Wu et al., 2011).

It has been suggested that LCAT-mediated esterification of cholesterol on HDLs is critical for RCT. Several lines of evidence, however, including the absence of major accumulation of cholesterol in LCAT-deficient subjects and failure of LCAT overexpression to enhance macrophage RCT in vivo (Tanigawa et al., 2009), suggest that LCAT is not the major driving force for the RCT pathway, but rather initial cholesterol efflux by ABCA1 is.

3.2 Cholesteryl Ester Transfer Protein

CETP and PLTP are members of the lipopolysaccharide binding/lipid transfer protein family, each containing 476 residues and both highly glycosylated. CETP mediates the transfer of CE from HDL to LDL and very-low-density lipoprotein (VLDL) and of TAG from VLDL to HDL and LDL. Approximately 60% of CE generated in HDLs by LCAT is transferred to LDLs and other apoB-containing lipoproteins, for eventual uptake by LDL receptors. TAG transfer by CETP from VLDLs results in TAG enrichment of HDLs and LDLs, particularly in situations in which plasma TAG levels are elevated. TAG-enriched HDLs are less stable and bind apoAI poorly. Furthermore, TAG-enriched HDLs are the preferred substrate for HL, reducing HDL particle size and further promoting apoAI release. TAG enrichment of LDLs also promotes their further hydrolysis by HL, producing smaller and denser LDLs, which are more atherogenic. This is likely due to easier transport across a damaged vascular endothelium and oxidation of the smaller LDL particles in the arterial intima.

CETP is expressed and secreted predominantly by the liver, adipose tissue and spleen and is upregulated in response to elevated cell cholesterol levels via LXR. The crystal structure of CETP indicates it contains a 6-nm-long tunnel that can accommodate two CE or TAG molecules, with the tunnel plugged at each end by a PC molecule (Qiu et al., 2007). It is suggested that the PLs in these pores merge with the PL surface monolayer of lipoproteins, allowing neutral lipids to enter and exit the tunnel. Association of CETP with lipoproteins appears to be lipid- rather than apolipoprotein-dependent, because CETP can also transfer lipids between liposomes lacking any apolipoproteins. Recent electron microscopy evidence suggests the hydrophobic N-terminal domain of CETP penetrates into the core of HDL particles and that the C-terminal domain binds to the surface of LDL and VLDL particles. This forms a ternary complex with a tunnel for transfer of neutral lipids from donor to acceptor lipoproteins ('tunnel hypothesis') (Zhang et al., 2012). Alternatively, the preference of CETP for HDLs and the absence of a similar ternary complex formation between VLDL and LDL suggest that HDL may act as a shuttle for transfer of TAG between VLDLs and LDLs ('shuttle hypothesis'). In this model, HDL is initially enriched with TAG by formation of a ternary complex between HDL, CETP and VLDL. The TAG-enriched HDL–CETP binary complex can then dissociate from VLDL and bind to LDL, resulting in transfer of VLDL-derived

TAG from HDL to LDL, in exchange for CE (Charles and Kane, 2012). Other CETP-mediated mechanisms of lipid exchange between lipoproteins may also exist; confirmation will require further direct visualisation methods.

3.3 Phospholipid Transfer Protein

PLTP has a primary structure that is approximately 25% identical to CETP and is predicted to have a similar tertiary structure (Day et al., 1994). PLTP is expressed in multiple tissues and like CETP is upregulated by elevated cell cholesterol content via LXR. The crystal structure of PLTP predicts a boomerang-shaped molecule with two domains, each with a hydrophobic pocket to bind the acyl chains of PLs. PLTP primarily transfers PC between lipoproteins, but can also transfer other PLs and α-tocopherol. PLTP mediates the transfer of PLs between HDL particles during the remodelling of HDL and facilitates transfer of surface lipid components from TAG-rich lipoproteins to HDLs. PLTP binds with highest affinity to TAG-rich HDLs (Settasatian et al., 2001) and transfers PL between HDL particles, promoting HDL particle fusion. This generates larger HDL species and also results in the dissociation of lipid-poor apoAI/PL complexes. During hydrolysis of VLDLs and chylomicrons by lipoprotein lipase, PLTP transfers excess surface PLs to HDLs, thereby contributing to the maturation of HDLs and the generation of VLDLs and chylomicron remnant lipoproteins and LDLs. PLTP contributes to RCT by both generation of HDL particles and their maturation during active and passive cholesterol removal from cells. PLTP has also been shown to directly promote ABCA1-dependent cholesterol removal from cells (Oram et al., 2008).

3.4 Hepatic Lipase

HL is a 65-kDa glycoprotein and serine hydrolase that releases fatty acids from TAG and PL of intermediate density lipoproteins (IDLs), HDLs and LDLs. This generates smaller, denser HDLs and LDLs and releases lipid-poor apoAI for further cycling in the RCT pathway (Brunzell et al., 2012). HL is expressed in the liver and functions as a homodimer on the endothelial surface of hepatocytes. TAG enrichment of HDLs by CETP enhances HL-dependent remodelling of HDLs. HL also appears to enhance SR-BI-dependent selective uptake of HDL–CE through both its hydrolytic and ligand-binding activities. Deficiency of HL leads to increased HDL-C levels, but also an increased risk of coronary heart disease, likely due to reduced regeneration of lipid-poor apoAI for RCT.

3.5 Endothelial Lipase

EL is a 68-kDa glycoprotein in the same family as HL, lipoprotein lipase (LPL) and pancreatic lipase. EL is synthesised and secreted by endothelial cells and a variety of tissues, as well as macrophages, and is bound to the endothelial cell surface by heparin sulphate proteoglycans, similar to HL and LPL. EL has phospholipase A_1 activity with substrate preference for PLs in HDLs and very little TAG hydrolase activity. Hydrolysis of HDL PLs by EL does not result in dissociation of apolipoproteins or the generation of lipid-poor apoAI particles (Jahangiri et al., 2005). While EL deficiency results in increased HDL-C levels in mice, there is no clear relationship between EL activity and atherosclerotic risk in humans or in mice.

3.6 Scavenger Receptor Class B, Type 1

SR-BI was the first HDL receptor recognised, mediating the transport of CE and cholesterol between HDL and cell membranes. SR-BI is an 82-kDa membrane glycoprotein, expressed in numerous tissues including the liver and adrenal glands, as well as endothelial cells and macrophages. In the late stage of RCT, HDL-CEs that are not already transferred to apoB-containing lipoproteins are delivered to the liver via interaction with SR-BI. The transfer of lipids from HDLs to the cell occurs in the absence of whole particle uptake. This 'selective lipid uptake' takes place via a two-step process involving initial apoAI-dependent binding of HDLs to SR-BI, followed by transfer of CE from bound HDLs to the plasma membrane. The CE is then internalised, by a nonendosomal pathway, and hydrolysed by neutral cholesterol esterases that have not yet been characterised. SR-BI-dependent selective lipid uptake in the liver is normally coupled to biliary excretion of cholesterol (Kozarsky et al., 1997) (Chapter 12).

SR-BI plays an important role in HDL-CE removal by the liver and in the delivery of cholesterol for steroid hormone synthesis in the adrenal glands, ovaries and testes. SR-BI deficiency, which results in increased diet-induced atherosclerosis in mice despite elevated HDL-C levels, has also recently been described in humans. Heterozygous carriers of a mutation affecting the extracellular loop of SR-BI showed a 28% increase in HDL-C levels, suggesting a role for SR-BI in removal of HDL-C by human liver. The subjects also had impaired adrenal steroidogenesis, but there was no obvious increase in atherosclerosis (Vergeer et al., 2011).

4. EXTREMES OF HIGH-DENSITY LIPOPROTEIN CHOLESTEROL LEVELS AND RELATIONSHIP TO ATHEROSCLEROSIS

In contrast to LDLs, in which mutations resulting in elevated LDL cholesterol (LDL-C) levels are clearly associated with a high risk of premature atherosclerotic heart disease and lowered LDL-C levels associated with marked reduction in CVD risk, neither rare nor common genetic variants associated with HDL-C levels have conclusively shown a link between HDL-C level and coronary heart disease. This is in stark divergence with the inverse correlation between HDL-C and atherosclerosis suggested by epidemiologic studies. While HDL likely has a number of protective actions in vivo, which in population studies could still be suggested by increasing HDL-C levels, it appears the protective actions of HDLs in a given individual are not predicted by simple measurement of HDL-C.

4.1 ApoAI Deficiency

Complete apoAI deficiency results from homozygous deletions of the apoAI gene or nonsense mutations early in the coding region of both alleles of the apoAI gene (Rader and Degoma, 2012). Affected individuals present with tuberoeruptive xanthomas, very low plasma HDL-C levels, and undetectable levels of apoAI. The weight of evidence indicates a high risk of early-onset CVD in individuals with homozygosity or with compound heterozygosity for apoAI deficiency. However, not all individuals with apoAI deficiency develop clinical atherosclerosis, suggesting other mechanisms influence their cardiovascular phenotype.

4.2 ABCA1 Deficiency

Complete deficiency of ABCA1 function, known as Tangier disease, is a rare cause of severe HDL deficiency due to mutations in both ABCA1 alleles. Levels of apoAI are extremely low due to rapid clearance of poorly lipidated apolipoproteins by the kidneys. ABCA1 deficiency is not clearly associated with increased risk of premature CVD, however, possibly because a reduction of LDL-C levels is also seen in many patients with Tangier disease. The largest analysis of 54 individuals with Tangier disease indicated, however, that there was a more than six-fold increase in CVD risk between ages 35 and 64 years in cases relative to controls, despite an average low LDL-C level (1.3 ± 1 mM) (Serfaty-Lacrosniere et al., 1994).

4.3 LCAT Deficiency

Complete deficiency of LCAT results in an inability of nascent HDL particles generated by ABCA1 to be converted to larger spherical HDLs with a neutral lipid core. The impaired maturation of nascent HDLs leads to their rapid clearance and low plasma levels of HDL-C and apoAI. Limited data from LCAT-deficient kindreds indicate that there is no clear increase in risk of premature atherosclerotic disease (Kunnen and Van Eck, 2012).

4.4 Cholesteryl Ester Transfer Protein Deficiency

Complete absence of CETP activity is associated with a marked increase in plasma HDL-C levels and a reduction in LDL-C levels, due to failure to transfer tissue-derived cholesterol from HDLs to LDLs. While it is not yet clear whether CETP deficiency is protective against atherosclerosis, the balance of available data from human gene polymorphisms and animal studies suggest that it is

(Barter and Kastelein, 2006). The net effect of inhibiting CETP on the uptake of tissue-derived cholesterol into the liver, and how this impacts atherosclerosis risk, remains to be determined.

5. PROTECTIVE ACTIONS OF HIGH-DENSITY LIPOPROTEINS

Despite the lack of consistency between levels of HDL-C and risk for CVD, population studies indicate that HDL particles have protective functions that indeed reduce atherosclerosis. The major protective action of HDL is presumed to be related to its ability to mediate the efflux of excess cholesterol from atherosclerotic lesion foam cells, the initial step of RCT. The rate-limiting mediator of this step is ABCA1.

Overexpression of ABCA1 in macrophages reduces progression of atherosclerotic plaque. Conversely, macrophage-specific deficiency of ABCA1 results in a marked increase in atherosclerotic plaque and reduction in RCT in mice. These manipulations do not change circulating HDL-C levels, because the contribution of foam cell cholesterol to total body cholesterol is minimal, despite it being the biochemical hallmark of atherosclerotic lesions. Animal studies using HDL infusion or transgenic expression of apoAI have consistently resulted in either prevention of atherosclerosis development or regression of established lesions (Rye and Barter, 2014). Human studies using infusions of apoAI and PL complexes also show regression of coronary atheromas. Other than the apoA-I transgenic studies, these treatments result in no persistent changes in circulating HDL-C level, suggesting it is the functional properties of newly formed or infused HDLs, not the quantity of HDL cholesterol, that are protective. In further support of this conclusion, HDL-C efflux capacity, but not HDL-C level, was independently predictive of incident cardiovascular events (Rohatgi et al., 2014).

While the promotion of cholesterol efflux and further steps in RCT are certainly a protective function of HDL apolipoproteins and particles, numerous additional protective actions of HDLs have been identified.

5.1 Anti-Inflammatory Effects

Cholesterol loading of cells, including macrophages, in the artery wall induces an inflammatory response, and removal of cell cholesterol by HDLs, apoAI or other cholesterol acceptors including cyclodextrin have been shown to reduce these responses (Yvan-Charvet et al., 2010b). Moreover, the presence of apoAI induces a shift in plaque macrophage phenotype from inflammatory to repair and allows migration of macrophages out of the plaque (Feig et al., 2011). Such effects may explain at least part of the ability of recombinant HDL infusions to induce plaque regression.

The anti-inflammatory effects of apoAI and HDLs have been shown to be mediated by ABCA1, ABCG1 and SR-BI and, in most cases, are related to changes in membrane cholesterol content or cellular cholesterol flux. These changes affect downstream signalling cascades that result in reduced expression of inflammatory cytokines.

Recent evidence suggests that HDLs are also a major carrier and delivery vehicle for microRNAs, including those that downregulate vascular cell adhesion molecule expression in endothelial cells, and thereby reduce the inflammatory response independent of alterations in cholesterol flux (Tabet et al., 2014).

5.2 Protection of Vascular Endothelium

HDLs have been shown to promote endothelial repair by stimulating endothelial cell proliferation and migration, as well as inducing endothelial progenitor cell differentiation (Mineo and Shaul, 2012). Plasma HDL-C levels have been shown to be an independent predictor of flow-mediated vasodilation, mediated by endothelial nitric oxide synthase (eNOS). HDL stimulates eNOS through an SR-BI-dependent interaction with caveolae on the cell surface, a response that requires cholesterol efflux. eNOS is also augmented by sphingosine-1-phosphate and paraoxonase (PON1), both of which are carried on HDLs (Mineo and Shaul, 2012).

5.3 Antioxidant Effects

HDL is thought to provide protection of LDL against oxidation. This may in part be due to it being more readily oxidisable than LDL, due to carrying lower levels of antioxidant molecules such as α-tocopherol than LDL (Bowry et al., 1992), and therefore acting as a preferential substrate for oxidation. In addition, oxidised PL hydroperoxides formed on LDL are transferred to HDL by CETP, thereby limiting oxidative modification of LDL. These hydroperoxides are subsequently reduced by methionine residues in apoAI, forming PL hydroxides and methionine sulphoxides. Enzymes carried on HDL, including PON1 and platelet-activating factor-acetylhydrolase (PAF-AH, also referred to as lipoprotein-associated phospholipase A2, LpPLA2), have also been considered to contribute to the anitoxidative properties of HDL. Evidence suggests that it is PAF-AH rather than PON1 that is the hydrolase for oxidised PL in HDL.

5.4 Antithrombotic Effects

HDL-C levels correlate inversely with platelet aggregability, perhaps directly, due to PL-dependent cholesterol efflux from platelets, or indirectly, through decreased secretion of platelet activating factor and increased activation of eNOS by endothelial cells on exposure to HDL (Mineo and Shaul, 2012). Further, HDL reduces endothelial cell thromboxane A_2 and upregulates prostacyclin expression, reducing platelet aggregation. HDL also suppresses endothelial cell tissue factor expression, a key initiator of thrombosis.

5.5 Other Protective Actions of HDLs

HDLs have been implicated in suppressing proliferation of haematopoietic stem cells through ABCA1- and ABCG1-dependent cholesterol efflux, providing a link between reduced HDL levels, leucocytosis and atherosclerotic CVD (Yvan-Charvet et al., 2010a). HDLs also appear to protect against diabetes mellitus by increasing insulin secretion from pancreatic β-cells, an effect stimulated by ABCA1-, ABCG1- and SR-BI-dependent cholesterol efflux (Fryirs et al., 2010). A further intriguing effect of HDLs is their ability to bind lipopolysaccharide and other microbial pathogen lipids in sepsis, thereby reducing toll-like receptor-mediated cytokine release and septic shock (Guo et al., 2013).

5.6 Loss of Protective Actions of HDLs

The heterogeneity of HDL particles makes prediction about the protective nature of HDLs in a given individual difficult. It is evident that HDL particles are highly susceptible to loss of their protective functions following oxidation, in diabetes mellitus (Chapter 19), and in acute phase responses such as sepsis and myocardial infarction (Chapter 18). The HDL proteome is highly variable, with more than 100 proteins now identified as potential cargo, and marked variation between HDL from healthy individuals and patients with coronary heart disease (Vaisar et al., 2007; Vaisar, 2012). The HDL lipidome is similarly diverse and also likely to be predictive of functional or dysfunctional HDL. The dynamic variability of HDL components and susceptibility of HDL to loss of function is not indicated by the simple measurement of plasma HDL-C. Further assays of HDL function are required, therefore, to predict the protective value of HDLs beyond HDL-C measurements.

6. HIGH-DENSITY LIPOPROTEIN-RAISING THERAPIES

Modest increases in HDL-C levels can be achieved with exercise through increased LPL activity in skeletal muscle, which then contributes surface components of TAG-rich lipoproteins to the HDL pool. Smoking cessation can also increase HDL-C levels through disinhibition of LPL activity. Reduced weight can increase HDL-C levels through a reduction of adipocyte CETP mass and activity. Niacin, through several potential mechanisms, and fibric acid derivatives, through stimulation of apoAI expression and reduction of TAG-rich particles, also raises HDL-C levels. However, the independent protective benefit of these treatments that is attributable to changes in HDL-C is unclear. Pharmacologic increases in HDL particle formation or HDL-C are being tested through increased transcription of apoA-I, increased ABCA1 expression using LXR agonists or ABCA1 microRNA inhibition, and inhibition of CETP. Full-length native and mutant forms of apoAI and peptides mimicking the structure of native apoAI, complexed with PLs in reconstituted HDL (rHDL), are being tested for their ability to promote removal of arterial wall cholesterol acutely for the prevention of cardiovascular events. The ability of any of these therapies to improve the current standard of care of lowering LDL-C levels in patients at risk of cardiovascular events remains to be determined.

7. SUMMARY AND FUTURE DIRECTIONS

The metabolism of HDL is a multifaceted pathway designed to promote the return of excess cholesterol from sterol-overloaded cells to the liver for excretion from the body. HDLs appear to have numerous other beneficial actions, including potent anti-inflammatory effects. The main measure of HDL clinically, HDL-C content, is clearly insufficient to determine whether an individual's HDL particles are protective or not. It is hoped that ongoing and further research will determine feasible methods of analysing the protective attributes of HDL and whether manipulations of the HDL production and metabolic cascade are independently and potently protective against atherosclerosis when measured against LDL-lowering treatments. An increased understanding of the complex biochemistry of HDL metabolism and of the various HDL particle subtypes can be expected to identify key functions and HDL subpopulations responsible for the protective effects of HDL and to direct future therapies to enhance or mimic these actions.

REFERENCES

Adorni, M.P., Zimetti, F., Billheimer, J.T., Wang, N., Rader, D.J., Phillips, M.C., Rothblat, G.H., 2007. The roles of different pathways in the release of cholesterol from macrophages. J. Lipid Res. 48, 2453–2462.

Asztalos, B.F., Brousseau, M.E., Mcnamara, J.R., Horvath, K.V., Roheim, P.S., Schaefer, E.J., 2001. Subpopulations of high density lipoproteins in homozygous and heterozygous Tangier disease. Atherosclerosis 156, 217–225.

Barr, D.P., Russ, E.M., Eder, H.A., 1951. Protein-lipid relationships in human plasma. II. In atherosclerosis and related conditions. Am. J. Med. 11, 480–493.

Barter, P.J., Kastelein, J.J., 2006. Targeting cholesteryl ester transfer protein for the prevention and management of cardiovascular disease. J. Am. Coll. Cardiol. 47, 492–499.

Bowden, K.L., Bilbey, N.J., Bilawchuk, L.M., Boadu, E., Sidhu, R., Ory, D.S., Du, H., Chan, T., Francis, G.A., 2011. Lysosomal acid lipase deficiency impairs regulation of ABCA1 gene and formation of high density lipoproteins in cholesteryl ester storage disease. J. Biol. Chem. 286, 30624–30635.

Bowry, V.W., Stanley, K.K., Stocker, R., 1992. High density lipoprotein is the major carrier of lipid hydroperoxides in human blood plasma from fasting donors. Proc. Natl. Acad. Sci. USA 89, 10316–10320.

Brunzell, J.D., Zambon, A., Deeb, S.S., 2012. The effect of hepatic lipase on coronary artery disease in humans is influenced by the underlying lipoprotein phenotype. Biochim. Biophys. Acta 1821, 365–372.

Charles, M.A., Kane, J.P., 2012. New molecular insights into CETP structure and function: a review. J. Lipid Res. 53, 1451–1458.

Day, J.R., Albers, J.J., Lofton-Day, C.E., Gilbert, T.L., Ching, A.F., Grant, F.J., O'hara, P.J., Marcovina, S.M., Adolphson, J.L., 1994. Complete cDNA encoding human phospholipid transfer protein from human endothelial cells. J. Biol. Chem. 269, 9388–9391.

Feig, J.E., Rong, J.X., Shamir, R., Sanson, M., Vengrenyuk, Y., Liu, J., Rayner, K., Moore, K., Garabedian, M., Fisher, E.A., 2011. HDL promotes rapid atherosclerosis regression in mice and alters inflammatory properties of plaque monocyte-derived cells. Proc. Natl. Acad. Sci. USA 108, 7166–7171.

Fielding, C.J., Fielding, P.E., 2008. Dynamics of lipoprotein transport in the circulatory system. In: Vance, D.E., Vance, J.E. (Eds.), Biochemistry of Lipids, Lipoproteins and Membranes. Elsevier, Amsterdam.

Francis, G.A., 2010. The complexity of HDL. Biochim. Biophys. Acta 1801, 1286–1293.

Francis, G.A., Knopp, R.H., Oram, J.F., 1995. Defective removal of cellular cholesterol and phospholipids by apolipoprotein A-I in Tangier Disease. J. Clin. Invest. 96, 78–87.

Fryirs, M.A., Barter, P.J., Appavoo, M., Tuch, B.E., Tabet, F., Heather, A.K., Rye, K.A., 2010. Effects of high-density lipoproteins on pancreatic beta-cell insulin secretion. Arterioscler Thromb. Vasc. Biol. 30, 1642–1648.

Gao, X., Yuan, S., Jayaraman, S., Gursky, O., 2012. Role of apolipoprotein A-II in the structure and remodeling of human high-density lipoprotein (HDL): protein conformational ensemble on HDL. Biochemistry 51, 4633–4641.

Guo, L., Ai, J., Zheng, Z., Howatt, D.A., Daugherty, A., Huang, B., Li, X.A., 2013. High density lipoprotein protects against polymicrobe-induced sepsis in mice. J. Biol. Chem. 288, 17947–17953.

Huang, R., Silva, R.A., Jerome, W.G., Kontush, A., Chapman, M.J., Curtiss, L.K., Hodges, T.J., Davidson, W.S., 2011. Apolipoprotein A-I structural organization in high-density lipoproteins isolated from human plasma. Nat. Struct. Mol. Biol. 18, 416–422.

Jahangiri, A., Rader, D.J., Marchadier, D., Curtiss, L.K., Bonnet, D.J., Rye, K.A., 2005. Evidence that endothelial lipase remodels high density lipoproteins without mediating the dissociation of apolipoprotein A-I. J. Lipid Res. 46, 896–903.

Jonas, A., Phillips, M.C., 2008. Lipoprotein structure. In: Vance, D.E., Vance, J.E. (Eds.), Biochemistry of Lipids, Lipoproteins and Membranes, fifth ed. Elsevier, New York.

Kozarsky, K.F., Donahee, M.H., Rigotti, A., Iqbal, S.N., Edelman, E.R., Krieger, M., 1997. Overexpression of the HDL receptor SR-BI alters plasma HDL and bile cholesterol levels. Nature 387, 414–417.

Kunnen, S., Van Eck, M., 2012. Lecithin:cholesterol acyltransferase: old friend or foe in atherosclerosis? J. Lipid Res. 53, 1783–1799.

Lund-Katz, S., Phillips, M.C., 2010. High density lipoprotein structure-function and role in reverse cholesterol transport. Subcell. Biochem. 51, 183–227.

Lyssenko, N.N., Nickel, M., Tang, C., Phillips, M.C., 2013. Factors controlling nascent high-density lipoprotein particle heterogeneity: ATP-binding cassette transporter A1 activity and cell lipid and apolipoprotein AI availability. FASEB J. 27, 2880–2892.

Mineo, C., Shaul, P.W., 2012. Novel biological functions of high-density lipoprotein cholesterol. Circ. Res. 111, 1079–1090.

Oram, J.F., Heinecke, J.W., 2005. ATP-binding cassette transporter A1: a cell cholesterol exporter that protects against cardiovascular disease. Physiol. Rev. 85, 1343–1372.

Oram, J.F., Wolfbauer, G., Tang, C., Davidson, W.S., Albers, J.J., 2008. An amphipathic helical region of the N-terminal barrel of phospholipid transfer protein is critical for ABCA1-dependent cholesterol efflux. J. Biol. Chem. 283, 11541–11549.

Ouimet, M., Franklin, V., Mak, E., Liao, X., Tabas, I., Marcel, Y.L., 2011. Autophagy regulates cholesterol efflux from macrophage foam cells via lysosomal acid lipase. Cell Metab. 13, 655–667.

Phillips, M.C., 2014. Molecular mechanisms of cellular cholesterol efflux. J. Biol. Chem. 289, 24020–24029.

Qiu, X., Mistry, A., Ammirati, M.J., Chrunyk, B.A., Clark, R.W., Cong, Y., Culp, J.S., Danley, D.E., Freeman, T.B., Geoghegan, K.F., Griffor, M.C., Hawrylik, S.J., Hayward, C.M., Hensley, P., Hoth, L.R., Karam, G.A., Lira, M.E., Lloyd, D.B., Mcgrath, K.M., Stutzman-Engwall, K.J., Subashi, A.K., Subashi, T.A., Thompson, J.F., Wang, I.K., Zhao, H., Seddon, A.P., 2007. Crystal structure of cholesteryl ester transfer protein reveals a long tunnel and four bound lipid molecules. Nat. Struct. Mol. Biol. 14, 106–113.

Rader, D.J., Degoma, E.M., 2012. Approach to the patient with extremely low HDL-cholesterol. J. Clin. Endocrinol. Metab. 97, 3399–3407.

Rohatgi, A., Khera, A., Berry, J.D., Givens, E.G., Ayers, C.R., Wedin, K.E., Neeland, I.J., Yuhanna, I.S., Rader, D.R., De Lemos, J.A., Shaul, P.W., 2014. HDL cholesterol efflux capacity and incident cardiovascular events. N. Engl. J. Med. 317, 2383–2393.

Rye, K.A., Barter, P.J., 2014. Cardioprotective functions of HDLs. J. Lipid Res. 55, 168–179.

Segrest, J.P., Jones, M.K., Klon, A.E., Sheldahl, C.J., Hellinger, M., De Loof, H., Harvey, S.C., 1999. A detailed molecular belt model for apolipoprotein A-I in discoidal high density lipoprotein. J. Biol. Chem. 274, 31755–31758.

Serfaty-Lacrosniere, C., Civeira, F., Lanzberg, A., Isaia, P., Berg, J., Janus, E.D., Smith Jr., M.P., Pritchard, P.H., Frohlich, J., Lees, R.S., et al., 1994. Homozygous Tangier disease and cardiovascular disease. Atherosclerosis 107, 85–98.

Settasatian, N., Duong, M., Curtiss, L.K., Ehnholm, C., Jauhiainen, M., Huuskonen, J., Rye, K.A., 2001. The mechanism of the remodeling of high density lipoproteins by phospholipid transfer protein. J. Biol. Chem. 276, 26898–26905.

Skipski, V.R., 1972. Lipid composition of lipoproteins in normal and diseased states. In: Nelson, G.J. (Ed.), Blood Lipids and Lipoproteins: Quantitation, Composition, and Metabolism. Wiley-Interscience, New York.

Tabet, F., Vickers, K.C., Cuesta Torres, L.F., Wiese, C.B., Shoucri, B.M., Lambert, G., Catherinet, C., Prado-Lourenco, L., Levin, M.G., Thacker, S., Sethupathy, P., Barter, P.J., Remaley, A.T., Rye, K.A., 2014. Hdl-transferred microRNA-223 regulates Icam-1 expression in endothelial cells. Nat. Commun. 5, 3292.

Tanigawa, H., Billheimer, J.T., Tohyama, J., Fuki, I.V., Ng, D.S., Rothblat, G.H., Rader, D.J., 2009. Lecithin: cholesterol acyltransferase expression has minimal effects on macrophage reverse cholesterol transport in vivo. Circulation 120, 160–169.

Vaisar, T., 2012. Proteomics investigations of HDL: challenges and promise. Curr. Vasc. Pharmacol. 10, 410–421.

Vaisar, T., Pennathur, S., Green, P.S., Gharib, S.A., Hoofnagle, A.N., Cheung, M.C., Byun, J., Vuletic, S., Kassim, S., Singh, P., Chea, H., Knopp, R.H., Brunzell, J., Geary, R., Chait, A., Zhao, X.Q., Elkon, K., Marcovina, S., Ridker, P., Oram, J.F., Heinecke, J.W., 2007. Shotgun proteomics implicates protease inhibition and complement activation in the antiinflammatory properties of HDL. J. Clin. Invest. 117, 746–756.

Vergeer, M., Korporaal, S.J., Franssen, R., Meurs, I., Out, R., Hovingh, G.K., Hoekstra, M., Sierts, J.A., Dallinga-Thie, G.M., Motazacker, M.M., Holleboom, A.G., Van Berkel, T.J., Kastelein, J.J., Van Eck, M., Kuivenhoven, J.A., 2011. Genetic variant of the scavenger receptor BI in humans. N. Engl. J. Med. 364, 136–145.

Wu, Z., Gogonea, V., Lee, X., May, R.P., Pipich, V., Wagner, M.A., Undurti, A., Tallant, T.C., Baleanu-Gogonea, C., Charlton, F., Ioffe, A., Didonato, J.A., Rye, K.A., Hazen, S.L., 2011. The low resolution structure of ApoA1 in spherical high density lipoprotein revealed by small angle neutron scattering. J. Biol. Chem. 286, 12495–12508.

Yvan-Charvet, L., Pagler, T., Gautier, E.L., Avagyan, S., Siry, R.L., Han, S., Welch, C.L., Wang, N., Randolph, G.J., Snoeck, H.W., Tall, A.R., 2010a. ATP-binding cassette transporters and HDL suppress hematopoietic stem cell proliferation. Science 328, 1689–1693.

Yvan-Charvet, L., Wang, N., Tall, A.R., 2010b. Role of HDL, ABCA1, and ABCG1 transporters in cholesterol efflux and immune responses. Arterioscler Thromb. Vasc. Biol. 30, 139–143.

Zhang, L., Yan, F., Zhang, S., Lei, D., Charles, M.A., Cavigiolio, G., Oda, M., Krauss, R.M., Weis-graber, K.H., Rye, K.A., Pownall, H.J., Qiu, X., Ren, G., 2012. Structural basis of transfer between lipoproteins by cholesteryl ester transfer protein. Nat. Chem. Biol. 8, 342–349.

Chapter 16

Assembly and Secretion of Triglyceride-Rich Lipoproteins

Roger S. McLeod[1], Zemin Yao[2]

[1]*Department of Biochemistry & Molecular Biology, Dalhousie University, Halifax, NS, Canada;*
[2]*Department of Biochemistry, Microbiology and Immunology, University of Ottawa, Ottawa, ON, Canada*

ABBREVIATIONS

ACAT Acyl-CoA: cholesterol acyltransferase
AFRP1 ADP-ribosylation factor related protein 1
Apo Apoliapoprotein
Arf1 ADP-ribosylation factor-1
CE Cholesteryl ester
CETP Cholesteryl ester transfer protein
CideB Cell death-inducing DFFA-like effector B
CM Chylomicron
DAG Diacylglycerol
DGAT Diacylglycerol acyltransferase
DMPC Dimyristoyl phosphatidylcholine
ER Endoplasmic reticulum
ERAD ER-associated degradation
ERK Extracellular signal-regulated kinase
HDL High density lipoprotein
IDL Intermediate density lipoprotein
LDL Low density lipoprotein
LXR Liver X receptor
MAG Monoacylglycerol
MAPK Mitogen-activated protein kinase
mTOR Mammalian target of rapamycin
MTP Microsomal triacylglycerol transfer protein
NAFLD Nonalcoholic fatty liver disease
PC Phosphatidylcholine
PCSK9 Proprotein convertase subtilisin/kexin type-9
PCTV Prechylomicron transport vesicles
PDI Protein disulphide isomerase
PE Phosphatidylethanolamine
PEMT PE methyltransferase
PERPP Post-ER presecretory proteolysis

Biochemistry of Lipids, Lipoproteins and Membranes. http://dx.doi.org/10.1016/B978-0-444-63438-2.00016-X

459

PL Phospholipid
PLTP Phospholipid transfer protein
PPAR Peroxisome proliferator activator receptor
PTP1B Protein tyrosine phosphatase 1B
PUFA Polyunsaturated fatty acid
RNAi RNA interference
SREBP1c Sterol regulatory element-binding protein 1c
TAG Triacylglycerol
TGH Triacylglycerol hydrolase
UPR Unfolded protein response
VLDL Very low density lipoprotein

1. OVERVIEW OF APOLIPOPROTEIN B-CONTAINING LIPOPROTEINS

Plasma lipoproteins are spherical microemulsions consisting of a neutral lipid core surrounded by a monolayer of phospholipids (PLs) and various proteins (termed apolipoproteins) (Figure 1). The neutral lipid core is composed mainly of triglycerides (TAG) and cholesteryl esters (CE), and the monolayer of PL is embedded with cholesterol. Under normal physiological conditions, the composition and the amount of lipids and apolipoproteins in each lipoprotein particle vary significantly.

The lipid-to-protein ratio determines buoyant density of lipoproteins, a property that has been used to separate lipoproteins by density gradient ultracentrifugation. Thus, lipoproteins are classified on the basis of their buoyant densities, such as very-low-density lipoprotein (VLDL) d < 1.006 g/ml; intermediate-density

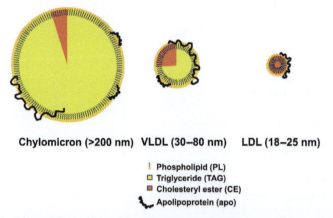

Chylomicron (>200 nm)   VLDL (30–80 nm)   LDL (18–25 nm)

} Phospholipid (PL)
□ Triglyceride (TAG)
▨ Cholesteryl ester (CE)
⟋ Apolipoprotein (apo)

FIGURE 1   **Lipoprotein structure.** Schematic structures of chylomicrons (CM), very-low-density lipoprotein (VLDL) and low-density lipoprotein (LDL). Lipoproteins are spherical emulsion particles composed of a neutral lipid core containing cholesteryl esters (CE) and triglycerides (TAG) and a surface monolayer of phospholipid (PL) and cholesterol with associated apolipoproteins (apo). Relative proportions of neutral lipid species in the core are indicated.

lipoprotein (IDL), d = 1.006–1.019 g/ml; low-density lipoprotein (LDL), d = 1.006–1.063 g/ml; and high-density lipoprotein (HDL), d = 1.063–1.21 g/ml. Another group of lipoproteins termed chylomicron (CM) has a buoyant density less than that of water, d < 0.990 g/ml because of the large quantities of TG associated with these particles. In addition, lipoproteins can be classified by their diameter, as determined by size exclusion chromatography, and by their electrophoretic mobility on agarose gel electrophoresis (alpha, pre-beta and beta). The latter technique separates particles based on their charge density. Lipoproteins within these broad classes also demonstrate significant heterogeneity, which can affect their metabolism and be indicative of pathological conditions.

Lipoproteins that contain apolipoprotein (apo) B are termed apoB-containing lipoproteins. With the exception of HDL, all plasma lipoproteins (VLDL, LDL, IDL and CM) are apoB-containing lipoproteins. The primary function of apoB-containing lipoproteins is to transport hydrophobic lipids, in the plasma through circulation, from the sites of synthesis to target tissues where they are used for various cellular functions. As described in other chapters in this volume, the metabolism of fatty acids and esters (Chapters 5 and 6), PLs (Chapter 7), and cholesterol (Chapter 11) needs to be integrated among tissues through the transport of these hydrophobic components in the blood compartment. For instance, TG is an energy-rich fatty acid ester that can be used to fuel metabolism in the cell or can be stored in fat tissue for later use in the same or other tissues.

Lipoproteins usually contain several apolipoproteins, the exception being LDL that contains only apoB100. apoB100 and apoB48 do not transfer between lipoprotein particles and are therefore termed 'non-exchangeable'. In contrast, the other apolipoproteins (apoA-I, A-II, A-IV, AV, C-I, C-II, C-III, D, E, etc.) move between lipoprotein particles as they are metabolised and are termed 'exchangeable' (see Chapter 15).

When produced in excess, or when removal mechanisms are impaired, apoB-containing lipoproteins can accumulate in the plasma. This represents an increased risk of developing cardiovascular diseases, including myocardial infarction, ischaemic stroke and peripheral vascular diseases (Chapter 18). Clinically, all the circulating apoB-containing lipoproteins have been shown to be risk factors for the development of atherosclerosis. Therefore, understanding the mechanisms that govern the production and clearance of apoB-containing lipoproteins is the subject of considerable medical interest. Modulation of the circulating levels of apoB-containing lipoproteins can be achieved by many dietary and pharmaceutical interventions, which affect either the production or the clearance of these particles.

Synthesis of apoB is achieved through protein translation mechanisms similar to those for other secretory proteins. Translation of the apoB messenger RNA (mRNA) occurs on the rough endoplasmic reticulum (ER), where the nascent polypeptide chain is translocated into the ER lumen. During and immediately after apoB translation, various lipid constituents of CM and VLDL, mainly TG, are recruited to assemble into a primordial lipoprotein particle. The intestinal enterocytes synthesise CM that carry TAG (and fat-soluble compounds such as vitamins) of dietary origin. On the other hand, the liver synthesises VLDL that carry TG derived from *de novo* lipogenesis and adipogenesis. Thus, metabolism of CM and VLDL are termed 'exogenous' and 'endogenous' pathways, respectively, to reflect the different origins of TAG.

Both CM and VLDL are metabolised in the circulation, where lipids, primarily TG, are removed by hydrolysis to deliver fatty acids to tissue sites. However, the apoB protein remains in association with the particle from which it originates until it is removed from the circulation. Here we describe how the apoB lipoproteins are assembled and secreted, with particular emphasis on recent observations that point to novel cell biological processes that regulate the assembly and secretion of apoB-containing lipoprotein.

2. STRUCTURE AND REGULATION OF THE APOLIPOPROTEIN B GENE

The human apoB gene is located on chromosome 2 and spans 43 kbp, with 29 exons and 28 introns. The apoB gene in mouse is found on chromosome 12 and has a similar basic architecture. Regulatory sequences involved in transcriptional control are mostly found in the 5' upstream region of the gene containing the promoter and in the intronic sequences. Promoter elements have been divided into distal (−128 to −86 from the transcription start site) and proximal (−86 to −70) elements. The promoter elements bind several liver-specific nuclear proteins, including BRF-2, AF-1, NF-BA1 and LIT-1, which restrict the tissue expression of the apoB gene. Closer to the transcription start site (−69 to −52) is a binding site for C/EBP that may also impact transcriptional activation. These promoter elements appear to govern the cell-specific expression of apoB proteins, but their role in regulation of apoB transcript levels under varying metabolic conditions has not been demonstrated. Additional regulatory sequences in introns 2 and 3 have enhancer activity and also bind to hepatic nuclear proteins involved in transcriptional regulation. Negative regulatory elements with repressor activity are found 2–3 kb upstream of the transcription start site that reduces apoB transcription in both hepatic and intestinal cells. These elements bind to ARP-1, EAR-2 and EAR-3 proteins.

The apoB transcript is detected not only in lipoprotein-producing tissues, such as liver and intestine, but also in the yolk sac of mice and in the heart. Production of apoB by the heart may be involved in the regulation of TAG energy stores in the organ. In the mouse, production of apoB is essential for embryonic development, but this is not the case in humans.

One of the most interesting aspects of apoB post-transcriptional regulation is the editing of the apoB mRNA in cells that produce apoB48 (reviewed in Davidson and Shelness (2000)). Intestinal apoB48 is essential for the formation of CM. ApoB48 is co-linear with the amino-terminal 48% of apoB100, and both proteins are generated from the same primary transcript. In the intestinal enterocytes of most animals, and also in the liver of some animals including the rat and mouse, specific editing of the apoB mRNA converts a cytosine at position 6666 of the transcript to a uracil. This deamination event introduces an inframe stop codon into the mRNA, which, when translated, produces apoB48. The cytidine deaminase that accomplishes the editing (APOBEC-1) is part of a multiprotein complex that recognises several areas of the apoB mRNA upstream and down-stream of the editing site for its assembly and activity. Many of these components have been characterised, and APOBEC-1 and components of the complex have homologues in bacteria and yeast. Moreover, editing activity is also present in human cell lines that do not normally express apoB. Thus, although mRNA editing is important for the tissue-specific production of apoB48, APOBEC-1 has other RNA editing functions that are still being revealed.

Expression of apoB is not primarily regulated at the level of gene transcription. Rather, the regulation of apoB synthesis and secretion occurs at the post-transcriptional level. This is remarkable, given that the apoB protein (4536 amino acids) and mRNA (14.1 kb) are large and therefore energetically expensive to synthesise constitutively if the protein is not required. Nevertheless, the level of apoB secretion from hepatic cells can vary considerably (as much as 10-fold) without change in the cellular level of apoB mRNA. As in the liver, apoB synthesis in the intestine also appears to be constitutive, despite the tremendous increase in CM production that occurs in the fed state.

Translational control of apoB synthesis has also been documented (Adeli, 2011). Insulin and microsomal triglyceride transfer protein (MTP) are known modulators of apoB translation. Sequences within the 5′ and 3′ untranslated regions of the mRNA appear to be involved in translation and these may adopt secondary structures that are recognised by *trans*-acting factors. The 5′ elements appear to be involved in translational control, whereas the 3′ elements may affect mRNA stability. Additionally, there appears to be sequestration of apoB mRNA within cytosolic granules that affects its availability for translation. The precise physiological and pathological roles of these processes in the regulation of apoB translation remain to be clarified.

3. STRUCTURAL FEATURES OF APOLIPOPROTEIN B

ApoB provides the structural framework for the assembly of TAG-rich lipoproteins in the liver and the intestine. Each apoB-containing lipoprotein particle contains a single molecule of apoB, either apoB100 or apoB48. The exact structure of the apoB protein on the lipoprotein particle has not been elucidated due to its hydrophobicity and size.

Mapping of antibody epitopes by electron microscopy suggested a ribbon and bow model for apoB on a spherical LDL particle (Chatterton et al., 1995). In this model, most of the apoB polypeptide sequence wraps once around the sphere, like a ribbon, and the C-terminal 11% of the sequence loops back to near the LDL receptor-binding site (~apoB75, a.a. ~3500), like a bow. This model suggested that the loop sequence might move away from the particle during lipolysis, exposing the receptor-binding site of apoB to the LDL receptor. A model for LDL as a sphere with apoB wrapping around the particle was also consistent with neutron scanning analysis and a three-dimensional reconstruction of LDL based on multiple cryoelectron microscopy images.

3.1 Computer Models

Computer modelling techniques, originally designed to search for sequence homology and internal repeat structures, when applied to structural analysis of apoB revealed proline-rich hydrophobic domains and repetitive α-helical regions within human apoB100. With these techniques, Segrest and colleagues developed a pentapartite model for apoB100, in which the amino terminus of the protein is a globular domain (termed β1 domain) followed by alternating β-strand and α-helix enriched domains (Figure 2(A)). Many studies have used the centile scale shown in Figure 2(A) to describe apoB polypeptides. For example, apoB100 denotes the full-length gene product, apoB48 is the N-terminal 48% of apoB100, and apoB37-42 represents the polypeptide sequence between the C-termini of apoB37 and apoB42. Other studies have identified the positions in the amino acid (a.a.) sequence of apoB100, which is also shown in Figure 2(A).

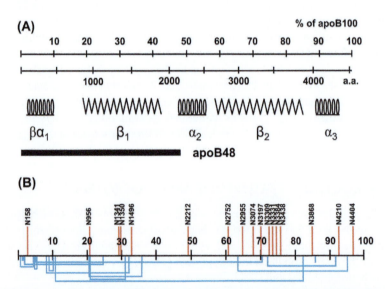

FIGURE 2 Apolipoprotein B100 primary structure. (A) Pentapartite Model–Schematic representation of the pentapartite model of apoB100 structure as described by Segrest and colleagues. Positions of the indicated domains in the model are on a centile scale, or by amino acid number (a.a.), and the location of apoB48 is shown for reference. (B) Sites of post-translational Modification–Locations of confirmed N-linked carbohydrate addition (*red*) and cysteine residues in disulphide bonds (*blue*) are shown on the centile scale map of apoB100. Positions of asparagine residues (N) are numbered according to amino acid in the 4536 residue apoB100.

The pentapartite model has also been applied to apoB100 sequences in eight other vertebrate species (chicken, frog, hamster, monkey, mouse, pig, rat and rabbit), although there is very little conservation in amino acid sequence between the individual species. Subsequently, a refined pentapartite model for apoB100 was developed to accommodate the presence of both α- and β-characteristics within the previously designated α1 domain. The new N-terminal domain is renamed βα1 and encompasses amino acid residues 1 through 1000 of apoB100 (Figure 2(A)). A lipid pocket model was proposed for the initiation of lipid recruitment by N-terminal sequences of the protein (Richardson et al., 2005). Correlation of the model with existing experimental evidence suggested that β1 and β2 domains may be in direct contact with the neutral lipid core of the particle and that these β-strand-enriched domains determine particle diameter. Furthermore, it suggested that as the particle increased in size, the contact of the α-domains with the particle surface increased, supporting a spherical model for LDL.

3.2 Experimental Evidence

The earliest attempts to study apoB structure predate the cloning of the apoB gene. Many studies were performed using delipidated, detergent-solubilised apoB protein or fragments generated by limited proteolysis of LDL. Importantly, in the absence of an amphipathic PL or detergent, the protein was insoluble. An increase in the α-helical content of the LDL protein was observed by circular dichroism on binding of the protein to PL or to monomeric detergent. These early studies established that apoB was unique among the apolipoproteins in its absolute requirement for lipids or detergent substitute for aqueous solubility.

Monoclonal antibodies raised against native or delipidated human apoB100 protein were used to show that each LDL particle contained a single molecule of apoB and that the lipid environment has a profound impact on apoB100 conformation on LDL. For instance, comparison of antibody interactions with intact LDL and VLDL showed differences in the conformation of several epitopes of apoB with lipidation. Moreover, blocking antibodies established that apoB on LDL was the ligand for the LDL receptor and that changes in the reactivity of some of the antibodies that recognise epitopes near the LDL receptor binding site of apoB may also reflect changes in apoB ligand:receptor interaction. Polyclonal antibodies against synthetic peptides established that apoB48 is the N-terminal 48% of apoB100. Multiple domains were also evident in electron microscopic images of LDL following lipid extraction, suggesting that the delipidated protein retains a multidomain structure (Johs et al., 2006).

When intact LDL was subjected to limited proteolysis and then reisolated by ultracentrifugation, approximately 80% of the protein content remained in association with the particle. The presence of these protease-resistant peptides suggests that most apoB100 polypeptides are tightly bound to lipids in LDL.

However, there was a decrease in the α-helical content following proteolysis, suggesting that α-helices in apoB are exposed on the surface of LDL and more susceptible to proteases. This was consistent with the observation that peptides released by proteolysis were able to change the physical properties of dimytristoyl phosphatidylcholine (DMPC) micelles, 'clearing' turbid solutions of these membrane mimetics. This observation is similar to those with exchangeable α-helical apolipoproteins (such as apoA-I and apoC-III), indicating lipid interactions with the released peptides. The picture that emerged was that α-helical regions of apoB were exposed on the LDL surface and the β-structures were firmly associated with lipids in the surface and core of the particle. This information was used to develop a model in which apoB100 contained five domains based on differential susceptibility to trypsin proteolysis. ApoB22 (a.a. 1–1000) contained trypsin-releasable peptides, apoB22-37 (a.a. 1001–1700) was a mixture of releasable and nonreleasable peptides, apoB37-68 (a.a. 1701–3070) was composed of nonreleasable peptides, apoB68-90 (a.a. 3071–4100) was releasable and apoB90-100 (a.a. 4101–4536) was mostly nonreleasable.

3.3 Homology Modelling

Amphipathic β-strands similar to the N-terminal domain of apoB100 were located by bioinformatics analysis in more primitive lipid transport proteins, such as vitellogenin and lipovitellin from frog, chicken, lamprey and *Caenorhabditis elegans*. MTP is also a member of this class of lipid transport proteins. Based on the known crystal structure of lipovitellin, the presence of amphipathic β-strands in the N-terminal domain of apoB and MTP suggested they form a complex that initiates lipoprotein assembly. Furthermore, the expansion of the core of the lipoprotein particle could involve the lipidation of β-strand elements from more distal sequences of apoB100. Further refinement of this model has suggested that two β-sheets in the apoB N-terminus (βA, apoB13-17, a.a 600-763; βB, apoB17-22, a.a. 780–1000) could form a lipid binding pocket with MTP to initiate lipid acquisition. Molecular modelling and mutagenesis studies show that apoB and MTP contain globular regions that are related to vitellogenin and that structural elements within apoB and MTP could form a stable complex for the transfer of lipids during lipoprotein assembly (Banaszak and Ranatunga, 2008).

Structural analysis of apoB17 revealed three independently folding lipovitellin-like domains (Jiang et al., 2005) such that a lipid nucleation site could be formed even without MTP. Analysis of peptides, some derived from apoB sequence, that adopt amphipathic β-strand characteristics has also been important in our understanding of apoB-lipid interactions (Wang and Small, 2004). Synthetic β-sheet peptides corresponding to authentic apoB sequences can associate with DMPC liposomes. These observations suggest that amphipathic β-peptides strongly associate with the outer PL layer of the LDL particle.

Wang and Small have examined the interfacial binding properties of 12 and 27 amino acid peptides from predicted β-strand sequences of apoB100 (from apoB21 to apoB41). These peptides bound tightly to a triolein:water interface and could not be desorbed under conditions in which amphipathic α-helical peptides were desorbed. These properties were also observed for an apoB polypeptide of 187 amino acids derived from the apoB37-42 region. ApoB37-42 was also able to recruit TAG into VLDL when expressed in rat hepatoma cells as a fusion protein with apoA-I as the N-terminus. These observations suggest that peptides of the amphipathic β-strand type have the functional properties to anchor the apoB polypeptide to the core lipids of the lipoprotein, making it nonexchangeable.

3.4 Post-Translational Modifications

Several post-translational modifications of apoB have been characterised, but the role of these modifications has not been fully elucidated (Figure 2(B)). The sites of N-linked glycosylation in apoB have been identified and their role in assembly has been explored, using chemical inhibition of glycosylation with tunicamycin and by site-specific mutagenesis. While the results of tunicamycin treatment can be considered equivocal because this compound also induces an ER stress response, the results of mutagenesis studies indicated that the addition of N-linked carbohydrate is essential for lipoprotein assembly and secretion. Expression in cells of apoB containing mutations in N-linked glycosylation sites can itself induce an ER stress response.

Two laboratories have examined the role of palmitoylation of apoB. However, the consequences of this modification and the outcome of its disruption are controversial and not clear at the present time. Similarly, serine phosphorylation of apoB has been reported in one laboratory, but not subsequently verified. A role for this modification in the diabetic state was suggested, but its role in apoB metabolism is uncertain.

The N-terminus of apoB is essential in the initiation of lipid recruitment during lipoprotein assembly. Inhibition of the formation of disulphide bonds in the N-terminal globular domain of apoB severely compromises the ability of the protein to assemble and secrete lipoproteins. This was shown both by dithiothreitol incubation of cells and by mutagenesis of selected cysteine residues. It is likely that correct disulphide bond formation is essential for the initiation of lipoprotein assembly, although it is also possible that interactions with MTP are impaired when the apoB N-terminal domain is incorrectly folded.

Truncated apoB proteins, as short as apoB20.1, contain solvent-exposed hydrophobic sequences that have high affinity for neutral lipids, and spontaneously form oligomers in the absence of lipids. This appears to be an important transitional region because the sequences between the C-termini of apoB19 and apoB20.1 can initiate not only the acquisition of neutral lipid, but also a change in the surface activity and elasticity of the polypeptide (Ledford et al., 2009). ApoB6.4–13 has interfacial properties indicating the capacity for lipid recruitment, although additional segments are necessary for the formation of a lipid pocket that is required for emulsion stabilisation. Thus, correct folding of the N-terminal domain of apoB, immediately following its translation, is essential for the early phases of VLDL assembly.

4. ASSEMBLY OF HEPATIC VERY LOW DENSITY LIPOPROTEINS

Assembly of VLDL is initiated in the hepatocyte as apoB protein translation begins. Like nearly all secreted proteins, the apoB primary amino acid sequence contains a signal peptide, which first arrests translation as it emerges from the ribosome. The cytosolic signal recognition particle recognises the apoB signal peptide and guides docking of the ribosome onto the Sec61 translocon on the ER membrane, allowing for transit of the signal peptide into ER lumen through an aqueous pore. Translation is then reinitiated and the polypeptide is translocated into the ER lumen during chain elongation.

Unlike other secretory proteins, however, secretion of apoB is dependent on acquisition of lipids, which begins early during the translation of the polypeptide. Lipid acquisition may be initiated only by apoB sequences or may be facilitated by MTP. Mitsche and colleagues (Mitsche et al., 2014) have described a model for the initiation of lipid recruitment by the βα1 domain of apoB. If these lipids are not available, for example due to absence or inhibition of MTP, lipoprotein assembly will be aborted and the apoB protein will be degraded (see below). Some studies have suggested that the entire VLDL assembly process is completed within the ER. Early work indicated that the rate of apoB and VLDL secretion from the cell was determined by the rate of exit from the ER. However, other work has suggested the addition of lipids, particularly TAG, occurs after the exit of immature VLDL particles from the ER.

Most of the available evidence supports a two-step model for the assembly of VLDL, by which a primordial TAG-poor particle is assembled by co-translational lipidation of the apoB protein, followed by post-translational addition of bulk TAG in a second step (Figure 3). Both steps appear to depend on the activity of MTP. The site of second step lipidation in the cell could be the ER or the Golgi apparatus. Although the differences in the cellular location of VLDL maturation may be attributable to the use of different model systems, more clarity on the sites of lipidation is still required.

Nevertheless, it is clear that the availability and transfer of TAG, PLs and CE from their sites of synthesis or storage, to the site of VLDL assembly, in the lumen of the secretory pathway, is critical to the successful secretion of the particle. Indeed, recent investigation has focused on the movement of lipids between cellular compartments. Developments using advanced lipidomic and live cell imaging techniques hold significant promise that this will be clarified very soon.

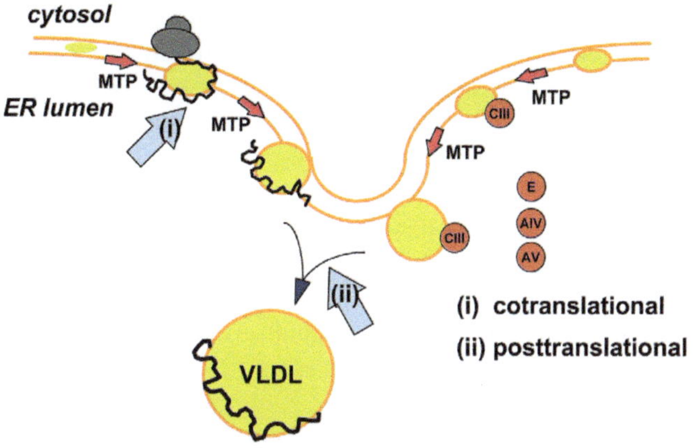

MTP = Microsomal TG transfer protein

FIGURE 3   Two-step model of hepatic VLDL assembly. Assembly of VLDL occurs in two steps, one cotranslational and a second that is post-translational. The two lipidation steps both appear to involve MTP. As the apoB polypeptide (*solid line*) is synthesised on the ribosome (*grey filled*), it acquires lipids from MTP. Independently, a lumenal lipid droplet is formed without apoB, involving the transfer of lipids by MTP and perhaps stabilised by exchangeable apolipoproteins, including apoC-III, E, A-IV, or A-V. The origin of the TAG is not yet established, but likely involves hydrolysis and re-esterification of TAG from cytosolic lipid droplets.

4.1 Role of Microsomal Triglyceride Transfer Protein

MTP is an obligatory factor for the production of apoB containing lipoproteins by the liver and intestine. Human subjects with abetalipoproteinaemia, characterised by the absence of circulating apoB lipoproteins, are unable to make either CM or VLDL due to the absence of functional MTP (Wetterau et al., 1992). MTP is a 97 kDa protein that forms a heterodimer with protein disulphide isomerase (PDI). The association with PDI maintains the lipid transfer function. MTP is present in the ER lumen and, to a lesser extent, in the Golgi apparatus, of both hepatocytes and enterocytes.

Expression of MTP and apoB in cells that are otherwise unable to produce lipoproteins endows them with the ability to assemble and secrete apoB-containing lipoproteins. Experiments in mouse and rat liver cells have established that the requirement for MTP in VLDL assembly is greater for apoB100 than it is for apoB48. Deletion of MTP from mouse liver eliminates apoB100 secretion, but has little effect on the secretion of apoB48. Moreover, inhibition of MTP activity with small molecule inhibitors or small interfering RNA blocks apoB100-containing lipoprotein secretion in rodent systems, but has much less of an effect on apoB48. Therefore, the requirement for MTP in VLDL assembly may be different between humans and rodents.

In vitro studies have shown that human MTP is able to transfer neutral lipids and PLs between vesicles, and the lack of this function would be expected to impair apoB lipoprotein assembly. The TAG and PL transfer activities of MTP are located in different regions of the MTP protein and may therefore be amenable to specific inhibition by some small molecule reagents. MTP is required for the acquisition of large amounts of TAG in the formation of mature VLDL and CM. This is consistent with a role for MTP in the assembly of apoB-free lipid droplets in the lumen of the ER, which then fuse with the nascent apoB-containing lipoprotein to form the mature particle in the second step of assembly (Figure 3). Importantly, mice that lack MTP also lack these lipid droplets (Raabe et al., 1999).

MTP may also have a chaperone activity to promote the early phases of co-translational folding of apoB. This could be either by direct protein:protein interaction or by providing lipids necessary for apoB folding. Interaction between MTP and the amino terminus of apoB has been detected, but not with sequences involved in the acquisition of TAG.

Low levels of MTP have also been detected in cells that do not make lipoproteins, such as cardiac myocytes and adipocytes. Expression of MTP in tissues that are not traditionally considered sources of lipoproteins suggests that MTP plays a more general role in intracellular TAG transport.

MTP orthologues have been identified in multiple organisms, but TAG transfer is the most evolutionarily advanced function, which first arose in fish. MTP in invertebrates is capable of PL transfer, but does not transfer TAG. Nevertheless, co-transfection experiments with truncated apoB proteins and the *Drosophila* orthologue of MTP show that the PL transfer activity is sufficient for the assembly and secretion of primordial apoB lipoproteins. Expression of *Drosophila* MTP in *Mttp*-null mice can alleviate hepatic steatosis (Khatun et al., 2012).

Overall, the evidence suggests that MTP is required for the movement of lipids into the secretory pathway, where the intrinsic lipid binding activity of apoB is responsible for recruitment. As a consequence, MTP is required for both the initiation and subsequent bulk lipidation steps in VLDL assembly.

In light of the central role of MTP in VLDL assembly, MTP inhibitors have been developed with the aim to reduce the hepatic output of atherogenic lipoproteins. However, the ability of MTP inhibitors to modulate liver lipoprotein production in human clinical trials has been disappointing. As might have been expected based on cell culture and animal studies, high-dose monotherapy with these agents produced hepatic steatosis in a number of subjects, and many of the early trials were discontinued. However, lower doses of MTP inhibitor, the addition of lipogenesis inhibitors (e.g. of DGAT), or developing inhibitors specific for the TAG transfer activity of MTP may be promising avenues for the future use of these agents.

Some natural products can impact on VLDL assembly via MTP. The citrus flavonoid naringenin decreases MTP expression in HepG2 cells by suppressing mitogen-activated protein kinase (MAPK) signalling. This compound also activates phosphoinositol-3-kinase in these cells but the mechanism of action is independent of the insulin receptor, despite its phenotypic similarities. In LDL receptor-deficient mice, these changes led to a decrease in atherosclerotic lesions in addition to the improvements in insulin sensitivity (Mulvihill et al., 2010).

MTP is regulated at transcriptional, post-transcriptional and post-translational levels by nutrient levels, hormones and other factors. Diurnal regulation of MTP expression in response to circadian clock genes that govern feeding behaviours has been reported (Pan et al., 2010).

4.2 VLDL Heterogeneity

VLDL particles are heterogeneous in composition, and the amount of TAG carried per particle can vary markedly. This heterogeneity is a reflection of the availability of TAG, but also influenced by other factors. VLDL can be separated into two forms based on flotation rate in the ultracentrifuge: $VLDL_1$ ($S_f > 100$), which is more TAG-rich and is recovered after 2.5 h centrifugation, and $VLDL_2$ (S_f 20–100) which contains less TAG and is recovered after 18 h centrifugation. In normal individuals, $VLDL_2$ is the major form secreted by the liver is, and only a minor portion is secreted as more TAG-rich $VLDL_1$. Assembly of $VLDL_1$ is achieved by recruiting large amount of TAG, a process requiring ADP ribosylation factor 1, phospholipase D1 and Extracellular signal-regulated kinase 2 (ERK2) (Asp et al., 2005). $VLDL_1$ secretion occurs under conditions of insulin resistance and type 2 diabetes. $VLDL_1$ are precursors for the production of small, dense LDL in the circulation, particles with well-recognised atherogenic potential.

4.3 Exchangeable Apolipoproteins

Several exchangeable apolipoproteins are found on circulating VLDL, and there is accumulating evidence that at least some of these apolipoproteins play intracellular roles in VLDL assembly. Overexpression of apoE in the mouse liver increased hepatic VLDL production, but it is not clear whether apoE promotes particle lipidation. ApoA-IV reduces the rate of VLDL precursor transit through hepatoma cells and enhances the TAG loading of the particles (Weinberg et al., 2012). Increased levels of plasma apoA-V have been associated with decreases in plasma TAG. Mechanisms for the effect may involve decreasing VLDL production, increasing lipolysis in the circulation, or a role for apoA-V as a ligand for receptor-mediated clearance of TAG-rich lipoproteins by the liver.

Plasma apoC-III levels have long been positively correlated with plasma TAG levels in human subjects with hypertriglyceridaemia. Yao and colleagues have proposed an intracellular role for apoC-III based on studies with hepatoma cells and mice. Overexpression of apoC-III enhanced the efficiency of TAG and apoB secretion in lipid-rich conditions, where increased MTP expression was also observed. The authors suggested that apoC-III is involved in the core expansion phase of $VLDL_1$ assembly. Importantly, mutant apoC-III that does not stimulate VLDL-TAG secretion also does not promote the accumulation of TAG in lumenal lipid droplets. Thus, it appears that apoC-III may be involved in TAG trafficking between the cytosolic lipid droplet and the ER lumenal pool, thereby promoting TAG loading of VLDL.

5. REGULATION OF HEPATIC VERY LOW DENSITY LIPOPROTEIN ASSEMBLY AND SECRETION

Under many metabolic conditions, apoB biosynthesis is constitutive, although apoB and/or TAG secretion is markedly altered. Therefore, the rate and quantity of VLDL secretion is a function of the hepatocyte capacity for VLDL assembly. Because many lipids make up VLDL, the cellular mass, extracellular sources and rate of biosynthesis of these compounds can all impact on assembly. Several transcription factors coordinate the biosynthesis of these lipids, including sterol regulatory element-binding proteins (SREBPs) and peroxisome proliferator activator receptors (PPARs). These can regulate the levels of cholesterol, CE, fatty acids and TAG in the cell.

5.1 Triglyceride Supply

Increasing TAG supply to the hepatocyte will enhance VLDL secretion. This has been shown in hepatoma cells and is also supported by observations in vivo, in which increased liver TAG stores (measured by magnetic resonance imaging) are associated with increased circulating VLDL. However, VLDL production is not governed merely by the amount of hepatic TAG synthesis or storage. Rather, location of the TAG within the cell is important, as microsome-associated lumenal lipid droplets are the major determinant of VLDL assembly and secretion. This has led to the concept of a 'secretion-coupled' pool of TAG in the liver. Most of the TAG that is secreted in VLDL is initially stored in the cytosolic lipid droplet and is then mobilised by hydrolysis and re-esterification for assembly into VLDL (see Chapter 5 for details). Re-esterified TAG is likely channelled into the secretion-coupled microsome-associated lumenal lipid droplet. Lumenal lipid droplet metabolism is an area of research of increasing interest, and similarities and differences between lumenal and cytosolic lipid droplet formation have been identified. Exchangeable apolipoproteins of VLDL appear to play a role in lipid droplet metabolism throughout the ER/Golgi secretory pathway.

Unesterified fatty acids, such as those released from adipose tissue lipolysis and arriving at the liver, can impact liver VLDL assembly. The effects of fatty acids on apoB100 secretion can be parabolic (Ota et al., 2008), with initial increases in secretion of apoB in VLDL as concentration of fatty acids increases. However, after a threshold fatty acid level is reached, there is a progressive decrease in VLDL output. While low levels of fatty acids can stimulate TAG synthesis, high levels may promote additional changes in the cell, including an ER stress response, leading to enhanced apoB degradation. This is particularly important, because reducing hepatic production of VLDL must be balanced to TAG levels to avoid TAG accumulation and hepatic steatosis (see below).

Some fatty acids, in particular the omega-3 fatty acids found in fish oils and perhaps other polyunsaturated long chain fatty acids (PUFA), decrease VLDL production, even at low concentrations. These effects may reflect the ability of these fatty acids to influence the fate of other fatty acids in the cell, accelerating their oxidation or decreasing their esterification into neutral lipids.

LIPIN proteins may play an important role in the provision of TAG for lipoprotein biosynthesis. These enzymes are phosphatidic acid phosphatases that generate diacylglycerol (DAG) for esterification into TAG (Csaki et al., 2013) (see Chapter 7). However, LIPINs are also transcriptional activators of TAG biosynthetic or catabolic enzymes. Overexpression of LIPIN1 in rat hepatoma cells increased the secretion of both TAG and apoB100 and decreased apoB degradation. It has also been suggested that these proteins may affect the secretory pathway itself (Bou Khalil et al., 2010).

CideB (cell death-inducing DFFA-like effector B) is a protein that is associated with both lipid droplets and the ER and is implicated in hepatic VLDL assembly. Knockout of CideB in mice caused TAG accumulation in the liver and reduced secretion of VLDL (Ye et al., 2009). The VLDL particles that were secreted had a reduced TAG:apoB ratio, and it was suggested that direct interaction between CideB and both the lipid droplet and apoB is necessary for TAG-enrichment of VLDL. Overexpression of peroxisome proliferator-activated receptor gamma coactivator 1-α (PGC1α) enhanced VLDL-TAG secretion from HepG2 cells, an effect that was dependent on CideB expression. Therefore, PGC1α may affect TAG partitioning for VLDL assembly via CideB.

The hepatic triglyceride hydrolase (TGH) has been studied extensively by Lehner and colleagues (see Chapter 5), and their work has shown that this enzyme is involved in the mobilisation of cytosolic lipid stores for assembly into VLDL. Isolated lipid droplets from the microsomal lumenal fraction of rat liver contain TGH, MTP and apoE. Other intracellular lipases have also been characterised and implicated in the mobilisation or disposition of TAG from the lipid droplet, affecting assembly into VLDL. However, the physiological significance of these lipases is not yet clear.

Some natural products can affect the mobilisation and storage of TAG in the liver. Epigallocatechin gallate (EGCG), a potent antioxidant derived from tea leaves, promotes cytosolic TAG storage and decreases the secretion of both TAG and VLDL by promoting the degradation of apoB. This suggests that agents that divert TAG away from the assembly pool could increase cytosolic TAG stores in the hepatocyte.

Liver TAG increases when there is an increase in circulating fatty acids. This can occur when adipose tissue function as a storage organ is compromised, as in type 2 diabetes, or in the absence of normal adipose tissue, which occurs in lipodystrophy. The response of the liver to the influx of fatty acids is the induction of PPARγ, which is normally expressed at very low levels. Storage of the fatty acids as TAG is enhanced under these conditions.

Diacylglycerol acyltransferases (DGAT) are the enzymes that catalyse the final step in the assembly of TAG, the addition of the third fatty acyl chain to DAG. Two enzymes, DGAT1 and DGAT2, are encoded by different genes (Liu et al., 2012). DGATs are expressed in both the liver and the intestine, but appear to have different roles in TAG metabolism. DGAT1 is expressed at low levels in the liver, but at high levels in enterocytes. The exact role of the two enzymes in liver VLDL assembly are not yet clear, although overexpression of DGAT1 in mouse liver increased the production of VLDL. In contrast, DGAT2 may direct its TAG products to the cytosolic lipid droplet.

A picture is emerging of a complex of proteins involved in the movement of TAG from its storage site in the cytosol to the ER lumen for VLDL assembly. Some of these proteins are enzymes involved in hydrolysis and re-esterification of the TAG, while others are transporters or stabilising factors. A major unresolved question in the assembly of hepatic VLDL is the cellular site and mechanism of the bulk TAG addition (second step) to the primordial apoB-containing lipoprotein particle. Studies of real time assembly of these particles with new microscopic techniques are likely to shed light on this issue in the near future.

5.2 Phospholipid Supply

Phosphatidylcholine (PC) is the major PL found in circulating VLDL (Cole et al., 2012). In the liver, PC can be made by one of two pathways (see Chapter 7). *De novo* biosynthesis, via the CDP-choline pathway, and methylation of phosphatidylethanolamine (PE) via PE methyltransferase (PEMT). The *de novo* biosynthetic pathway is essential for VLDL assembly, because knockout of the key enzyme of CDP-choline pathway (*Pcyt1a*) in the liver decreases VLDL secretion and causes accumulation of hepatic TAG. There is evidence to suggest that the methylation pathway is also important for provision of PC for VLDL assembly, because liver-specific knockout of PEMT reduces VLDL output.

Impaired PC biosynthesis leads to nascent VLDL particles that are recognised as defective and removed from the secretory pathway, likely by post-ER autophagy (see below). Phospholipid transfer protein (PLTP) may work in a coordinated manner with MTP for delivery of PL substrates for VLDL assembly. Changes in lysophosphatidylcholine metabolism may also affect particle lipidation.

5.3 Cholesterol Synthesis

Early studies using the statin inhibitors of hydroxymethylglutaryl-coenzyme A reductase gave ambiguous results on the role of cholesterol in hepatic VLDL production. Although statins decrease the output of VLDL, they also increase LDL receptor levels, making it difficult to parse the individual contributions of decreased output or increased removal, or both. Some studies have used knockdown of the ATP-binding cassette transporter ABCA1 in rat hepatoma cells, which allows accumulation of cellular cholesterol. This manipulation increased the secretion of $VLDL_1$, suggesting that cellular cholesterol accumulation may enhance assembly of TAG-rich lipoproteins. Simvastatin decreased both the production of VLDL and the direct hepatic production of LDL, based on kinetic studies of apoB in human subjects, without affecting the clearance of LDL apoB (Myerson et al., 2005). Overall, the evidence suggests that decreasing cholesterol in the liver can reduce circulating VLDL at least in part by reducing VLDL production.

The role of CE supply in VLDL production is similar to the role of TAG. CE is a minor component of VLDL, generated via esterification of cholesterol by acyl-CoA:cholesterol acyltransferases (ACAT). ACAT1 and ACAT2, encoded by separate genes, catalyse the same enzymatic reaction, but differ in that ACAT2 expression can be modulated in both hepatocytes and enterocytes, whereas ACAT1 expression is constitutive. ACAT inhibition has been shown to decrease VLDL production, suggesting that CE supply regulates VLDL production. However, this was not observed in other studies, reflecting the variable contribution of the two ACAT enzymes on lipoprotein production.

MTP may also play an indirect role in modulating CE production, because MTP can remove CE from the cell by assembly into VLDL. This could release product inhibition of ACATs, and may promote further CE formation and cellular cholesterol utilisation. The core esterified lipid composition of secreted apoB-containing lipoproteins is a reflection of the activities of DGAT1, ACAT1 and ACAT2 in the rat hepatoma cell, suggesting that these enzymes provide the TAG and CE necessary for assembly.

5.4 Cellular Trafficking

Cellular trafficking plays an integral role during the assembly and secretion (or degradation) of nascent VLDL. Although most of the evidence suggests that VLDL are assembled and transit through the ER/Golgi secretory pathway, $VLDL_1$ is a particularly large cargo and may require cellular specialisation for its transport and secretion. It would appear likely that there are trafficking events or pathways that can accommodate this special cargo, although evidence for this is not yet at hand. ADP-ribosylation factor-1(Arf1) is required for vesicular transport in the ER to Golgi secretory pathway, and its overexpression in rat hepatoma cells increases the production of TAG-rich $VLDL_1$, without affecting the production of $VLDL_2$ (Asp et al., 2005). This suggests that acquisition of substantial TAG content during $VLDL_1$ formation requires transport out of the ER. Currently, whether or not formation $VLDL_1$ shares the same assembly mechanism as for $VLDL_2$ is unclear.

5.4.1 Sortilin

Sortilins are highly conserved, membrane proteins that function as receptors to regulate protein trafficking in the exocytic and endocytic pathways (Willnow et al., 2011). Polymorphisms in the vicinity of the sortilin gene locus are associated with hypercholesterolaemia and myocardial infarction. Elevated sortilin expression in the liver is correlated with substantial reductions of plasma LDL-C and reduced risk for myocardial infarction in humans.

Sortilin is postulated to act as an intracellular sorting receptor for apoB in the Golgi apparatus, where it influences intracellular transport and secretion of apoB-containing lipoproteins. A small amount of cellular sortilin (10%)

is found on the liver plasma membrane, where it may also act as an alternative receptor for LDL. Increased sortilin expression could provide an additional means for LDL reduction via increased hepatic uptake and degradation of LDL. In mice, overexpression of liver sortilin decreased plasma levels of LDL cholesterol, whereas liver-specific sortilin knockdown increased LDL cholesterol (Strong et al., 2012). The reductions in LDL were the result of both decreased apoB secretion and enhanced LDL catabolism. Conversely, however, increasing hepatic sortilin expression in a global sortilin knockout mouse had the opposite effect, increasing the secretion of apoB-containing lipoproteins from the liver (Kjolby et al., 2010). These observations suggest that extrahepatic sortilin may have additional effects on lipoprotein metabolism.

Some evidence suggests that regulation of apoB and VLDL secretion by sortilin may be of pathological relevance. Sortilin expression is decreased in obese mice and restoration of expression reduced hepatic TAG and apoB secretion. Thus, dysregulation of sortilin may underlie some changes in apoB secretion in disease states. Furthermore, inhibition of mammalian target of rapamycin (mTOR) increased sortilin expression concomitant with reduced ER stress markers, suggesting that induction of ER stress through the mTOR signalling pathway may contribute to dyslipidaemia in obese states via sortilin.

5.4.2 Proprotein Convertase Subtilisin/Kexin Type-9 and LDL Receptor

The LDL receptor has been reported to regulate apoB secretion (Blasiole et al., 2008). Reducing the expression of LDL receptors in hepatocytes and hepatoma cells increased the secretion of VLDL particles and decreased the intracellular degradation of apoB in a post-ER compartment of the cell. Interaction of the LDL receptor with either apoB or apoE within the late secretory pathway is required for degradation to occur by this mechanism.

The proprotein convertase subtilisin/kexin type-9 (PCSK9) is a liver-derived protease that plays a significant role in the targeting of the LDL receptor for lysosomal degradation (Lagace et al., 2006) (see Chapter 17). PCSK9 reduces the level of cell surface LDL receptors, thus decreasing the rate of LDL removal and elevating plasma LDL. PCSK9 can also interfere with the LDL receptor interaction with apoB in the secretory pathway and allow the lipoprotein to escape lysosomal degradation. However, in mice without LDL receptors, overexpression of PCSK9 increased plasma apoB, LDL and VLDL, indicating that the effect of PCSK9 could be independent of interaction with the LDL receptor (Sun et al., 2012). This indicates that a direct interaction between PCSK9 and apoB in the hepatocyte could prevent the removal of apoB from the secretory pathway. PCSK9 could alternatively associate with the LDL receptor or with apoB, thereby coordinately regulating lipoprotein import and export. These potential mechanisms of regulation require further exploration.

5.5 Pharmacologic Agents That Regulate VLDL Secretion

Many pharmacologic agents and natural products have been identified that decrease the production of hepatic VLDL. Central among these are those that modulate TAG addition into the VLDL particle. Selective inhibition of TAG transfer activity of MTP may be useful in reducing hepatic VLDL output without promoting hepatic steatosis.

PPARα agonists, such as the hypolipidemic fibrates, enhance fatty acid oxidation at the expense of TAG synthesis in the liver. ApoC-III production is also decreased by these agents. These effects contribute to the reduction in VLDL production and reduced plasma TAG levels in humans. However, PPARα activation also increases uptake of fatty acids by the liver, and some studies have indicated increased apoB production. At least part of this paradoxical observation is because some PPARα agonists increase the expression of MTP. The use of PPARα agonists is further complicated by the fact that they also promote TAG storage in the liver by increasing the levels of perilipin-2. Changes in perilipin-2 expression will affect whether TAG is stored in cytosolic lipid droplets or secreted in VLDL.

MTP is downregulated by insulin, and the appropriate reductions in MTP can be disrupted in the insulin-resistant state. Insulin sensitisers repress transcription of the human MTP gene, but the negative promoter element responsible for this repression may not be present in the rodent MTP gene. Thus, mechanisms regulating human MTP expression may be difficult to extrapolate from studies with rodents.

Liver X receptor α (LXRα) agonists increase plasma TAG and VLDL in hamsters via activation of SREBP1c target genes. ApoB stability was enhanced under these conditions, but MTP expression was unaffected. LXR activation during dysregulation of insulin signalling may enhance VLDL production in insulin resistance states (see Chapter 19).

6. INTRACELLULAR DEGRADATION OF APOLIPOPROTEIN B

Availability of sufficient lipids at the site of apoB polypeptide translation is requisite for initiation of VLDL assembly. Under conditions where this criterion cannot be met, either elongation of the apoB polypeptide will be aborted, or if translation is completed, the nascent lipid-poor lipoprotein will be degraded intracellularly (reviewed by Olofsson and Boren (2012)). Several levels of quality control mechanisms exist in the cell to ensure disposal of the partially assembled lipoproteins. It is clear that the structural elements within the apoB sequence that mediate lipid recruitment co-localise with elements that mediate degradation of the protein, particularly the ER-associated degradation (ERAD) mediated by the proteasome. The interdependence between apoB-mediated lipid recruitment and apoB ERAD highlights the intricate nature of the regulatory mechanisms for apoB-containing lipoprotein assembly and secretion.

6.1 Ubiquitin-Proteasome System

Intracellular degradation of apoB was first identified in rat hepatocytes and subsequently characterised as a proteasomal pathway in HepG2 cells. In this pathway, apoB is recognised when in a topologically ambiguous state due to the failure to acquire sufficient lipidation during its translation. This causes an uncoupling of the normally coordinated translation and translocation steps of polypeptide transport across the ER bilayer. Consequently, part of the poly-peptide becomes exposed to the cytosol, which is recognised as an abnormal condition by the cellular quality control machinery. This initiates the addition of polyubiquitin chains to the exposed portion of apoB, and its subsequent removal from the ER for degradation in the cytosolic proteasome.

Several of the proteins involved in the recognition and degradation of apoB have been identified (Figure 4(A)). The ER lumenal chaperone BiP engages apoB polypeptides that have not assembled efficiently, the cytosolic E3 ubiq-uitin ligase gp78 catalyses polyubiquitination of cytosolically exposed regions and the AAA-ATPase p97 mediates delivery of these species to the proteasome. Reduction of gp78 increases the amount of apoB and TAG that can be assem-bled into VLDL in HepG2 (Fisher et al., 2011), and overexpression of gp78 decreases apoB secretion from these cells (Liang et al., 2003). This suggests that ubiquitination of the protein can regulate its secretion and that this may be one

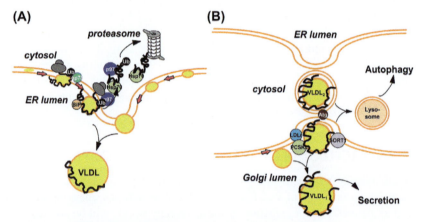

FIGURE 4 **Models of apoB degradation.** (A) Proteasomal degradation—If the transfer of lipids to the apoB polypeptide is insufficient to sustain the initial step of VLDL assembly, translocation of the growing polypeptide will slow or cease and a portion of the apoB will become exposed to the cytoplasm. This will trigger the ubiquitination (*Ub*) of the apoB by the E3 ubiquitin ligase (*gp78*) and the association with the ER lumenal chaperone (*BiP*) with the lumenal sequences of apoB. Removal of the ubiquitinated apoB from the ER involves the AAA-ATPase p97 (*p97*) and cytosolic chaperone proteins (*Hsp70*). Finally, the protein is delivered to the proteasome for degradation. (B) Nonproteasomal degradation—VLDL particles in the secretory pathway (*VLDL₂*) can be subject to degradative quality control after leaving the ER. Association of the particle with the LDL receptor (*LDLr*), sortilin (*SORT1*), or PCSK9 (*PCSK9*) may initiate budding of autophagosomes from secre-tory membrane compartments, via recruitment of autophagy proteins (*Atg*). Once formed, these can fuse with lysosomes to initiate degradation.

mechanism by which the cell can modulate apoB levels in response to lipid avail-ability. This pathway is particularly prominent in HepG2 cells, but is also present in primary hepatocytes and may serve as a surveillance mechanism *in vivo*.

Increasing lipid availability decreases proteasomal degradation of apoB con-comitant with an increase in VLDL assembly. The type of lipids also exerts an effect on assembly and secretion of VLDL, as some PUFA decrease VLDL assembly. However, it is not clear whether or not all PUFA affect proteasomal degradation of apoB. In some studies, n-3 fatty acids were shown not to affect proteasomal apoB degradation or influence VLDL secretion. Some fatty acids are more efficiently oxidised than incorporated into TG, and therefore do not increase the liver TAG resources for assembly.

The unfolded protein response (UPR) is activated by misfolded proteins in the ER to maintain cellular homoeostasis, in part by mediating ERAD. Indeed, agents that induce UPR, such as glucosamine, cause changes in apoB confor-mation that lead primarily to co-translational degradation of apoB by ERAD. Mechanisms other than ERAD are also likely involved. The UPR is composed of three arms: the serine/threonine-protein kinase/endoribonuclease IRE1α arm; the PERK (protein kinase RNA-like endoplasmic reticulum kinase) arm; and the ATF6 (activating transcription factor six) arm. Failure to respond to mis-folded protein and remove it by degradation can result in the death of the cell by apoptosis. When IRE1α is deleted, hepatocytes undergo apoptosis when placed under an ER stress condition. The normal role of IRE1α in the hepatocyte is to maintain lipid homoeostasis and lipoprotein secretion and to prevent lipid accumulation, which is also achieved by repression of several transcriptional regulators including C/EBPβ, C/EBPδ and PPARγ. The IRE1α arm of the UPR appears to be involved in the partitioning of TAG into the lumenal lipid droplets for VLDL assembly (Wang et al., 2012).

Despite substantial investigation of the ER stress response and proteasomal degradation, the details of how apoB polypeptides are selected for degradation by the proteasome remain elusive. Some studies suggest that translocation arrested apoB is targeted, as described above, but it has been difficult to reproducibly iden-tify the incomplete peptide chains that would be expected to arise by this mecha-nism. Another model suggests that the apoB polypeptide is completely translated and then the transbilayer movement of the complete polypeptide is reversed (retrotranslocation) to remove and degrade apoB in the cytosol. Although this has been shown for some model apoB proteins, it is difficult to imagine how apoB100, in a partially assembled lipoprotein, would be extracted or how it could be 'dis-assembled' in the ER lumen to allow for retrotranslocation.

6.2 Nonproteasomal Mechanisms

6.2.1 Macroautophagy

Macroautophagy is a highly conserved and ubiquitous cellular process that involves the engulfment of cellular organelles by a double-membrane struc-ture termed the autophagosome. Autophagosomes subsequently fuse with lysosomes to form autophagolysosomes, in which degradation of proteins and lipids takes place (Figure 4(B)). For our purposes, the term 'macroautophagy' is abbreviated as 'autophagy'. Autophagy occurs continuously, but can also be upregulated, for example during starvation, to provide energy sources for cellular metabolism. ApoB and cytosolic lipid droplets are subject to degradation by autophagy, which may act as a late-stage quality control mechanism for the removal of incompletely assembled apoB-containing lipoproteins. It has been further suggested that lipophagy, the autophagy of lipid droplets, is involved in mobilisation of TAG from the cellular lipid storage pool for VLDL assembly (Christian et al., 2013).

There is substantial experimental evidence for autophagy of apoB, and this process may be the predominant mode of apoB degradation, particularly in primary hepatocytes. Intracellular apoB colocalised with proteasomes, autophagosomes and lipid droplets (Ohsaki et al., 2006). These observations suggest that for apoB, autophagy and proteasomal degradation both occur near lipid droplets and that the two processes are part of a quality control continuum, with the proteasome as an early stage mechanism and autophagy a late stage event. Autophagic degradation of apoB also occurs in primary hepatocytes.

6.2.2 Reuptake or Redirection

Post-ER presecretory proteolysis (PERPP) (Pan et al., 2004) is a mechanism by which n-3 fatty acids reduce production of VLDL. In this degradation process, oxidation of the apoB100 protein and aggregation of the damaged protein is reported to trigger removal from the cell by autophagy, although this is not yet firmly established.

As described earlier, the LDL receptor can mediate reuptake of newly secreted apoB-containing lipoproteins. Interaction between apoB and LDL receptor in the hepatocyte has been proposed as a mechanism for regulation of VLDL secretion (Twisk et al., 2000). Although this pathway may redirect apoB lipoproteins in the secretory pathway, the physiological relevance is not clear. As well, PCSK9 may also impact interactions between the LDL receptor and apoB in the secretory pathway, potentially regulating presecretory degradation.

7. DYSREGULATION OF VERY LOW DENSITY LIPOPROTEIN ASSEMBLY AND SECRETION

7.1 Insulin Resistance

Dysregulation of VLDL metabolism often accompanies the insulin resistant state in which increased VLDL output is a feature. However, the mechanism underlying the process is not yet clear. Increased circulating fatty acids occur when adipocyte uptake and storage pathways are compromised during insulin resistance. Hepatic insulin resistance does not prevent excess fatty acids from being stored as TAG. As a consequence, assembly of larger, more TAG-rich

VLDL$_1$ is often observed in type 2 diabetes. Knockout of protein tyrosine phosphatase 1B (PTP1B), which promotes insulin signalling, prevents the increase of VLDL secretion that is observed when mice are fed a fructose diet to induce insulin resistance (Qiu et al., 2004). Conversely, overexpression of PTP1B in HepG2 cells increased apoB secretion. Thus, the changes in the activity of this protein phosphatase of the insulin signalling pathway may govern some of the hepatic changes associated with diabetic dyslipidaemia (see Chapter 19).

7.2 Nonalcoholic Fatty Liver Disease

Although we have discussed the potential consequences of VLDL overproduction on metabolism and disease, it should be appreciated that the failure to assemble and secrete VLDL also has its consequences. Accumulation of TAG in the liver is termed hepatic steatosis and is the essential feature of nonalcoholic fatty liver disease (NAFLD). This relatively benign condition can progress to steatohepatitis and to cirrhosis, if unresolved. Autophagy is involved in the resolution of the steatosis and may prevent the progression to steatohepatitis. It should be noted, however, that steatosis is not simply a direct consequence of the failure to assemble VLDL. As described in Section 5.1, the initial response of the liver to increases in fatty acid flux is to increase VLDL output. However, if the concentration exceeds a threshold fatty acid level, or if the exposure is prolonged, there is a progressive decrease in VLDL output. Thus, decreased plasma TAG can occur even under severe hepatosteatosis conditions.

Familial hypobetalipoproteinaemia is a rare condition that is characterised by the failure to assemble VLDL normally, usually due to a defect in the coding region of the apoB100 gene. In addition to abnormalities associated with the absence of VLDL, subjects with this disorder are also prone to hepatic steatosis, steatohepatitis, hepatic cirrhosis and hepatocarcinoma. Thus, the clinical outcome of VLDL assembly failure can be more serious than VLDL overproduction.

8. ASSEMBLY AND SECRETION OF CHYLOMICRONS

CM assembly is essential for TAG absorption and acquisition of fat-soluble vitamins from the gut. TAG are hydrolysed in the gut and absorbed primarily as free fatty acids and monoacylglycerol, which are then efficiently re-esterified to TAG in enterocytes. The resulting TAG is incorporated into CM, initially as precursor lipoproteins in the ER of the enterocyte (Figure 5). Pre-CM are transported from ER in prechylomicron transport vesicles (PCTV) for final lipidation in the Golgi apparatus. Mature CM are secreted across the basolateral membranes of the cells into the lymph. Similar to VLDL assembly in the liver, MTP is essential for CM assembly. ApoB48 is the structural protein for CM assembly although some apoB100 containing CM are also produced by the enterocytes. The accumulation of CM remnant particles in the plasma is a risk for atherosclerosis, particularly in insulin resistance.

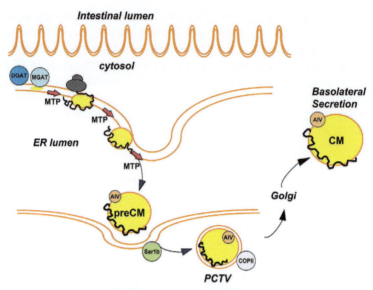

FIGURE 5 **Assembly of chylomicrons.** Chylomicron assembly in intestinal enterocytes utilises fatty acids and monoacylglycerol absorbed from the intestinal lumen and converted to TAG by mon oacylglycerol:acyltransferase *(MGAT)* and diacylglycerol:acyltransferase *(DGAT)*. In the ER, TAG is transferred to the lumen by MTP. Pre-CM acquire apoA-IV *(A-IV)* during assembly and are then recruited into pre-CM transport vesicles *(PCTV)*, initiated by components of the vesicle transport pathway *(Sar1b* and *COPII)*. Additional acquisition of lipid components occurs, and the mature CM is secreted into the lymph from the basolateral membrane.

8.1 Similarities to VLDL Assembly

Assembly of CM in the intestinal enterocyte and of VLDL in the hepatocyte is regulated by many of the same factors. As described for VLDL assembly, the ER and lipid droplet associated protein CideB plays a role in CM assembly, likely in the delivery of TAG to the nascent particle. The ADP-ribosylation factor related protein 1, ARFRP1, is involved in CM and VLDL assembly, likely at the maturation stage at which bulk TAG is added. Exchangeable apolipoproteins, such as apoA-IV, are associated with cytosolic lipid droplets in Caco-2 cells, suggesting a potential role in lipidation similar to the proposed role for apoA-IV and apoC-III in VLDL assembly.

There is evidence for proteasomal degradation of apoB48, which could modulate CM assembly in a manner similar to degradation of apoB100 in the liver. However, evidence for proteasomal degradation of apoB48 is controversial and more work is needed to support this form of regulation (see Chapter 19 for details of CM production during insulin resistance).

8.2 Distinguishing Features

The roles of Sar1 and PCTV are key distinguishing features of CM assembly (Figure 5). Sar1b promotes intracellular transport of pre-CM in PCTV and upregulation of lipid biosynthetic genes necessary for CM assembly in enterocytes

(Levy et al., 2011). Defective Sar1 is the proximal cause for CM retention disease, in which CM are assembled but fail to be secreted from the enterocyte. CM assembly is driven by the amount of fat in the diet, and intestinal enterocytes are specialised for the assembly and transport of dietary TAG into circulation as CM.

MTP is perhaps more important for CM assembly than for assembly of VLDL in the liver. Deletion of MTP in mouse liver had little effect on the secretion of apoB48 in hepatic VLDL, whereas intestinal deletion reduced the assembly and secretion of apoB48 in CM.

ApoA-IV may play a critical role in CM assembly; overexpression of apoA-IV in an intestinal cell line increased the size of secreted CM and also increased MTP expression. These changes may reflect increased partitioning of TAG into lumenal lipid droplets.

9. HEPATOCYTE AND ENTEROCYTE MODELS – STRENGTHS AND LIMITATIONS

Much of the current knowledge concerning assembly and secretion of apoB-containing lipoproteins has been derived from studies with cell culture models. For the liver, this includes hepatoma cell models of human, mouse, or rat origin and primary hepatocytes isolated from mice, rats, or hamsters. A smaller number of studies have been performed in hepatocytes from larger mammals, including pigs, cows and nonhuman primates.

Hepatoma cells are often the models of choice because of the ease with which they can be cultured, the lack of specialised expertise required for isolation, and the relatively low cost of culture maintenance. An additional advantage of hepatoma cells is that they are readily amenable to genetic manipulation, including transfection with plasmid DNA containing transgenes or RNA interfering (RNAi) elements to study specific genes of interest. However, it must be appreciated that the transformation of these cells has altered the cellular phenotype, and often some of the metabolic characteristics of normal cells are not reliably represented in the hepatoma cell line. As such, some of the results based on these model studies must be interpreted with caution and verified in whole animals.

One prominent model is the human hepatocarcinoma cell line HepG2, which has been widely used for studies of apoB-containing lipoprotein assembly. Other human hepatoma cells (such as Huh7) appear to have no particular advantages over HepG2 cells. It is known that HepG2 cells are relatively inefficient, compared to primary hepatocytes, in the assembly of VLDL, such that many of the lipoprotein particles secreted into the culture medium are of IDL or LDL size and density. HepG2 cells constitutively express the Ras gene as part of the transformation phenotype leading to constitutive activation of the downstream MEK/ERK pathway in the cell. This defect in VLDL formation can be overcome by inhibition of the MEK/ERK pathway (Tsai et al., 2007), making it possible to model VLDL assembly in a human hepatoma cell line.

The rat hepatoma cell line McA-RH7777 has also been used extensively for studies of VLDL assembly. These cells divide prolifically and are readily amenable to transfection. In contrast to HepG2 cells, they are able to assemble and secrete VLDL of authentic size and density. However, because they are derived from rat hepatocytes, McA-RH7777 cells contain apoB mRNA editing activity and therefore secrete both apoB100 and apoB48 lipoproteins. Nevertheless, this cell line has been invaluable in studies of apoB structure and function.

Primary hepatocytes from hamster, mouse and rat have been utilised as a more physiological cell model in comparison to hepatoma cells. However, primary hepatocyte preparation is technically demanding, and often complicated by irreproducibility between preparations. Moreover, primary hepatocytes have a short lifespan and, until recently, were difficult to manipulate with transgenes or RNAi. The development of viral vectors, including adenovirus and lentivirus, have made the manipulation of specific genes in primary hepatocytes much more feasible. However, the viable window for working with these cells in culture is still a limitation.

Rat hepatocytes are perhaps the easiest to prepare in large quantities, but suffer the same drawback as rat hepatoma cells; they make both apoB100 and apoB48. Hamster hepatocytes are an excellent alternative, because these cells, like human liver cells, make only apoB100. Although there are practical and ethical hurdles to the harvest and use of hepatocytes from humans and nonhuman primates, some experiments have been reported using these cells.

There are distinct advantages to the use of primary mouse hepatocytes, particularly because the mouse genome has been studied extensively, and transgenic and knockout mouse models for many genes of interest are commercially available (Breslow, 1996). Several mouse models have 'humanised' lipoprotein metabolism, generated by the introduction or deletion of specific genes, including the introduction of cholesteryl ester transfer protein (CETP), deletion of hepatic apoB mRNA editing such that the liver only produces apoB100, and the complete replacement of the mouse apoB gene with the human apoB100.

Rodent models are particularly useful for dietary manipulation experiments, which are relatively straightforward and inexpensive. Additionally, hepatocyte and whole animal lipoprotein metabolism can be studied under the same conditions. The inhibitors poloxomer and Triton WR1339 have been used to prevent intravascular lipolysis of TAG and can allow measurement of the output of hepatic VLDL in the absence of circulatory catabolism.

Cell culture models for intestinal lipoprotein assembly and secretion have been much more challenging than hepatocytes. Most studies use Caco-2, one of the few established cell culture models of enterocytes. These cells are derived from colonic epithelium, but retain some of the characteristics of absorptive epithelium, including the ability to assemble and secrete intestinal lipoproteins. Importantly, these cells can be maintained in polarised cell cultures, such that both the apical (intestinal lumen) and basolateral (secretory) surfaces of the cell can be sampled and manipulated independently. Unfortunately, long culture times (on the order of 14 days or longer) are required before the cells express the enterocyte phenotype. Primary entero-cytes from mouse, hamster, and rat intestine have also been used, but prepa-ration of these cells is technically challenging, and they are short-lived in culture. Nevertheless, significant information has been obtained from these experimental models.

10. FUTURE DIRECTIONS

The future discoveries in TAG-rich lipoprotein assembly will be guided by recent improvements in technology. Advances in cell imaging techniques will allow the dissection of TAG movement within cells and characterisation of the roles of enzymes and exchangeable apolipoproteins. Additionally, the molecular charac-terisation of the proteome of liver and intestinal microsomes, lipid droplets from the cytosol and subcellular organelles will also enhance our understanding of the mechanisms involved. Moreover, we anticipate that improvements in the ability to define protein structures at atomic resolution will allow greater insight into apoB structure, although intact apoB proteins are unlikely to yield to these tech-niques in the foreseeable future. Exploration of roles that lipoproteins play in host defence and inflammation will likely experience substantial progress, because of increasing appreciation of the inter-relationship among inflammation, obesity, insulin resistance and dyslipidaemia. Mechanisms involved in the regulation of lipid and lipoprotein production in fatty liver disease should be revealed, and the interplay between the hepatitis C virus propagation and the role of VLDL assem-bly pathway (reviewed in Felmlee et al., (2013) will also provide fertile research ground, beyond the traditional boundaries of lipid and lipoprotein metabolism. We look forward to the contributions of the next generation of researchers.

REFERENCES

Adeli, K., 2011. Translational control mechanisms in metabolic regulation: critical role of RNA binding proteins, microRNAs, and cytoplasmic RNA granules. Am. J. Physiol. Endocrinol. Metab. 301, E1051–E1064.

Asp, L., Magnusson, B., Rutberg, M., Li, L., Boren, J., Olofsson, S.O., 2005. Role of ADP ribosyl-ation factor 1 in the assembly and secretion of ApoB-100-containing lipoproteins. Arterioscler. Thromb. Vasc. Biol. 25, 566–570.

Banaszak, L.J., Ranatunga, W.K., 2008. The assembly of apoB-containing lipoproteins: a structural biology point of view. Ann. Med. 40, 253–267.

Blasiole, D.A., Oler, A.T., Attie, A.D., 2008. Regulation of ApoB secretion by the low density lipo-protein receptor requires exit from the endoplasmic reticulum and interaction with ApoE or ApoB. J. Biol. Chem. 283, 11374–11381.

Bou Khalil, M., Blais, A., Figeys, D., Yao, Z., 2010. Lipin - the bridge between hepatic glycerolipid biosynthesis and lipoprotein metabolism. Biochim. Biophys. Acta 1801, 1249–1259.

Breslow, J.L., 1996. Mouse models of atherosclerosis. Science 272, 685–688.

Chatterton, J.E., Phillips, M.L., Curtiss, L.K., Milne, R., Fruchart, J.C., Schumaker, V.N., 1995. Immunoelectron microscopy of low density lipoproteins yields a ribbon and bow model for the conformation of apolipoprotein B on the lipoprotein surface. J. Lipid Res. 36, 2027–2037.

Christian, P., Sacco, J., Adeli, K., 2013. Autophagy: emerging roles in lipid homeostasis and metabolic control. Biochim. Biophys. Acta 1831, 819–824.

Cole, L.K., Vance, J.E., Vance, D.E., 2012. Phosphatidylcholine biosynthesis and lipoprotein metabolism. Biochim. Biophys. Acta 1821, 754–761.

Csaki, L.S., Dwyer, J.R., Fong, L.G., Tontonoz, P., Young, S.G., Reue, K., 2013. Lipins, lipinopathies, and the modulation of cellular lipid storage and signaling. Prog. Lipid Res. 52, 305–316.

Davidson, N.O., Shelness, G.S., 2000. Apolipoprotein B: mRNA editing, lipoprotein assembly, and presecretory degradation. Annu. Rev. Nutr. 20, 169–193.

Felmlee, D.J., Hafirassou, M.L., Lefevre, M., Baumert, T.F., Schuster, C., 2013. Hepatitis C virus, cholesterol and lipoproteins–impact for the viral life cycle and pathogenesis of liver disease. Viruses 5, 1292–1324.

Fisher, E.A., Khanna, N.A., McLeod, R.S., 2011. Ubiquitination regulates the assembly of VLDL in HepG2 cells and is the committing step of the apoB-100 ERAD pathway. J. Lipid Res. 52, 1170–1180.

Jiang, Z.G., Carraway, M., McKnight, C.J., 2005. Limited proteolysis and biophysical characterization of the lipovitellin homology region in apolipoprotein B. Biochemistry (NY) 44, 1163–1173.

Johs, A., Hammel, M., Waldner, I., May, R.P., Laggner, P., Prassl, R., 2006. Modular structure of solubilized human apolipoprotein B-100. Low resolution model revealed by small angle neutron scattering. J. Biol. Chem. 281, 19732–19739.

Khatun, I., Zeissig, S., Iqbal, J., Wang, M., Curiel, D., Shelness, G.S., Blumberg, R.S., Hussain, M.M., 2012. Phospholipid transfer activity of microsomal triglyceride transfer protein produces apolipoprotein B and reduces hepatosteatosis while maintaining low plasma lipids in mice. Hepatology 55, 1356–1368.

Kjolby, M., Andersen, O.M., Breiderhoff, T., Fjorback, A.W., Pedersen, K.M., Madsen, P., Jansen, P., Heeren, J., Willnow, T.E., Nykjaer, A., 2010. Sort1, encoded by the cardiovascular risk locus 1p13.3, is a regulator of hepatic lipoprotein export. Cell. Metab. 12, 213–223.

Lagace, T.A., Curtis, D.E., Garuti, R., McNutt, M.C., Park, S.W., Prather, H.B., Anderson, N.N., Ho, Y.K., Hammer, R.E., Horton, J.D., 2006. Secreted PCSK9 decreases the number of LDL receptors in hepatocytes and in livers of parabiotic mice. J. Clin. Invest 116, 2995–3005.

Ledford, A.S., Cook, V.A., Shelness, G.S., Weinberg, R.B., 2009. Structural and dynamic interfacial properties of the lipoprotein initiating domain of apolipoprotein B. J. Lipid Res. 50, 108–115.

Levy, E., Harmel, E., Laville, M., Sanchez, R., Emonnot, L., Sinnett, D., Ziv, E., Delvin, E., Couture, P., Marcil, V., Sane, A.T., 2011. Expression of Sar1b enhances chylomicron assembly and key components of the coat protein complex II system driving vesicle budding. Arterioscler. Thromb. Vasc. Biol. 31, 2692–2699.

Liang, J.S., Kim, T., Fang, S., Yamaguchi, J., Weissman, A.M., Fisher, E.A., Ginsberg, H.N., 2003. Overexpression of the tumor autocrine motility factor receptor Gp78, a ubiquitin protein ligase, results in increased ubiquitinylation and decreased secretion of apolipoprotein B100 in HepG2 cells. J. Biol. Chem. 278, 23984–23988.

Liu, Q., Siloto, R.M., Lehner, R., Stone, S.J., Weselake, R.J., 2012. Acyl-CoA:diacylglycerol acyltransferase: molecular biology, biochemistry and biotechnology. Prog. Lipid Res. 51, 350–377.

Mitsche, M.A., Packer, L.E., Brown, J.W., Jiang, Z.G., Small, D.M., McKnight, C.J., 2014. Surface tensiometry of apolipoprotein B domains at lipid interfaces suggests a new model for the initial steps in triglyceride-rich lipoprotein assembly. J. Biol. Chem. 289, 9000–9012.

Mulvihill, E.E., Assini, J.M., Sutherland, B.G., DiMattia, A.S., Khami, M., Koppes, J.B., Sawyez, C.G., Whitman, S.C., Huff, M.W., 2010. Naringenin decreases progression of atherosclerosis by improving dyslipidemia in high-fat-fed low-density lipoprotein receptor-null mice. Arterioscler. Thromb. Vasc. Biol. 30, 742–748.

Myerson, M., Ngai, C., Jones, J., Holleran, S., Ramakrishnan, R., Berglund, L., Ginsberg, H.N., 2005. Treatment with high-dose simvastatin reduces secretion of apolipoprotein B-lipoproteins in patients with diabetic dyslipidemia. J. Lipid Res. 46, 2735–2744.

Ohsaki, Y., Cheng, J., Fujita, A., Tokumoto, T., Fujimoto, T., 2006. Cytoplasmic lipid droplets are sites of convergence of proteasomal and autophagic degradation of apolipoprotein B. Mol. Biol. Cell. 17, 2674–2683.

Olofsson, S.O., Boren, J., 2012. Apolipoprotein B secretory regulation by degradation. Arterioscler. Thromb. Vasc. Biol. 32, 1334–1338.

Ota, T., Gayet, C., Ginsberg, H.N., 2008. Inhibition of apolipoprotein B100 secretion by lipid-induced hepatic endoplasmic reticulum stress in rodents. J. Clin. Invest. 118, 316–332.

Pan, M., Cederbaum, A.I., Zhang, Y.L., Ginsberg, H.N., Williams, K.J., Fisher, E.A., 2004. Lipid peroxidation and oxidant stress regulate hepatic apolipoprotein B degradation and VLDL production. J. Clin. Invest. 113, 1277–1287.

Pan, X., Zhang, Y., Wang, L., Hussain, M.M., 2010. Diurnal regulation of MTP and plasma triglyceride by CLOCK is mediated by SHP. Cell Metab. 12, 174–186.

Qiu, W., Avramoglu, R.K., Dube, N., Chong, T.M., Naples, M., Au, C., Sidiropoulos, K.G., Lewis, G.F., Cohn, J.S., Tremblay, M.L., Adeli, K., 2004. Hepatic PTP-1B expression regulates the assembly and secretion of apolipoprotein B-containing lipoproteins: evidence from protein tyrosine phosphatase-1B overexpression, knockout, and RNAi studies. Diabetes 53, 3057–3066.

Raabe, M., Veniant, M.M., Sullivan, M.A., Zlot, C.H., Bjorkegren, J., Nielsen, L.B., Wong, J.S., Hamilton, R.L., Young, S.G., 1999. Analysis of the role of microsomal triglyceride transfer protein in the liver of tissue-specific knockout mice. J. Clin. Invest. 103, 1287–1298.

Richardson, P.E., Manchekar, M., Dashti, N., Jones, M.K., Beigneux, A., Young, S.G., Harvey, S.C., Segrest, J.P., 2005. Assembly of lipoprotein particles containing apolipoprotein-B: structural model for the nascent lipoprotein particle. Biophys. J. 88, 2789–2800.

Strong, A., Ding, Q., Edmondson, A.C., Millar, J.S., Sachs, K.V., Li, X., Kumaravel, A., Wang, M.Y., Ai, D., Guo, L., Alexander, E.T., Nguyen, D., Lund-Katz, S., Phillips, M.C., Morales, C.R., Tall, A.R., Kathiresan, S., Fisher, E.A., Musunuru, K., Rader, D.J., 2012. Hepatic sortilin regulates both apolipoprotein B secretion and LDL catabolism. J. Clin. Invest. 122, 2807–2816.

Sun, H., Samarghandi, A., Zhang, N., Yao, Z., Xiong, M., Teng, B.B., 2012. Proprotein convertase subtilisin/kexin type 9 interacts with apolipoprotein B and prevents its intracellular degradation, irrespective of the low-density lipoprotein receptor. Arterioscler. Thromb. Vasc. Biol. 32, 1585–1595.

Tsai, J., Qiu, W., Kohen-Avramoglu, R., Adeli, K., 2007. MEK-ERK inhibition corrects the defect in VLDL assembly in HepG2 cells: potential role of ERK in VLDL-ApoB100 particle assembly. Arterioscler. Thromb. Vasc. Biol. 27, 211–218.

Twisk, J., Gillian-Daniel, D.L., Tebon, A., Wang, L., Barrett, P.H., Attie, A.D., 2000. The role of the LDL receptor in apolipoprotein B secretion. J. Clin. Invest. 105, 521–532.

Wang, L., Small, D.M., 2004. Interfacial properties of amphipathic beta strand consensus peptides of apolipoprotein B at oil/water interfaces. J. Lipid Res. 45, 1704–1715.

Wang, S., Chen, Z., Lam, V., Han, J., Hassler, J., Finck, B., Davidson, N., Kaufman, R., 2012. IRE1α-XBP1s induces PDI expression to increase MTP activity for hepatic VLDL assembly and lipid homeostasis. Cell Metab. 16, 473–486.

Weinberg, R.B., Gallagher, J.W., Fabritius, M.A., Shelness, G.S., 2012. ApoA-IV modulates the secretory trafficking of apoB and the size of triglyceride-rich lipoproteins. J. Lipid Res. 53, 736–743.

Wetterau, J.R., Aggerbeck, L.P., Bouma, M.E., Eisenberg, C., Munck, A., Hermier, M., Schmitz, J., Gay, G., Rader, D.J., Gregg, R.E., 1992. Absence of microsomal triglyceride transfer protein in individuals with abetalipoproteinemia. Science 258, 999–1001 (New York, NY).

Willnow, T.E., Kjolby, M., Nykjaer, A., 2011. Sortilins: new players in lipoprotein metabolism. Curr. Opin. Lipidol. 22, 79–85.

Ye, J., Li, J.Z., Liu, Y., Li, X., Yang, T., Ma, X., Li, Q., Yao, Z., Li, P., 2009. Cideb, an ER- and lipid droplet-associated protein, mediates VLDL lipidation and maturation by interacting with apolipoprotein B. Cell Metab. 9, 177–190.

Chapter 17

Lipoprotein Receptors

Wolfgang J. Schneider

Department of Medical Biochemistry, Medical University of Vienna, Vienna, Austria

ABBREVIATIONS

(V)LDL (very) low density lipoprotein
Apo Apolipoprotein
ApoER2 Apo E receptor type 2
ARH Autosomal recessive hypercholesterolaemia
Dab1 Disabled-1
EGF Epidermal growth factor
FH Familial hypercholesterolaemia
LCAT Lecithin-cholesterol acyltransferase
HDL High density lipoprotein
PCSK9 Proprotein convertase subtilisin-like kexin-type 9
LOX-1 Lectin-like oxidised LDL receptor
LRP LDL receptor-related protein
PDGF Platelet derived growth factor
SR Scavenger receptor
SR-BI Scavenger receptor class B type I

1. INTRODUCTION: RECEPTOR-MEDIATED LIPOPROTEIN METABOLISM

The physiologically most important task of lipoprotein receptors in systemic lipid metabolism is the clearance of lipoproteins from the circulation, body fluids and interstitial spaces. There are several reasons why lipoproteins need to be redirected from extracellular fluids into cellular compartments, for instance: (1) lipoprotein particles are transport vehicles for components that are vital to the target cells; (2) their uptake serves signalling and/or regulatory roles in processes controlling cellular and organ metabolism; (3) following completion of their job, they have become dispensable; and (4) they might have deleterious effects if allowed to remain in extracellular compartments for prolonged time. Chapters 15 and 16 have described principles of assembly, secretion and interconversion of lipoprotein particles in the circulation, as well as structural aspects of these highly diverse macromolecules. Here, the players in mechanisms of lipoprotein transport from the plasma compartment to various types of cells of the body, that is the

Biochemistry of Lipids, Lipoproteins and Membranes. http://dx.doi.org/10.1016/B978-0-444-63438-2.00017-1

489

roles of bona-fide lipoprotein receptors, are described. In addition, newly discovered functions of receptors thus far thought to be specialised exclusively for lipoprotein transport are summarised.

When considering lipoprotein transport and targeting via cell surface receptors, an important aspect is that, in many cases, the two major lipid components of lipoproteins, triacylglycerols and cholesterol (unesterified and/or esterified), have different fates. Triacylglycerols are delivered primarily to adipose tissue and muscle, where their fatty acids are stored or oxidised for production of energy, respectively (see Chapter 5). The major pool of cholesterol, in contrast, is continuously shuttled among the liver, intestine and other extra-hepatic tissues. The major transport form of cholesterol is fatty acyl esters of the sterol; within cells these cholesteryl esters are hydrolysed, and the unesterified sterol has multiple uses. Among their many functions, sterols serve as structural components of cellular membranes, as substrates for the synthesis of steroid hormones and bile acids, and they perform several regulatory functions. A classical example of cholesterol-mediated regulation is the low density lipoprotein (LDL) receptor pathway (Section 2.2). For correct targeting of lipoproteins to sites of metabolism and removal, the lipoproteins rely heavily on the apolipoproteins (apos) associated with their surface coat. Apos mediate and/or regulate the interaction of lipoprotein particles with enzymes, transfer proteins and transmembrane transporters (see Chapter 15) and with cell surface receptors, the topic of this chapter.

Key features of human lipoprotein metabolic pathways are schematically summarised in Figure 1; this outline necessarily omits many of the details less significant for aspects of receptor-mediated removal of lipoproteins. The net of complex pathways can be divided into exogenous and endogenous branches, concerned with the transport of dietary and liver-derived lipids, respectively. Both metabolic sequences are initiated by the production and secretion of triacylglycerol-rich lipoproteins (see Chapter 16). In the exogenous branch, intestinally derived chylomicrons (Figure 1) containing the principal apos apoB48 and apoE are secreted into the lymph and subsequently enter the bloodstream, where they function as energy carriers by providing triacylglycerol-derived fatty acids to peripheral tissues. This lipolytic extraction of fatty acids from the triacylglycerol-core of the lipoprotein particles ('lipolysis' in Figure 1) is achieved mainly, but not exclusively, by the enzyme lipoprotein lipase, which is bound to the lumenal surface of the endothelial cells lining the capillary bed. Removal of triacylglycerol in extrahepatic tissues decreases the size of the chylomicrons and produces cholesteryl ester-rich lipoprotein particles termed chylomicron remnants. During this conversion, apos of the C-type are lost from the surface of the particles; the remnants, having completed their task, are destined for rapid uptake and catabolism by the liver, which occurs almost exclusively by receptor-mediated processes. Both the so-called LDL receptor-related protein (LRP/LRP1; see Section 4.1) and the LDL receptor mediate their rapid removal via recognition of apo E. ApoB48, the intestine-specific isoform of apoB, which resides on chylomicrons throughout their life span, is not recognised by these receptors.

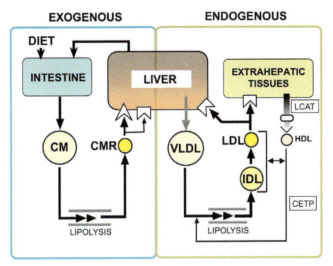

FIGURE 1   **Simplified schematic summary of the essential pathways for receptor-mediated human lipoprotein metabolism.** The liver is the crossing point between the exogenous pathway (left-hand side), which deals with dietary lipids, and the endogenous pathway (right-hand side) that starts with the hepatic synthesis of VLDL. The exogenous metabolic branch starts with the production of chylomicrons (CM) in the intestine, which are converted to chylomicron remnants (CMR). Very low density lipoprotein (VLDL) particles are metabolised to LDL particles, which bind to the LDL receptor. Abbreviations: IDL, intermediate density lipoproteins; LDL, low density lipoproteins; HDL, high density lipoproteins; LCAT, lecithin:cholesterol acyltransferase; CETP, cholesteryl ester transfer protein; ⋀, LDL receptor-related protein (LRP1); and ⋂, LDL receptor; 'Lipolysis' denotes lipoprotein lipase-catalysed triacylglycerol lipolysis in the capillary bed.

In analogy to the exogenous lipid transport branch, the endogenous pathway begins with the production and secretion of triacylglycerol-rich lipoprotein particles by the liver; these newly synthesised lipoprotein particles are termed very low density lipoprotein (VLDL). In significant contrast to chylomicrons, VLDL contains the isoform B100 of apoB, and on maturation in the circulation, the particles acquire apoC and apoE. Lipoprotein lipase in the capillary bed hydrolyses triacylglycerol of secreted VLDL, but less efficiently than from chylomicrons, which likely is one of the reasons for slower plasma clearance of VLDL ($t_{1/2}$ is days) compared to chylomicrons ($t_{1/2}$ is minutes to a few hours). Lipolysis during the prolonged residency of VLDL particles in the plasma compartment generates intermediate density lipoproteins (IDL) (Figure 1) and finally, LDL, which can be considered a class of VLDL-remnants analogous to chylomicron remnants. In parallel, apos and surface components (mostly phospholipids and unesterified cholesterol), but also cholesteryl esters and triacylglycerol, are subject to transfer and exchange between particles in the VLDL lipolysis pathway and certain species of high density lipoproteins (HDLs). In addition, the so-called VLDL receptor (Section 4.2) appears to deliver fatty acids to a limited set of peripheral tissues. Enzymes involved in cholesterol loading and esterification and in interparticle-transfer reactions are lecithin-cholesterol acyl-transferase (LCAT), cholesteryl ester transfer protein (CETP) and phospholipid transfer protein; certain aspects of these three proteins described in Chapters 15 and 19. IDL particles (which still harbour some apoE), to a variable degree, and LDL as the end product of VLDL catabolism in the plasma are finally cleared from the blood by uptake via the LDL receptor. This receptor is found in the liver (which harbours 60–70% of all the LDL receptors in the body) as well as in extrahepatic tissues, and is a key regulatory element of systemic cholesterol homoeostasis (Section 2.2). Thus, steady-state plasma LDL levels are not only the result of lipoprotein receptor numbers and activities but are also influenced by the rate of VLDL synthesis, the activity of lipoprotein lipase and auxiliary lipases, and many other metabolic processes.

As far as HDL levels and metabolism are concerned, one result of LCAT- and transfer protein-catalysed reactions is the production of a dynamic spectrum of particles with a wide range of sizes and lipid compositions (Chapter 15). Nascent HDL particles contain mostly, but not exclusively, apoA1 and phospholipids and undergo modulation and maturation in the circulation. For instance, the unesterified cholesterol incorporated into plasma HDL is converted to cholesteryl esters by LCAT, creating a concentration gradient of cholesterol between HDL and cell membranes that is required for efficient cholesterol efflux from cells to HDL. In addition, CETP transfers a significant amount of HDL cholesteryl ester to VLDL, IDL and LDL for further transport, primarily to the liver. Thus, a substantial fraction of cell-derived cholesterol is transferred as part of HDL indirectly to the liver via hepatic endocytic receptors for IDL and LDL; this process is termed 'reverse cholesterol transport' (Chapter 15). However, receptor-mediated delivery of HDL cholesteryl to cells is fundamentally different from the classic LDL receptor-mediated endocytic pathway (see Section 7.3.2 and Chapter 15).

In addition to receptor-mediated metabolism of lipoproteins, that is the predominant mechanism for removal of intact lipoproteins, individual components of lipoproteins, particularly unesterified cholesterol and fatty acids, can diffuse into cells across the plasma membrane. Other minor uptake processes include so-called fluid-phase endocytosis, which does not involve high-affinity binding of lipoproteins to specific cell surface proteins, and phagocytosis, in which lipoproteins attach to the cell surface via more or less specific forces to subsequently become engulfed by the plasma membrane.

2. REMOVAL OF LOW-DENSITY LIPOPROTEIN FROM THE CIRCULATION

The supply of cells with cholesterol via receptor-mediated endocytosis of LDL is one of the best-characterised processes of macromolecular transport across the plasma membrane of eukaryotic cells. The following sections describe this process, provide an overview of the biochemical and physiological properties of the LDL receptor, and discuss mutations in the gene for the LDL receptor as the molecular basis for the genetic disease termed autosomal dominant familial hypercholesterolaemia (FH). This genetic disease belongs to the group of autosomal dominant hypercholesterolaemias (ADH), known causes of which are genetic defects in the LDL receptor, PCSK9 (see Section 3.2), or apoB.

2.1 Receptor-Mediated Endocytosis

This multistep process, originally defined as a distinct mechanism for the cellular uptake of macromolecules, emerged from studies to elucidate the normal function of LDL by M.S. Brown, J.L. Goldstein and their colleagues (Goldstein et al., 1985), who discovered that the pathway operates in cultured human fibroblasts, which thus became a convenient and powerful experimental system to study lipoprotein metabolism. The salient features of the itinerary of an LDL particle (mean diameter ~22 nm) from the plasma into a normal human fibroblast are summarised in Figure 2. First, the lipoprotein particle binds to one of the approximately 15,000 LDL receptors exposed on the surface of the cell. LDL receptors are not evenly distributed on the cell surface; rather, up to 80% are localised to specialised regions of the plasma membrane comprising only 2% of the cell surface. These regions form pits and are lined on their cytoplasmic side with an arrangement of proteins that in electron micrographs has the appearance of a fuzzy coat. Each of these so-called 'coated pits' contains several kinds of endocytic receptors in addition to LDL receptors, but LDL particles exclusively bind to 'their' receptors due to the extremely high affinity and specificity for each other. Next, the receptor/LDL complex undergoes rapid invagination of the coated pit, which eventually culminates in the release of the coated pit into the interior of the cell. At this point, the coated pit has been transformed into an endocytic 'coated vesicle', a membrane-enclosed organelle that is coated on its exterior (cytoplasmic) surface with a polygonal network of fibrous protein(s), the main structural and functional component of which is a fascinating protein called clathrin (Goldstein et al., 1985). Subsequently, the coat is rapidly removed in concert with acidification of the vesicles' interior and their fusion with other uncoated endocytic vesicles; commonly, vesicular compartments of this kind are termed 'endosomes'. Transiently, LDL and the receptor are found in smooth vesicles, in which the lipoprotein particles dissociate from the receptor due to the acidic environment. LDL is then delivered to lysosomes, where the particle is disassembled to liberate individual components, while the receptor escapes this fate and recycles back to the cell surface, homes in on a coated pit, and is ready to bind and internalise new ligand molecules (Goldstein et al., 1985). It has been calculated that one LDL receptor can be reused approximately 100 times before it is removed from the recycling pathway and directed toward degradation.

There are variations to the LDL receptor-LDL path, because ligand degradation and receptor recycling are not always coupled. However, all these systems have in common the initial steps leading to the formation of endosomes. Subsequently, the receptors are either degraded, recycled back to the cell surface, or are transported (for example, across polarised cells); their respective ligands can follow the same or divergent routes (Goldstein et al., 1985). The possibility to reuse the LDL receptor via recycling constitutes an elegant and economical way to regulate and achieve efficient removal of LDL from the extracellular space.

2.2 The Low-Density Lipoprotein Receptor Pathway

The LDL receptor is a key component in the mechanism ensuring maintenance of cholesterol homoeostasis (Goldstein et al., 1985). In fact, as an active interface between extracellular and intracellular cholesterol pools, it is itself subject to regulation at the cellular level (Figure 2). LDL-derived cholesterol

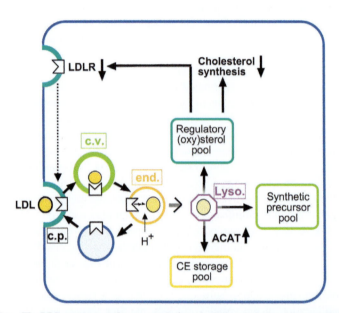

FIGURE 2   **The LDL receptor pathway: regulation of cellular cholesterol homoeostasis.** LDL receptors (LDLR) are synthesised in the endoplasmic reticulum, undergo post-translational modification in the Golgi compartment, and travel to the cell surface, where they collect in coated pits (c.p.). LDL particles bound to LDL receptors (⋈) are internalised in coated vesicles (c.v.), which become uncoated and acidified by proton (H+) transport into their lumen, resulting in endosomes (end.) in which the LDLs dissociate from the receptors due to the low pH. From there, the LDLs are delivered to lysosomes (Lyso.), but almost all the receptors travel back to the cell surface (where they become incorporated into c.p. again) within a recycling vesicle. Lysosomal degradation of LDL results in the complete breakdown of apoB100 and liberation of cholesteryl via hydrolysis of cholesteryl esters. The LDL-derived cholesteryl has three main fates: (a) it is reconverted to cholesteryl esters via stimulation of acyl-CoA:cholesteryl acyltransferase (ACAT) for storage in droplets (CE storage pool; bottom); (b) it is used as biosynthetic precursor for bile acids, steroid hormones, membranes (synthetic precursor pool, right-hand side); and (c) it serves, especially if converted to oxysterols (top), several regulatory functions. The most important of these are suppression of cholesteryl synthetic enzymes and decrease in the production of LDL receptors.

(generated by hydrolysis of LDL-borne cholesteryl esters) and its intracellularly generated oxidised derivatives mediate a complex series of feedback control mechanisms that protect the cell from overaccumulation of cholesterol. First, sterols suppress the activities of key enzymes that determine the rate of cellular cholesterol biosynthesis. Second, the cholesterol activates the cytoplasmic enzyme acyl-CoA: cholesterol acyltransferase, which allows the cells to store excess cholesterol in re-esterified form. Third, the synthesis of new LDL receptors is suppressed, preventing further cellular entry of LDL, and thus cholesterol overloading (Figure 2). The coordinated regulation of LDL receptors and cholesterol synthetic enzymes relies on the sterol-modulated rate of proteolytic cleavage of membrane-bound transcription factor precursors to produce soluble sterol-regulatory element-binding proteins (SREBPs). This elegant mechanism encompassing sterol-level sensing, induced formation of a transcription activation complex, and several protein translocation events is described in detail in Chapter 11.

The overall benefits from, and consequences of, this LDL receptor-mediated regulatory system are the coordinated use of intracellular and extracellular sources of cholesterol at the systemic level. Mammalian cells are able to subsist in the absence of lipoproteins, because they can synthesise cholesterol from acetyl-CoA. When LDL is available, however, most cells primarily import LDL cholesterol and keep their own synthetic activity suppressed (cholesterol biosynthesis requires considerable amounts of energy and intermediate molecules). Thus, while the external supply of cholesterol in the form of lipoproteins can undergo large fluctuation, a constant level of cholesterol is maintained within the cell at all times.

Most of these concepts have arisen from detailed studies in cultured fibroblasts from normal subjects and from patients with the heritable disease, familial hypercholesterolaemia (FH). Lack of the above described regulatory features in FH fibroblasts led to the conclusion that the abnormal phenotype is caused by lack of LDL receptor function and the disruption of the LDL receptor pathway. In particular, the balance between extracellular and intracellular cholesterol pools is severely disturbed in patients with FH. Clinically, the most important effect of LDL receptor deficiency is hypercholesterolaemia, with ensuing accelerated development of atherosclerosis and its complications (Chapter 18). The following sections provide a detailed description of LDL receptor biology, emphasising the impact of mutations in the gene specifying the receptor on its structure and function.

2.3 Relationships between Structure and Function of the Low-Density Lipoprotein Receptor

Studies at the levels of protein chemistry, molecular biology and cell biology have led to a detailed understanding of the biological properties of the LDL receptor. The mature human receptor is a highly conserved integral membrane glycoprotein of 839 residues consisting of five domains (Figure 3). In order of appearance from the amino terminus, these domains are: (1) the ligand binding domain; (2) a domain that has a high degree of homology with the precursor to epidermal growth factor (EGF); (3) a domain that contains a cluster of O-linked carbohydrate chains; (4) a single transmembrane domain; and (5) a short cytoplasmic region. A highly schematised model of the arrangement of these domains (which also applies to the VLDL receptor, Section 4.2) is presented in Figure 3.

The Ligand Binding Domain: This domain mediates the interaction between the receptor and lipoproteins containing apoB100 and/or apoE (Esser et al., 1988). As stated in Section 1, the domain does not interact with apoB48, which is specific for lipoproteins of intestinal origin. The domain,

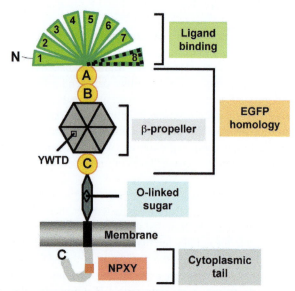

FIGURE 3 **Domain model of the LDL and VLDL receptor.** The five domains of the mature proteins, from the amino-terminus (N) to the carboxy-terminus (C) are: (i) the ligand binding domain, characterised by seven (LDL receptor) or eight (VLDL receptor) cysteine-rich repeats containing clusters of negatively charged amino acids, the core of which consists of Ser-Asp-Glu; repeats 2–7 and 2–8, respectively, cooperatively bind apoB100 and apoE; (ii) the epidermal growth factor-precursor region (EGFP-Homology), consisting of approximately 400 amino acid residues; adjacent to the ligand binding domain and at the carboxy-terminus of this region, respectively, are located three repeats with high homology to repeat motifs found in the precursor to epidermal growth factor (encircled letters A, B and C). The remaining portion of this domain, termed β-propeller, consists of six internally homologous stretches of approximately 50 amino acid residues, each of which contains the sequence Tyr-Trp-Thr-Asp (YWTD); (iii) the O-linked sugar domain, consisting of 58 amino acids with 18 serine and threonine residues containing O-linked carbohydrate chains; (iv) a single membrane-spanning domain; and (v) the cytoplasmic tail with 50 amino acid residues containing the internalisation sequence Asn-Pro-Val-Tyr ('NPXY'; the Val is not absolutely conserved in all species).

located at the amino terminus of the receptor, comprises seven repeats of approximately 40 residues each, called ligand binding repeats. Each of the repeats contains six cysteines, which presumably mediate the folding of each repeat into a compact structure by forming three disulfide bridges with clusters of negatively charged residues with the signature tripeptide Ser-Asp-Glu (SDE) on its surface. The linkers between the individual repeats provide flexibility to the domain to accommodate lipoprotein ligands of different sizes, mainly via positively charged residues on apoB100 or apoE (Jeon and Blacklow, 2005).

The Epidermal Growth Factor Precursor Homology Domain: This region of the LDL receptor lies adjacent to the ligand binding site and comprises approximately 400 amino acids; the outstanding feature is the sequence similarity of this region to parts of the EGF precursor, that is three regions termed 'growth factor repeats' (also called type B repeats). Two of these repeats are located in tandem at the amino-terminus (A and B in Figure 3), and the third (C in Figure 1) is at the carboxy-terminus of the precursor homology region of the LDL receptor. The remainder consists of modules of about 40 residues containing a consensus tetrapeptide, Tyr-Trp-Thr-Asp (YWTD; see Figure 3); six of these tandemly arranged modules form a so-called six-bladed β-propeller. In turn, the β-propeller and the type B repeats constitute the EGF precursor (EGFP) homology domain. X-ray crystallographic studies suggest that the acidic conditions in the endosome (pH 5.0) cause a conformational change that expels LDL from the binding domain in the endosome (see Figure 2, *end.*) via close apposition of binding repeats four and five and the β-propeller (Beglova and Blacklow, 2005).

The O-linked Sugar Domain: This domain of the human LDL receptor is a 58-amino acid stretch enriched in serine and threonine residues, located just outside the plasma membrane. Most, if not all, of the 18 hydroxylated amino acid side chains are glycosylated. The O-linked oligosaccharides undergo elongation in the course of receptor synthesis and maturation. When leaving the endoplasmic reticulum, *N*-acetylgalactosamine is the sole O-linked sugar present, and on processing in the Golgi, galactosyl- and sialyl-residues are added. Possibly, the presence of this domain protects the receptor from degradation by extracellular proteases; however, despite the detailed knowledge about the structure of this region, its physiological importance remains unclear (May et al., 2003).

The Membrane Anchoring Domain: This domain lies carboxy-terminally to the O-linked carbohydrate cluster and consists of ~20 hydrophobic amino acids. As expected, deletion of this domain in certain naturally occurring mutations, or by site-directed mutagenesis, leads to secretion of truncated receptors from the cells.

The Cytoplasmic Tail: This domain of the LDL receptor constitutes a short stretch of 50 amino acid residues involved in the targeting of LDL receptors to coated pits. Naturally occurring mutations and site-specific mutagenesis have identified an 'internalisation signal', Asn-Pro-Xxx-Tyr

(NPXY in Figure 3, in which X denotes any amino acid). Recently, the cytoplasmic domains of the LDL receptor and structural relatives have come into new focus, because they play important roles in the involvement of these receptors in regulation of their activity (Section 3.1) and in signal transduction (Sections 5.2, 5.3 and 6.1). For further details on these aspects, see (Dieckmann et al., 2010).

2.4 The Human Low-Density Lipoprotein Receptor Gene: Organisation and Naturally Occurring Mutations

The ~48 kb human LDL receptor gene contains 18 exons and is localised on the distal short arm of chromosome 19. There is a strong correlation between the functional domains of the protein and the exon organisation of the gene. For instance, the seven ligand binding repeats of the domain are encoded by exons 2 (repeat 1), 3 (repeat 2), 4 (repeats 3, 4 and 5), 5 (repeat six) and 6 (repeat 7). The EGF precursor homology domain is encoded by eight exons, organised in a manner similar to the gene for the EGF precursor itself. The O-linked sugar domain is translated from a single exon between introns 14 and 15. Thus, the LDL receptor gene is a compound of shared coding sequences; in fact, many more molecules containing all or some of these elements have been discovered (see Sections 4 and 5). Membrane proteins with clusters of ligand binding repeats in their extracellular domains are now recognised as relatives of the so-called LDL receptor gene family. Figure 4 depicts a summary of the simplified schematised structures of these membrane proteins.

Molecular genetic studies in patients with autosomal dominant familial FH, which is caused by mutations in the LDL receptor, have identified more than 1100 unique allelic variants in the LDL receptor gene. Comprehensive information about LDL receptor mutations in patients with FH is available at a Website maintained by the Leiden Open Source Variation Database (LOVD) platform, http://www.ucl.ac.uk/ldlr/LOVDv.1.1.0/, where the current status can be retrieved (Leigh et al., 2008). It is important to realise that the primary clinical features of a patient with classical FH (i.e. elevated plasma LDL cholesterol levels) may arise from carrying one or two mutant alleles at the receptor locus. Patients with one mutant allele are termed FH heterozygotes (with mild to significant hypercholesterolaemia), and those with two identical mutant alleles are termed FH homozygotes (with dramatically elevated LDL cholesterol). Most commonly, clinically defined so-called homozygous FH patients in fact have inherited two different mutant alleles from their parents, and therefore their genotype is termed compound heterozygosity.

Based on insights into the molecular nature of these disease-causing mutations, they can be grouped into five classes according to their effects on the LDL receptor protein as follows.

2.4.1 Class 1: Null Alleles—No Detectable Receptor

These mutant alleles fail to produce receptor proteins; cells from subjects carrying two alleles with such mutations do not bind any LDL in saturable fashion. This phenotype arises from point mutations causing premature termination codons early in the protein coding region, mutations in the promoter region(s) that block transcription, mutations that lead to abnormal splicing, and/or instability of the messenger RNA (mRNA), and large deletions.

2.4.2 Class 2: Slow or Absent Processing of the Precursor

These alleles, accounting for more than 50% of all mutant LDL receptor alleles, specify transport-deficient receptor precursors that fail to move with normal rates from the endoplasmic reticulum to and through the Golgi apparatus and on to the cell surface. While some mutations merely attenuate post-translational processing of the LDL receptor, most of these mutations are complete in that transport from the endoplasmic reticulum fails and the mutant receptors never reach the cell surface, that is are unable to bind LDL extracellularly.

2.4.3 Class 3: Defective Ligand Binding

These receptors in general reach the cell surface at normal rates, but are unable to bind LDL efficiently due to subtle structural changes close to, or only mildly affecting, the ligand binding domain. By definition, these mutant receptors undergo the normal posttranslational maturation process.

2.4.4 Class 4: Internalisation-Defective

One of the prerequisites for effective ligand internalisation—the localisation of LDL receptors to coated pits—is not met by the products of these mutant alleles. The failure of 'internalisation-defective' receptors to localise to coated structures results from mutations that alter the carboxy-terminal domain of the receptor. Variants of class 4 mutations have large deletions that lead to a lack of both the cytoplasmic and transmembrane domains. Most of these truncated LDL receptor variants are, as expected, secreted.

2.4.5 Class 5: Recycling-Defective

The classification of these mutations into a separate class is based on the observation that deletion of the first two EGF precursor domains (A and B in Figure 3) of the human LDL receptor allows the truncated receptor to bind and internalise ligand. However, as demonstrated by biochemical means (Beglova and Blacklow, 2005), the release of LDL in the acidic environment of the endosome is blocked, the mutant receptor cannot recycle to the cell surface, and consequently is rapidly degraded intracellularly. All class 5 mutations affect the EGF precursor homology domain, and most often the YWTD modules thereof.

In summary, to a large extent through the delineation of natural mutations in the LDL receptor gene, structure–function relationships and regulatory features of receptor-mediated metabolism of the major cholesterol-carrying lipoprotein in human plasma are now thought to be thoroughly understood. This knowledge has led to the development of the most prescribed lipid-lowering group of pharmaceuticals, the well-known statins. These drugs inhibit cellular cholesterol biosynthesis, which reduces the intracellular cholesterol concentration and consequently increases the number of LDL receptor, a particularly efficient way to accelerate LDL removal by the liver.

Despite these thorough insights from studies of the LDL receptor, recent studies have revealed additional mechanisms for control of LDL receptor activity; three of these modulatory mechanisms and their key components are outlined in the next section. As will be described, genetic defects affecting these regulatory mechanisms also cause heritable diseases, the key feature of which is hypercholesterolaemia.

3. POST-TRANSLATIONAL MODULATORS OF LOW-DENSITY LIPOPROTEIN RECEPTOR ACTIVITY

All molecules and/or mechanisms described below have been identified in studies of clinical FH proven not to be caused by mutations in the LDL receptor gene itself.

3.1 Autosomal Recessive Hypercholesterolaemia

Autosomal recessive hypercholesterolaemia (ARH) is characterised by defects in LDL receptor endocytosis and/or recycling that manifest in specific cell types, such as lymphocytes and liver cells, but not in fibroblasts (this may be one reason why it was discovered late). The clinical phenotype, characterised by xanthomatosis and premature coronary artery disease akin to classical FH, is caused by homozygosity of mutant alleles specifying ARH. This 308-residue cytoplasmic adaptor protein interacts with the cytoplasmic domain of the LDL receptor (the NPXY motif, see Section 2.3), phosphoinositide lipids, and two important components of the coated pit/coated vesicle compartment, that is clathrin and adaptor protein-2 (AP-2) (Cohen et al., 2003). Binding of ARH via its phosphotyrosine binding (PTB) domain to the LDL receptor tail and to either clathrin or AP-2 facilitates normal LDL receptor clustering and internalisation. Interestingly, in fibroblasts, ARH deficiency may be compensated by a mechanism mediated by the PTB-protein Disabled-2 that appears not to operate in lymphocytes and hepatocytes. The affected cells of subjects with ARH display an increased number of LDL receptors on the surface, but because internalisation and recycling are impaired, LDL degradation is greatly reduced. Consequently, ARH homozygotes have LDL cholesterol levels that in general are in the range of receptor-defective FH homozygotes.

3.2 Proprotein Convertase Subtilisin-Like Kexin-Type 9

Proprotein convertase subtilisin-like kexin-type 9 (PCSK9) accelerates the degradation of hepatic LDL receptors. The surprising initial finding was that gain-of-function mutations cause hypercholesterolaemia (like classical FH, this is an autosomal dominant hypercholesterolaemia, or ADH), while loss-of-function mutations in PCSK9 lead to hypocholesterolaemia and protect from heart disease. The current model for PCSK9 action (Horton et al., 2007) holds that the protein, following maturation and secretion from cells, binds to the LDL receptor on the hepatocyte surface. Subsequently, PCSK9 is internalised and mediates rerouting of the LDL receptor from endosomes to lysosomes, where the receptor is degraded (Zhang et al., 2007). This property of PCSK9 is based on its sequence-specific interaction with the receptor's first EGF precursor repeat (see Section 2.3), which apparently interferes with the acid-dependent conformational change essential for receptor recycling (see Section 2.4.5). Furthermore, it explains the effects of both loss- and gain-of-function mutations in this LDL receptor modulator gene.

Because loss-of-function mutations in PCSK9 apparently have no deleterious effects, PCSK9 inhibition has become an attractive new strategy for lowering LDL-cholesterol concentrations. Possible approaches to inhibiting or ablating PCSK9 activity are application of inhibitory antibodies, small molecules including PCSK9-derived peptides and gene silencing. Clinical studies to demonstrate the beneficial effects of these PCSK9-targeted therapies, which may act in addition to currently available drug regimens, show promise.

3.3 Inducible Degrader of the Low-Density Lipoprotein Receptor

Inducible degrader of the LDL receptor (IDOL) is a post-translational modulator of LDL receptor levels. The activity of IDOL is that of an ubiquitin ligase; ubiquitination is associated with accelerated degradation of substrate proteins. The key regulatory system for IDOL is the nuclear liver X receptor (LXR), a sterol-responsive transcription factor (Calkin and Tontonoz, 2012). Thus, the LXR-Idol-LDLR axis defines a pathway antagonistic to the SREBPs (see Section 2.2) for sterol regulation of cholesterol uptake. IDOL overexpression reduces LDLR protein levels in vitro and in vivo and inhibits LDL uptake, whereas knockdown of IDOL increases LDLR protein and LDL uptake (Hong et al., 2010). Analogous to its action on the LDL receptor, IDOL also affects the activities of two closely related receptors, that is the VLDL receptor and apoE receptor type 2 (see Sections 4.2 and 5.1, respectively) via ubiqitination of a sequence in the cytoplasmic domains that is highly conserved in all three receptors.

Current research based on these and future insights into the complex underlying pathways is aimed at determining whether IDOL could be an attractive and feasible therapeutic target. In this respect, studies with known drugs, for example statins, indicate parallel or opposite actions on IDOL and PCSK9 with an unpredictable net effect (Dong et al., 2011). Also, IDOL expression is higher in peripheral cells such as macrophages compared with the liver, suggesting that the primary biological role of IDOL may be to protect the periphery from excess cholesterol uptake (Calkin and Tontonoz, 2012).

Interestingly, the pharmacologic activation of LXR led to regression of established atherosclerotic lesions in mice, but, on the other hand, LXR activation in the liver raises plasma levels of triacylglycerides, an independent risk factor for cardiovascular disease. As hypertriglyceridaemia is the result of an imbalance between production of triacylglycerol-rich lipoproteins (see Chapters 16 and 19) and their catabolism, in keeping with the topic of this chapter, our current, still incomplete knowledge about mechanisms for receptor-mediated plasma clearance of triacylglycerol-rich lipoproteins is outlined in the following sections.

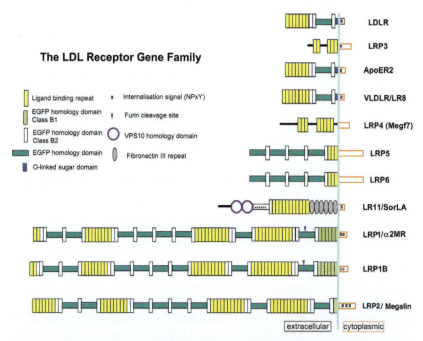

FIGURE 4   The LDL receptor gene family. The structural building blocks making up these proteins are listed in the legend at the top left (for more details, see Figure 3). Presumed extracellular domains are depicted to the left of the plasma membrane (grey vertical line). The standard modules are: negatively charged ligand binding repeats with six cysteines each; epidermal growth factor precursor (EGFP) homology repeats (in the entire family, two subclasses with slightly different consensus sequences are distinguished, termed B1 and B2; these repeats also contain six cysteines each); the 'YWTD' motifs within the β-propeller structure of EGFP homology domains; the O-linked sugar domains, just outside the plasma membrane, typical for LDL receptor, apoER2, and VLDL receptor/LR8; and the consensus or presumed internalisation signals, (FD)NPXY. Several large members of the gene family harbour consensus furin cleavage sites; LR11 (Section 6.2) contains two VPS10 domains and six fibronectin type III repeats, domains not found in other relatives.

4. RECEPTOR-MEDIATED REMOVAL OF TRIACYLGLYCEROL-RICH LIPOPROTEINS FROM THE PLASMA

4.1 Catabolism of Chylomicrons by Low Desity Lipoprotein Receptor-Related Protein 1

Chylomicrons are too large to cross the endothelial barrier; thus, their prior lipolysis to remnants serves a dual function: transport of energy to tissues, and decrease in size to facilitate terminal catabolism. The triacylglycerides in chylomicrons are hydrolysed by lipoprotein lipase along the luminal surface of the capillary endothelial cells, a process that is dependent on molecules that anchor the enzyme on the endothelial surface. A protein termed glycosylphosphatidylinositol-anchored high-density lipoprotein-binding protein 1 (GPIHBP1) has been proposed by S.G. Young to play a critical role in the lipolytic processing of triglyceride-rich lipoprotein particles, possibly enhanced by interaction with apoA-V, an apo independently presumed to be involved in triacylglycerol homoeostasis. GPIHBP1 is highly expressed in heart and adipose tissue, the same tissues that express high levels of lipoprotein lipase, in agreement with GPIHBP1 being an important platform for the processing of chylomicrons in capillaries (Young and Zechner, 2013). The role of hepatic heparan sulfates in triacylglyceride-rich lipoprotein metabolism was studied by inactivating, in murine hepatocytes, the gene for GlcNAc *N*-deacetylase/*N*-sulfotransferase 1, an important enzyme in heparin-sulphate biosynthesis. These mice were viable and healthy, but they accumulated triglyceride-rich lipoprotein particles containing apoE, apoB100, apoB48 and apoC-I to IV, suggesting that heparan sulphate proteoglycan(s) participate in the clearance of both intestinally derived and hepatic lipoprotein particles (Dallinga-Thie et al., 2010).

While these studies indicate the requirement for intermediary steps in the overall clearance pathways for triacylglyceride-rich lipoproteins, the ultimate uptake of the particles across the plasma membrane is believed to occur by endocytotic mechanisms mediated by lipoprotein-specific receptors. Studies in a variety of experimental systems predicted that the high-affinity removal of chylomicron remnants occurs independent of the LDL receptor, despite the presence of apoE on these particles. Accordingly, individuals with homozygous FH, who lack functional LDL receptors, show no signs of delayed clearance of chylomicron remnants. It took a decade after discovery of the LDL receptor to identify the receptor mediating this clearance process.

Because the LDL receptor and the chylomicron remnant receptor must share at least the ability to bind apoE, attempts to isolate this receptor were based on the anticipated similarity of its ligand binding region to that of the LDL receptor. Indeed, homology cloning resulted in the characterisation of an unusually large membrane protein, composed exclusively of structural elements found in the LDL receptor molecule; it therefore has been termed LDL receptor-related protein, or LRP, later designated LRP1 (Herz et al., 1988). As shown in Figure 4, LRP1 (LRP/α_2MR), a 4,526-amino acid integral membrane glycoprotein, contains 31 ligand binding repeats and 22 repeats of the growth factor type. LRP1 clearly binds lipoproteins in an apoE-dependent fashion. In addition, a role of LRP1 in cellular retention of apoE for subsequent use in formation and secretion of HDL particles has been suggested (Laatsch et al., 2012).

Interestingly, soon after its cloning, LRP1 was shown to be identical to the receptor for α_2-macroglobulin, a major plasma protein that functions in 'trapping', and thereby inactivating, cellular proteinases that have entered the plasma compartment. Since then, many more plasma proteins and protein complexes have been identified that, at least in vitro, bind to LRP1 (Lillis et al., 2008). Importantly, α_2-macroglobulin-proteinase complexes are cleared by the liver with the same kinetics as chylomicron remnants, indicating that LRP1 may indeed perform multiple functions in the removal of spent vehicles of intestinal lipid transport and of potentially harmful proteinases. In addition, recent findings also support roles of LRP1 in signal transduction (described in Section 6.2). Another LDL receptor relative with an even broader range of functions is introduced in the following section.

4.2 The So-Called Very Low-Density Lipoprotein Receptor: A Role in Catabolism of Very Low Density Lipoprotein?

The name VLDL receptor was coined for a protein discovered with elegant cloning methodology by T. Yamamoto and colleagues. The overall modular structure of the VLDL receptor is virtually superimposable with that of the LDL receptor, except that the ligand recognition domain contains an additional (eighth) binding repeat located at the amino terminus (Figures 3 and 4). The VLDL receptor shows an amazing degree of sequence conservation among different species; there is 95% identity between the corresponding mammalian proteins. Even the VLDL receptor homologues of more distant species such as the chicken and frog show 84% and 73%, respectively, identity with the human VLDL receptor. In addition, VLDL receptors exist in variant forms, arising from differential splicing of exon 16, which specifies an O-linked sugar domain. However, uptake of significant amounts of VLDLs into cells by this receptor has not been shown conclusively despite the fact that its tissue distribution is highly suggestive of a role in triacylglycerol transport into metabolically active tissues. However, apoJ (clusterin), a multifunctional protein often found associated with HDL, is internalised via the VLDL receptor (Leeb et al., 2014). In contrast to the LDL receptor, and as expected from a receptor implicated in triacylglycerol transport, the VLDL receptor is not regulated by cellular sterols, but its level appears to be influenced by hormones such as oestrogen and thyroid hormone. On the other hand, its expression pattern is different from that of lipoprotein lipase, with which the VLDL receptor would be expected to act in concert. Nevertheless, numerous studies suggest that the VLDL receptor is, at least in part, involved in the delivery of fatty acids derived from VLDL-triacylglycerols to peripheral tissues, mainly adipose tissue. Such delivery may also be mediated by binding of lipoprotein lipase to both the VLDL receptor and LRP1 and may promote lipid accumulation in, and inflammation of, adipose tissue (Nguyen et al., 2014).

Subsequent to the realisation that the VLDL receptor likely has only a limited role in lipoprotein metabolism, novel VLDL receptor functions have been uncovered. Elegant experimentation showed that it, together with apoE receptor type 2 (apoER2) (see Section 5.1), is involved in neuronal migration in the developing brain via binding of reelin, a ligand distinct from lipoproteins (Honda et al., 2011). These important results are described in Section 6. However, there is a VLDL receptor, described below, which deserves this name because it has a well-defined function in lipoprotein metabolism.

4.3 A Multifunctional Very Low Density Lipoprotein Receptor in the Chicken

The functions of the VLDL receptor homologue of the chicken (termed LDL receptor relative with eight binding repeats, or LR8; Figure 4) are documented by both biochemical and genetic evidence. LR8 mediates a key step in the reproductive effort of the hen, that is, oocyte growth via deposition of yolk lipoproteins. This conclusion is based on studies of a nonlaying chicken strain carrying a single mutation at the *lr8* locus that disrupts LR8 function (the 'restricted ovulator', R/O, strain; reviewed in (Elkin et al., 2012)). As a consequence of the mutation, the hens fail to deposit into their oocytes VLDL and the lipophosphoglycoprotein vitellogenin, which are produced at normal levels in the liver, and the mutant females develop severe hyperlipidaemia and features of atherosclerosis. The phenotypic consequences of the single-gene mutation in R/O hens revealed the extraordinary multifunctionality of the avian VLDL receptor, which recognises more than 95% of all the plasma precursors making up the yolk mass of the fully grown oocyte. R/O hens, which represent a unique animal model for an oocyte-specific receptor defect leading to FH (Section 2.2), are sterile because they do not lay eggs (Elkin et al., 2012).

LR8 exists in isoforms that arise by differential exon splicing; the larger form is termed LR8+, and the smaller one is termed LR8-. The somatic cells of chicken tissues, in particular the granulosa cells surrounding the oocytes, heart, and skeletal muscle, express predominantly LR8+ (albeit at very low levels), while the oocyte is by far the major site of LR8- expression. In the male gonad, the same expression dichotomy exists: somatic cells express the larger, and spermatocytes the shorter, form of LR8.

The properties of LR8 and its central role in reproduction strengthen the hypothesis that the avian receptor is the product of an ancient gene with the ability to interact with many, if not all, ligands of more recent additions to the LDL receptor gene family (Figure 4). In this context, vitellogenin, which is absent from mammals, and apoE, which is not found in birds, possess certain common biochemical properties and regions of sequence similarities and have been suggested to be functional analogues (Schneider and Nimpf, 2003). Even lipophorin, an abundant lipoprotein in the circulatory compartment of insects, is endocytosed in a variety of tissues via an LR8 homologue with very high similarity to the VLDL receptor (Schneider and Nimpf, 2003). Presumably, binding of lipophorin to this receptor is mediated by apolipophorins-I and -II, which share sequence homology with mammalian apoB, and thus may behave similarly to the major vertebrate yolk precursor proteins.

5. OTHER RELATIVES OF THE LOW-DENSITY LIPOPROTEIN RECEPTOR FAMILY

In addition to the VLDL receptor, apoER2, and LRP1, several additional LDL receptor gene family members have been identified at the molecular level in recent years. These receptor proteins are often referred to as 'LDL receptor relatives'. Because several of them are better known under their originally proposed names, these are included here when appropriate. The most prominent members of the family are, listed in the order of their discovery, LDL receptor; LRP1; megalin (also called LRP2, and originally named gp330); VLDL receptor (in chicken termed LR8); LR11 (also named SorLA); apoE receptor type 2 (apoER2); LRPs 3, 4, 5, and 6; and LRP1B (also termed LRP1-DIT) (see Figure 4).

Common features of these proteins are the structurally and functionally defined domains described in Section 2.3 for the human LDL receptor (i.e. the ligand binding domain, the EGF-precursor domain, a facultatively present O-linked sugar domain, a single membrane-spanning stretch and the cytoplasmic domain). Because all LDL receptor relatives, by definition, contain LDL receptor ligand-binding repeat clusters and may therefore play roles in lipid-related metabolism, they are also described in this chapter.

5.1 ApoER2—A Close Relative of the Very Low Density Lipoprotein Receptor

The structure of apoER2 (also known as LRP8) is highly reminiscent of that of the VLDL receptor (Figure 4). However, proteins produced from differentially spliced apoER2 mRNA harbour clusters of either 3, 4, 5, 7 or 8 binding repeats, dependent on the species and organ expressing the gene (Schneider et al., 1997). ApoER2 is predominantly found in the brain, placenta and testis, in contrast to other members of the LDL receptor family, which are expressed to a small extent in the brain, but most prominently in other organs. Besides the liver, the most prevalent site of mammalian apoE expression is the brain, where apoE serves a role in local lipid transport and lipoprotein metabolism, as shown in groundbreaking studies by R.W. Mahley in the 1990s. For instance, ligand-binding studies with human apoER2 demonstrated that the receptor likely is involved in apoE-mediated transport processes in the brain. In addition, the apoER2 splice variant containing eight binding repeats can act as receptor for α_2-macroglobulin (also a ligand of LRP1, Section 4.1) in the brain, which suggests a role in the clearance of α_2M/proteinase complexes from the cerebrospinal fluid and from the surface of neurons. However, as described in Section 6.1, the exciting discovery of the involvement of apoER2 and the VLDL receptor in signal transduction has quickly refocused interest in these receptors.

5.2 Small and Midsize Low-Density Lipoprotein Receptor Relatives: LRP 3, 4, 5, and 6

These members of the gene family (Figure 4) were discovered more or less serendipitously. Degenerative DNA probes corresponding to the highly conserved amino acid sequence WRCDGD, found in LDL receptor ligand-binding repeats, were used to screen a rat liver complementary DNA (cDNA) library, resulting first in the cloning of LRP3, and then of its human homologue from a HepG2 cDNA library. The same approach, using a murine heart cDNA library, resulted in the cloning of murine LRP4 (also called Megf7) and of human LRP4. Mature LRP3 is a 733-residue membrane protein with clusters of two and three binding repeats. Human LRP4 (mature protein, 1885 residues) contains two clusters of LDL receptor binding repeats with three and five modules. Defects in LRP4 lead to malformations of the kidney and limbs (hallmarks of Cenani-Lenz syndrome). Recent studies have shown that many of these receptors interact with ligands other than apoE and amyloid precursor peptide known to interact with other LDL receptor relatives, are endocytically competent, and have, at least in part, delineated their spectrum of in vivo functions. For more details, see (Dieckmann et al., 2010).

Two other members of the LDL receptor family, LRP5 and LRP6 (reviewed in (Dieckmann et al., 2010)) have been discovered in the course of attempts to identify the nature of the insulin-dependent diabetes mellitus locus IDDM4 on chromosome 11q13. Human and mouse LRP5 and LRP6 are very similar type I membrane proteins, approximately 1600 residues long (about twice as large as the LDL receptor), and their extracellular domains are organised exactly as the central portion of LRP1 (Figure 4). The cytoplasmic domains of LRP5/LRP6 contain motifs (dileucine; and aromatic-X-X-aromatic/large hydrophobic) similar to those known to function in endocytosis of other receptors. Furthermore, they harbour serine- and proline-rich stretches that can interact with Src homology 3 and WW (a variant of Src homology 3) domains, properties that relate these receptors to signal transduction pathways different from those involving apoER2 and the VLDL receptor. For instance, LRP5/6 (the orthologs of *Drosophila arrow*) have been shown to act as co-receptors for Wnt proteins, which trigger signalling pathways important for correct development of anterior structures. LRP6 also interacts with Dickkopf, which inhibits Wnt-signalling by releasing receptor-bound Wnt, and with Axin, a component in the cascade that regulates the activation of gene expression in the nucleus of target cells (Dieckmann et al., 2010). The interplay of Wnt-, Dickkopf- and Axin-binding to LRP5/6 apparently holds the key to important developmental signals, similar to the role of VLDL receptor and apoER2 in neuronal migration (Section 6.1). Indeed, following reports that mutations in LRP5 are causally linked to alterations in human bone mass, considerable progress has been made toward understanding the molecular links between Wnt-signalling and bone development and remodelling (Kim et al., 2013).

5.3 The Unusual One: LR11

This membrane protein, a structurally complex member of both the LDL receptor (Figure 4) and sortilin-related receptor gene family, was discovered first in rabbit and subsequently in chicken (Morwald et al., 1997), man, and mouse. Significantly, overall sequence identities between the ~250 kDa proteins range from 80% (man versus chicken) to 94% (man versus rabbit). LR11 (also termed SorLA) is made up of seven distinct domains (Figure 4), including a cluster of 11 LDL receptor ligand repeats. Unusual features are a ~70 kDa domain highly homologous to a yeast receptor for sorting of proteins to the vacuole, Vps10p, and six tandem fibronectin type III repeats. The membrane-spanning and cytoplasmic domains are extremely highly conserved. Interestingly, as in several related Vps10p-containing membrane proteins, the extracellular domain of LR11 can be proteolytically released from the cell surface, producing a soluble form, sLR11, which is found in the circulation. The plasma levels of sLR11 are positively associated with elevated risk for increased intima-media thickness via increasing the migration of smooth muscle cells (Jiang et al., 2008) and appears to be involved in regulation of myeloid cell mobilisation (Shimizu et al., 2014).

LR11 is found mainly in the nervous system, smooth muscle cells and, depending on the species, also in testis, ovary, adrenal glands and kidney. LR11 levels are increased during proliferation, but decrease following differentiation of neuroblastoma cells. A genetic association of LR11/SorLA with Alzheimer's disease has been shown, in agreement with the finding that LR11 binds apoE and that decreased expression of LR11 in the brain increases β-amyloid production, a hallmark of the degenerative disease (Dieckmann et al., 2010). Thus, to date, available information suggests that LR11 and sLR11 are involved in cellular proliferation during development and haematopoiesis, as well as in the pathological processes underlying atherosclerosis and Alzheimer's disease. Notably, despite its demonstrated ability to bind apoE-containing ligands, a direct involvement of LR11 in the metabolism of lipoproteins harbouring apoE has not been demonstrated unequivocally to date.

5.4 Large Low-Density Lipoprotein Receptor Relatives: LRP2 and LRP1B

5.4.1 LRP2/Megalin, a True Transport Receptor

This 600-kDa protein is another large member of the LDL receptor gene family containing four clusters of LDL receptor ligand-binding repeats. Although many proteins that bind to LRP1 are also ligands of LRP2, its expression pattern and specificity for certain ligands account for physiological roles distinct from those of LRP1. LRP2 is expressed mainly in polarised epithelial cells of the kidney, lung, eye, intestine, uterus, oviduct and male reproductive tract. Its roles in embryogenesis are well established; for instance, LRP2 is required for development of the forebrain by taking up apoB-containing lipoproteins into the embryonic neuroepithelium and removing Bmp4 (bone morphogenetic protein-4) from the extracellular space (Spoelgen et al., 2005). Another important function of LRP2 is its role in the metabolism of certain lipophilic vitamins. For instance, in the kidney, vitamin B_{12}/transcobalamin complexes are recaptured from the ultrafiltrate directly by binding to megalin expressed on proximal tubule cells (reviewed in (Dieckmann et al., 2010)). Furthermore, megalin mediates the reabsorption from the proximal tubules of 25-(OH) vitamin D_3/ vitamin D-binding protein complexes, which constitutes a key step in converting the precursor into active vitamin D_3 in the kidney (Fisher and Howie, 2006).

5.4.2 LRP1B

This 4599-residue type I membrane protein contains 32 LDL receptor ligand-binding repeats in its extracellular portion (Figure 4). Among all its relatives, LRP1B shows the highest homology to LRP1. Compared to LRP1, it contains one additional ligand-binding repeat and an insertion of 33 amino acids in the cytoplasmic domain, translated from two additional exons. These structural properties suggest roles of LRP1B in lipoprotein metabolism, which have not yet been demonstrated directly, however. Binding and uptake studies suggested that LRP1B endocytoses ligands much slower than LRP1, suggesting that these receptors may act antagonistically by competing for common ligands (Marzolo and Bu, 2009). LRP1B is frequently deleted in non-small cell lung cancer cell lines and is normally expressed mainly in brain skeletal muscle and smooth muscle cells. Most likely, LRP1B has a nonredundant role in the control of cell cycle and proliferation in the context of cancer. In regard to the receptor's function(s) in smooth muscle cells, its regulation has been studied in cells derived from rabbit arteries and in an established smooth muscle cell line. In both systems, LRP1B expression is induced during the exponential phases of cellular proliferation, similar to the expression pattern of LR11 (Section 5.3) (Bujo and Saito, 2006). However, peak levels seem to occur at later points than those observed for LR11 induction, consistent with different physiological roles of the two proteins. In any case, LRP1B deficiency in smooth muscle cells causes atherosclerosis via modulation of intracellular PDGF signalling, as shown for other LDL receptor family members (Dieckmann et al., 2010). In the following section, signal transduction activities of certain lipoprotein receptors are described in more, but not exhaustive, detail.

6. ROLES OF LIPOPROTEIN RECEPTORS IN SIGNAL TRANSDUCTION

6.1 Genetic Models Reveal New Roles for apoER2 and Very Low Density Lipoprotein Receptor in Signal Transduction

Surprisingly, targeted disruption of both the VLDL receptor and the apoER2 genes in mice (double-knockout mice) elicits a dramatic phenotype, which is essentially identical to that of mice lacking the extracellular matrix glycoprotein reelin (Levenson et al., 2008); single-knockout mice of either receptor gene show only subtle phenotypes. Reelin, secreted by cells in the outermost layer of the developing cerebral cortex, orchestrates the migration of neurons along radial fibres, thus forming distinct cortical layers in the cerebrum.

The reason for the grossly abnormal phenotype of the double-knockout mice (disturbed or inverted foliation of the cortical layers) is that reelin normally interacts with the extracellular domains of both the VLDL receptor and apoER2 (Dieckmann et al., 2010), but the ensuing vital signal cascade remains inactivated when the receptors are missing. This cascade is triggered by reelin binding to the VLDL receptor and apoER2, leading to receptor clustering and phosphorylation of the cytoplasmic adaptor protein Disabled-1 (Dab-1) bound to the NPxY-motifs present in the receptors' tails (see Section 2.3). In turn, receptor clustering leads to transphosphorylation of tyrosine kinases and their recruitment to the tails by binding to Dab1. Reelin-stimulated tyrosine-phosphorylation of Dab-1 then starts kinase cascade(s) controlling cell motility and shape by acting on the neuronal cytoskeleton (most notably actin and microtubules), motor proteins and ion channels, specifically the N-methyl-D-aspartate (NMDA) receptors (reviewed in (Dieckmann et al., 2010)). Because Dab1 binds to the intracellular domains of VLDL receptor, apoER2, LDL receptor, and LRP1, it is possible that besides endocytosis of macromolecules, signal transduction might be a general function common to many members of the LDL receptor family. In such a scenario, adaptor molecules like Dab1 might be part of machineries that define the specific function of a particular receptor family member. Hence, in the case of VLDL receptor and apoER2, the specificity of the reelin signalling is achieved by the selective binding of reelin to these two receptors, but not to other members of the LDL receptor family.

6.2 Signalling through LRP1

As described in Section 4.1, where its role in lipoprotein transport is described, LRP1 is one of the largest members of the LDL receptor family. In contrast to the main function of apoER2, endocytosis of a broad range of ligands that bind to the extracellular domain of LRP1 is undisputed and probably the most important task performed by this receptor (Lillis et al., 2008). LRP1 has an intracellular domain that is significantly larger than those of LDL receptor, VLDL receptor, and apoER2, and mediates ligand uptake with very high endocytosis rates. Interestingly, an LRP1-tail-specific YXXL motif, but not the also present NPXY motif, appears to be the dominant determinant for internalisation (Marzolo and Bu, 2009). In addition, cAMP-dependent phosphorylation of a serine residue within the cytoplasmic tail of LRP1 modulates endocytosis efficiency, indicating a possible regulation of ligand uptake by external signals.

Phosphorylation of tyrosine(s) in the intracellular domain of LRP1 by the kinase v-Src, however, might be involved in cellular transformation. Through generating a binding site on phospho-LRP1 for the PTB-domain of Shc, Shc comes in close proximity to v-Src, which also becomes tyrosine-phosphorylated. Thus, in this case, LRP1 acts as an anchor rather than as bona fide signalling receptor. On the other hand, platelet-derived growth factor (PDGF) binds to LRP1 and with the PDGF-receptor induces the tyrosine-phosphorylation of the intracellular domain of LRP1. Interestingly, PDGF induces phosphorylation only of LRP1 that is present in caveolae, suggesting that PDGF heterodimerises LRP1 and PDGF-receptor in the caveolae fraction of the plasma membrane. This phosphorylation event is prevented by the binding of apoE-containing lipoproteins to LRP1, possibly explaining the powerful effect of apoE on preventing PDGF-induced smooth muscle cell migration in vitro. Identification of LRP1 as part of a signalling pathway involving the PDGF-receptor confirms results that show that apoE inhibits PDGF-induced cell migration in an LRP1-dependent manner.

Another signalling system based on LRP1 heterodimerisation appears to involve the NMDA receptor, which mediates calcium flux in cultured primary neurons. Postsynaptic density protein-95 (PSD95), which binds to the LRP1 tail, might be a candidate adaptor protein mediating the interaction of LRP1 with the NMDA receptor (Dieckmann et al., 2010). Finally, LRP1 can be proteolytically processed by a γ-secretase-like activity within the transmembrane domain leading to the release of the entire cytoplasmic domain of the receptor into the cytosol. The question whether the released LRP tail acts as co-factor of a putative transcription factor complex and/or as a modulator for the localisation of LRP1-targeted adaptor proteins is not yet answered.

7. SCAVENGER RECEPTORS: LIPID UPTAKE AND BEYOND

In addition to the type of scavenger receptors (SRs) described in Chapter 18, there is a growing list of hepatic and extrahepatic SRs with potential disease-related functions. For many of these, the criterion for being classified as SR is their broad spectrum of ligands, the first identified of which included diverse polyanionic compounds and modified lipoproteins such as oxidised LDL. Some researchers have proposed the fitting name 'pattern recognition receptors'. Thus far, the SRs have been classified into eight classes (classes A–H). In March 2014, a consortium of 15 researchers in the field proposed a unified and standardised SR nomenclature, mainly to avoid redundancies and miscommunication about the identity of currently known and yet to be discovered receptors (Prabhudas et al., 2014). This newly proposed nomenclature is indicated here. There are now 10 classes of SRs, that is SR-A to SR-J. The most prominent and probably best understood human SRs in the context of lipoprotein metabolism are the SR-As (in particular, SR-A1 and its alternatively spliced isoform, SR-A1.2), SR-Bs (SR-B1 and SR-B2), and the class E SR, SR-E1 (Figure 5). These SRs will be described in the following section, because they are under study for their presumed roles in lipoprotein metabolism, atherosclerosis, and/ or co-regulating the balance between lipid accumulation and inflammation in the arterial wall. The more familiar previous names are also provided.

7.1 Class A Scavenger Receptors

The human class A receptors SR-A1 (previously named SR-AI or SCARA1) and SR-A1.2 (previously SR-AIII) are trimeric membrane proteins characterised structurally by a small amino-terminal intracellular region, an extracellular coiled-coil collagen-like stalk and a cysteine-rich carboxy-terminal domain. Four more receptors are classified as A-types, among which the previously termed MARCO (now SR-A6) and SR with C-type lectin (SRCL; now SR-A4) (Figure 5) may also be involved in the biology of modified lipoproteins. The two human SR-A1 isoforms are produced from the same gene by differential

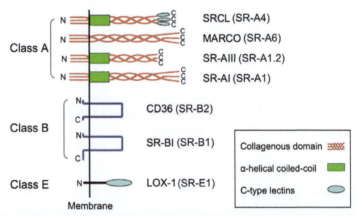

FIGURE 5 **Scavenger receptors.** Structural features of SRs with proposed roles in atherosclerosis (SR-As, SR-Bs and LOX/SR-E1) that are described in this chapter. The plasma membrane is indicated by the vertical line, with the cytoplasmic domains to the left. The individual domains are described in the box. The classical designations are indicated next to the receptor structures, and the proposed new names (see Section 7) are in parentheses. Not drawn to scale.

splicing and are expressed at different levels in tissue macrophages, Kupffer cells and various extrahepatic endothelial cells. SR-A1 expression is induced by some of its ligands, which include, in addition to modified lipoproteins and polyanions, Gram-positive bacteria, heparin, lipoteichoic acid and a precursor of lipid A from lipopolysaccharide of Gram-negative bacteria.

A potential role of the splice forms of SR-A1s in atherosclerotic plaque development was initially reported in a study on apoE- and SR-A1 double-knockout mice. Mice deficient in apoE developed severe plaques, but simultaneous absence of the SR-A1s led to a reduction in plaque size by 58%. This reduction may be related to the greatly reduced uptake of acetylated LDL and oxidised LDL that can be observed in in vitro uptake studies using macrophages and liver cells of SR-A1 knockout mice. However, these early results could not be confirmed. The discrepancy arising from conflicting results may be reconciled by SR-A1 having different roles in early versus advanced atherosclerotic lesions or by different genetic backgrounds of the mice used (Prabhudas et al., 2014).

7.2 Lectin-Like Oxidised Low-Density Lipoprotein Receptor

This 50-kDa transmembrane protein shows no structural similarity to other SRs (Figure 5). The amino-terminal cytoplasmic tail of the lectin-like molecule contains several potential phosphorylation sites. As shown in the laboratory of T. Kita and others, lectin-like oxidised low-density lipoprotein receptor (LOX-1, now SR-E1) can act as endocytic receptor for atherogenic oxidised LDL, but in contrast to the SR-A1s, it interacts only weakly, if at all, with negatively charged LDL (so-called acetylated LDL), which can be produced in vitro by extensive acetylation of lysine residues of apoB. Also, LOX-1 differs from other SRs in that binding of oxidised LDL is inhibited by polyinosinic acid and delipidated oxidised LDL, but not by acetylated LDL, maleylated bovine serum albumin, or fucoidin. LOX-1 is found in thoracic and carotid vessels and highly vascularised tissues such as placenta, lungs, brain and liver, where the cell types expressing it include vascular endothelial cells, platelets, smooth muscle cells, adipocytes and macrophages. Its expression is apparently not constitutive, but can be induced by pro-inflammatory stimuli; it is then detectable in cultured macrophages and in activated smooth muscle cells. LOX-1 also mediates the recognition of aged red blood cells and apoptotic cells. However, little is known about whether LOX-1 indeed functions in clearance of damaged cells in vivo.

7.3 Class B Scavenger Receptors

7.3.1 CD36 or Fatty Acid Translocase (now SR-B2): Role in Lipid Uptake

Also termed fatty acid translocase (FAT), SR-B2 is the prototype of a gene family of proteins with two internal hydrophobic, putative membrane-spanning domains adjacent to short cytoplasmic amino- and carboxy-terminal tails.

This molecule was originally identified as a platelet receptor for throm bospondin, and only later recognised as a macrophage receptor for moderately oxidised LDL, but not for extensively oxidised LDL and acetylated LDL. It can also bind native LDL, HDL, VLDL, collagen and fatty acids and has been proposed to mediate the uptake of fatty acids from the circulation (Ehehalt et al., 2006). It also binds erythrocytes infected with the malaria parasite *Plasmodium falciparum*. The most significant role of CD36 in lipoprotein metabolism is the clearance of oxidised LDL, contributing more than 60% of the cholesteryl esters accumulating in macrophages exposed to oxidised LDL. In agreement with these in vitro data, apoE$^{-/-}$/CD36$^{-/-}$ double knockout mice show marked reductions in atherosclerotic lesion area compared to apoE$^{-/-}$ mice. CD36 has also been reported to promote proinflammatory signalling that may lead to chronic inflammation in the artery wall, and also in Alzheimer's disease. Yet, the signalling pathway(s) triggered by CD36 activation have not been identified beyond doubt.

7.3.2 Removal of High-Density Lipoprotein: SR-BI/II (Proposed Designations, SR-B1/B1.1)

HDL binding to SR class B type I (SR-BI) on the cell surface is now believed to mediate the transfer of its cholesteryl esters to the cell and subsequent release of the lipid-depleted HDL into the extracellular fluid; a small contribution may also be mediated via uptake and lysosomal degradation of the particles. The novel cellular mechanism not involving holoparticle uptake, called selective lipid uptake, can extract cholesteryl esters from LDL and HDL. The mechanism is based on the reversible incorporation of cholesteryl esters into the plasma membrane that may be followed by irreversible internalisation and subsequent hydrolysis through a nonlysosomal pathway (Ashraf and Gupta, 2011). In rats, the liver removes 60–70% of HDL cholesteryl esters from plasma via the selective lipid uptake pathway. Selective uptake of HDL cholesterol by steroidogenic tissues can also provide cholesteryl as a substrate for steroid hormone synthesis. For example, in the rodent adrenal gland, selective uptake accounts for 90% or more of the cholesterol destined for steroid hormone production (Ashraf and Gupta, 2011). The main pathway for cholesterol removal via HDL into the liver is essential for cholesterol homoeostasis, because peripheral cells are unable to degrade cholesterol. This pathway, known as reverse cholesterol transport, is considered to be partly responsible for the antiatherogenic properties of HDL and is discussed in Chapter 15.

The predicted sequences of SR-B1 proteins (509 amino acids) from different mammalian species share approximately 80% sequence identity. As for CD36/SR-B2, the bulk of the protein lies between the two hydrophobic plasma membrane-anchoring domains on the extracellular aspect of cells and contains a set of conserved cysteines. An alternatively spliced mRNA of SR-BI and its corresponding protein have been identified and designated

SR-BII, now proposed to be named SR-B1.1 (Prabhudas et al., 2014). The difference from SR-B1 is that the last 42 amino acids in the C-terminal cytoplasmic domain of SR-B1 are replaced by 39 residues encoded by an alternatively spliced exon, and that SR-B1.1's efficiency in selective lipid uptake is about 25% of that of SR-B1.

In summary, SRs are a widely expressed and highly diverse group of membrane proteins that are appropriately named for their recognition of a broad array (or 'pattern') of ligands. At least some members of this intriguing group of receptors, as indicated here, may play roles in the metabolism of modified lipoproteins and consequently have been implicated in lipid disorders, atherosclerosis and inflammatory processes.

8. OUTLOOK

It is now clear that the increasingly better understood multifunctionality of LDL receptor gene family members is not limited to their extracellular moiety. The original concepts regarding their functions in lipoprotein metabolism remain valid, but have been extended significantly by the discovery that these membrane proteins also play roles in important signal transduction pathways. Given the wealth of signalling pathways, future efforts will have to develop concepts that define the contributions of the individual intracellular domains of LDL receptor relatives. A significant challenge is that, at least in the LDL receptor family, the cytoplasmic domains and the ligand-binding structures of its members are highly similar. Therefore, the molecular basis for the specificity of ligand binding and signal transduction must be identified for each receptor. These ongoing efforts are also expected to add alternative aspects to the ongoing quest to delineate the evolutionary history of the LDL receptor family. In fact, based on the most recently gained knowledge, an evolutionary theory must consider the combinatorial events arising from the multitude of both intracellular and extracellular domains of these proteins in the generation of signalling and/or transport platforms. At the gene level, the different modes of transcriptional, post-transcriptional, post-translational, and epigenetic regulation pose an even greater challenge and offer chances to delineate novel mechanisms of general applicability.

Another general point to take into account is the apparent functional redundancy of receptors involved in a multitude of metabolic pathways and events. This aspect has been addressed here not only for the LDL receptor gene family, but also for the group of SRs. which show broad overlapping ligand specificities (Section 7). In the case of the well-understood LDL receptor gene family, when need arises (e.g. when one or more receptors are dysfunctional), certain members can substitute for others, or are at least sufficiently active for preserving life. For example, VLDL receptor$^{-/-}$ mice, and apoER2$^{-/-}$ mice, have phenotypes that are at first sight indistinguishable from normal. However, the double-knockout mice show gross neurological abnormalities (Section 6.1).

Thus, functional redundancy of LDL receptor relatives can be due to their simultaneous expression in the same cells in a given organ. In turn, different receptor functions despite similar ligand spectra may arise from their expression in different cell types within a tissue or organism. In vivo, receptors presumably have access to different ligands in different environments, and in addition, their cytoplasmic domains would interact with cell-type specific adaptor proteins to mediate a spectrum of signal transduction pathways.

Many of these considerations likely apply to the SRs as well. However, an added complication is the difficulty of determining the in vivo ligand spectrum for these receptors, because there may be enormous overlaps under physiological and/or pathophysiological conditions, and the affinities of SRs for ligands may vary greatly in different settings. As a consequence, specific functions of lipoprotein receptor families must be defined at two levels: the cellular level, to delineate molecular events, and the physiological level, where state-of-the-art genetic manipulations will continue to reveal the functional relevance of receptor redundancy.

Finally, increasing our knowledge about already and yet-to-be identified modulators of the activity of the LDL receptor family (see Section 3) and, possibly, of SRs (Section 7) will be important for the characterisation of modifier genes of lipoprotein metabolic pathways in the general population.

REFERENCES

Ashraf, M.Z., Gupta, N., 2011. Scavenger receptors: Implications in atherothrombotic disorders. Int. J. Biochem. Cell Biol. 43 (5), 697–700.

Beglova, N., Blacklow, S.C., 2005. The LDL receptor: how acid pulls the trigger. Trends Biochem. Sci. 30 (6), 309–317.

Bujo, H., Saito, Y., 2006. Modulation of smooth muscle cell migration by members of the low-density lipoprotein receptor family. Arterioscler Thromb. Vasc. Biol. 26 (6), 1246–1252.

Calkin, A.C., Tontonoz, P., 2012. Transcriptional integration of metabolism by the nuclear sterol-activated receptors LXR and FXR. Nat. Rev. Mol. Cell Biol. 13 (4), 213–224.

Cohen, J.C., Kimmel, M., et al., 2003. Molecular mechanisms of autosomal recessive hypercholesterolemia. Curr. Opin. Lipidol. 14 (2), 121–127.

Dallinga-Thie, G.M., Franssen, R., et al., 2010. The metabolism of triglyceride-rich lipoproteins revisited: new players, new insight. Atherosclerosis 211 (1), 1–8.

Dieckmann, M., Dietrich, M.F., et al., 2010. Lipoprotein receptors–an evolutionarily ancient multifunctional receptor family. Biol. Chem. 391 (11), 1341–1363.

Dong, B., Wu, M., et al., 2011. Suppression of idol expression is an additional mechanism underlying statin-induced up-regulation of hepatic LDL receptor expression. Int. J. Mol. Med. 27 (1), 103–110.

Ehehalt, R., Fullekrug, J., et al., 2006. Translocation of long chain fatty acids across the plasma membrane–lipid rafts and fatty acid transport proteins. Mol. Cell Biochem. 284 (1–2), 135–140.

Elkin, R.G., Bauer, R., et al., 2012. The restricted ovulator chicken strain: an oviparous vertebrate model of reproductive dysfunction caused by a gene defect affecting an oocyte-specific receptor. Anim. Reprod. Sci. 136 (1–2), 1–13.

Esser, V., Limbird, L.E., et al., 1988. Mutational analysis of the ligand binding domain of the low density lipoprotein receptor. J. Biol. Chem. 263 (26), 13282–13290.

Fisher, C.E., Howie, S.E., 2006. The role of megalin (LRP-2/Gp330) during development. Dev. Biol. 296 (2), 279–297.

Goldstein, J.L., Brown, M.S., et al., 1985. Receptor-mediated endocytosis: concepts emerging from the LDL receptor system. Annu. Rev. Cell Biol. 1, 1–39.

Herz, J., Hamann, U., et al., 1988. Surface location and high affinity for calcium of a 500-kd liver membrane protein closely related to the LDL-receptor suggest a physiological role as lipoprotein receptor. EMBO J. 7 (13), 4119–4127.

Honda, T., Kobayashi, K., et al., 2011. Regulation of cortical neuron migration by the Reelin signaling pathway. Neurochem. Res. 36 (7), 1270–1279.

Hong, C., Duit, S., et al., 2010. The E3 ubiquitin ligase IDOL induces the degradation of the low density lipoprotein receptor family members VLDLR and ApoER2. J. Biol. Chem. 285 (26), 19720–19726.

Horton, J.D., Cohen, J.C., et al., 2007. Molecular biology of PCSK9: its role in LDL metabolism. Trends Biochem. Sci. 32 (2), 71–77.

Jeon, H., Blacklow, S.C., 2005. Structure and physiologic function of the low-density lipoprotein receptor. Annu. Rev. Biochem. 74, 535–562.

Jiang, M., Bujo, H., et al., 2008. Ang II-stimulated migration of vascular smooth muscle cells is dependent on LR11 in mice. J. Clin. Invest 118 (8), 2733–2746.

Kim, W., Kim, M., et al., 2013. Wnt/beta-catenin signalling: from plasma membrane to nucleus. Biochem. J. 450 (1), 9–21.

Laatsch, A., Panteli, M., et al., 2012. Low density lipoprotein receptor-related protein 1 dependent endosomal trapping and recycling of apolipoprotein E. PLoS One 7 (1), e29385.

Leeb, C., Eresheim, C., et al., 2014. Clusterin is a ligand for apolipoprotein E receptor 2 (ApoER2) and very low density lipoprotein receptor (VLDLR) and signals via the Reelin-signaling pathway. J. Biol. Chem. 289 (7), 4161–4172.

Leigh, S.E., Foster, A.H., et al., 2008. Update and analysis of the University College London low density lipoprotein receptor familial hypercholesterolemia database. Ann. Hum. Genet. 72 (Pt 4), 485–498.

Levenson, J.M., Qiu, S., et al., 2008. The role of reelin in adult synaptic function and the genetic and epigenetic regulation of the reelin gene. Biochim. Biophys. Acta 1779 (8), 422–431.

Lillis, A.P., Van Duyn, L.B., et al., 2008. LDL receptor-related protein 1: unique tissue-specific functions revealed by selective gene knockout studies. Physiol. Rev. 88 (3), 887–918.

Marzolo, M.P., Bu, G., 2009. Lipoprotein receptors and cholesterol in APP trafficking and proteolytic processing, implications for Alzheimer's disease. Semin. Cell Dev. Biol. 20 (2), 191–200.

May, P., Bock, H.H., et al., 2003. Differential glycosylation regulates processing of lipoprotein receptors by gamma-secretase. J. Biol. Chem. 278 (39), 37386–37392.

Morwald, S., Yamazaki, H., et al., 1997. A novel mosaic protein containing LDL receptor elements is highly conserved in humans and chickens. Arterioscler Thromb. Vasc. Biol. 17 (5), 996–1002.

Nguyen, A., Tao, H., et al., 2014. Very low density lipoprotein receptor (VLDLR) expression is a determinant factor in adipose tissue inflammation and adipocyte-macrophage interaction. J. Biol. Chem. 289 (3), 1688–1703.

Prabhudas, M., Bowdish, D., et al., 2014. Standardizing scavenger receptor nomenclature. J. Immunol. 192 (5), 1997–2006.

Schneider, W.J., Nimpf, J., 2003. LDL receptor relatives at the crossroad of endocytosis and signaling. Cell Mol. Life Sci. 60 (5), 892–903.

Schneider, W.J., Nimpf, J., et al., 1997. Novel members of the low density lipoprotein receptor superfamily and their potential roles in lipid metabolism. Curr. Opin. Lipidol. 8 (5), 315–319.

Shimizu, N., Nakaseko, C., et al., 2014. G-CSF induces the release of the soluble form of LR11, a regulator of myeloid cell mobilization in bone marrow. Ann. Hematol. 93 (7), 1111–1122.

Spoelgen, R., Hammes, A., et al., 2005. LRP2/megalin is required for patterning of the ventral telencephalon. Development 132 (2), 405–414.

Young, S.G., Zechner, R., 2013. Biochemistry and pathophysiology of intravascular and intracellular lipolysis. Genes Dev. 27 (5), 459–484.

Zhang, D.W., Lagace, T.A., et al., 2007. Binding of PCSK9 to EGF-A repeat of LDL receptor decreases receptor recycling and increases degradation. J. Biol. Chem. 282 (25), 18602–18612.

Chapter 18

Atherosclerosis

Murray W. Huff[1], Alan Daugherty[2], Hong Lu[2]

[1]*Departments of Medicine and Biochemistry, Robarts Research Institute, The University of Western Ontario, London, ON, Canada;* [2]*Saha Cardiovascular Research Center, University of Kentucky, Lexington, KY, USA*

ABBREVIATIONS

ABC ATP-binding cassette
ACAT Acyl-CoA:cholesterol acyltransferase
antagomiR Antagonist of a microRNA
apo(a) Apolipoprotein(a)
ApoE$^{-/-}$ Apolipoprotein E-deficient mice
ASO Antisense oligonucleotide
CC β-chemokine
CD36 Cluster of differentiation 36
CD68 Cluster of differentiation 68
CE Cholesteryl ester
CETP Cholesteryl ester transfer protein
CXC α-Chemokine
CXCL, CCL Chemokine ligand
CXCR, CCR Chemokine receptor
DGAT Diacylglycerol acyltransferase
DHA Docosahexenoic acid
eNOS Endothelial cell nitric oxide synthase
EPA Eicosapentenoic acid
ER Endoplasmic reticulum
FH Familial hypercholesterolaemia
GCKR Glucokinase (hexokinase 4) regulator
HDL High-density lipoprotein
HTGL Hepatic triglyceride lipase
ICAM Intracellular adhesion molecule
IDL Intermediate-density lipoprotein
IFN Interferon
IL Interleukin
JNK c-Jun N-terminal kinases
LDL Low-density lipoprotein
LDLR Low-density lipoprotein receptor
Ldlr$^{-/-}$ Low-density lipoprotein receptor-deficient mice
LOX1 Lectin-like oxidised low-density lipoprotein receptor-1

Biochemistry of Lipids, Lipoproteins and Membranes. http://dx.doi.org/10.1016/B978-0-444-63438-2.00018-3

Lp(a) Lipoprotein(a)
LPL Lipoprotein lipase
LRP Low-density lipoprotein-related protein
LXR Liver X receptor
M1 Classically activated macrophages
M2 Alternatively activated macrophages
MCP-1, CCL2 Monocyte chemoattractant protein-1
MERTK Tyrosine protein kinase MER
miR MicroRNA
MyD Myeloid differentiation factor
NLRP3 NOD-, leucine-rich repeat and pyrin domain-containing 3
NOD Nucleotide-binding oligomerisation domain
NPC Niemann–Pick type C
NPC1L1 Niemann–Pick 1-Like 1 protein
Nrf2 Nuclear factor-erythroid-derived 2-related factor 2
oxLDL Oxidised LDL
PCSK9 Proprotein convertase subtilisin/kexin type 9
PECAM Platelet endothelial cell adhesion molecule
PLA2G Secretory phospholipase A2 group
PON-1 Paraoxanase-1
PPAR Peroxisome proliferator-activated receptor
SR-A Scavenger receptor class A
SR-B Scavenger receptor class B
SR-PSOX; CXCL16 Scavenger receptor for phosphatidylserine and oxLDL
SREBF, SREBP Sterol regulatory element-binding factor
TAG Triacylglycerol
TLR Toll-like receptor
TMA Trimethylamine
TMAO Trimethylamine-N-oxide
TNF Tumor necrosis factor
TRIB1 Tribbles homolog 1
VCAM Vascular cell adhesion molecule 1
VLDL Very-low-density lipoprotein
VSMC Vascular smooth muscle cell

1. ATHEROSCLEROSIS

Atherosclerotic diseases, particularly coronary artery disease and stroke, are leading causes of death (Mozaffarian et al., 2015). Atherosclerosis is a chronic process that progresses in a clinically silent manner for decades. The devastating consequence of lesion evolution is the abrupt formation of thrombi, due to lesion rupture or erosion that occludes blood flow, with the attendant overt clinical symptoms (Bentzon et al., 2014) (Figure 1). The slow progression of human atherosclerosis, combined with tissue inaccessibility and the absence of effective imaging techniques, provides challenges for characterisation of lesion evolution. Hence, the understanding of mechanisms underlying the evolution of human atherosclerotic lesion progression is incomplete.

Atherosclerotic Plaque

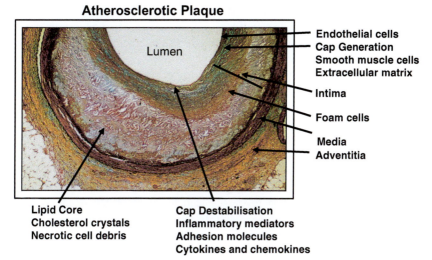

Lumen

Endothelial cells
Cap Generation
Smooth muscle cells
Extracellular matrix

Intima

Foam cells

Media
Adventitia

Lipid Core
Cholesterol crystals
Necrotic cell debris

Cap Destabilisation
Inflammatory mediators
Adhesion molecules
Cytokines and chemokines

FIGURE 1 **A section of a human atherosclerotic coronary artery.** The lesion contains a large lipid-rich core, containing cholesterol crystals, derived from apoptotic and/or necrotic foam cells. A thin fibrous cap, consisting primarily of collagen underlies a compromised endothelial cell layer. Smooth muscle cells have migrated from the media into the intima where they proliferate and can become lipid-rich foam cells. Macrophage-derived and smooth muscle cell-derived foam cells populate the edges of the growing lesion and the fibrous cap. The lesion reduces the artery lumen diameter. A lesion with a thin fibrous cap covering a large lipid core is prone to rupture and rapid thrombus formation resulting in an acute coronary event.

As a result, there has been a dependence on information gleaned from animal models of the disease (Getz and Reardon, 2012). One feature common to all animal models is the need to induce hypercholesterolaemia for lesion development, which is achieved by dietary and/or genetic manipulations. The mechanisms described in this chapter are derived predominantly from studies in apolipoprotein E ($ApoE^{-/-}$) or low-density lipoprotein (LDL) receptor ($Ldlr^{-/-}$)-deficient mice. While many factors influence the initiation, severity and nature of lesions in hypercholesterolaemic animals, the basic tenet of this chapter is that aberrant lipoprotein metabolism is a requirement for atherosclerosis development.

Atherosclerosis develops in specific regions of the arterial tree that are characterised by turbulent blood flow. The initial morphological change is the subendothelial (intimal) deposition of aggregated lipoproteins (Tabas et al., 2007). These aggregates promote chemotaxis of leukocytes through several mechanisms, including oxidation, and enzymatic and nonenzymatic proteolysis of apolipoproteins. In addition, monocyte/macrophage adhesion to the arterial endothelium occurs in the initial stages of experimental atherosclerosis, followed by progressive subendothelial accumulation of macrophages in both experimental and human lesions (Moore and Tabas, 2011). In addition to recruiting more macrophages, these cells avidly take up native and modified lipoproteins resulting in lipid droplets filled predominantly with cholesteryl esters (CE), a process known as 'foam cell' formation. Macrophage uptake and clearance of subendothelial lipoproteins are likely beneficial during the initial stages of atherosclerosis.

The resulting dysregulation of lipid metabolism alters macrophage phenotype and compromises crucial functions of the cell (Moore et al., 2013). Foam cells that accumulate in lesions have a decreased ability to migrate from the lesion, which contributes to chronic inflammation and lesion progression. More complex plaques involve other immune cells and vascular smooth muscle cells (VSMCs). In advanced lesions, macrophages continue as major contributors to lesion growth and the inflammatory response through secretion of proinflammatory mediators and reactive oxygen and nitrogen species, as well as matrix-degrading proteases. These processes lead to eventual cell death by either necrosis or apoptosis. Dead macrophages release their lipid contents and tissue factors, resulting in the formation of a prothrombotic necrotic core, which is a critical component of unstable plaques. Instability contributes to the rupture of a thin fibrous cap that covers the plaque and leads to subsequent thrombus formation that underlies an acute myocardial infarction or stroke (Figure 1).

Although many cell types, including monocytes, dendritic cells, lymphocytes, eosinophils, mast cells, endothelial cells and smooth muscle cells, contribute to lesion formation and growth, foam cells are central to the pathophysiology of atherosclerosis. Therefore, this chapter will focus on the relationship of altered lipid and lipoprotein metabolism to the mechanisms underlying monocyte recruitment, macrophage foam cell formation, the inflammatory response, cell death, macrophage emigration and the resolution of inflammation (Moore et al., 2013). Initially, we will review critical components and pathways in lipid metabolism that contribute to atherosclerosis, followed by a review of emerging mechanisms underlying the influence of lipid metabolism in atherosclerosis at various stages of lesion development or regression. We also review the current development of therapeutic strategies for treatment of atherosclerotic diseases.

2. LIPOPROTEIN TRANSPORT IN ATHEROSCLEROSIS

2.1 Low-Density Lipoprotein

The most atherogenic dyslipidaemia is hypercholesterolaemia, especially familial hypercholesterolaemia (FH) (Sahebkar and Watts, 2013a,b). FH is characterised by greatly increased plasma concentrations of LDL cholesterol, primarily due to downregulation, deficiency or dysfunction of LDL receptors (Chapter 17). Pathogenic mutations causing FH frequently occur in the LDL receptor gene and less frequently in apolipoprotein B (apoB) and proprotein convertase subtilisin/kexin type 9 (PCSK9) genes. As a consequence, plasma LDL cholesterol concentrations increase, due to an inability of hepatic LDL receptors to clear LDL particles. Overproduction of LDL also contributes to elevated LDL. LDL penetrates the subendothelium and accumulates within

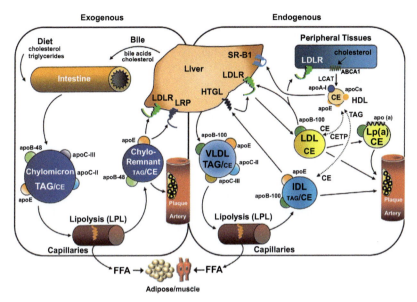

FIGURE 2 A schematic of exogenous and endogenous lipoproteins involved in atherogenesis. Elevated low-density lipoproteins (LDL) promote atherosclerosis as typified by patients with familial hypercholesterolaemia. Very-low-density lipoproteins (VLDL), their remnants (IDL) and chylomicron remnants also amplify atherogenesis. Lp(a) readily enters the arterial intima and promotes atherosclerosis. Low levels of HDL or dysfunctional HDL are unable to promote reverse cholesterol transport from lesion macrophages, contributing to the formation of foam cells.

the intima of susceptible arteries (Figure 2). LDL then aggregates and/or is readily oxidised, resulting in an inflammatory response as well as foam cell formation in both macrophages and smooth muscle cells. This process leads to premature atherosclerosis in patients with hypercholesterolaemia, particularly those with FH.

2.2 Very-Low-Density Lipoprotein

Another dyslipidaemia that predisposes to premature atherosclerosis is familial combined hyperlipidaemia (FCH) (Sahebkar and Watts, 2013a). FCH is the most common form of inherited dyslipidaemia and is genetically heterogeneous, resulting from variations in multiple genes involved in the metabolism and clearance of plasma lipoproteins (Chapter 15). Combined hyperlipidaemia is often accompanied by additional metabolic abnormalities, including adipose tissue dysfunction, increased fatty acid flux to the liver and insulin resistance (Chapter 19). The metabolic phenotype of this dyslipidaemia is hepatic overproduction of very-low-density lipoprotein (VLDL), elevated plasma VLDL triacylglycerol (TAG), apoB and small dense LDL (Figure 2). These lipoproteins also readily accumulate within the arterial intima, eliciting inflammatory and other cellular responses linked to atherogenesis.

Hypertriglyceridaemia is a dyslipidaemia characterised by increased plasma VLDL TAG concentrations and in the more severe form, elevated chylomicrons and chylomicron remnants (Hegele et al., 2014) (Chapter 16). Hypertriglyceridaemia is associated with increased risk of coronary heart disease. Heritability accounts for more than 50% of the individual variations in plasma TAG concentrations. Genome-wide association studies identified gene variants, including *APOC3*, *LPL*, *TRIB1*, *GCKR* and *APOA5* that modulate plasma TAG and the risk for atherosclerosis. Hypertriglyceridaemia results from hepatic overproduction and/or decreased catabolism of TAG-rich lipoproteins, accumulation of small dense LDL particles and low concentrations of high-density lipoprotein (HDL) (Figure 2). This dyslipidaemia is often linked to obesity, metabolic syndrome and type 2 diabetes (Chapter 19). Although small dense LDL particles and dysfunctional HDL contribute, atherosclerosis associated with this dyslipidaemia is primarily due to retention of TAG-rich lipoprotein remnants in the artery wall. Remnants are retained by intimal extracellular matrix proteins, and can be oxidised and readily induce foam cell formation. Fatty acids from these particles readily elicit an inflammatory response and promote lipotoxicity within cells involved in atherosclerosis.

2.3 Remnants of VLDL and Chylomicrons

The most compelling evidence that remnants of TAG-rich lipoproteins are atherogenic comes from patients with dysbetalipoproteinaemia who have a very high incidence of premature atherosclerosis (Brahm and Hegele, 2015). In this rare disorder, remnants of VLDL and chylomicrons accumulate in plasma due to a genetic variant of *APOE* (*APOE2*) bound to the surface of these particles (Chapter 15). Normally, apoE is the ligand responsible for the clearance of chylomicron remnants and VLDL remnants by hepatic LDL receptors; however, apoE2 binds poorly to these receptors, resulting in delayed remnant clearance. When coupled to an overproduction of intestinal or hepatic TAGs (Chapter 16), remnants attain high plasma concentrations, thereby increasing their exposure to the arterial intima (Figure 2).

2.4 Lipoprotein(a)

Elevated plasma concentrations of lipoprotein(a) (Lp(a)) are strongly associated with increased risk of atherosclerosis, although the definitive function of this lipoprotein is still unclear (Koschinsky and Boffa, 2014) (Figure 2). Lp(a) is composed of LDL with an additional single molecule of apo(a) that is bound covalently to apoB-100 (Chapter 16). Plasma Lp(a) concentrations are not modulated by diet, age, sex or physical activity, but are largely determined by variation at the *LPA* gene locus, making elevated Lp(a) highly heritable.

The apo(a) gene contains a variable number of tandem repeats that encode a kringle moiety identical to kringle IV of plasminogen. This leads to Lp(a) isoforms that differ in molecular mass. There is a strong and inverse association between plasma Lp(a) concentration and molecular mass of apo(a). Smaller apo(a) isoforms possess greater atherogenic potential. Similar to LDL, Lp(a) is retained within the arterial intima through extracellular matrix interactions, and induces chemokine secretion, inflammation and foam cell formation. Plasma Lp(a) carries oxidised phospholipids, and its atherogenicity may be associated with delivery of these proinflammatory lipids to lesions. Lp(a) can also induce apoptosis in endoplasmic reticulum (ER)-stressed macrophages, thereby contributing to lesion vulnerability and thrombosis. In addition, apo(a) homology to plasminogen may inhibit the fibrinolytic pathway, making Lp(a) a thrombogenic lipoprotein.

2.5 High-Density Lipoprotein

HDL cholesterol concentrations are inversely related to the risk of cardiovascular disease and death (Luscher et al., 2014). However, there is increasing awareness that HDL function plays a more important role in protection from atherosclerotic disease, a feature not captured in measures of plasma HDL concentration. Many dyslipidaemias are associated with low plasma concentrations and/or dysfunctional HDL. Causes of low HDL cholesterol are not well understood, but involve enhanced catabolism of the cholesteryl ester-depleted HDL particle (Chapter 15).

HDL possesses anti-inflammatory and antioxidant properties, but its principal antiatherogenic function is due to its ability to mediate cholesterol export from cells involved in atherosclerosis and subsequent elimination of HDL cholesterol through reverse cholesterol transport (Chapter 15) (Figure 2). In experimental models of atherosclerosis, promotion of macrophage cholesterol export attenuates progression or induces lesion regression. Other functions of HDL impact atherogenesis. HDL-associated paraoxanase-1 (PON-1) normally prevents HDL from oxidative modification. Reduced levels of HDL-associated PON-1 activity lead to the generation of modified HDL that inhibits endothelial cell nitric oxide synthase (eNOS) activation, thereby losing its anti-inflammatory properties. These HDL are also defective in cholesterol export.

3. LIPOPROTEIN RECEPTORS AND LIPID TRANSPORTERS

Lipoproteins are complex macromolecules that are recognised by a variety of receptors (Chapter 17) and function to facilitate physiological and pathological processes. There is direct in vivo evidence for a role of lipoprotein receptors in the development of atherosclerosis as summarised in Table 1.

3.1 LDL Receptors

LDL receptors were defined by the classic studies of Goldstein and Brown in which they described a process for transporting large lipoprotein particles

TABLE 1 Selected Lipoprotein Receptors with Proposed Roles in Atherosclerosis

Receptor	Full and Alternative Names	Selected Lipoprotein Ligands	Effects of Deletion on Atherosclerosis
LDLR	Low-density lipoprotein receptor	LDL	↑ (Ishibashi et al., 1994)
SR-A (I and II)	Scavenger receptor class A	Acetylated and oxidised forms of LDL	↓ (Suzuki et al., 1999) ↔ (Moore et al., 2005) ↑(de Winther et al., 1999)
CD36	Thrombospondin receptor, fatty acid translocase	Modified LDL	↓ (Febbraio et al., 2000) ↔ (Moore et al., 2005)
LOX1	Lectin-like oxidised LDL receptor	Oxidised LDL	↓ (Inoue et al., 2005; Mehta et al., 2007)
SR-B1	Scavenger receptor class B	Acetylated and oxidised forms of LDL, HDL	↑ (Covey et al., 2003)
LRP1	Low-density lipoprotein receptor-related protein, alpha 2 macroglobulin receptor	apoE-containing lipoproteins	↑ (Overton et al., 2007; Hu et al., 2006; Boucher et al., 2003)

(~24 nm) across the cell membrane. The first step in this process is the interaction of the apoB of LDL with the cysteine-rich receptor-binding domain of LDL receptors. Engagement of apoB with LDL receptors results in endocytosis of the entire LDL particle with subsequent delivery to the lysosome where the particle is degraded to its components. Delivery of free (unesterified) cholesterol to the ER induces homeostatic mechanisms that inhibit continuous exogenous delivery or endogenous cholesterol synthesis (Chapters 17 and 11). Deficiency of LDL receptors increases plasma cholesterol and accelerates atherosclerosis. Humans with homozygous FH develop severe atherosclerotic disease within two decades of life if untreated. Deficiency of LDL receptors in rabbits also leads to pronounced hypercholesterolaemia and accelerated atherosclerosis. Genetic depletion of LDL receptors in mice leads to a modest hypercholesterolaemia, which is greatly augmented by feeding diets that are enriched in saturated fat and/or cholesterol. LDL receptors can also be negatively regulated by PCSK9. Expression of gain-of-function PCSK9 mutants leads to depletion of LDL receptors, hyper-cholesterolaemia and atherosclerosis (Dadu and Ballantyne, 2014).

3.2 Scavenger Receptors

LDL receptors are downregulated by increased cellular cholesterol to prevent sterol engorgement, which is a characteristic of cells in atherosclerotic lesions. Therefore, it is unlikely that LDL promotes atherosclerosis through this receptor. As an alternative, it was proposed that in hypercholesterolaemic states, an increased residence time of LDL particles in the subendothelial space leads to modifications, resulting in enhancement of macrophage uptake that is not regulated by cellular sterol content. Acetylation of LDL was the first modification identified that enhanced uptake in macrophages. Binding of acetylated LDL to macrophages is saturable, consistent with a receptor-mediated interaction. These recognitions led to the concept that hypercholesterolaemia promotes atherosclerosis through modified LDL uptake by 'scavenger' receptors (Chapter 17).

The subsequent effort to identify scavenger receptors led to the cloning of a trimeric membrane glycoprotein that was expressed as two major splice variants. This receptor is now referred to as scavenger receptor class A or SR-A (Canton et al., 2013). Although expression of the cloned protein demonstrated the ability of acetylated LDL to bind to SR-A in cultured cells, studies in which SR-A was genetically manipulated in atherosclerosis-susceptible mice demonstrated a full spectrum of effects on lesion size from being increased, unaffected or decreased (Canton et al., 2013) (Table 1). SR-A is now recognised as the receptor for a wide array of ligands that influence atherosclerosis, and is also an integrin-independent adhesion molecule (Chapter 17). Therefore, it is unclear whether the multiple ligands of SR-A contribute to the variable effects observed on atherosclerotic lesion formation.

There are now a large number of other receptors that are collectively classed as scavenger receptors. Most of these receptors bind to modified LDL, in which acetylation or oxidation is the most common modification. In addition to SR-A, the major members of this class that have been studied for effects on experimental atherosclerosis are CD36, SR-B1 and LOX1 (Canton et al., 2013) (Chapter 17). CD36 was originally described as a fatty acid transporter and was subsequently identified as a receptor for copper-oxidised LDL. It is highly expressed in cultured macrophages and atherosclerotic lesions. Like SR-A, the data on the role of CD36 in atherosclerosis in either $ApoE^{-/-}$ or $Ldlr^{-/-}$ mice have been inconsistent (Table 1).

SR-B1 was identified as a structural variant of CD36 with similar binding characteristics for acetylated LDL. While many of the same ligands bind both CD36 and SR-B1, LDL only binds to SR-B1. It was subsequently demonstrated that SR-B1 is an HDL receptor that facilitates bidirectional movement of cholesterol between cells and HDL particles (Canton et al., 2013) (Chapter 15).

Mice with deficiencies of both SR-B1 and apoE exhibit an array of cardiovascular pathologies at an early age, including accelerated atherosclerosis. Lesions in these mice are present in coronary arteries, which is uncommon in most small animal models of atherosclerosis. Overexpression of SR-B1 in $Ldlr^{-/-}$ mice decreases atherosclerosis with concomitant decreases in plasma HDL cholesterol concentrations. Deleting SR-B1 increases atherosclerosis in $Ldlr^{-/-}$ mice with increases in plasma cholesterol concentrations and very large HDL particles (Table 1). The effects of SR-B1 on plasma lipoprotein concentrations and characteristics do not appear to affect atherosclerosis since deletion of SR-B1 in bone marrow-derived cells augments lesion size in mice, without discernible effects on plasma cholesterol.

Although macrophages have been the major focus of expression for most scavenger receptors, endothelial dysfunction is also crucial in atherosclerotic lesions. A screen of an endothelial cell cDNA library identified a protein designated as lectin-like oxidised LDL (oxLDL) receptor or LOX-1 that binds oxLDL (Chapter 17). LOX-1 is also expressed in macrophages and smooth muscle cells of atherosclerotic lesions. LOX1 deficiency reduces atherosclerosis, whereas its overexpression in endothelial cells increases atherosclerosis in hypercholesterolaemic mice, suggesting that this protein contributes to lesion development (Table 1).

3.3 Low-Density Lipoprotein Receptor-Related Protein

LDL receptor-related protein (LRP1), a multifunctional protein in the LDL receptor superfamily, is present on the surface of multiple cell types (Chapter 17). Since its discovery by Herz and colleagues in 1988, more than 40 ligands have been identified to interact with LRP1, implicating complex functions of this protein (Chapter 17). This large protein contains a 515-kDa α chain (amino-terminal part) and an 85-kDa β chain (membrane-bound) carboxyl terminal fragment. Deficiency of LRP1 in macrophages reduces VLDL uptake, and enhances the secretion of inflammatory mediators including monocyte chemoattractant protein-1 (MCP-1, CCL2), interleukin (IL)-1, IL-6 and TNFα. Deficiency of LRP1 in macrophages does not affect plasma cholesterol concentrations or lipoprotein–cholesterol profiles, but augments atherosclerosis in hypercholesterolaemic mice (Table 1). In addition to macrophages, LRP1 is abundant in VSMCs. Deficiency of LRP1 in VSMCs augments atherosclerosis in hypercholesterolaemic mice (Gonias and Campana, 2014).

3.4 ATP-Binding Cassette Subfamily (ABCs)

Cells normally respond to excessive lipid accumulation by upregulation of pathways that promote export of cholesterol and other lipids (Chapter 15). Although passive diffusion of cholesterol from the plasma membrane occurs, macrophage foam cells possess several transporters that export lipids, including members of the ATP-binding cassette subfamily (ABCs) ABCA1 and ABCG1. ABCA1 facilitates cholesterol export to lipid-poor apoA-I within pre-β-HDL, whereas ABCG1 facilitates export to mature HDL particles (Figure 2). Macrophage expression of ABCA1 and ABCG1 is upregulated in response to increased cellular cholesterol concentrations by liver X receptors (LXR). This ligand-activated nuclear receptor acts as a sterol sensor and is transcriptionally activated by metabolites of lipoprotein-derived cholesterol, including the oxysterols 27-OH-cholesterol, 24-OH-cholesterol, 7-keto-cholesterol and desmosterol, a cholesterol precursor. All of these metabolites accumulate in macrophage foam cells. Activation of LXR also inhibits inflammatory responses. In experimental models, strategies that increase activation or expression of LXR or ABC transporters attenuate atherosclerosis, whereas inhibition of their expression promotes atherosclerosis.

4. CONTRIBUTIONS OF LIPOPROTEIN-MEDIATED INFLAMMATION TO ATHEROSCLEROSIS

Multiple cell types are involved in the development of atherosclerosis. One well-recognised mechanistic theory is that endothelial dysfunction of the arterial wall is the initial step in provoking monocyte adhesion, followed by macrophage infiltration into the subendothelial space (intima) to form lipid-laden foam cells. Resident cells of the arterial wall together with leukocytes recruited from peripheral blood orchestrate a series of inflammatory responses in the arterial wall, thereby promoting the progression of atherosclerosis. The 'response-to-retention' hypothesis (Tabas et al., 2007) proposes that lipoproteins normally diffuse across the endothelial barrier, become trapped in the subendothelium, stimulate inflammatory responses and promote accumulation of lipid-laden macrophages. It is generally accepted that accumulated lipids within macrophages are derived from plasma lipoproteins through uptake by endocytic, phagocytic, receptor-involved or nonreceptor-mediated processes. This section updates the current knowledge regarding lipid and lipoprotein-stimulated cellular changes and cell-specific manipulations of components that have been studied in atherosclerotic animal models.

4.1 Cell Types Involved in Atherosclerotic Lesions

Associations between inflammation and atherosclerosis are complex, they involve multiple cell types and a network of signalling pathways. A subtle change in this complex system may lead to substantial changes in a spectrum of signalling pathways. Hypercholesterolaemia induces leukocytosis (Murphy et al., 2014), thereby increasing their likelihood of recruitment into the subendothelial space. Modified lipoproteins can activate endothelial cells to secrete adhesion molecules such as vascular cell adhesion molecule 1 (VCAM-1), intracellular adhesion molecule 1 (ICAM-1) and selectins. These molecules, together with chemoattractant mediators including complement factors and MCP-1, promote monocyte adhesion to the impaired endothelial barrier and emigration into the intima. Monocytes become activated to macrophages, which take up native and modified lipoproteins to form lipid-laden foam cells. Among all leukocyte types, macrophages are the predominant cell type in atherosclerosis and are major contributors to the development of lesions. Therefore, the contributions of this cell type to atherosclerosis will be reviewed in detail in Section 4.2. Lipoproteins accumulating in the intimal space (intima) and their uptake by macrophages contribute to the secretion of multiple proatherogenic cytokines such as IL-1, IL-6 and TNFα (Figure 3).

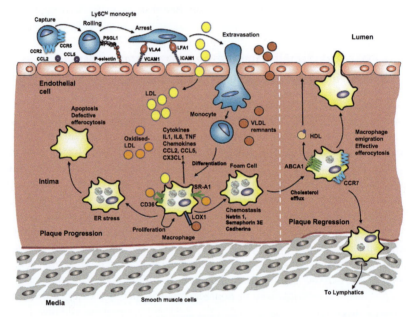

FIGURE 3 **Pathways for monocyte recruitment and macrophage retention, foam cell formation, apoptosis, lipid export and emigration in plaques.** GR1+LY6C^{hi} monocytes are recruited to mouse atherosclerotic lesions using chemokine–chemokine receptor pairs (CCR2:CCL2 and CCR5:CCL5) to infiltrate the intima, which is facilitated by endothelial adhesion molecules, including P-selectin and P-selectin glycoprotein ligand 1 (PSGL1), intercellular adhesion molecule 1 (ICAM1) and vascular adhesion molecule 1 (VCAM1). Recruited monocytes differentiate into macrophages in the intima, where they interact with atherogenic lipoproteins (see text) resulting in foam cell formation. Foam cells secrete proinflammatory cytokines, chemokines and retention factors that induce macrophage proliferation, promote foam cell accumulation within lesions and amplify the inflammatory response. The inflammatory milieu and excess lipid accumulation induce macrophage ER stress and apoptosis. Cell death, together with defective efferocytosis, results in the formation of a necrotic core, which is characteristic of advanced lesions. The promotion of lipid unloading by ABCA1 and cholesterol export to lipid-poor apoA-I reverses cellular lipid accumulation and induces plaque regression. Events that stimulate efferocytosis lead to reverse transmigration to the lumen and macrophage emigration to adventitial lymphatics. LFA1, lymphocyte function-associated antigen 1; VLA4, very late antigen 4. *Adapted from Moore et al. (2013).*

In addition to macrophages, T lymphocytes are present in atherosclerotic lesions and, like macrophages, accumulate in early lesions (Libby, 2012). LDL stimulates T lymphocytes to produce interferon-γ (IFN-γ), which contributes to further recruitment of T lymphocytes. In contrast to T lymphocytes, atherosclerotic lesions contain minimal B lymphocytes. oxLDL stimulates antibody production from B lymphocytes that accumulate in atherosclerotic lesions. It is generally accepted that T lymphocytes augment atherosclerosis, whereas antibodies produced by a subclass of B lymphocytes, termed B1 cells, protect against formation and development of atherosclerosis.

Neutrophils are the most numerous leukocyte type in blood. This cell type is present in blood for only 1–2 days until activated, which promotes migration into tissues. In the innate immune system, neutrophils are the first line in acute inflammatory responses. The role of neutrophils in atherosclerosis has not been appreciated until recently. Neutrophils become activated and are recruited to sites of atherosclerotic lesions, but the mechanisms by which neutrophils contribute to atherosclerosis are unclear (Weber and Noels, 2011).

VSMCs are the predominant cell type in the normal arterial medial layer. Vascular inflammation leads to VSMC proliferation, and their migration into atherosclerotic lesions. Recent studies revealed that progenitor cells present in the adventitia of the arterial wall have the capability to become VSMCs and contribute to the development of atherosclerosis. VSMCs secrete extracellular matrix that stabilises lesions through the formation of a fibrous cap. VSMCs also have the capacity to transform into foam cells by accumulating lipoprotein-derived lipids; however, VSMC foam cells lose their capacity to secrete extracellular matrix. Lipoproteins, such as oxLDL, induce the proliferation of smooth muscle cells. In human coronary lesions, recent studies found that over 50% of foam cells were derived from smooth muscle cells, and these cells expressed the macrophage marker CD68 (Allahverdian et al., 2014). These discoveries have made research on vascular smooth muscle-related mechanisms of atherosclerosis more challenging (Tellides and Pober, 2015; Zhang and Xu, 2014).

4.2 Foam Cells

4.2.1 Recruitment of Circulating Monocytes

Activation of endothelium promotes recruitment of circulating monocytes (Moore et al., 2013) (Figure 3). Of significance, hypercholesterolaemia is associated with increased numbers of circulating monocytes. Their precursors, hematopoietic stem and progenitor cells, become enriched in cholesterol and deficient in cholesterol export, leading to cell proliferation. Circulating monocytes in mice consist of at least two major subsets, Ly6C^{hi} and Ly6C^{low} monocytes. Ly6C^{hi} cells constitute the majority of cells recruited to the intima, and represent precursors of M1 classically activated macrophages and thus are key participants in foam cell formation and inflammatory responses.

Recruitment of circulating monocytes into atherosclerotic lesions requires integration of at least three discrete processes. These are capture, rolling and transmigration and each step is regulated by several molecular factors (Figure 3). Molecules on the surface of activated endothelial cells mediate capture and rolling of monocytes. This depends on their immobilisation by endothelial cell chemokines, P-selectin, VCAM1 and ICAM1, which interact with chemokine receptors and integrins on monocytes. Extravasation of monocytes across the endothelium and into plaques is mediated by the interaction of monocyte chemokine receptors with the chemokines PECAM1 or VCAM1 that are secreted by endothelial cells, macrophages and smooth muscle cells. In addition, neuronal guidance cues expressed by endothelial cells, including ephrin B2, netrin 1 and semaphorin 3A, can increase monocyte recruitment or inhibit the chemokine-directed monocyte migration (Figure 3).

4.2.2 Foam Cell Formation

Macrophage foam cell formation represents one of the earliest stages of lesion development, and continues throughout lesion evolution (Moore and Tabas, 2011) (Figure 4). Although macrophages can take up apoB-containing lipoproteins through the LDL receptor, this receptor is downregulated during foam cell formation by increased cellular cholesterol. Modified lipoprotein uptake by scavenger receptors, which is not regulated by intracellular cholesterol, contributes to foam cell formation. Multiple means of lipoprotein modification have been described; however, the relevant pathways that promote foam cell formation in vivo remain to be fully elucidated.

Oxidative stress in the arterial intima promotes lipoprotein modification that generates 'damage' signals that are recognised by scavenger receptors on macrophages and other cells of the innate immune system. oxLDL has been identified in both human and mouse lesions and antibodies that recognise oxidation-specific epitopes of LDL have also been discovered. Mediators of lipoprotein oxidation, including 12/15-lipoxygenase, myeloperoxidase, free radicals including superoxide, hydrogen peroxide and nitric oxide, have been identified in the artery wall. Lipoproteins modified through these mechanisms are readily endocytosed by scavenger receptors in vitro (Figure 4). Scavenger receptors, including SR-A1 (encoded by *MSR1*), macrophage receptor with collagenous structure (MARCO; also known as SR-A2), CD36 (also known as platelet glycoprotein 4), SR-B1, LOX1, scavenger receptor expressed by endothelial cells 1 (SREC1) and scavenger receptor for phosphatidylserine and oxidised LDL (SR-PSOX; also known as CXCL16) all can bind oxidised lipoproteins and promote unimpeded foam cell formation (Section 3.2). SR-A1 and CD36 mediate the majority of uptake of oxidised lipoproteins by macrophages in vitro. These receptors internalise lipoproteins into late endosomes and lysosomes, where lysosomal acid lipase mediates the hydrolysis of lipoprotein CE to cholesterol and fatty acids (Figure 4). Endolysosomal cholesterol is trafficked to the ER or plasma membrane by NPC1 and NPC2. In the ER, cholesterol undergoes

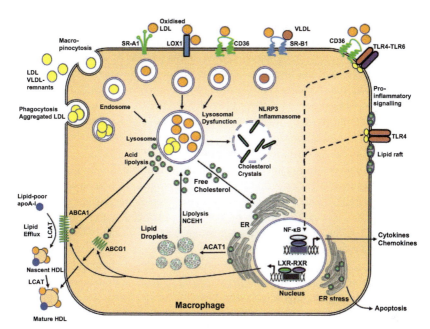

FIGURE 4 **Mechanisms controlling macrophage lipoprotein uptake, lipid export and intracellular cholesterol trafficking.** Macrophages internalise native LDL, VLDL and oxidised lipoproteins within the lesion by scavenger receptors outlined in Table 1. Internalised lipoproteins and their lipids are digested in the lysosome. Cholesterol is transferred to the plasma membrane for export from the cell or to the endoplasmic reticulum for esterified by acyl-coenzyme A:cholesterol acyltransferase 1 (ACAT1) and ultimately stored as CE in cytosolic lipid droplets. Stored lipids are released for export by either neutral cholesteryl ester hydrolase 1 (NCEH1)-mediated lipolysis or via lipophagy (not shown). Cellular cholesterol accumulation increases oxysterol synthesis and activation of LXR–retinoid X receptor (RXR) transcription factor complex that upregulates expression of ABCA1 and ABCG1. These transporters transfer cholesterol to lipid-poor apoA-I to form nascent HDL or to lipidate mature HDL particles. Exported cholesterol is esterified by lecithin cholesterol acyltransferase (LCAT). Excessive cellular cholesterol can interfere with the function of the ER as well as induce cholesterol crystal formation in the lysosome, resulting in activation of the inflammasome. If prolonged, this results in cell death by apoptosis. Increased cholesterol content of membrane-associated lipid rafts increases proinflammatory Toll-like receptor 4 (TLR4) signalling that activates nuclear factor-κB (NF-κB), resulting in the production of proinflammatory cytokines and chemokines. *Adapted from Moore et al. (2013).*

reesterification by acyl-CoA:cholesterol acyltransferase 1 (ACAT1) to CE that are stored within neutral lipid droplets that appear 'foamy' under the microscope (Figure 4).

Although *ApoE*$^{-/-}$ mice deficient in both SR-A1 and CD36 have reduced lesion inflammation, macrophage apoptosis and plaque necrosis, loss of these receptors did not reduce foam cell formation or lesion size (Moore et al., 2005). Together with the results of human clinical trials of antioxidant vitamins E and C that failed to show a reduction of cardiovascular events, these studies have prompted the field to consider alternative mechanisms for foam cell formation.

Proteases and lipases within the arterial intima also mediate lipoprotein modifications, particularly LDL aggregation (Moore et al., 2013). Extracellular matrix glycoproteins within the intima contribute to this process by retaining apoB-containing lipoproteins and by modulating the activity of enzymes, including group IIA secretory phospholipase A2 (PLA2G2A), PLA2G5 and PLA2G10, as well as secretory sphingomyelinase. These lipolytic enzymes produce modified forms of LDL that are taken up by macrophages independent of scavenger receptors. Evidence from mouse studies supports a function for PLA2 in atherosclerosis progression, and circulating PLA2 levels in humans correlate with risk for atherosclerosis. Although foam cell formation by native apoB-containing lipoproteins was not originally considered a major pathway, recent studies have documented macrophage uptake of LDL by pinocytosis in vitro and in the arterial intima leading to foam cell formation (Figure 4). This receptor-independent endocytic pathway also delivers cholesterol to the endolysosomal compartment and stimulates ACAT1-mediated cholesterol esterification. Thus, in vivo, it is likely that atherogenic LDL and other apoB-containing lipoproteins induce foam cell formation by multiple pathways.

Excessive uptake of lipoproteins eventually results in defective lipid metabolism within macrophages, ultimately leading to increased lesion complexity. When retained within lipid droplets, CE is metabolically inert, whereas cholesterol within cell membranes can be toxic. Enrichment of macrophage ER membranes with cholesterol, which occurs with excessive lipoprotein uptake, eventually suppresses its esterification by ACAT1, resulting in further cholesterol accumulation. Furthermore, plasma membranes become cholesterol enriched, leading to amplified inflammatory signalling through Toll-like receptors (TLR; see Section 4.6) (Figure 4). In addition, trafficking of cholesterol from lysosomes becomes compromised, which inhibits macrophage cholesterol export and further enhances the inflammatory response. This dysregulated lipid metabolism contributes to macrophage ER stress, and if prolonged, can ultimately lead to apoptosis and cell death (Figure 4). Efficient clearance of apoptotic cells from lesions by surrounding macrophages (known as efferocytosis) requires the engulfing cells to metabolise lipids derived from the apoptotic cells (Figure 3). Therefore, excess lipoprotein uptake by macrophages compromises cellular lipid metabolism, promotes apoptosis and suppresses efferocytosis, which leads to secondary necrosis and a necrotic core (Figure 4). A necrotic core, together with a thinned fibrous cap, is characteristic of plaques more vulnerable to rupture.

4.3 Macrophage Polarisation

On entry into the arterial intima, monocytes are polarised to M1 macrophages, known as classically activated inflammatory cells, or M2 macrophages, known as alternately activated cells that resolve inflammation (Moore et al., 2013). It is thought that M1 macrophages are derived from $Ly6C^{hi}$ monocytes, whereas M2 macrophages are from $Lys6C^{low}$ monocyte precursors. In vivo, it is likely that macrophage phenotype is more complex. Furthermore, the factors in lesion microenvironments that promote macrophage polarisation remain incompletely defined. M1 macrophages represent foam cell precursors and in human plaques they are enriched in lipids and localised to areas distinct from the less inflammatory M2 macrophages. In mouse models, M1 macrophages become predominant in complex lesions, whereas enrichment of M2 macrophages occurs in lesions in which regression of atherosclerosis has been induced. Administration of the M2-polarising cytokine IL-13 to *Ldlr*−/− mice drives lesion macrophages to M2-like cells and inhibits lesion progression. Recent evidence indicates that oxLDL induces a distinct macrophage phenotype that has been termed Mox, characterised by the increased expression of NRF2, although the function of these cells in atherosclerosis has not been defined.

4.4 Inflammatory Responses

Innate immune activation is considered a central feature of the pathogenesis of atherosclerosis and is largely a consequence of dysregulated lipid metabolism within developing lesions (Moore et al., 2013). Lipids, oxidised lipids and other ligands that accumulate within lesions trigger macrophage receptors including scavenger receptors, TLRs and nucleotide-binding oligomerisation domain (NOD)-like receptors that in turn activate inflammatory responses (Figure 4). Cholesterol crystals are detected in extracellular spaces and within macrophages in both early and more advanced lesions. Cholesterol crystals induce the NLRP3 inflammasome, which leads to the processing and secretion of several proinflammatory cytokines (Figure 4). The exposure of macrophages to fatty acids derived from VLDL or VLDL remnants also activates the expression of proinflammatory cytokines, independent of TLRs.

TLRs are a class of membrane-spanning, noncatalytic receptors that play critical roles in innate immune responses. Multiple ligands, including oxLDL and oxidised phospholipids, activate TLR2- and/or TLR4-related signalling cascades to promote inflammation. In the presence of lipid ligands, CD36 and SRA interact with TLRs, resulting in activation of myeloid differentiation factor (MyD) 88 and c-Jun N-terminal kinases (JNK), which promote apoptosis (Figure 4). Their role in atherosclerosis is supported by studies in *ApoE*−/− mice and *Ldlr*−/− mice, whereby genetic deficiency of TLR2, TLR 4 or the TLR adapter protein MyD88 reduces lesion development. However, despite consistent effects of global deficiency of TLRs and MyD88 in reducing atherosclerosis in mice, their absence in bone marrow-derived cells has been conflicting. These studies suggest that the primary impact of TLR signalling on atherosclerosis is not through hematopoietic cells, including macrophages.

4.5 Atherosclerotic Lesion Macrophage Retention and Emigration

Atherogenesis is related to macrophage accumulation within lesions that is determined by monocyte recruitment, macrophage proliferation, foam cell formation and macrophage emigration and death (Moore et al., 2013) (Figure 3). Atherogenic lipoproteins in the intima are primary determinants of macrophage recruitment. Macrophage emigration occurs in early atherosclerotic plaques, but the rate of egress decreases with lesion progression. In mouse models, emigration is an important factor linked to attenuation of progression or the induction of lesion regression. Lesional macrophages are subject to both retention and emigration signals. Macrophage foam cells exhibit increased expression of the neuro-immune guidance cues, netrin 1 and semaphorin 3E, both of which induce macrophage chemostasis (Figure 3). Deficiency of netrin 1 attenuates lesion progression and increases macrophage emigration. The signals that guide macrophages to exit lesions, by transmigration through the endothelium to the lumen or by migrating through the media to the adventitial lymphatics, are not well defined, but likely involve macrophage expression of the CC-chemokine receptor CCR7 and the CC-chemokine ligands CCL19 and CCL21 that regulate cell homing the lymph. As lesions progress, the continued presence of foam cells in a lipid-rich lesion environment leads to cholesterol- and saturated fatty acid-induced cytotoxicity, and ER stress, thereby increasing apoptosis (Figure 4). Efferocytosis, the ability of macrophages to clear apoptotic cells through receptors including tyrosine protein kinase MER (MERTK) and LRP1, becomes compromised by cholesterol accumulation in the engulfing macrophage. Defective efferocytosis contributes to secondary necrosis as well as to formation and expansion of lipid-rich necrotic cores, which, in turn, contribute to lesion rupture.

4.6 Atherosclerotic Lesion Regression

With the exception of early lesions, which are dominated by foam cells, atherosclerosis was originally considered irreversible (Moore et al., 2013). Recent studies demonstrating macrophage emigration from plaques in mice, and that inflammation-resolving and tissue-remodelling M2 macrophages are present in human and animal lesions, suggest that regression in humans may be achievable. Some of the determinants of lesion regression have recently emerged (Figure 3). Common to all mouse models of regression is the requirement for aggressive lowering of plasma lipids and the observation that within the regressing lesion there is a marked decrease in macrophage number. Of those remaining, a majority display an M2 phenotype. In macrophages from regressing lesions, the expression of genes including netrin 1, semaphorin 3E and members of the cadherin family is downregulated, whereas CCR7 and cell motility factors are upregulated, thereby enhancing the migratory ability of macrophages. Although other key factors that promote regression remain to be identified, improved lipid metabolism, including marked attenuation of foam cell formation, is critical for regression to be achieved.

5. NEW EMERGING MECHANISMS OF LIPID METABOLISM INFLUENCING ATHEROSCLEROSIS

5.1 MicroRNAs

MicroRNAs (miRs) are a class of small noncoding RNAs that participate in a diversity of biological and pathophysiological processes through posttranscriptional regulation of gene expression. Many miRs regulate lipoprotein metabolism in the circulation and/or local tissues (Novak et al., 2014). A major focus of recent research is regulation of HDL metabolism by miRs. HDL particles are carriers of miRs in plasma. Conversely, miRs also regulate expression of many genes associated with HDL biosynthesis, uptake and metabolism.

miR-33 is the most intensely studied miR that regulates HDL metabolism. miR-33 is embedded in intron 16 of the sterol regulatory element-binding factor (SREBP) gene, a transcription factor involved in cholesterol homeostasis (Chapter 11). Although many studies reported that inhibition of miR-33 by antisense oligonucleotides (ASO) in mice and monkeys or genetic deficiency of miR-33 in mice increased plasma HDL cholesterol concentrations (Figure 5), other studies have demonstrated that inhibition of miR-33 did not increase, or only increased transiently, plasma HDL cholesterol concentrations. As with conflicting findings in regulation of HDL cholesterol, effects of miR-33 on atherosclerosis are also inconsistent. In hypercholesterolaemic mice, an ASO against miR-33 or its genetic deficiency prevented, or had no effect on, the development of atherosclerosis, or regressed preexisting atherosclerosis. These conflicting findings admonish further studies before regulation of miR-33 can be used in human studies. In addition to conflicting findings in mouse models, multiple differences between mice and humans warrant careful interpretation. For example, miR-33 has differential effects in mice and humans on the expression of Niemann–Pick C1 (NPC1) and ABCG1. These differences may be partially attributed to the finding that mice only have miR-33a, whereas humans and primates have both miR-33a and miR-33b (Fernandez-Hernando and Moore, 2011; Naar, 2013).

Several other miRs have also been shown to regulate lipoprotein metabolism. We provide a brief review of those that have been studied in atherosclerotic animal models. HDL carries more miRs than LDL particles. miR-155 is one of the few exceptions and exhibits higher abundance in LDL than in HDL. miR-155 regulates inflammatory responses. In addition, enhancement of miR-155 increases oxLDL uptake into cultured macrophages and impairs macrophage cholesterol export to apoA-I, and suppression of miR-155 leads to increases of cholesterol export to apoA-I in the absence of changes in the expression of ABCA1 and ABCG1. Inhibition of miR-155 by an antagonist of a microRNA (antagomiR) system reduced atherosclerotic lesion size in $ApoE^{-/-}$ mice fed a high-fat, high-cholesterol diet without affecting plasma total cholesterol concentrations. miR-155 deficiency in bone marrow-derived cells reduced atherosclerotic lesions in $ApoE^{-/-}$ mice, but augmented atherosclerotic lesion size in $Ldlr^{-/-}$ mice (Nazari-Jahantigh et al., 2015). It is unclear whether miR-155 has

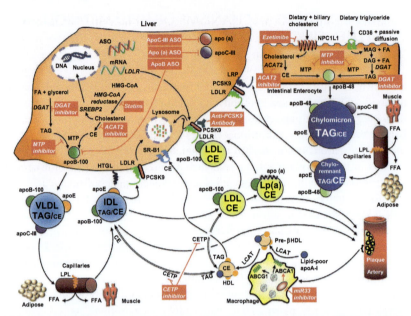

FIGURE 5 **Existing and potential therapeutic targets for the treatment of lipid metabolism and prevention or treatment of atherosclerosis.** The therapeutic objective is to decrease the concentration of atherogenic lipoproteins or increase antiatherogenic lipoproteins in the circulation, thereby improving lesion pathology or plaque regression. Statins inhibit the hepatic enzyme HMG-CoA reductase, leading to increased expression of the LDLR and stimulation of the hepatic clearance of LDL and VLDL remnants. Cholesterol-absorption inhibitors (ezetimibe) reduce the absorption of cholesterol from the intestine by inhibiting the transporter NPC1L1 and are very effective when combined with statins. Inhibition of PCSK9 with injected monoclonal antibodies blocks LDLR degradation and increases hepatic cell surface LDLR, which results in an increased hepatic clearance of LDL, VLDL remnants and Lp(a). Inhibition of microsomal TAG transfer protein (MTP) decreases the assembly of VLDL in the liver and chylomicrons in the intestine by blocking the transfer of TAG and CE to apoB-100 or apoB-48. CETP inhibitors increase HDL cholesterol and decrease LDL cholesterol levels as well as Lp(a). CETP normally transfers CE from HDL to VLDL, IDL and LDL, where theoretically this can promote the atherogenicity of these lipoproteins. ApoB antisense oligonucleotides (ASO) inhibit the translation and synthesis of apoB and therefore decrease the secretion of apoB-containing lipoproteins (VLDL) into plasma. An apoC-III ASO inhibits the synthesis and secretion of apoC-III, a protein that normally delays the catabolism of TAG-rich lipoproteins. An apo(a) ASO inhibits the hepatic synthesis of apo(a), therefore inhibiting the formation of Lp(a). Inhibition of ACAT-2 with ASOs in mice inhibits the secretion of both chylomicrons and VLDL into plasma and attenuates atherosclerosis in mice. Specific DGAT-2 inhibitors decrease hepatic VLDL secretion and LDL cholesterol in experimental animals and are currently under development for clinical use. MicroRNA-33 (miR-33) inhibits the expression of ABCA1 and ABCG1 and decreases HDL formation. miR-33 ASOs or anti-miR-33 in mice increases cellular ABCA1 expression and plasma HDL cholesterol concentrations and attenuate atherosclerosis in mice.

differential effects between $Ldlr^{-/-}$ mice and $ApoE^{-/-}$ mice, leading to different contributions of miR-155 to atherosclerosis in these two mouse strains.

miR-30c targets the 3′ untranslated region of microsomal triacylglyceride transfer protein (MTP) mRNA. In addition to reducing MTP activity and apoB secretion, miR-30c diminishes lipid synthesis in an MTP-independent manner.

Lentiviral transduction of miR-30c reduced, and anti-miR-30c increased, plasma cholesterol concentrations in C57BL/6 or *ApoE^{-/-}* mice fed a Western diet, an effect attributed to changes in secretion of plasma apoB-containing lipoproteins. Consistent with modulation of plasma apoB-containing lipoproteins, transduction of miR-30c in female *ApoE^{-/-}* mice fed a Western diet attenuated atherosclerosis, whereas inhibition of this miR augmented lesion development.

Given their multitargeting features, some circulating miRs might serve as biomarkers to predict the development of atherosclerosis, and inhibiting or enhancing certain miRs might be a good therapeutic strategy. However, the most comprehensively studied miRs, such as miR-33, still require further characterisation and thorough preclinical testing prior to applying therapeutic strategies to humans (Naar, 2013).

5.2 Inflammasomes

Cholesterol crystal accumulation in macrophages is a distinguishing feature of atherosclerosis from initiation through advanced stages of lesion development, and is both a consequence of lipid overloading in cells and a critical contributor to the progression of atherosclerosis (Figures 1 and 4). Cholesterol crystals activate immune cells and induce inflammation through multiple mechanisms. One proposed mechanism is that cholesterol crystals activate nucleotide-binding oligomerisation domain receptors (NLRP)3 inflammasomes (Figure 4) (Lu and Kakkar, 2014).

The inflammasome is a cytosolic caspase-activating molecular complex that contributes to inflammatory responses. Uptake of oxLDL leads to activation of NLRP3 inflammasomes in cultured macrophages, and increases IL-1β release. As a consequence, NLRP3 deficiency in bone marrow-derived cells reduces hypercholesterolaemia-induced atherosclerosis in mice. This is a leukocyte-specific effect of NLRP3 inflammasomes because whole body deficiency of NLRP3 inflammasome components such as NLRP3, apoptosis-associated speck-like protein or caspase-1 have no effect on atherosclerosis development in hypercholesterolaemic mice.

oxLDL uptake increases IL-1β secretion in macrophages, an important cytokine representing NLRP3 inflammasome activation. However, global deficiency of IL-1β, but not its deficiency in bone marrow-derived cells, affects atherosclerosis in hypercholesterolaemic mice, implicating macrophage NLRP3 inflammasomes in promotion of atherosclerosis through an IL-1β-independent mechanism. Activation of NLRP3 inflammasomes also results in release of IL-1α. Comparable to effects of IL-1β, global deficiency of IL-1α reduces hypercholesterolaemia-induced atherosclerosis in mice. Fatty acids, especially oleic acid, increase secretion of IL-1α in macrophages independent of NLRP3 inflammasome activation. Deficiency of IL-1α in bone marrow-derived cells reduces atherosclerosis. Therefore, although activation of NLRP3 inflammasomes increases both IL-1α and IL-1β, these two cytokines contribute to atherosclerosis through different mechanisms.

Currently, a clinical trial, CANTOS (NCT01327846), is ongoing to determine whether inhibition of IL-1β has beneficial effects on coronary artery disease. Despite the conflicting findings in animal models, completion of this clinical trial will provide insights into the effects and mechanisms of this IL-1 subtype in the development of atherosclerosis in humans (Lu and Kakkar, 2014).

5.3 Trimethylamine and Trimethylamine-N-oxide

For decades, physicians and research investigators have noted that consumption of food abundant in saturated fat and cholesterol increases risk for atherosclerosis. These foods are also rich in trimethylamine (TMA). Recently, potential mechanisms by which a TMA-enriched diet promotes atherosclerosis have been examined. In addition to food, TMA is produced by intestinal microbiota digestion of choline and phosphatidylcholine. On absorption, TMA is oxidised by hepatic flavin monooxygenases to generate trimethylamine-N-oxide (TMAO). TMAO has been shown to augment atherosclerosis in mice through the regulation of multiple components of lipid metabolism. For example, TMAO decreases mRNA abundance of ABCA1 and ABCG1 in cultured mouse peritoneal macrophages, resulting in attenuated cholesterol export to apoA-I (Chapter 15). TMAO increases CD36 and SR-A in macrophages, modulates bile acid metabolism and sterol transporters in both the liver and the intestine, and promotes cholesterol delivery but diminishes reverse cholesterol transport in vivo (Tang and Hazen, 2014). However, TMAO does not influence mRNA levels of LDL receptor or cholesterol synthesis genes (Chapters 11 and 17). Effects of TMAO on atherosclerosis development are also supported by findings in human cohort studies. Plasma concentrations of TMAO are associated with common events of atherosclerosis such as myocardial infarction, stroke or death (Tang and Hazen, 2014; Micha et al., 2010).

6. TRADITIONAL AND EVOLVING LIPID-LOWERING THERAPIES FOR THE TREATMENT OF ATHEROSCLEROSIS

In addition to lifestyle changes, lipid-regulating agents are widely used to improve dyslipidaemia in high-risk patients (Chan et al., 2014). Current guidelines recommend statins as first-line lipid regulators. Inhibitors of cholesterol absorption or bile acid sequestrants represent additional or alternative agents for patients who are intolerant to statins or require optimal reduction in plasma LDL cholesterol. However, many statin-treated patients, including those with metabolic syndrome or type 2 diabetes, have significant residual atherosclerosis risk, due in part to persistent abnormalities in TAG-rich lipoproteins and HDL (Chapter 19). Treatment approaches involve the additional use of other lipid-regulating agents to treat atherogenic dyslipidaemia by harnessing their complementary mechanisms of action. Increased understanding of the biological and molecular mechanisms underlying the atherosclerotic process at its interface with lipoprotein metabolism, together with progress in our comprehension of the molecular genetics of lipid disorders, has highlighted several new targets for therapeutic intervention, as well as novel treatment options (Figure 5).

6.1 Statins

Inhibition of HMG-CoA reductase, a rate-limiting enzyme in hepatic cholesterol synthesis by statins, results in the reduction of intracellular cholesterol content that in turn induces an increase in SREBP-2-mediated hepatic LDL receptor synthesis (Sahebkar and Watts, 2013b) (Chapter 11). This increases the clearance of atherogenic lipoproteins, particularly LDL, as well as chylomicron remnants and VLDL remnants (Figure 5). Statins are the most efficacious agents for lowering the plasma concentration of LDL cholesterol and apoB-100. Statins also decrease chylomicron remnants and VLDL TAGs, reduce small dense LDL particles and modestly increase HDL cholesterol. Statins decrease hepatic apoB-100 production, although results are inconsistent. Divergent results have also been reported on the effects of statins on HDL metabolism. However, there is no evidence that statin-induced HDL effects contribute to cardiovascular benefit. Irrespective of its effects, clinical trials have consistently demonstrated that statin therapy reduces cardiovascular events.

6.2 Fibrates

Several trials demonstrate that fibrates decrease atherosclerotic events in patients with the metabolic syndrome and type 2 diabetes, which is characterised by increased plasma VLDL TAG and reduced plasma HDL cholesterol (Do et al., 2014). Fibrates are agonists of the nuclear hormone receptor, peroxisome proliferator-activated receptor-α (PPAR-α). Fibrates decrease plasma TAG up to 50%, LDL cholesterol up to 20%, increase HDL cholesterol up to 20% and decrease small dense LDL particles. The mechanistic action of fibrates is not fully understood, but they decrease TAG substrate availability to liver by stimulating fatty acid oxidation, thereby decreasing hepatic VLDL secretion. Fibrates also promote VLDL lipolysis by activating lipoprotein lipase (LPL) and reducing apoC-III gene expression. Fibrates increase the expression of apoA-I and ABCA1 by activating LXR, thereby promoting reverse cholesterol transport.

6.3 Niacin

Niacin has the theoretical potential to be an optimal drug for the treatment of atherosclerotic diseases (Do et al., 2014). Niacin lowers plasma TAG concentrations by up to 30%, LDL cholesterol up to 15%, and increases HDL cholesterol by up to 25% and causes a shift of small dense LDL to large buoyant LDL particles. Niacin is one of the few agents that can decrease plasma Lp(a). The mechanism by which niacin modulates lipid metabolism has not been elucidated. Niacin decreases hepatic secretion of VLDL TAG by inhibiting fatty acid mobilisation from peripheral adipocytes, thus decreasing the hepatic synthesis and secretion of VLDL and the concentrations of its products, intermediate-density lipoprotein (IDL) and LDL. Niacin also increases plasma HDL cholesterol and apoA-I concentrations by decreasing clearance of HDL apoA-I. Despite this desirable profile of action on plasma lipoproteins, and positive results from early clinical trials, two recent clinical trials failed to demonstrate significant benefits of niacin on cardiovascular events in statin-treated patients (Rached et al., 2014).

6.4 Cholesterol Absorption Inhibitors

Ezetimibe selectively lowers plasma LDL cholesterol concentrations (Do et al., 2014). Ezetimibe inhibits the function of Niemann–Pick 1-Like 1 protein (NPC1L1) resulting in reduced intestinal cholesterol absorption, attenuated cholesterol transport in chylomicron remnants and reduced hepatic cholesterol content (Figure 5). SREBP-2-mediated upregulation of hepatic LDL receptors stimulates hepatic uptake of plasma apoB-100-containing lipoproteins and accounts for up to 20% decreases in plasma LDL cholesterol. Ezetimibe also significantly decreases LDL particle number, but has an inconsistent effect on LDL particle size. Ezetimibe decreases intrahepatic TAGs, independent of body weight, visceral fat and insulin sensitivity, although the mechanism remains unclear. Inhibition of cholesterol absorption and hepatic cholesterol synthesis with ezetimibe in combination with a statin, has complementary effects on the fractional catabolism of apoB-containing lipoproteins. Accordingly, a combination of ezetimibe and a low-dose statin can more effectively decrease LDL cholesterol than higher doses of statin alone. Although some short-term imaging studies have reported little or no benefit of ezetimibe treatment, the recent Improved Reduction of Outcomes: Vytorin Efficacy International Trial (IMPROVE-IT) study demonstrated significant benefits of ezetimibe for the reduction of cardiovascular events.

6.5 ω-3 Polyunsaturated Fatty Acids

Fish oils are a rich source of ω-3 polyunsaturated fatty acids (PUFAs), mainly eicosapentenoic acid (EPA) and docosahexenoic acid (DHA). Fish oils are recommended as an adjunct to diet for the reduction of VLDL–TAG and prevention of acute pancreatitis (Rached et al., 2014). Human studies demonstrate that ω-3 PUFA supplementation decreases hepatic VLDL production and accordingly reduces VLDL–TAG. There are no consistent effects on plasma concentrations of LDL cholesterol, HDL cholesterol or HDL–apoA-I. The mechanistic action of ω-3 PUFAs on plasma TAGs includes inhibition of diacylglycerol acyltransferase (DGAT), fatty acid synthase and acyl-CoA carboxylase enzyme activities, resulting in decreased TAG and fatty acid synthesis. ω-3 PUFAs also enhance fatty acid β-oxidation by stimulating PPAR-α. However, they are much weaker PPAR-α agonists than fibrates. Although some early trials have shown cardiovascular benefits, a recent meta-analysis has cast doubts on the therapeutic efficacy of ω-3 PUFA and concluded that its supplementation is not associated with decreased risk of atherosclerosis-related events.

6.6 Proprotein Convertase Subtilisin/Kexin Type 9 Inhibitors

PCSK9 targets the LDL receptor for degradation and was discovered as the third gene involved in autosomal dominant hypercholesterolaemia. Some mutations of PCSK9 enhance its function and cause hypercholesterolaemia, whereas loss-of-function mutations are associated with hypocholesterolaemia and reduced cardiovascular risk. By enhancing hepatic LDL receptor degradation, PCSK9 increases plasma LDL cholesterol. PCSK9 is primarily regulated transcriptionally by SREBP2 and the secreted protein binds the epidermal growth factor-like repeat A domain of the LDLR, which results in attenuated LDLR recycling and its targeting to lysosomes for degradation. While statins effectively reduce plasma LDL cholesterol in many patients, statins also upregulate PCSK9 via SREBP2, thereby limiting statin's efficacy in reducing plasma LDL cholesterol.

Antibodies against PCSK9 have been developed that increase the numbers of LDL receptors available at the cell surface and reduce plasma LDL cholesterol (Figure 5) (Rached et al., 2014). These include human mononclonal antibodies or humanised antibodies that are injected intravenously or subcutaneously. In phase II and III trials, anti-PCSK9 therapy of patients lowers LDL cholesterol by 40–65% in both heterozygous FH and non-FH patients, with or without statin and ezetimibe treatment. Lp(a) concentrations were also decreased significantly. Other strategies to inhibit PCSK9, including ASO therapies, small interfering RNAs and recombinant adnectins are under development. PCSK9 inhibition is a promising new approach for treatment of hypercholesterolaemia and potentially atherosclerosis.

6.7 ASO Therapies

ASOs are short, single-stranded, synthetic analogues of natural nucleic acids designed to bind to a target mRNA in a sequence-specific manner. Injection of a short complementary ASO sequence binds to mRNA and induces either selective degradation of the complex by endogenous nucleases or inhibition of mRNA processing and/or function. The most advanced ASO is mipomersen that is designed to inhibit hepatic apoB-100 synthesis (Rached et al., 2014; Do et al., 2014) (Figure 5). Phase II and III trials revealed that mipomersen reduced plasma concentrations of apoB, LDL cholesterol, non-HDL cholesterol and TAGs by 20–65% in patients with different forms of hyperlipidaemia, including heterozygous and homozygous FH, and in patients treated with statins and ezetimibe. Mipomersen reduced Lp(a) levels by 25% by inhibiting its synthesis. The main side effects involve injection-site reactions and hepatic steatosis.

Beyond apoB and PCSK9, several other potential ASO targets for treatment of hyperlipidaemia have been identified, including apoC-III and Lp(a) (Figure 5). ApoC-III normally inhibits LPL-mediated lipolysis of chylomicrons and VLDL, and attenuates hepatic clearance of their remnants. Human genetic studies have associated decreased plasma concentrations of apoC-III with lower TAG concentrations and diminished cardiovascular events. In animal models and Phase I human trials, an apoC-III ASO inhibitor produced potent, selective reductions in plasma apoC-III and TAG concentrations as well as marked increases in HDL-C. ASOs with the potential to lower PCSK9 and Lp(a) plasma concentrations are in early stages of development.

6.8 Microsomal Triacylglyceride Transfer Protein Inhibitors

MTP is critical for the formation and secretion of apoB-containing lipoproteins from the liver and intestine (Do et al., 2014) (Chapter 16). MTP transfers TAG, CE and phospholipid to apoB within the cell during the lipoprotein assembly process. Mutations in the MTP gene lead to a rare condition known as abetalipoproteinaemia, in which plasma apoB-containing lipoproteins are undetectable. In animal models, MTP inhibition results in profound reductions in plasma TAG and cholesterol concentrations (Figure 5). Although early MTP inhibitors reduced LDL cholesterol, further development of most inhibitors has been discontinued due primarily to hepatic fat accumulation. Lomitapide is the only systemic MTP inhibitor currently in development. Lomitapide substantially reduced levels of LDL cholesterol in homozygous FH and in clinical trials proved very effective in reducing LDL cholesterol in patients with homozygous FH as well as in patients with moderate hypercholesterolaemia, either as monotherapy or when combined with ezetimibe. However, variable gastrointestinal side effects, minor elevations in liver transaminase levels and increases in hepatic fat were reported. Lomitapide was recently approved for treatment of homozygous FH, as this drug was considered to provide the benefits of cardiovascular protection that outweighed risks of increased hepatic fat. An intestine-targeted MTP inhibitor has been shown to decrease both VLDL and chylomicron production and resulted in weight loss without elevations of liver enzymes or increases in hepatic fat, suggesting that this approach may be more applicable to treatment of a broader range of lipid disorders (Rached et al., 2014).

6.9 ACAT and DGAT Inhibitors

One of the important steps in atherogenesis and cholesterol accumulation in the arterial wall is esterification of cholesterol in macrophages, which promotes foam cell formation. This provided a rationale for the development of ACAT inhibitors; however, early versions inhibited both ACAT isoforms (ACAT-1 and -2) (Rached et al., 2014). Imaging based clinical trials failed to demonstrate reduced atherosclerosis progression in patients with FH. By contrast, an ACAT-1-selective inhibitor K-604 is under development by Kowa Pharmaceuticals. ACAT-2 is the isoform responsible for cholesterol esterification in the liver and intestine and the provision of CE for lipoprotein synthesis (Figure 5). Genetic deletion of ACAT-2 or pharmacological inhibition of ACAT-2 with ASOs in mice inhibits the secretion of both chylomicrons and VLDL into plasma and attenuates atherosclerosis in $Ldlr^{-/-}$ mice. As a result, specific inhibitors of ACAT-2 are currently under development.

DGATs esterify the third fatty acid to DAG to form TAG in adipose tissue, the intestine and the liver. Studies in mice with DGAT-1 deficiency or mice treated with DGAT-1 inhibitors demonstrated reduced plasma TAGs, hepatic steatosis and obesity, which were paralleled by improved insulin resistance. In patients with severe hypertriglyceridaemia, with LPL deficiency, a DGAT-1 inhibitor decreased fasting plasma TAG levels. Clinical trials are ongoing in patients with coronary artery disease to further substantiate these findings. DGAT-2 is the isoform responsible for TAG synthesis in liver and intestine and the provision of CE for lipoprotein synthesis (Figure 5). Specific DGAT-2 inhibitors decrease plasma and liver TAG in experimental animals and are currently under development for clinical use.

6.10 High-Density Lipoprotein Modulating Drugs

Epidemiological evidence demonstrates that low HDL cholesterol concentrations predict cardiovascular events across multiple populations. However, therapeutic strategies to target HDL have not yet achieved success (Rached et al., 2014). Three HDL cholesterol-raising drugs, two early cholesteryl ester transfer protein (CETP) inhibitors and niacin, have failed in prospective intervention trials due to off-target side effects or lack of efficacy. Current therapeutic approaches are focussed on improved HDL metabolism, increased HDL particle number and enhanced HDL function. Inhibition of CETP lipid transfer activity elevates plasma HDL cholesterol concentrations by 30–140% and reduces concentrations of LDL cholesterol and Lp(a) by up to 40% (Figure 5). Large cardiovascular end-point trials of two recently developed CETP inhibitors are ongoing. Other approaches that transiently increase HDL particle numbers and enhance HDL functionality include infusions of reconstituted HDL, administration of apoA-I mimetic peptides, antagonists of miR33 (antimiR33) (Figure 5) and injection of delipidated HDL. These strategies involve structural remodelling of endogenous HDL particles and are targeted towards export of cholesterol from the atherosclerotic plaque (Figure 5). The impact of these agents on cardiovascular events remains to be evaluated.

7. FUTURE DIRECTIONS

The pathogenesis and mechanisms of atherosclerosis are complex. In the past several decades, lipid and lipoprotein-related mechanisms have been studied extensively, and medical treatments of atherosclerosis targeting lipid metabolism have been improved. However, many questions remain unresolved: (1) Although inflammation is recognised as a critical mechanism in hypercholesterolaemia-induced atherosclerosis, the therapeutic effects of directly targeting inflammation have not been translated into human use. (2) There is substantial experimental evidence that HDL reduces atherosclerosis; however, efforts to increase HDL cholesterol were hampered by either off-target effects or lack of beneficial effects on atherosclerotic diseases. (3) Currently, there is only weak evidence that drugs regress preexisting atherosclerosis in humans. Several animal studies have demonstrated that therapeutic intervention can induce regression of atherosclerosis, but only in the setting of significant lowering of plasma cholesterol concentrations. (4) Therapeutic approaches that directly influence mechanisms of atherogenesis within the arterial wall, including those that specifically target cellular lipid metabolism, have yet to be developed or evaluated. In future studies, in addition to continuously exploring the complex mechanisms of lipid/lipoprotein-related atherosclerosis in animal models, it will be important to apply convincing findings from animal models into human clinical trials.

REFERENCES

Allahverdian, S., Chehroudi, A.C., McManus, B.M., Abraham, T., Francis, G.A., 2014. Contribution of intimal smooth muscle cells to cholesterol accumulation and macrophage-like cells in human atherosclerosis. Circulation 129, 1551–1559.

Bentzon, J.F., Otsuka, F., Virmani, R., Falk, E., 2014. Mechanisms of plaque formation and rupture. Circ. Res. 114, 1852–1866.

Boucher, P., Gotthardt, M., Li, W.P., Anderson, R.G., Herz, J., 2003. LRP: role in vascular wall integrity and protection from atherosclerosis. Science 300, 329–332.

Brahm, A.J., Hegele, R.A., 2015. Chylomicronaemia-current diagnosis and future therapies. Nat. Rev. Endocrinol.. 11. http://dx.doi.org/10.1038/nrendo.2015.26

Canton, J., Neculai, D., Grinstein, S., 2013. Scavenger receptors in homeostasis and immunity. Nat. Rev. Immunol. 13, 621–634.

Chan, D.C., Barrett, P.H., Watts, G.F., 2014. The metabolic and pharmacologic bases for treating atherogenic dyslipidaemia. Best Pract. Res. Clin. Endocrinol. Metab. 28, 369–385.

Covey, S.D., Krieger, M., Wang, W., Penman, M., Trigatti, B.L., 2003. Scavenger receptor class B type I-mediated protection against atherosclerosis in LDL receptor-negative mice involves its expression in bone marrow-derived cells. Arterioscler. Thromb. Vasc. Biol. 23, 1589–1594.

Dadu, R.T., Ballantyne, C.M., 2014. Lipid lowering with PCSK9 inhibitors. Nat. Rev. Cardiol. 11, 563–575.

Do, R.Q., Nicholls, S.J., Schwartz, G.G., 2014. Evolving targets for lipid-modifying therapy. EMBO Mol. Med. 6, 1215–1230.

Febbraio, M., Podrez, E.A., Smith, J.D., Hajjar, D.P., Hazen, S.L., Hoff, H.F., Sharma, K., Silverstein, R.L., 2000. Targeted disruption of the class B scavenger receptor CD36 protects against atherosclerotic lesion development in mice. J. Clin. Invest. 105, 1049–1056.

Fernandez-Hernando, C., Moore, K.J., 2011. MicroRNA modulation of cholesterol homeostasis. Arterioscler. Thromb. Vasc. Biol. 31, 2378–2382.

Getz, G.S., Reardon, C.A., 2012. Animal models of atherosclerosis. Arterioscler. Thromb. Vasc. Biol. 32, 1104–1115.

Gonias, S.L., Campana, W.M., 2014. LDL receptor-related protein-1: a regulator of inflammation in atherosclerosis, cancer, and injury to the nervous system. Am. J. Pathol. 184, 18–27.

Hegele, R.A., Ginsberg, H.N., Chapman, M.J., Nordestgaard, B.G., Kuivenhoven, J.A., Averna, M., Boren, J., Bruckert, E., Catapano, A.L., Descamps, O.S., Hovingh, G.K., Humphries, S.E., Kovanen, P.T., Masana, L., Pajukanta, P., Parhofer, K.G., Raal, F.J., Ray, K.K., Santos, R.D., Stalenhoef, A.F., Stroes, E., Taskinen, M.R., Tybjaerg-Hansen, A., Watts, G.F., Wiklund, O., 2014. The polygenic nature of hypertriglyceridaemia: implications for definition, diagnosis, and management. Lancet Diabetes Endocrinol. 2, 655–666.

Hu, L., Boesten, L.S., May, P., Herz, J., Bovenschen, N., Huisman, M.V., Berbee, J.F., Havekes, L.M., van Vlijmen, B.J., Tamsma, J.T., 2006. Macrophage low-density lipoprotein receptor-related protein deficiency enhances atherosclerosis in ApoE/LDLR double knockout mice. Arterioscler. Thromb. Vasc. Biol. 26, 2710–2715.

Inoue, K., Arai, Y., Kurihara, H., Kita, T., Sawamura, T., 2005. Overexpression of lectin-like oxidized low-density lipoprotein receptor-1 induces intramyocardial vasculopathy in apolipoprotein E-null mice. Circ. Res. 97, 176–184.

Ishibashi, S., Herz, J., Maeda, N., Goldstein, J.L., Brown, M.S., 1994. The two-receptor model of lipoprotein clearance: tests of the hypothesis in 'knockout' mice lacking the low density lipoprotein receptor, apolipoprotein E, or both proteins. Proc. Natl. Acad. Sci. U.S.A. 91, 4431–4435.

Koschinsky, M.L., Boffa, M.B., 2014. Lipoprotein(a): an important cardiovascular risk factor and a clinical conundrum. Endocrinol. Metab. Clin. North Am. 43, 949–962.

Libby, P., 2012. Inflammation in atherosclerosis. Arterioscler. Thromb. Vasc. Biol. 32, 2045–2051.

Lu, X., Kakkar, V., 2014. Inflammasome and atherogenesis. Curr. Pharm. Des. 20, 108–124.

Luscher, T.F., Landmesser, U., von Eckardstein, A., Fogelman, A.M., 2014. High-density lipoprotein: vascular protective effects, dysfunction, and potential as therapeutic target. Circ. Res. 114, 171–182.

Mehta, J.L., Sanada, N., Hu, C.P., Chen, J., Dandapat, A., Sugawara, F., Satoh, H., Inoue, K., Kawase, Y., Jishage, K., Suzuki, H., Takeya, M., Schnackenberg, L., Beger, R., Hermonat, P.L., Thomas, M., Sawamura, T., 2007. Deletion of LOX-1 reduces atherogenesis in LDLR knockout mice fed high cholesterol diet. Circ. Res. 100, 1634–1642.

Micha, R., Wallace, S.K., Mozaffarian, D., 2010. Red and processed meat consumption and risk of incident coronary heart disease, stroke, and diabetes mellitus: a systematic review and meta-analysis. Circulation 121, 2271–2283.

Moore, K.J., Kunjathoor, V.V., Koehn, S.L., Manning, J.J., Tseng, A.A., Silver, J.M., McKee, M., Freeman, M.W., 2005. Loss of receptor-mediated lipid uptake via scavenger receptor A or CD36 pathways does not ameliorate atherosclerosis in hyperlipidemic mice. J. Clin. Invest. 115, 2192–2201.

Moore, K.J., Sheedy, F.J., Fisher, E.A., 2013. Macrophages in atherosclerosis: a dynamic balance. Nat. Rev. Immunol. 13, 709–721.

Moore, K.J., Tabas, I., 2011. Macrophages in the pathogenesis of atherosclerosis. Cell 145, 341–355.

Mozaffarian, D., Benjamin, E.J., Go, A.S., Arnett, D.K., Blaha, M.J., Cushman, M., de Ferranti, S., Despres, J.P., Fullerton, H.J., Howard, V.J., Huffman, M.D., Judd, S.E., Kissela, B.M., Lackland, D.T., Lichtman, J.H., Lisabeth, L.D., Liu, S., Mackey, R.H., Matchar, D.B., McGuire, D.K., Mohler 3rd, E.R., Moy, C.S., Muntner, P., Mussolino, M.E., Nasir, K., Neumar, R.W., Nichol, G., Palaniappan, L., Pandey, D.K., Reeves, M.J., Rodriguez, C.J., Sorlie, P.D., Stein, J., Towfighi, A., Turan, T.N., Virani, S.S., Willey, J.Z., Woo, D., Yeh, R.W., Turner, M.B., 2015. Heart disease and stroke statistics–2015 update: a report from the American Heart Association. Circulation 131, e29–322.

Murphy, A.J., Dragoljevic, D., Tall, A.R., 2014. Cholesterol efflux pathways regulate myelopoiesis: a potential link to altered macrophage function in atherosclerosis. Front. Immunol. 5, 490.

Naar, A.M., 2013. Anti-atherosclerosis or no anti-atherosclerosis: that is the miR-33 question. Arterioscler. Thromb. Vasc. Biol. 33, 447–448.

Nazari-Jahantigh, M., Egea, V., Schober, A., Weber, C., 2015. MicroRNA-specific regulatory mechanisms in atherosclerosis. J. Mol. Cell. Cardiol. 84. http://dx.doi.org/10.1016/j.yjmcc.2014.10.021

Novak, J., Bienertova-Vasku, J., Kara, T., Novak, M., 2014. MicroRNAs involved in the lipid metabolism and their possible implications for atherosclerosis development and treatment. Mediators Inflamm. 2014, 275867.

Overton, C.D., Yancey, P.G., Major, A.S., Linton, M.F., Fazio, S., 2007. Deletion of macrophage LDL receptor-related protein increases atherogenesis in the mouse. Circ. Res. 100, 670–677.

Rached, F.H., Chapman, M.J., Kontush, A., 2014. An overview of the new frontiers in the treatment of atherogenic dyslipidemias. Clin. Pharmacol. Ther. 96, 57–63.

Sahebkar, A., Watts, G.F., 2013a. Fibrate therapy and circulating adiponectin concentrations: a systematic review and meta-analysis of randomized placebo-controlled trials. Atherosclerosis 230, 110–120.

Sahebkar, A., Watts, G.F., 2013b. New therapies targeting apoB metabolism for high-risk patients with inherited dyslipidaemias: what can the clinician expect? Cardiovasc. Drugs Ther. 27, 559–567.

Suzuki, K., Doi, T., Imanishi, T., Kodama, T., Tanaka, T., 1999. Oligonucleotide aggregates bind to the macrophage scavenger receptor. Eur. J. Biochem. 260, 855–860.

Tabas, I., Williams, K.J., Boren, J., 2007. Subendothelial lipoprotein retention as the initiating process in atherosclerosis: update and therapeutic implications. Circulation 116, 1832–1844.

Tang, W.H., Hazen, S.L., 2014. The contributory role of gut microbiota in cardiovascular disease. J. Clin. Invest. 124, 4204–4211.

Tellides, G., Pober, J.S., 2015. Inflammatory and immune responses in the arterial media. Circ. Res. 116, 312–322.

de Winther, M.P., Gijbels, M.J., van Dijk, K.W., van Gorp, P.J., Suzuki, H., Kodama, T., Frants, R.R., Havekes, L.M., Hofker, M.H., 1999. Scavenger receptor deficiency leads to more complex atherosclerotic lesions in APOE3Leiden transgenic mice. Atherosclerosis 144, 315–321.

Weber, C., Noels, H., 2011. Atherosclerosis: current pathogenesis and therapeutic options. Nat. Med. 17, 1410–1422.

Zhang, L., Xu, Q., 2014. Stem/progenitor cells in vascular regeneration. Arterioscler. Thromb. Vasc. Biol. 34, 1114–1119.

Chapter 19

Diabetic Dyslipidaemia

Khosrow Adeli,[1] Jennifer Taher,[1] Sarah Farr,[1] Changting Xiao,[2]
Gary F. Lewis[2]

[1]*The Hospital for Sick Children, University of Toronto, Toronto, ON, Canada;* [2]*University Health
Network, University of Toronto, Toronto, ON, Canada*

ABBREVIATIONS

CE Cholesteryl ester
CETP Cholesteryl ester transfer protein
CVD Cardiovascular disease
DGAT Diacylglycerol acyltransferase
DPP-4 Dipeptidyl peptidase-4
EL Endothelial lipase
FFA Free fatty acid
GLP-1 Glucagon-like peptide-1
GWAS Genome-wide association study
HDL High-density lipoprotein
HDL-C HDL-cholesterol
HL Hepatic lipase
IL-6 Interleukin-6
IR Insulin receptor
IRS Insulin receptor substrate
LDL Low-density lipoprotein
LPL Lipoprotein lipase
MGAT Monoacylglycerol acyltransferase
MTP Microsomal transfer protein
NAFLD Nonalcoholic fatty liver disease
NASH Nonalcoholic steatohepatitis
PIP$_3$ Phosphatidylinositol 3,4,5-trisphosphate
PTEN Phosphatase and tensin homologue
PTP-1B Protein tyrosine phosphatase-1B
PUFA Polyunsaturated fatty acid
SAA Serum amyloid A
sdLDL Small, dense LDL
T1D Type 1 diabetes
T2D Type 2 diabetes
TG Triglyceride
TNF-α Tumour necrosis factor-α
TRL Triglyceride-rich lipoprotein
VLDL Very-low-density lipoprotein

Biochemistry of Lipids, Lipoproteins and Membranes. http://dx.doi.org/10.1016/B978-0-444-63438-2.00019-5
549

1. INTRODUCTION TO THE TYPICAL DYSLIPIDAEMIA OF INSULIN-RESISTANT STATES

The metabolic syndrome, which is linked to obesity and can lead to type 2 diabetes (T2D), is increasing in prevalence and poses one of the major public health challenges worldwide. At least one-quarter of the North American population has evidence of the metabolic syndrome and the prevalence is increasing. Metabolic syndrome describes a cluster of closely related risk factors for diabetes and cardiovascular disease (CVD), including abdominal obesity, dyslipidaemia, hyperglycaemia and hypertension. Despite the absence of a universally accepted definition, dyslipidaemia is a common feature of the metabolic syndrome. In an attempt to standardise the diagnostic criteria of the metabolic syndrome, a joint statement from several expert groups described the dyslipidaemia as increased triglycerides (TG; ≥150 mg/dL or 1.7 mmol/L) and reduced high-density lipoprotein (HDL)-cholesterol (HDL-C; <40 mg/dL or 1.0 mmol/L in men; <50 mg/dL or 1.3 mmol/L in women) (Alberti et al., 2005). Insulin-deficient states on the other hand, as exemplified by type 1 diabetes (T1D), are also characterised by lipid abnormalities but these are highly dependent on the degree of metabolic control and insulin replacement, as will be discussed briefly below.

1.1 Major Dyslipidaemia of Insulin-Resistant States: Hypertriglyceridaemia, Low HDL and Qualitative Changes in Low-Density Lipoprotein (Small, Dense Low-Density Lipoprotein)

The typical dyslipidaemia of the insulin-resistant state (also referred to as 'atherogenic dyslipidaemia') is characterised by a cluster of quantitative and qualitative lipid and lipoprotein abnormalities. This includes increased plasma concentrations of fasting and postprandial apolipoprotein B (apoB)-containing triglyceride-rich lipoproteins (TRL), including very-low-density lipoprotein (VLDL) and chylomicrons (CMs) (see Chapter 16). Also evident is reduced HDL particle number and cholesterol content, as assessed by plasma apoA-I and HDL-C, respectively (see Chapter 15), and a predominance of small, dense LDL (sdLDL) particles (Chahil and Ginsberg, 2006). Altered metabolism of TRL, both overproduction and impaired clearance (Figure 1), is central to the pathophysiology of the atherogenic dyslipidaemia.

In humans, there are significant correlations between fasting and postprandial TG concentrations, both of which are inversely correlated with HDL-C and apoA-I plasma concentrations, suggesting a close link between TRL and HDL metabolism. In fact, an elevated TG:HDL-C ratio may be the single most characteristic biomarker of the metabolic syndrome, even more predictive than the presence of abdominal obesity. Plasma LDL-C concentrations are similar to, or only modestly higher than, those in insulin-sensitive, nondiabetic individuals. In subjects with the metabolic syndrome and/or T2D, the higher

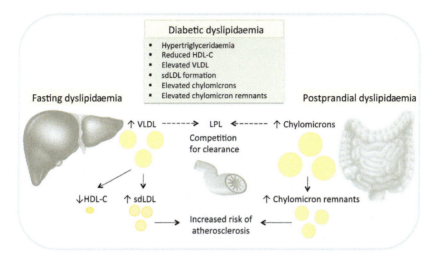

FIGURE 1 **Fasting and postprandial diabetic dyslipidaemia.** Both hepatic and intestinal lipo-protein overproduction contribute to diabetic dyslipidaemia. In the fasting state, increased produc-tion of hepatic VLDL results in hypertriglyceridaemia, which then contributes to the formation of sdLDL and decreased levels of HDL. In the postprandial state, in addition to VLDL, release of intestinally derived chylomicron particles results in postprandial hypertriglyceridaemia. There is also an accumulation of chylomicron remnants in those with type 2 diabetes. VLDL pool size is also a major determinant of postprandial lipaemia since VLDL and chylomicrons compete for clearance. Finally, increases in sdLDL, chylomicron remnants and other smaller triglyceride-rich lipoprotein particles, and reduced HDL-C jointly contribute to increased atherosclerotic risk. Abbreviations: HDL-C, high-density lipoprotein cholesterol; LPL, lipoprotein lipase; sdLDL, small dense LDL; VLDL, very-low-density lipoprotein.

concentration of plasma TRL particles (referred to as increased TRL pool size) promotes exchange of neutral lipids, by mass action, between HDL and TRL and between LDL and TRL. This leads to the formation of TG-rich HDL particles, which are more rapidly catabolised, and sdLDL particles, which are catabolised more slowly than their normal counterparts. sdLDL are also considered more atherogenic. Therefore, abnormalities in TRL metabolism quantitatively and qualitatively affect the metabolism of both HDL and LDL, leading to increased plasma TRL remnants, low HDL (see below Section 4) and increased sdLDL, all of which are strongly associated with increased CVD risk.

1.2 Lipid Profile of Individuals with T1D

Marked elevations in TRL may rarely be seen in insulin-deficient T1D. This occurs in newly diagnosed individuals prior to initiation of insulin treatment or in treated individuals who omit insulin therapy or who have severe, intercurrent illness. TRL accumulates due to reduced activity of lipoprotein lipase (LPL), an insulin-regulated enzyme that plays a key role in TRL catabolism. Removal of CM and VLDL remnants is also impaired in T1D, since insulin regulates their uptake by the liver. LDL-C and apoB-100 may or may not be increased in patients with insulin deficiency, since LDL receptor expression is regulated but to a limited extent by insulin. In metabolically controlled, insulin-treated T1D patients, plasma lipid and lipoprotein levels can be lower than nondiabetic individuals (i.e. HDL-C higher and TG lower). This could be due to the hyperinsulinaemia that often results from systemic insulin administration, which stimulates LPL. In cases where individuals with T1D also develop insulin resistance (often associated with obesity), their lipid abnormality resembles that of T2D patients, with hypertriglyceridaemia and low HDL-C (Chahil and Ginsberg, 2006).

1.3 Role of Diabetic Dyslipidaemia in Atherosclerosis and CVD

LDL has been firmly established as the primary atherogenic lipoprotein (see Chapter 18) and LDL-C lowering, primarily with statin therapy, has been well established as an effective therapeutic strategy to prevent atherosclerotic cardiovascular events. However, even at low LDL-C concentrations, CVD risks are not completely abrogated. The residual CVD risk suggests that additional interventions, besides LDL-C lowering, are required to more significantly reduce CVD events. Because the dyslipidaemia and other features of the insulin-resistant state (e.g. metabolic, inflammatory and procoagulant) are so intricately linked, genetically and metabolically, it has been impossible to distinguish which of the individual components of the metabolic syndrome or diabetes are primarily causative in premature atherosclerosis. A consensus is developing that the atherosclerosis is multifactorial.

In epidemiological analyses, low HDL-C is associated with CVD independent of other lipid parameters. In T2D patients, the UK Prospective Diabetes Study identified HDL-C as the second most important coronary risk factor, following LDL-C. Hypertriglyceridaemia, the common form of dyslipidaemia in T2D, is frequently associated with increased risk of CVD (Do et al., 2013), although the direct atherogenic role of TRL particles continues to be debated. In addition, hypertriglyceridaemia correlates strongly with low HDL-C and the presence of sdLDL particles, and mechanistic studies have shown a causative role between the elevation of TRL and the other features of the dyslipidaemia. Furthermore, the abnormality in TRL metabolism results in plasma accumulation of TRL remnants, that is, intermediate-density lipoprotein (IDL) and CM remnants, which are cholesterol enriched and atherogenic in humans and animal models. Both elevated TRL remnants and reduced HDL play a role in the pathophysiology of atherosclerosis and contribute to CVD risk, especially in insulin resistance (Chapman et al., 2011). Many studies have also demonstrated a strong association between LDL size or density and CVD, although the direct role of sdLDL particles in the atherosclerotic process continues to be debated.

1.4 Aetiology of the Dyslipidaemia: Genetic and Environmental Factors

The majority of cases of the typical dyslipidaemia of insulin-resistant states and T2D result from both genetic and nongenetic factors. Genetic variants can increase the propensity to develop environmentally induced lipid abnormalities, whereas environmental and lifestyle factors can force expression of a clinically significant dyslipidaemic phenotype on a genetically susceptible background. The typical dyslipidaemia of insulin-resistant states is therefore best viewed as a complex phenotype resulting from the cumulative interaction of multiple susceptibility genes and environmental stressors. Physiologically, both the genetic and the nongenetic components can contribute to the defective metabolism of TRL and their remnants, that is, impaired clearance or increased production, or both, causing the abnormal levels of circulating lipids (Lewis et al., 2015).

Genetic variations play an important role in the development of dyslipidaemia, especially hypertriglyceridaemia. Hypertriglyceridaemia can develop with mutations in genes that encode proteins needed for the efficient metabolism of TRL, including *LPL, APOC2, APOA5, LMF1, GPIHBP1* and *GPD1*. Variation in these gene products can, directly or indirectly, decrease LPL function, reduce TRL catabolism and result in increased fasting plasma TG. Although monogenic forms of hypertriglyceridaemia do occur, the majority of cases are polygenic and often coexist with nongenetic conditions. Plasma TG levels are modulated by the cumulative effects of both common and rare variations in multiple genes, together with the influence of secondary, nongenetic factors (Hegele et al., 2013). As a result, more than 95% of patients with hypertriglyceridaemia have a multigenic susceptibility component, which interacts with nongenetic factors and perturbs the production and catabolism of TRL particles. Clinically relevant abnormalities of plasma TG appear to require a polygenic foundation of common variants. These create a background state of predisposition that can interact with additional rare heterozygous variants or nongenetic factors, then forcing expression of a more extreme TG phenotype. Genome-wide association studies (GWAS) have identified common variants in at least 45 loci associated with variation in plasma TG concentrations (Willer et al., 2013). When rare heterozygous variants are present in the genome of an individual who has a high burden of common predisposing variants, the phenotype can be even more severe.

The nongenetic factors include, but are not limited to, lifestyle and environmental factors such as abdominal obesity, physical inactivity, poor diet and excessive alcohol intake. The presence of these secondary factors can exert sufficient perturbation to overwhelm normal physiology and lipid metabolism. Combined with a predisposing genetic condition, these factors eventually may push the risks above a threshold for clinically significant dyslipidaemia. Many of these factors are associated with defective TRL metabolism; thus TRL overproduction and impaired clearance occur in obesity, insulin resistance and T2D.

2. DYSLIPIDAEMIA OF INSULIN-RESISTANT STATES: KEY FACTORS AND MECHANISMS, WITH A FOCUS ON HEPATIC LIPOPROTEIN OVERPRODUCTION

2.1 Apolipoprotein B-Containing Lipoproteins: Alterations in Insulin Resistance and T2D

ApoB100 is a 550 kDa amphipathic lipid-binding protein synthesised in the liver. Unlike other apolipoproteins, apoB100 is not exchanged between lipoproteins and is the structural protein component of VLDL, IDL and LDL particles, remaining an integral part of the secreted particle as it moves down the lipolytic cascade from VLDL to IDL and finally LDL (see Chapter 16).

Individuals with insulin resistance or T2D can have normal or elevated LDL particle number, with a smaller and denser LDL particle profile, even though plasma LDL-C concentration is not particularly elevated by the insulin-resistant or diabetic state. In general, VLDL particles are larger and LDL particles smaller in insulin-resistant states. sdLDL particles have a longer residence time in the circulation, are more glycated and oxidised and can cross the endothelial surface more readily, with greater retention in the arterial intima, contributing to plaque formation (see Chapter 18). In hypertriglyceridaemia, neutral lipid (TG and cholesteryl esters (CE)) exchange between VLDL and LDL particles, facilitated by cholesteryl ester transfer protein (CETP), favours TG enrichment of LDL. The TG-rich LDL particles are then hydrolysed by either LPL or hepatic lipase (HL) to sdLDL particles (Figure 1). A similar lipid exchange occurs between VLDL and HDL particles, additionally affecting plasma HDL composition, as will be discussed below.

2.2 Molecular Mechanisms Underlying Hepatic Insulin Resistance and Increased VLDL Secretion

VLDL overproduction is a fundamental and consistent abnormality of insulin-resistant states and T2D and is the primary cause of the fasting hypertriglyceridaemia in these patients. Insulin plays a major role in regulation of hepatic lipid homeostasis. Following a meal, pancreatic insulin secretion into the portal vein acutely inhibits VLDL secretion. While the mechanism is complex and incompletely understood, insulin acutely inhibits apoB synthesis, reduces microsomal transfer protein (MTP) expression, and enhances apoB100 degradation (Figure 2). Insulin suppression of both apoB synthesis and secretion has been demonstrated in isolated human and rat hepatocytes, and is associated with increased apoB100 degradation. The mechanism is thought to be PI3-kinase dependent, given that acute inhibition of PI3-kinase prevents and reverses the action of insulin and results in increased VLDL-apoB and TG in vivo (Chirieac et al., 2006).

In contrast to the acute actions of insulin in the liver, chronic exposure of rat hepatocytes to insulin results in increased apoB100 secretion. Chronic

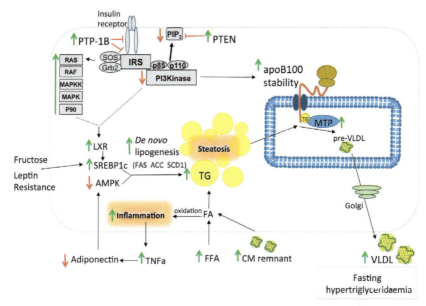

FIGURE 2 **Mechanisms of hepatic steatosis and VLDL overproduction in insulin resistance.** In hepatic insulin resistance, which is commonly associated with hyperinsulinaemia, upregulation of the phosphatases PTEN and PTP-1B leads to disruption of the insulin signalling pathway via dephosphorylation of PIP_3 and IR/IRS, respectively. This decreases signalling in the PI3K/PKB arm of the insulin signalling pathway, which mediates the metabolic responses to insulin. The PI3K-dependent induction of apoB100 degradation by insulin is impaired, thereby increasing apoB100 stability and enhancing VLDL particle biosynthesis. The combination of increased MAPK and decreased PI3K/PKB signalling results in increased de novo lipogenesis through increases in the transcription factor LXR and the downstream target SREBP-1c. Enhanced fructose consumption and/or leptin resistance also increases hepatic de novo lipogenesis by stimulating SREBP-1c, thereby contributing to increased hepatic steatosis. In addition to enhanced de novo lipogenesis, increased FFA flux (from adipocyte lipolysis and reduced adipose tissue fatty acid trapping) and fatty acids released from the intracellular degradation of chylomicron remnants taken up by the liver also contribute to hepatic steatosis. The oxidation of FFA in preference to cytosolic TG increases hepatic inflammation via increased TNF-α. This is associated with reductions in adiponectin and hepatic AMPK, and enhanced de novo lipogenesis. The increased lipid substrate leads to stabilisation of newly synthesised apoB100 due to its reduced posttranslational degradation, and thus increased VLDL particle production, ultimately resulting in hypertriglyceridaemia. Abbreviations: ACC, Acetyl-CoA carboxylase; AMPK, AMP-activated protein kinase; apoB, apolipoproteinB; CM, chylomicron; FAS, fatty acid synthase; FFA, free fatty acid; IRS, insulin receptor substrate, Lep, Leptin; LXR, liver X receptor; MTP, microsomal transfer protein; PI3kinase, phosphoinositide 3-kinase; PIP_3, phosphatidylinositol 3,4,5-trisphosphate; PTEN, phosphatase and tensin homologue; PTP-1B, protein tyrosine phosphatase 1B; SCD1, stearoyl desaturase-1; SREBP-1c, sterol regulatory element-binding protein-1c; TG, triglyceride; TNF-α, tumour necrosis factor-α; VLDL, very-low-density lipoprotein.

hyperinsulinaemia, which inevitably accompanies insulin resistance, has stimulatory effects on VLDL secretion in humans. This may be explained by partial insulin resistance in hepatocytes. The acute inhibitory effect of insulin on VLDL particle assembly and secretion is blunted in the insulin-resistant state, whereas its stimulatory effects on hepatic de novo lipogenesis remain, contributing to the increased production of VLDL particles. The lipogenic actions of insulin in the liver are mediated through the activation of the transcription factor SREBP-1c (Figure 2). Insulin activates SREBP-1c, which then enhances transcription of both acetyl-CoA carboxylase (ACC) and fatty acid synthase (FAS) (see Chapter 5), thereby increasing hepatic production of TG that are then secreted from the liver in VLDL particles. The lipogenic SREBP-1c pathway remains insulin sensitive in T2D, resulting in increased hepatic TG and VLDL production (Brown and Goldstein, 2008).

In addition to increased hepatic de novo lipogenesis, animal models of T2D and/or insulin resistance also display increased hepatic MTP mass, mRNA and activity with enhanced VLDL production and fasting hypertriglyceridaemia (Fisher et al., 2014). The inhibitory effects of insulin on MTP are prevented in insulin-resistant states. Thus, the increase in MTP activity in insulin resistance would be expected to further stimulate hepatic apoB100 production and secretion. The MTP promoter contains negative sterol response elements, indicating regulatory control of MTP by SREBP-1c. Thus, MTP expression would be expected to decrease with increased SREBP-1c levels; however, genetic and diet-induced animal models of insulin resistance show a simultaneous upregulation of both SREBP-1c and MTP gene expression in the liver. This may be the result of unknown factors that block SREBP-1c-mediated inhibition of MTP and may also be due to increased hepatic TG stores that contribute to MTP stimulation.

Taken together, enhanced VLDL secretion observed in insulin-resistant states is due to a combination of both impaired insulin action on apoB100 degradation, thereby increasing apoB stability, and increased stimulation of SREBP-1c, thereby enhancing hepatic de novo lipogenesis. MTP upregulation observed in insulin-resistant states is likely a key factor in enhanced VLDL secretion.

Dysregulation of hepatic lipid and lipoprotein metabolism in insulin resistance has also been linked to specific defects in the hepatic insulin signalling cascade (Figure 2). Insulin signalling is initiated with insulin binding to the cell surface insulin receptor (IR). On binding, receptor autophosphorylation is initiated and results in tyrosine phosphorylation of intracellular insulin receptor substrates (IRS). These effectors then bind to a signalling protein complex (Grb-2–SOS) to activate the MAP kinase pathway that mediates the mitogenic effects of insulin. IRS can also bind to the regulatory subunit of PI-3K, which activates the PI-3K/PKB pathway that mediates the metabolic responses to insulin. This signalling arm results in the production of phosphatidylinositol 3,4,5-trisphosphate (PIP_3) required for the inhibitory effects of insulin on apoB production.

The metabolic arm of insulin signalling is reversibly modulated by dephosphorylation of the signal mediators by downstream regulatory phosphatases, particularly phosphatase and tensin homologue (PTEN), which dephosphorylates PIP_3, and protein tyrosine phosphatase-1B (PTP-1B), which dephosphorylates IR and IRS. Increased PTP-1B and PTEN expression is linked to hepatic insulin resistance and VLDL overproduction as shown in studies of knockout mice. PTP-1B knockout mice are resistant to diet-induced obesity and have decreased fasting plasma TG levels, while PTEN liver-specific knockout mice display hypersensitivity to insulin and a reduction in both hepatic apoB and MTP mass (Qiu et al., 2008). A high-fructose diet can interrupt hepatic insulin signalling by increasing PTP-1B, leading to decreased tyrosine phosphorylation of IR, IRS1 and IRS2 and ultimately resulting in VLDL overproduction (Taghibiglou et al., 2002).

2.3 Mechanisms of Hepatic VLDL Overproduction in Insulin Resistance: Multiorgan Cross Talk, Hormones and Dietary Factors

Nutrients and hormones are critical regulators of hepatic lipid metabolism. Hyperglycaemia, for instance, enhances hepatic de novo lipogenesis through ChREBP, a key transcription factor and regulator of both glycolysis and lipogenesis. The activation of ChREBP by glucose acts synergistically with SREBP-1c (activated by insulin) to increase FAS and ACC expression and increase TG synthesis. In this way, hyperglycaemia may contribute to the onset of hepatic steatosis and VLDL hypersecretion. In fact, inhibition of ChREBP in the liver of *ob/ob* mice decreases both hepatic steatosis and insulin resistance (Dentin et al., 2006), but has not been directly linked to changes in VLDL production.

Extrahepatic insulin-sensitive tissues, particularly adipose tissue, play an indirect role in hepatic lipoprotein overproduction. The inhibitory effects of insulin on adipocyte TG lipolysis and its ability to enhance adipocyte fatty acid trapping are blunted in insulin resistance, thereby leading to increased flux of free fatty acids (FFA) from adipose tissue and increased exposure of the liver and other insulin-sensitive tissues to elevated FFA flux (Lewis et al., 2002). Up to 60% of hepatic lipid content is attributed to FFA influx and thus increased FFA levels can contribute to hepatic steatosis and VLDL hypersecretion.

Many adipokines released from adipose tissue, such as the satiety hormone leptin, are critical regulators of both hepatic lipid and lipoprotein metabolism. *Ob/ob* mice lack leptin and are a genetic model of T2D and hepatic steatosis. In the *ob/ob* mouse, increases in hepatic SREBP-1c, ACC, FAS and SCD1 result in enhanced hepatic de novo lipogenesis (Perfield et al., 2013). Additionally, increased adipocyte TG lipolysis results in enhanced FFA delivery to the liver, further exacerbating the hepatic lipid load. Leptin also indirectly modulates hepatic apoB100 levels by acutely lowering plasma insulin levels. Selective hepatic leptin resistance results in increased VLDL particle size due to increased hepatic lipid levels and therefore increased lipid substrate for VLDL particles (Huynh et al., 2013). Thus, the absence of a functional leptin receptor in *db/db* mice is associated with dyslipidaemia, particularly enhanced LDL-C levels and hypercholesterolaemia. However, in the absence of leptin resistance, leptin has been shown to improve insulin sensitivity, dyslipidaemia and hepatic steatosis in diabetic states.

In addition to leptin, the adipokine adiponectin also plays a beneficial role in regulation of lipid metabolism. Adiponectin levels are negatively correlated with adiposity and are decreased in patients with T2D. Peripheral adiponectin administration reduces body mass and visceral adiposity via the activation of AMP-activated protein kinase (AMPK), which results in decreased hepatic de novo lipogenesis due to phosphorylation and inhibition of ACC as well as decreased SREBP-1c mRNA and protein expression. Insulin resistance contributes to decreased adiponectin levels and increased leptin resistance, thus preventing the hormonally regulated lipid-lowering effects of these adipokines.

The gut peptide glucagon-like peptide (GLP)-1 does not acutely regulate hepatic VLDL secretion in humans (Xiao et al., 2012), but prolonged exposure to the peptide may indirectly inhibit VLDL secretion. GLP-1 stimulates insulin secretion and inhibits glucagon secretion from the pancreas, and both have been shown to regulate VLDL secretion. T2D patients treated with GLP-1 receptor agonists or dipeptidyl peptidase-4 (DPP-4; the enzyme that degrades GLP-1) inhibitors have improved lipid profiles, including reduced plasma TG, LDL-C and total cholesterol levels, indicating potential CVD benefits (Xiao et al., 2015). GLP-1 receptor agonism prevents fructose-induced dyslipidaemia and hepatic VLDL overproduction in insulin resistance through an indirect mechanism involving altered energy utilisation, decreased hepatic lipid synthesis and parasympathetic signalling (Taher et al., 2014). Interestingly, GLP-1 has also been suggested as a potential treatment for nonalcoholic fatty liver disease (NAFLD), given its ability to reduce hepatic lipid stores, as shown in various murine models including high-fat fed, hyperglycaemic and *ob/ob* mice (reviewed in Farr et al. (2014)). In addition to GLP-1, the pancreatic hormone glucagon has also been linked to regulation of hepatic lipid and lipoprotein metabolism (Longuet et al., 2008). A study in humans showed that physiological hyperglucagonaemia reduced VLDL production and clearance simultaneously without a net effect on VLDL apoB-100 concentrations (Xiao et al., 2011b).

A major determinant of hepatic lipid and lipoprotein synthesis and export is dietary intake. Diets that are rich in fat and refined carbohydrates stimulate hepatic lipogenesis, thereby playing a pivotal role in hepatocyte lipid accumulation and increased export as VLDL. Excess food intake or diet-induced obesity results in fat overload and ectopic fat accumulation in the liver. Diets that are rich in omega-6 polyunsaturated fatty acids (PUFA), mono-unsaturated fatty acids and saturated fatty acids induce hepatic steatosis, whereas diets containing omega-3 PUFA decrease hepatic lipid deposition and were found to decrease multivariate-adjusted risk for T2D by 33% (Virtanen et al., 2014). Interestingly, decreased dietary fat consumption in the United States from 1994 to 2014 did not correlate with a decrease in obesity, which may be related to the increase in per capita fructose consumption. Fructose bypasses the tightly regulated phosphofructokinase rate-limiting enzyme for glycolysis, thus permitting fructose to continuously enter into the glycolytic pathway, leading to rapid catabolism and the formation of glycerol-3-phosphate and FFA, and ultimately in TG accumulation. Thus fructose, but not glucose, consumption increases postprandial TG and apoB-containing lipoproteins, in association with visceral adiposity, insulin resistance and increased hepatic de novo lipogenesis in humans, which further leads to increased levels of circulating TG in comparison to glucose consumption alone (Stanhope et al., 2009). Fructose is primarily metabolised in the liver, where it activates hepatic SREBP-1c, and increases NF-kB involved in hepatic inflammation (Dekker et al., 2010). Animals fed a high-fructose diet display fasting dyslipidaemia due to VLDL overproduction. Increased dietary fructose consumption over the past few decades is suggested to be a factor contributing to the rising prevalence of obesity and T2D in children and adults. In summary, a number of dietary, hormonal and other regulatory factors play pivotal roles in the regulation of VLDL secretion and contribute to the hypersecretion of hepatic VLDL in insulin resistance and T2D.

2.4 Association of Fatty Liver/Inflammation and Diabetic Dyslipidaemia

Fatty liver or NAFLD is commonly observed in insulin-resistant states. NAFLD is a continuum of disease states, ranging from simple steatosis to nonalcoholic steatohepatitis (NASH), to fibrosis and finally to hepatic cirrhosis. NAFLD is currently the most common liver disorder in Western society, with 80% of NAFLD patients considered to be overweight and hyperlipidaemic. NAFLD also has a high prevalence among individuals with T2D and has been shown to be an independent risk factor, increasing mortality by 2.4-fold. Recently, insulin-sensitising agents have been employed to treat hepatic steatosis since liver lipid levels, inflammation and fibrosis are all improved with enhanced insulin sensitivity (Farr et al., 2014). In hepatic insulin resistance, the lipogenic effects of insulin are not compromised, and thus insulin continues to stimulate hepatic de novo lipogenesis by stimulating LXR, a transcription factor which regulates *SREBP-1c* and *FAS* gene transcription (Figure 2). This is considered to be the 'first hit' according to a two-hit hypothesis resulting in hepatic lipid accumulation. The second hit is required for the progression to inflammation and fibrosis. In cases of enhanced FFA flux to the liver, TG synthesis does not account for disposal of all FFA, with a considerable amount being oxidised. The oxidation of FFA results in the formation of reactive oxidation species, leading to mitochondrial damage as well as proinflammatory cytokine release. This is accompanied by enhanced tumour necrosis factor-α (TNF-α) release from adipocytes, which is associated with dyslipidaemia, oxidative stress, insulin resistance and hepatic inflammation. With the onset of hepatic inflammation and necrosis, simple steatosis progresses to NASH. Both hepatic injury and insulin resistance can then further promote myofibroblastic stellate cell proliferation, which eventually results in cirrhosis. The effect on hepatic VLDL production depends on the stage of NAFLD along this continuum. In patients with simple steatosis, VLDL overproduction results as a compensatory mechanism for the hepatic lipid overload, which ultimately increases fasting dyslipidaemia and atherosclerotic risk. Alternatively, progression to NASH results in impaired VLDL synthesis and export, causing all sources of hepatic lipid to be retained within the liver which can result in liver failure.

3. POSTPRANDIAL DYSLIPIDAEMIA AND INTESTINAL CHYLOMICRON HYPERSECRETION IN INSULIN-RESISTANT STATES

Abnormalities in circulating lipids and lipoproteins after a meal are also characteristic of insulin-resistant states and T2D, and elevated nonfasting TG levels are considered to be an independent predictor of CVD. Postprandial lipaemia is determined by both the production rate and the clearance of triglyceride-rich lipoproteins (TRL) from the circulation. Since hepatic VLDL and intestinally derived CM compete for removal by common, saturable lipolytic clearance mechanisms, the fasting VLDL pool size is a major determinant of postprandial lipaemia and the postprandial TG excursion correlates directly with fasting TG concentration. Not only do CMs rise postprandially, but the major contributor to postprandial lipaemia is an elevation of VLDL (apoB-100-containing) particles, secondary to an increase in secretion and a reduction in clearance.

In the past, postprandial lipaemia in insulin-resistant states was solely attributed to defective particle clearance from the circulation. This was supported by observations of reduced catabolism of CM preparations in patients with T2D. However, increased production of VLDL particles, their competition with CM for clearance, and reduced postprandial stimulation of LPL in insulin-resistant states, all contribute to the observed changes. Moreover, intestinal lipoprotein production was thought to occur exclusively in response to the amount of lipid ingested. Although ingested fat is undoubtedly the major determinant and regulator of intestinal lipoprotein secretion, since 2004 we have come to appreciate that intestinal lipoprotein production is highly regulated even in the fasted state by paracrine and endocrine factors and circulating metabolites, analogous to hepatic lipoprotein secretion (Xiao et al., 2011a). The secretion rate of apoB48-containing lipoproteins increased in obese and insulin-resistant individuals, as well as in patients with T2D. Furthermore, the rate of apoB48 production is inversely correlated with the degree of insulin sensitivity, even in nondiabetic individuals, and improved metabolic control in patients with T2D can help to lower postprandial levels of apoB48-containing particles.

3.1 Mechanisms of Intestinal Lipoprotein Overproduction in Insulin-Resistant States

Insulin acutely inhibits intestinal lipoprotein particle secretion, as it does with hepatic lipoproteins, in animals (Federico et al., 2006) and humans (Pavlic et al., 2010). Molecular evidence indicates that the intestine itself can become insulin

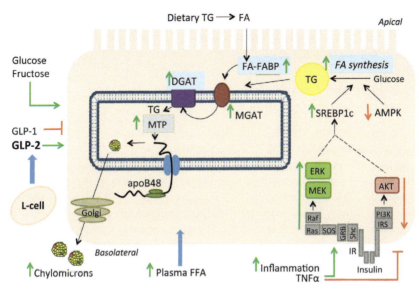

FIGURE 3 **Mechanisms of chylomicron overproduction in insulin resistance.** Insulin resistance is associated with impairment in signalling at the level of the enterocyte, increased enterocyte lipid availability, altered activity of the cellular machinery of lipoprotein biogenesis and secretion, and changes in circulating endocrine and inflammatory factors that contribute to the hypersecretion of chylomicrons. Insulin typically acutely inhibits chylomicron secretion, but in insulin-resistant/ chronic hyperinsulinaemia there is downregulation of the PI3K/Akt arm and chronic upregulation of the MAPK arm of the insulin receptor signalling pathway, changes that are amplified by inflammatory factors such as TNF-α. Aberrant insulin signalling contributes to enhanced expression of SREBP-1c and increased de novo lipogenesis in the enterocyte, thereby increasing lipid availability. Reduced AMPK activity can also contribute to enhanced de novo lipogenesis. In addition, insulin resistance is further associated with elevated levels of circulating FFA, which are taken up by enterocytes at their basolateral surface. This increase in lipid substrate is handled by elevations in cytosolic FABP, and increased activity of MTP at the endoplasmic reticulum for TG resynthesis and lipidation of apoB48. Finally, increasing consumption of monosaccharides (glucose and fructose) and imbalance in GLP-1 (inhibitory) and GLP-2 (stimulatory) activity can contribute to chylomicron overproduction in insulin-resistant states. Abbreviations: AMPK, AMP-activated protein kinase; apoB48, apolipoprotein B48; DGAT, diacylglycerol acyltransferase; FA, fatty acid; ERK, extracellular signal-regulated kinase; FABP, fatty acid-binding protein; FFA, free fatty acid; GLP-1, glucagon-like peptide-1; GLP-2, glucagon-like peptide-2; IR, insulin receptor; MTP, microsomal triglyceride transfer protein; MAPK, mitogen-activated protein kinases; MEK, mitogen/extracellular signal-regulated kinase; MGAT, monoacylglycerol acyltransferse; PI3K, phosphoinositide 3-kinase; SREBP-1c, sterol regulatory element-binding protein-1c; TG, triglyceride; TNF-α, tumour necrosis factor-α.

resistant in rodents and obese humans. In insulin-resistant enterocytes, perturbations in the PI3K/Akt metabolic arm of the insulin signalling pathway are evident, while signalling through the mitogenic MAPK arm is increased (Veilleux et al., 2014) (Figure 3). These changes can contribute to apoB48 hypersecretion. In addition to changes in insulin signalling at the level of the enterocyte, increased lipid availability in the intestine, altered expression of intestinal lipid and lipoprotein-related genes (see Chapters 5 and 16), and changes in endocrine factors also contribute to CM overproduction in insulin-resistant states, as summarised in Figure 3.

3.2 Lipid and Carbohydrate Regulation of Intestinal Lipoprotein Secretion

The main stimulus for intestinal lipoprotein production is luminal availability of ingested lipids. CM lipidation increases with the amount of fat consumed, and meals rich in monounsaturated, polyunsaturated and saturated fatty acids can differentially affect the number, size and clearance of CM particles (Xiao and Lewis, 2012). In addition to the lipid content of the diet, dietary carbohydrates influence intestinal lipid handling. Simple sugars like glucose and fructose raise CM production in healthy humans during intestinal lipid infusion. The effects of glucose may involve mobilisation of TG storage pools within the enterocyte, while fructose itself is directly lipogenic. Alternatively, these monosaccharides increase intestinal lipid absorption, as the efficiency of fat absorption in humans is improved when carbohydrates are co-infused into the small intestine. Given that the increasing prevalence of T2D and the metabolic syndrome in North America closely mirrors the increased consumption of refined carbohydrates and simple sugars over the twentieth century, both lipid and carbohydrate ingestion appear to be important contributors to the hypertriglyceridaemia and CM overproduction observed in these states (Morgantini et al., 2014).

In the fasting state where dietary lipids are no longer present at the apical brush border of the enterocyte, the intestine is still able to secrete small, poorly lipidated CM particles. The lipid required for generating these small particles can be acquired in part from circulating FFAs. Acutely raising circulating FFA levels enhances apoB48 production in rodents and healthy humans. This excess nondietary source of fat at the basolateral surface of the enterocyte can increase intracellular TG synthesis or mobilisation for apoB48 lipidation and stability, supported by the recovery of FFAs in TG from mesenteric lymph where CMs are secreted. In this way, increased FFA flux to the intestine contributes to the CM overproduction observed in insulin-resistant states (Figure 3).

Though often overlooked, the small intestine not only absorbs dietary lipids but is a contributor to the de novo synthesis of TG. De novo lipogenesis generates fatty acids from acetyl-CoA, which is largely derived from glucose (see Chapter 5). Various animal models of insulin resistance have increased intestinal fatty acid and cholesterol synthesis, and decreased intestinal fatty acid oxidation. Increased fatty acid synthesis has also been observed in the intestine of obese insulin-resistant subjects compared to those who have remained insulin sensitive (Veilleux et al., 2014). This provides more lipid substrate for CM assembly. The mechanisms underlying these observations remain poorly understood, but may involve increased activity of SREBP-1c, a transcriptional regulator of fatty acid biosynthetic enzymes, and perturbations in intestinal AMPK signalling to promote fatty acid oxidation while reducing fatty acid synthesis (Figure 3).

3.3 Alterations in Other Pathways Involved in Lipoprotein Assembly and Secretion

In addition to elevated dietary, circulating and de novo sources of fatty acids for TG synthesis and CM lipidation, several other molecular changes are reported at the level of the enterocyte in insulin resistance. Insulin-resistant, obese individuals display increased intestinal expression of the liver and intestinal fatty acid-binding proteins L-FABP and I-FABP (Veilleux et al., 2014). L-FABP plays an important role in sequestering fatty acids in the cytoplasm of the enterocyte, while I-FABP appears to target fatty acids away from β-oxidation and toward TG synthesis. Increased monoacylglycerol acyltransferse (MGAT) and diacylglycerol acyltransferase (DGAT) expression and activity have also been observed in insulin-resistant rodent models, which may serve to convert fatty acids in enterocytes from diacylglycerol to TG (see Chapter 5). Finally, expression and activity of MTP, the protein responsible for CM lipidation (see Chapter 16), are also increased in the intestine of insulin-resistant obese subjects (Figure 3). The increased capacity for apoB48 lipidation prevents its degradation, resulting in enhanced apoB48 stability for CM production.

3.4 Gut Peptides and Inflammatory Factors Affect Intestinal Lipoprotein Secretion

As discussed above, defects in insulin signalling contribute to postprandial dyslipidaemia. In addition to insulin, changes in other hormones also affect CM secretion, particularly the glucagon-like peptides (GLP-1 and GLP-2), secreted from the intestine following nutrient ingestion (Xiao et al., 2015). In recent years, studies in several animal models and in both healthy humans and patients with T2D have shown that GLP-1 regulates postprandial dyslipidaemia by reducing intestinal apoB secretion and preventing the postprandial increase in circulating TG levels (Figure 3). The ability of GLP-1 to improve insulin sensitivity, inhibit gastric emptying, promote satiety and lower body weight may contribute to its beneficial metabolic effects in a chronic setting. However, given that GLP-1 reduces CM production even after a single dose, its mechanism of action may involve a more immediate and direct effect on the intestine (Xiao et al., 2015), or the activation of a rapid neuronal brain-gut axis (Farr et al., 2015). GLP-1 secretion appears to be reduced in patients with T2D, which may contribute to the CM overproduction observed in these individuals. The intestine also co-secretes its sister peptide GLP-2, which has a well-known role in maintaining the integrity of the intestinal mucosa and promoting bowel growth. GLP-2 has been found to promote postprandial CM release and to increase postprandial TG, FFA and apoB48 levels in rodent models as well as healthy humans. Though GLP-1 and GLP-2 are co-secreted in equimolar quantities, in an insulin-resistant state the action of GLP-2 appears to be enhanced while the action of GLP-1 is impaired. This imbalance may contribute to the CM overproduction observed in insulin resistance, and correcting this imbalance by pharmacologically raising endogenous levels of GLP-1 may serve as a treatment for postprandial diabetic dyslipidaemia (Hein et al., 2013).

Finally, in insulin-resistant individuals, the intestine expresses elevated levels of the pro-inflammatory factors TNF-α and interleukin-6 (IL-6), and displays increased NFκB pathway activation, which amplifies inflammation. TNF-α impairs intestinal insulin signalling and increases CM secretion (Figure 3). Therefore, the low-grade inflammation often observed in obese, insulin-resistant individuals likely contributes to CM overproduction.

4. LOW HIGH-DENSITY LIPOPROTEIN IN INSULIN RESISTANCE AND TYPE 2 DIABETES

Plasma HDL-C concentration and its most abundant apolipoprotein, apoA-I, are inversely correlated with atherosclerotic CVD risk (Boden, 2000). Both HDL-C and apoA-I levels are commonly reduced in insulin-resistant individuals and features of insulin resistance can be detected in a very large number of individuals who come to medical attention primarily for low HDL-C.

4.1 HDL Lowering Due to Increased Catabolism in Hypertriglyceridaemia and Insulin Resistance

The inverse correlation between plasma TG and HDL-C reflects the complex, inter-linked metabolism of TRL and HDL. LPL lipolysis of TRL results in the formation of redundant surface materials such as phospholipids, cholesterol and apolipoproteins that are transferred to HDL, contributing to HDL maturation and stability, and ultimately plasma concentrations of HDL (see Chapter 15). Normal insulin-mediated stimulation of LPL that occurs postprandially is blunted in the presence of insulin resistance. In addition, CETP-mediated hetero-exchange of core neutral lipids between TRL and HDL leads to mass transfer of TG from TRL to HDL in exchange for CE. The loss of CE from HDL is magnified in hypertriglyceridaemic states, and produces HDL particles that are TG enriched and CE depleted. This contributes to the lower concentration of HDL-C measured in the plasma. Although apoA-I production may also be affected, kinetic studies have demonstrated that the HDL-apoA-I fractional clearance rate is increased in hypertriglyceridaemic and insulin-resistant individuals, without a significant reduction in apoA-I production rates (Pont et al., 2002).

4.2 Increased Apolipoprotein A-I Catabolism Due to TG Enrichment, Combined with Increased HL Activity

TG enrichment of HDL has an additional effect on HDL metabolism that ultimately determines the circulating HDL concentration. It enhances the remodelling of HDL particles in the circulation, with shedding of lipid-poor apoA-I and generation of HDL remnants, both contributing to increased HDL catabolism (Lewis and Rader, 2005). The CETP-mediated transfer of neutral core lipids is affected by the pool size of TRL in the circulation and the TG enrichment of HDL particles enhances apoA-I catabolism in hypertriglyceridaemia.

Compared with normal HDL, HDL enriched with TG after CETP-mediated exchange with TRL are preferred substrates for HL. HL lipolysis of TG-rich HDL, but not TG-poor HDL, dissociates apoA-I from the particle and generates HDL remnants that are small and core lipid depleted. Both the lipid-poor apoA-I and HDL remnants undergo more rapid catabolism. In humans, HDL isolated from individuals receiving intravenous infusion of intralipid are TG enriched in vivo and, when administered intravenously to healthy individuals, have significantly increased fractional catabolic rate compared with HDL that is not TG enriched (Lamarche et al., 1999). In addition, HDL apoA-I catabolism correlates strongly with HDL TG content, but not with other HDL components.

Compositional changes of HDL particles alone are not sufficient to destabilise the particle. Lipolytic modifications are necessary to promote HDL catabolism. A number of lines of evidence suggest that HL plays an important role in HDL metabolism. Postheparin HL activities are elevated in insulin-resistant, obese and T2D patients. HL activity is correlated with low HDL-C levels in these conditions and plays an important role in the increased catabolism of TG-enriched HDL. Lipolysis of TG-enriched HDL particles enhances apoA-I catabolism, thus lipolytically modified, TG-enriched HDL particles are cleared faster from the circulation than those not hydrolysed by HL. In isolated, perfused rabbit kidney, catabolism of apoA-I in TG-enriched HDL is not increased unless the HDL particles are treated with partially purified HL. In New Zealand white rabbits, an animal model that lacks HL, TG-enriched HDL do not show increased apoA-I catabolism in vivo. In contrast, apoA-I catabolism is increased when the TG-rich HDL particles are lipolysed by HL, either by ex vivo incubation with purified human HL or by adenovirus-mediated transfer of the human HL gene into the rabbits (Rashid et al., 2003). Insulin resistance increases the activity of HL, which then hydrolyses phospholipids in LDL and HDL, contributing to the formation of sdLDL and low HDL. Collectively, both TG enrichment and HL activity are needed to promote catabolism of HDL.

4.3 Role of Inflammation and Endothelial Lipase in HDL Metabolism

Insulin-resistant states are characterised by chronic, low-grade inflammation. Both acute and chronic inflammation are associated with low HDL-C. HDL undergoes significant changes in composition and structure during the acute phase of inflammation (van der Westhuyzen et al., 2007). Acute-phase HDL particles are larger, depleted of CE and enriched in cholesterol, TG and FFA. ApoA-I content is decreased and serum amyloid A (SAA) increases and can become the major apolipoprotein of HDL. SAA may lower HDL by displacing apoA-I from the particle and inhibiting CE uptake. These changes also cause the HDL particles to lose their anti-inflammatory and antioxidant properties, promoting LDL oxidation. Increased SAA and TG content in HDL particles and circulating levels of group II sPLA$_2$, an acute phase protein that associates with HDL under inflammatory conditions and hydrolyses HDL phospholipids, may promote HDL remodelling and increase catabolism. Inflammatory cytokines also modulate HDL metabolism. For instance, IL-6 may lower HDL production through inhibition of LPL activity. Inflammation also increases VLDL production and plasma TG, which contribute to lowered HDL and apoA-I levels by promoting catabolism, as discussed above.

Endothelial lipase (EL), a member of the LPL gene family, is synthesised by endothelial cells and functions at the vascular endothelial surface. In contrast to LPL and HL, EL has primarily phospholipase A1 activity with little TG hydrolase activity. EL acts on HDL more efficiently than other lipoprotein fractions, plays an important role in HDL metabolism and is a major determinant of plasma HDL-C. By hydrolysing phospholipids, EL reduces HDL size and generates small, structurally compromised, surface lipid-depleted HDL particles that are rapidly cleared by the kidney and liver. Overexpression of EL reduces, while inhibition with antibody or gene knockout increases, HDL-C and apoA-I in mice. In addition to its catalytic activity, EL acts as a bridge between HDL particles and endothelial cells by binding to heparan sulphte proteoglycans to promote binding and cellular metabolism of apoB- and apoA-I-containing lipoproteins. Inhibition of EL could be a new approach to raising HDL-C. In humans, polymorphism in the gene encoding EL results in loss of function and increased HDL-C, although the effects on CVD remain uncertain (Yasuda et al., 2010). In humans, plasma EL levels are directly correlated with inflammatory markers and are increased with experimental endotoxaemia. This mechanism may partially explain the low levels of HDL-C seen in insulin-resistant states (Badellino et al., 2008).

5. TREATMENT OF THE DYSLIPIDAEMIA OF INSULIN-RESISTANT STATES

Guidelines for the management of the typical dyslipidaemia of insulin-resistant states vary from country to country and differ according to various professional organisations, even within individual countries. The lipid levels that would initiate pharmacotherapy and on-treatment lipid targets generally depend on the presence of CVD or the global risk for CVD, estimated using validated risk algorithms. At least one prominent professional organisation has recently advocated drug treatment without consideration of specific lipid targets (Stone et al., 2014). Adults who have the metabolic syndrome and T2D are seldom if ever at low risk for CVD and are generally at least at moderate to high risk. Their metabolic risk factors should be assessed in relation to traditional risk factors such as age, gender, smoking, hypertension, presence of CVD, etc. Because treatment guidelines are not consistent, and are subject to frequent revisions, discussion of specific lipid levels that would initiate treatment and on-treatment targets is beyond the scope of this chapter. Interested readers are referred to the original, updated documents of authoritative professional organisation's guidelines. We will, however, discuss some of the general principles of treatment of the dyslipidaemia of insulin-resistant states here, in light of the present body of clinical trial evidence.

5.1 Lifestyle Modification

Lifestyle intervention addressing aggravating causes of dyslipidaemia constitutes the first step in management. Dietary changes, weight reduction and physical exercise are modifications for improving the underlying insulin resistance as well as the resulting dyslipidaemia and this should be the first approach to management. Lifestyle modification is also an essential aspect of the treatment of hyperglycaemia in those who have T2D. Optimal recommended food consumption is outlined in the various lipid management guideline recommendations and will not be discussed in detail here. In patients who have T2D, treatment of hyperglycaemia can also have a major effect in improving the typical dyslipidaemia of this condition. Unless there is very marked hyperlipidaemia or established CVD, it is prudent to first treat the poorly controlled hyperglycaemia with lifestyle and glucose-lowering therapies (the latter if indicated) prior to starting pharmacotherapy for the dyslipidaemia. Finally, dyslipidaemia is only one of a number of treatable CVD risk factors that frequently coexist in individuals with insulin resistance, and additional risk factors (such as hypertension, smoking and hyperglycaemia) should be addressed in the individual.

5.2 Pharmacotherapies

Statins should be the first lipid-modifying therapies for individuals who are deemed to be at sufficiently high CVD risk and who have failed 4–6 months of lifestyle intervention measures. Although statins are predominantly LDL-lowering agents and do not adequately correct the hypertriglyceridaemia and low HDL typical of insulin-resistant states, they have been shown in a number of large, prospective, double blind, placebo-controlled, randomised clinical trials to effectively prevent CVD in those with the metabolic syndrome and T2D. In patients with established CVD and those with diabetes plus multiple additional CVD risk factors, regardless of LDL-C levels, pharmacotherapy is indicated, concurrent with lifestyle intervention. For patients who cannot tolerate statins or those who do not meet LDL-C treatment goals with maximum statin doses, other LDL-C-lowering drugs, such as the cholesterol absorption inhibitor ezetimibe and bile acid sequestrants may be used, either as monotherapy or in combination with lower doses of statins. A number of new cholesterol-lowering therapies (PCSK9 antibodies, CETP inhibitors, apoB antisense, MTP inhibitors and others) will very likely become available over the next few years. It is too early to comment on the role that these new therapies will play in lipid management in T2D.

There is currently no consensus with respect to pharmacotherapy for hyper-triglyceridaemia, except in the case of marked hypertriglyceridaemia (plasma TG > approximately 10 mmol/L), in which TG-lowering therapy is indicated for the prevention of acute pancreatitis. Disappointing clinical trial results from the fibric acid derivative therapeutic agents and niacin and its derivatives have resulted in a general lack of enthusiasm for these classes of lipid-modifying drug, even though their therapeutic effect favourably alters the typical dyslipi-daemia of insulin-resistant states. Although clinical trials do not currently sup-port the first line use of fibrates or their addition to statins in cases of mild hypertriglyceridaemia (TG < 200 mg/dL or 2.3 mmol/L), there is some evidence that they may be effective in patients with more marked hypertriglyceridaemia. There is clinical trial evidence for the effectiveness of fibrates (mainly gemfi-brozil) as monotherapy for CVD risk reduction but clinical trials have generally not tested the effectiveness of fibrates in preventing CVD in those with moder-ate hypertriglyceridaemia (TG ≥ 2 mmol/L). Therefore, at the discretion of the treating physician and in the absence of firm clinical trial evidence favouring effective CVD prevention, fibrates may be used for CVD risk reduction, either alone or added to statins (but not gemfibrozil, because of a specific interaction with statins), in those with moderate hypertriglyceridaemia and an increased risk for CVD.

Niacin and extended release niacin are not favoured at the present time in light of two recent, large clinical trials that have showed no benefit when extended release niacin was added to statin therapy. Novel HDL-C raising thera-pies have not shown promise thus far in clinical trials, despite their impres-sive HDL raising effect. The effectiveness of CETP inhibition may need to be evaluated with the next generation of CETP inhibitors that also show promis-ing LDL and apoB-lowering effects. In the future, effects of HDL-modifying drugs on CVD risk reduction may be focused more on their ability to enhance macrophage-specific reverse cholesterol transport and other putative antiathero-sclerotic properties rather than raising plasma HDL-C concentrations (Yasuda et al., 2010).

6. CONCLUSIONS

Individuals with the insulin-resistant syndromes such as the metabolic syn-drome, obesity and T2D are at high risk for CVD, and the typical dyslipidae-mia that accompanies these conditions likely contributes to their risk, although other factors are also felt to play an important role. Insulin resistance plays a central role in the development of lipid abnormalities in these conditions. TRL overproduction, both by the liver and in the intestine, together with impaired clearance of TRL particles and their remnants, accounts for the elevated fasting and postprandial TG, and contributes to the lowering of HDL-C and the forma-tion of sdLDL particles, which collectively present an atherogenic lipid profile. Interactions between genes and environmental factors lead to cumulative risks for development of dyslipidaemia. Lifestyle modification and pharmacological approaches are needed for treatment of the dyslipidaemia that is so prevalent in insulin-resistant states.

REFERENCES

Alberti, K.G., Zimmet, P., Shaw, J., 2005. The metabolic syndrome–a new worldwide definition. Lancet 366 (9491), 1059–1062.

Badellino, K.O., Wolfe, M.L., Reilly, M.P., Rader, D.J., 2008. Endothelial lipase is increased in vivo by inflammation in humans. Circulation 117 (5), 678–685.

Boden, W.E., 2000. High-density lipoprotein cholesterol as an independent risk factor in cardiovascular disease: assessing the data from Framingham to the Veterans Affairs High–Density Lipoprotein Intervention Trial. Am. J. Cardiol. 86 (12A), 19L–22L.

Brown, M.S., Goldstein, J.L., 2008. Selective versus total insulin resistance: a pathogenic paradox. Cell Metab. 7 (2), 95–96.

Chahil, T.J., Ginsberg, H.N., 2006. Diabetic dyslipidemia. Endocrinol. Metab. Clin. North Am. 35 (3), 491-viii.

Chapman, M.J., Ginsberg, H.N., Amarenco, P., Andreotti, F., Boren, J., Catapano, A.L., Descamps, O.S., Fisher, E., Kovanen, P.T., Kuivenhoven, J.A., Lesnik, P., Masana, L., Nordestgaard, B.G., Ray, K.K., Reiner, Z., Taskinen, M.R., Tokgozoglu, L., Tybjaerg-Hansen, A., Watts, G.F., 2011. Triglyceride-rich lipoproteins and high-density lipoprotein cholesterol in patients at high risk of cardiovascular disease: evidence and guidance for management. Eur. Heart J. 32 (11), 1345–1361.

Chirieac, D.V., Davidson, N.O., Sparks, C.E., Sparks, J.D., 2006. PI3-kinase activity modulates apo B available for hepatic VLDL production in apobec-1-/- mice. Am. J. Physiol. Gastrointest. Liver Physiol. 291 (3), G382–G388.

Dekker, M.J., Su, Q., Baker, C., Rutledge, A.C., Adeli, K., 2010. Fructose: a highly lipogenic nutrient implicated in insulin resistance, hepatic steatosis, and the metabolic syndrome. Am. J. Physiol. Endocrinol. Metab. 299 (5), E685–E694.

Dentin, R., Benhamed, F., Hainault, I., Fauveau, V., Foufelle, F., Dyck, J.R., Girard, J., Postic, C., 2006. Liver-specific inhibition of ChREBP improves hepatic steatosis and insulin resistance in ob/ob mice. Diabetes 55 (8), 2159–2170.

Do, R., Willer, C.J., Schmidt, E.M., Sengupta, S., Gao, C., Peloso, G.M., Gustafsson, S., Kanoni, S., Ganna, A., Chen, J., Buchkovich, M.L., Mora, S., Beckmann, J.S., Bragg-Gresham, J.L., Chang, H.Y., Demirkan, A., Den Hertog, H.M., Donnelly, L.A., Ehret, G.B., Esko, T., Feitosa, M.F., Ferreira, T., Fischer, K., Fontanillas, P., Fraser, R.M., Freitag, D.F., Gurdasani, D., Heikkila, K., Hypponen, E., Isaacs, A., Jackson, A.U., Johansson, A., Johnson, T., Kaakinen, M., Kettunen, J., Kleber, M.E., Li, X., Luan, J., Lyytikainen, L.P., Magnusson, P.K., Mangino, M., Mihailov, E., Montasser, M.E., Muller-Nurasyid, M., Nolte, I.M., O'Connell, J.R., Palmer, C.D., Perola, M., Petersen, A.K., Sanna, S., Saxena, R., Service, S.K., Shah, S., Shungin, D., Sidore, C., Song, C., Strawbridge, R.J., Surakka, I., Tanaka, T., Teslovich, T.M., Thorleifsson, G., Van den Herik, E.G., Voight, B.F., Volcik, K.A., Waite, L.L., Wong, A., Wu, Y., Zhang, W., Absher, D., Asiki, G., Barroso, I., Been, L.F., Bolton, J.L., Bonnycastle, L.L., Brambilla, P., Burnett, M.S., Cesana, G., Dimitriou, M., Doney, A.S., Doring, A., Elliott, P., Epstein, S.E., Eyjolfsson, G.I., Gigante, B., Goodarzi, M.O., Grallert, H., Gravito, M.L., Groves, C.J., Hallmans, G., Hartikainen, A.L., Hayward, C., Hernandez, D., Hicks, A.A., Holm, H., Hung, Y.J., Illig, T., Jones, M.R., Kaleebu, P., Kastelein, J.J., Khaw, K.T., Kim, E., Klopp, N., Komulainen, P., Kumari, M., Langenberg, C., Lehtimaki, T., Lin, S.Y., Lindstrom, J., Loos, R.J., Mach, F., McArdle, W.L., Meisinger, C., Mitchell, B.D., Muller, G., Nagaraja, R., Narisu, N., Nieminen, T.V.,

Nsubuga, R.N., Olafsson, I., Ong, K.K., Palotie, A., Papamarkou, T., Pomilla, C., Pouta, A., Rader, D.J., Reilly, M.P., Ridker, P.M., Rivadeneira, F., Rudan, I., Ruokonen, A., Samani, N., Scharnagl, H., Seeley, J., Silander, K., Stancakova, A., Stirrups, K., Swift, A.J., Tiret, L., Uitterlinden, A.G., van Pelt, L.J., Vedantam, S., Wainwright, N., Wijmenga, C., Wild, S.H., Willemsen, G., Wilsgaard, T., Wilson, J.F., Young, E.H., Zhao, J.H., Adair, L.S., Arveiler, D., Assimes, T.L., Bandinelli, S., Bennett, F., Bochud, M., Boehm, B.O., Boomsma, D.I., Borecki, I.B., Bornstein, S.R., Bovet, P., Burnier, M., Campbell, H., Chakravarti, A., Chambers, J.C., Chen, Y.D., Collins, F.S., Cooper, R.S., Danesh, J., Dedoussis, G., de, F.U., Feranil, A.B., Ferrieres, J., Ferrucci, L., Freimer, N.B., Gieger, C., Groop, L.C., Gudnason, V., Gyllensten, U., Hamsten, A., Harris, T.B., Hingorani, A., Hirschhorn, J.N., Hofman, A., Hovingh, G.K., Hsiung, C.A., Humphries, S.E., Hunt, S.C., Hveem, K., Iribarren, C., Jarvelin, M.R., Jula, A., Kahonen, M., Kaprio, J., Kesaniemi, A., Kivimaki, M., Kooner, J.S., Koudstaal, P.J., Krauss, R.M., Kuh, D., Kuusisto, J., Kyvik, K.O., Laakso, M., Lakka, T.A., Lind, L., Lindgren, C.M., Martin, N.G., Marz, W., McCarthy, M.I., McKenzie, C.A., Meneton, P., Metspalu, A., Moilanen, L., Morris, A.D., Munroe, P.B., Njolstad, I., Pedersen, N.L., Power, C., Pramstaller, P.P., Price, J.F., Psaty, B.M., Quertermous, T., Rauramaa, R., Saleheen, D., Salomaa, V., Sanghera, D.K., Saramies, J., Schwarz, P.E., Sheu, W.H., Shuldiner, A.R., Siegbahn, A., Spector, T.D., Stefansson, K., Strachan, D.P., Tayo, B.O., Tremoli, E., Tuomilehto, J., Uusitupa, M., van Duijn, C.M., Vollenweider, P., Wallentin, L., Wareham, N.J., Whitfield, J.B., Wolffenbuttel, B.H., Altshuler, D., Ordovas, J.M., Boerwinkle, E., Palmer, C.N., Thorsteinsdottir, U., Chasman, D.I., Rotter, J.I., Franks, P.W., Ripatti, S., Cupples, L.A., Sandhu, M.S., Rich, S.S., 2013. Common variants associated with plasma triglycerides and risk for coronary artery disease. Nat.Genet. 45 (11), 1345–1352.

Farr, S., Taher, J., Adeli, K., 2014. Glucagon-like peptide-1 as a key regulator of lipid and lipoprotein metabolism in fasting and postprandial states. Cardiovasc. Hematol. Disord. Drug Targets 14 (2), 126–136.

Farr, S., Baker, C., Naples, M., Taher, J., Iqbal, J., Hussain, M., Adeli, K., 2015. Central nervous system regulation of intestinal lipoprotein metabolism by glucagon-like peptide-1 via a brain-gut axis. Arterioscler. Thromb. Vasc. Biol. ATVBAHA-114.

Federico, L.M., Naples, M., Taylor, D., Adeli, K., 2006. Intestinal insulin resistance and aberrant production of apolipoprotein B48 lipoproteins in an animal model of insulin resistance and metabolic dyslipidemia: evidence for activation of protein tyrosine phosphatase-1B, extracellular signal-related kinase, and sterol regulatory element-binding protein-1c in the fructose-fed hamster intestine. Diabetes 55 (5), 1316–1326.

Fisher, E., Lake, E., McLeod, R.S., 2014. Apolipoprotein B100 quality control and the regulation of hepatic very low density lipoprotein secretion. J. Biomed. Res. 28 (3), 178–193.

Hegele, R.A., Ginsberg, H.N., Chapman, M.J., Nordestgaard, B.G., Kuivenhoven, J.A., Averna, M., Boren, J., Bruckert, E., Catapano, A.L., Descamps, O.S., Hovingh, G.K., Humphries, S.E., Kovanen, P.T., Masana, L., Pajukanta, P., Parhofer, K.G., Raal, F.J., Ray, K.K., Santos, R.D., Stalenhoef, A.F., Stroes, E., Taskinen, M.R., Tybjaerg-Hansen, A., Watts, G.F., Wiklund, O., 2013. The polygenic nature of hypertriglyceridaemia: implications for definition, diagnosis, and management. Lancet Diabetes Endocrinol. 2 (8), 655–666.

Hein, G.J., Baker, C., Hsieh, J., Farr, S., Adeli, K., 2013. GLP-1 and GLP-2 as yin and yang of intestinal lipoprotein production: evidence for predominance of GLP-2-stimulated postprandial lipemia in normal and insulin-resistant states. Diabetes 62 (2), 373–381.

Huynh, F.K., Neumann, U.H., Wang, Y., Rodrigues, B., Kieffer, T.J., Covey, S.D., 2013. A role for hepatic leptin signaling in lipid metabolism via altered very low density lipoprotein composition and liver lipase activity in mice. Hepatology 57 (2), 543–554.

Lamarche, B., Uffelman, K.D., Carpentier, A., Cohn, J.S., Steiner, G., Barrett, P.H., Lewis, G.F., 1999. Triglyceride enrichment of HDL enhances in vivo metabolic clearance of HDL apo A-I in healthy men. J. Clin. Invest. 103 (8), 1191–1199.

Lewis, G.F., Carpentier, A., Adeli, K., Giacca, A., 2002. Disordered fat storage and mobilization in the pathogenesis of insulin resistance and type 2 diabetes. Endocr. Rev. 23 (2), 201–229.

Lewis, G.F., Rader, D.J., 2005. New insights into the regulation of HDL metabolism and reverse cholesterol transport. Circ. Res. 96 (12), 1221–1232.

Lewis, G.F., Xiao, C., Hegele, R.A., 2015. Hypertriglyceridemia in the genomic era: a new paradigm. Endocr. Rev. 36 (1), 131–147. Epub 2015 Jan 2.

Longuet, C., Sinclair, E.M., Maida, A., Baggio, L.L., Maziarz, M., Charron, M.J., Drucker, D.J., 2008. The glucagon receptor is required for the adaptive metabolic response to fasting. Cell Metab. 8 (5), 359–371.

Morgantini, C., Xiao, C., Dash, S., Lewis, G.F., 2014. Dietary carbohydrates and intestinal lipoprotein production. Curr. Opin. Clin. Nutr. Metab. Care 17 (4), 355–359.

Pavlic, M., Xiao, C., Szeto, L., Patterson, B.W., Lewis, G.F., 2010. Insulin acutely inhibits intestinal lipoprotein secretion in humans in part by suppressing plasma free fatty acids. Diabetes 59 (3), 580–587.

Perfield, J.W., Ortinau, L.C., Pickering, R.T., Ruebel, M.L., Meers, G.M., Rector, R.S., 2013. Altered hepatic lipid metabolism contributes to nonalcoholic fatty liver disease in leptin-deficient Ob/Ob mice. J. Obes. 2013, 296537.

Pont, F., Duvillard, L., Florentin, E., Gambert, P., Verges, B., 2002. High-density lipoprotein apolipoprotein A-I kinetics in obese insulin resistant patients. An in vivo stable isotope study. Int. J. Obes. Relat. Metab. Disord. 26 (9), 1151–1158.

Qiu, W., Federico, L., Naples, M., Avramoglu, R.K., Meshkani, R., Zhang, J., Tsai, J., Hussain, M., Dai, K., Iqbal, J., Kontos, C.D., Horie, Y., Suzuki, A., Adeli, K., 2008. Phosphatase and tensin homolog (PTEN) regulates hepatic lipogenesis, microsomal triglyceride transfer protein, and the secretion of apolipoprotein B-containing lipoproteins. Hepatology 48 (6), 1799–1809.

Rashid, S., Trinh, D.K., Uffelman, K.D., Cohn, J.S., Rader, D.J., Lewis, G.F., 2003. Expression of human hepatic lipase in the rabbit model preferentially enhances the clearance of triglyceride-enriched versus native high-density lipoprotein apolipoprotein A-I. Circulation 107 (24), 3066–3072.

Stanhope, K.L., Schwarz, J.M., Keim, N.L., Griffen, S.C., Bremer, A.A., Graham, J.L., Hatcher, B., Cox, C.L., Dyachenko, A., Zhang, W., McGahan, J.P., Seibert, A., Krauss, R.M., Chiu, S., Schaefer, E.J., Ai, M., Otokozawa, S., Nakajima, K., Nakano, T., Beysen, C., Hellerstein, M.K., Berglund, L., Havel, P.J., 2009. Consuming fructose-sweetened, not glucose-sweetened, beverages increases visceral adiposity and lipids and decreases insulin sensitivity in overweight/obese humans. J. Clin. Invest. 119 (5), 1322–1334.

Stone, N.J., Robinson, J.G., Lichtenstein, A.H., Bairey Merz, C.N., Blum, C.B., Eckel, R.H., Goldberg, A.C., Gordon, D., Levy, D., Lloyd-Jones, D.M., McBride, P., Schwartz, J.S., Shero, S.T., Smith Jr., S.C., Watson, K., Wilson, P.W., Eddleman, K.M., Jarrett, N.M., LaBresh, K., Nevo, L., Wnek, J., Anderson, J.L., Halperin, J.L., Albert, N.M., Bozkurt, B., Brindis, R.G., Curtis, L.H., DeMets, D., Hochman, J.S., Kovacs, R.J., Ohman, E.M., Pressler, S.J., Sellke, F.W., Shen, W.K., Smith Jr., S.C., Tomaselli, G.F., 2014. 2013 ACC/AHA guideline on the treatment of blood cholesterol to reduce atherosclerotic cardiovascular risk in adults: a report of the American College of Cardiology/American Heart Association Task Force on Practice Guidelines. Circulation 129 (25 Suppl. 2), S1–S45.

Taghibiglou, C., Rashid-Kolvear, F., Van Iderstine, S.C., Le-Tien, H., Fantus, I.G., Lewis, G.F., Adeli, K., 2002. Hepatic very low density lipoprotein-ApoB overproduction is associated with attenuated hepatic insulin signaling and overexpression of protein-tyrosine phosphatase 1B in a fructose-fed hamster model of insulin resistance. J. Biol. Chem. 277 (1), 793–803.

Taher, J., Baker, C.L., Cuizon, C., Masoudpour, H., Zhang, R., Farr, S., Naples, M., Bourdon, C., Pausova, Z., Adeli, K., 2014. GLP-1 receptor agonism ameliorates hepatic VLDL overproduction and de novo lipogenesis in insulin resistance. Mol. Metab. 3 (9), 823–833.

van der Westhuyzen, D.R., de Beer, F.C., Webb, N.R., 2007. HDL cholesterol transport during inflammation. Curr. Opin. Lipidol. 18 (2), 147–151.

Veilleux, A., Grenier, E., Marceau, P., Carpentier, A.C., Richard, D., Levy, E., 2014. Intestinal lipid handling: evidence and implication of insulin signaling abnormalities in human obese subjects. Arterioscler. Thromb. Vasc. Biol. 34 (3), 644–653.

Virtanen, J.K., Mursu, J., Voutilainen, S., Uusitupa, M., Tuomainen, T.P., 2014. Serum omega-3 polyunsaturated fatty acids and risk of incident type 2 diabetes in men: the Kuopio Ischemic Heart Disease Risk Factor study. Diabetes Care 37 (1), 189–196.

Willer, C.J., Schmidt, E.M., Sengupta, S., Peloso, G.M., Gustafsson, S., Kanoni, S., Ganna, A., Chen, J., Buchkovich, M.L., Mora, S., Beckmann, J.S., Bragg-Gresham, J.L., Chang, H.Y., Demirkan, A., Den Hertog, H.M., Do, R., Donnelly, L.A., Ehret, G.B., Esko, T., Feitosa, M.F., Ferreira, T., Fischer, K., Fontanillas, P., Fraser, R.M., Freitag, D.F., Gurdasani, D., Heikkila, K., Hypponen, E., Isaacs, A., Jackson, A.U., Johansson, A., Johnson, T., Kaakinen, M., Kettunen, J., Kleber, M.E., Li, X., Luan, J., Lyytikainen, L.P., Magnusson, P.K., Mangino, M., Mihailov, E., Montasser, M.E., Muller-Nurasyid, M., Nolte, I.M., O'Connell, J.R., Palmer, C.D., Perola, M., Petersen, A.K., Sanna, S., Saxena, R., Service, S.K., Shah, S., Shungin, D., Sidore, C., Song, C., Strawbridge, R.J., Surakka, I., Tanaka, T., Teslovich, T.M., Thorleifsson, G., Van den Herik, E.G., Voight, B.F., Volcik, K.A., Waite, L.L., Wong, A., Wu, Y., Zhang, W., Absher, D., Asiki, G., Barroso, I., Been, L.F., Bolton, J.L., Bonnycastle, L.L., Brambilla, P., Burnett, M.S., Cesana, G., Dimitriou, M., Doney, A.S., Doring, A., Elliott, P., Epstein, S.E., Eyjolfsson, G.I., Gigante, B., Goodarzi, M.O., Grallert, H., Gravito, M.L., Groves, C.J., Hallmans, G., Hartikainen, A.L., Hayward, C., Hernandez, D., Hicks, A.A., Holm, H., Hung, Y.J., Illig, T., Jones, M.R., Kaleebu, P., Kastelein, J.J., Khaw, K.T., Kim, E., Klopp, N., Komulainen, P., Kumari, M., Langenberg, C., Lehtimaki, T., Lin, S.Y., Lindstrom, J., Loos, R.J., Mach, F., McArdle, W.L., Meisinger, C., Mitchell, B.D., Muller, G., Nagaraja, R., Narisu, N., Nieminen, T.V., Nsubuga, R.N., Olafsson, I., Ong, K.K., Palotie, A., Papamarkou, T., Pomilla, C., Pouta, A., Rader, D.J., Reilly, M.P., Ridker, P.M., Rivadeneira, F., Rudan, I., Ruokonen, A., Samani, N., Scharnagl, H., Seeley, J., Silander, K., Stancakova, A., Stirrups, K., Swift, A.J., Tiret, L., Uitterlinden, A.G., van Pelt, L.J., Vedantam, S., Wainwright, N., Wijmenga, C., Wild, S.H., Willemsen, G., Wilsgaard, T., Wilson, J.F., Young, E.H., Zhao, J.H., Adair, L.S., Arveiler, D., Assimes, T.L., Bandinelli, S., Bennett, F., Bochud, M., Boehm, B.O., Boomsma, D.I., Borecki, I.B., Bornstein, S.R., Bovet, P., Burnier, M., Campbell, H., Chakravarti, A., Chambers, J.C., Chen, Y.D., Collins, F.S., Cooper, R.S., Danesh, J., Dedoussis, G., de, F.U., Feranil, A.B., Ferrieres, J., Ferrucci, L., Freimer, N.B., Gieger, C., Groop, L.C., Gudnason, V., Gyllensten, U., Hamsten, A., Harris, T.B., Hingorani, A., Hirschhorn, J.N., Hofman, A., Hovingh, G.K., Hsiung, C.A., Humphries, S.E., Hunt, S.C., Hveem, K., Iribarren, C., Jarvelin, M.R., Jula, A., Kahonen, M., Kaprio, J., Kesaniemi, A., Kivimaki, M., Kooner, J.S., Koudstaal, P.J., Krauss, R.M., Kuh, D., Kuusisto, J., Kyvik, K.O., Laakso, M., Lakka, T.A., Lind, L., Lindgren, C.M., Martin, N.G., Marz, W., McCarthy, M.I., McKenzie, C.A., Meneton, P., Metspalu, A., Moilanen, L., Morris, A.D., Munroe, P.B., Njolstad, I., Pedersen, N.L., Power, C., Pramstaller, P.P., Price, J.F., Psaty, B.M., Quertermous, T., Rauramaa, R., Saleheen, D., Salomaa, V., Sanghera, D.K., Saramies, J., Schwarz, P.E., Sheu, W.H., Shuldiner, A.R., Siegbahn, A., Spector, T.D., Stefansson, K., Strachan, D.P., Tayo, B.O., Tremoli, E., Tuomilehto, J., Uusitupa, M., van Duijn, C.M., Vollenweider, P., Wallentin, L., Wareham, N.J., Whitfield, J.B., Wolffenbuttel, B.H., Ordovas, J.M., Boerwinkle, E., Palmer, C.N., Thorsteinsdottir, U., Chasman, D.I., Rotter, J.I., Franks, P.W.,

Ripatti, S., Cupples, L.A., Sandhu, M.S., Rich, S.S., Boehnke, M., Deloukas, P., 2013. Discovery and refinement of loci associated with lipid levels. Nat. Genet. 45 (11), 1274–1283.

Xiao, C., Bandsma, R.H., Dash, S., Szeto, L., Lewis, G.F., 2012. Exenatide, a glucagon-like peptide-1 receptor agonist, acutely inhibits intestinal lipoprotein production in healthy humans. Arterioscler. Thromb. Vasc. Biol. 32 (6), 1513–1519.

Xiao, C., Dash, S., Morgantini, C., Adeli, K., Lewis, G.F., 2015. Gut peptides are novel regulators of intestinal lipoprotein secretion: implications for diabetic dyslipidemia. Diabetes (In Press).

Xiao, C., Hsieh, J., Adeli, K., Lewis, G.F., 2011a. Gut-liver interaction in triglyceride-rich lipoprotein metabolism. Am. J. Physiol. Endocrinol. Metab. 301 (3), E429–E446.

Xiao, C., Lewis, G.F., 2012. Regulation of chylomicron production in humans. Biochim. Biophys. Acta 1821 (5), 736–746.

Xiao, C., Pavlic, M., Szeto, L., Patterson, B.W., Lewis, G.F., 2011b. Effects of acute hyperglucagonemia on hepatic and intestinal lipoprotein production and clearance in healthy humans. Diabetes 60 (2), 383–390.

Yasuda, T., Ishida, T., Rader, D.J., 2010. Update on the role of endothelial lipase in high-density lipoprotein metabolism, reverse cholesterol transport, and atherosclerosis. Circ. J. 74 (11), 2263–2270.

Index

Note: Page numbers followed by "f" or "t" indicates figures and tables respectively.

E